T0332772

Handbook of Research on Advanced Applications of Graph Theory in Modern Society

Madhumangal Pal
Vidyasagar University, India

Sovan Samanta
Tamralipta Mahavidyalaya, India

Anita Pal
National Institute of Technology Durgapur, India

A volume in the Advances in Computer and
Electrical Engineering (ACEE) Book Series

Published in the United States of America by
IGI Global
Engineering Science Reference (an imprint of IGI Global)
701 E. Chocolate Avenue
Hershey PA, USA 17033
Tel: 717-533-8845
Fax: 717-533-8661
E-mail: cust@igi-global.com
Web site: http://www.igi-global.com

Library of Congress Cataloging-in-Publication Data

Names: Pal, Madhumangal, editor. | Samanta, Sovan, 1985- editor. | Pal,
   Anita, 1976- editor.
Title: Handbook of research on advanced applications of graph theory
in modern society / Madhumangal   Pal, Sovan Samanta, and Anita Pal, editors.
Description: Hershey, PA : Engineering Science Reference, [2020]
Identifiers: LCCN 2019001909| ISBN 9781522593805 (h/c) | ISBN 9781522593829
   (eISBN) | ISBN 9781522593812 (s/c)
Subjects: LCSH: Graph theory.
Classification: LCC QA166 .A34 2020 | DDC 511/.5--dc23 LC record available at https://lccn.loc.gov/2019001909

This book is published in the IGI Global book series Advances in Computer and Electrical Engineering (ACEE) (ISSN: 2327-039X; eISSN: 2327-0403)

British Cataloguing in Publication Data
A Cataloguing in Publication record for this book is available from the British Library.

All work contributed to this book is new, previously-unpublished material. The views expressed in this book are those of the authors, but not necessarily of the publisher.

For electronic access to this publication, please contact: eresources@igi-global.com.

Advances in Computer and Electrical Engineering (ACEE) Book Series

Srikanta Patnaik
SOA University, India

ISSN:2327-039X
EISSN:2327-0403

MISSION

The fields of computer engineering and electrical engineering encompass a broad range of interdisciplinary topics allowing for expansive research developments across multiple fields. Research in these areas continues to develop and become increasingly important as computer and electrical systems have become an integral part of everyday life.

The **Advances in Computer and Electrical Engineering (ACEE) Book Series** aims to publish research on diverse topics pertaining to computer engineering and electrical engineering. **ACEE** encourages scholarly discourse on the latest applications, tools, and methodologies being implemented in the field for the design and development of computer and electrical systems.

COVERAGE

- VLSI Design
- Microprocessor Design
- Power Electronics
- Computer Architecture
- Computer Science
- Optical Electronics
- Computer Hardware
- Sensor Technologies
- Qualitative Methods
- VLSI Fabrication

IGI Global is currently accepting manuscripts for publication within this series. To submit a proposal for a volume in this series, please contact our Acquisition Editors at Acquisitions@igi-global.com or visit: http://www.igi-global.com/publish/.

Titles in this Series

For a list of additional titles in this series, please visit:
https://www.igi-global.com/book-series/advances-computer-electrical-engineering/73675

Novel Practices and Trends in Grid and Cloud Computing
Pethuru Raj (Reliance Jio Infocomm Ltd. (RJIL), India) and S. Koteeswaran (Vel Tech, India)
Engineering Science Reference • © 2019 • 374pp • H/C (ISBN: 9781522590231) • US $255.00 (our price)

Blockchain Technology for Global Social Change
Jane Thomason (University College London, UK) Sonja Bernhardt (ThoughtWare, Australia) Tia Kansara (Replenish Earth Ltd, UK) and Nichola Cooper (Blockchain Quantum Impact, Australia)
Engineering Science Reference • © 2019 • 230pp • H/C (ISBN: 9781522595786) • US $195.00 (our price)

Contemporary Developments in High-Frequency Photonic Devices
Siddhartha Bhattacharyya (RCC Institute of Information Technology, India) Pampa Debnath (RCC Institute of Information Technology, India) Arpan Deyasi (RCC Institute of Information Technology, India) and Nilanjan Dey (Techno India College of Technology, India)
Engineering Science Reference • © 2019 • 369pp • H/C (ISBN: 9781522585312) • US $225.00 (our price)

Applying Integration Techniques and Methods in Distributed Systems and Technologies
Gabor Kecskemeti (Liverpool John Moores University, UK)
Engineering Science Reference • © 2019 • 351pp • H/C (ISBN: 9781522582953) • US $245.00 (our price)

Handbook of Research on Cloud Computing and Big Data Applications in IoT
B. B. Gupta (National Institute of Technology Kurukshetra, India) and Dharma P. Agrawal (University of Cincinnati, USA)
Engineering Science Reference • © 2019 • 609pp • H/C (ISBN: 9781522584070) • US $295.00 (our price)

Multi-Objective Stochastic Programming in Fuzzy Environments
Animesh Biswas (University of Kalyani, India) and Arnab Kumar De (Government College of Engineering and Textile Technology Serampore, India)
Engineering Science Reference • © 2019 • 420pp • H/C (ISBN: 9781522583011) • US $215.00 (our price)

Renewable Energy and Power Supply Challenges for Rural Regions
Valeriy Kharchenko (Federal Scientific Agroengineering Center VIM, Russia) and Pandian Vasant (Universiti Teknologi PETRONAS, Malaysia)
Engineering Science Reference • © 2019 • 432pp • H/C (ISBN: 9781522591795) • US $205.00 (our price)

701 East Chocolate Avenue, Hershey, PA 17033, USA
Tel: 717-533-8845 x100 • Fax: 717-533-8661
E-Mail: cust@igi-global.com • www.igi-global.com

List of Contributors

Table of Contents

Detailed Table of Contents

Chapter 1
    E. Sampathkumar, University of Mysore, India
    L. Pushpalatha, Yuvaraja's College, India

The study of domination in graphs originated around 1850 with the problems of placing minimum number of queens or other chess pieces on an n x n chess board so as to cover/dominate every square. The rules of chess specify that in one move a queen can advance any number of squares horizontally, vertically, or diagonally as long as there are no other chess pieces in its way. In 1850 enthusiasts who studied the problem came to the correct conclusion that all the squares in an 8 x 8 chessboard can be dominated by five queens and five is the minimum such number. With very few exceptions (Rooks, Bishops), these problems still remain unsolved today. Let $G = (V,E)$ be a graph. A set $S \subset V$ is a dominating set of G if every vertex in V–S is adjacent to some vertex in D. The domination number $\gamma(G)$ of G is the minimum cardinality of a dominating set.

Chapter 2
    Madhumangal Pal, Vidyasagar University, India

In this chapter, a very important class of graphs called intersection graph is introduced. Based on the geometrical representation, many different types of intersection graphs can be defined with interesting properties. Some of them—interval graphs, circular-arc graphs, permutation graphs, trapezoid graphs, chordal graphs, line graphs, disk graphs, string graphs—are presented here. A brief introduction of each of these intersection graphs along with some basic properties and algorithmic status are investigated.

Chapter 3
    V. R. Kulli, Gulbarga University, India

A molecular graph is a finite simple graph representing the carbon-atom skeleton of an organic molecule of a hydrocarbon. Studying molecular graphs is a constant focus in chemical graph theory: an effort to better understand molecular structure. Many types of graph indices such as degree-based graph indices, distance-based graph indices, and counting-related graph indices have been explored recently. Among degree-based graph indices, Zagreb indices are the oldest and studied well. In the last few years, many new graph indices were proposed. The present survey of these graph indices outlines their mathematical properties and also provides an exhaustive bibliography.

The Zagreb indices are the oldest among all degree-based topological indices. For a connected graph G, the first Zagreb index M1(G) is the sum of the term dG(u)+dG(v) corresponding to each edge uv in G, that is, M1 , where dG(u) is degree of the vertex u in G. In this chapter, the authors propose a weighted first Zagreb index and calculate its values for some standard graphs. Also, the authors study its correlations with various physico-chemical properties of octane isomers. It is found that this novel index has strong correlation with acentric factor and entropy of octane isomers as compared to other existing topological indices.

The inverse sum indeg (ISI) index of a graph G is defined as the sum of the weights dG(u)dG(v)/ dG(u)+dG(v) of all edges uv in G, where dG(u) is the degree of the vertex u in G. This index is found to be a significant predictor of total surface area of octane isomers. In this chapter, the authors present some lower and upper bounds for ISI index of subdivision graphs, t-subdivision graphs, s-sum and st -sum of graphs in terms of some graph parameters such as order, size, maximum degree, minimum degree, and the first Zagreb index. The extremal graphs are also characterized for their sharpness.

Let G = (V(G), E(G)) be a graph. The complement of G is denoted by Gc. The forgotten topological index of G, denoted F(G), is defined as the sum of the cubes of the degrees of all the vertices in G. The second Zagreb index of G, denoted M2(G), is defined as the sum of the products of the degrees of pairs of adjacent vertices in G. A graph G is k-Hamiltonian if for all X ⊂V(G) with |X| ≤ k, the subgraph induced by V(G) – X is Hamiltonian. Clearly, G is 0-Hamiltonian if and only if G is Hamiltonian. A graph G is k-path-coverable if V(G) can be covered by k or fewer vertex-disjoint paths. Using F(Gc) and M2(Gc), Li obtained several sufficient conditions for Hamiltonian and traceable graphs (Li, The Hyper-Zagreb Index and Some Properties of Graphs). In this chapter, the author presents sufficient conditions based upon F(Gc) and M2(Gc) for k-Hamiltonian, k-edge-Hamiltonian, k-path-coverable, k-connected, and k-edge-connected graphs.

One important problem in graph theory is graph coloring or graph labeling. Labeling problem is a well-studied problem due to its wide applications, especially in frequency assignment in (mobile) communication system, coding theory, ray crystallography, radar, circuit design, etc. For two non-negative integers, labeling of a graph is a function from the node set to the set of non-negative integers such that if and if, where it represents the distance between the nodes. Intersection graph is a very important subclass of graph. Unit disc graph, chordal graph, interval graph, circular-arc graph, permutation graph, trapezoid graph, etc. are the important subclasses of intersection graphs. In this chapter, the authors discuss labeling for intersection graphs, specially for interval graphs, circular-arc graphs, permutation graphs, trapezoid graphs, etc., and have presented a lot of results for this problem.

A set-valuation of a graph G=(V,E) assigns to the vertices or edges of G elements of the power set of a given nonempty set X subject to certain conditions. A set-indexer of G is an injective set-valuation f:V(G)→2x such that the induced set-valuation f⊕:E(G)→2X on the edges of G defined by f⊕(uv)=f(u)⊕f(v) ∀uv∈E(G) is also injective, where ⊕ denotes the symmetric difference of the subsets of X. Set-valued graphs such as set-graceful graphs, topological set-graceful graphs, set-sequential graphs, set-magic graphs are discussed. Set-valuations with a metric, associated with each pair of vertices is defined as distance pattern distinguishing (DPD) set of a graph (open-distance pattern distinguishing set of a graph (ODPU)) is ∅≠M⊆V(G) and for each u∈V(G), fM(u)={d(u,v): v ∈ M} be the distance-pattern of u with respect to the marker set M. If fM is injective (uniform) then the set M is a DPD (ODPU) set of G and G is a DPD (ODPU)-graph. This chapter briefly reports the existing results, new challenges, open problems, and conjectures that are abound in this topic.

Graph labelling is an assignment of labels to the vertices and/or edges of a graph with respect to certain restrictions and in accordance with certain predefined rules. The sumset of two non-empty sets A and B, denoted by A+B, is defined by A+B=\{a=b: a\inA, b\inB\}. Let X be a non-empty subset of the set \Z and \sP(X) be its power set. An \textit{sumset labelling} of a given graph G is an injective set-valued function f: V(G)\to\sP_0(X), which induces a function f+: E(G)\to\sP_0(X) defined by f+(uv)=f(u)+f(v), where f(u)+f(v) is the sumset of the set-labels of the vertices u and v. This chapter discusses different types of sumset labeling of graphs and their structural characterizations. The properties and characterizations of certain hypergraphs and signed graphs, which are induced by the sumset-labeling of given graphs, are also done in this chapter.

A popular area of graph theory is based on a paper written in 1930 by F. P. Ramsey titled "On a Problem on Formal Logic." A theorem which was proved in his paper triggered the study of modern Ramsey theory. However, his premature death at the young age of 26 hindered the development of this area of study at the initial stages. The balanced size multipartite Ramsey number mj (H,G) is defined as the smallest positive number s such that Kj×s→ (H,G). There are 36 pairs of (H, G), when H, G represent connected graphs on four vertices (as there are only 6 non-isomorphic connected graphs on four vertices). In this chapter, the authors find mj (H, G) exhaustively for all such pairs in the tripartite case j=3, and in the quadpartite case j=4, excluding the case m4 (K4,K4). In this case, the only known result is that m4 (K4,K4) is greater than or equal to 4, since no upper bound has been found as yet.

Chapter 11

Harishchandra S. Ramane, Karnatak University, India

The energy of a graph G is defined as the sum of the absolute values of the eigenvalues of its adjacency matrix. The graph energy has close correlation with the total pi-electron energy of molecules calculated with Huckel molecular orbital method in chemistry. A graph whose energy is greater than the energy of complete graph of same order is called hyperenergetic graph. A non-complete graph having energy equal to the energy of complete graph is called borderenergetic graph. Two non-cospectral graphs are said to be equienergetic graphs if they have same energy. In this chapter, the results on graph energy are reported. Various bounds for graph energy and its characterization are summarized. Construction of hyperenergetic, borderenergetic, and equienergetic graphs are reported.

Chapter 12

Natalia G. Miloslavskaya, National Research Nuclear University MEPhI (Moscow Engineering Physics Institute), Russia
Andrey Nikiforov, National Research Nuclear University MEPhI (Moscow Engineering Physics Institute), Russia
Kirill Plaksiy, National Research Nuclear University MEPhI (Moscow Engineering Physics Institute), Russia
Alexander Tolstoy, National Research Nuclear University MEPhI (Moscow Engineering Physics Institute), Russia

A technique to automate the generation of criminal cases for money laundering and financing of terrorism (ML/FT) based on typologies is proposed. That will help an automated system from making a decision about the exact coincidence when comparing the case objects and their links with those in the typologies. Several types of subgraph changes (mutations) are examined. The main goal to apply these mutations is to consider other possible ML/FT variants that do not correspond explicitly to the typologies but have a similar scenario. Visualization methods like the graph theory are used to order perception of data and to reduce its volumes. This work also uses the foundations of information and financial security. The research demonstrates possibilities of applying the graph theory and big data tools in investigating information security incidents. A program has been written to verify the technique proposed. It was tested on case graphs built on the typologies under consideration.

Radi Romansky, Technical University of Sofia, Bulgaria

Globalization is an important characteristic of the digital age which is based on the informatization of the
society as a social-economical and science-technical process for changing the information environment
while keeping the rights of citizens and organizations. The key features of the digital age are knowledge
orientation, digital representation, virtual and innovative nature, integration and inter-network interactions,
remote access to the information resources, economic and social cohesion, dynamic development, etc.
The graph theory is a suitable apparatus for discrete presentation, formalization, and model investigation
of the processes in the modern society because each state of a process could be presented as a node in
a discrete graph with connections to other states. The chapter discusses application of the graph theory
for a discrete formalization of the communication infrastructure and processes for remote access to
information and network resources. An extension of the graph theory like apparatus of Petri nets is
discussed and some examples for objects investigation are presented.

Seethalakshmi R., SASTRA University (Deemed), India

Mathematics acts an important and essential need in different fields. One of the significant roles in
mathematics is played by graph theory that is used in structural models and innovative methods, models
in various disciplines for better strategic decisions. In mathematics, graph theory is the study through
graphs by which the structural relationship studied with a pair wise relationship between different objects.
The different types of network theory or models or model of the network are called graphs. These graphs
do not form a part of analytical geometry, but they are called graph theory, which is points connected
by lines. The various concepts of graph theory have varied applications in diverse fields. The chapter
will deal with graph theory and its application in various financial market decisions. The topological
properties of the network of stocks will provide a deeper understanding and a good conclusion to the
market structure and connectivity. The chapter is very useful for academicians, market researchers,
financial analysts, and economists.

Kousik Das, Vidyasagar University, India
Rupkumar Mahapatra, Vidyasagar University, India
Sovan Samanta, Tamralipta Mahavidyalaya, India
Anita Pal, National Institute of Technology Durgapur, India

Social network is the perfect place for connecting people. The social network is a social structure formed
by a set of nodes (persons, organizations, etc.) and a set of links (connection between nodes). People
feel very comfortable to share news and information through a social network. This chapter measures
the influential persons in different types of online and offline social networks.

Manoj Kumar Srivastav, Champdani Adarsh Sharmik Vidyamandir, India
Robin Singh Bhadoria, Indian Institute of Information Technology Bhopal, India
Tarasankar Pramanik, Khanpur Gangche High School, India

The internet plays important role in the modern society. With the passage of time, internet consumers are increasing. Therefore, the traffic loads during communication between client and its associated server are getting complex. Various networking systems are available to send the information or to receive messages via the internet. Some networking systems are so expensive that they cannot be used for the regular purpose. A user always tries to use that networking system that works on expansion of optimizing the cost. A content delivery network (CDN) also called as content distribution network has been developed to manage better performance between client and list of available servers. This chapter presents the mathematical model to find optimization among client and cache server during delivery of content based on fuzzy logic.

Alejandro Vega-Muñoz, Universidad San Sebastián, Chile
Juan Manuel Arjona-Fuentes, Universidad Loyola Andalucia, Spain

This chapter presents how the analysis of social networks supported in graph theory contributes to the search for "distant knowledge" in the field of industrial engineering, discipline of engineering that in its current form began in the early 20th century when the first engineers began to apply scientific theory to manufacturing. In particular, the case of Chilean documented scientific production in this area of engineering is analyzed as a category of the web of science distinguishing its degree of connection with the great knowledge, generating organizations worldwide, determining its high dissociation with the great contemporary theoretical referents, and recommending the way to reduce these problems in the future.

Ganesh Ghorai, Vidyasagar University, India
Kavikumar Jacob, Universiti Tun Hussein Onn Malaysia, Malaysia

In this chapter, the authors introduce some basic definitions related to fuzzy graphs like directed and undirected fuzzy graph, walk, path and circuit of a fuzzy graph, complete and strong fuzzy graph, bipartite fuzzy graph, degree of a vertex in fuzzy graphs, fuzzy subgraph, etc. These concepts are illustrated with some examples. The recently developed concepts like fuzzy planar graphs are discussed where the crossing of two edges are considered. Finally, the concepts of fuzzy threshold graphs and fuzzy competitions graphs are also given as a generalization of threshold and competition graphs.

Relationship is the core building block of a network, and today's world advances through the complex networks. Graph theory deals with such problems more efficiently. But whenever vagueness or imprecision arises in such relationships, fuzzy graph theory helps. However, fuzzy hypergraphs are more advanced generalization of fuzzy graphs. Whenever there is a need to define multiary relationship rather than binary relationship, one can use fuzzy hypergraphs. In this chapter, interval-valued fuzzy hypergraph is discussed which is a generalization of fuzzy hypergraph. Several approaches to find shortest path between two given nodes in an interval-valued fuzzy graphs is described here. Many researchers have focused on fuzzy shortest path problem in a network due to its importance to many applications such as communications, routing, transportation, etc.

In this chapter, firstly some basic definitions like fuzzy graph, its adjacency matrix, eigenvalues, and its different types of energies are presented. Some upper bound and lower bound for the energy of this graph are also obtained. Then certain notions, including energy of m-polar fuzzy digraphs, Laplacian energy of m-polar fuzzy digraphs and signless Laplacian energy of m-polar fuzzy digraphs are presented. These concepts are illustrated with several example, and some of their properties are investigated.

The authors introduce neutrosophic cubic graphs and single-valued netrosophic Cubic graphs in bipolar setting and discuss some of their algebraic properties such as Cartesian product, composition, m-union, n-union, m-join, n-join. They also present a real time application of the defined model which depicts the main advantage of the same. Finally, the authors define a score function and present minimum spanning tree algorithm of an undirected bipolar single valued neutrosophic cubic graph with a numerical example.

Preface

In mathematics and computer science, graph theory is a very widely studied topic. A graph is a mathematical structure used to model pairwise relations between objects. A graph in this context is made up of vertices and nodes which are connected by edges, or lines. A graph may be undirected, meaning that there is no distinction between the two vertices associated with each edge, or its edges may be directed from one vertex to another. Graphs are one of the prime objects of study in discrete mathematics.

The paper written by Leonhard Euler on the Seven Bridges of Königsberg and published in 1736 is regarded as the first paper in the history of graph theory. Euler's formula relating the number of edges, vertices, and faces of a convex polyhedron was studied and generalized by Cauchy and L'Huilier.

In particular, the term "graph" was introduced by Sylvester in a paper published in 1878 in Nature, where he draws an analogy between "quantic invariants" and "co-variants" of algebra and molecular diagrams. Dénes Kőnig wrote the first textbook on graph theory and published in 1936. Another book by Frank Harary, published in 1969, was "considered the world over to be the definitive textbook on the subject", and enabled mathematicians, chemists, electrical engineers and social scientists to talk to each other.

This book has interesting co-related topics like social networks, graph colouring, domination, decomposition and graph drawings. Social network analysis (SNA) is the process of investigating social structures through the use of networks and graph theory. It characterizes network structures in terms of nodes (individual actors, people, or things within the network) and the ties, edges, or links (relationships or interactions) that connect them. Examples of social structures commonly visualized through social network analysis include social media networks, friendship and acquaintance networks, collaboration graphs, kinship, disease transmission, and sexual relationships. These networks are often visualized through sociograms in which nodes are represented as points and ties are represented as lines. Social network analysis has emerged as a key technique in modern sociology.

Graph colouring has many applications in real life like traffic light problems, map colouring and many more. This book will find suitable applications of graph covering, dominations, decompositions, graph drawing, labelling, social network, fuzzy graph, etc.

On the other hand, the real world is full of uncertainties. To handle such uncertainties, two theories are developed, viz. probability theory and fuzzy theory. Many real-life problems can be modelled using graph theory with uncertain vertices, edges, or both vertices and edges. Sometimes it happens that the weights of the vertices and/or edges may be uncertain. Such types of graph is known as a fuzzy graph. Rosenfeld defined fuzzy relations on fuzzy sets and developed the theory of fuzzy graphs in 1975. After that, many researchers enrich the fuzzy graph theory.

This book is a description of such updated theories and applications of graph theory. The summaries of all 22 chapters are presented below.

Chapter 1

The study of domination in graphs originated around 1850 with the problems of placing a minimum number of queens or other chess pieces on a $n \times n$ chess board to cover/dominate every square. The rules of chess specify that in one move, a queen can advance any number of squares horizontally, vertically or diagonally as long as there are no other chess pieces in its way. In 1850 enthusiasts who studied the problem came to the correct conclusion that five queens and five can dominate all the squares in an 8×8 chessboard is the minimum such number. With very few exceptions (Rooks, Bishops), these problems remain unsolved today. Let $G = (V,E)$ be a graph. A set $S \subset V$ is a dominating set of G if every vertex in $V - S$ is adjacent to some vertex in D. The domination number $\gamma(G)$ of G is the minimum cardinality of a dominating set. This chapter is a study of the domination theory of graphs.

Chapter 2

In this chapter, an essential class of graphs, called the intersection graph, is introduced. Based on the geometrical representation, many different types of intersection graphs can be defined with exciting properties. Some of them, viz. interval graphs, circular-arc graphs, permutation graphs, trapezoid graphs, chordal graphs, line graphs, disk graphs, string graphs are presented here. A brief introduction of each of these intersection graphs along with some basic properties and algorithmic status is investigated.

Chapter 3

A molecular graph is a finite simple graph, representing the carbon-atom skeleton of an organic molecule of a hydrocarbon studying molecular graphs is a constant focus in Chemical Graph Theory: an effort to understand the molecular structure better. Many types of graph indices such as degree based graph indices, distance-based graph indices and counting related graph indices are explored during the past recent years. Among degree based graph indices, Zagreb indices are the oldest one and studied well. In the last few years, many new graph indices were proposed. The present survey of these graph indices outlines their some mathematical properties and also provides an exhaustive bibliography. This chapter is an extensive discussion of different graph indices.

Chapter 4

The Zagreb indices are the oldest among all degree based topological indices. For a connected graph G, the first Zagreb index $M_1(G)$ is the sum of the term $d_G(u) + d_G(v)$ corresponding to each edge uv in G, i.e.,

$$M_1\left(G\right) = \sum \_ \left(uv \in E\left(G\right)\right) \lfloor \left(d_G\left(u\right) + d_G\left(v\right)\right],$$

where $d_G(u)$ is a degree of the vertex u in G. In this communication, the authors propose a weighted first Zagreb index and calculate its values for some standard graphs. Also, the authors study its correlations with various physicochemical properties of octane isomers. It is found that this new index has a strong correlation with Acentric factor and Entropy of octane isomers as compared to other existing topological indices. In this chapter, the Zagreb index is described.

Chapter 5

Another index, the inverse sum in-deg (ISI) index of a graph G, is defined in this chapter. The index is the sum of the weights

$$dG\left(u\right)dG\left(v\right)/dG\left(u\right)+dG\left(v\right)$$

of all edges uv in G, where $dG(u)$ is the degree of the vertex u in G. This index is found to be a significant predictor of total surface area of octane isomers. In this chapter, the authors present some lower and upper bounds for ISI index of subdivision graphs, t-subdivision graphs, s-sum and st -the sum of graphs in terms of some graph parameters such as order, size, maximum degree, minimum degree and the first Zagreb index. Their sharpness also characterises the extremal graphs.

Chapter 6

Let $G = \left(V\left(G\right), E\left(G\right)\right)$ be a graph. G^c denotes the complement of G. The forgotten topological index of G, denoted $F(G)$, is defined as the sum of the cubes of the degrees of all the vertices in G. The second Zagreb index of G, denoted $M_2(G)$, is defined as the sum of the products of the degrees of pairs of adjacent vertices in G. A graph Gisk-Hamiltonian if for all $X \subset V\left(G\right)$ with $|X|{\leq}k$, the subgraph induced by $V(G)$ - X is Hamiltonian. G is 0-Hamiltonian if and only if G is Hamiltonian. A graph G is k-path-coverable if $V(G)$ can be covered by k or fewer vertex-disjoint paths. Using $F(G^c)$ and $M_2(G^c)$, Li obtained several sufficient conditions for Hamiltonian and traceable graphs (Rao Li, Topological Indexes and Some Hamiltonian Properties of Graphs). In this chapter, the authors will present sufficient conditions based upon $F(G^c)$ and $M_2(G^c)$ for k-Hamiltonian, k-edge-Hamiltonian, k-path-coverable, k-connected, and k-edge-connected graphs.

Chapter 7

A critical problem in graph theory is graph colouring or graph labelling. L(h,k)-Labelling problem is a well-studied problem due to its full applications, especially in frequency assignment in a (mobile) communication system, coding theory, x-ray crystallography, radar, circuit design, etc. For two non-negative integers and, an L(h,k)-labelling of a graph is a function from the node set to the set of non-negative integers such that if and if, where represents the distance between the nodes and. Intersection graph is a very important subclass of a graph. Unit disc graph, chordal graph, interval graph, circular-arc graph, permutation graph, trapezoid graph etc. are the essential subclasses of intersection graphs. In this chapter, the authors discuss L(h,k)-labelling for intersection graphs, especially, for interval graphs, circular-arc graphs, permutation graphs, trapezoid graphs etc. and have presented many results for this problem.

Chapter 8

A set-valuation of a graph $G = (V,E)$ assigns to the vertices or edges of G elements of the power set of a given nonempty set X subject to certain conditions. A set-indexer of G is an injective set-valuation f: $V(G) \rightarrow 2X$ such that the induced set-valuation $f\oplus$: $E(G) \rightarrow 2X$ on the edges of G defined by

$$f \oplus (uv) = f(u) \oplus f(v) \forall uv \in E(G)$$

is also injective, where $\oplus$ denotes the symmetric difference of the subsets of X. Set-valued graphs such as set-graceful graphs, topological set-graceful graphs, set-sequential graphs, set-magic graphs are discussed. Set-valuations with a metric, associated with each pair of vertices is defined as distance pattern distinguishing (DPD) set of a graph (open-distance pattern distinguishing set of a graph (ODPU)) of a graph is, $\varnothing \neq M \subseteq V(G)$ and for each

$$u \in V(G), fM(u) = \{d(u,v) : v \in M\}$$

be the distance-pattern of u with respect to the marker set M. If $f M$ is injective (uniform) then the set M is a DPD (ODPU) set of G and G is a DPD (ODPU)-graph. This, briefly reports the existing results, new challenges, open problems and conjectures that abound in this topic.

Chapter 9

Graph labelling is an assignment of labels to the vertices and edges of a graph concerning certain restrictions and by certain predefined rules. The sumset of two non-empty sets A and B, denoted by $A+B$, is defined by

$$A + B = \{a + b : a \in A, b \in B\}.$$

Let X be a non-empty subset of the set of integers Z, and P(X) be its power set. A *sumset labelling* of a given graph G is an injective set-valued function f: $V(G) \rightarrow P_0(X)$, which induces a function f^+: $E(G) \rightarrow P_0(X)$ defined by $f^+(uv) = f(u)+f(v)$, where $f(u)+f(v)$ is the sumset of the set-labels of the vertices u and v. This chapter discusses different types of sumset labellings of graphs and their structural characterisations. The properties and characterisations of certain hypergraphs and signed graphs, which are induced by the sumset-labellings of given graphs have also done in this chapter.

Chapter 10

A popular area of Graph theory is based on a paper written in 1930 by Ramsey titled "On a problem on formal logic". A theorem which was proved in his paper triggered the study of modern Ramsey Theory. However, his premature death at the young age of 26, hindered the development of this area of study at the initial stages. The balanced size multipartite Ramsey number $m_j(H,G)$ is defined as the smallest positive number s such that $K_{j\times s} \rightarrow (H,G)$. There are 36 pairs of (H, G), when H, G represent connected graphs on 4 vertices (as there are only 6 non-isomorphic connected graphs on 4 vertices). In this chapter, the authors find $m_j(H,G)$ exhaustively for all such pairs in the tripartite case $j=3$, and in the quadripartite case $j=4$, excluding the case $m_4(K_4,K_4)$. In this case, the only known result is that $m_4(K_4,K_4)$ is greater than or equal to 4 since no upper bound has been found as yet.

Chapter 11

The energy of a graph G is defined as the sum of the absolute values of the eigenvalues of its adjacency matrix. The graph energy has a close correlation with the total pi-electron energy of molecules calculated with Huckel molecular orbital method in chemistry. A graph whose energy is higher than the energy of the complete graph of the same order is called a hyperenergetic graph. A non-complete graph having energy equal to the energy of a complete graph is called border-energetic graph. Two non-cospectral graphs are said to be equi-energetic graphs if they have same energy. In this chapter, the results of graph energy are reported. Various bounds for graph energy and its characterization are summarized. Construction of hyperenergetic, border-energetic and equi-energetic graphs are reported.

Chapter 12

A technique to automate the generation of criminal cases for money laundering and financing of terrorism (ML/FT) based on typologies is proposed. That will help an automated system from deciding the exact coincidence when comparing the case objects and their links with those in the typologies. Several types of subgraph changes (mutations) are examined. The main goal to apply these mutations is to consider other possible ML/FT variants that do not correspond explicitly to the typologies but have a similar scenario. Visualization methods like the graph theory are used to order perception of data and to reduce its volumes. This work also uses the foundations of information and financial security. The research demonstrates the possibilities of applying the graph theory and big data tools in investigating information security incidents. A program has been written to verify the technique proposed. It was tested on case graphs built on the typologies under consideration.

Chapter 13

Digital age has important key features – knowledge orientation, virtual nature, remote access to resources, integration and inter-network interactions, dynamic development, etc. All these features determine many digital services as e-access, e-society, e-policy, e-democracy, e-voting, e-governance/e-government, e-learning, e-health, e-business/e-commerce, e-consultation, e-inclusion, etc. The graph theory is a suitable apparatus for discrete presentation, formalization and model investigation of different processes in the modern society because each state of a process (or object) could be presented as a node in a discrete graph with connections (arcs) to other states (objects). In this reason, the chapter discusses the application of the graph theory for a discrete formalization of the communication infrastructure and processes for remote access to information and network resources. An extension of the graph theory like apparatus of Petri Nets is discussed, and some examples for objects investigation are presented.

Chapter 14

Mathematics acts as an essential and essential need in different fields. One of the significant roles in Mathematics is played by Graph theory that is used in structural models and innovative methods models in various disciplines for better strategic decisions. In mathematics, graph theory is the study through graphs by which the structural relationship studied with a pairwise relationship between different objects. The different types of network theory or models or model of the network are called as graphs. These graphs do not form a part of analytical geometry, but they called graph theory, which is ' points connected by lines'. The various concepts of Graph theory have diverse applications in diverse fields. The present chapter will deal with graph theory and its application in the various financial market decision. The topological properties of the network of stocks will provide a deeper understanding and a good conclusion to the market structure and connectivity. The current chapter is beneficial for academicians, market researcher, financial analyst and economist

Chapter 15

Humans are a social person and best decision maker. They cannot stay alone; they want to make friends, share news, information etc. Social network is the perfect place for connecting people. The social network is a social structure formed by a set of nodes (persons, organizations, etc.) and a set of links (the connection between nodes). People feel very comfortable to share news, information through a social network. This study measures the influential persons in different types of online and offline social networks.

Chapter 16

The internet plays an essential role in modern society. With time, internet consumers are increasing. Therefore, the traffic loads during communication between the client and its associated server are getting complex. Various networking systems are available to send the information or to receive messages via the internet. Some networking system is more so expensive that it cannot be used for the regular purpose. A user always tries to use that networking system which works on expansion of optimizing the cost. A Content Delivery Networks (CDN) also called as Content Distribution Networks, which has been developed to manage better performance between client and list of available servers. This chapter presents the mathematical model to find optimization among client and cache server during delivery of content based on fuzzy logic.

Chapter 17

This chapter presents how the analysis of social networks supported in the graph theory contributes to the search for 'distant knowledge' in the field of industrial engineering, discipline of engineering that in its current form, began in the early 20th century, when the first engineers began to apply scientific theory to manufacturing. In particular, the case of Chilean documented scientific production in this area of engineering is analyzed, as a category of the Web of Science. It is distinguishing its degree of connection with the excellent knowledge generating organizations worldwide and determining its high dissociation with the tremendous contemporary theoretical referents, recommending the way to reduce these problems in the future.

Chapter 18

In this chapter, the authors introduce some basic definitions related to fuzzy graphs like a directed and undirected fuzzy graph, walk, path and circuit of a fuzzy graph, complete and robust fuzzy graph, bipartite fuzzy graph, degree of a vertex in fuzzy graphs, fuzzy subgraph etc. These concepts are illustrated with some examples. The recently developed concepts like fuzzy planar graphs are discussed where the crossing of two edges is considered. Finally, the concepts of fuzzy threshold graphs and fuzzy competitions graphs are also given as a generalization of threshold and competition graphs.

Chapter 19

Relationship is the core building block of a network, and today's world advances through the complex and more complex networks. Graph theory deals with such problems more efficiently. However, whenever vagueness or imprecision arises in such relationships, the fuzzy graph theory helps in a better way. However, fuzzy hypergraphs are a more advanced generalization of fuzzy graphs. Whenever there is a need to define military relationship rather than a binary relationship, one can use fuzzy hypergraphs. In this chapter, interval-valued fuzzy hypergraph is discussed, which is a generalization of the fuzzy hypergraph. Several approaches to find the shortest path between two given nodes in an interval-valued fuzzy graph is described here. Many researchers have focused on fuzzy shortest path problem in a network due to its importance to many applications such as communications, routing, transportation, etc.

Chapter 20

In this chapter, firstly some basic definitions like a fuzzy graph, its adjacency matrix, eigenvalues and its different types of energies are presented. Some upper bound and lower bound for the energy of this graph are also obtained. Then certain notions, including the energy of m-polar fuzzy digraphs, Laplacian energy of m-polar fuzzy digraphs and signless Laplacian energy of m-polar fuzzy digraphs, are presented. These concepts are illustrated with several examples. Also, some of their properties are investigated.

Chapter 21

In this framework, the authors introduce Neutrosophic Cubic graphs and single-valued Neutrosophic cubic graphs in bipolar setting and discuss some of their algebraic properties such as Cartesian Product, Composition, M-Union, N-Union, M-Join, N-Join. We also present a real-time application of the defined model, which depicts the main advantage of the same. Finally, we define a score function and present a minimum spanning tree algorithm of an undirected bipolar single-valued neutrosophic cubic graph with a numerical example.

This book covers advanced research area of graph theory and its applications. This book is designed for researchers those who are working on (crisp) graph theory (in contrast of fuzzy graph, the conventional graph is sometimes called as a crisp graph), fuzzy graph and related topics. Many applications of graph theory and fuzzy graph theory are discussed in this book.

We are highly grateful to all the contributors for writing the chapters by spending their valuable time and expertise. Also, we are very thankful to the reviewers for their time and suggestion to review the chapters within the stipulated period. Without the support of contributors and reviewers, it is not possible for us to bring this book to graph theory lovers. We are also thankful to our colleagues and friends for their continuous support and encouragement to prepare the book.

We shall feel great to receive constructive criticisms for the improvement of the book from the experts as well as the learners.

We thank the IGI Global Publisher for their sincere care and suggestion in the publication of the book.

Madhumangal Pal
Vidyasagar University, India

Sovan Samanta
Tamralipta Mahavidyalaya, India

Anita Pal
National Institute of Technology Durgapur, India

Chapter 1
Domination Theory in Graphs

E. Sampathkumar
University of Mysore, India

L. Pushpalatha
Yuvaraja's College, India

ABSTRACT

The study of domination in graphs originated around 1850 with the problems of placing minimum number of queens or other chess pieces on an n x n chess board so as to cover/dominate every square. The rules of chess specify that in one move a queen can advance any number of squares horizontally, vertically, or diagonally as long as there are no other chess pieces in its way. In 1850 enthusiasts who studied the problem came to the correct conclusion that all the squares in an 8 x 8 chessboard can be dominated by five queens and five is the minimum such number. With very few exceptions (Rooks, Bishops), these problems still remain unsolved today. Let G = (V,E) be a graph. A set S ⊂ V is a dominating set of G if every vertex in V–S is adjacent to some vertex in D. The domination number γ(G) of G is the minimum cardinality of a dominating set.

INTRODUCTION

The study of Domination in Graphs originated around 1850 with the problems of placing minimum number of queens or other chess pieces on an $n \times n$ chess board so as to cover/dominate every square. The rules of chess specify that in one move, a queen can advance any number of squares horizontally, vertically or diagonally as long as there are no other chess pieces in its way. In 1850 enthusiasts who studied the problem came to the correct conclusion that all the squares in an *8 x 8* chessboard can be dominated by five queens and five is the minimum such number. With very few exceptions (Rooks, Bishops), these problems still remain unsolved today.

DOI: 10.4018/978-1-5225-9380-5.ch001

Figure 1.

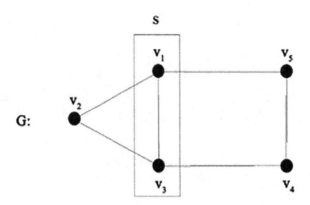

Let $G = (V, E)$ be a graph. A set $S \subset V$ is a dominating set of G if every vertex in $V - S$ is adjacent to some vertex in D. The domination number $\gamma(G)$ of G is the minimum cardinality of a dominating set.

$S = \{v_1, v_3\}$ is a dominating set and $\gamma(G) = 2$.

APPLICATIONS

1. Berge (1973) in his book "Graphs and Hypergraphs" mentions the problem of keeping all points in a network under the surveillance of a set of radar stations. A number of strategic locations $v_1, v_2, \ldots$ called cells are kept under the surveillance of radar. Radar in cell v_2 can survey the locations v_1, v_2 or v_5. Similarly, v_3 can be surveyed by radar located at v_2 or v_3.

What is the minimum number of radar stations needed to survey all locations? It is the domination number of the network.

2. In a similar vein, Liu (1968) in his Book "Introduction to combinatorial Math," discusses the application of dominance to communication network, where a dominating set represents a set of cities which, acting as transmitting stations, can transmit messages to every city in the network.
3. Berge (1972) also discusses relationship between Kernels in graphs, i.e. dominating sets which are also independent and solutions to game theory.
4. The notion of dominance is used in coding theory. If one defines a graph whose vertices are n - dimensional vectors with coordinates chosen from (1, 2, ..., p) and two vertices are adjacent if they differ in one coordinate, the sets of vectors which are (n, p) covering sets, or simple error correcting codes, or perfect covering sets are all dominating sets of the graph with certain additional properties. See for example, Kalbfleish, Stanton and Horton (1971), on covering sets and error correcting codes.

5.   Ore (1982) in his Book "Theory of Graphs," mentions the problem of placing minimum number of Queens on a chess board so that each square is controlled by at least one queen.

Open Problem

What is the minimum number of Queens that can be placed on an $n \times n$ chess board so as to cover/dominate every square?

6.   The notion of dominance is also related to the Theory of matchings. A matching is a maximal set of independent edges in a graph. Clearly, any matching in a graph G corresponds to an independent dominating set in the line graph $L(G)$ of G.

7.   Dominating sets also occur in applications on clutters and blockers of clutters. A clutter is a non-empty family C of subsets of a set A, none of which properly contains any other. The blocker of a clutter C consists of all minimal subsets of A which contain at least one element of each subset in C. For details see

a.   Edmonds, J. & Fulkerson, D.R. (1970) Bottleneck extrema, J. Comb. Theory 8, pp. 299-306.

b.   Billera, L.J. (1971) On the composition and decomposition of clutters. Journal of Combinatorial Theory 11B, pp. 243-245,

BASIC PROPERTIES

Proposition 1

A dominating set D is a minimal dominating set if for each $d \in D$, one of the following holds (Ore, 1982).

1.   d is not adjacent to any vertex in D
2.   There is a vertex $c \notin D$ such that

$$N(c) \cap D = \{d\}$$

where $N(c) = \{v \in V : vc \in E\}$

Proof

In order that $D - d$ not be a dominating set it is necessary and sufficient that at least one of the two conditions hold.

Proposition 2

If G is a graph without isolated vertices, and D is a minimal dominating set, then V-D contains a minimal dominating set (Ore, 1982).

A dominating set D is an independent dominating set if D is independent. The independent domination number i(G) of G is the minimum cardinality of such a set.

Proposition 3

If D is an independent dominating set, then D is both a minimal dominating set and a maximal independent set (Berge, 1973). Conversely, if D is a maximal independent set then D is an independent dominating set.

BOUNDS FOR THE DOMINATION NUMBER

Proposition 4

If G is a graph having p vertices, q edges and maximum degree Δ, then (Berge, 1973):

$$p - q \leq \gamma\left(G\right) \leq p - \Delta$$

Proof

Let D be a γ-set and $\left|V - D\right| = r$. Then there are at least r edges from $V - D$ to D and $\left|D\right| = p - r$.

Since $r \leq q$, we have $p - q \leq \left|D\right| = \gamma$

To establish the upper bound, suppose $deg\, v = \Delta$. Clearly, the set $V - N\left(v\right)$ is a dominating set of G, and hence $\gamma\left(G\right) \leq p - \Delta$.

Corollary 4.1

$\gamma\left(G\right) \leq p - k\left(G\right)$, where $k\left(G\right)$ is the connectivity of G.

For a real number r, let [r] be the greatest integer not greater than r, and {r} the least integer not less than r.

Proposition 5

For any (p, q) graph G without isolates

$$\left[\frac{p}{\Delta + 1}\right] \leq \gamma\left(G\right) \leq p - \beta_0$$

where β_0 is the independence number.

$\underline{\text{Proof:}}$ let $|D| = r$ and $D = \{v_1, v_2, \ldots v_r\}$. Since every vertex in $V - D$ is adjacent to some vertex in D, we have

$$|V - D| \leq \sum_{i=1}^{r} d(V_i) \leq \gamma(G).\Delta$$

Thus, $p - \gamma(G) \leq \gamma(G).\Delta$, and the lower bound follows. To establish the upper bound, let S be a maximum independent set of vertices. Then |S| = β_0. Clearly, every vertex in S is adjacent to some vertex in $V - S$. Hence S is a dominating set and the upper bound follows.

Corollary 5.1

In any connected graph of order $p \geq 2$, any vertex cover is a dominating set, and $\gamma(G) \leq \alpha_0$ where α_0 is the vertex covering number.

Proposition 6

For a (p, q) graph G

$$q \leq \left\lceil \left(p - \gamma(G)\right)\left(p - \gamma(G) + 2\right) / 2 \right\rceil$$

Proposition 7

Let I(G) be the intersection graph on the set of all independent sets in a graph (Cockayne & Hedetneimi, 1974). Then:

$$i(I(G)) = \chi(G)$$

Proposition 8

Let (V, E) be a graph which does not have an induced subgraph isomorphic to $K_{1,3}$, then $\gamma(G) = i(G)$.

Proof

Clearly $\gamma(G) \leq i(G)$ for any graph. We now show that $i(G) \leq \gamma(G)$.

Let

$$m = \gamma(G) \text{ and } D_{-1} = \{w_0, w_1, \ldots, w_{m-1}\} \subset V$$

a minimum dominating set. For a non-empty $V' \subset V$, let $a(V')$ denote the number of edges in the subgraph induced by V'. Clearly,

$$0 \leq a(D_{-1}) \leq \binom{m}{2}.$$

If $a(D_{-1}) = 0$, then D_{-1} is independent, and since D_{-1} is a maximal independent set. D_{-1} is an independent dominating set by Proposition 3.

Hence

$$i(G) \leq m = \gamma(G).$$

Now assume $a(D_{-1}) \neq 0$ and without loss of generality, assume $w_0 w_1 \in E$.

Now, by Proposition 1, the set

$$N_0 = \left\{ u \in V - D_{-1} : N(u) \cap D_{-1} = \{w_0\} \right\}$$

is not empty.

Let u and w be any two distinct vertices of N_0 and consider $\{w_0, w_1, u, w\} \subset V$. The subgraph induced by this set certainly contains $\{w_0 w_1, w_0 u, w_0 w\}$. By hypothesis,

$$\{w_1 u, w_1 w, uw\} \cap E \neq \varnothing,$$

Otherwise, $K_{1,3}$ will be induced by $\{u, w_0, w, w_1\}$

But since

$$N(u) \cap D_{-1} = \{w_0\} = N(w) \cap D_{-1}$$

Figure 2.

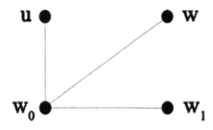

it follows that, $uw \notin E$, $ww_1 \notin E$. Hence, $uv \in E$. Now, it follows that any two distinct elements of $N_0 \cup \{w_0\}$ are adjacent.

Take $u_0 \in N_0$ and consider

$$D_0 = \{u_0, w_1, w_2, \ldots, w_{m-1}\}.$$

Let

$$z \in V - D_0 = M \cup K.$$

Where

$$M = \left(N_0 - \{u_0\}\right) \cup \{w_0\} \text{ and } K = V - \left(N_0 \cup D_{-1}\right).$$

If $z \in M$, then $z = w_0$ or $z ~~ {N}_{0} ~~ \left\{ {w}_{0} \right\}$. If $z = w_0$, then $z{u}_{0} = ~{w}_{0}{u}_{0} ~~ E$. If $z \in N_0 - \{w_0\}$, then zu_0 is an edge since,

$$u_0 \in L = N_0 v \{w_0\}$$

and any two vertices in L are adjacent. If $z \in K,$ then

$$N(z) \cap D_{-1} \supseteq \{W_i\},$$

where $1 \leq i \leq m - 1$, which says that $w_i \in E$. (Note: Since D_{-1} is a dominating set, every vertex in $V - D_{-1}$ is adjacent to some vertex in D_{-1}). This implies that D_0 is a dominating set such that $|D_0| = m$. Now, $N(u_0) \cap D_0 = \varnothing$, since $u_0 \in N_0$ and u_0 is adjacent to only w_0 in D_{-1}. Hence

$$0 \leq a(D_0) \leq \binom{m-1}{2}.$$

If $a(D_0) = 0$, then

$$i(G) \leq m = \gamma(G)$$

as before. If $a(D_0) > 0$, we can repeat the process used to obtain D_0 to obtain, without loss of generality, a dominating set

$$D_1 = \left\{ u_0, u_1, u_2, \ldots, u_{m-1} \right\}$$

such that

$$N\left(u_i\right) \cap D_1 = \varnothing$$

for i = 0, 1. We then have

$$0 \leq a\left(D_1\right) \leq \binom{m-2}{2}.$$

Again if $a\left(D_1\right) = 0$, then we are done and if $a\left(D_1\right) > 0$ then we repeat the process used to obtain D_0 and then D_1.

Clearly the repetitions of the process must terminate in at most m-1 steps with a dominating set

$$D_k, \; -1 \leq k \leq m-2,$$

such that

$$\left|D_k\right| = m = \gamma\left(G\right) \text{ and } a\left(D_k\right) = 0.$$

Hence D_k is an independent dominating set, which by proposition 2 implies that D_k is a maximal independent set. Hence,

$$i\left(G\right) \leq \left|D_k\right| = m = \gamma\left(G\right)$$

and this completes the proof.

Since

$$\gamma\left(K_{1,3}\right) = 1 = i\left(K_{1,3}\right),$$

we see that the hypothesis of the proposition is not a necessary condition. However, we do have the following:

Corollary 8.1

For any graph G,

$$\gamma\big(L\big(G\big)\big) = i\big(L\big(G\big)\big),$$

where L(G) denotes the line graph.

Proof

This follows from Proposition 8, since L(G) does not have an induced subgraph isomorphic to $K_{1,3}$.

DOMATIC NUMBER OF A GRAPH

A D-partition of G is a partition of $V\big(G\big)$ into dominating sets. The domatic number $d\big(G\big)$ of G is the maximum order of a D-partition of G.

The indomatic number $id\big(G\big)$ of G is the maximum order of a partition over $V\big(G\big)$ into sets which are both independent and dominating. (We note that such a partition may not exist, in which case $id\big(G\big) = 0$).

ELEMENTARY PROPERTIES OF DOMATIC NUMBER

Clearly, a D-partition is a partition of the vertices of G into n subsets such that every vertex is adjacent to at least one vertex in every subset other that its own. One way of obtaining a D-partition of a graph is to assign colors to the vertices of G in such a way that every vertex is adjacent to a vertex of every color different from its own.

Thus, if a graph has domatic number k, then it is evident that every vertex must be adjacent to at least k-1 vertices, one in each dominating subset of a D-partition of order k. Hence, the following upper bound for the domatic number is obtained.

Proposition 9

For any graph G (Cockayne & Hedetniemi, 1978, 1977):

$$d\big(G\big) \le \delta\big(G\big) + 1$$

A graph G is (domatically) full if

$$d\big(G\big) = \delta\big(G\big) + 1$$

The following proposition determines $d\big(G\big)$ for special classes of graphs. The proofs of these results are simple and are omitted. Asterisks indicate that every graph in the class is full.

Proposition 10 (Cockayne & Hedetniemi, 1978, 1977)

$$d\left(K_{n}+G\right)=n+d\left(G\right)$$

1. $d\left(K_{n}\right)=n$; $d\left(\overline{K_{n}}\right)=1$
2. $d\left(G\right)\geq 2$ if and only if has no isolated (Ore, 1982)
3. *For any tree with $p\geq 2$ vertices, $d\left(T\right)=2$.
4. For any $n\geq 1$, * $d\left(C_{3n}\right)=3$, and $d\left(C_{3n+1}\right)=d\left(C_{3n+2}\right)=2$
5. For any $2\leq m\leq n, d\left(K_{m,n}\right)=m$.
6. If G is uniquely n-colorable, then $d\left(G\right)\geq n.$
7. *If G is maximal outer planar then $d\left(G\right)=3.$

Let F_{6} be the graph obtained from K_{6} with a 1-factor removed. Then F_{6} is a maximal planar graph. But $\delta\left(F_{6}\right)=3\,and\,d(F_{6})=3$, hence is not full. Thus, maximal planar graphs are not all full. But maximal outer planar graphs are full.

We now relate $d\left(G\right)$ and $d\left(\overline{G}\right).$

Proposition 11

For any graph having p vertices (Cockayne & Hedetniemi, 1978, 1977):

$$d\left(G\right)+d\left(\overline{G}\right)\leq p+1.$$

Proof: By proposition 9,

$$d\left(G\right)\leq\delta\left(G\right)+1$$

and

$$d\left(\overline{G}\right)\leq\delta\left(G\right)+1\leq\Delta\left(\overline{G}\right)+1.$$

Therefore,

$$d\left(G\right)+d\left(\overline{G}\right)\leq\delta\left(G\right)+\Delta\left(\overline{G}\right)+2=\left(p-1\right)+2=p+1.$$

The next result sharpens Proposition 11.

Proposition 12

For a graph G of order p,

$$d(G) + d(\bar{G}) = p + 1$$

if, and only if, $G = K_p$ or $\overline{k_p}$.

Proof

Clearly,

$$d(K_p) + d(\overline{K_p}) = p + 1.$$

Conversely, suppose $G \neq K_p$ or $\overline{k_p}$, and

$$d(G) + d(\bar{G}) = p + 1.$$

Case 1

If $\delta(G) = 0$, then $d(G) = 1$, G has an isolated vertex and $\bar{G}$ has a vertex of degree $p - 1$. Hence $\bar{G} = K_1 + F$, where F has p-1 vertices and $F \neq K_{p-1}$. Since $d(F) < p - 1$.

$$d(G) + d(\bar{G}) = d(G) + 1 + d(F)$$

by proposition 10 (i)

$$= 2 + d(F)$$

$$< 2 + p - 1 = p + 1,$$

a contraction

Case 2

$$0 < \delta(G) < \frac{p}{2}.$$

By assumption,

$$d(G) + d(\bar{G}) = p + 1$$

Therefore,

$$d(\bar{G}) = p + 1 - d(G)$$

$$\geq (p+1) - (\delta(G) + 1)$$

$$= p - \delta(G)$$

If all dominating sets in a maximal D- partition of $\bar{G}$ have at least two vertices then

$$p \geq 2d(\bar{G}) \geq 2(p - \delta(G))$$

using (2)

$$> 2p - p = p,$$

a contradiction.

Hence, some vertex v dominates $\bar{G}$.

Therefore, $deg_{\bar{G}}^{v} = p - 1$ and $deg_{G}^{v} = 0$

Hence, $\delta(G) = 0$ Contrary to (1)

Case 3

$$\frac{p}{2} \leq \delta(G) < p - 1 \tag{3}$$

In this case, since for any graph G of p vertices

$$\delta(G) + \delta(\bar{G}) \leq p - 1$$

$$\delta(\bar{G}) \leq p - 1 - \delta(G) \leq p - 1 - \frac{p}{2}$$

$$= \frac{p}{2} - 1$$

If $\delta(\bar{G}) = 0$, apply case 1 to $\bar{G}$, otherwise apply case 2 to $\bar{G}$. This completes the proof.

Proposition 13

For any graph G of order p (Jaeger & Payen, 1972):

$$\min\{3, p+1\} \leq \gamma(G) + \gamma(\bar{G}) \leq p + 1$$

$$\min\{2, p\} \leq \gamma(G).\gamma(\bar{G}) \leq p$$

Proposition 14

For any graph G, (Cockayne & Hedetniemi, 1978, 1977):

$$\gamma(G) \leq d(\bar{G})$$

Proposition 15

For any graph G of order $p, \gamma(G).\gamma(G) = p$ if and only if, the following two conditions hold simultaneously (Sampathkumar & Walikar, 1979):

$$\gamma(G).d(G) = p$$

$$\gamma(\bar{G}) = d(G)$$

Proof: If the conditions hold, then we have

$$\gamma(G).\gamma(\bar{G}) = \gamma(G).d(G) = P.$$

To prove the converse, assume that $\gamma(G).\gamma(\bar{G}) = P$.

Then by proposition 14, we get

$$\gamma\bar{G} = \frac{P}{\gamma(G)} \leq d(G), \text{ so that}$$

$$\gamma(G).d(G) \geq p.$$

To establish that inequality, let P = { $v_1, v_2, \ldots, v_d$ } be a domatic partition of G such that $d = d(G)$. Then

(A) P = $\displaystyle\sum_{i=1}^{d} |V_I| \geq \gamma(G)d(G)$

Thus, (i) follows. By (A),

$$\gamma(G) \leq \frac{P}{d(G)}$$

Since

$$\gamma(G).\gamma(\overline{G}) = P$$

by hypothesis, this implies

$$\frac{P}{d(\overline{G})} = \gamma(G) \leq \frac{P}{d(G)},$$

so that $\gamma(\overline{G}) \geq d(G)$. But, $\gamma(\overline{G}) \leq d(G)$. Hence $\gamma(\overline{G}) = d(G)$.

DOMINATION RELATED PARAMETERS

Let $G = (V, E)$ be a graph. Two vertices u, v in G dominate each other if u and v are adjacent. Thus, a set $D \subset V$ is a domination set if every vertex in $V - D$ is dominated by some vertex in D.

A set $D \subset V$ is a total dominating set if every vertex in V is dominated by some vertex in D. Thus, a dominating set D is a total dominating set if and only if the subgraph $\langle D \rangle$ induced by D has no isolates. The total domination number γ_t of G is the minimum cardinality of a total dominating set. (See Allan, Laskar & Hedetniemi, 1984, Cockayne, Dawqes & Hedetniemi, 1980).

Note that this parameter is defined only for graphs without isolated vertices.

Let G be graph without isolates. Then the complement $V - D$ of a minimal set D is itself a dominating set. Let $d \in D$. If a is not adjacent to a vertex in $V - D$, then $D - \{d\}$ would be a dominating set of G, a contradiction to the minimality of D. This implies, $V - D$ is dominating set, and $\gamma(G) \leq p/2$, where p is the order of G. This bound is too small for the total domination number. In fact, $\gamma_t(G) = |V|$, for $G = mK_2$. It is easy to see that mK_2 is the only graph having this property. The paths P_3 and P_6 are examples of connected graphs for which $\gamma_t(G) = 2p / 3$. The following result shows that this is a best possible bound for γ_t.

Proposition 1

If G is a connected graph with $p \geq 3$ vertices, then $\gamma_t(G) = 2p / 3$.

Proof: See (Cockayne, Dawqes & Hedetniemi, 1980).

Proposition 2

1.   1. If G has p vertices and no isolates then

$$\gamma_t(G) \leq p - \Delta(G) + 1.$$

2.   If G is connected and

$$\Delta(G) \leq p - 1,$$

then

$$\gamma_t(G) \leq p - \Delta(G).$$

Proof. (i) Let vertex v have degree $\Delta(G)$ and let

$$X = V - \left(\{v\} \cup N(v)\right).$$

If

$$X = \varnothing, \Delta(G) = p - 1$$

and

$$\gamma_t = 2 = p - \Delta + 1.$$

Now suppose $X \neq \varnothing$ let S be the set of isolates of $\langle X \rangle$. If $S = \varnothing$ then $X \cup \{u\}$, where $u \in N(v)$ is a total dominating set of cardinality $p - \Delta(G) + 1$. If $S \neq \varnothing$, by hypothesis each $s \in S$ is adjacent to some vertex in $N(v)$ and we define $M(S)$ to be a set of smallest cardinality contained in $N(v)$ such that each $s \in S$ is adjacent to an element of $M(S)$, and we note that $|M(S)| \le |S|$. Then

$$\{v\} \cup M(S) \cup X - S$$

is total dominating set and

$$\gamma_t(G) \le 1 + |S| + |X - S| = 1 + |X| = p - \Delta(G).$$

This completes the proof of (i).

(ii) Let $v, X \ and \ S$ be defined as in the proof of (i) since

$$\Delta(G) < p - 1, X \ne \varnothing.$$

If $S = \varnothing$, by connectivity, some $x \in X$ is adjacent to some $y \in N(v)$. Let C, Δ' be the vertex set and maximum degree of the complement of $\langle X \rangle$ which contains y. By (i), $\langle C \rangle$ has a total dominating set Y of cardinality atmost $|C| - \Delta' + 1$. If $\Delta' = 1$ then $|C| = 2$ and the set

$$\langle v, y, x \rangle \cup (X - C)$$

is the total dominating set in G and

$$\gamma_t(G) \le 3 + (p - \Delta(G) - 1) - 2 = p - \Delta(G).$$

If $\Delta' > 1$, then the set

$$\langle v, y \rangle \cup Y \cup (X - C)$$

is total dominating in G. Hence,

$$\gamma_t \le 2 + (|C| - \Delta' + 1) + p - \Delta(G) - 1 - |C|$$

$$= p - \Delta(G) + (2 - \Delta') \le p - \Delta(G).$$

The remaining case $(S \ne \varnothing)$ is proved in a manner identical to that used in the proof of (i).\

Proposition 3

If G has p vertices, no isolates, and $\Delta(G) < p - 1$, then $\gamma_t + \overline{\gamma_t} \le p + 2$, with equality if, and only if, G or $\overline{G} = mk_2$.

Proof. See (Cockayne, Dawqes & Hedetniemi, 1980).

Problem: Prove: (i) $\gamma \le \gamma_t \le 2\gamma$

and (ii) if $\gamma_t = 2\gamma$, then $\gamma = 1$.

(iii) $\gamma \le 1 \le \beta_0.$

(iv) $\gamma \le \alpha_0$

Proposition 4

For any graph without isolates (i) $\gamma + 1 \le p$ (ii) $\gamma + \gamma_t \le p$ (iii) $1 + \gamma_t \le p$

Proof. See "A note on total domination" (Allan, Laskar & Hedetniemi, 1984)

CONNECTED DOMINATION

A dominating set $D \subset V$ is a connected dominating set if the subgraph $\langle D \rangle$ induced by D is connected. The connected domination number $\gamma_c(G)$ of G is the minimum cardinality of such a set. (See Hedetniemi & Laskar, 1984; Sampathkumar & Walikar, 1979).

One can easily establish the following:

1. $\gamma_c(K_p) = 1$
2. $\gamma_c(K_p + G) = 1$ for any graph G.

$$\gamma_c(K_{m,n}) = \begin{cases} 1 \text{ if either } m \text{ or } n = 1 \\ \quad 0 \text{ otherwise} \end{cases}$$

$$\gamma_c(C_p) = p - 2$$

3. For any tree T of order p, $\gamma_c = p - e$ where e is the number of pendant vertices in T. Clearly, for any connected graph $G, \gamma \le \gamma_c.$ one can easily establish the following.

Proposition 5

Let G be any graph and H a spanning subgraph of G. Then every dominating set of H is also a dominating set of G, and hence

$$\gamma\big(G\big) \le \gamma\big(H\big).$$

Corollary 5.1

Let G be a connected graph and H a connected spanning subgraph of G. then every connected dominating set of H is also a connected dominating set of G, and hence

$$\gamma_c\big(G\big) \le \gamma_c\big(H\big).$$

Proposition 6

For any connected graph of order

$$p \ge 3,\; \gamma_c\big(G\big) \le p - 2$$

and the bound is sharp.

    Proof. Let T be a spanning tree of G. Then

$$\gamma_c\big(G\big) \le \gamma_c\big(T\big) = p - e\,(\mathbf{B})$$

Where e is the number of pendant vertices in T. Since $e \ge 2$, the result is follows:

Proposition 7

For any connected $\big(p, q\big) -$ graph G with maximum degree Δ

$$\frac{p}{\Delta + 1} \le \gamma_c\big(G\big) \le 2q - p.$$

    Proof. The lower bound is obvious since it is also a lower bound for γ.
Now

$$\gamma_c \le p - 2 = 2\big(p - 1\big) \le 2q - p.$$

since G is connected.
    We now show that $\gamma_c = 2q - p$ iff G is a path. If G is a path, then

$$\gamma_c = p - 2 = 2\big(p - 1\big) - p = 2q - p.$$

Conversely, let $\gamma_c = 2q - p$. Then by (B), $2q - p \le p - 2$, which implies $q \le p - 1$. Since G is connected, $q = p - 1$, and hence G must be a tree. So, $\gamma_c = p - e$. If $e > 2$, then

$$\gamma_c \left(G \right) = p - e = 2q - p$$

as above, a contradiction. Thus $e = 2$, and G is a path.

Let $\epsilon_T \left(\epsilon_T \right)$ denote the maximum number of pendant edges in any spanning forest (tree) of G.

Proposition 8

For any connected graph G with p vertices $\gamma + \epsilon_T = p$.

Proposition 9

For any connected graph with p vertices $\gamma_c + \epsilon_T = p$. (Hedetniemi & Laskar, 1984)

Proof. Let T be any spanning tree of G, with ϵ_T end vertices and let F denote the set of end vertices. Then $T - F$ is connected dominating set with $p - \epsilon_T$ vertices.

Thus, $\gamma_c \le p - \epsilon_T$.

In order to prove $\gamma_c \ge p - \epsilon_T$, let D be a connected dominating set with γ_c vertices. Since $\left\langle D \right\rangle$ is a connected subgraph of G, let T_D be any spanning tree of $\left\langle D \right\rangle$. we can now form a spanning tree T of G by adding the remaining $p - \gamma_c$ vertices of G to T_D, joining each of these vertices to one vertex of D to which it is adjacent. In this way, T will have at least $p - \gamma_c$ vertices, i.e. $\epsilon_T \ge p - \gamma_c$ or $\gamma_c \ge p - \epsilon_T$.

Proposition 10

$\gamma_c \le p - \Delta$ for any connected graph of order p.

Proof. Let deg $v = \Delta$ in G. then a spanning tree T of G can be formed in which v is adjacent to each of its neighbours. i.e. T has a vertex v of degree Δ and hence has at least Δ end vertices. Thus, by Proposition 9, $\gamma_c \le p - \Delta$.

Proposition 11

For any tree $T, \gamma_c = p - \Delta$ if, and only if, T is a spider, (i.e. a tree having at most one vertex opf degree not less than 3).

Proposition 12

For any connected graph G, diam $G - 1 < \gamma_c$.

Proof

Let diam $G = k$, and let $d(u,v) = k$ for $u, v \in V(G)$. Let D be a connecting dominating set and consider whether u and/or v are in D.

Case 1

$u, v \in D$. Then since $\langle D \rangle$ is a connected subgraph, $|D| \geq k+1$.

Case 2

$u \in D, v \notin D$. Then since u must be adjacent to at least one vertex in D. It follows $|D| \geq k$.

Case 3

$u \notin D, v \notin D$. In this case, it is easy to see that $|D| \geq k-1$.

Let $\beta_1(G)$ be the edge independence number (matching number) of G. It is easy to see that for any connected graph $G, \gamma(G) \leq 2\beta_1$ i.e. the set of all vertices contained in any maximum matching forms a dominating set of G. This result can be strengthened, however, by selecxting a maximum matching the vertices of which form a connected subgraph.

Proposition 13

Every connected graph G contains at least one β_1- set M such that $\langle V(M) \rangle$ is connected sungraph.

Proof. Let M be a maximum matching of G that maximizes the cardinality of a largest connected component C of $\langle V(M) \rangle$. Suppose $\langle V(M) \rangle$ is not connected. If T is a spanning tree of G there exists another connected component C' of $\langle V(M) \rangle$, which is connected to C by a pair of $T-$edges, uv and uv' with $v \in C, v' \in C'$. We have $u \notin \langle V(M) \rangle$. Replacing in M, the edge containing v' by the edge uv', we obtain a matching $M', < C \cup \langle u, v' \rangle$ is connected, hence $\langle V(M') \rangle$ has a connected component larger than C; contradiction with our choice of M and C.

Corollary 13.1

For any connected graph , $\gamma_c \leq 2\beta_1$.

Proposition 14

For any graph G such that both G and $\overline{G}$ are connected $\gamma_c + \overline{\gamma_c} \leq p+1$.

Proof. From proposition 10, $\gamma_c \leq p - \Delta$, and $\overline{\gamma_c} \leq p - \overline{\Delta}$

Therefore

$$\gamma_c + \overline{\gamma_c} \le \left(p - \Delta\right) + \left(p - \overline{\Delta}\right)$$

$$= 2p - \left(\Delta + \overline{\Delta}\right)$$

$$= 2p - \left(\Delta + p - 1 - \overline{\Delta}\right)$$

$$= p + 1 + 6 - \Delta \le p + 1, \text{ since } 6 - \Delta \le 0.$$

Note that for

$$C_5, \gamma_c + \overline{\gamma_c} = 3 + 3 = p + 1.$$

Thus, the upper bound is sharp.

GLOBAL DOMINATION

A dominating set D of a graph $G = (V, E)$ is said to be a global dominating set if it dominates both G and its complement $\bar{G}$. The minimum cardinality of a global dominating set is called the global domination number $\gamma_g \left(G\right)$.

This concept was introduced and studied by Sampathkumar (1989) and Brigham and Dutton (1990).

STRONG AND WEAK DOMINATION

Let G = (V, E) be a graph and u,v ∈ V. Then *u strongly dominates v* and *v weakly dominates u* if (i) *uv* ∈ E and (ii) *deg(u)* ≥ *deg(v)*. A set $S \subseteq V$ is a *strong dominating set* of G if every vertex in $V - S$ is strongly dominated by at least one vertex in S. The *strong domination number* γ_s of G is the minimum cardinality of such a set. A *weak dominating set* and the *weak domination number* γ_w are defined similarly. The concept was introduced by Sampathkumar and Pushpalatha (1996).

The following proposition presents properties of vertices in a minimal strong (weak) dominating set.

Proposition 16

Let S be a minimal strong (weak) dominating set of a graph $G = (V, E)$. Then for each $v \in S$, one of the following holds:

1. No vertex in S strongly (weakly) dominates v, or
2. There exists a vertex $u \in V - S$ such that v is the only vertex in S which strongly (weakly) dominates u.

The related concept of independent strong (weak) domination has also been defined. The independent strong domination number γ_{is} of G is the minimum cardinality of a strong dominating set which is also independent. The independent weak domination number γ_{iw} of G is similarly defined.

A result similar to that of Allan and Laskar regarding $K_{1,3}$–free graphs holds in the case of the strong and weak domination numbers also.

Proposition 17

If a graph does not contain $K_{1,3}$ as an induced subgraph then,

1. $\gamma_s = \gamma_{is}$ and
2. $\gamma_w = \gamma_{iw}$.

Several generalizations of the domination parameter have been studied. Set domination and Point-set domination represent a few such. . Global aspects of these parameters have also been studied.

REFERENCES

Allan, R. B., Laskar, R., & Hedetniemi, S. T. (1984). A note on total domination. *Discrete Mathematics*, *49*(1), 7–13. doi:10.1016/0012-365X(84)90145-6

Berge, C. (1973). *Graphs and Hypergraphs*. Amsterdam: North Holland.

Billera, L. J. (1971). On the composition and decomposition of clutters. *Journal of Combinatorial Theory*, *11B*, 243–245.

Brigham, R. C., & Dutton, R. D. (1990). Factor domination in graphs. *Discrete Mathematics*, *86*(1-3), 127–136. doi:10.1016/0012-365X(90)90355-L

Cockayne, E. J., Dawqes, R. M., & Hedetniemi, S. T. (1980). Total domination in graphs. *Networks*, *10*(3), 211–219. doi:10.1002/net.3230100304

Cockayne, E. J., & Hedetniemi, S. T. (1974). Independence graphs. *Proc. Fifth. S.E. conf. on Combinotories, Graph Theory and Computing, Utilities Mathematics*, 471-479.

Edmonds, J., & Fulkerson, D. R. (1970). Bottleneck extrema. *J. Comb. Theory*, *8*(3), 299–306. doi:10.1016/S0021-9800(70)80083-7

Hedetniemi, S. T., & Laskar, R. (1984). *Connected domination in graphs*. In B. Bollobas (Ed.), *Graph theory and Combinatories* (pp. 209–218). London: Academic Press.

Jaeger, F., & Payen, C. (1972). Relations du type Nordhaus Gaddum pour le nombre d' absorption d'un graphs simple. *C.R. Acad. Sci. Paris*, *274*, 728–730.

Kalbfleisch, J. G., Stanton, R. G., & Horton, J. D. (1971). On covering sets and error-correcting codes. *Journal of Combinatorial Theory Series A*, *11*(3), 233–250. doi:10.1016/0097-3165(71)90051-3

Liu, C. L. (1968). *Introduction to combinatorial mathematics*. McGraw-Hill.

Ore, O. (1982). Theory of Graphs. *Amer. Math. Scc. Collog. Publ., 38*, 206-212.

Sampathkumar, E. (1989). The global domination number of a graph. *J. Math. Phys. Sci., 23*, 377–385.

Sampathkumar, E., & Pushpalatha, L. (1996). Strong, weak domination and domination balance in a graph. *Discrete Mathematics, 161*(1-3), 235–242. doi:10.1016/0012-365X(95)00231-K

Sampathkumar, E., & Walikar, H. B. (1979). The connected domination number of a graph. *J. Math. Phys. Sci, 13*(6).

Chapter 2
An Introduction to Intersection Graphs

Madhumangal Pal
Vidyasagar University, India

ABSTRACT

In this chapter, a very important class of graphs called intersection graph is introduced. Based on the geometrical representation, many different types of intersection graphs can be defined with interesting properties. Some of them—interval graphs, circular-arc graphs, permutation graphs, trapezoid graphs, chordal graphs, line graphs, disk graphs, string graphs—are presented here. A brief introduction of each of these intersection graphs along with some basic properties and algorithmic status are investigated.

INTRODUCTION

The graphs are very useful tool to model a huge number of real life problems starting from science, technology, medical science, social science and many other areas. The geometrical and topological stuctures of any communication system such as Internet, Facebook, Whatsapp, ResearchGate, Twiter, etc. are based on graph. So graph theory is an old as well as young topic of research as till today graphs are used to solve several problems. Depending on the geometrical structures and properties different type of graphs are defined, viz. path, cycle, tree, complete graph, planar graph, perfect graph, chordal graph, tolarence graph, intersection graph, etc.

In this chapter, diferent types of intersection graphs (IntGs) are defined and investigated their properties.

Let $S = \{S_1, S_2, \ldots\}$ be a finite or infinite set of sets. For each set S_i we consider a vertex (v_i) and there is an edge between the vertices v_i and v_j if the corresponding sets $S_i \cap S_j$ have a non-empty intersection. That is, the set of vertices $V = \{v_1, v_2, \ldots\}$ and the set of edges

$$E = \{(v_i, v_j) : S_i \cap S_j \neq \varphi\}.$$

The resultant graph is called intersection graph.

DOI: 10.4018/978-1-5225-9380-5.ch002

Table 1. Table of abbreviations

Abbreviation	Description
IntG	Intersection graph
InvG	Interval graph
CirG	Circular-arc graph
PerG	Permutation graph
TraG	Trapezoidal graph
TolG	Tolerance graph
PCA	Proper circular-arc
UCA	Unit circular arc
GIG	Grid intersection graphs

One important class of graph is perfect graph defined as follows:

A graph (undirected) $G = (V, E)$ is called χ-perfect if for all

$$A \subseteq V, \omega(G(A)) = \chi(G(A)),$$

and G is called α-perfect if for all

$$A \subseteq V, \alpha(G(A)) = \kappa(G(A)),$$

where $G(A)$ is the induced subgraph by the subset A. A graph is said to be *perfect* if it is either χ-perfect or α-perfect. In the famous *perfect graph theorem* is it proved that a graph is χ-perfect if and only if it is α-perfect.

An undirected graph G is said to be *triangulated* if every cycle of length four or more has a chord. This type of graphs are also known as *chordal graphs, monotone transitive graphs, rigid-circuit graphs* and *perfect elimination graphs.*

A clique of a graph $G = (V, E)$ is a set of vertices $S \subseteq V$ in which all the vertices of S are adjacent, i.e. $(v_i, v_j) \in E$ for all $v_i, v_j \in V$. A *maximal clique* is a special type of clique for which no further vertex can be added so that it remains a clique. That is, if C is a maximal clique of the graph $G = (V, E)$, then $C \cup \{v\}$ is not a clique for $v \notin C$ and any $v \in V$. A clique is said to be *maximum* if the cardinality is maximum among all others cliques of the graph.

The *clique graph* $C(G)$ of a graph G is the IntG of the family of all cliques of G. *Cographs* are defined as the graphs which can be reduced to single vertices by recursively complementing all connected subgraphs. This graphs are known as complement reducible graphs.

Let $N[v](N(v))$ be the closed (open) neighbour of the vertex v. A vertex v is called *simplicial* if and only if $N[v]$ is a clique. The ordering $(v_1, v_2, ..., v_n)$ of V is called a perfect elimination ordering if and only if for all $i \in \{1, 2, ..., n\}$ the vertex v_i is simplicial in G_i, where

$$G_i = G(\{v_i, v_{i+1}, ..., v_n\}).$$

The graph G is called *chordal* if and only if G possesses a perfect elimination ordering. The simplicial ordering $(v_1, v_2, ..., v_n)$ is said to be a strong elimination ordering if and only if for all

$$i \in \{1, 2, ..., n\}, \ N_i[v_j] \subseteq N_i[v_k]$$

when $v_j, v_k \in N_i[v_i]$ and $i < k$. The graph G is called *strongly chordal* if and only if G possesses a strong elimination ordering. A vertex $u \in N[v]$ is said to be a maximum neighbourhood of v if and only if for all $w \in N[v]$ $N[w] \subseteq N[u]$ holds. Here u and v may be same. The simplicial ordering $(v_1, v_2, ..., v_n)$ is said to be a maximum neighbourhood ordering if for all $i \in \{1, 2, ..., n\}$ there is a maximum neighbour $u_i \in N_i[v_i]$, for all

$$w \in N_i[v_i], N_i[w] \subseteq N_i[u_i]$$

holds. The graph G is said to be *dually chordal* if and only if G has a maximum neighbourhood ordering. The graph G is called *doubly chordal* if and only if G is both chordal and dually chordal.

A set $A \subseteq V$ is said to be an *asteroidal set* if for every $u \in A$, $A - \{u\}$ is contained in one component of $G - N[u]$. An asteroidal set with three vertices is said to be an *asteroidal triple (AT)*. A graph is said to be AT-free if it has no asteroidal triple. It cqan be shown that every asteroidal set is an independent set. A triple $\{u, v, w\}$, $u, v, w \in V$ is an asteroidal triple if and only if for every two of these three vertices there is a path between them avoiding the closed neighbourhood of the third.

A graph is called a *comparability graph* if its edges can be given a transitive orientation. A *cocomparability* graph is the complement of a comparability graph.

With these terminologies one can define the intersection graphs.

INTERSECTION GRAPHS (INTGS)

Due to the rich theories and wide applications of IntGs, along with others, the IntGs have a seperate mathematics classification number 05C62. A huge number of properties of IntGs are available in (McKee & McMorris, 1999). The IntG is defined as follows:

A graph $G = (V, E)$ is called an *IntG* for a finite family $S = \{S_1, S_2, ...\}$ of a non-empty set if there is a one-to-one correspondence between S and V such that two sets in S have at least one common element if and only if their corresponding vertices in V are connected by an edge. The family of sets S is called an intersection model of G. For an intersection model S, $G(S)$ to represents the intersection graph for S.

Depending on the geometric representation or mathematical nature of the sets $S_1, S_2, ...$ different types of IntGs are defined below:

- Interval graphs (when S is the set of intervals on a real line)
- Circular-arc graphs (when S is the set of arcs on a circle)
- Permutation graphs (when S is the set of line segments between two parallem line segments)
- Trapezoid graphs (when S is the set of trapeziums between two line segments)
- Disk graphs (when S is the set of circles on a plane)
- Circle graphs (when S s the set of chords within a (unit) circle)
- Chordal graphs (when S is the set of connected subgraphs of a tree)
- String graphs (when S is the set of curves in a plane)
- Graphs with boxicity k (when S is the set of boxes of dimension k)
- Line graphs (when S is the set of edges of a graph).

It can be easily verified that all the above mentioned graphs are IntGs.

INTERVAL GRAPHS (INVG)

Let $I = \left\{ I_1, I_2, \ldots, I_n \right\}$ be a set of intervals on the real line.

An undirected graph $G = (V, E)$ is called an *interval graph* (InvG) if for each interval I_i there is a vertex v_i of G and two vertices v_i and v_j are adjacent in G if and only if their corresponding intervals I_i and I_j have non-empty intersection. This means there is a bijection from V to I.

The set of intervals I is called an interval representation of the graph G and conversely G is said to be the InvG of the intervals I (Gulumbic, 1980).

InvGs occur in the process of modelling many real life problems, particularly time dependent situations or other restrictions that are linear in nature. InvG and its subclasses occur in many daily life problems in the areas of genetics, archeology, psychology, sociology, molecular biology, circuit routing, traffic planning, scheduling, VLSI design, etc.

InvG is a very interesting and useful graph class of IntGs. This graph has numerous applications and has rich mathematical theories.

The InvG obey the *hereditary property*.

Lemma 2.1. (Gulumbic, 1980) *"An induced subgraph of an interval graph is an interval graph."*

The following useful property is also satisfied for InvG.

"Every simple cycle of length strictly greater than 3 possesses a chord."

This property is known as tringulated graph property.

The following property is called the transitive orientation property of graph, which is also satisfied for an InvG.

Let G be a graph. Each edge of G can be assigned a one-way direction in such a way that the oriented graph (V, E) satisfies the following condition:

$$(u, v) \in E \, and \, (v, w) \in E \Rightarrow (u, w) \in E, u, v, w \in V.$$

Lemma 2.3. (Ghouilo-Houri, 1962) *"The complement of an interval graph satisfies the transitive orientation property."*

Gilmore and Hoffman proved the following theorem and shown that InvG is a perfect graph.

Theorem 2.1. (Gilmore & Hoffman, 1964) *"Let G be an undirected graph. The following statements are equivalent.*

1.   *G is an IG.*
2.   *G contains no chordless cycle of length 4 and its complement $\overline{G}$ is a comparability graph.*
3.   *The maximal cliques of G can be linearly ordered such that, for every vertex u of G, the maximal cliques containing u occur consecutively."*

The statement (c) is a power result. From this result we can construct a matrix whose rows are vertices and columns are maximal cliques or vice-versa. If the vertex v_i belongs to the clique C_j then the entry of the matrix is 1 otherwise 0. So the entries of this matrix are either zero and one. This incidence matrix is called *clique matrix*.

For an InvG, this matrix has important properties called *consecutive 1's property for columns* if its rows can be permuted in such way that the 1's in each column occur consecutively.

From the clique matrix one can recognize the interval graph.

Theorem 2.2. (Fulkerson & Gross, 1965) *"An undirected graph G is an interval graph if and only if its clique matrix has the consecutive 1's property for columns."*

Theorem 2.3. (Lekkerkerker & Boland, 1962) *"An undirected graph G is an InvG if and only if the following two conditions are satisfied:*

1.   *G is a triangulated graph and*
2.   *Any three vertices of G can be ordered in such a way that every path from the first vertex to the third vertex passes through a neighbour of the second vertex."*

Following theorem is also useful to recognize an InvG, but computationally it is hard to recognize an InvG.

Theorem 2.4. (Lekkerkerker & Boland, 1962) *"A graph is an interval graph if and only if it contains none of the graphs shown in Figure 1 as an induced subgraph."*

Let $G = (V, E)$ be a simple connected InvG. For simplicty the vertices are numbered as 1, 2, ..., n, where n is the number of vertices and m is the number of edges of the graph. It is mentioned that the intersection model of InvG is a set of intervals on a real line. That is, for every vertex there is an interval. Suppose the interval representation of the InvG be $I = \{I_1, I_2, ..., I_n\}$, where $I_j = \left[a_j, b_j\right]$, and b_j are the left and right endpoints of the interval $I_j, j = 1, 2, ..., n$.

To design the algorithms for solving a problem modeled by InvG, following are the obvious assumption.

• The set of intervals are indexed by increasing values of the right endpoints, i.e. $b_1 < b_2 < \cdots < b_n$
• All the intervals are closed. That is, both of its endpoints are included with the interval and that no two intervals share a common endpoint.
• The vertices and the intervals are samething.

Figure 1. Forbidden structures for interval graphs

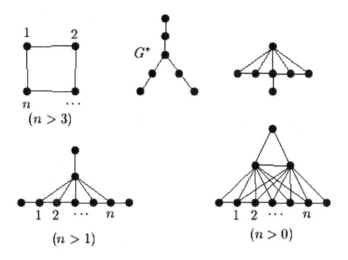

Figure 2. An interval graph and its interval representation.

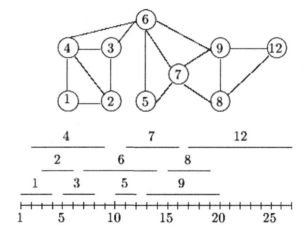

- The InvG is connected.
- The list of sorted endpoints is given.

The set intervals of any InvG can be ordered in a non-decreasing order as per right endpoints. This ordering is called as IG ordering.

For algorithmic purpose, it is assumed that the vertices of an InvG are labelled as IG ordering. The IG ordering is unique for the given set of intervals.

For the above assumption the following result is highly useful to design algoithms on interval graph.

Lemma 2.4. (Ramalingam & Pandu Rangan, 1988) "Let u, v, w be three vertices of an IG and $u < v < w$ in the IG ordering. If there is an edge between the vertices u and w, then there must be an edge between v and w."

This result is also known as unbrela properties.

For illustration, the InvG of Fig. 2 is considered.

29

A large number of algorithms are degined for InvGs using different techniques and data structures. Here, a data structure called *interval tree* (IT) is described which is defined in (Pal & Bhattacharjee, 1997a).

This data structure is used to solve several problems on InvG.

INTERVAL TREE

Let $H(v)$ and $L(v)$ be two functions, $v \in V$ and they are defined below:

$$H(v) = \max\{u : (u,v) \in E, u \geq v\},$$

and

$$L(v) = \min\{u : (u,v) \in E, u \leq v\}.$$

Note that that H is monotonic non-decreasing function, i.e. $H\left(u\right) \leq H\left(v\right)$ for $u < v$ where $u, v \in V$. Let $G' = \left(V, E'\right)$ be a spanning subgraph defined as

$$E' = \{(u,v) : u \in V \text{ and } v = H(u), u \neq n\}.$$

It can be verified that the subgraph G' is nathing but a spanning tree of G. This tree is said to be *interval tree (IT)* and let it be denoted by $T_I(G)$. Note that IT is a rooted tree with root node n. The IT exists for every connected InvG and it is unique for a giveninterval representation. The IT $T_I(G)$ of the InvG of Fig. 2 is shown in Fig. 3.

Let us defined the level for each vertex $u \in V$ in the IT as follows:

The lavel of the root node n is assigned to 0. The lavel of each node which are adjacent with the node n in the IT is 1 and so on. The level of the vertex u in IT is denoted by $level_I(u)$.

Let N_i be the set of vertices which are at a distance i in IT from the vertex n, i.e. N_i is the set of vertices which are at level i in IT, i.e.

$$N_i = \{u : \delta_G(u,n) = i\}$$

and N_0 is the singlaton set $\{n\}$. According to the definiton, if $u \in N_I$ then $level_I(u) = i$. Let k be the ecentricity of the vertex n. It can easily be vertified that the set N_k is non-empty while the set N_{k+1} is empty.

PROPERTIES OF INTERVAL TREE

Note that each set N_i is a finite set of integers. So it has both minimum and maximum elements.

Let $\max(N_i)$ and $\min(N_i)$ be the maximum and minimum vertices of the set N_i respectively. That is,

$$\max(N_i) = \max\{x : x \in N_i\}$$

and

$$\min(N_i) = \min\{x : x \in N_i\}.$$

Another observation for the set N_i is that the elements of N_i are consecutive integers and also the maximum value of the set N_{i+1} is one less than the minimum value of N_i, i.e.

$$\max(N_{i+1}) = \min(N_i) - 1$$

for all i.

It is obvious that if $level_I(u) < level_I(v)$ then $u > v$.

Figure 3. The IT for the graph of Fig. 2.

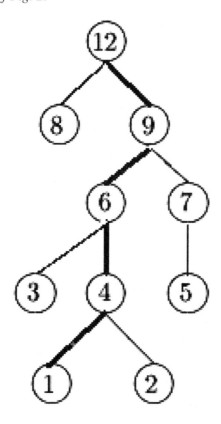

The maximum and minimum lavel of the IT are respectively the height of the IT $h(T_I(G))$ and 0. Thus

$$0 \leq level_I(u) \leq h(T_I(G)).$$

This minimum level 0 is assigned to the vertex n only. Note that $\delta_G(v,n)$ is maximum when $level_I(v) = h(T_I(G))$ and the maximum value of $\delta_G(v,n)$ is $h(T_I(G)) + 1$. As per definition of lavel, $level_I(1) = h(T_I(G))$ and $level_I(n) = 0$.

It may be noted that the path from the vertex 1 to the vertex n in the IT is the longest path among all other paths ending at n. This longest path is marked as *main path*. The main path of the graph of Fig. 2 is depicted in Fig. 3 by thick lines.

If two vertices have same level in IT then the distance between them in G is either 1 or 2 proved below.

Lemma 2.5. (Pal & Bhattacharjee, 1997b) *For any vertices* $x, y \in V$ *if* $level_I(x) = level_I(y)$ *then*

$$\delta_G(x,y) = \begin{cases} 1, & if\ (x,y) \in E(G) \\ 2, & otherwise. \end{cases}$$

But, the result is different for the IT. That is, if

$$level_I(x) = level_I(y), x, y \in V$$

then the distance between the vertices x and y, i.e. $\delta_{T_I(G)}(x,y)$ is not necessarily 1 or 2. It may be more than 3 for some cases. For example, for the IT of Fig. 3,

$$level_I(3) = level_I(5) = 3$$

and

$$\delta_{T_I(G)}(3,5) = 4.$$

One important observation about the connectivity between the vertices on IT is stated below:

Let the level of the vertex x be j and if it is adjacent to the vertex y in G then y must be either at level $j-1$ or j or $j+1$. Again, if $x, y \in V$ and

$$|\ level_I(x) - level_I(y)\ | > 1$$

then $(x,y) \notin E(G)$ (Pal & Bhattacharjee, 1997b).

By utilizing the properties of IT many of problems of InvG have been solved. Here, some of them are discussed.

APPLICATIONS OF INTERVAL TREE

Construction of Tree 3-Spanner

Spanning sub graph or spanning tree has many applications in planning purpose.

A spanning subgraph $K(G)$ is said to be a *t-spanner* of a graph G if the distance between every pair of vertices is at most t times their distance in G. That is, if the condition

$$\delta_H(x, y) \leq t\, \delta_G(x, y)$$

holds for all $x, y \in V$ then K is the spanning subgraph of G. The parameter t is called the *stretch factor*.

The *minimum t-spanner* problem is sated below:

For a fixed t and the graph G, find a t-spanner K with the fewest possible edges. The spanning subgraph K is called a minimum t-spanner of G and it is denoted by $K_t(G)$.

A *spanning tree* of a connected graph G is a cycle free connected spanning subgraph of G containing all vertices. A *tree spanner* of a graph G is a spanning tree that approximates the distance between the vertices in the original graph. In particular, a spanning tree T is said to be a *tree t-spanner* of a graph G if the distance between any two vertices in T is at most t times their distance in G, i.e.

$$\delta_T(x, y) \leq t\, \delta_G(x, y)$$

for all $x, y \in V$. It is obvious that if G is connected, then the number of edges of $K_t(G)$ is at least $n - 1$, the equality holds if and only if G has a tree t-spanner.

It can be verify that the IT for a given InvG may or may not be a tree 3-spanner.

Lemma 2.6. (Saha et al., 2005a) *"The IT of an InvG G may or may not be a tree 3-spanner."*

The tree 3-spanner, $T_{3S}(G)$, of G can be constructed from the interval tree by rearranging the parent vertex.

For a given interval representation of an InvG, the IT exists and is unique. Also, the tree 3-spanner of an InvG is obtained by rearranging the parents of the vertices of IT. Therefore, for every connected InvG there is a tree 3-spanner. Moreover, it is unique for a given interval representation.

Theorem 2.5. (Saha et al., 2005a) *A tree 3-spanner for an InvG can be determined using O(n) time, wher n is the number of vertices of the graph.*

Computation of Diameter

Let $G = (V, E)$ be a graph. The eccentricity of a vertex $x \in V$ is denoted by $ecen(x)$ and which is defined as

$$ecen(x) = \max_{y \in V}\{\delta_G(x,y)\}.$$

The radius ($\rho(G)$) and diameter ($diam(G)$) of the graph G are defined as

$$\rho(G) = \min_{x \in V}\{ecen(x)\}, \quad diam(G) = \max_{x \in V}\{ecen(x)\}.$$

In geometry, the radius is one half of the diameter. But, it is not true, in general, for a graph. It is true for InvG. There is a nice relation between the diameter of an InvG and the height of the corresponding IT, which is stated below.

Lemma 2.7. (Pal, 1995) *"Let $x_1^* \in N_1$ be the vertex on the main path of the IT at level 1. If all the vertices of level 1 of IT are adjacent to x_1^* in G, then*

$$diam(G) = h(T_I(G)), \text{ otherwise } diam(G) = h(T_I(G)) + 1. \text{ "}$$

Theorem 2.6. (Pal & Bhattacharjee, 1995c; Pal & Bhattacharjee, 1997a) *"The diameter, radius and centre of an* InvG *can be determined in linear time."*

All-Pairs Shortest Distances

From IT it follows that if the level of two vertics x and y are same then the shortest distance between them is either 1 or 2. But, when the vertices are in different levels then the shortest distance between them may be 1 or 2 or more. It is mentioned that two vertices x and y of IT are adjacent in G if there level difference is at most two. Keeping this result in mind we can find the distance between two vertices x and y, when there levels in IT are different by using the following method.

To find the distance between the vertices x and y, $x < y$, check the adjacency of the vertex y with the vertices at levels $level(y) + 1$, $level(y)$ and $level(y) - 1$. The distance $\delta_G(x,y)$ between the vertices x and y is determined by the following result.

Lemma 2.8. (Pal & Bhattacharjee, 1997b) *Let $x,y \in V$. Let u_1 be a vertex at level $level(u_1) + 1$ on the path marked $\min(x)$ and $u_2 = H(u_1)$. If $level(x) > level(y)$, then*

$$\delta_G(x,y) = \begin{cases} level(x) - level(y), & if \ (u_1, y) \in E \\ level(x) - level(y) + 1, & if \ (u_1, y) \notin E \ and \ (u_2, y) \in E \\ level(x) - level(y) + 2, & otherwise \end{cases}$$

Using this result one can calculate all-pairs shortest distances for an InvG. The time complexity of the algorithm to find all-pairs shortest distances using this method is stated below.

Theorem 2.7. (Pal & Bhattacharjee, 1997b) *"The all-pairs shortest distances of an* InvG *can be determined in $O\left(n^2\right)$, where n is the number of vertices."*

After introduction of InvG, lot of varieties of it are proposed by several scholars. For example, unit InvG, proper InvG, dotted InvG, tolerance graphs, etc. Some of them are discussed here.

DOTTED INTERVAL GRAPHS

Let

$$\{a, a+d, a+2d, \ldots, t\}$$

be an arithmetic progression, where a, t and d are positive inetegers and d is the common difference called it jump and t is the terminal term. This arithmetic progression is called *dotted interval* and it is denoted by $I(a, t, d)$. When $d = 1$ the dotted interval $I(a, t, d)$ becomes the conventional interval $[a, t]$ on the positive integer line. From the set of dotted intervals one can defined dotted interval graph. A *dotted interval graph* is an IntG of dotted intervals. The dotted interval graph is defined as follows:

Let $\{I_1, I_2, \ldots, I_k\}$ be a set of k dotted intervals. For each dotted interval $I_i, i = 1, 2, \ldots, k$ there is a vertex i and two vertices i and j are connected by an edge in G if $I_i \cap I_j \neq \varphi$. If the jumps of all the intervals are at most d, then the graph is called the d-dotted-interval (d-DIG). Let us consider five dotted intervals

$$I_a = I(1, 5, 2) = \{1, 3, 5\},$$

$$I_b = I(2, 3, 1) = \{2, 3\},$$

$$I_c = I(1, 7, 2) = \{1, 3, 5, 7\},$$

$$I_d = I(4, 6, 2) = \{4, 6\},$$

$$I_e = I(6, 8, 2) = \{6, 8\}.$$

For this case, the vertices of the graph are $\{a, b, c, d, e\}$. Since $I_a \cap I_b \neq \phi$, so there is an edge between the vertices a, b. But, $I_a \cap I_d = \phi$ and hence there is no edge between the vetices a and d. The corresponding 2-dotted interval graph is depicted in Fig. 4.

Aumann et al. (2005, 2012) introduced the dInvGs in the context of high throughput genotyping. They used dotted intervals to model microsatellite polymorphisms which are used in a genotyping technique called microsatellite genotyping. The respective genotyping problem translates to minimum coloring in d-DIGs for small d. They have shown that minimum coloring in d-DIGs is NP-hard even for $d = 2$.

Figure 4. A set of dotted intervals and corresponding 2-dotted interval graph

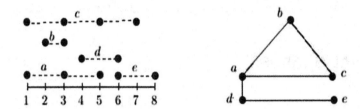

Theorem 2.15. (Aumann et al. 2005) *"Every graph with a countable number of vertices is a dotted interval graph."*

Theorem 2.16. (Aumann et al. 2005) *"For all $d \geq 1$, d-DIG $\nsubseteq (d+1)$-DIG."*

Theorem 2.17. (Hermelin et al., 2012) *"The minimum vertex cover and maximum independent set of any d-DIG graphs can be determined in $O(dn^d)$-time."*

The minimum dominating set of *d*-DIG graphs can be determined by using the approach which is used to solve maximum independent set problem (Hermelin et al., 2012).

Theorem 2.18. (Hermelin et al., 2012) *"For a d-DIG graph with n vertices, the minimum dominating set can be computed in $O(d^2 n^{O(d^2)})$ time."*

Definition 2.3. *A clique C in a d-DIG graph with dotted interval representation I is a point clique if there exists a point $p \in N$ which is included in every $I_v \in I$ with $v \in C$.*

It can be shown that, if *k* is the maximum clique size of an InvG, then its pathwidth is *k*-1. But, for a d-DIG graph with maximum point clique size k has pathwidth at most $dk - 1$.

An interesting corollary of the above lemma is stated below.

Theorem 2.19. (Aumann et al. 2005) *For any $d \geq 2$, the coloring problem for d-DIGs is NP-complete.*

It is easy to observed that when $d = 1$, *d-DIGs* are InvGs.

TOLERANCE GRAPHS

Let $I = \{I_u : u \in V\}$ be the sets of intervals and $t = \{t_u : u \in V\}$ the set of tolerances. The tolerances are positive real numbers. Suppose for each vertex $u \in V$, there is a closed interval I_v and a tolerance $t_v \in R^+$. A graph $G = (V, E)$ is called a tolerance graph (TolG) (Gulumbic et al., 1984) with vertex set *V* and there is an between the vertices *u* and *v* if and only if

$$| I_u \cap I_v | \geq \min\{t_u, t_v\}.$$

The collection $\langle I, t \rangle$ of intervals and tolerances is called a tolerance representation of the TolG *G*. If $t_v \leq | I_v |$ for all $v \in V$ of the TolG *G*, then G is called a bounded TolG and such representation is called a bounded tolerance representation.

For illustration let

$$I = \{I_a = [0,3], I_b = [1,4], I_c = [3,9], I_d = [8,11], I_e = [9,11], I_f = [4,8], I_g = [5,7]\}$$

be a set of intervals and

$$t = \{t_a = 1, t_b = 1, t_c = 4, t_d = 1, t_e = 1, t_f = 1, t_g = \infty\}$$

be a set of tolerances. The InvG for the interval representation

$$I = \{I_a = [0,3], I_b = [1,4], I_c = [3,9], I_d = [8,11], I_e = [9,11], I_f = [4,8], I_g = [5,7]\}$$

is drawn in Fig. 5.

But, if incorporate the tolerances along the intervals, i.e. $\langle I, t \rangle$ then the corresponding graph becomes TolG depicted in Fig. 6.

Not that, the graph of Fig. 6(b) is a TolG but not an InvG.

It can be easily verified that an induced subgraph of a TolG is a TolG and induced subgraph of a TolG and induced subgraph of a bounded TolG is also a bounded TolG. If $\langle I, t \rangle$ is a tolerance representation of a TolG G and I_v is a point, then the corresponding vertex v is an isolated vertex of G. Again, if $\langle I, t \rangle$ is bounded tolerance representation of G, then no interval I_v is a point.

Figure 5. Interval representation and its corresponding interval graph

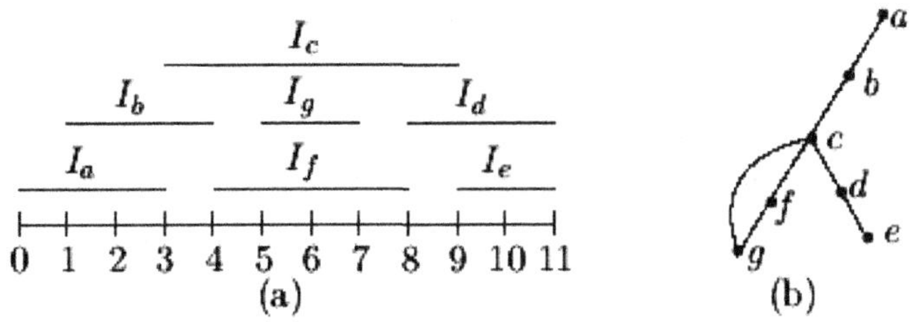

Figure 6. Tolerance representation and its corresponding tolerance graph

Note that there is no edge between the vertices c and g in tolerance graph (though $I_c \cap I_g \neq \varphi$) since $|I_c \cap I_g| = 2 \not\geq \{t_c, t_g\}$.

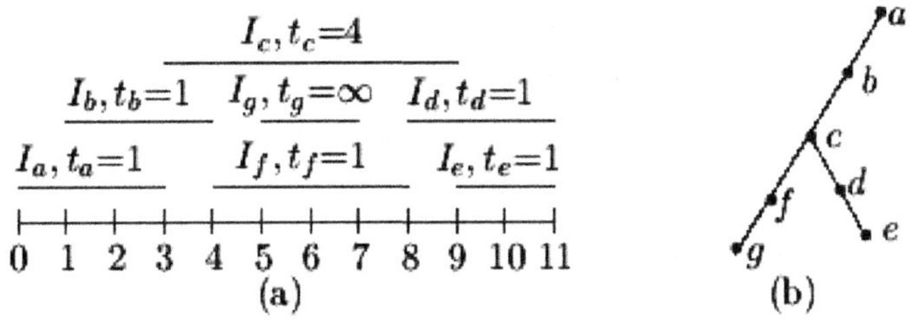

A tolerance representation is called regular (Gulumbic & Trenk, 2004) if

1. all tolerances are distinct.
2. any tolerance large than the length of its corresponding interval is set to infinity.
3. no two different intervals share an endpoint.

Lemma 3.1. (Gulumbic & Trenk, 2004) *"For every TolG there is a tolerance representation."*
If the tolerances for all the vertices are equal then the tolerances are called constant. The following theorem is valid in case of constant tolerance.

Theorem 3.1. (Gulumbic et al., 1984; Gulumbic & Trenk, 2004) *"The following are equivalent statements about a graph G.*

1. *G is an interval graph.*
2. *G is a tolerance graph with constant tolerance.*
3. *G is a bounded tolerance graph with constant tolerances."*

Theorem 3.3. (Golumbic, Monma & Trotter 1984) *"Tolerance graphs are weakly chordal."*
They also proved the following theorem.

Theorem 3.4. (Gulumbic & Trenk, 2004) *"The tolerance graphs are perfect."*
A TolG is called unit TolG if all intervals have the same length. A graph is called 50% tolerance graph if it has a tolerance representation $\langle I, t \rangle$ so that $t_v = 0.5 \mid I_v \mid$ for all $v \in V$. In general, if G has a tolerance representation for which there is a constant c with $\mid I_v \mid - 2t_v = c$ for all $v \in V$, then the tolerance representation has constant cores.

Theorem 3.5. (Gulumbic & Trenk, 2004) *"The following statements are equivalent.*

1. *G is a unit tolerance graph.*
2. *G is a 50% tolerance graph.*
3. *G has a bounded tolerance representation with constant cores."*

The following theorem establishes the relation between InvG and TolG.

Theorem 3.6. (Bogart et al., 2001) *"Interval graphs are unit tolerance graph."*

Theorem 3.7. (Gulumbic & Trenk, 2004) *"Let $G = (X, Y, E)$ be a bipartite graph. The following conditions are equivalent*

1. *G is a bounded tolerance graph.*
2. *G is a trapezoid graph.*
3. *G is a cocomparability graph.*
4. *G is AT-free.*
5. *G is a permutation graph."*

CIRCULAR-ARC GRAPHS

Let $C = \{C_1, C_2, \ldots, C_n\}$ be a family of arcs around a circle.

A graph $G = (V, E)$ is a *circular-arc* graph (CirG) if there is a one-one correspondence between the arcs in C and vertices of G, such that two vertices x and y are adjacent in G if and only if the corresponding arcs C_x and C_y intersect in C. The family of arcs C is called an *arc representation* for G or intersection model for G. A CirC and its circular-arc representation are shown in Fig. 7.

A graph CirG G is called a *proper circular-arc (PCA)* graph if there exists a family of arc C for G such that no arc properly included in another. An efficient recognition algorithm is designed for recognizing PCA graphs by Tucker (Tucker, 1971) using matrix characterizations. Another recognition algorithm is presented by Deng *et al.* (1996) to recognized PCA graph, which also produces a PCA model. The time complexity of this algorithm is linear.

If the length of all arcs are same, then the corresponding CirG is called a *unit circular-arc (UCA)* graph. Tucker (1974) characterized this class of graphs by forbidden subgraphs. Using this characterization Tucker shown that UCA graphs are proper subclass of PCA graphs. These graphs are useful in traffic signaling, when it is necessary that the green lights for each lane at a street intersection are on for the same amount of time (Gulumbic, 1980).

Let C be a family C of subsets. The family C satisfied the *Helly property* if for every subfamily of C consisting of pairwise intersecting subsets has a common element. A graph $G = (V, E)$ is called a *Helly circular-arc (HCA)* graph if there exists an arc model C for G which satisfy the Helly property.

CirGs have many applications in several areas, viz. traffic control, genetic research, statistics, computer compiler design, etc.. McConnell (2003) has designed an algorithm for recognizing a CirG. The running time of this algorithm is $O(n + m)$.

Let $A = \{a_1, a_2, \ldots, a_n\}$ be a family of n arcs on a circle C. The each endpoint of every are is a positive integer, called *coordinate*. The endpoints of the arcs are assigned on the circumference of the circle C in the ascending order of the values of the coordinates in the clockwise direction. In mathematical representation each arc $a_i, i = 1, 2, \ldots, n$, has two endpoints (s_i, f_i), where s_i and f_i are the *starting point and* the *finishing point* of the arc a_i when it is traversed in clockwise direction, starting from any an arbitrary point on the circle C. For a set of n arcs there are $2n$ endpoints.

Figure 7. Example of a circular-arc graph and its circular arc representation.

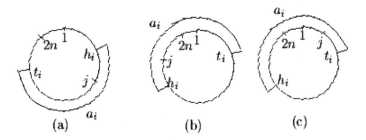

(a)   (b)   (c)

Some easy observations for circular arcs are presented below:

1. No single arc in A covers the whole circle C by itself. In this case, the shortest path problem becomes trivial, i.e. the shortest distance between any two (vertices) arcs is either 1 or 2 unit.
2. The endpoints of the arcs distinct.
3. The union of all arcs cover the entire circle, otherwise, the corresponding graph becomes an interval graph.
4. The arcs are sorted in increasing values of s_i's, i.e. $s_i > s_j$ for $i > j$.

The family of arcs A is called *canonical* if

1. For $i = 1, 2, \ldots, n$, s_i's and f_i's are all distinct integers between 1 and $2n$, and
2. The point 1 is the starting point of the arc a_1.

If the set of n arcs A is not canonical, then by sorting one can convert a canonical family of arcs using $O(n \log n)$ time. Here, it is assumed that the family of arcs A is canonical.

Let A be a set of arcs. A *path* from the arc $x \in S$ to the arc $y \in S$ is a sequence $\sigma = (a_1, a_2, \ldots, a_k)$ of arcs in A, where $a_1 = x$ and $a_k = y$, and a_j and a_{j+1} have some common portion for every $j = 1, 2, \ldots, k-1$. The *length* of the path σ is the number of arcs on σ, and it will be shortest if it length is minimum among all possible paths between x and y in A.

If the some portion or a single point of the circle do not cover by circular-arcs of a CirG G, then the graph is topologically same as an interval representation of G. In this situation, we can cut the circle at an uncover position and straighten it out to a straight line, the set of arcs become set of intervals. From this topological point of view we can say that every InvG is a CirG, but, the converse is not true.

Let $a_i \in A$ and an endpoint j of another arc in A. The arc a_i *contains point j* if any one of the following three conditions hold.

1. $1 \leq s_i < j < f_i \leq 2n$ (Figure 8(a)).
2. $1 \leq f_i < s_i < j \leq 2n$ (Figure 8(b)).
3. $1 \leq j < f_i < s_i \leq 2n$ (Figure 8(c)).

Figure 8. Position of the point j within the are a_i (a) $1 \leq s_i < j < f_i \leq 2n$, (b) $1 \leq f_i < s_i < j \leq 2n$, (c) $1 \leq j < f_i < s_i \leq 2n$

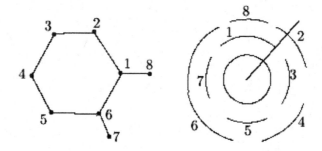

Theorem 4.3. (Gulumbic, 1980; Tucker, 1974) *"If G is a proper circular-arc graph, then G has a proper circular-arc model in which no pair of arcs covers the circle."*

Lemma 4.1. (Tucker, 1974) *"If G is a unit circular-arc graph, then it is also a proper circular-arc graph."*

Two arcs a_i and a_j in S are said to be *intersect* with each other if one of them contains at least one of the endpoints of the other arc, otherwise, a_i and a_j are said to be *independent* from each other. If a_i contains both endpoints of a_j, then a_i *contains* a_j.

For CirG, the arc a_i and the corresponding vertex v_i or i are one and the same thing.

Corollary 4.1. *"A tree is a circular-arc graph if and only if it is an interval graph."*

Proof: Let G be a CirG which is a tree and it is assumed that G is not an InvG. Therefore, by Theorem 2.4, G contain some of the graphs depicted in Fig. 1. Since G is a tree, this induced subgraph can only be G^*. But, this it is not a CirG, which is a contradiction.

The converse is true, as InvGs are a subclass of CirGs. **Lemma 4.3.** *Let P be the maximum weight independent set of A and Q be any clique, then $\mid P \cap Q \mid = 1$.*

Proof: Any two vertices of Q are connected by at least one edge. So, at most one vertex can take from Q into P, otherwise P does not remain an independent set. Thus, $\mid P \cap Q \mid \leq 1$. Also, if $\mid P \cap Q \mid = \varphi$ then inclusion of any vertex of Q into P, it remains independent and weight will be more than the previous set. Hence, $\mid P \cap Q \mid = 1$. The following result is very interesting.

Lemma 4.4. *"The graph $G \setminus N[i]$ is an InvG for any i, where $N[i]$ is the close neighbour of the vertex i."*

Let S_i be the maximum weight independent set of the graph $G \setminus N[i]$. Then it can easily be verified that the set $S_i \cup \{i\}$ is a maximal independent set of the CirG G.

Theorem 4.4. (Mandal & Pal, 2006a) *"The time complexity to find the maximum weight independent set of a CirG with n vertices is $O\left(n^2\right)$ time."*

Theorem 4.5. (Mandal & Pal, 2007) *The time to find the set of hinge vertices of a CirG can be determined using $O\left(n\right)$ time, where n represents the number of vertices of the graph.*

One of the another important problem of graph theory is covering problem. The 2-neighbourhood covering problem on CirG can be solved by converting the problem to an appropriate InvG. The main reason for this conversion is that the InvG can easily be take up with its good data structure IT. Pal and Bhattachajee (1997b) have developed this data structure. Many problems on CirG can be solved by converting it to an InvG. Thus, to solve a problem on CirG, we first transfer the set of arcs of CirG to an equivalent set of intervals on a real line.

Theorem 4.6. *"The 2-neighbourhood covering set of a CirG can be determined using $O\left(n\right)$ time."*

To compute all-pairs shortest distances of a CirG can be obtained by converting it to two equivalent InvGs. Thus, to solve this problem on CirG, at first we transfer CirG into two InvGs. For these two InvGs, two ITs are drawn. Next, find the distances between two vertices x and y from two ITs. The minimum of these two values is the shortest distance between the vertices x and y in CirG.

Theorem 4.7. (Saha et al., 2005b) *"The time complexity to find all-pair shortest paths on CirG is $O\left(n^2\right)$."*

The next-to-shortest path from the vertex x to the vertex y is the shortest path from between the vertices x and y amongst those the distances strictly greater than the shortest distance. If no such path exist, then the distance of next-to-shortest path is considered as ∞. The length of the next-to-shortest path is the *next-to-shortest distance*.

Theorem 4.8. (Mandal & Pal, 2006b) *"The next-to-shortest paths on CirG can be determined in* $O\left(n^2\right).$ *"*

CHORDAL GRAPHS

A graph $G = \left(V, E\right)$ is called *chordal* if each of its cycles of length four or more has a chord, (this is an edge joining two vertices which are not adjacent in the cycle). In other words, a chordal graph is a graph with no induced cycles of length more than three.

From the above definition, it follows that an undirected graph is chordal if it does not contain an induced subgraph isomorphic to C_n for $n > 3$. Thefefore, InvGs are chordal.

The chordal graphs are a subclass of perfect graphs (Gulumbic, 1980). These graphs are also known as *triangulated graphs* or *rigid circuit graphs*.

The vertices of a chordal graph can be ordered in a special way which is known as *perfect elimination ordering*. This ordering is very useful to design algorithms on chordal graphs. A vertex is said to be *simplicial* if its adjacency set induces a complete sub graph. A permutation $\sigma = [v_1, v_2, \ldots, v_n]$ of the vertices of an undirected graph G, or a bijection $\sigma : V \rightarrow \{1, 2, \ldots, n\}$, is called a perfect elimination order if each v_i is a simplicial vertex of the subgraph of G induced by $\{v_1, v_2, \ldots, v_n\}$.

Theorem 5.1. (Fulkerson & Gross, 1965) *"A graph is chordal if and only if it has a perfect elimination ordering."*

Theorem 5.2. *"For a graph G with n vertices, the following statements are equivalent:*

1. *G has a perfect elimination ordering.*
2. *G is chordal.*
3. *If H is any induced subgraph of G and S is a vertex separator of H of minimal size then vertices of S induce a clique."*

Chordal graphs can be recognized in linear time.

Lemma 5.1. *"A chordal graph has at most n maximal cliques."*

Chordal graphs are a subclass of the perfect graphs. Other superclasses of chordal graphs are even-hole-free graphs, odd-hole-free graphs and weakly chordal graphs. In fact, chordal graphs are precisely the graphs that are both odd-hole-free and even-hole-free.

Using results of Dirac (1961), Fulkerson and Gross (1965), Buneman (1974), Gavril (1974) and Rose et al. (1976), we have:

Theorem 5.3. *"The following statements are equivalent and characterize chordal graphs.*

1. *G has a simplicial elimination scheme.*
2. Every minimal separator is a clique.

3.   *G admits a maximal clique tree.*

4.   *G is the intersection graph of subtrees in a tree.*

5.   Any LexBFS provides a simplicial elimination scheme."

Given two non-adjacent vertices x and y of the undirected graph $G = (V, E)$. A subset $S \subset V$ is called x,y-separator if the removal of S separates x and y in distinct connected components. If no proper subset of S is an x,y-separator then S is a minimal x,y-separator. A minimal separator is a set of vertices S for which there exist non-adjacent vertices x and y such that S is a minimal x,y-separator.

The minimal separators of a chordal graph are complete subgraphs (Dirac, 1961). Let $G = (V, E)$ be a chordal graph. The clique-graph of G, denoted by $C(G) = (V_c, E_c, \mu)$ with weight function $\mu : E_c \to N$, is defined by

1.   The set of vertices V_c is the set of maximal cliques of the graph G;

2.   There is an edge between the vertices C_1 and C_2 if and only if the intersection $C_1 \cap C_2$ is a minimal x,y-separator for each $x \in (C_1 - C_2)$ and each $y \in (C_2 - C_1)$;

3.   The weight of the edge $(C_i, C_j) \in E_c$ is the cardinality of the corresponding minimal separator.

Let $S_{ij} = C_i \cap C_j$ if and only if S_{ij} is a minimal x,y-separator for each $x \in C_i - C_j$ and each $y \in C_j - C_i$.

Lemma 5.2. *"Let (C_1, C_2, C_3) be a 3-cycle in $C(G)$ and let S_{12}, S_{13}, S_{23} be the associated minimal separators of G. Then two of these three minimal separators are equal and included in the third."*

But, the converse of this result false.

Lemma 5.3. (Galinier et al., 1995) *"Let $C(G)$ be the clique graph of the chordal graph G. Let C_1, C_2, C_3 be three maximal cliques such that $(C_1, C_2) \in E_c$ and $(C_1, C_3) \in E_c$, then $S_{12} \subset S_{13} \Rightarrow (C_2, C_3) \in E_c$."*

PERMUTATION GRAPHS

The permutation graph (PerG) can be defined in two different ways, one is algebraic way and another is geometrical way. An undirected graph $G = (V, E)$ with set of vertices $V = \{1, 2, \ldots, n\}$ is called a PerG if there exists a permutation π on the set of natural numbers $N = \{1, 2, \ldots, n\}$ such that for all $x, y \in N$,

$$(x - y)(\pi^{-1}(x) - \pi^{-1}(y)) < 0$$

if and only if x and y are joined by an edge in G (Gulumbic, 1980). Geometrically, the integers $1, 2, \ldots, n$ are drawn in order on a real line called as *upper line* and the permutations $\pi(1), \pi(2), \ldots, \pi(n)$ are drawn on another line parallel to this line called as *lower line* such that for each $x \in N$, x is directly below $\pi(x)$. For this geometrical representation, the intergers 1, 2, ..., n are written on the upper line in ascending order and the same set of intergers are also writen on the lower line, but they are not in any order.

Next, we drawn a line segment from the integer x on the lower line to the same integer x on the upper line and this line segment is denoted by $l(x)$. Then from definition from PerG it follows that there is an edge between the vertices x and y in G if and only if the line segment $l(x)$ for the vertex x intersects the line segment $l(y)$ for y.

A PerG with its permutation representation is drawn in Fig. 9.

A huge amount of works have been done on PerGs (Mandal et al., 1999; Mandal et al., 2002; Mandal et al., 2003b; Mandal et al., 2003c; Pal, 1998a; Pal, 1998b; Saha & Pal, 2003; Yu & Chen, 1993).

Like InvG and CirG, many useful characterizations are available on PerGs, some of them are discussed below.

Let π be a permutation on the set of natural numbers and $G\big[\pi\big]$ be the corresponding PerG. If we reverse the permutation π then we obtain another permutation say π'. For this permutation there is another PerG $G\big[\pi'\big]$. But, interestingly $G\big[\pi'\big]$ is the complement of the PerG $G\big[\pi\big]$. In other words, $G[\pi'] = G'[\pi]$ (Gulumbic, 1980).

This analogy leads to the following result.

Lemma 6.1. (Gulumbic, 1980) *"The complement of a permutation graph is also a permutation graph."* The PerG $G\big[\pi\big]$ is transitively orientable.

Theorem 6.1. (Pnueli et al., 1971) *"An undirected graph G is a PerG if and only if G and its complement graph G' are comparability graph."*

Following is the relation among cliques, stable sets and subsequences of π.

Lemma 6.2. *"The decreasing subsequences of π and the cliques of $G[\pi]$ are in one-to-one correspondence. The increasing subsequence of π and the stable sets of $G[\pi]$ are in one-to-one correspondence."*

Let P^2 be a two dimensional plane whose axes are indexed by x and $y = \pi^{-1}(x)$. For each vertex

$$x \in V = \big\{1,2,...,n\big\}$$

a point

Figure 9. (a) A permutation π, (b) The corresponding PerG.

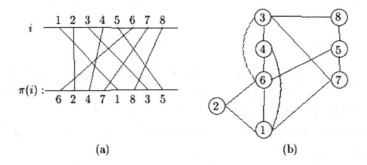

$$p_{xy} = (x,y), y = \pi^{-1}(x)$$

is defined in P^2. The origin of the plane is the point $p_{00} = (0,0)$. Let

$$S(G) = \left\{ p_{xy} : y = \pi^{-1}(x), x = 1,2,...,n \right\}$$

be the set of all such points for the graph G in the plane P^2. Thus there is a one-one correspondence between the the set of points $S(G)$ and the set of vertices V of the PerG. This plane representation of the PerG of Fig. 9 is depicted in Fig. 10.

Two points $p_{xy}, p_{zu} \in S(G)$ are said to be *non-connected* if ine of the following condistions is satisfied:

1. $x < z$ and $y < u$
2. $x > z$ and $z > u$.

Again, p_{xy} is called *directly non-connected* with p_{zu} if p_{xy} is non-connected with p_{zu} and there exists no other point

$$p_{vw} \in S(G), x < v < n, y < w < u$$

such that p_{xy} is non-connected with p_{vw} and p_{vw} is non-connected with p_{zu}.

For example, the points p_{22} and p_{74} of Fig. 10 are non-connected as 2 is less than 7 and 4. The points p_{15} and p_{22} are not non-connected as 1 is less than 2 but 5 is greater than 2. The points p_{22} and p_{43} are directly non-connected, but, p_{22} and p_{74} are not directly non-connected because there exists a point p_{43} between the points p_{22} and p_{74} satisfying $2 < 4 < 7$ and $2 < 3 < 4$. Two points are said to be *connected* if they are not non-connected.

Lemma 6.3. *"Two points p_{xy}, $p_{zu} \in S(G)$ are connected if* (a) $x < z$ and $y > u$ *or* (b) $x > z$ and $y < u$*."*

The following result establishes a relation between connected points of $S(G)$ and corresponding edges of G.

Lemma 6.4. *"Let p_{xy}, $p_{zu} \in S(G)$ be two points corresponding to two vertices $x, z \in V$. The vertices x and z are adjacent in G if and only if p_{xy} and p_{zu} are connected."*

A sequence of points in $S(G)$ is called a *chain* if any point of it is directly non-connected with the next one. A chain of $S(G)$ is said to be *maximal* if it is not contained in any other chain of $S(G)$. A chain of $S(G)$ is said to be *maximum* if its length (the number of points in the chain) is maximum. Suppose the given PerG is weighted and each vertex has a weight $w(x), x \in V$. For this weighted PerG, we can assign weight to each point p_{xy} as

Figure 10. Plane representation of the PerG of Fig. 9 in P^2.

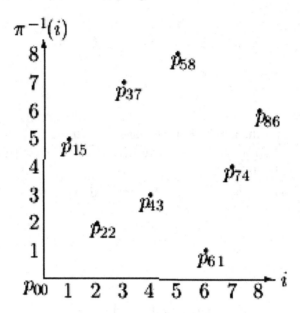

$$w\left(p_{xy}\right) = w\left(x\right), x \in V.$$

A chain of $S(G)$ is said to be of *maximum weight* if the sum of the weights of the points of that chain is maximum.

From this point representation, one can construct a tree $T(G)$ for any PerG G, which is described below:

The root of $T(G)$ is $p_{00} = \left(0,0\right)$. The children of p_{00} are all the points which are directly non-connected with p_{00}. Suppose $D_x\left(G\right)$ be the set of all directly non-connected points for the point p_{xy}. If p_{xy} is a vertex of the tree $T(G)$ then its children are the members of $D_x\left(G\right)$. It is easily verifed that the set of vertices of $T(G)$ is $\{p_{00}\} \cup S(G)$, but, the number of nodes is not the cardinality of $\{p_{00}\} \cup S(G)$. From the definition of $T(G)$, it is obvious that, if p_{xy} is the parent of p_{zu} then p_{xy} is directly non-connected with p_{zu}. A vertex may occurs more than once in the tree $T(G)$. But, no vertices appear more than once in a path starting from a leaf to the root of $T(G)$. Thus the number of vertices of any path starting from a leaf to the root of $T(G)$ is not more than n.

From this discussion it is obvious that if the PerG G has n maximal independent sets then the number of vertices of the tree $T(G)$ is $O(n^2)$. This case happens for most of the PerGs. But, a PerG may have more than $O(n)$ maximal independent sets. For example, let $G = (V, E)$ where $V = \{1, 2, ..., n\}$ and

$$E = \{(x, x+1) : x = 1, 2, ..., n-1\}$$

be the sets of vertices and edges of a PerG. This graph has more than $O(n)$ maximal independent sets. But, this graph is a chain, a tree, an interval graph, a trapezoid graph, etc.

For simplicity, let the numbers of vertices of the tree $T(G)$ be *N*.

The tree $T(G)$ for the graph of Fig. 9 is depicted in Fig. 11.

This tree can be used to solve many problems. Here, we discuss a method to find all maximal independent sets of a PerG. By the construction of the tree $T(G)$ from *G* it follows that if any two vertices lie on the same path in the tree $T(G)$ then the corresponding vertices are non-connected in the PerG *G*. This, the set of points of any path from a leaf to the root of $T(G)$ forms a chain.

Lemma 6.5. (Saha et al., 2005c) *"The vertices corresponding to the points of any path from a leaf to the root of $T(G)$ form a maximal independent set in G ."*

Theorem 6.2. (Pal, 1996; Saha et al., 2005c) *"A maximum weight independent set of a PerG can be determined in $O(n^2)$ time, provided the graph has at most $O(n)$ maximal independent sets, where n is the number of vertices of the graph."*

Another version of independent set is maximum weight *k*-independent set, which is defined below. Let *G* be a weighted PerG and $w(x)$ is the positive weight for the vertex $x \in V$. The maximum weight k -independent (MWkI) set problem on G is to find a set of k disjoint partitions $P_1, P_2, \ldots, P_k$ of *G* such that the weight where, is maximum.

Theorem 6.3. (Pal & Bhattacharjee, 1995c; Pal & Bhattacharjee, 1996; Saha & Pal, 2003) *"The maximum weight k -independent set problem can be solved by using $O\left(kn\sqrt{\log c} + m\right)$ time on InvGs and using $O(kn^2)$ time on PerGs, where c and m represent the weight of the longest path and the number of edges in the graph respectively."*

Figure 11. The tree $T(G)$ for the graph of Fig. 9.

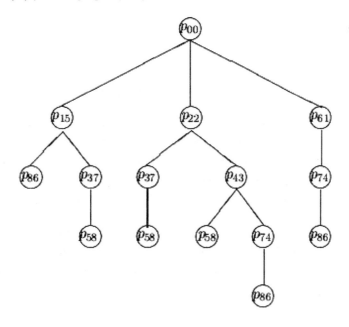

The depth-first search (DFS) is one of the fundamental searching techniques of graph. By exploiring the properties of PerG many algorithms are design in this graph. A new approach is used to performed a DFS on PerGs (Mandal et al., 1999). The proposed method also constucts a depth-first tree.

Theorem 6.4. (Mandal et al., 1999) *"The O(n) time needed to construct the depth-first tree of a PerG."*

Theorem 6.5. (Rana et al., 2011a) *"Minimum cardinality 2-neighbourhood covering problem on PerGs can be solved $O(n + \overline{m})$ time, where $\overline{m}$ is the number of edges in the complement graph.*

Let us define k-dominating set, where $k > 1$ (a positive integer). A vertex $x \in V$ is said to be k-*dominates* the vertex $y \in V$ if $d(x,y) \leq k$. A subset $D \subseteq V$ is called a k-*dominating set* in G if for every $x \in V$ there is at least one $y \in D$ with $d(x,y) \leq k$. The set D is called a *total k-dominating set* of G if every vertex $x \in V$ is k-dominated by a vertex in D and for $x \in D$ there is at least one vertex $z(\neq x) \in D$ with $d(x,z) \leq k$. The subgraph induced by a total k-dominating set may produces a set of isolated vertices for $k > 1$.

Theorem 6.6. (Rana et al., 2011c) *"The minimum cardinality $\underline{k}$-domination set and a minimum cardinality total k-domination set for a PerG determine in $O(n + \overline{m})$ time."*

A subset D of V is called a distance k-dominating set of G if each $x \in V \setminus D$ is within distance k from some vertices of D. The minimum cardinality of a distance k-dominating set in G is the distance k-domination number of G, denoted by $\gamma_k(G)$. The distance k-*domination problem* is to find a $\gamma_k(G)$ in G. The distance 1-domination number $\gamma_1(G)$ is the usual domination number $\gamma(G)$. In general, determining $\gamma_k(G)$ is NP-complete (Garey & Jhonson, 1979).

Theorem 6.7. (Rana et al., 2011c) *"The time to find minimum cardinality distance k-domination set on PerGs is $O\left(n^2\right)$."*

Like interval graph, the all-pair shortest path problem on permutation graph can also be solved in optimal time as stated below.

Theorem 6.8. (Mandal et al., 2002) *"For PerGs, the all-pair shortest paths problem can be solved in $O\left(n^2\right)$ time. Also, the average distance beyween the vertices can be computed using $O\left(n^2\right)$ time."*

Theorem 6.9. (Barman et al., 2009) *"The time to find the next-to-shortest path between any two given vertices in PerG is $O\left(n^2\right)$."*

TRAPEZOID GRAPHS

It is another important subclass of IntG. It's geometrical structure is more complicated than both PerG and InvG. The trapezoid graph (TraG) is the superclass of both InvG and PerG. First we define what is trapezoid between two parallel lines. A trapezoid T_i is defined by four corner points $[a_i, b_i, c_i, d_i]$, where $a_i < b_i$ and $c_i < d_i$ with a_i, b_i lying on top line and c_i, d_i lying on bottom line of a rectangular channel. An undirected graph $G = (V, E)$ with vertex set $V = \{x_1, x_2, \ldots, x_n\}$ and edge set $E = \{e_1, e_2, \ldots, e_m\}$ is said to be a TraG if a trapezoid representation can be obtained such that each vertex $x_i \in V$ corresponds to a trapezoid T_i and two vertices x_i and x_j are adjacent if and only if the $T_i \cap T_j \neq \phi$. For simplicity, most of the time the vertices $x_1, x_2, \ldots, x_n$ are numbered as 1, 2,…, *n*. Fig. 12 illustrates a TraG with seven trapezoids/vertices and its corresponding trapezoid representation. Like other IntG,

Figure 12. A TraG G and its corresponding trapezoid representation

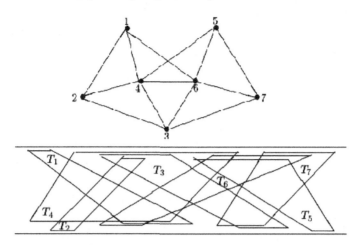

trapezoids and vertices are same thing. It is interesting to note that if $a_i = b_i$ and $c_i = d_i$ then the corresponding trapezoid T_i becomes a straight line. So, in this manner, if all the trapezoids reduce to straight lines the corresponding TraG becomes a PerG. For simplicity, we assume that the corner points of the trapezoids are all distinct. Therefore, they can be numbered as $1, 2, \ldots, 2n$ from left to right on both channels. In addition to this we may label these n trapezoids in increasing order of their right corner points on top channel, i.e. for two trapezoids T_i and $T_j, i < j$ if and only if b_i lies on the left of b_j.

Ma and Spinrad's have designed an algorithm to recognized TraG in $O\left(n^2\right)$ time (Ma & Spinrad, 1994).

Hota et al. (Hota et al., 1999) shown that TraGs are used to model *channel routing* problem in VLSI, in a single-layer-per-net model. A *routing* is a connection of every net by wires inside the channel such that no two wires from different nets overlap (see Fig. 13). The channel routing problem and minimum colouring problem of a TraG are equivalent, where each net is represented by a trapezoid.

A TraG with *n* vertices can be represented geometrically either by,

1. a set of *n* trapezoids drawn inside a rectangular channel or by,
2. a set of *n* segments drawn on a two dimensional plane or by,
3. a set of *n* boxes drawn on a two dimensional plane or by,
4. a permutation diagram π of $2n$ lines drawn inside a channel.

TRAPEZOID REPRESENTATION

The following points are assumed during the design of algorithm on TraGs:

1. each trapezoid has four distinct corner points and that no two trapezoids share a common end point,
2. The vertices in the trapezoid graph and trapezoids in the trapezoid representation and are one and same thing,

Figure 13. (a) A routing instance, where the routing region is a channel. (b) A routing instance, where the routing region is bounded by a straight line and a module.

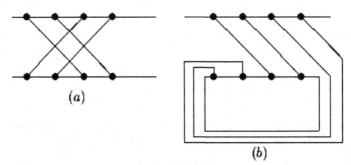

(a)

(b)

3.   the trapezoids in the trapezoid representation T are indexed by increasing right end points on the top line, i.e. for any two trapezoids T_i and T_j in the trapezoid representation $i < j$ if and only if $b_i < b_j$.

This kind of ordering gives the following result which is quite useful in designing efficient algorithms.

Let i, j, k be three vertices of a TraG $G = (V, E)$ such that $i < j < k$. If $(i, k) \in E$ then either $(i, j) \in E$ or $(j, k) \in E$.

This type of ordering is called as *cocomparability ordering*. This ordering can be done in an $O(n^2)$ time for general graph (Spinrad, 1985). But, for a TraG, this ordering can be done in $O(n)$ time only, if the trapezoid representation is given.

The adjacency relation between any two vertices can be tested using the following result:

Let i and j be two vertices of a TraG. Then two vertices i and j are not adjacent if and only if either (i) $b_i < a_j$ and $d_i < c_j$ or (ii) $b_j < a_i$ and $d_j < c_i$.

Otherwise, the vertices i and j are adjacent.

A TraG can be stored by storing its four corner points. For a TraG with n vertices there are $4n$ corner points. So, a trapezoid graph can be stored using only $4n$ memory units. The adjacency relation between two vertices can be tested using $O(1)$ time.

Time complexity of some useful algorithms on TraGs are stated below:

Theorem 7.3. (Mandal et al., 2002) *"The all pairs shortest distances of a trapezoid graph can be computed using $O(n^2)$ time."*

Theorem 7.4. (Barman et al., 2007) *"The next-to-shortest path between any two vertices u and v in TraG is can be determined in $O(n^2)$ time."*

Theorem 7.5. (Bera et al., 2002) *"The time complexity to generate all maximal cliques of a TraG is $O(n^2 + \gamma n)$ where γ is the output size and n is the number of vertices."*

Theorem 7.7. (Hota et al., 2001) *"The maximum weight k-independent set of a TraG can be determined in $O(kn^2)$ time. In particular, the time complxity for $k = 2$ is $O(n^2)$."*

A *clique cover* of the graph $G = (V, E)$ is a partition of the vertices $V_1, V_2, ..., V_k$ of V such that each $V_i, i = 1, 2, ..., k$ is a clique. A clique cover with the minimum cardinality $c(c \leq n)$ is known as a *minimum clique cover* (MCC). The number c is called the *clique cover number*.

The MCC is a well known NP-complete problem on general graphs (Garey & Jhonson, 1979, Gulumbic, 1980). But, the problem can be solved in polynomial time for some particular type of graphs, like interval graphs, chordal graphs, circular-graphs, circular permutation graph, etc.

Theorem 7.8. (Hota, 2005) *"A MCC of the TraG can be determined in $O(n^2)$ time."*

Theorem 7.9. (Hota, 2005) *"The time complexity to find the diameter and center of a TraG is $O(dn)$, where d is the degree of vertex 1."*

It is proved in (Corneil et al., 1997) that every connected AT-free graph contains a dominating pair. The TraG is a subclass of AT-free graphs and follow the following results (Hota, 2005).

Every TraG has at least one dominating pair and the vertices u, v of a TraG form a dominating pair if $\delta(u, v) = diam(G)$.

A *diameter path* in a graph is a shortest path whose length is equal to the diameter of the graph.

Theorem 7.10. (Hota, 2005) *"If $p(u, v)$ is a diameter path in a TraG G, then $p(u, v)$ is minimum connected dominating path."*

Theorem 7.11. (Hota, 2005) *"The time to find out the dominating pairs and minimum connected dominating paths of a TraG is $O(dn)$."*

Several problems on TraGs have been solve in polynomial or polynomial-log time. Some of them are presented below:

Theorem 7.12. (Bera et al., 2003a) *"The time to compute of all hinge vertices of a TraG is $O(n \log n)$ and space is $O(n)$."*

Theorem 7.13. (Bera et al., 2003b) *"A spanning tree of a TraG can be constructed in linear time."*

Theorem 7.14. (Barman et al., 2010a) *"A tree 4-spanner of a TraG can be constructed in linear time, however a tree 3-spanner on the same graph can be determined using $O(n^2)$ time."*

INTERVAL BIGRAPH AND INTERVAL DIGRAPH

Let *H* be a bipartite graph, with the partition of vertex set (*X, Y*). The graph *H* is called an interval bigraph if the vertices of *H* can be represented by a family of intervals

$$I_u, u \in V, where\, V = X \cup Y,$$

and there is an edge between u and v if and only if there corresponding intervals I_u and I_v intrsect. That is, an interval bigraph is an InvG and it is bipartite. Interval bigraphs were introduced in (Horary, 1982). This graph can be recognized in $O(n^5 m^6 \log n)$ time (Müller, 1997). An asteroidal triple of edges form by three edges where any two are connected by a path avoiding vertices and neighbors of the third edge. It is proved that *no interval bigraph contains an asteroidal triple of edges* (Müller, 1997).

Let $B(G)$ be the bigraph of the graph $G = (V, E)$. Then the vertex set of $B(G)$ is $V \cup \{u' : u \in V\}$ and there is an edge between the vertices u and v' where $u \in V$ and $v \in V$, whenever $uv \in E$ or $u = v$.

Müller (1997) has proved the following results.

Theorem 9.1. *"Interval bigraphs do not contain induced cycles of length 6 or more."*

Theorem 9.2. *"A graph G is an interval graph if and only if B(G) is an interval bigraph."*

***Proposition* 9.1.** (Prisner, 1998) *"If G is an IntG of P-sets, then B(G) is an intersection bigraph of sets that are union of two intersecting P-sets."*

There is a nice relation between interval bigraphs and CirGs.

Theorem 9.3. *"If H is an interval bigraph, then the complement $\bar{H}$ of H is a CirG."*

Let us define insect of a bipartite graph. This is a bipartite graph with twelve vertices containing a copy of $K_{3,3}$ a matching from those vertices to the remaining vertices (called feet), and possible additional edges among the feet. The graph without additional edges is the minimal insect *I*, and the maximal insect *I** is the Cartesian product between $K_{3,3}$ and K_2, i.e. $K_{3,3} \square K_2$. It can be proved that a bipartite graph *G* is an insect if and only if $I \subseteq G \subseteq I^*$.

Lemma 9.1. (Steiner, 1996) "An interval bigraph does not contain an induced subgraph isomorphic to an insect."

Theorem 9.4. (Steiner, 1996) *"A bipartite graph is a proper interval bigraph if and only if it is a PerG."*

A proper interval bigraph can be recognized in $O(n^2)$ time.

Let

$$R = \left\{ (S_u, T_u), u \in V \right\}$$

be a family of a pairs of sets for an intersection graph *G=(V,E)*. The intersection digraph for the family *R* is defined as: *V* is the vertex set and there is an edge between *u* and *v* whenever $S_u \cap T_u \neq \phi$.

Now we define interval digraph below:

Let D be a digraph. D is called an interval digraph, if there exists a family of ordered pairs of intervals $(I_v, J_v), v \in V$ of D, such that *uv* is an edge of *D* if and only if I_v intersects J_v. Let D* be the associated bipartite graph obtained from *D* by replacing each vertex *v* by two vertices v', v'', and each arc *uv* of *D* by the edge $v'v''$. Then it is clear from the definitions that *D* is an interval digraph if and only if D^* is an interval bigraph.

The concept of digraph is old but the concept of interval bigraph is more general.

Here we introducing a new type of clique called exobiclique. This is a particular type of biclique (a complete bipartite subgraph). Let H be a bipartite graph with the vertex set (X, Y). An exobiclique in *H* is a biclique with nonempty subsets $M \subseteq X$ and $N \subseteq Y$ such that each of $X - M$ and $Y - N$ contain three vertices with incomparable neighborhoods in the biclique.

Proposition 9.2. (Hell and Huang, 2004) "If the bipartite graph *H* contains an exobiclique, then in any representation of $E(\bar{H})$ by circular arcs, there are two arcs that together cover the whole circle."

The following result follows from the above theorem.

Corollary 9.1. (Hell and Huang, 2004) "A bipartite graph with an exobiclique is not an interval bigraph."

Given a graph H, the associated graph $H*$ of H is a bipartite graph with vertex set $\left\{ v', v'' : v \in V\left(H\right) \right\}$ and the edge set

$$\left\{ v'v'' : v \in V\left(H\right) \cup \left\{ u'v'' : uv \in E\left(H\right) \right\} \right\}.$$

Theorem 9.5. (Hell and Huang, 2004) "For a graph H, the following statements are equivalent

1. H is an InvG;
2. $H*$ is an interval bigraph;
3. $H*$ is the complement of a CirG."

GRAPH WITH BOXICITY k

An interval on the real line can be extended to a k-box in R^k. Let R_i be a closed interval on the real line. A k-box $B = (R_1, R_2, \ldots, R_k)$ is defined as the Cartesian product of k real lines $R_i, i = 1, 2, \ldots, k$, i.e. $R_1 \times R_2 \times \cdots \times R_k$. If each R_i is an interval of length one, then the graph B is called k-cube. Thus, 1-boxes, i.e. $B = \left(R_1\right)$ are nothing but closed intervals on the real line, whereas 2-boxes, i.e. $B = \left(R_1, R_2\right)$ are axis-parallel rectangles in two dimensional plane. The parameter boxicity of a graph G, denoted as $box(G)$, is the minimum integer k such that G is an intersection graph of k-boxes. Similarly, the cubicity of G, denoted as $cub(G)$, is the minimum integer k such that G is an intersection graph of k-cubes. The boxicity of InvGs are at most 1 and that of unit InvGs are at most 1. These graph parameters were first defined by Roberts (1969).

Any graph can be represented as the intersection graph of a set of rectangles in d dimensional plane where $d \geq 1$. The graphs with boxicity-2 are called the rectangle IntGs. Followings are some results available in literature.

1. A graph has boxicity one if and only if it is an InvG.
2. The boxicity of outerplanar graph is at most two.
3. The boxicity of planar graph is at most three.
4. Every bipartite graph with boxicity two can be represented as an IntG of axis-parallel line segments in the plane.
5. Let τ be the tree-width of a graph G. Then the upper bound of the boxicity is $\tau + 1$.

A graph which is not an InvG can be represented by 2-boxes. For example, the graph C_4, can be represented as an intersection graph of 2-boxes (see Figure 14).

The following lemma due to Roberts (1969) is very interesting.

Figure 14. A 2-box representation for C_4

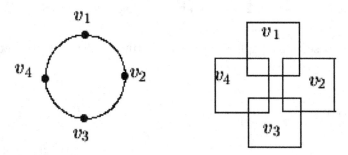

Lemma 10.1. (Roberts, 1969) *For any graph G, $box(G) \leq k$ if and only if there exists k interval graphs $I_1, \ldots, I_k$ such that $G = I_1 \cap I_2 \cap \cdots \cap I_k$.*

Roberts also proved the following theorem.

Theorem 10.1. (Roberts, 1969) *Let G be a graph G with n vertices. Then $box(G) \leq n$.*

The problem to represent a graph into boxes is NP-complete. The decision version of the problem, i.e. to test whether the boxicity of a given graph is at most some given value k, even for $k=2$, is NP-complete. An efficient solution or approximation algorithm is available for lower values of k to represent a graph into boxes. Chandran et al. (2006) described algorithms for finding representations of arbitrary graphs as IntGs of boxes, with a dimension that is within a logarithmic factor of the maximum degree of the graph.

Theorem 10.2. (Chandran and Sivadasan, 2007) *"Let G be a graph with n vertices and maximum degree Δ. Then $box(G) \leq \lceil (\Delta + 2) \ln n \rceil$."*

POWER OF INTERSECTION GRAPHS

Let G be a graph. Then k-th power of the graph $G = (V, E)$ is denoted by $G^k = (V', E')$ and is defined as $V' = V$ and

$$E = \{(u, v) : d(u, v) \leq k, d(u, v)$$

is the distance between the vertices u and v in G}.

It is interesting that many intersection graphs are closed under power. That is, if G is the intersection class and if $G \in$ G then $G^k \in$ G for all $k \geq 1$. An intersection class is called strongly closed under power, if $G^k \in$ G implies that $G^{k+1} \in$ G for every $k \geq 1$.

This property is stated for interval graph.

Theorem 11.1. (Raychaudhuri, 1987): *"The intersection classes of InvGs, proper InvGs and asteroidal triple-free graphs are strongly closed under power".*

For the circular-arc graphs, Prisner and Flotow stated the following results.

Theorem 11.2. (Prisner, 1997): *"The intersection class of proper CirGs is strongly closed under powers".*

Theorem 11.3. (Flotow, 1995): "*The intersection class of CirGs is closed under powers and for all $k \geq 2$, G^k is a CirG implies that G^{k+2} is a CirG*".

But, the result is different for chordal graphs discussed below:

The square of a chordal graph is not chordal. But, Wallis and Wu stated the following theorem when square of chordal graph is chordal.

Theorem 11.4. "*Let G be a chordal graph. Then G^2 is chordal iff the clique graph K(G) of G is closed. Also, for any graph G, if K(G) is chordal then G^2 is chordal*".

The following theorems are related to the closer property of power of chordal graph.

Theorem 11.5. (Duchet, 1984): "*If G^k is chordal then G^{k+2} is chordal. If G and G^2 are both chordal, then all powers of G are chordal*".

Theorem 11.6. (Balakrishnan & Paulraja, 1983) "*If G is chordal and k is odd, then G^k is chordal. If G is chordal and G^{2k} is not chordal, then none of the edges of any chordless cycle of G^{2k} is an edge of G^p, $p < 2k$* ".

Theorem 11.7. (Flotow, 1995): "*If G contains no induced A-graph or $K_{1,3}$ or C_n with $n \geq 2k + 2$, then G^k is chordal. If G contains no induced $K_{1,3}$ or C_n with $n \geq 4$, then every power of G is chordal. If G contains no induced $K_{1,3}$ or C_n with $n \geq 6$, then every odd power G^k, $k \geq 3$ of G is chordal*".

For the strongly chordal graphs Raychaudhuri gives closed property of power of graphs.

Theorem 11.8. (Raychaudhuri, 1987): "*The intersection class of strongly chordal graphs is strongly closed under powers*".

Also, every power of a bock graph is chordal.

Let us define *m*-trapezoid graph. For a positive integer *m*, let l_1, l_2, ..., l_{m+1} be (*m+1*) parallel lines in R^2, where l_i are indexed according to their ordering. Let [a_i, b_i] be a closed interval on l_i. Then the interior of the closed polygon a_1, a_2,..., a_{m+1}, b_{m+1}, b_m,...,b_2, b_1, a_1 form an m-trapezoid.

Definition 11.1. (*m*-trapezoid graph): Let l_1, l_2, ..., l_{m+1} be (*m+1*) parallel lines on R^2 and they are indexed according to their ordering. Let { T_1, T_2,..., T_n } be a set of n number of m-trapeziods on l_1, l_2, ..., l_{m+1} . A graph $G = (V, E)$ is called an *m*-trapezoid graph, where $V = \{T_1, T_2, ..., T_n\}$ and there is an edge $(i, j) \in E$ iff $T_i \cap T_j \neq \varnothing$.

Remark 11.1. The class of 0-trapezoid graphs is exactly the class of InvG and class of 1-trapezoid graph is exactly the class of TraG.

Flotow has proved several results on *m*-trapezoid graph.

Theorem 11.9. (Flotow, 1995) Let *G* be a co comparability graph. Then there exists a positive integer *m*, such that *G* is an *m*-trapezoid graph.

The closer property of *m*-trapezoid graph on power is stated below.

Theorem 11.10. (Flotow, 1995) "*If G^{k-1} is an m-trapezoid graph, then G^k is an m-trapezoid graph.*"

As corollary, the following results are obvious.

Corollary 11.1. (Flotow, 1995) "*If G^{k-1} is a trapezoid graph, then G^k is a trapezoid graph.*"

Corollary 11.2. (Flotow, 1995) "*If G^{k-1} is a cocomparability graph, then G^k is a cocomparability graph.*"

Thus, combining these entire results one can conclude that

"Power of an InvGs are InvGs, power of cocomparability graphs are cocomparability graphs and power of TraGs are TraGs".

HYPERGRAPH

Let I be the index and $\{A_u, u \in I\}$ be a family of sets. The family of sets $\{A_u, u \in I\}$ are called the hypergraphs. The sets A_u are called hyperedges. A hypergraph is denoted by $H = (V, A_u, u \in I\}$. The IntG $F(H)$ of a hypergraph $H = (V, A_u, u \in I\}$ is defined as follows:

The set of vertices of H is I and there is an edge between two vertices u and v if $A_u \cap A_v \neq \varnothing$.

Two edges of a hypergraph are l-intersecting if they have at least l common vertices i.e. A_u and A_v are l-intersecting if $|A_u \cap A_v| \geq l$. A hypergraph H is called k-uniform if all edges of H have exactly k vertices. A hypergraph is linear if any two edges have at most one common vertex. A 2-uniform linear hypergraph is the graph.

Many different restrictions may be imposed on the set A_u. Suppose P be a set property and all the hyperedges (sets) $\{A_u, u \in I\}$ of H obey this set property. Then the class of all such finite IntGs is denoted by G(P).

If the hypergraph has no multiple sets, then it is called simple hypergraph.

The class of all IntGs of simple hypergraphs is denoted by G0(P), where all hypergraphs obey the property ρ.

The graph classes G(P) and G0(P) are called Scheinerman classes. The graph class G(P) or G0(P) is only important for discrete intersection models.

Lemma 12.1. (Scheinerman, 1985) *"For every Scheinerrman class G(P), there is a countable subset P' of P such that G(P) =G$(P)'$."*

The IntG of k-uniform simple hypergraphs is NP-complete for all $k \geq 3$, but it can be done in linear time for $k = 2$.

INTERSECTION GRAPH

Let $H = (V, A_u, u \in I)$ be a hypergraph. The l-intersection graph $F_l(H)$ of H is defined as: V is the set of vertices of $F_l(H)$ and two distinct vertices u and v have an edge if $|A_u \cap A_v| \geq l$. This kind of intersection occurs only in the discrete intersection models.

A t-linear hypergraph is a hypergraph in which every two hyper edges have at most t points in common.

It can be proved that an l-intersection graphs of k-uniform t-linear hypergraphs are intersection graphs of certain $\binom{k}{l}$–uniform, $\binom{t}{l}$-linear hypergraphs.

LINE GRAPHS

One of the interesting graph classes is line graph. Let G be a graph. The *line graph* of G is generally denoted by $L(G)$ and defined as follows:

Given a graph G, its line graph $L(G)$ is a graph such that

1.   For each edge of G there is a vertex of $L(G)$; and
2.   Two vertices of $L(G)$ are adjacent if and only if corresponding edges share a common endpoint in G.

That is, line graph is the IntG of the edges of G, representing each edge by the set of its two endpoints.

The line is also known as *the covering graph, the derivative, the conjugate, theta-obrazom, the edge-to-vertex dual, and the representative graph, the interchange graph, the adjoint graph, as well as the edge graph, and the derived graph.*

Let us consider a graph G shown in Figure 15(a). The edges of this graph are

$$e_1 = (1,2),\; e_2 = (1,5),\; e_3 = (1,6),\; e_4 = (2,3),\; e_5 = (3,4),\; e_6 = (4,5),\; e_7 = (4,6).$$

Seven vertices are drawn (see Figure 15(b)) for seven edges. Since e_1 and e_2 have a common vertex 1, so there is an edge between e_1 and e_2. Again, there is no common vertex between the edges e_2 and e_5 so there is no edge between e_2 and e_5. In this way the entire line graph $L(G)$ (Figure 15(b)) is constructed from G.

The line graph of the complete graph K_n is the triangular graph. Every line graph is a claw-free graph. Some of the properties of claw-free graphs are generalizations of those of line graphs. The line graph of a bipartite graph is perfect. The line graphs of bipartite graphs form one of the key building blocks of perfect graphs.

Figure 15. (a) A graph G, (b) Line graph of G

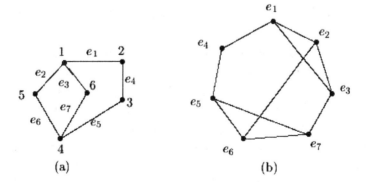

Some important results of line graph are presented below:

1. The line graph of a connected graph is connected.
2. A maximum independent set in a line graph corresponds to maximum matching in the original graph.
3. The edge chromatic number of a graph G is equal to the vertex chromatic number of its line graph $L(G)$.
4. The line graph of an edge-transitive graph is vertex-transitive.
5. If a graph G has an Euler cycle, that is, if G is connected and has an even number of edges at each vertex, then the line graph of G is Hamiltonian.
6. The line graphs of trees are exactly the claw-free block graph.

It is very interesting that line graph of line graph is not the original graph, it gives some other graph. L can be treated as an operator. Rooij and Wilf (1965) consider the sequence of graphs

$$G, L(G), L(L(G)), L(L(L(G))), \dots$$

They proved the following behaviours of this sequence.

1. If G is a cycle graph then $L(G)$ and each subsequent graph in this sequence is isomorphic to G itself. These are the only connected graphs for which $L(G)$ is isomorphic to G.
2. If G is a claw $K_{1,3}$, then $L(G)$ and all subsequent graphs in the sequence are triangles.
3. If G is a path graph then each subsequent graph in the sequence is a shorter path until eventually the sequence terminates with a graph having only one vertex.
4. In all remaining cases, the sizes of the graphs in this sequence eventually increase.

STRING GRAPHS

A *string graph* is an IntG of curves in the plane and each curve is called a "string". Let S be a collection of strings in a plane such that no three strings intersect at a single point. One can draw an IntG for a set of such strings on plane. Draw a vertex for each string and an edge for each intersecting pair of strings. This IntG is called *string graph*. The mathematical study of string graphs began with work of Ehrlich et al. (1976). But, recognition of string graphs is NP-complete (Pach & T´oth, 2002).

A problem on string graph occurs naturally in the context of layout problem RC circuits (Ehrlich et al., 1976). Nowadays the circuit layout setting is modeled as paths (wires) intersecting on a grid. In the knock-knee layout models, all the wires may bend (turn) at a grid point, but intersect with another wire is not allowed. A layout may have multiple layers and in a legal layout, on each layer, the vertex IntG of paths on a grid is an independent set. The minimum coloring problem of VPG graphs defines the knock-knee multiple layout with minimum number of layers.

It is easy to verify that every planar graph is a string graph. The string representation is described as follows: draw a string for each vertex that loops around the vertex and around the midpoint of each adjacent edge. For any edge (u, v) of the graph, the strings for u and v cross each other twice near the midpoint of the edge (u, v), and there are no other crossings, so the pairs of strings that cross represent exactly the adjacent pairs of vertices of the original planar graph. Chalopin et al. (2007) proved that every planar graph has a string representation in which each pair of strings has at most one crossing point, unlike the representations described above.

Let P be a set of simple paths on a (rectangular) grid. The vertex IntG VPG(P) of P has vertex set V, where every vertex $v \in V$ corresponds to a path $P_v \in P$, and edge set E, where $(u, v) \in E$ if and only if the corresponding paths P_u and P_v intersect, i.e.

$$E = \{(u, v) \mid u, v \in V, u \neq v, P_u \cap P_v \neq \varphi\}.$$

We call a graph G a VPG graph if $G \cong \text{VPG(P)}$, for some P. If P is a set of simple paths on a grid, where each path has at most k bends ($90°$ turns), then the graph G is called $B_k - VPG$.

$B_k - VPG$ graphs are related to several other families of IntG s that have been studied in the literature. It is rather simple to prove that the string graphs are equivalent to the VPG graphs when there is no restriction on the number of bends per path in the grid. InvG s and trees are both subfamilies of $B_0 - VPG$, and the so called grid IntGs (Hartman et al., 1991) are equivalent to the bipartite $B_0 - VPG$ graphs.

Circle graphs are a subfamily of string graphs, and it is easy to show that they are contained in the class $B_1 - VPG$. This immediately implies that the coloring problem is NP-complete on $B_1 - VPG$ graphs. We prove the stronger result that the coloring problem is NP-complete for $B_0 - VPG$ graphs. It is proved that every planar graph is a $B_3 - VPG$ graph (Asinowski et al., 2012). Another connection between planar graphs and string graphs began when Scheinerman and West (1983) conjectured that planar graphs are contained in the family of segment graphs (SEG), the intersection graphs of straight-line segments in the plane with an arbitrary number of directions. Recently, Chalopin et al. (2007) proved Scheinerman's conjecture.

Theorem 14.1. (Asinowski et al., 2012) *"The family of VPG graphs is equivalent to the family of string graphs."*

The Grid Intersection Graphs (GIG) are the IntG s of horizontal and vertical line segments on the plane, such that no two horizontal segments or two vertical segments intersect.

Theorem 14.2. (Asinowski et al., 2012) *"The family of bipartite $B_0 - VPG$ graphs is equivalent to the family of GIG."*

The GIGs are the bipartite graphs with boxicity two. It can easily be proved that every tree is a bipartite $B_0 - VPG$ graph.

Another type of IntG is defined below.

The family of IntG s of straight-line segments parallel with at most d directions is called the d-DIR graphs.

When $d = 2$, without loss of generality, it is assume that the two directions are orthogonal to each other.

Theorem 14.3. (Asinowski et al., 2012) *"The $B_0 - VPG$ graphs are exactly the 2-DIR graphs."*

REFERENCES

Asinowski, A., Cohen, E., Golumbic, M. C., Limouzy, V., Lipshteyn, M., & Stern, M. (2012). Vertex intersection graphs of paths on a grid. *Journal of Graph Algorithms and Applications*, *16*(2), 129–150. doi:10.7155/jgaa.00253

Aumann, Y., Lewenstein, M., Melamud, O., Pinter, R. Y., & Yakhini, Z. (2005). Dotted interval graphs and high throughput genotyping. *Proceedings of the Sixteenth Annual ACM-SIAM Symposium on Discrete Algorithms*, 339–348.

Aumann, Y., Lewenstein, M., Melamud, O., Pinter, R. Y., & Yakhini, Z. (2012). Dotted interval graphs. *ACM Transactions on Algorithms*, *8*(2).

Balakrishnan, R., & Paulraja, P. (1983). Powers of chordal graphs. *Journal of the Australian Mathematical Society*, *35*(2), 211–217. doi:10.1017/S1446788700025696

Barman, S. C., Mondal, S., & Pal, M. (2009). An efficient algorithm to find next-to-shortest path on permutation graphs. *Journal of Applied Mathematics and Computing*, *31*(1), 369–384. doi:10.100712190-008-0218-1

Barman, S. C., Mondal, S., & Pal, M. (2010a). A linear time algorithm to construct a tree 4-spanner on trapezoid graphs. *International Journal of Computer Mathematics*, *87*(4), 743–755. doi:10.1080/00207160802037880

Barman, S. C., Pal, M., & Mondal, S. (2007). An efficient algorithm to find next-to-shortest path on trapezoid graphs. *Advances in Applied Mathematical Analysis*, *2*(2), 97–107.

Barman, S. C., Pal, M., & Mondal, S. (2010b). The k-neighbourhood-covering problem on interval graphs. *International Journal of Computer Mathematics*, *87*(9), 1918–1935. doi:10.1080/00207160802676570

Bera, D., Pal, M., & Pal, T. K. (2002). An efficient algorithm to generate all maximal cliques on trapezoid graphs. *International Journal of Computer Mathematics*, *79*(10), 1057–1065. doi:10.1080/00207160212707

Bera, D., Pal, M., & Pal, T. K. (2003a). An efficient algorithm for finding all hinge vertices on trapezoid graphs. *Theory of Computing Systems*, *36*(1), 17–27. doi:10.100700224-002-1004-3

Bera, D., Pal, M., & Pal, T. K. (2003b). An optimal PRAM algorithm for a spanning tree on trapezoid graphs. *Journal of Applied Mathematics and Computing*, *12*(1–2), 21–29. doi:10.1007/BF02936178

Bera, D., Pal, M., & Pal, T. K. (2001). A parallel algorithm for computing all hinge vertices on interval graphs. *Korean Journal of Computational & Applied Mathematics*, *8*(2), 295–309.

Bogart, K. P., Jacobson, M. S., Langley, L. J., & Mcmorris, F. R. (2001). Tolerance orders and bipartite unit tolerance graphs. *Discrete Mathematics*, *226*(1), 35–50. doi:10.1016/S0012-365X(00)00124-2

Buneman, P. (1974). A characterisation of rigid circuit graphs. *Discrete Mathematics*, *9*(3), 205–212. doi:10.1016/0012-365X(74)90002-8

Carlisle, M. C., & Lloyd, E. L. (1995). On the k-coloring of intervals. *Discrete Applied Mathematics*, *59*(3), 225–235. doi:10.1016/0166-218X(95)80003-M

Chalopin, J., Gonçalves, D., & Ochem, P. (2007). Planar graphs are in 1-STRING. *Proceedings of the Eighteenth Annual ACM-SIAM Symposium on Discrete Algorithms*, 609–617.

Chandran, L. S., & Sivadasan, N. (2006, August 15). *Geometric representation of graphs in low dimension*. Academic Press. doi:10.1007/11809678_42

Chandran, L. S., & Sivadasan, N. (2007). Boxicity and treewidth. *Journal of Combinatorial Theory Series B*, *97*(5), 733–744. doi:10.1016/j.jctb.2006.12.004

Cheah, F., & Corneil, D. G. (1996). On the structure of trapezoid graphs. *Discrete Applied Mathematics*, *66*(2), 109–133. doi:10.1016/0166-218X(94)00158-A

Chen, E., Yang, L., & Yuan, H. (2007). Improved algorithms for largest cardinality 2-interval pattern problem. *Journal of Combinatorial Optimization*, *13*(3), 263–275. doi:10.100710878-006-9030-8

Corneil, D., Olariu, S., & Stewart, L. (1997). Asteroidal triple-free graphs. *SIAM Journal on Discrete Mathematics*, *10*(3), 399–430. doi:10.1137/S0895480193250125

Crochemore, M., Hermelin, D., Landau, G. M., Rawitz, D., & Vialette, S. (2008). Approximating the 2-interval pattern problem. *Theoretical Computer Science*, *395*(2), 283–297. doi:10.1016/j.tcs.2008.01.007

Dagan, I., Golumbic, M. C., & Pinter, R. Y. (1988). Trapezoid graphs and their coloring. *Discrete Applied Mathematics*, *21*(1), 35–46. doi:10.1016/0166-218X(88)90032-7

Deng, X., Hell, P., & Huang, J. (1996). Linear-time representation algorithms for proper circular-arc graphs and proper interval graphs. *SIAM Journal on Computing*, *25*(2), 390–403. doi:10.1137/S0097539792269095

Dirac, G. A. (1961). On rigid circuit graphs. *Abhandlungen aus dem Mathematischen Seminar der Universität Hamburg*, *25*(1), 71–76. doi:10.1007/BF02992776

Duchet, P. (1984). Classical perfect graphs: an introduction with emphasis on triangulated and interval graphs. In *North-Holland mathematics studies* (Vol. 88, pp. 67–96). North-Holland.

Ehrlich, G., Even, S., & Tarjan, R. E. (1976). Intersection graphs of curves in the plane. *Journal of Combinatorial Theory Series B*, *21*(1), 8–20. doi:10.1016/0095-8956(76)90022-8

Fabri, J. (1982). *Automatic storage optimization*. Academic Press.

Flotow, C. (1995). On powers of m-trapezoid graphs. *Discrete Applied Mathematics*, *63*(2), 187–192. doi:10.1016/0166-218X(95)00062-V

Fomin, F. V., Gaspers, S., Golovach, P., Suchan, K., Szeider, S., van Leeuwen, E. J., … Villanger, Y. (2012). k-gap interval graphs. In D. Fernández-Baca (Ed.), LATIN 2012: Theoretical Informatics (pp. 350–361). Springer Berlin Heidelberg.

Fulkerson, D., & Gross, O. (1965). Incidence matrices and interval graphs. *Pacific Journal of Mathematics*, *15*(3), 835–855. doi:10.2140/pjm.1965.15.835

Galinier, P., Habib, M., & Paul, C. (1995). Chordal graphs and their clique graphs. In M. Nagl (Ed.), *Graph-Theoretic Concepts in Computer Science* (pp. 358–371). Springer Berlin Heidelberg.

Garey, M. R., & Johnson, D. S. (1990). *Computers and Intractability; A Guide to the Theory of NP-Completeness*. New York, NY: W. H. Freeman & Co.

Gavril, F. (1974). The intersection graphs of subtrees in trees are exactly the chordal graphs. *Journal of Combinatorial Theory Series B*, *16*(1), 47–56. doi:10.1016/0095-8956(74)90094-X

Ghouilo-Houri, A. (1962). *Characterisation des graphs non orientes dont on peut orienter les arretes de maniere a obtenir le graphe d'une relation d'ordre* (Vol. 254). Paris: C. R. Acad. Sci.

Gilmore, P. C., & Hoffman, A. J. (1964). A characterization of comparability graphs and of interval graphs. *Canadian Journal of Mathematics*, *16*, 539–548. doi:10.4153/CJM-1964-055-5

Golumbic, M. C. (1980). *Algorithmic graph theory and perfect graphs* (Vol. 57). Academic Press.

Golumbic, M. C., Monma, C. L., & Trotter, W. T. (1984). Tolerance graphs. *Discrete Applied Mathematics*, *9*(2), 157–170. doi:10.1016/0166-218X(84)90016-7

Golumbic, M. C., & Trenk, A. N. (2004). *Tolerance Graphs*. Cambridge, UK: Cambridge University Press.

Griggs, J., & West, D. (1980). Extremal Values of the Interval Number of a Graph. *SIAM Journal on Algebraic Discrete Methods*, *1*(1), 1–7. doi:10.1137/0601001

Harary, F., Kabell, J. A., & McMorris, F. R. (1982). Bipartite intersection graphs. *Commentationes Mathematicae Universitatis Carolinae*, *23*(4), 739–745.

Hartman, I. B.-A., Newman, I., & Ziv, R. (1991). On grid intersection graphs. *Discrete Mathematics*, *87*(1), 41–52. doi:10.1016/0012-365X(91)90069-E

Hell, P., & Huang, J. (2004). Interval bigraphs and circular arc graphs. *Journal of Graph Theory*, *46*(4), 313–327. doi:10.1002/jgt.20006

Hermelin, D., Mestre, J., & Rawitz, D. (2014). Optimization problems in dotted interval graphs. *Discrete Applied Mathematics*, *174*, 66–72. doi:10.1016/j.dam.2014.04.014

Horne, J. A., & Smith, J. C. (2005). Dynamic programming algorithms for the conditional covering problem on path and extended star graphs. *Networks*, *46*(4), 177–185. doi:10.1002/net.20086

Hota, M., Pal, M., & Pal, T. K. (1999). An efficient algorithm to generate all maximal independent sets on trapezoid graphs. *International Journal of Computer Mathematics*, *70*(4), 587–599. doi:10.1080/00207169908804777

Hota, M., Pal, M., & Pal, T. K. (2001). An efficient algorithm for finding a maximum weight k-independent set on trapezoid graphs. *Computational Optimization and Applications*, *18*(1), 49–62. doi:10.1023/A:1008791627588

Hota, M., Pal, M., & Pal, T. K. (2004). Optimal sequential and parallel algorithms to compute all cut vertices on trapezoid graphs. *Computational Optimization and Applications*, *27*(1), 95–113. doi:10.1023/B:COAP.0000004982.13444.bc

Hota, M. (2005). *Sequential and parallel algorithms on some problems of trapezoid graph*. Midnapore, India: Vidyasagar University.

Lekkeikerker, C., & Boland, J. (1962). Representation of a finite graph by a set of intervals on the real line. *Fundamenta Mathematicae, 51*, 45–64. doi:10.4064/fm-51-1-45-64

Ma, T. H., & Spinrad, J. P. (1994). On the 2-chain subgraph cover and related problems. *Journal of Algorithms, 17*(2), 251–268. doi:10.1006/jagm.1994.1034

Mandal, S., & Pal, M. (2006a). A sequential algorithm to solve next-to-shortest path problem on circular-arc graphs. *The Journal of Physiological Sciences; JPS, 10*, 201–217.

Mandal, S., & Pal, M. (2006b). Maximum weight independent set of circular-arc graph and its application. *Journal of Applied Mathematics and Computing, 22*(3), 161–174. doi:10.1007/BF02832044

Mandal, S., & Pal, M. (2007). An optimal algorithm to compute all hinge vertices on circular-arc graphs. *Arab Journal of Mathematics and Mathematical Sciences, 1*(1), 16–27.

McConnell, R. M. (2003). Linear-time recognition of circular-arc graphs. *Algorithmica, 37*(2), 93–147. doi:10.100700453-003-1032-7

McKee, T., & McMorris, F. (1999). *Topics in intersection graph theory*. Retrieved from https://epubs.siam.org/doi/abs/10.1137/1.9780898719802

McKee, T. A., & McMorris, F. R. (1999). *Topics in intersection graph theory*. SIAM.

Mondal, S., Pal, M., & Pal, T. K. (1999). An optimal algorithm for finding depth-first spanning tree on permutation graphs. *Korean Journal of Computational & Applied Mathematics, 6*(3), 493–500. doi:10.1007/BF03009943

Mondal, S., Pal, M., & Pal, T. K. (2002a). An optimal algorithm for solving all-pairs shortest paths on trapezoid graphs. *International Journal of Computational Engineering Science, 03*(02), 103–116. doi:10.1142/S1465876302000575

Mondal, S., Pal, M., & Pal, T. K. (2002b). An optimal algorithm to solve 2-neighbourhood covering problem on interval graphs. *International Journal of Computer Mathematics, 79*(2), 189–204. doi:10.1080/00207160211921

Mondal, S., Pal, M., & Pal, T. K. (2003a). Optimal sequential and parallel algorithms to compute a Steiner tree on permutation graphs. *International Journal of Computer Mathematics, 80*(8), 937–943. doi:10.1080/0020716031000112330

Mondal, S., Pal, M., & Pal, T. K. (2003b). An optimal algorithm to solve the all-pairs shortest paths problem on permutation graph. *Journal of Mathematical Modelling and Algorithms, 2*(1), 57–65. doi:10.1023/A:1023695531209

Müller, H. (1997). Recognizing interval digraphs and interval bigraphs in polynomial time. *Discrete Applied Mathematics, 78*(1), 189–205. doi:10.1016/S0166-218X(97)00027-9

Pach & Tóth. (2002). Recognizing string graphs is decidable. *Discrete & Computational Geometry, 28*(4), 593–606. doi:10.100700454-002-2891-4

Pal, M. (1995a). *Some sequential and parallel algorithms on interval graphs* (Ph. D Thesis). Indian Institute of Technology, Kharagpur, India.

Pal, M., & Bhattacharjee, G. P. (1995b). Optimal sequential and parallel algorithms for computing the diameter and the center of an interval graph. *International Journal of Computer Mathematics, 59*(1–2), 1–13. doi:10.1080/00207169508804449

Pal, M. (1996a). An efficient parallel algorithm for computing a maximum-weight independent set of a permutation graph. *Proc.: 6th National Seminar on Theoretical Computer Science, Banasthali Vidyapith, Rajasthan, India*, 276–285.

Pal, M., & Bhattacharjee, G. (1996b). An optimal parallel algorithm to color an interval graph. *Parallel Processing Letters, 6*(4), 439–449.

Pal, M., & Bhattacharjee, G. P. (1996c). A sequential algorithm for finding a maximum weight k-independent set on interval graphs. *International Journal of Computer Mathematics, 60*(3–4), 205–214. doi:10.1080/00207169608804486

Pal, M., & Bhattacharjee, G. (1997a). An optimal parallel algorithm for all-pairs shortest paths on un-weighted interval graphs. *Nordic Journal of Computing, 4*(4), 342–356.

Pal, M., & Bhattacharjee, G. P. (1997b). A data structure on interval graphs and its applications. *Journal of Circuits, Systems, and Computers, 07*(03), 165–175. doi:10.1142/S0218126697000127

Pal, M. (1998a). A parallel algorithm to generate all maximal independent sets on permutation graphs. *International Journal of Computer Mathematics, 67*(3–4), 261–274. doi:10.1080/00207169808804664

Pal, M. (1998b). Efficient algorithms to compute all articulation points of a permutation graph. *Korean Journal of Computational & Applied Mathematics, 5*(1), 141–152. doi:10.1007/BF03008943

Pnueli, A., Lempel, A., & Even, S. (1971). Transitive orientation of graphs and identification of per-mutation graphs. *Canadian Journal of Mathematics, 23*(1), 160–175. doi:10.4153/CJM-1971-016-5

Ramalingam, G., & Rangan, C. P. (1988). A unified approach to domination problems on interval graphs. *Information Processing Letters, 27*(5), 271–274.

Rana, A. (2011). Pal, A., & Pal, M. (2011a). *The 2-neighbourhood covering problem on permutation graphs. Advanced Modelling and Optimization, 13*(3), 463–476.

Rana, A., Pal, A., & Pal, M. (2011c). Efficient algorithms to solve k-domination problem on permutation graphs. In Y. Wu (Ed.), High Performance Networking, Computing, and Communication Systems (pp. 327–334). Springer Berlin Heidelberg.

Raychaudhuri, A. (1987). On powers of interval and unit interval graphs. *Congr. Numer., 4*, 235–242.

Roberts, F. S. (1969). On the boxicity and cubicity of a graph. *Recent Progresses in Combinatorics*, 301–310.

Rose, D., Tarjan, R., & Lueker, G. (1976). Algorithmic aspects of vertex elimination on graphs. *SIAM Journal on Computing, 5*(2), 266–283. doi:10.1137/0205021

Saha, A., & Pal, M. (2003). Maximum weight k-independent set problem on permutation graphs. *International Journal of Computer Mathematics, 80*(12), 1477–1487. doi:10.1080/00207160310001614972

Saha, A., Pal, M., & Pal, T. K. (2005a). An optimal parallel algorithm to construct a tree 3-spanner on interval graphs. *International Journal of Computer Mathematics, 82*(3), 259–274. doi:10.1080/00207160412331286851

Saha, A., Pal, M., & Pal, T. K. (2005b). An optimal parallel algorithm for solving all-pairs shortest paths problem on circular-arc graphs. *Journal of Applied Mathematics and Computing, 17*(1), 1–23.

Saha, A., Pal, M., & Pal, T. K. (2005c). An efficient PRAM algorithm for maximum-weight independent set on permutation graphs. *Journal of Applied Mathematics and Computing, 19*(1–2), 77–92. doi:10.1007/BF02935789

Saha, A., Pal, M., & Pal, T. K. (2007). Selection of programme slots of television channels for giving advertisement: A graph theoretic approach. *Information Sciences, 177*(12), 2480–2492. doi:10.1016/j.ins.2007.01.015

Scheinerman, E. R. (1985). Characterizing intersection classes of graphs. *Discrete Mathematics, 55*(2), 185–193. doi:10.1016/0012-365X(85)90047-0

Scheinerman, E. R., & West, D. B. (1983). The interval number of a planar graph: Three intervals suffice. *Journal of Combinatorial Theory Series B, 35*(3), 224–239. doi:10.1016/0095-8956(83)90050-3

Spinrad, J. (1985). On comparability and permutation graphs. *SIAM Journal on Computing, 14*(3), 658–670. doi:10.1137/0214048

Steiner, G. (1996). The recognition of indifference digraphs and generalized semiorders. *Journal of Graph Theory, 21*(2), 235–241.

Tiskin, A. (2015). Fast distance multiplication of unit-Monge matrices. *Algorithmica, 71*(4), 859–888. doi:10.100700453-013-9830-z

Tucker, A. (1971). Matrix characterizations of circular-arc graphs. *Pacific Journal of Mathematics, 39*(2), 535–545.

Tucker, A. (1974). Structure theorems for some circular-arc graphs. *Discrete Mathematics, 7*(1), 167–195. doi:10.1016/S0012-365X(74)80027-0

van Rooij, A. C. M., & Wilf, H. (1965). The interchange graph of a finite graph. *Acta Mathematica Hungarica, 16*(3–4), 263–269. doi:10.1007/BF01904834

Yu, C.-W., & Chen, G.-H. (1993). Generate all maximal independent sets in permutation graphs. *International Journal of Computer Mathematics, 47*(1–2), 1–8. doi:10.1080/00207169308804157

Chapter 3
Graph Indices

V. R. Kulli
Gulbarga University, India

ABSTRACT

A molecular graph is a finite simple graph representing the carbon-atom skeleton of an organic molecule of a hydrocarbon. Studying molecular graphs is a constant focus in chemical graph theory: an effort to better understand molecular structure. Many types of graph indices such as degree-based graph indices, distance-based graph indices, and counting-related graph indices have been explored recently. Among degree-based graph indices, Zagreb indices are the oldest and studied well. In the last few years, many new graph indices were proposed. The present survey of these graph indices outlines their mathematical properties and also provides an exhaustive bibliography.

ZAGREB INDICES

Let G be a simple graph with vertex set $V(G)$ and edge set $E(G)$. Let $|V(G)| = n$ and $|E(G)| = m$. The edge connecting the vertices u and v will be denoted by uv. The degree $d_G(u)$ of u is the number of vertices adjacent to u.

A graph invariant is a function on a graph which is not depending on the labeling of its vertices. Such quantity is called a topological index or molecular descriptor or graph index. Graph indices play a significant role in Theoretical Chemistry, especially in, structure property relationships and structure activity relationships (Gutman, 2013, 1986; Todeschini & Consonni, 2009). In Chemical Science, graph indices have been found to be useful in chemical documentation, isomer discrimination and pharmaceutical drug design. There has been considerable interest in the general problem of determining graph indices. Wiener index (Wiener, 1947) is the best-known index and it is defined as the sum of the distances between all pairs of vertices of a graph G

$$W(G) = \sum_{\{u,v \in V(G)\}} d(u,v).$$

DOI: 10.4018/978-1-5225-9380-5.ch003

This index is used for the calculation of the boiling points of alkenes. In Gutman and Trinajstić (1972), Gutman and Trinajstić (1972) introduced the first and second Zagreb indices, defined as

$$M_1(G) = \sum_{u \in V(G)} d_G(u)^2,$$

$$M_2(G) = \sum_{uv \in E(G)} d_G(u) d_G(v).$$

The first Zagreb index is defined as (Nikolic et al., 2003).

$$M_1(G) = \sum_{uv \in E(G)} \Big[d_G(u) + d_G(v) \Big].$$

Many properties and chemical applications of these indices can be found in Caparossi, Hansen, and Vukicevic (2010), Das and Gutman (2004), Das, Gutman, and Horoldagya (2012), Das, Gutman,a nd Zhou (2009), Das, Xu, and Gutman (2013), Doslic et al. (2011), Fath-Tabar (2011), Gutman and Das (2004), Hosamani and Gutman (2014), Ilic and Stevanovic (2011), Khalifeh, Yousefi-Azari, and Ashrafi (2009), Liu and Yoi (2011), Ranjini, Lokesha, and Cangul (2011), Reti (2012), Trinajstić et al. (2010), Xu et al. (2015), and Zhou (2004, 2007). Several results on Zagreb indices were established and many modifications of Zagreb indices were introduced.

Shirdel et al. proposed the first hyper Zagreb index (Shirdel, Rezapour & Sayadi, 2013)

$$HM_1(G) = \sum_{uv \in E(G)} \Big[d_G(u) + d_G(v) \Big]^2.$$

Some properties of the first hyper Zagreb index were obtained (Basavanagoud & Patil, 2016; Falahati, Nezhad & Azari, 2016; Gao, Jamil & Farahani, 2017; Gutman, 2017; Luo, 2016; Pattabiraman & Vijayaragayan, 2017).

The second hyper Zagreb index was defined as (Gao et al., 2016):

$$HM_2(G) = \sum_{uv \in E(G)} \Big[d_G(u) d_G(v) \Big]^2.$$

The simple modification of these indices is to propose on their definition a variable parameter. In Li and Zhao (2004) and Li and Zheng (2005), the generalization of Zagreb indices were introduced, defined as

$$M_1^a(G) = \sum_{u \in V(G)} d_G(u)^a = \sum_{uv \in E(G)} \Big[d_G(u)^{a-1} + d_G(v)^{a-1} \Big]$$

$$M_2^a(G) = \sum_{uv \in E(G)} \left[d_G(u)d_G(v) \right]^a.$$

where a is a real number.

The general sum connectivity index is defined as (Zhou & Trinajstić, 2009):

$$M_1^{*a}(G) = \sum_{uv \in E(G)} \left[d_G(u) + d_G(v) \right]^a.$$

Several publications on these indices can be found in Britto et al. (2014), Kulli (2017a, 2017b, 2017c), and Liu and Liu (2010).

$M_1^a(G)$ for $a = 2$ is the first Zagreb index. Also $M_2^a(G)$ for $a = 1$ is the second Zagreb index.

$M_1^{*a}(G)$ for $a = 2$ is the first hyper Zagreb index. Also $M_2^a(G)$ for $a = 2$ is the second hyper Zagreb index.

$M_1^a(G)$ for $a = 3$ is the forgotten topological index or F-index, which is defined as (Furtula & Gutman, 2015):

$$F(G) = \sum_{u \in V(G)} d_G(u)^3 = \sum_{uv \in E(G)} \left[d_G(u)^2 + d_G(v)^2 \right].$$

Followed by many papers (Basavanagoud & Desai, 2016; Che & Chen, 2016; De, 2016; Gao et al., 2016; Kulli, 2017d; Ramane & Jummannaver, 2016).

$M_1^{*a}(G)$ for $a = -\dfrac{1}{2}$ is the sum connectivity index

$$S(G) = \sum_{uv \in E(G)} \frac{1}{\sqrt{d_G(u) + d_G(v)}}$$

was proposed by Zhou and Trinajstić (2009). Many results on $S(G)$ can be found in Das, Das, and Zhou (2016), Wang, Zhou, and Trinajstić (2011), and Xing, Zhou, and Trinajstić (2010).

$M_2^a(G)$ for $a = -\dfrac{1}{2}$ is the product connectivity or Randić index

$$P(G) = \sum_{uv \in E(G)} \frac{1}{\sqrt{d_G(u)d_G(v)}}$$

was introduced by Randić (1975). More information on $P(G)$ can be found in Li and Gutman (2006), and Gutman and Furtula (2008). This index is most useful graph index in Chemistry.

Estrada et al. (1998) introduced the atom bond connectivity index

$$ABC(G) = \sum_{uv \in E(G)} \sqrt{\frac{d_G(u) + d_G(v) - 2}{d_G(u)d_G(v)}}.$$

It has several chemical applications (Das, 2010; Das, Gutman & Furtula, 2011; Furtula, Graovac & Vukicevic, 2009; Gam, Hou & liu, 2011; Xing, Zhou & Du, 2010).

The geometric-arithmetic index was proposed by Vukičević and Furtula (2009), as

$$GA(G) = \sum_{uv \in E(G)} \frac{2\sqrt{d_G(u)d_G(v)}}{d_G(u) + d_G(v)}$$

Many results on *GA(G)* can be found in Das (2010) and Mogharrab and Fath-Tabar (2011).

FIFTH ZAGREB INDICES

Let $S_G(u)$ denote the sum of degrees of all vertices adjacent to u. Graovac et al. (2011) introduced fifth version of Zagreb indices as:

$$M_1G_5(G) = \sum_{uv \in E(G)} \left[S_G(u) + S_G(v) \right],$$

$$M_2G_5(G) = \sum_{uv \in E(G)} \left[S_G(u)S_G(v) \right].$$

The fifth version of hyper Zagreb indices were defined as [66]

$$HM_1G_5(G) = \sum_{uv \in E(G)} \left[S_G(u) + S_G(v) \right]^2,$$

$$HM_2G_5(G) = \sum_{uv \in E(G)} \left[S_G(u)S_G(v) \right]^2.$$

In [66], Kulli defined the general fifth version of Zagreb indices as,

$$M_1^a G_5(G) = \sum_{uv \in E(G)} \left[S_G(u) + S_G(v) \right]^a,$$

$$M_2^a G_5(G) = \sum_{uv \in E(G)} \left[S_G(u) S_G(v) \right]^a.$$

Several publications on indices using $S_G(u)$ can be found in the literature.

Recently using $S_G(u)$, the first neighborhood Zagreb index (may be called as fifth version of first Zagreb index) is defined by Mondal et al. (2018) and Basvanagoud et al. (2018) as:

$$NM_1(G) = \sum_{u \in V(G)} S_G(u)^2.$$

KV INDICES

Let $M_G(u)$ denote the product of degrees of all vertices adjacent to u. The first and second *KV* indices

$$KV_1(G) = \sum_{uv \in E(G)} \left[M_G(u) + M_G(v) \right],$$

$$KV_2(G) = \sum_{uv \in E(G)} M_G(u) M_G(v)$$

introduced by Kulli (2018a).

The first vertex *KV* index of *G* is defined by Kulli (2018a) as

$$KV_{01}(G) = \sum_{u \in V(G)} M_G(u)^2.$$

Kulli proposed the first and second hyper KV indices as (Kulli, 2019a):

$$HKV_1(G) = \sum_{uv \in E(G)} \left[M_G(u) + M_G(v) \right]^2,$$

$$HKV_2(G) = \sum_{uv \in E(G)} \left[M_G(u) M_G(v) \right]^2,$$

followed by the papers (Kulli, 2019b, 2018b).

DAKSHAYANI INDICES

The complement $\overline{G}$ of G is the graph with vertex set $V(G)$ in which two vertices are adjacent if they are not adjacent in G. Then $d_{\overline{G}}(u) = n - 1 - d_G(u)$.

Miličević et al. defined the modified first Zagreb index (Milicevic & Nikolic, 2004)

$$^m M_1(G) = \sum_{u \in V(G)} \frac{1}{d_G(u)^2}.$$

The inverse degree has attracted through a conjecture of Grafitti (Fajtolowicz, 1988), defined as (Zhang, Zhang & Lu, 2005):

$$ID(G) = \sum_{u \in V(G)} \frac{1}{d_G(u)}.$$

In Dakshavayani Indices, Kulli (2018c) elaborated the idea of using the degree of a vertex in $\overline{G}$. This leads to the Dakshayani indices, defined as

$$DK_1(G) = \sum_{u \in V(G)} d_{\overline{G}}(u) \frac{1}{d_G(u)}$$

$$DK_2(G) = \sum_{u \in V(G)} d_{\overline{G}}(u) \frac{1}{d_G(u)^2}.$$

In [77], the Lanzhou index was defined as

$$Lz(G) = \sum_{u \in V(G)} d_{\overline{G}}(u) d_G(u)^2.$$

The following identities (Kulli, 2018c; Vukicevic et al., 2018) are immediate:

$$DK_1(G) = (n-1)ID(G) - n.$$

$$DK_2(G) = (n-1)^m M_1(G) - ID(G).$$

$$Lz(G) = (n-1)M_1(G) - F(G).$$

The general Dakshayani index was introduced by Kulli (2018c) as

$$DK^a(G) = \sum_{u \in V(G)} d_{\bar{G}}(u)d_G(u)^a.$$

Clearly,

$$DK^{-1}(G) = DK_1(G), \quad DK^{-2}(G) = DK_2(G), \quad DK^2(G) = Lz(G).$$

GOURAVA INDICES

Motivated by the definitions of the Zagreb indices and their applications, Kulli (2017) proposed the first Gourava index

$$GO_1(G) = \sum_{uv \in E(G)} \left[d_G(u) + d_G(v) + d_G(u)d_G(v) \right].$$

In [79], the redefined third Zagreb index was defined as

$$GO_2(G) = \operatorname{Re} ZG_3(G) = \sum_{uv \in E(G)} \left[d_G(u) + d_G(v) \right]\left[d_G(u)d_G(v) \right].$$

In Kulli (2017), the same index was proposed under the name second Gourava index.

Based on $GO_1(G), \ GO_2(G)$, the general first and second Gourava indices were defined as (Kulli, 2017):

$$GO_1^a(G) = \sum_{uv \in E(G)} \left[d_G(u) + d_G(v)d_G(u)d_G(v) \right]^a,$$

$$GO_2^a(G) = \sum_{uv \in E(G)} \left[\left(d_G(u) + d_G(v) \right) d_G(u)d_G(v) \right]^a.$$

$GO_1^a(G)$ for $a = 2$ is the first hyper Gourava index and $GO_2(G)$ for $a = 2$ is the second hyper Gourava index. These are defined in Kulli (2017).

GO_1^a for $a = -\dfrac{1}{2}$ is the sum connectivity Gourava index, defined in Kulli (2017).

GO_2^a for $a = \dfrac{1}{2}$ is the product connectivity Gourava index, defined in Kulli (2017).

Properties of these indices may be established in the future.

REDUCED ZAGREB INDICES

The reduced first (Ediz, 2016) and second (Furtula, Gutman & Ediz, 2014) Zagreb indices were defined as:

$$RM_1(G) = \sum_{u \in V(G)} \left(d_G(u) - 1\right)^2,$$

$$RM_2(G) = \sum_{uv \in E(G)} \left(d_G(u) - 1\right)\left(d_G(v) - 1\right).$$

The following identity is immediate (Gutman, Furtula & Ephick, 2014):

$$RM_2(G) = M_2(G) - M_1(G) + m.$$

The straightforward modification of the reduced Zagreb indices are to introduce

$$RM_1^a(G) = \sum_{u \in V(G)} \left(d_G(u) - 1\right)^a,$$

$$RM_2^a(G) = \sum_{uv \in E(G)} \left[\left(d_G(u) - 1\right)\left(d_G(v) - 1\right)\right]^a.$$

These generalizations of the reduced Zagreb indices were introduced by Kulli (2018d, 2018e). $RM_1^a(G)$ for $a = 3$ is the reduced F-index, defined as (Kulli, 2018d):

$$RF(G) = \sum_{u \in V(G)} \left(d_G(u) - 1\right)^3.$$

$RM_2^a(G)$ for $a = 2$ is the reduced second hyper Zagreb index which is defined as (Kulli, 2018f)

$$RHM_2(G) = \sum_{uv \in E(G)} \left[\left(d_G(u) - 1\right)\left(d_G(v) - 1\right)\right]^2.$$

$RM_2^a(G)$ for $a = -\dfrac{1}{2}$ is the reduced product connectivity index (Kulli, 2018g)

$$RP(G) = \sum_{uv \in E(G)} \frac{1}{\sqrt{\left(d_G(u) - 1\right)\left(d_G(v) - 1\right)}}.$$

For further research on reduced Zagreb indices see (An & Xiong, 2015; Horoldagya, Buyantoglok & Dorisembe, 2017).

REVERSE ZAGREB INDICES

Let $\Delta(G)$ denote the largest of all degrees of G. The reverse vertex degree of u in G is $c_u = \Delta(G) - d_G(u) + 1$ (Ediz, 2018). Ediz (2018) introduced the reverse Zagreb indices, as

$$CM_1(G) = \sum_{uv \in E(G)} \left(c_u + c_v\right),$$

$$CM_2(G) = \sum_{uv \in E(G)} c_u c_v.$$

The general first and second reverse Zagreb indices of G are defined as

$$CM_1^a(G) = \sum_{uv \in E(G)} \left(c_u + c_v\right)^a,$$

$$CM_2^a(G) = \sum_{uv \in E(G)} \left(c_u c_v\right)^2.$$

$CM_1^a(G)$ and $CM_2^a(G)$ for $a=2$ are first and second reverse Zagreb indices respectively as (Kulli, 2018h)

$$HCM_1(G) = \sum_{uv \in E(G)} \left(c_u + c_v\right)^2,$$

$$HCM_2(G) = \sum_{uv \in E(G)} \left(c_u c_v\right)^2.$$

$CM_1^a(G)$ for $a = -\dfrac{1}{2}$ is the sum connectivity reverse index (Kulli, 2017)

$$SC(G) = \sum_{uv \in E(G)} \frac{1}{\sqrt{c_u + c_v}}.$$

$CM_2^a(G)$ for $a = -\dfrac{1}{2}$ is the product connectivity reverse index (Kulli, 2017)

$$PC(G) = \sum_{uv \in E(G)} \frac{1}{\sqrt{c_u c_v}}.$$

Further research on reverse Zagreb indices can be found in Ediz (2015, 2016), Gao et al. (2018), Kandan and Kennedy (2018), Kulli (2017, 2018).

REVAN INDICES

Let $\delta(G)$ denote the smallest of all degrees of G. The Revan vertex degree of u in G (Kulli, 2017) is

$$r_G(u) = \Delta(G) + \delta(G) - d_G(u).$$

Note that if $\delta(G) = 1$, then $r_G(u) = c_u$ and if $\delta(G) = \Delta(G)$, then $r_G(u) = d_G(u)$.
The first and second Revan indices (Kulli, 2017) are defined as

$$R_1(G) = \sum_{uv \in E(G)} \left[r_G(u) + r_G(v) \right],$$

$$R_2(G) = \sum_{uv \in E(G)} r_G(u) r_G(v).$$

In Hyper Revan indices, Kulli (2018) introduced the first and second hyper Revan indices

$$HR_1(G) = \sum_{uv \in E(G)} \left[r_G(u) + r_G(v) \right]^2,$$

$$HR_2(G) = \sum_{uv \in E(G)} \left[r_G(u) r_G(v) \right]^2.$$

The general first and second Revan indices are defined as

$$R_1^a(G) = \sum_{uv \in E(G)} \left[r_G(u) + r_G(v) \right]^a,$$

$$R_2^a(G) = \sum_{uv \in E(G)} \left[r_G(u) r_G(v) \right]^a.$$

$R_1^a(G)$ for $a = -\frac{1}{2}$ is the sum connectivity Revan index (Kulli, 2017)

$$SR(G) = \sum_{uv \in E(G)} \frac{1}{\sqrt{r_G(u) + r_G(v)}}.$$

$R_2^a(G)$ for $a = -\frac{1}{2}$ is the product connectivity Revan index (Kulli, 2017)

$$PR(G) = \sum_{uv \in E(G)} \frac{1}{\sqrt{r_G(u) r_G(v)}}.$$

More results on Revan Zagreb indices can be found in Baig, Nadeem, and Gao (2018), Kandan, Chandrasekaran, and Priyadharshini (2018), and Kulli (2018).

LEAP ZAGREB INDICES

The 2-distance $d_2(u)$ in G is defined the number of second neighbors of u in G. Naji et al. (2017) introduced the first and second leap Zagreb indices as

$$LM_1(G) = \sum_{u \in V(G)} d_2(u)^2,$$

$$LM_2(G) = \sum_{uv \in E(G)} d_2(u) d_2(v).$$

More on these indices can be found in Basavanagoud and Chitra, 2018), Basavanagoud and Jakkannayer (2018), Naji and Soner (2018), and Shiladhar, Naji, and Soner (2018, n.d.).

A new version of the first leap Zagreb index is defined as (Kulli, 2018).

$$LM_1^*(G) = \sum_{uv \in E(G)} \left[d_2(u) + d_2(v) \right].$$

In [119], Kulli proposed the general first and second leap Zagreb indices, defined as

$$LM_1^a(G) = \sum_{uv \in E(G)} \left[d_2(u) + d_2(v) \right]^a,$$

$$LM_2^a(G) = \sum_{uv \in E(G)} \left[d_2(u) d_2(v) \right]^a.$$

$LM_1^a(G)$ for $a = 2$ is the first leap hyper Zagreb index and $LM_2^a(G)$ for $a = 2$ is the second hyper Zagreb index. These leap hyper Zagreb indices are defined in Kulli (2018).

$LM_1^a(G)$ for $a = -\dfrac{1}{2}$ is the sum connectivity leap index and $LM_2^a(G)$ for $a = -\dfrac{1}{2}$ is the product connectivity leap index. These indices are introduced in Kulli (2018) respectively.

VE-DEGREE INDICES

The set of all vertices which adjacent to u is called the open neighborhood of u and denoted by $N(u)$. The closed neighborhood set of u is the set $N[u] = N(u) \cup \{u\}$. The *ve*-degree $d_{ve}(u)$ of u in G is the number of different edges that incident to any vertex from the closed neighborhood of u *(Chellali et al., 2017)*. In Ediz (2017), the author defined the first and second ve-degree Zagreb indices as

$$Ve_1(G) = \sum_{uv \in E(G)} \left[d_{ve}(u) + d_{ve}(v) \right].$$

$$Ve_2(G) = \sum_{uv \in E(G)} d_{ve}(u) d_{ve}(v).$$

More results on *ve*-degree indices can be found in Ediz (2018, 2017), Sahin and Ediz (2018), and Cancan and Aldemir (2017).

The first and second hyper ve-degree indices (Kulli, 2018) are defined as

$$HVe_1(G) = \sum_{uv \in E(G)} \left[d_{ve}(u) + d_{ve}(v) \right]^2$$

$$HVe_2(G) = \sum_{uv \in E(G)} \left[d_{ve}(u) d_{ve}(v) \right]^2.$$

The general first and second ve-degree indices are defined as

$$Ve_1^a(G) = \sum_{uv \in E(G)} \left[d_{ve}(u) + d_{ve}(v) \right]^a,$$

$$Ve_2^a(G) = \sum_{uv \in E(G)} \left[d_{ve}(u) d_{ve}(v) \right]^a.$$

The *F-ve*-degree index was introduced by Kulli (2018), defined as

$$F_{ve}(G) = \sum_{uv \in E(G)} \left[d_{ve}(u)^2 + d_{ve}(v)^2 \right].$$

Different type of *F*-indices were studied in Kulli (2018).

MINUS INDICES

In Albertson (1997), the author introduced the irregularity index or minus index as

$$Alb\left(G\right) = \sum_{uv\in E\left(G\right)} \left|d_G\left(u\right) - d_G\left(v\right)\right|.$$

The general minus index was introduced by Kulli (2018) and defined as

$$M_i^a\left(G\right) = \sum_{uv\in E\left(G\right)} \left[\left|d_{ve}\left(u\right) - d_{ve}\left(v\right)\right|\right]^a.$$

$M_i^a\left(G\right)$ for $a = 1$ is the minus index.

$M_i^a\left(G\right)$ for $a = 2$ is the sigma index, which is defined by Gutman et al. (2018) and Gutman, Milovanovic, and Milovanovic (2018).

$M_i^a\left(G\right)$ for $a = \dfrac{1}{2}$ is the reciprocal minus index.

$M_i^a\left(G\right)$ for $a = -\dfrac{1}{2}$ is the minus connectivity index.

$M_i^a\left(G\right)$ for $a = -1$ is the modified minus index.

In Kulli (2018), the square ve-degree index was introduced, defined as

$$Q_{ve}\left(G\right) = \sum_{uv\in E\left(G\right)} \left[d_{ve}\left(u\right) - d_{ve}\left(v\right)\right]^2.$$

For other square indices see (Kulli, 2018).

KULLI - BASAVA INDICES

Let $S_e(u)$ denote the sum of the degrees of all edges incident to u. In Basavanagoud and Jakkannavar (2018), the authors proposed four Kulli-Basava indices, defined as

- first Kulli-Basava index: $KB_1\left(G\right) = \sum_{uv\in E\left(G\right)} \left[S_e\left(u\right) + S_e\left(v\right)\right].$

- second Kulli-Basava index: $KB_2\left(G\right) = \sum_{uv\in E\left(G\right)} S_e\left(u\right) S_e\left(v\right)$

- third Kulli-Basava index: $KB_3\left(G\right) = \sum_{uv\in E\left(G\right)} \left|S_e\left(u\right) - S_e\left(v\right)\right|$

- modified first Kulli-Basava index: $KB_1^*(G) = \sum\limits_{u \in V(G)} S_e(u)^2$.

In Basavanagoud and Jakkannayer (2018), the identity

$$KB_1(G) = F(G) - 2M_1(G) + M_1G_5(G)$$

was obtained.

BANHATTI INDICES

In Kulli (2018), the author proposed the first and second K Banhatti indices, intending to take into account the contributions of pairs of incident elements. The first and second K Banhatti indices are defined as

$$B_1(G) = \sum\limits_{ue} \left[d_G(u) + d_G(e)\right],$$

$$B_2(G) = \sum\limits_{ue} d_G(u) d_G(e).$$

where ue means that the vertex u and edge e are incident in G.

The first and second K hyper Banhatti indices were introduced by Kulli (2018), defined as

$$HB_1(G) = \sum\limits_{ue} \left[d_G(u) + d_G(e)\right]^2,$$

$$HB_2(G) = \sum\limits_{ue} \left[d_G(u) d_G(v)\right]^2.$$

In Gutman et al. (2017), the identities

$$B_1(G) = 3M_1(G) - 4m,$$

$$B_2(G) = HM_1(G) - 2M_1(G)$$

were established.

The general first and second K Banhatti indices were proposed by Kulli (2018), defined as

$$B_1^a(G) = \sum\limits_{ue} \left[d_G(u) + d_G(e)\right]^a,$$

$$B_2^2(G) = \sum_{ue} \left[d_G(u) d_G(e) \right]^a.$$

$B_1^a(G)$ for $a = -\dfrac{1}{2}$ is the sum connectivity Banhatti index (Kulli et al., n.d.)

$$SB(G) = \sum_{ue} \frac{1}{\sqrt{d_G(u) + d_G(e)}}.$$

$B_2^a(G)$ for $a = -\dfrac{1}{2}$ is the product connectivity Banhatti index (Kulli et al., 2019)

$$PB(G) = \sum_{ue} \frac{1}{\sqrt{d_G(u) d_G(e)}}.$$

Some papers have been devoted to the study of the Banhatti indices (Firdous, Neezer & Farahani, 2018; Gao et al., 2017; Kulli, 2017; Kulli et al., 2017; Kulli, Chaluvaraju & Boregowda, 2017).

MULTIPLICATIVE VERSIONS OF GRAPH INDICES

Todeschini et al. (2010) and Todeschini and Consonni (2010) introduced the multiplicative versions of first and second Zagreb indices, defined as

$$II_1(G) = \prod_{u \in V(G)} d_G(u)^2,$$

$$II_2(G) = \prod_{uv \in E(G)} d_G(u) d_G(v).$$

Narumi et al. (1984) considered the product

$$NK(G) = \prod_{u \in V(G)} d_G(u)$$

which now a days is referred to as the Narumi-Katayama index. Evidently,

$$II_1(G) = \left[NK(G) \right]^2.$$

Eliasi at al. (2012) introduced a multiplicative version of the first Zagreb index

$$II_1^*(G) = \prod_{uv \in E(G)} \left[d_G(u) + d_G(v) \right].$$

Based on the successful consideration of multiplicative Zagreb indices, Kulli (2016) continued to define the first and second multiplicative hyper Zagreb indices as

$$HII_1(G) = \prod_{uv \in E(G)} \left[d_G(u) + d_G(v) \right]^2,$$

$$HII_2(G) = \prod_{uv \in E(G)} \left[d_G(u) d_G(v) \right]^2.$$

Kulli et al. (2017) continued this generalization and defined the general first and second multiplicative Zagreb indices as

$$II_1^a(G) = \prod_{uv \in E(G)} \left[d_G(u) + d_G(v) \right]^a,$$

$$II_2^a(G) = \prod_{uv \in E(G)} \left[d_G(u) d_G(v) \right]^a,$$

$II_1^a(G)$ for $a = -\dfrac{1}{2}$ is the multiplicative sum connectivity index (Kulli, 2016) and $II_2^a(G)$ for $a = -\dfrac{1}{2}$ is the multiplicative product connectivity index (Kulli, 2016).

Kulli (2016) introduced the multiplicative atom bond connectivity and geometric-arithmetic indices as

$$ABCII(G) = \prod_{uv \in E(G)} \sqrt{\frac{d_G(u) + d_G(v) - 2}{d_G(u) d_G(v)}},$$

$$GAII(G) = \prod_{uv \in E(G)} \frac{2\sqrt{d_G(u) d_G(v)}}{d_G(u) + d_G(v)}.$$

More properties on the multiplicative connectivity indices see (Kulli, 2018).

Multiplicative versions of other graph indices were proposed and studied in the subject Chemical Graph Theory. Details of these multiplicative graph indices and their properties go beyond the scope of this survey.

REFERENCES

Albertson, M. O. (1997). The irregularity of a graph. *Ars Combinatoria, 46*, 219–225.

An, M., & Xiong, L. (2015). Some results on the difference of the Zagreb indices of a graph. *Bulletin of the Australian Mathematical Society, 92*(2), 117–186. doi:10.1017/S0004972715000386

Baig, A. Q., Nadeem, M., & Gao, W. (2018). Revan and hyper Revan indices of octahedral and icosahedral networks. *Applied Mathematics and Nonlinear Sciences, 3*(1), 33–40. doi:10.21042/AMNS.2018.1.00004

Basavanagoud, B., Barangi, A. P., & Hosamani, S. M. (2018). First neighbourhood Zagreb index of some nanostructures. *Proceedings of IAM, 7*(2), 178-193.

Basavanagoud, B., & Chitra, E. (2018). On leap Zagreb indices of some nanostructures. *Malaya Journal of Matematik, 6*(4), 816–822.

Basavanagoud, B., & Desai, V. R. (2016). Forgotten topological index and hyper Zagreb index of generalized transformation graphs. *Bull. Math. Sci. Appl., 14*, 1–6.

Basavanagoud, B., & Jakkannavar, P. (2018). Computing first leap Zagreb index of some nanostructures. *Int. J. Math. And. Appl., 6*(2-B), 141–150.

Basavanagoud, B., & Jakkannavar, P. (2018). Kulli-Basava indices of graphs. *Inter. J. Appl. Engg. Research, 14*(1), 325–342.

Basavanagoud, B., & Patil, S. (2016). A note on hyper Zagreb index of graph operation. *Iran. J. Math. Chem., 7*, 89–92.

Britto Antony Xavier, G., Suresh, E., & Gutman, I. (2014). Counting relations for general Zagreb indices, Kragujevac, J. *Math., 38*, 95–103.

Cancan & Aldemir. (2017). *On ve-degree and ev-degree Zagreb indices of titania nanotubes.* Academic Press.

Caparossi, G., Hansen, P., & Vukičević, D. (2010). Comparing Zagreb indices of cyclic graphs. *MATCH Commun. Math. Comput. Chem, 63*, 44–451.

Che, Z., & Chen, Z. (2016). Lower and upper bounds of the forgotten topological index. *MATCH Commun. Math. Comput. Chem., 76*, 635–648.

Chellali, M., Hynes, T. W., Hedetniemi, S. T., & Lewis, T. W. (2017). On ve-degrees and ev-degrees in graphs. *Discrete Mathematics, 340*(2), 31–38. doi:10.1016/j.disc.2016.07.008

Das, K. C. (2010). Atom bond connectivity index of graphs. *Discrete Applied Mathematics, 158*(11), 1181–1188. doi:10.1016/j.dam.2010.03.006

Das, K. C. (2010). On geometric-arithmetic index of graphs. *MATCH Commun. Math. Comput. Chem., 64*, 619–630.

Das, K. C., Das, S., & Zhou, B. (2016). Sum connectivity index of a graph. *Frontiers of Mathematics in China*, *11*(1), 47–54. doi:10.100711464-015-0470-2

Das, K. C., & Gutman, I. (2004). Some properties of the second Zagreb index. *MATCH Commun. Math. Comput. Chem.*, *52*, 103–112.

Das, K. C., Gutman, I., & Furtula, B. (2011). On atom bond connectivity index. *Chemical Physics Letters*, *511*(4-6), 452–454. doi:10.1016/j.cplett.2011.06.049

Das, K. C., Gutman, I., & Horoldagva, B. (2012). Comparing Zagreb indices and coindices of trees. *MATCH Commun. Math. Comput. Chem.*, *67*, 189–198.

Das, K. C., Gutman, I., & Zhou, B. (2009). New upper bounds on Zagreb indices. *Journal of Mathematical Chemistry*, *46*(2), 514–521. doi:10.100710910-008-9475-3

Das, K. C., Xu, K., & Gutman, I. (2013). On Zagreb and Harary indices. *MATCH Commun. Math. Comput. Chem.*, *70*, 301–314.

De, Nayeem, & Pal. (2016). F-index of some graph operations. *Discrete Math. Algorithms Appl., 8*.

De, N. (2016). F-index of bridge and chain graphs. *Malay. J. Fund. Appl. Sci.*, *12*, 109–113.

Doslic, T., Furtula, B., Graovać, A., Gutman, I., Moradi, S., & Yarahmadi, Z. (2011). On vertex degree based molecular structure descriptors. *MATCH Commun. Math. Comput. Chem.*, *66*, 613–626.

Ediz, S. (2015). Maximum chemical trees of the second reverse Zagreb index. *Pacific J. Appl. Math.*, *7*(4), 291–295.

Ediz, S. (2017). Predicting some physicochemical properties of octane isomers: A topological approach using ve-degree and ve-degree Zagreb indices. *International Journal of System Science and Applied Mathematics*, *2*, 87–92.

Ediz, S. (2017). A new tool for QSPR researches: Ve-degree Randić index. *Celal Bayar University Journal of Science*, *13*(3), 615–618.

Ediz, S. (2018). Maximal graphs of the first reverse Zagreb beta index. *TWMS J. Appl. Eng. Math*, *8*, 306–310.

Ediz, S. (2018). On ve-degree molecular topological properties of silicate and oxygen networks, Int. J. *Computing Science and Mathematics*, *9*(1), 1–12.

Ediz, S., & Cancan, M. (2016). Reverse Zagreb indices of Cartesian product of graphs. *International Journal of Mathematics and Computer Science*, *11*(1), 51–58.

(2016). Ediz, On the reduced first Zagreb index of graphs. *Pacific J. Appl. Math.*, *8*(2), 99–102.

Eliasi, M., Irammanesh, A., & Gutman, I. (2012). Multiplicative versions of first Zagreb index. *MATCH Commun. Math. Comput. Chem.*, *68*, 217–230.

Estrada, E., Torres, L., Rodriguez, L., & Gutman, I. (1998). An atom bond connectivity index: Modeling the enthalpy of formation of alkanes. *Indian Journal of Chemistry*, *37A*, 849–855.

Fajtolowicz, S. (1988). On conjectures of Grafitti. *Discrete Mathematics*, *72*(1-3), 113–118. doi:10.1016/0012-365X(88)90199-9

Falahati Nezhad, F., & Azari, M. (2016). Bounds on the hyper Zagreb index. *J. Appl. Math. Inform.*, *34*(3_4), 319–330. doi:10.14317/jami.2016.319

Fath-Tabar, G. H. (2011). Old and new Zagreb indices of graphs. *MATCH Commun. Math. Comput. Chem.*, *65*, 79–84.

Firdous, S., Nazeer, W., & Farahani, M. R. (2018). Mathematical properties and computations of Banhatti indices for a nanostructure "Toroidal Polyhex Network". *Asian Journal of Nanoscience and Materials*, *1*, 43–47.

Furtula, B., Graovac, A., & Vukičević, D. (2009). Atom bond connectivity index of trees. *Discrete Applied Mathematics*, *157*(13), 2828–2835. doi:10.1016/j.dam.2009.03.004

Furtula, B., & Gutman, I. (2015). A forgotten topological index. *Journal of Mathematical Chemistry*, *53*(4), 1184–1190. doi:10.100710910-015-0480-z

Furtula, B., Gutman, I., & Ediz, S. (2014). On difference of Zagreb indices. *Discrete Applied Mathematics*, *178*, 83–88. doi:10.1016/j.dam.2014.06.011

Gan, L., Hou, H., & Liu, B. (2011). Some results on atom bond connectivity index of graphs. *MATCH Commun. Math. Comput. Chem.*, *66*, 669–680.

Gao, W., Farahani, M. R., Siddiqui, M. K., & Jamil, M. K. (2016). On the first and second Zagreb and first and second hyper Zagreb indices of carbon nanocones CNC_k [n]. *Journal of Computational and Theoretical Nanoscience*, *13*(10), 7475–7482. doi:10.1166/jctn.2016.5742

Gao, W., Jamil, M. K., & Farahani, M. R. (2017). The hyper-Zagreb index and some graph operations. *J. Appl. Math. Comput.*, *54*(1-2), 263–275. doi:10.100712190-016-1008-9

Gao, W., Muzaffar, B., Nazeer, W., & Banhatti, K. (2017). K-Banhatti and K-hyper Banhatti Indices of Dominating David Derived Network. *Open J. Math. Anal.*, *1*(1), 13–24. doi:10.30538/psrp-oma2017.0002

Gao, W., Siddiqui, M. K., Imran, M., Jamil, M. K., & Farahani, M. R. (2016). Forgotten topological index of chemical structure in drugs. *Saudi Pharmaceutical Journal*, *24*(3), 258–264. doi:10.1016/j.jsps.2016.04.012 PMID:27275112

Gao, W., Yonuas, M., Farooq, A., Virk, A. R., & Nazeer, W. (2018). Some reverse degree based topological indices and polynomials of dendrimers. *Mathematics*, *6*(10), 214. doi:10.3390/math6100214

Graovac, A., Ghorbani, M., & Hosseinzadeh, M. A. (2011). Computing fifth geometric-arithmetic index of nanostar dendrimers. *Journal of Mathematical Nanoscience*, *1*(1), 33–42.

Gutman, Milovanović, & Milovanović. (2018). Beyond the Zagreb indices. *AKCE International Journal of Graphs and Combinatorics*. .2018.05.002 doi:10.1016/jakcej

Gutman, I. (2013). Degree based topological indices. *Croatica Chemica Acta, 86*(4), 351–361. doi:10.5562/cca2294

Gutman, I. (2017). On hyper Zagreb index and coindex, Bull. Acad. Sebre Sci. Arts. *Cl. Sci. Math. Natur., 150*, 1–8.

Gutman, I., & Das, K. C. (2004). The first Zagreb index 30 years after. *MATCH Commun. Math. Comput. Chem., 50*, 83–92.

Gutman, I., & Furtula, B. (Eds.). (2008). *Recent Results in the theory of Randić index*. Kragujevac: University Kragujevac.

Gutman, I., Furtula, B., & Elphick, C. (2014). Three new/old vertex degree based topological indices. *MATCH Commun. Math. Comput. Chem., 72*, 617–682.

Gutman, I., Kulli, V. R., Chaluvaraju, B., & Boregowda, H. S. (2017). On Banhatti and Zagreb indices. *Journal of the International Mathematical Virtual Institute, 7*, 53–67.

Gutman, I., & Polansky, O. E. (1986). *Mathematical Concepts in Organic Chemistry*. Berlin: Springer. doi:10.1007/978-3-642-70982-1

Gutman, I., Togan, M., Yurtlas, A., Cavik, A. S., & Cangul, I. N. (2018). Inverse problem for sigma index. *MATCH Commun. Math. Comput. Chem., 79*, 491–508.

Gutman, I., & Trinajstić, N. (1972). Graph theory and molecular orbitals. Total π-electron energy of alternant hydrocarbons. *Chemical Physics Letters, 17*(4), 535–538. doi:10.1016/0009-2614(72)85099-1

Horoldagva, B., Buyantoglok, L., & Dorjsembe, S. (2017). Difference of Zagreb indices and reduced second Zagreb index of cyclic graphs with cut edges. *MATCH Commun. Math. Comput. Chem., 78*, 337–350.

Hosamani, S. M., & Gutman, I. (2014). Zagreb indices of transformation graphs and total transformation graphs. *Applied Mathematics and Computation, 247*, 1156–1160. doi:10.1016/j.amc.2014.09.080

Ilić, A., & Stevanović, D. (2011). On Comparing Zagreb indices. *MATCH Commun. Math. Comput. Chem., 66*, 681–687.

Kandan, P., Chandrasekaran, E., & Priyadharshini, M. (2018). The Revan weighted szeged index of graphs. *Journal of Emerging Technologies and Innovative Research, 5*(9), 358–366.

Kandan, P., & Joseph Kennedy, A. (2018). Reverse Zagreb indices of corona product of graphs. *Malaya Journal of Matematik, 6*(4), 720–714. doi:10.26637/MJM0604/0003

Khalifeh, M. H., Yousefi-Azari, H., & Ashrafi, A. R. (2009). The first and second Zagreb indices of some graph operations. *Discrete Applied Mathematics, 157*(4), 804–811. doi:10.1016/j.dam.2008.06.015

Kulli. (2017a). Computing topological indices of dendrimer nanostars. *International Journal of Mathematics and Its Applications, 5*(3-A), 163-169.

Kulli. (2017). On the product connectivity reverse index of silicate and hexagonal networks. *International Journal of Mathematics and its Applications, 5*(4-B), 175-179.

Kulli. (2017). Geometric-arithmetic reverse and sum connectivity reverse indices of silicate and hexagonal networks. *International Journal of current Research in Science and Technology, 3*(10), 29-33.

Kulli. (2017). Revan indices of oxide and honeycomb networks. *International Journal of Mathematics and its Applications, 5*(4-E), 663-667.

Kulli. (2018). Hyper Revan indices and their polynomials of silicate networks. *International Journal of Mathematics and its Applications, 4*(3), 17-21.

Kulli, Chaluvaraju, & Boregowda. (n.d.). *On sum connectivity Banhatti index of some nanostructures.* (submitted)

Kulli, Stone, Wang, & Wei. (2017). Generalized multiplicative indices of polycyclic aromatic hydrocarbons and benzenoid systems. *Z. Naturforsch, 72*(6), 573-576.

Kulli, V. R. (2016). On K hyper-Banhatti indices and coindices of graphs. *International Research Journal of Pure Algebra, 6*(5), 300–304.

Kulli, V. R. (2016). On K hyper-Banhatti indices and coindices of graphs. *International Research Journal of Pure Algebra, 6*(5), 300–304.

Kulli, V. R. (2016). Multiplicative connectivity indices of certain nanotubes. *Annals of Pure and Applied Mathematics, 12*(2), 169–176. doi:10.22457/apam.v12n2a8

Kulli, V. R. (2017). General fifth M-Zagreb indices and fifth M-Zagreb polynomials of PAMAM dendrimers. *Intern. J. Fuzzy Mathematical Archive, 13*(1), 99–103.

Kulli, V. R. (2017). The Gourava indices and coindices of graphs. *Annals of Pure and Applied Mathematics, 14*(1), 33–38. doi:10.22457/apam.v14n1a4

Kulli, V. R. (2017). Computation of some Gourava indices of titania nanotubes. *Intern. J. Fuzzy Mathematical Archive, 12*(2), 75–81.

Kulli, V. R. (2017). On hyper-Gourava indices and coindices. *International Journal of Mathematical Archive, 8*(12), 116–120.

Kulli, V. R. (2017). On the sum connectivity Gourava index. *International Journal of Mathematical Archive, 8*(6), 211–217.

Kulli, V. R. (2017). The product connectivity Gourava index. *J. Comp. and Math. Sci., 8*(6), 235–242.

Kulli, V. R. (2017). On the sum connectivity reverse index of oxide and honeycomb networks. *J. Comp. and Math. Sci., 8*(9), 408–413.

Kulli, V. R. (2017). The sum connectivity Revan index of silicate and hexagonal networks. *Annals of Pure and Applied Mathematics*, *14*(3), 401–406. doi:10.22457/apam.v14n3a6

Kulli, V. R. (2017). On the product connectivity Revan index of certain nanotubes. *J. Comp. and Math. Sci.*, *8*(10), 562–567.

Kulli, V. R. (2017b). General topological indices of tetrameric 1,3-adamantane. *International Journal of Current Research in Science and Technology*, *3*(8), 26–33.

Kulli, V. R. (2017c). General topological indices of some dendrimer nanostars. *Journal of Global Research in Mathematical Archives*, *14*(11), 83–90.

Kulli, V. R. (2017d). Edge version of F-index, general sum connectivity index of certain nanotubes. *Annals of Pure and Applied Mathematics*, *14*(3), 449–455. doi:10.22457/apam.v14n3a11

Kulli, V. R. (2018). Atom bond connectivity reverse and product connectivity reverse indices of oxide and honeycomb networks. *Intern. J. Fuzzy Mathematical Archive*, *15*(1), 1–5.

Kulli, V. R. (2018). Square reverse index and its polynomial of certain networks. *International Journal of Mathematical Archive*, *9*(10), 27–33.

Kulli, V. R. (2018). Computation of F-reverse and modified reverse indices of some nanostructures. *Annals of Pure and Applied Mathematics*, *18*(1), 37–43.

Kulli, V. R. (2018). Revan indices and their polynomials of certain rhombus networks. *International Journal of Current Research in Life Sciences*, *7*(5), 2110–2116.

Kulli, V. R. (2018). Computing the F-Revan and modified Revan indices of certain nanostructures. *J. Comp. and Math. Sci.*, *9*(10), 1326–1333.

Kulli, V. R. (2018). Leap hyper-Zagreb indices and their polynomials of certain graphs. *International Journal of Current Research in Life Sciences*, *7*(10), 2783–2791.

Kulli, V. R. (2018). Product connectivity leap index and ABC leap index of helm graphs. *Annals of Pure and Applied Mathematics*, *18*(2), 183–193.

Kulli, V. R. (2018). On ve-degree indices and their polynomials of dominating oxide networks. *Annals of Pure and Applied Mathematics*, *18*(1), 1–7. doi:10.22457/apam.v13n1a1

Kulli, V. R. (2018). Computing the F-ve-degree index and its polynomial of dominating oxide and regular triangulate oxide networks. *Intern. J. Fuzzy Mathematical Archive*, *16*(1), 1–6.

Kulli, V. R. (2018). Computation of F-reverse and modified reverse indices of some nanostructures. *Annals of Pure and Applied Mathematics*, *18*(1), 37–43.

Kulli, V. R. (2018). Computing the F-Revan and modified Revan indices of certain nanostructures. *J. Comp. and Math. Sci.*, *9*(10), 1326–1333.

Kulli, V. R. (2018). Computation of some minus indices of titania nanotubes. *International Journal of Current Research in Science and Technology*, *4*(12), 9–13.

Kulli, V. R. (2018). On the square ve-degree index and its polynomial of certain network. *Journal of Global Research in Mathematical Archives*, *5*(10), 1–4.

Kulli, V. R. (2018). Square reverse index and its polynomial of certain networks. *International Journal of Mathematical Archive*, *9*(10), 22–33.

Kulli, V. R. (2018). Minus leap and square leap indices and their polynomials of some special graphs. *Int. Res. J. Pure Algebra*, *8*(11), 54–60.

Kulli, V. R. (2018). Computing Banhatti indices of networks. *International Journal of Advances in Mathematics*, *2018*(1), 31–40.

Kulli, V. R. (2018). *Multiplicative Connectivity Indices of Nanostructures*. LAP Lambert Academic Publishing.

Kulli, V. R. (2018a). On KV indices and their polynomials of two families of dendrimers. *International Journal of Current Research in Life Sciences*, *7*(9), 2739–2744.

Kulli, V. R. (2018b). Multiplicative connectivity indices of dendrimers. *Journal of Mathematics and Informatics*, *15*, 1–7.

Kulli, V. R. (2018c). Dakshayani indices. *Annals of Pure and Applied Mathematics*, *18*(2), 139–146.

Kulli, V. R. (2018d). On reduced Zagreb indices of polycyclic aromatic hydrocarbons and benzenoid systems. *Annals of Pure and Applied Mathematics*, *18*(1), 73–78.

Kulli, V. R. (2018f). Reduced second hyper-Zagreb index and its polynomial of certain silicate networks. *Journal of Mathematics and Informatics*, *14*, 11–16. doi:10.22457/jmi.v14a2

Kulli, V. R. (2018g). Computing reduced connectivity indices of certain nanotubes. *Journal of Chemistry and Chemical Sciences*, *8*(11), 1174–1180. doi:10.29055/jccs/688

Kulli, V. R. (2018h). Reverse Zagreb and reverse hyper-Zagreb indices and their polynomials of rhombus silicate networks. *Annals of Pure and Applied Mathematics*, *16*(1), 47–51. doi:10.22457/apam.v16n1a6

Kulli, V. R. (2019). Sum connectivity leap index and geometric-arithmetic leap index of certain windmill graphs. *J. Global Research in Mathematical Archives*, *6*(1), 15–20.

Kulli, V. R. (2019a). On hyper KV and square KV indices and their polynomials of certain families of dendrimers. *J. Comp. and Math. Sci.*, *10*(2), 279–186.

Kulli, V. R. (2019b). On connectivity KV indices of certain families of dendrimers. *International Journal of Mathematical Archive*, *10*(2), 14–17.

Kulli, V. R., Chaluvaraju, B., & Baregowda, H. S. (2017). K-Banhatti and K hyper-Banhatti indices of windmill graphs. *South East Asian J. of Math. and Math. Sci*, *13*(1), 11–18.

Kulli, V. R., Chaluvaraju, B., & Boregowda, H. S. (2017). Connectivity Banhatti indices for certain families of benzenoid systems. *Journal of Ultra Chemistry*, *13*(4), 81–87. doi:10.22147/juc/130402

Kulli, V. R., Chaluvaraju, B., & Boregowda, H. S. (2019). On sum connectivity Banhatti index of a graph. *Discussiones Mathematicae. Graph Theory*, *39*, 205–217.

Kulli, V. R., & New, K. (2017). Banhatti topological indices. *Intern.J.Fuzzy Mathematical Archive*, *12*(1), 29–37.

Kulli, V. R., & On, K. (2016). Banhatti indices of graphs. *J. Comp. and Math. Sci.*, *7*, 213–218.

Li, X., & Gutman, I. (2006). *Mathematical Aspects of Randić-type Molecular Structure Descriptors*. Kragujevac: University Kragujevac.

Li, X., & Zhao, H. (2004). Trees with the first three smallest and largest generalised topological indices. *MATCH Commun. Math. Comput. Chem.*, *50*, 57–62.

Li, X., & Zheng, J. (2005). A unified approach to the external trees for different indices. *MATCH Commun. Math. Comput. Chem.*, *54*, 195–208.

Liu, B., & You, Z. (2011). A survey on comparing Zagreb indices, A. Ilić, D. Stevanović, On Comparing Zagreb indices. *MATCH Commun. Math. Comput. Chem.*, *66*, 581–593.

Liu, M., & Liu, B. (2010). Some properties of the first general Zagreb index. *Australas J. Combin.*, *47*, 285–294.

Luo, Z. (2016). Applications on hyper Zagreb index of generalized hierarchical product graphs. *Journal of Computational and Theoretical Nanoscience*, *13*(10), 7355–7361. doi:10.1166/jctn.2016.5726

Milićević, A., & Nikolić, S. (2004). On variable Zagreb indices. *Croatica Chemica Acta*, *77*, 97–101.

Mogharrab, M., & Fath-Tabar, G. H. (2011). Some bound on GA_1 index of graphs. *MATCH Commun. Math. Comput. Chem*, *65*, 33–38.

Mondal, De, & Pal. (2018). *On neighbourhood index of product of graphs*. arXivi1805.05273v1

Naji, A. M., & Soner, N. D. (2018). The first leap Zagreb index of some graph operations. *Int. J. Appl. Graph Theory*, *2*(1), 7–18.

Naji, A. M., Soner, N. D., & Gutman, I. (2017). On leap Zagreb indices of graphs. *Commun. Comb. Optim.*, *2*(2), 99–107.

Narumi, H., & Katayama, M. (1984). Simple topological index, A newly devised index characterizing the topological nature of structural isomers of saturated hydrocarbons. *Mem. Fac. Engin. Hokkaido Univ.*, *16*, 209–214.

Nikolić, S., Kovaćević, G., Milićević, A., & Trinajstić, N. (2003). The Zagreb indices 30 years after. *Croat. Chem. Acta CCACAA*, *76*(2), 113–124.

Pattabiraman, K., & Vijayaragavan, M. (2017). Hyper Zagreb indices and its coindices of graphs. *Bull. Int. Math. Virt. Inst.*, *7*, 31–41.

Ramane, H. S., & Jummannaver, R. B. (2016). Note on forgotten topological index of chemical structure in drugs. *Applied Mathematics and Nonlinear Sciences*, *1*(2), 369–374. doi:10.21042/AMNS.2016.2.00032

Randić, M. (1975). On characterization of molecular branching. *Journal of the American Chemical Society*, *97*(23), 6609–6615. doi:10.1021/ja00856a001

Ranjini, P. S., Lokesha, V., & Cangül, I. N. (2011). On the Zagreb indices of the line graphs of the subdivision graphs. *Applied Mathematics and Computation*, *218*(3), 699–702. doi:10.1016/j.amc.2011.03.125

Ranjini, P. S., Lokesha, V., & Usha, A. (2013). Relation between phenylene and hexagonal squeeze using harmonic index. *Int. J. Graph Theory*, *1*, 116–121.

Réti, T. (2012). On the relationships between the first and second Zagreb indices. *MATCH Commun. Math. Comput. Chem.*, *68*, 169–188.

Sahin, B., & Ediz, S. (2018). On ev-degree and ve-degree topological indices. *Iranian J. Math. Chem.*, *9*(4), 263–277.

Shiladhar, Naji, & Soner. (n.d.). Computation of leap Zagreb indices of some windmill graphs. *Int. J. Math. And. Appl.,* *6*(2-B, 183-191.

Shiladhar, P., Naji, A. M., & Soner, N. D. (2018). Leap Zagreb indices of some wheel related graphs. *J. Comp. and Math. Sci.*, *9*(3), 221–231.

Shirdel, G. H., Rezapour, H., & Sayadi, A. M. (2013). The hyper-Zagreb index of graph operations. *Iranian J. Math. Chem.*, *4*(2), 213–220.

Todeschini, R., Ballabio, D., & Consonni, V. (2010). Noval molecular descriptors based on functions of new vertex degrees. In I. Gutman & B. Furtula (Eds.), *Novel molecular structure descriptors – Theory and Applications I* (pp. 73–100). Kragujevac: Univ. Kragujevac.

Todeschini, R., & Consonni, V. (2009). *Molecular Descriptors for Chemoinformatics*. Weinheim: Wiley-VCH. doi:10.1002/9783527628766

Todeschini, R., & Consonni, V. (2010). New local vertex invariants and descriptors based on functions of vertex degrees. *MATCH Commun. Math. Comput. Chem.*, *64*, 359–372.

Trinajstić, N., Nikolić, S., Miličvić, A., & Gutman, I. (2010). On Zagreb indices. *Kem. Ind.*, *59*, 577–589.

(2018e). V.R. Kulli General reduced second Zagreb index of certain networks. *International Journal of Current Research in Life Sciences*, *7*(11), 2827–2833.

Vukičević, D., & Furtula, B. (2009). Topological index on the ratios of geometrical and arithmetical means of end-vertex degrees of edges. *Journal of Mathematical Chemistry*, *46*(4), 1369–1376. doi:10.100710910-009-9520-x

Vukičević, D., Li, Q., Sedlar, J., & Došlić, T. (2018). Lanzhou index. *MATCH Commun. Math. Comput. Chem.*, *8*, 863–876.

Wang, S., Zhou, B., & Trinajstić, N. (2011). On the sum connectivity index. *Filomat*, *25*(3), 29–42. doi:10.2298/FIL1103029W

Wiener, H. (1947). Structural determination of parattin boiling points. *Journal of the American Chemical Society*, *69*(1), 17–20. doi:10.1021/ja01193a005 PMID:20291038

Xing, R., Zhou, B., & Du, Z. (2010). Further results on atom bond connectivity index of trees, discrete. *Applications of Mathematics*, *158*, 1536–1545.

Xing, R., Zhou, B., & Tranajstić, N. (2010). Sum connectivity index of molecular trees. *Journal of Mathematical Chemistry*, *48*(3), 583–591. doi:10.100710910-010-9693-3

Xu, K., Tang, K., Liu, H., & Wang, J. (2015). The Zagreb indices of bipartite graphs with more edges. *J. Appl. Math. and Informatics*, *33*(3), 365–377. doi:10.14317/jami.2015.365

Zhang, Z., Zhang, J., & Lu, X. (2005). The relation of matching with inverse degree of a graph. *Discrete Mathematics*, *301*(2-3), 243–246. doi:10.1016/j.disc.2003.01.001

Zhou, B. (2004). Zagreb indices. *MATCH Commun. Math. Comput. Chem.*, *52*, 113–118.

Zhou, B. (2007). Remarks on Zagreb indices. *MATCH Commun. Math. Comput. Chem.*, *57*, 597–616.

Zhou, B., & Gutman, I. (2005). Further properties of Zagreb indices. *MATCH Commun. Math. Comput. Chem.*, *54*, 233–239.

Zhou, B., & Trinajstic, N. (2009). On a novel connectivity index. *Journal of Mathematical Chemistry*, *46*(4), 1252–1270. doi:10.100710910-008-9515-z

Chapter 4
A Novel Weighted First Zagreb Index of Graph

Jibonjyoti Buragohain
Dibrugarh University, India

A. Bharali
Dibrugarh University, India

ABSTRACT

The Zagreb indices are the oldest among all degree-based topological indices. For a connected graph G, the first Zagreb index $M_1(G)$ is the sum of the term $d_G(u)+d_G(v)$ corresponding to each edge uv in G, that is, $M_1(G) = \sum_{UV \in E(G)}[(d_{G(u)} + d_{G(v)})]$, where $d_G(u)$ is degree of the vertex u in G. In this chapter, the authors propose a weighted first Zagreb index and calculate its values for some standard graphs. Also, the authors study its correlations with various physico-chemical properties of octane isomers. It is found that this novel index has strong correlation with acentric factor and entropy of octane isomers as compared to other existing topological indices.

INTRODUCTION

The introduction of topological index or molecular structure descriptor in literature is a great success as it can correlate various physical properties, biological reactivity or chemical activity of molecules without undergoing actual experimentation. A topological index is a real number that can be associated to a molecule based on its molecular graph. A molecular graph is a simple graph corresponding to a molecule in which vertices represents atoms and edges represents various chemical bonds between them. For historical background of the topological index interested reader can go through (Trinajstić, 2011). The success of this simple mathematical quantity is because of the fact that the properties of a chemical compound is highly related to its molecular structure and we calculate the indices directly from the molecular graph of a molecule using various graph theoretic notions (e.g. degree, distance, etc). Some of the topological indices may be found in (Das & Tinajstić, 2010; Gutman, 2013; Estrada, 2000; Gutman, Milovanović & et al., 2018) and the references therein.

DOI: 10.4018/978-1-5225-9380-5.ch004

Throughout the chapter only undirected, finite and simple connected graphs are considered. The degree of a vertex u in a graph G is the number of vertices incident on u and it is denoted as $d_G(u)$ or simply as $d(u)$ if there is no scope for confusion. The notation $uv \in E(G)$ represents any edge between two vertices u and v in a graph G. $d(u,v)$ is used to denote the distance between vertices u and v which is nothing but the length of the shortest path between u and v. The status of a vertex u in a graph G is denoted by $\sigma_G(u)$ or simply $\sigma(u)$ and is defined as the sum of distances of all other vertices from u in G.

Among all the degree based topological indices that exist in literature, the Zagreb indices are the oldest (Gutman, 2013). The Zagreb indices are defined by Gutman and Trinajstić in 1972 during their study of the total π electron energy of alternant hydrocarbons (Gutmam & Trinajstić, 1972). They have defined these indices as

$$M_1(G) = \sum_{uv \in E(G)} \left[d_G(u) + d_G(v) \right] \text{ and } M_2(G) = \sum_{uv \in E(G)} d_G(u) d_G(v),$$

where M_1 is called first Zagreb index and M_2 is called the second Zagreb index. Some of the research works on these indices can be seen in (Deng, Sarala & et al., 2016; Da Fonseca & Stevanović, 2014; Gutman & et al., 2015; Khalifeh, Yousefi-Azari & et al., 2009) and references cited therein. In 2016 Ramane & et al. defined a status based topological index and called it as Harmonic status index (HS). For a connected graph G, HS is given as

$$HS(G) = \sum_{uv \in E(G)} \frac{2}{\sigma(u) + \sigma(v)}.$$

In the next section, the authors propose a new topological index which is basically a weighted version of first Zagreb index. Then, the values of this newly proposed index for some of the popular graphs are provided. Further a study of the correlation of this novel index with various physicochemical properties of octane isomers is also reported in this chapter and at the end of the chapter, the conclusions are made.

WEIGHTED FIRST ZAGREB INDEX

In Llić & Milosavljević (2017) the authors introduced a new topological index and named it as *Weighted Szeged index*. This index is defined as

$$wSz(G) = \sum_{uv \in E(G)} \left[d(u) + d(v) \right] . n_u(e) . n_v(e)$$

where $n_u(e)$ is cardinality of the set

$$N_u = \{x \in V(G) : d(x,u) < d(x,v)\}.$$

Motivated by this the authors of this chapter propose a new topological index and called it as *Weighted First Zagreb Index*. This new index is denoted as wM_1.

Definition 2.1: (Weighted First Zagreb Index): For a graph G, wM_1 index is defined as

$$wM_1(G) = \sum_{uv \in E(G)} W_{uv} \left[d_G(u) + d_G(v) \right],$$

where $W_{uv} = \dfrac{1}{\sigma(u) + \sigma(v)}$ and $\sigma(u)$ is the status of the vertex u and so on.

Clearly, for any connected graph G, $d(u) \geq 1$ and $\sigma(u) \geq n-1 \ \forall u \in V(G)$ where n is the number of vertices in G. Thus, we have

$$\frac{2}{\sigma(u) + \sigma(v)} \leq \frac{d(u) + d(v)}{\sigma(u) + \sigma(v)} \leq \frac{d(u) + d(v)}{2(n-1)}$$

$$\Rightarrow \sum_{uv \in E(G)} \frac{2}{\sigma(u) + \sigma(v)} \leq \sum_{uv \in E(G)} \frac{d(u) + d(v)}{\sigma(u) + \sigma(v)} \leq \sum_{uv \in E(G)} \frac{\left[d(u) + d(v) \right]}{2(n-1)}$$

$$\Rightarrow HS(G) \leq wM_1(G) \leq \frac{M_1(G)}{2(n-1)}.$$

Again, for any vertex u in a connected graph G we have $\sigma(u) \geq 2n - 2 - d(u)$. Therefore,

$$d(u) + d(v) \geq 4n - 4 - \left(\sigma(u) + \sigma(v) \right)$$

$$\Rightarrow \frac{d(u) + d(v)}{\sigma(u) + \sigma(v)} \geq \frac{4n - 4 - \left(\sigma(u) + \sigma(v) \right)}{\sigma(u) + \sigma(v)}$$

$$\Rightarrow \sum_{uv \in E(G)} \frac{d(u) + d(v)}{\sigma(u) + \sigma(v)} \geq \sum_{uv \in E(G)} \left[\frac{4n - 4 - \left(\sigma(u) + \sigma(v) \right)}{\sigma(u) + \sigma(v)} \right]$$

$$\Rightarrow wM_1(G) \geq 2(n-1)HS - \left| E(G) \right|.$$

WEIGHTED FIRST ZAGREB INDEX OF SOME STANDARD GRAPHS

In this section, the Weighted first Zagreb index of some standard graphs are calculated.

Proposition 3.1: *Let* K_n *be a complete graph on n vertices. Then,* $wM_1\left(K_n\right) = n\left(n-1\right)/2$.

Proof. For a complete graph K_n , $d\left(u\right) = n-1$ and $\sigma\left(u\right) = n-1$ for all vertices u in $V\left(K_n\right)$. Therefore,

$$W_{uv} = \frac{1}{\left(n-1\right)+\left(n-1\right)} = \frac{1}{2\left(n-1\right)} \forall uv \in E\left(K_n\right).$$

Now,

$$wM_1\left(K_n\right) = \sum_{uv \in E\left(K_n\right)} \left(d_{K_n}\left(u\right) + d_{K_n}\left(v\right)\right)W_{uv} = \sum_{uv \in E\left(K_n\right)} \left(2\left(n-1\right)\right).\frac{1}{2\left(n-1\right)} = \frac{n\left(n-1\right)}{2}.$$

Hence the result.

Proposition 3.2: *Let* $K_{p,q}$ *be a complete bipartite graph. Then,*

$$wM_1\left(K_{p,q}\right) = pq\left(p+q\right)/\left(3\left(P+q\right)-4\right).$$

Proof. Let V_1 and V_2 be the two independent partitions of vertices of $K_{p,q}$ such that

$$V_1 = \{u \in V\left(K_{p,q}\right) \mid d\left(u\right) = p\}$$

and

$$V_2 = \{u \in V\left(K_{p,q}\right) \mid d\left(u\right) = q\} .$$

$K_{p,q}$ has $p+q$ vertices and pq edges. The edges uv in $K_{p,q}$ are of the type such that one vertex is in V_1 and other in V_2. Without loss of generality let $u \in V_1$ and $v \in V_2$. Then,

$$\sigma\left(u\right) = p + 2\left(q-1\right)$$

and

$$\sigma\left(v\right) = q + 2\left(p-1\right).$$

Therefore,

$$\sigma(u) + \sigma(v) = 3(p+q) - 4 \ \forall \ uv \in E(K_{p,q}).$$

$$\therefore \ wM_1(K_{p,q}) = \sum_{uv \in E(K_{p,q})} (p+q) \cdot \frac{1}{3(p+q)-4} = \frac{pq(p+q)}{3(p+q)-4}.$$

Hence the result.

Proposition 3.3: *For a cycle C_n on n vertices, $n \geq 3$, we have*

$$wM_1(C_n) = \begin{cases} \dfrac{8}{n} & \text{if } n \text{ is even} \\ \dfrac{8n}{n^2-1} & \text{if } n \text{ is odd.} \end{cases}$$

Proof. For a cycle every vertex is of degree 2 i.e., $d(u) = 2$ for all vertices $u \in V(C_n)$. Now, if n is even then

$$\sigma(u) = n^2 / 4 \ \forall \ u \in C_n.$$

Therefore,

$$wM_1(C_n) = \sum_{uv \in E(C_n)} (2+2) \frac{2}{n^2} = \frac{8}{n^2} n = \frac{8}{n}.$$

Again, if n is odd then

$$\sigma(u) = (n^2 - 1) / 4 \ \forall \ u \in C_n.$$

Therefore,

$$wM_1(C_n) = \sum_{uv \in E(C_n)} (2+2) \frac{2}{n^2-1} n = \frac{8n}{n^2-1}.$$

Hence the proposition.

Figure 1. A wheel graph W_9.

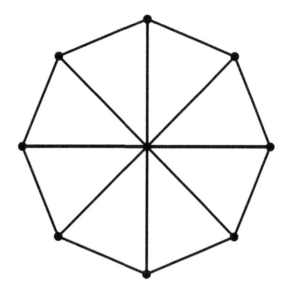

Definition 3.4: (Wheel graph) (Ramane, Basavanagoud et al., 2016) *A wheel graph* W_{n+1} *is a simple connected graph obtained from the cycle* C_n, $n \geq 3$ *by adding a new vertex and connecting it to all the vertices of* C_n.

Proposition 3.5: *Let* W_{n+1} *be wheel graph on* $n+1$ *vertices. Then,*

$$wM_1\left(W_{n+1}\right) = \frac{n\left(11n^2 - 6n - 9\right)}{3\left(n-1\right)\left(2n-3\right)}.$$

Proof. Partition the edge set of W_{n+1} into two parts

$$E_1 = \{uv \in E\left(W_{n+1}\right) \mid d\left(u\right) = n, d\left(v\right) = 3\}$$

and

$$E_2 = \{uv \in E\left(W_{n+1}\right) \mid d\left(u\right) = d\left(v\right) = 3\}.$$

Clearly, $\left|E_1\right| = \left|E_2\right| = n$. Now, let $uv \in E_1$ then,

$$\sigma\left(u\right) = n, \text{and } \sigma\left(v\right) = 2n - 3.$$

Therefore,

$$\sigma\left(u\right) + \sigma\left(v\right) = 3n - 3.$$

Again, let $uv \in E_2$ then,

$$\sigma\left(u\right) = 2n - 3, \text{and } \sigma\left(v\right) = 2n - 3.$$

This implies that $\sigma\left(u\right) + \sigma\left(v\right) = 4n - 6$. Thus,

$$wM_1\left(W_{n+1}\right) = \sum_{E_1}\left(d\left(u\right) + d\left(v\right)\right)W_{uv} + \sum_{E_2}\left(d\left(u\right) + d\left(v\right)\right)W_{uv}$$

$$= \sum_{E_1}\left(n + 3\right)\frac{1}{3n - 3} + \sum_{E_2}\left(3 + 3\right)\frac{1}{4n - 6}$$

$$= \frac{n\left(n + 3\right)}{3\left(n - 1\right)} + \frac{3n}{2n - 3}$$

$$= \frac{n\left(11n^2 - 6n - 9\right)}{3\left(n - 1\right)\left(2n - 3\right)}.$$

Hence proved.

Definition 3.6: (Friendship graph) (Ramane, Basavanagoud & et al., 2016) *A friendship graph F_n, $n \geq 2$, is a simple connected graph that can be constructed by joining n copies of the cycle C_3 of length 3 with a common vertex. It has $2n + 1$ vertices and $3n$ edges.*

Figure 2. A friendship graph F_4.

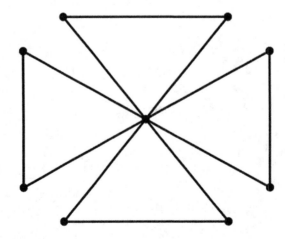

Proposition 3.7: *For a friendship graph* F_n, $n \geq 2$, *we have*

$$wM_1\left(F_n\right) = \frac{n\left(4n^2 + 5n - 3\right)}{\left(3n - 1\right)\left(2n - 1\right)}.$$

Proof. Partition the edge set of F_n as

$$E_1 = \{uv \in E\left(F_n\right) \mid d\left(u\right) = 2n,\, d\left(v\right) = 2\}$$

and

$$E_2 = \{uv \in E\left(F_n\right) \mid d\left(u\right) = d\left(v\right) = 2\}.$$

$$\left|E_1\right| = 2n \text{ and } \left|E_2\right| = n. \text{ If } uv \in E_1 \text{ then}$$

$$\sigma\left(u\right) = 2n, \sigma\left(v\right) = 4n - 2$$

and if $uv \in E_2$ then

$$\sigma\left(u\right) = \sigma\left(v\right) = 4n - 2.$$

Therefore,

$$wM_1\left(F_n\right) = \sum_{E_1}\left(d\left(u\right) + d\left(v\right)\right)W_{uv} + \sum_{E_2}\left(d\left(u\right) + d\left(v\right)\right)W_{uv}$$

$$= \sum_{E_1}\left(2n + 2\right)\frac{1}{6n - 2} + \sum_{E_2}\left(2 + 2\right)\frac{1}{2\left(4n - 2\right)}$$

$$= \frac{n + 1}{3n - 1}2n + \frac{1}{2n - 1}n$$

$$= \frac{n\left(4n^2 + 5n - 3\right)}{\left(3n - 1\right)\left(2n - 1\right)}.$$

Hence the proposition.

REGRESSION ANALYSIS BETWEEN wM_1 AND OCTANE ISOMERS

Topological indices are highly useful in *Quantitative Structure Property Relationships* (QSPR) and *Quantitative Structure Activity Relationships* (QSAR) of molecules. In this section the study of correlation of the newly proposed topological index with various properties of isomers of octane molecule $\left(C_8H_{18}\right)$ is carried out. In Figure 3 the hydrogen-suppressed molecular graphs of octane isomers is provided. The hydrogen atoms have no contribution in determining the various physical as well as chemical properties of molecular compound because of its smaller size. Thus in the study of chemical correlations of topological index only the hydrogen suppressed molecular graph are considered.

We observe that the Weighted first Zagreb index is strongly correlated with many properties of octane isomers, especially with acentric factor(AcentFac) and entopy(S). The R^2 value between wM_1 index and AcentFac of octane isomers is 0.9906. See the scatter plot in Figure 4. As compared to M_1 i.e., the fisrt Zagreb index and HS index, this novel index has better correlation with AcentFac of octane isomers. The R^2 value between M_1 and AcentFac of octane isomers is 0.9469 and that between HS index and AcentFac of octane isomers is 0.9430. The wM_1 index also has good correlation with the entropy(S) of octane isomers. For this R^2 value is 0.8996. A scatter plot between S and wM_1 index is depicted in Figure 5.

Figure 3. Hydrogen suppressed molecular graphs of octane isomers $\left(C_8H_{18}\right)$.

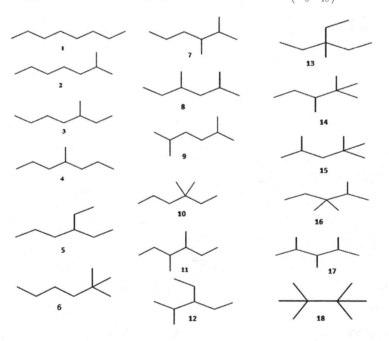

Table 1. Some properties of octane isomers, M_1, wM_1 and HS index

S. No.	Octane Isomers	AcentFac	Entropy	M_1	HS	wM_1
1	n-octane	0.3979	111.67	26	0.3601	0.6803
2	2-methylheptane	0.3779	109.84	28	0.386	0.7823
3	3-methylheptane	0.3710	111.26	28	0.4061	0.8336
4	4-methylheptane	0.3715	109.32	28	0.4139	0.8538
5	3-ethylhexane	0.3625	109.43	28	0.4352	0.9038
6	2,2-dimethylhexane	0.3394	103.42	32	0.4401	1.0282
7	2,3-dimethylhexane	0.3482	108.02	30	0.4483	0.9881
8	2,4-dimethylhexane	0.3442	106.98	30	0.4386	0.9598
9	2,5-dimethylhexane	0.3568	105.72	30	0.4153	1.8972
10	3,3-dimethylhexane	0.3226	104.74	32	0.4753	1.1296
11	3,4-dimethylhexane	0.3403	106.59	30	0.4648	1.0317
12	3-ethyl, 2-methylhexane	0.3324	106.06	30	0.4733	1.0513
13	3-ethyl, 3-metylpentane	0.3069	101.48	32	0.5034	1.207
14	2,2,3-trimethylpentane	0.3008	101.31	34	0.5091	1.2724
15	2,2,4-trimethylpentane	0.3054	104.09	34	0.4779	1.1776
16	2,3,3-trimethylpentane	0.2932	102.06	34	0.5202	1.3077
17	2,3,4-trimethylpentane	0.3174	102.39	32	0.488	1.1429
18	2,2,3,3-tetramethylbutane	0.2553	93.06	38	0.5615	1.5538

Figure 4. Regression line between Acentric factor(AcentFac) and wM_1 index of octane isomers

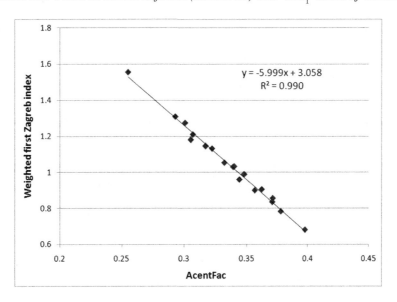

Figure 5. Regression line between entropy(S) and wM_1 index of octane isomers

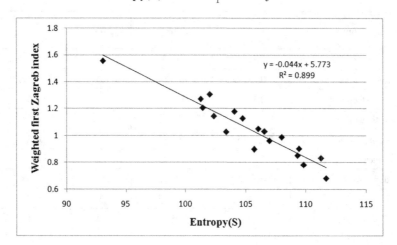

CONCLUSION

In this work, it is tried to mix two parameters of vertices viz, degree and status to propose a novel topological index. A regression analysis is also carried out to establish the usefulness of this extension. Further the authors has done a comparision study of the newly proposed index with two other classical indices viz, M_1 and HS. M_1 is based on degree and the other is based on status. As we know that a vertex with higher degree ends up with lower status. Degree based topological indices like first Zagreb index is more efficient to predict properties where terminal edges play leading role, while status based indices are found to be efficient in capturing the properties mostly rely on interior edges. So, this extension is an attempt to capture those properties, where both terminal and internal edges play almost equal roles.

REFERENCES

Da Fonseca, C. M., & Stevanović, D. (2014). Further properties of the second Zagreb index. *MATCH Commun. Math. Comput. Chem.*, *72*, 655–668.

Das, K. C., & Trinajstić, N. (2010). Comparison between first geometric-arithmetic index and atom-bond connectivity index. *Chemical Physics Letters*, *497*(1-3), 149–151. doi:10.1016/j.cplett.2010.07.097

Deng, H., Saralab, D., Ayyaswamy, S. K., & Balachandran, S. (2016). The Zagreb indices of four operations on graphs. *Applied Mathematics and Computation*, *275*, 422–431. doi:10.1016/j.amc.2015.11.058

Estrada, E. (2000). Characterization of 3D molecular structure. *Chemical Physics Letters*, *319*(5-6), 713–718. doi:10.1016/S0009-2614(00)00158-5

Gutman, I. (2013). Degree-based topological indices. *Croatica Chemica Acta*, *86*(4), 315–361. doi:10.5562/cca2294

Gutman, I., Furtula, B., Vukićević, Ž. K., & Popivoda, G. (2015). On Zagreb indices and Coindices. *MATCH Commun. Math. Comput. Chem.*, *74*, 5–16.

Gutman, I., Milovanović, E., & Milovanović, I. (2018). Beyond the Zagreb indices. *AKCE International Journal of Graph and Combinatorics.*

Gutman, I., & Trinajstić, N. (1972). Graph theory and molecular orbitals. Total π. *Chemical Physics Letters, 17*(4), 535–538. doi:10.1016/0009-2614(72)85099-1

Khalifeh, M. H., Yousefi-Azari, H., & Ashrafi, A. R. (2009). The first and second Zagreb indices of some graph operations. *Discrete Applied Mathematics, 157*(4), 804–811. doi:10.1016/j.dam.2008.06.015

Llić, A., & Milosavljević, N. (2017). The weighted vertex PI index. *Mathematical and Computer Modelling, 57*, 393–406.

Ramane, H. S., Basavanagoud, B., & Yalnaik, A. S. (2016). Harmonic Status Index og Graphs. *Bulletin of Mathematical Sciences and Applications, 17*, 24–32. doi:10.18052/www.scipress.com/BMSA.17.24

Trinajstić, N. (2011). *Chemical graph theory* (2nd ed.). CRC Press.

Chapter 5

Inverse Sum Indeg Index of Subdivision, t–Subdivision Graphs, and Related Sums

Amitav Doley
DHSK College, India

Jibonjyoti Buragohain
Dibrugarh University, India

A. Bharali
Dibrugarh University, India

ABSTRACT

The inverse sum indeg (ISI) index of a graph G is defined as the sum of the weights dG(u)dG(v)/ dG(u)+dG(v) of all edges uv in G, where dG(u) is the degree of the vertex u in G. This index is found to be a significant predictor of total surface area of octane isomers. In this chapter, the authors present some lower and upper bounds for ISI index of subdivision graphs, t-subdivision graphs, s-sum and st -sum of graphs in terms of some graph parameters such as order, size, maximum degree, minimum degree, and the first Zagreb index. The extremal graphs are also characterized for their sharpness.

INTRODUCTION

A topological index (TI), also known as graph-based molecular descriptor, is a mapping $T : \Im \to \mathbb{R}$ from the collection of graphs $\Im$ to the set $\mathbb{R}$ of real numbers which characterizes numerically a topological structure of a molecular graph numerically. It is a graph invariant which does not depend on the labelling or the pictorial representation of a graph, i.e., $T\left(G\right) = T\left(H\right)$ iff the graphs G and H are isomorphic. The values of these indices are very helpful in predicting various physical and chemical properties of a molecular compound, among them boiling point, melting point, strain energy, stability, surface tension of isomers are to name a few. It studies quantitative structure–activity relationship (QSAR) and

DOI: 10.4018/978-1-5225-9380-5.ch005

quantitative structure–property relationship (QSPR) which are used in predicting the biological activities and properties of the chemical compounds.

The first TI was defined by Wiener in the year 1947 (Wiener, 1947), named after his name as Wiener index which is a distance-based TI. Historically Zagreb indices can be considered as the first degree-based topological indices, which came into picture during the study of total π-electron energy of alternant hydrocarbons by Gutman and Trinajstić in 1972 (Gutman & Trinajstić, 1972). Since then many topological indices are proposed and studied based on degree, distance and other parameters of graph. Inverse sum indeg or simply ISI index is relatively a new addition to this wide list of degree based topological indices.

Formally the ISI index can be defined as

$$ISI\left(G\right) = \sum_{uv \in E\left(G\right)} \frac{d_G\left(u\right)d_G\left(v\right)}{d_G\left(u\right) + d_G\left(v\right)}.$$

ISI index was selected by Vukičević and Gašperov (Vukičević & Gašperov, 2010) as a significant predictor of total surface area of octane isomers, and included in the list of 20 (twenty) indices which were selected as significant predictors of physico-chemical properties of a molecular compound out of a list of 148 discrete Adriatic indices considered in their study.

The study of ISI index for various operations of graphs is found to be limited in literature. Sedlar and Stevanović et al. determined extremal values of ISI index across several graph classes, including connected graphs, chemical graphs, trees and chemical trees (Sedlar, Stevanović, & Vasilyev, 2015). Some exact formulas for the inverse sum indeg index of some nanotubes is computed in (Falahati-Nezhad & Azari, 2016). In 2017, Falahati-Nezhad and Azari, et al. again published a paper (Falahati-Nezhad, Azari & Došlić, 2017) on ISI index in which several sharp upper and lower bounds for ISI index in terms of the order, size, radius, number of pendant vertices, minimal and maximal vertex degrees, and minimal non-pendent vertex degree are presented, and linked this index to various well-known TIs such as the Zagreb indices, Randić index, sum-connectivity index, modified Zagreb index, harmonic index, forgotten index, and eccentric connectivity index. Some extremal molecular graphs with the minimum and the maximum value of ISI index in $MG\left(i, n\right)$, where $MG\left(i, n\right)$ is the class of all n-vertex molecular graphs with minimum degree i, is proposed in Hasani (2017). In Pattabiraman (2018), several upper and lower bounds on the ISI index is presented in terms of some molecular structural parameters and linked this index to various well-known molecular descriptors. Some sharp bounds for ISI index of graphs with given matching number, independence number and vertex-connectivity are presented and characterized the extremal graphs for which the bounds are obtained in An and Xiong (2018). Chen and Deng derived some bounds for ISI index in terms of vertex (edge) connectivity, chromatic number, vertex bipartiteness etc. in Chen and Deng (2018). Gao and Jamil et al. computed ISI index of some chemical graphs (Gao, Jamil, Javed, Farahani, & Imran, 2018) by using the line graphs of subdivision of the graphs. In Gutman, Matejić, Milovanović, and Milovanović (2020) some lower bounds of ISI index reported in the literature are analysed and determined some new lower bounds for ISI index. Very recently, Gutman and Rodríguez et al. has established some linear and non-linear inequalities of ISI index in terms maximum degree, minimum degree, vertices and edges, and in terms of some well-known topological index in which some inequalities are also generalised (Gutman, Rodríguez, & Sigarreta 2019).

The graph operations namely, subdivision graph, vertex-semitotal graph, edge-semitotal graph and total graph and the four sums based on these operations, called F-sums, were defined and studied in connection with wiener index in 2009 by Eliasi and Taeri (2009). These operations seem to be important, as they can be used to model the atom-bond connectivity alongwith atom-atom connectivity in a molecular graph. Since then these operations were studied for various TIs. Some exact expressions of the reverse-Wiener index of the F-sum graphs are presented in Metsidik, Zhang, and Duan (2010). In 2016, Deng and Sarala et al. studied the first and second Zagreb indices for these four operations on graphs (Deng, Sarala, Ayyaswamy & Balachandran, 2016). The hyper-Zagreb index of F-sums on graphs is studied in Basavanagoud and Patil (2017). In Onagh (2017) the harmonic index of subdivision graphs, t-subdivision graphs and also, S-sum and S_t-sum of graphs is studied. In 2018, Onagh again, published a paper on harmonic index of edge-semitotal graphs, total graphs and related sums of graphs (Onagh, 2018). Sarala and Deng et al. studied the F index of strong product of two connected graphs in which one of the graphs is obtained by using F-sums of graphs and the other is any connected graph (Sarala, Deng, Natarajan, & Ayyaswamy, 2019). But there is limited work for ISI index for subdivision graphs, vertex semi-total graphs, edge semi-total graphs and four operations of graphs in the literature. This is why this research work is done to evaluate ISI index for of subdivision graphs, t-subdivision graphs, S-sum and S_t-sum of graphs.

In this work, the authors present some bounds for ISI index of subdivision graphs, t-subdivision graphs, S-sum and S_t-sum of graphs and characterize the extremal graphs.

PRELIMINARIES

Graph theory is the branch of mathematics concerned with networks of points connected by lines. It has been independently discovered many times from various recreational mathematical problems. Indeed, the earliest recorded mention of the subject occurs in the work of *Euler,* the father of graph theory, when in 1736 he settled a famous unsolved problem of his day called the Königsberg Bridge Problem. In this section, the authors discuss on some basic concepts of graphs and graph operations and present some definitions which will be explored in the rest of the chapter.

- **Degree of Vertex:** The degree of a vertex v in an undirected graph G is the number of edges in G that are incident to the vertex v.
- **Subgraph:** A subgraph H of a graph G is a graph whose vertices and edges are also in G and each edge of H has the same end vertices in H as in G.
- **Walk:** A walk of a graph G is a subgraph of G which is a finite alternating sequence of vertices and edges, beginng and ending with vertices, such that each edge is incident with the vertices preceding and following it. No edge in a walk apears more than once.
- **Path:** A path is an open walk in which no vertex appears more than once. The number of edges in a path is the length of that path. A path of n veritces is usually denoted by P_n.
- **Cycle:** A closed walk in which no vertex, except the initial and the terminal vertices, appears more than once is called a cycle. Clearly, every vertex in a cycle is of degree two. A cycle of n vertices is usually denoted by C_n.

- **Connected Graph:** A graph G is called connected if there exists at least one path between every pair of vertices in G.
- **Regular Graph:** A graph in which all vertices are of equal degree is called a regular graph. A graph G is called r-regular if every vertex of G is of dgree r.
- **Biregular Graph:** A graph G is called a (δ, Δ) -biregular graph if the degree of any vertex in G is either δ or Δ.
- **Complete Graph:** A simple graph in which there exists an edge between every pair of vertices is called a complete graph. The complete graph of n vertices is denoted by K_n.
- **Molecular Graph:** A molecular graph $G = (V, E)$ is a simple graph having $|V|$ vertices and $|E|$ edges. The vertices $v_i \in V$ represent non-hydrogen atoms and the edges $(v_i, v_j) \in E$ represent covalent bonds between the corresponding atoms. In particular, hydrocarbons are formed only by carbon and hydrogen atom and their molecular graphs represent the carbon skeleton of the molecule.

Definition 2.1. *The first Zagreb index (Gutman & Trinajstić, 1972)* $M_1(G)$ *of a graph G is defined as*

$$M_1(G) = \sum_{u \in V(G)} d_G^2(u) = \sum_{uv \in E(G)} \left[d_G(u) + d_G(v) \right].$$

Definition 2.2. *The subdivision graph (Eliasi &Taeri, 2009)* $S(G)$ *of a graph G is a graph which is obtained from G by inserting an additional vertex on each edge of G. The t-subdivision graph* $S_t(G)$ *of G is obtain from G by replacing each edge of G by a path of length* $t + 1$.
Clearly,

$$d_{S(G)}(u) = \begin{cases} d_G(u), u \in V(G) \\ 2, \ u \notin V(G) \end{cases}$$

An example of $S(P_3)$ and $S_4(P_2)$ is depicted in Figure 1.

Definition 2.3. *The S-sum (Eliasi & Taeri, 2009)* $G_1 +_S G_2$ *of the graphs* G_1 *and* G_2 *is a graph with vertex set*

$$\left(V(G_1) \cup E(G_1) \right) \times V(G_2)$$

in which two vertices (u_1, v_1) *and* (u_2, v_2) *are adjacent iff* $[u_1 = u_2 \in V(G_1)$ *and*

$v_1 v_2 \in E(G_2)]$ or $[v_1 = v_2 \in V(G_2)$ and $u_1 u_2 \in E(S(G_1))]$.

Clearly,

$$d_{G_1 +_S G_2}(u,v) = \begin{cases} d_{G_1}(u) + d_{G_2}(v), u \in V(G_1), v \in V(G_2) \\ 2, u \in E(G), v \in V(G_2). \end{cases}$$

Definition 2.4. *The S_t-sum (Onagh, 2017) $G_1 +_{S_t} G_2$ of the graphs G_1 and G_2 is a graph with vertex set $V(S_t(G_1)) \times V(G_2)$ in which two vertices (u_1, v_1) and (u_2, v_2) are adjacent iff*

$$\left[u_1 = u_2 \in V(G_1) \, and \, v_1 v_2 \in E(G_2) \right]$$

or

$$\left[v_1 = v_2 \in V(G_2) \, and \, u_1 u_2 \in E(S_t(G_1)) \right].$$

Clearly,

$$d_{G_1 +_{S_t} G_2}(u,v) = \begin{cases} d_{G_1}(u) + d_{G_2}(v), u \in V(G_1), v \in V(G_2) \\ 2, otherwise. \end{cases}$$

An example of $P_3 +_S P_2$ and $P_3 +_{S_4} P_2$ is depicted in Figure 1.

Throughout this chapter all graphs are finite, simple, non-trivial and connected. The vertex set and the edge set of the graph G will be denoted by $V(G)$ and $E(G)$ respectively. Two vertices u and v of a graph G are said to be adjacent if there is an edge connecting them. The connecting edge is usually denoted by uv. The notations and terminologies used but not clearly stated in this article may be found in West (2002).

MAIN RESULTS

The main results of ISI index for subdivision graph, t-subdivision graph, S-sum and S_t-sum of graphs are presented in this section. The extremal graphs are also characterized.

Let G be a graph with m edges. Then by the definition of ISI index,

$$ISI(S(G)) = \sum_{uv \in E(S(G))} \frac{d_{S(G)}(u) d_{S(G)}(v)}{d_{S(G)}(u) + d_{S(G)}(v)} = \sum_{uv \in E(G)} \left(\frac{2d_G(u)}{d_G(u)+2} + \frac{2d_G(v)}{d_G(v)+2} \right) \quad (1)$$

Again,

Figure 1. The graph operations

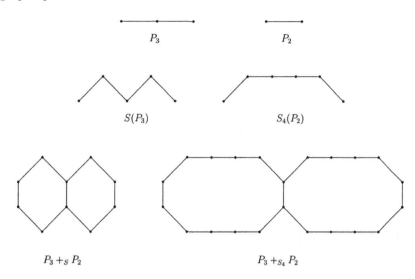

$$ISI\left(S_{t}\left(G\right)\right) = \sum_{uv\in E\left(G\right)} \left| \frac{2d_{G}\left(u\right)}{d_{G}\left(u\right)+2} + \overbrace{\frac{2.2}{2+2} + ... + \frac{2.2}{2+2}}^{(t-1)times} + \frac{2d_{G}\left(v\right)}{d_{G}\left(v\right)+2} \right|$$

$$= \sum_{uv\in E\left(G\right)} \left| \frac{2d_{G}\left(u\right)}{d_{G}\left(u\right)+2} + \frac{2d_{G}\left(v\right)}{d_{G}\left(v\right)+2} \right| + m\left(t-1\right) = ISI\left(S\left(G\right)\right) + m\left(t-1\right)$$

(2)

Clearly,

$$ISI\left(S_{t}\left(G\right)\right) > ISI\left(S\left(G\right)\right),$$

for $t \geq 2$.

In the following two examples we evaluate the ISI index for subdivision graphs and *t*-subdivision graphs of a path, cycle, complete graph and *k*-regular graph of n vertices using equation (1) and (2).

Example 1. a*) For*

$$n \geq 2, \ ISI\left(S\left(P_{n}\right)\right) = 2n - \frac{8}{n}$$

and

$$ISI\left(S_{t}\left(P_{n}\right)\right) = 2n - \frac{8}{n} + \left(t-1\right)\left(n-1\right).$$

b) For

$$n \geq 3, \ ISI\left(S\left(C_n\right)\right) = 2n$$

and

$$ISI\left(S_t\left(C_n\right)\right) = 2n + \left(t-1\right)n.$$

c) $ISI\left(S\left(K_n\right)\right) = \dfrac{2n(n-1)^2}{n+1}$ and

$$ISI\left(S_t\left(K_n\right)\right) = \dfrac{2n(n-1)^2}{n+1} + \left(t-1\right)\dfrac{n\left(n-1\right)}{2}.$$

Example 2. *If G is a k-regular graph of n vertices, then*

$$ISI\left(S\left(G\right)\right) = \dfrac{2nk^2}{k+2}$$

and

$$ISI\left(S_t\left(G\right)\right) = \dfrac{2nk^2}{k+2} + \left(t-1\right)\dfrac{nk}{2}.$$

Lower Bounds for ISI Index of Subdivision Graphs

Some lower bounds for ISI index of subdivision graphs and *t*-subdivision graphs are established in this section which are obtained in terms of vertices and edges. The following two well known inequalities are used to obtain the bounds.

Jensen Inequality: (Dragomir, 2003) If f is convex on the interval I and $x_1, x_2, ..., x_n \in I$, then

$$f\left(\dfrac{x_1 + x_2 + ... + x_n}{n}\right) \leq \dfrac{f\left(x_1\right) + f\left(x_2\right) + ... + f\left(x_n\right)}{n}$$

with equality iff $x_1 = x_2 = ... = x_n$.

Cauchy Schwarz' Inequality: (Dragomir, 2003) Let a_i and b_i $\left(i = 1, 2, ..., n\right)$, be real numbers. Then

$$\left(\sum_{i=1}^{n} a_i b_i\right)^2 \le \sum_{i=1}^{n} a_i^2 \sum_{i=1}^{n} b_i^2$$

with equality iff $\dfrac{a_i}{b_i}$ is constant for $i = 1, 2, \ldots, n$.

Theorem 3.1. *Let G be a graph with n vertices and m edges. Then*

$$ISI\left(S\left(G\right)\right) = \sum_{u \in V(G)} \frac{2d_G^2\left(u\right)}{d_G\left(u\right) + 2} = 4m - 4n + 8 \sum_{u \in V(G)} \frac{1}{d_G\left(u\right) + 2}.$$

Proof. For every neighbor of the vertex u, the expression $\dfrac{2d_G\left(u\right)}{d_G\left(u\right) + 2}$ appears exactly once in the sum

$$\sum_{u \in E(G)} \left(\frac{2d_G\left(u\right)}{d_G\left(u\right) + 2} + \frac{2d_G\left(v\right)}{d_G\left(v\right) + 2}\right).$$

Therefore,

$$ISI\left(S\left(G\right)\right) = \sum_{u \in V(G)} \left(\overbrace{\frac{2d_G\left(u\right)}{d_G\left(u\right) + 2} + \ldots + \frac{2d_G\left(u\right)}{d_G\left(u\right) + 2}}^{d_G(u) times}\right).$$

By which we have

$$ISI\left(S\left(G\right)\right) = \sum_{u \in V(G)} \frac{2d_G^2\left(u\right)}{d_G\left(u\right) + 2} = 4m - 4n + 8 \sum_{u \in V(G)} \frac{1}{d_G\left(u\right) + 2}.$$

This completes the proof.

Corollary 3.2. *Let G be a graph with n vertices and m edges. Then*

$$ISI\left(S_t\left(G\right)\right) = m\left(t + 3\right) - 4n + 8 \sum_{u \in V(G)} \frac{1}{d_G\left(u\right) + 2}.$$

Proof. It can be easily obtained from the fact that

$$ISI\left(S_t\left(G\right)\right) = ISI\left(S\left(G\right)\right) + m\left(t - 1\right).$$

Theorem 3.3. *For the graph G with n vertices and m edges,* $ISI\left(S\left(G\right)\right) \geq 3m - n$ *with equality iff* $G = C_{n}$.

Proof. As

$$ISI\left(S\left(G\right)\right) = \sum_{uv \in E(G)} \left(\frac{2d_G\left(u\right)}{d_G\left(u\right)+2} + \frac{2d_G\left(v\right)}{2+d_G\left(v\right)}\right)$$

$$= 2 \sum_{uv \in E(G)} \left[1 - \frac{2}{d_G\left(u\right)+2} + 1 - \frac{2}{2+d_G\left(v\right)}\right] = 4m - 4 \sum_{u \in V(G)} \frac{d_G\left(u\right)}{d_G\left(u\right)+2}$$

for any vertex u in $V\left(G\right)$, therefore by Jensen's inequality we obtain

$$ISI\left(S\left(G\right)\right) \geq 4m - \sum_{u \in V(G)} \left[1 + \frac{1}{2}d_G\left(u\right)\right] = 3m - n$$

with equality iff $\dfrac{2}{d_G\left(u\right)} = 1$, for all $u \in V\left(G\right)$, i.e., iff G is a 2-regular graph i.e., iff $G = C_{n}$ as G is connected.

Corollary 3.4. *For the graph G with n vertices and m edges,*

$$ISI\left(S_{t}\left(G\right)\right) \geq m\left(t+2\right) - n$$

with equality iff $G = C_{n}$.

A better lower bound can be obtained by using Cauchy-Schwarz' inequality whose equality condition is satisfied by any regular graph.

Theorem 3.5. *For the graph G with n vertices and m edges,*

$$ISI\left(S\left(G\right)\right) \geq \frac{4m^2}{m+n}$$

with equality iff G is regular.

Proof. By Cauchy-Schwarz' inequality, for any vertex u in $V\left(G\right)$, we have

$$\sum_{u \in V(G)} \left(\sqrt{d_G\left(u\right)+2}\right)^2 \sum_{u \in V(G)} \left(\frac{1}{\sqrt{d_G\left(u\right)+2}}\right)^2 \geq \left(\sum_{u \in V(G)} \sqrt{d_G\left(u\right)+2} \frac{1}{\sqrt{d_G\left(u\right)+2}}\right)^2,$$

whence

$$\sum_{u \in V(G)} \frac{1}{d_G(u) + 2} \geq \frac{n^2}{2(m+n)}.$$

Now the desired result follows from Theorem 3.1 with equality iff $d_G(u) + 2 = a$ constant, $\forall\, u \in V(G)$ i.e., iff G is regular.

Corollary 3.6. *For the graph G with n vertices and m edges,*

$$ISI\left(S_t(G)\right) \geq \frac{4m^2}{m+n} + m(t-1)$$

with equality iff G is regular.

Upper Bounds for ISI Index of Subdivision Graphs

In this section, the authors present some upper bounds for ISI index of subdivision graphs and t-subdivision graphs in terms of number of vertices, edges, first Zagreb index and maximum degree and minimum degree of any vertex in a graph using Jensen's inequality and the following inequality.

Schweitzer's Inequality: (Dragomir, 2003) Let $x_1, x_2, ..., x_n$ be positive real numbers and let

$$m \leq x_i \leq M, i = 1, 2, ..., n.$$

Then

$$\sum_{i=1}^{n} x_i \sum_{i=1}^{n} \frac{1}{x_i} \leq \frac{n^2(m+M)^2}{4mM},$$

with equality iff

$$x_1 = ... = x_n = m = M$$

or if *n* is even,

$$x_1 = ... = x_{n/2} = m$$

and

$$x_{n/2+1} = ... = x_n = M,$$

where $m < M$ and $x_1 \leq x_2 \leq ... \leq x_n$.

Theorem 3.7. *Let G be a graph with n vertices and m edges. Then,*

$$ISI\left(S\left(G\right)\right) \leq m + \frac{1}{4}M_1\left(G\right)$$

with equality iff $G = C_n$.

Proof. From Theorem 3.1.,

$$ISI\left(S\left(G\right)\right) = \sum_{u \in V(G)} \frac{2d_G^2\left(u\right)}{d_G\left(u\right) + 2}.$$

Now by Jensen's inequality,

$$\sum_{u \in V(G)} \frac{2d_G^2\left(u\right)}{d_G\left(u\right) + 2} \leq \frac{1}{2} \sum_{u \in V(G)} \left[d_G\left(u\right) + \frac{1}{2}d_G^2\left(u\right)\right] = \frac{1}{2}\left[2m + \frac{1}{2}M_1\left(G\right)\right]$$

with equality iff

$$\frac{1}{d_G\left(u\right)} = \frac{2}{d_G^2\left(u\right)} \quad \forall u \in V\left(G\right),$$

i.e., iff $G = C_n$ as G is connected graph, completing the proof.

Corollary 3.8. *Let G be a graph with n vertices and m edges. Then,*

$$ISI\left(S_t\left(G\right)\right) = tm + \frac{1}{4}M_1\left(G\right)$$

with equality iff $G = C_n$.

Theorem 3.9. *Let G be graph with n vertices and m edges. Then*

$$ISI\left(S\left(G\right)\right) \leq 4\left(m - n\right) + \frac{n^2(\delta + \Delta + 4)^2}{\left(m + n\right)\left(\delta + 2\right)\left(\Delta + 2\right)}$$

with equality iff G is regular or if n is even, G is $\left(\delta, \Delta\right)$-biregular where δ and Δ are the minimum and maximum degree of any vertex in G.

Proof. For any vertex u in $V\left(G\right)$,

$$\delta + 2 \leq d_G(u) + 2 \leq \Delta + 2.$$

Now by Schweitzer's inequality

$$\sum_{u \in V(G)} \big(d_G(u) + 2\big) \sum_{u \in V(G)} \frac{1}{d_G(u) + 2} \leq \frac{n^2 (\delta + 2 + \Delta + 2)^2}{4(\delta + 2)(\Delta + 2)} \Rightarrow \sum_{u \in V(G)} \frac{1}{d_G(u) + 2} \leq \frac{n^2 (\delta + \Delta + 4)^2}{8(m + n)(\delta + 2)(\Delta + 2)}.$$

The desired inequality follows from Theorem 3.1 with equality iff G is a regular or if n is even, $\dfrac{n}{2}$ vertices of G have degree δ and remaining $\dfrac{n}{2}$ vertices have degree Δ i.e., iff G is regular or if n is even, G is (δ, Δ)-biregular.

Corollary 3.10. *Let G be a graph with n vertices and m edges. Then,*

$$ISI\big(S_t(G)\big) \leq m(t + 3) - 4n + \frac{n^2(\delta + \Delta + 4)^2}{(m + n)(\delta + 2)(\Delta + 2)}$$

with equality iff G is regular or if n is even, G is (δ, Δ)-biregular.

The ISI Index for S-Sum and S_t-Sum of Graphs

In this section, the authors present an upper bound for ISI index of S-sum and S_t-sum of graphs in terms of number of vertices, edges and first Zagreb index.

Theorem 3.11. *Let G_1 and G_2 be two graphs with*

$$\big|V(G_1)\big| = n_1, \ \big|V(G_2)\big| = n_2$$

and

$$\big|E(G_1)\big| = m_1, \ \big|E(G_2)\big| = m_2.$$

Then,

$$ISI\big(G_1 +_S G_2\big) \leq 2m_1 m_2 + m_1 n_2 + \frac{1}{4}\Big[n_1 M_1\big(G_2\big) + n_2 M_1\big(G_1\big)\Big]$$

with equality iff $G_1 = G_2 = K_2$.

Proof. Partition the edge set $E(G)$ of $G = G_1 +_S G_2$ into two sets E_1 and E_2, where

$$E_1 = \left\{ \left(u_i, v_k\right)\left(u_j, v_l\right) \mid u_i, u_j \in V\left(G_1\right), v_k, v_l \in V\left(G_2\right) \right\}$$

and

$$E_2 = \left\{ \left(u_i, v_k\right)\left(e_j, v_k\right) \mid u_i \in V\left(G_1\right), v_k \in V\left(G_2\right), e_j \in E\left(G_1\right) \right\}.$$

Now,

$$ISI\left(G\right) = \sum_{\left(u_i, v_k\right)\left(u_j, v_l\right) \in E_1} \frac{d_G\left(u_i, v_k\right) d_G\left(u_j, v_l\right)}{d_G\left(u_i, v_k\right) + d_G\left(u_j, v_l\right)} + \sum_{\left(u_i, v_k\right)\left(e_j, v_k\right) \in E_2} \frac{d_G\left(u_i, v_k\right) d_G\left(e_j, v_k\right)}{d_G\left(u_i, v_k\right) + d_G\left(e_j, v_k\right)} := \sum 1 + \sum 2$$

By Jensen's inequality, for any edge $\left(u_i, v_k\right)\left(u_j, v_l\right) \in E_1$, we obtain

$$\sum 1 = \sum_{\left(u_i, v_k\right)\left(u_j, v_l\right) \in E_1} \frac{\left[d_{G_1}\left(u_i\right) + d_{G_2}\left(v_k\right)\right]\left[d_{G_1}\left(u_j\right) + d_{G_2}\left(v_l\right)\right]}{\left[d_{G_1}\left(u_i\right) + d_{G_2}\left(v_k\right)\right] + \left[d_{G_1}\left(u_j\right) + d_{G_2}\left(v_l\right)\right]}$$

$$\leq \frac{1}{4} \sum_{\left(u_i, v_k\right)\left(u_j, v_l\right) \in E_1} \left[\left(d_{G_1}\left(u_j\right) + d_{G_2}\left(v_l\right)\right) + \left(d_{G_1}\left(u_i\right) + d_{G_2}\left(v_k\right)\right)\right]$$

$$= \frac{1}{4} \sum_{u_i \in V\left(G_1\right)} \sum_{v_k v_l \in E\left(G_2\right)} \left[2d_{G_1}\left(u_i\right) + \left(d_{G_2}\left(v_k\right) + d_{G_2}\left(v_l\right)\right)\right] = m_1 m_2 + \frac{n_1}{4} M_1\left(G_2\right)$$

with equality iff

$$d_{G_1}\left(u_i\right) + d_{G_2}\left(v_l\right) = d_{G_1}\left(u_i\right) + d_{G_2}\left(v_k\right), \forall u_i, u_j \in V\left(G_1\right); v_k, v_l \in V\left(G_2\right)$$

i.e., iff both G_1 and G_2 are regular. Similarly,

$$\sum 2 = \sum_{\left(u_i, v_k\right)\left(e_j, v_k\right) \in E_2} \frac{\left[d_{G_1}\left(u_i\right) + d_{G_2}\left(v_k\right)\right]2}{\left[d_{G_1}\left(u_i\right) + d_{G_2}\left(v_k\right)\right] + 2} \leq \frac{1}{4} \sum_{\left(u_i, v_k\right)\left(e_j, v_k\right) \in E_2} \left[2 + \left(d_{G_1}\left(u_i\right) + d_{G_2}\left(v_k\right)\right)\right]$$

$$= \frac{1}{4} \sum_{v_k \in V\left(G_2\right)} \sum_{u_i \in E\left(S\left(G_1\right)\right)} \left[2 + \left(d_{G_1}\left(u_i\right) + d_{G_2}\left(v_k\right)\right)\right] = m_1 n_2 + m_1 m_2 + \frac{1}{4} n_2 M_1\left(G_1\right),$$

with equality iff

$$d_{G_1}\left(u_i\right) + d_{G_2}\left(v_k\right) = 2 \ \forall u_i \in V\left(G_1\right), v_k \in V\left(G_2\right),$$

i.e., iff

$$d_{G_1}\left(u_i\right) = 2 - d_{G_2}\left(v_k\right) \; \forall u_i \in V\left(G_1\right), v_k \in V\left(G_2\right),$$

i.e., iff $G_2 = K_2$ and $G_1 = K_2$, as both G_1 and G_2 are non-trivial graphs.
    Hence,

$$ISI\left(G_1 +_S G_2\right) \leq 2m_1m_2 + m_1n_2 + \frac{1}{4}\left[n_1M_1\left(G_2\right) + n_2M_1\left(G_1\right)\right]$$

with equality iff $G_1 = G_2 = K_2$.
    Corollary 3.12. *Let G_1 and G_2 be two graphs with*

$$\left|V\left(G_1\right)\right| = n_1, \; \left|V\left(G_2\right)\right| = n_2, \; \left|E\left(G_1\right)\right| = m_1 \text{ and } \left|E\left(G_2\right)\right| = m_2.$$

Then

$$ISI\left(G_1 +_{S_t} G_2\right) \leq 2m_1m_2 + m_1n_2 + \frac{1}{4}\left[n_1M_1\left(G_2\right) + n_2M_1\left(G_1\right)\right] + \left(t-1\right)m_1n_2$$

with equality iff $G_1 = G_2 = K_2$.
    Proof. Partition the edge set $E\left(G\right)$ of $G = G_1 +_{S_t} G_2$ into three sets E_1, E_2 and E_3 where

$$E_1 = \left\{\left(u_i, v_k\right)\left(u_j, v_l\right) | u_i, u_j \in V\left(G_1\right), v_k, v_l \in V\left(G_2\right)\right\},$$

$$E_2 = \left\{\left(u_i, v_k\right)\left(e_j, v_k\right) | u_i \in V\left(G_1\right), v_k \in V\left(G_2\right), e_j \in E\left(G_1\right)\right\}$$

and

$$E_3 = \left\{\left(e_i, v_k\right)\left(e_j, v_k\right) | e_i, e_j \in V\left(S\left(G_1\right)\right), v_k \in V\left(G_2\right)\right\}.$$

Clearly, $\left|E_3\right| = \left(t-1\right)m_1n_2$. The rest of the proof follows from Theorem 3.11.

CONCLUSION

In recent times, the ISI index draws attention of various researchers and various results are reported for different graphs and operations of graphs. In this work the bounds of ISI index for subdivision graphs, *t*-subdivision graphs, *S*-sum and S_t-sum of graphs are established. The external cases of the bounds are also studied for their sharpness. An immediate extension may be the computation of ISI index for vertex-semitotal graphs, edge-semitotal graphs and total graphs and the related sums. These graph operations, especially *t*-subdivision can also be cosidered for other topological indices like ABC index, augmented Zagreb index for these graph operations.

REFERENCES

An, M., & Xiong, L. (2018). Some results on the inverse sum indeg index of a graph. *Information Processing Letters*, *134*, 42–46. doi:10.1016/j.ipl.2018.02.006

Basavanagoud, B., & Patil, S. (2017). The Hyper-Zagreb index of four operations on graphs. *Mathematical Sciences Letters*, *6*(2), 193–198. doi:10.18576/msl/060212

Chen, H., & Deng, H. (2018). The inverse sum indeg index of graphs with some given parameters. *Discrete Mathematics, Algorithms, and Applications*, *10*(1). doi:10.1142/S1793830918500064

Deng, H., Sarala, D., Ayyaswamy, S. K., & Balachandran, S. (2016). The Zagreb indices of four operations on graphs. *Applied Mathematics and Computation*, *275*, 422–431. doi:10.1016/j.amc.2015.11.058

Dragomir, S. S. (2003). A survey on Cauchy-Buyakorsky-Schwarz type discrete inequalities. *Journal of Inequalities in Pure and Applied Mathematics*, *4*(3), 63.

Eliasi, M., & Taeri, B. (2009). Four new sums of graphs and their Wiener indices. *Discrete Applied Mathematics*, *157*(4), 794–803. doi:10.1016/j.dam.2008.07.001

Falahati-Nezhad, F., & Azari, M. (2016). The inverse sum indeg index of some nanotubes. *Studia Ubb Chemia*, *LXI*(1), 63–70.

Falahati-Nezhad, F., Azari, M., & Došlić, T. (2017). Sharp bounds on the inverse sum indeg index. *Discrete Applied Mathematics*, *217*(2), 185–195. doi:10.1016/j.dam.2016.09.014

Gao, W., Jamil, M. K., Javed, A., Farahani, M. R., & Imran, M. (2018). Inverse sum indeg index of the line graphs of subdivision graphs of some chemical structures. *U. P. B. Sci. Bull. Series B*, *80*(3), 97–104.

Gutman, I., Matejić, M., Milovanović, E., & Milovanović, I. (2020). Lower bounds for inverse sum indeg index of graphs. *Kragujevac J. Math.*, *44*(4), 551–562.

Gutman, I., & Rodr. (2019). Linear and non-linear inequalities on the inverse sum indeg index. *Discrete Applied Mathematics*, *258*, 123–134. doi:10.1016/j.dam.2018.10.041

Gutman, I., & Trinajstić, N. (1972). Graph Theory and Molecular orbitals. Total π . *Chemical Physics Letters*, *17*(4), 535–538. doi:10.1016/0009-2614(72)85099-1

Hasani, M. (2017). Study of inverse sum indeg index. *Journal of Mathematical Nanoscience*, *7*(2), 103–109.

Metsidik, M., Zhang, W., & Duan, F. (2010). Hyper- and reverse-Wiener indices of F-sums of graphs. *Discrete Applied Mathematics*, *158*(13), 1433–1440. doi:10.1016/j.dam.2010.04.003

Onagh, B. N. (2017). The Harmonic index of subdivision graphs. *Transactions on Combinatorics*, *6*(4), 15–27.

Onagh, B. N. (2018). The harmonic index of edge-semitotal graphs, total graphs and related sums. *Kragujevac J. Math.*, *42*(2), 217–228. doi:10.5937/KgJMath1802217O

Pattabiraman, K. (2018). Inverse Sum Indeg index of graphs. *AKCE International Journal of Graph and Combinatorics, 15*(2), 155-167.

Sarala, D., Deng, H., Natarajan, C., & Ayyaswamy, S.K. (2019). F index of graphs based on four new operations related to the strong product. *AKCE International Journal of Graphs and Combinatorics*. doi:10.1016/j.akcej.2018.07.003

Sedlar, J., Stevanović, D., & Vasilyev, A. (2015). On the inverse sum indeg index. *Discrete Applied Mathematics*, *184*, 202–212. doi:10.1016/j.dam.2014.11.013

Vukičević, D., & Gašperov, M. (2010). Bond additive Modeling 1. Adriatic indices. *Croatica Chemica Acta*, *83*, 243–260.

West, D. B. (2002). *Introduction to graph Theory*. Prentice Hall India.

Wiener, H. (1947). Structural Determination of Paraffin Boiling Points. *Journal of the American Chemical Society*, *69*(1), 17–20. doi:10.1021/ja01193a005 PMID:20291038

Chapter 6
The Hyper–Zagreb Index and Some Properties of Graphs

Rao Li

University of South Carolina at Aiken, USA

ABSTRACT

Let G = (V(G), E(G)) be a graph. The complement of G is denoted by Gc. The forgotten topological index of G, denoted F(G), is defined as the sum of the cubes of the degrees of all the vertices in G. The second Zagreb index of G, denoted M2(G), is defined as the sum of the products of the degrees of pairs of adjacent vertices in G. A graph G is k-Hamiltonian if for all X ⊂ V(G) with |X| ≤ k, the subgraph induced by V(G) – X is Hamiltonian. Clearly, G is 0-Hamiltonian if and only if G is Hamiltonian. A graph G is k-path-coverable if V(G) can be covered by k or fewer vertex-disjoint paths. Using F(Gᶜ) and $M_2(G^c)$, Li obtained several sufficient conditions for Hamiltonian and traceable graphs (Li, The Hyper-Zagreb Index and Some Properties of Graphs). In this chapter, the author presents sufficient conditions based upon F(Gᶜ) and $M_2(G^c)$ for k-Hamiltonian, k-edge-Hamiltonian, k-path-coverable, k-connected, and k-edge-connected graphs.

INTRODUCTION

We consider only finite undirected graphs without loops or multiple edges. Notation and terminology not defined here follow that in Bondy and Murty (1976). Let $G = \left(V\left(G \right), E\left(G \right) \right)$ be a graph. We use n, e, δ, and κ to denote the order, size, minimum degree, and connectivity of G, respectively. The complement of G is denoted by G^c. We also use K_n and E_n to denote the complete graph and the empty graph of order n. The forgotten topological index of G, denoted $F\left(G \right)$, is defined as $\sum_{v \in V(G)} \left(d\left(v \right) \right)^3$

(see Furtala & Gutman, 2015). The second Zagreb index of G, denoted $M_2\left(G \right)$, is defined as $\sum_{uv \in E(G)} d\left(u \right) d\left(v \right)$

(see Gutman et al., 1975). The hyper-Zagreb index of G, denoted $HZ\left(G \right)$, is defined as $F\left(G \right) + 2M_2\left(G \right)$

(see Milovanovic, Matejic & Milovanovic, 2019). We use $\mu_n\left(G \right)$ to denote the largest eigenvalue of

DOI: 10.4018/978-1-5225-9380-5.ch006

the adjacency matrix of a graph G of order n. For two disjoint graphs G_1 and G_2, we use $G_1 + G_2$ and $G_1 \vee G_2$ to denote respectively the union and join of G_1 and G_2. The concept of closure of a graph G was introduced by Bondy and Chvátal in Bondy and Chvatal (1976). The k-closure of a graph G, denoted $cl_k(G)$, is a graph obtained from G by recursively joining two nonadjacent vertices such that their degree sum is at least k until no such pair remains. We use $C(n, r)$ to denote the number of r-combinations of a set with n distinct elements. A cycle C in a graph G is called a Hamiltonian cycle of G if C contains all the vertices of G. A graph G is called Hamiltonian if G has a Hamiltonian cycle. A path P in a graph G is called a Hamiltonian path of G if P contains all the vertices of G. A graph G is called traceable if G has a Hamiltonian path. A graph G is k-Hamiltonian if for all $X \subset V(G)$ with $|X| \leq k$, the subgraph induced by $V(G) - X$ is Hamiltonian. Clearly, G is 0-Hamiltonian if and only if G is Hamiltonian. A graph G is k-edge-Hamiltonian if any collection of vertex-disjoint paths with at most k edges is in a Hamiltonian cycle in G. Clearly, G is 0-edge-Hamiltonian if and only if G is Hamiltonian. A graph G is k-path-coverable if $V(G)$ can be covered by k or fewer vertex-disjoint paths. Clearly, G is 1-path-coverable if and only G is a traceable. A graph G is k-connected if it has more than k vertices and G is still connected whenever fewer than k vertices are removed from G. A graph G is k-edge-connected if it has at least two vertices and G is still connected whenever fewer than k edges are removed from G.

The following results were obtained by Fiedler and Nikiforov.

Theorem 1: (Fiedler & Nikiforov, 2010) Let G be a graph of order n.

1. If $\mu_n(G^c) \leq \sqrt{n-1}$, then G contains a Hamiltonian path unless $G = K_{n-1} + v$.

2. If $\mu_n(G^c) \leq \sqrt{n-2}$, then G contains a Hamiltonian cycle unless $G = K_{n-1} + e$.

Using the ideas and techniques developed by Fiedler and Nikiforov (2010), Li (2019) obtained sufficient conditions which involve the hyper-Zagreb indexes of the complements of the graphs for the Hamiltonian and traceable graphs. It is found that the ideas and techniques in Li (2019) can also be utilized to obtain sufficient conditions based upon hyper-Zagreb indexes for the additional properties of graphs. The aim of this paper is to present those conditions for k-Hamiltonian, k-edge-Hamiltonian, k-path-coverable, k-connected, and k-edge-connected graphs. The main results of this paper are as follows.

Theorem 2: Let G be a graph of order $n \geq k + 6$, where k is an integer and $k \geq 1$. If

$$HZ(G^c) = F(G^c) + 2M_2(G^c) \leq (n-1)^2(n-k-2),$$

then G is k-Hamiltonian or

$$G = K_{k+1} \vee (K_1 + K_{n-k-2}).$$

Theorem 3: Let G be a graph of order $n \geq k + 6$, where k is an integer and $k \geq 1$. If

$$HZ\left(G^c\right) = F\left(G^c\right) + 2M_2\left(G^c\right) \le (n-1)^2 \left(n-k-2\right),$$

then G is k-edge-Hamiltonian or

$$G = K_{k+1} \vee \left(K_1 + K_{n-k-2}\right).$$

Theorem 4: Let G be a graph of order $n \ge 5k+6$, where k is an integer and $k \ge 2$. If

$$2HZ\left(G^c\right) = 2\left(F\left(G^c\right) + 2M_2\left(G^c\right)\right) \le (n+k-1)^2 \left(2\left(k+1\right)n - k^2 - 5k - 4\right),$$

then G is k-path-coverable.

Theorem 5: Let G be a graph of order $n \ge k+1$, where k is an integer and $k \ge 1$. If

$$HZ\left(G^c\right) = F\left(G^c\right) + 2M_2\left(G^c\right) \le (n-1)^2 \left(n-k\right),$$

then G is k-connected or

$$G = K_{k-1} \vee \left(K_1 + K_{n-k}\right).$$

Theorem 6: Let G be a graph of order $n \ge k+1$, where k is an integer and $k \ge 2$. If

$$HZ\left(G^c\right) = F\left(G^c\right) + 2M_2\left(G^c\right) < (n-k+1)^2 \left(\left(k+3\right)n - 2k^2 - 2k - 3\right),$$

then G is k-edge-connected.

PREVIOUS RESULTS

We need the following previous results as lemmas to prove our theorems. The first five results are from Bondy and Chvatal (1976) and the second five results are from Feng et al. (2017).

Lemma 1: (2.11 on Page 113 in Bondy and Chvatal, 1976) A graph G of order n is k-Hamiltonian if and only if $cl_{n+k}\left(G\right)$ is k-Hamiltonian.

Lemma 2: (2.12 on Page 113 in Bondy and Chvatal, 1976) A graph G of order n is k-edge-Hamiltonian if and only if $cl_{n+k}\left(G\right)$ is k-edge-Hamiltonian.

Lemma 3: (2.14 on Page 113 in Bondy and Chvatal, 1976) A graph G of order n is k-path-coverable if and only if $cl_{n-k}\left(G\right)$ is k-path-coverable.

Lemma 4: (2.8 on Page 113 in Bondy and Chvatal, 1976) A graph G of order n is k-connected if and only if $cl_{n+k-2}\left(G\right)$ is k-connected.

Lemma 5: (2.9 on Page 113 in Bondy and Chvatal, 1976) A graph G of order n is k-edge-connected if and only if $cl_{n+k-2}(G)$ is k-edge-connected.

Lemma 6: (Claim in the Proof of Theorem 3.3 on Page 190 in Feng et al., 2017) Let G be a graph of order $n \geq k+6$, where k is an integer and $k \geq 1$. If

$$e(G) \geq C(n,2) - n + k + 2,$$

then G is k-Hamiltonian or

$$e(G) = C(n,2) - n + k + 2$$

and

$$G = K_{k+1} \vee \left(K_1 + K_{n-k-2}\right).$$

Lemma 7: (Claim in the Proof of Theorem 3.4 on Page 192 in Feng et al., 2017) Let G be a graph of order $n \geq k+6$, where k is an integer and $k \geq 1$. If

$$e(G) \geq C(n,2) - n + k + 2,$$

then G is k-edge-Hamiltonian or

$$e(G) = C(n,2) - n + k + 2$$

and

$$G = K_{k+1} \vee \left(K_1 + K_{n-k-2}\right).$$

Lemma 8: (Claim in the Proof of Theorem 3.6 on Page 194 in Feng et al., 2017) Let G be a graph of order $n \geq 5k+6$, where k is an integer and $k \geq 2$. If

$$e(G) \geq \frac{1}{2}\left(n^2 - (2k+3)n + k^2 + 5k + 4\right),$$

then G is K-path-coverable or

$$e(G) = \frac{1}{2}\left(n^2 - (2k+3)n + k^2 + 5k + 4\right)$$

and

$$G = K_1 \vee \left(E_{k+1} + K_{n-k-2} \right).$$

Lemma 9: (Claim in the Proof of Theorem 3.1 on Page 187 in Feng et al., 2017) Let G be a graph of order $n \geq k+1$, where k is an integer and $k \geq 1$. If

$$e(G) \geq C(n,2) - n + k,$$

then G is k-connected or

$$e(G) = C(n,2) - n + k$$

and

$$G = K_{k-1} \vee \left(K_1 + K_{n-k} \right).$$

Lemma 10: (Claim in the Proof of Theorem 3.2 on Page 189 in Feng et al., 2017) Let g be a graph of order $n \geq k+1$, where K is an integer and $k \geq 2$. If

$$e(G) \geq \frac{1}{2}\left(n^2 - (k+4)n + 2k^2 + 2k + 4 \right),$$

then G is k-edge-connected.

PROOFS

Proof of Theorem 2. Let G be a graph satisfying the conditions in Theorem 2 and G is not k-Hamiltonian. Then Lemma 1 implies that $H := cl_{n+k}(G)$ is not k-Hamiltonian and therefore H is not K_n. Thus there exist two vertices x and y in $V(H)$ such that $xy \notin E(H)$ and for any pair of nonadjacent vertices u and v in $V(H)$ we have

$$d_H(u) + d_H(v) \leq n + k - 1.$$

Hence for any pair of adjacent vertices u and v in $V(H^c)$ we have that

$$d_{H^c}(u) + d_{H^c}(v) = n - 1 - d_H(u) + n - 1 - d_H(v) \geq n - k - 1.$$

Thus

$$(d_{H^c}(u) + d_{H^c}(v))^2 \geq (n - k - 1)^2.$$

Therefore

$$\sum_{uv \in E(H^c)} (d_{H^c}(u) + d_{H^c}(v))^2 \geq (n - k - 1)^2 e(H^c).$$

Hence

$$\sum_{uv \in E(H^c)} \left(d_{H^c}^2(u) + d_{H^c}^2(v)\right) + 2 \sum_{uv \in E(H^c)} d_{H^c}(u) d_{H^c}(v) \geq (n - k - 1)^2 e(H^c).$$

Notice that

$$F(H^c) = \sum_{x \in V(H^c)} (d_{H^c}(x))^3 = \sum_{uv \in E(H^c)} \left(d_{H^c}^2(u) + d_{H^c}^2(v)\right).$$

So we have that

$$HZ(H^c) = F(H^c) + 2M_2(H^c) \geq (n - k - 1)^2 e(H^c).$$

Notice that

$$d_{H^c}(u) \leq d_{G^c}(u)$$

for each

$$u \in V(H^c) = V(G^c)$$

and

$$E(H^c) \subseteq E(G^c).$$

We have that

$$HZ\left(G^c\right) = F\left(G^c\right) + 2M_2\left(G^c\right) \geq F\left(H^c\right) + 2M_2\left(H^c\right) \geq (n-k-1)^2 e\left(H^c\right).$$

Notice that H is not k-Hamiltonian. Lemma 6 implies that

$$e\left(H\right) \leq C\left(n,2\right) - n + k + 1$$

or

$$e\left(H\right) = C\left(n,2\right) - n + k + 2$$

and

$$H = K_{k+1} \vee \left(K_1 + K_{n-k-2}\right).$$

If

$$e\left(H\right) \leq C\left(n,2\right) - n + k + 1\,,$$

then

$$(n-1)^2\left(n-k-2\right) \geq HZ\left(G^c\right) = F\left(G^c\right) + 2M_2\left(G^c\right) \geq F\left(H^c\right) + 2M_2\left(H^c\right)$$
$$\geq (n-1)^2 e\left(H^c\right) \geq (n-1)^2\left(C\left(n,2\right) - C\left(n,2\right) + n - k - 1\right) = (n-1)^2\left(n-k-1\right),$$

a contradiction.

If

$$H = K_{k+1} \vee \left(K_1 + K_{n-k-2}\right),$$

then

$$(n-1)^2\left(n-k-2\right) \geq HZ\left(G^c\right) = F\left(G^c\right) + 2M_2\left(G^c\right) \geq F\left(H^c\right) + 2M_2\left(H^c\right)$$
$$\geq (n-1)^2 e\left(H^c\right) \geq (n-1)^2\left(C\left(n,2\right) - C\left(n,2\right) + n - k - 2\right) = (n-1)^2\left(n-k-2\right).$$

Thus

$$F\left(G^c\right) = F\left(H^c\right),\ d_{G^c}\left(u\right) = d_{H^c}\left(u\right)$$

for each

$$u \in V\left(H^c\right) = V\left(G^c\right), \ M_2\left(G^c\right) = M_2\left(H^c\right), \ d_{G^c}\left(u\right)d_{G^c}\left(v\right) = d_{H^c}\left(u\right)d_{H^c}\left(v\right)$$

for each $uv \in E\left(H^c\right)$, and

$$d_{H^c}\left(u\right) + d_{H^c}\left(v\right) = n - k - 1$$

for each $uv \in E\left(H^c\right)$. Therefore $G=H$.

Proof of Theorem 3. The proof of Theorem 3 is almost the same as the proof of Theorem 2. Replacing respectively k-Hamiltonian, Lemma 1, and Lemma 6 by k-edge-Hamiltonian, Lemma 2, and Lemma 7 in the proof of Theorem 2, we can complete the proof of Theorem 3.

Proof of Theorem 4. Let G be a graph satisfying the conditions in Theorem 4 and G is not k-path-coverable. Then Lemma 3 implies that $H := cl_{n-k}\left(G\right)$ is not k-path-coverable and therefore H is not K_n. Thus there exist two vertices x and y in $V\left(H\right)$ such that $xy \notin E\left(H\right)$ and for any pair of nonadjacent vertices u and v in $V\left(H\right)$ we have

$$d_H\left(u\right) + d_H\left(v\right) \leq n - k - 1 .$$

Hence for any pair of adjacent vertices u and v in $V\left(H^c\right)$ we have that

$$d_{H^c}\left(u\right) + d_{H^c}\left(v\right) = n - 1 - d_H\left(u\right) + n - 1 - d_H\left(v\right) \geq n + k - 1 .$$

Thus

$$(d_{H^c}\left(u\right) + d_{H^c}\left(v\right))^2 \geq (n + k - 1)^2 .$$

Therefore

$$\sum_{uv \in E\left(H^c\right)} (d_{H^c}\left(u\right) + d_{H^c}\left(v\right))^2 \geq (n + k - 1)^2 e\left(H^c\right) .$$

Hence

$$\sum_{uv \in E\left(H^c\right)} \left(d_{H^c}^2\left(u\right) + d_{H^c}^2\left(v\right)\right) + 2\sum_{uv \in E\left(H^c\right)} d_{H^c}\left(u\right)d_{H^c}\left(v\right) \geq (n + k - 1)^2 e\left(H^c\right) .$$

Notice that

$$F\left(H^c\right) = \sum_{x \in V\left(H^c\right)} \left(d_{H^c}\left(x\right)\right)^3 = \sum_{uv \in E\left(H^c\right)} \left(d_{H^c}^2\left(u\right) + d_{H^c}^2\left(v\right)\right).$$

So we have that

$$HZ\left(H^c\right) = F\left(H^c\right) + 2M_2\left(H^c\right) \geq (n + k - 1)^2 e\left(H^c\right).$$

Notice that

$$d_{H^c}\left(u\right) \leq d_{G^c}\left(u\right)$$

for each

$$u \in V\left(H^c\right) = V\left(G^c\right)$$

and

$$E\left(H^c\right) \subseteq E\left(G^c\right).$$

We have that

$$HZ\left(G^c\right) = F\left(G^c\right) + 2M_2\left(G^c\right) \geq F\left(H^c\right) + 2M_2\left(H^c\right) \geq (n + k - 1)^2 e\left(H^c\right).$$

Notice that H is not k-path-coverable. Lemma 8 implies that

$$2e\left(H\right) \leq n^2 - \left(2k + 3\right)n + k^2 + 5k + 3$$

or

$$2e\left(H\right) = n^2 - \left(2k + 3\right)n + k^2 + 5k + 4$$

and

$$H = K_1 \vee \left(E_{k+1} + K_{n-k-2}\right).$$

If

$$2e\left(H\right) \le n^2 - \left(2k + 3\right)n + k^2 + 5k + 3 \,,$$

then

$$
\begin{aligned}
\left(n + k - 1\right)^2 &\left(2\left(k + 1\right)n - k^2 - 5k - 4\right) \ge 2HZ\left(G^c\right) \\
&= 2\left(F\left(G^c\right) + 2M_2\left(G^c\right)\right) \ge 2\left(F\left(H^c\right) + 2M_2\left(H^c\right)\right) \ge 2(n + k - 1)^2 e\left(H^c\right) \\
&\ge \left(n + k - 1\right)^2 \left(2C\left(n, 2\right) - n^2 + \left(2k + 3\right)n - k^2 - 5k - 3\right) \\
&= \left(n + k - 1\right)^2 \left(2\left(k + 1\right)n - k^2 - 5k - 3\right),
\end{aligned}
$$

a contradiction.

If

$$H = K_1 \vee \left(E_{k+1} + K_{n-k-2}\right),$$

then

$$
\begin{aligned}
\left(n + k - 1\right)^2 &\left(2\left(k + 1\right)n - k^2 - 5k - 4\right) \ge 2HZ\left(G^c\right) \\
&= 2\left(F\left(G^c\right) + 2M_2\left(G^c\right)\right) \ge 2\left(F\left(H^c\right) + 2M_2\left(H^c\right)\right) \ge 2(n + k - 1)^2 e\left(H^c\right) \\
&\ge \left(n + k - 1\right)^2 \left(2C\left(n, 2\right) - n^2 + \left(2k + 3\right)n - k^2 - 5k - 4\right) \\
&= \left(n + k - 1\right)^2 \left(2\left(k + 1\right)n - k^2 - 5k - 4\right).
\end{aligned}
$$

Thus

$$F\left(G^c\right) = F\left(H^c\right),\ d_{G^c}\left(u\right) = d_{H^c}\left(u\right)$$

for each

$$u \in V\left(H^c\right) = V\left(G^c\right),\ M_2\left(G^c\right) = M_2\left(H^c\right),\ d_{G^c}\left(u\right)d_{G^c}\left(v\right) = d_{H^c}\left(u\right)d_{H^c}\left(v\right)$$

for each $uv \in E\left(H^c\right)$, and

$$d_{H^c}\left(u\right) + d_{H^c}\left(v\right) = n + k - 1$$

for each $uv \in E\left(H^c\right)$, a contradiction.

Proof of Theorem 5. Let G be a graph satisfying the conditions in Theorem 5 and G is not k-connected. Then Lemma 4 implies that $H := cl_{n+k-2}(G)$ is not k-connected and therefore H is not K_n. Thus there exist two vertices x and y in $V(H)$ such that $xy \notin E(H)$ and for any pair of nonadjacent vertices u and v in $V(H)$ we have

$$d_H(u) + d_H(v) \leq n + k - 3.$$

Hence for any pair of adjacent vertices u and v in $V(H^c)$ we have that

$$d_{H^c}(u) + d_{H^c}(v) = n - 1 - d_H(u) + n - 1 - d_H(v) \geq n - k + 1.$$

Thus

$$(d_{H^c}(u) + d_{H^c}(v))^2 \geq (n - k + 1)^2.$$

Therefore

$$\sum_{uv \in E(H^c)} (d_{H^c}(u) + d_{H^c}(v))^2 \geq (n - k + 1)^2 e(H^c).$$

Hence

$$\sum_{uv \in E(H^c)} \left(d_{H^c}^2(u) + d_{H^c}^2(v)\right) + 2\sum_{uv \in E(H^c)} d_{H^c}(u) d_{H^c}(v) \geq (n - k - 1)^2 e(H^c).$$

Notice that

$$F(H^c) = \sum_{x \in V(H^c)} (d_{H^c}(x))^3 = \sum_{uv \in E(H^c)} \left(d_{H^c}^2(u) + d_{H^c}^2(v)\right).$$

So we have that

$$HZ(H^c) = F(H^c) + 2M_2(H^c) \geq (n - k + 1)^2 e(H^c).$$

Notice that

$$d_{H^c}(u) \leq d_{G^c}(u)$$

for each

$$u \in V\left(H^c\right) = V\left(G^c\right)$$

and

$$E\left(H^c\right) \subseteq E\left(G^c\right).$$

We have that

$$HZ\left(G^c\right) = F\left(G^c\right) + 2M_2\left(G^c\right) \geq F\left(H^c\right) + 2M_2\left(H^c\right) \geq (n-k+1)^2 e\left(H^c\right).$$

Notice that H is not k-connected. Lemma 9 implies that

$$e\left(H\right) \leq C\left(n,2\right) - n + k - 1$$

or

$$e\left(H\right) = C\left(n,2\right) - n + k$$

and

$$H = K_{k-1} \vee \left(K_1 + K_{n-k}\right).$$

If

$$e\left(H\right) \leq C\left(n,2\right) - n + k - 1,$$

then

$$(n-1)^2\left(n-k\right) \geq HZ\left(G^c\right) = F\left(G^c\right) + 2M_2\left(G^c\right) \geq F\left(H^c\right) + 2M_2\left(H^c\right)$$
$$\geq (n-1)^2 e\left(H^c\right) \geq (n-1)^2\left(C\left(n,2\right) - C\left(n,2\right) + n - k + 1\right) = (n-1)^2\left(n-k+1\right),$$

a contradiction.
If

$$H = K_{k-1} \vee \left(K_1 + K_{n-k}\right),$$

then

$$(n-1)^2\left(n-k\right) \geq HZ\left(G^c\right) = F\left(G^c\right) + 2M_2\left(G^c\right) \geq F\left(H^c\right) + 2M_2\left(H^c\right)$$
$$\geq (n-1)^2 e\left(H^c\right) \geq (n-1)^2\left(C\left(n,2\right) - C\left(n,2\right) + n - k\right) = (n-1)^2\left(n-k\right).$$

Thus

$$F\left(G^c\right) = F\left(H^c\right), \; d_{G^c}\left(u\right) = d_{H^c}\left(u\right)$$

for each

$$u \in V\left(H^c\right) = V\left(G^c\right), \; M_2\left(G^c\right) = M_2\left(H^c\right), \; d_{G^c}\left(u\right)d_{G^c}\left(v\right) = d_{H^c}\left(u\right)d_{H^c}\left(v\right)$$

for each $uv \in E\left(H^c\right)$, and

$$d_{H^c}\left(u\right) + d_{H^c}\left(v\right) = n - k + 1$$

for each $uv \in E\left(H^c\right)$. Therefore $G=H$.

Proof of Theorem 6. Let G be a graph satisfying the conditions in Theorem 6 and G is not k-edge-connected. Then Lemma 5 implies that $H := cl_{n+k-2}\left(G\right)$ is not k-edge-connected and therefore H is not K_n. Thus there exist two vertices x and y in $V\left(H\right)$ such that $xy \notin E\left(H\right)$ and for any pair of nonadjacent vertices u and v in $V\left(H\right)$ we have

$$d_H\left(u\right) + d_H\left(v\right) \leq n + k - 3.$$

Hence for any pair of adjacent vertices u and v in $V\left(H^c\right)$ we have that

$$d_{H^c}\left(u\right) + d_{H^c}\left(v\right) = n - 1 - d_H\left(u\right) + n - 1 - d_H\left(v\right) \geq n - k + 1.$$

Thus

$$(d_{H^c}\left(u\right) + d_{H^c}\left(v\right))^2 \geq (n - k + 1)^2.$$

Therefore

$$\sum_{uv \in E(H^c)} (d_{H^c}(u) + d_{H^c}(v))^2 \geq (n - k + 1)^2 e(H^c).$$

Hence

$$\sum_{uv \in E(H^c)} \left(d_{H^c}^2(u) + d_{H^c}^2(v)\right) + 2 \sum_{uv \in E(H^c)} d_{H^c}(u) d_{H^c}(v) \geq (n - k + 1)^2 e(H^c).$$

Notice that

$$F(H^c) = \sum_{x \in V(H^c)} (d_{H^c}(x))^3 = \sum_{uv \in E(H^c)} \left(d_{H^c}^2(u) + d_{H^c}^2(v)\right).$$

So we have that

$$HZ(H^c) = F(H^c) + 2M_2(H^c) \geq (n + k - 1)^2 e(H^c).$$

Notice that

$$d_{H^c}(u) \leq d_{G^c}(u)$$

for each

$$u \in V(H^c) = V(G^c)$$

and

$$E(H^c) \subseteq E(G^c).$$

We have that

$$HZ(G^c) = F(G^c) + 2M_2(G^c) \geq F(H^c) + 2M_2(H^c) \geq (n - k + 1)^2 e(H^c).$$

Notice that H is not k-edge-connected. Lemma 10 implies that

$$2e(H) \leq n^2 - (k + 4)n + 2k^2 + 2k + 3.$$

Thus

$$(n - k + 1)^2 \left(\left(k + 3 \right) n - 2k^2 - 2k - 3 \right) > 2HZ \left(G^c \right)$$
$$= 2 \left(F \left(G^c \right) + 2M_2 \left(G^c \right) \right) \geq 2 \left(F \left(H^c \right) + 2M_2 \left(H^c \right) \right) \geq 2(n - k + 1)^2 e \left(H^c \right)$$
$$\geq (n - k + 1)^2 \left(2C \left(n, 2 \right) - n^2 + \left(k + 4 \right) n - 2k^2 - 2k - 3 \right) = (n - k + 1)^2 \left(\left(k + 3 \right) n - 2k^2 - 2k - 3 \right),$$

a contradiction.

REFERENCES

Bondy, J. A., & Chvátal, V. (1976). A method in graph theory. *Discrete Mathematics*, *15*(2), 111–135. doi:10.1016/0012-365X(76)90078-9

Bondy, J. A., & Murty, U. S. R. (1976). Graph Theory with Applications. Macmillan.

Feng, L., Zhang, P., Liu, H., Liu, W., Liu, M., & Hu, Y. (2017). Spectral conditions for some graphical properties. *Linear Algebra and Its Applications*, *524*, 182–198. doi:10.1016/j.laa.2017.03.006

Fiedler, M., & Nikiforov, V. (2010). Spectral radius and Hamiltonicity of graphs. *Linear Algebra and Its Applications*, *432*(9), 2170–2173. doi:10.1016/j.laa.2009.01.005

Furtula, B., & Gutman, I. (2015). A forgotten topological index. *Journal of Mathematical Chemistry*, *53*(4), 1184–1190. doi:10.100710910-015-0480-z

Gutman, I., Ruščić, B., Trinajstić, N., & Wilcox, C. F. Jr. (1975). Graph theory and molecular orbitals. XII. Acyclic polyenes. *The Journal of Chemical Physics*, *62*(9), 3399–3405. doi:10.1063/1.430994

Li, R. (2019). The hyper-Zagreb index and some Hamiltonian properties of graphs. *Discrete Mathematics Letters*, *1*, 54–58.

Milovanović, E., Matejić, M., & Milovanović, I. (2019). Some new upper bounds for the hyper-Zagreb index. *Discrete Mathematics Letters*, *1*, 30–35.

Chapter 7
L(h,k)–Labeling of Intersection Graphs

Sk. Amanathulla
Vidyasagar University, India

Madhumangal Pal
Vidyasagar University, India

ABSTRACT

One important problem in graph theory is graph coloring or graph labeling. Labeling problem is a well-studied problem due to its wide applications, especially in frequency assignment in (mobile) communication system, coding theory, ray crystallography, radar, circuit design, etc. For two non-negative integers, labeling of a graph is a function from the node set to the set of non-negative integers such that if and if, where it represents the distance between the nodes. Intersection graph is a very important subclass of graph. Unit disc graph, chordal graph, interval graph, circular-arc graph, permutation graph, trapezoid graph, etc. are the important subclasses of intersection graphs. In this chapter, the authors discuss labeling for intersection graphs, specially for interval graphs, circular-arc graphs, permutation graphs, trapezoid graphs, etc., and have presented a lot of results for this problem.

INTRODUCTION

Almost all problems in the world can be solve by designing graphs. So, during long period graph theory is being researched. In engineering, physical science, mathematical science, *graph* has lot of applications. One important problem in graph theory is graph coloring or graph labeling. *L(h,k)*-labeling problem is a well studied problem due to its wide applications, specially in frequency assignment in (mobile) communication system, coding theory, X-ray crystallography, radar, circuit design, etc. For two non-negative integers h and k, an *L(h,k)*-labeling of a graph $G = (V, E)$ is a function f from the node set V to the set of non-negative integers such that $|f(x) - f(y)| \geq h$ if $d(x, y) = 1$ and $|f(x) - f(y)| \geq k$ if $d(x, y) = 2$, where $d(x, y)$ represents the distance between the nodes x and y.

DOI: 10.4018/978-1-5225-9380-5.ch007

Intersection graph is a very important subclasses of graph. Unit disc graph, chordal graph, interval graph, circular-arc graph, permutation graph, trapezoid graph etc. are the important subclasses of intersection graphs.

In this chapter, we discuss *L(h,k)*-labeling for intersection graphs, specially, for interval graphs, circular-arc graphs, permutation graphs, trapezoid graphs etc. and have presented a lot of results for this problem.

BASIC CONCEPT OF *L(h,k)*-LABELING

In this section, the definition and span of *L(h,k)*-labeling is presented. Different variations of *L(h,k)*-labeling is also highlighted in this section. The definition of *L(h,k)*-labeling is as follows.

Definition 1 *L(h,k)*-labeling: Given a graph $G = (V, E)$ and two nonnegative integers h and k, an *L(h,k)*-labeling is an assignment of non-negative integers to the nodes of G such that adjacent nodes are labelled using colours at least h apart, and nodes having a common neighbour are labelled using colours at least k apart. The difference between largest and smallest labels is called the span. The aim of the $L(h,k)$-labeling problem is to minimize the span. The minimum span over all possible labeling functions is denoted by $\lambda_{h,k}(G)$ and is called $\lambda_{h,k}$-number of G.

In other words, if $f(x)$ is the label assigned to the node x then

$$| f(x) - f(y) |\geq h \text{ if } d(x,y) = 1$$

and

$$| f(x) - f(y) |\geq k \text{ if } d(x,y) = 2,$$

where $d(x,y)$ is the distance (i.e. number of edges) between x and y.

The *L(h,k)*-labeling problem can also be referred to as:

- Distance-2-coloring and $D2$-node coloring problem (when $h = k = 1$);
- Radiocoloring problem and λ-coloring problem (when $h = 2$ and $k = 1$);
- Frequency assignment problem;
- Distance two labeling, etc.

For different values of h and k different *L(h,k)*-labeling problems are addressed by the researchers, specially $L(2,1)$, $L(0,1)$ and $L(1,1)$-labeling problems. For general graphs, the lower bound for $\lambda_{2,1}(G)$ is $\Delta + 1$. But the upper bound has gradually improved. Griggs and Yeh (1992) proved that $\lambda_{2,1}(G) \leq \Delta^2 + 2\Delta$ and have proposed the following conjecture.

Griggs and Yeh Conjecture

For a graph G with maximum degree $\Delta \geq 2$, $\lambda_{2,1}(G) \leq \Delta^2$. In 1993, Jonas (1993) has shown that

$$\lambda_{2,1}(G) \le \Delta^2 + 2\Delta - 4 \,.$$

Later Chang and Kuo (1996) have proved that

$$\lambda_{2,1}(G) \le \Delta^2 + \Delta$$

in 1996. This is a significant improvement. In 2003, Král' and Skrekovski (2003) showed that

$$\lambda_{2,1}(G) \le \Delta^2 + \Delta - 1$$

and it is further improved to

$$\lambda_{2,1}(G) \le \Delta^2 + \Delta - 2$$

(Goncalves, 2008). In 2008, Havet et al. (2008) obtained new upper bound, i.e.

$$\lambda_{2,1}(G) \le \Delta^2 + c \,,$$

for some constant c. Obviously, the bound of $\Delta^2 + c$ is better than $\Delta^2 + \Delta - 2$ as Δ increases enough, but c is unfortunately a rather huge number.

For some classes of graphs $\lambda_{2,1}(G)$ can be calculated efficiently. Such types of graphs are paths (Griggs & Yeh, 1992), cycles (Griggs & Yeh, 1992), wheels (Griggs & Yeh, 1992), trees (Chang & Kuo, 1996), cographs (Chang & Kuo, 1996), cactus graphs (Khan, Pal & Pal, 2012a, 2012b), etc. For path, the result is as follows:

$$\lambda_{2,1}(P_2) = 2, \lambda_{2,1}(P_3) = \lambda_{2,1}(P_4) = 3 \, and \, \lambda_{2,1}(P_n) = 4, for \, n \ge 5.$$

For a cycle C_n of length n, $\lambda_{2,1}(C_n) = 4$, for any n. Let W_n be a wheel of n nodes then $\lambda_{2,1}(W_n) = n + 1$. Tree is another graph classes on which $L(2,1)$-labeling problem has been extensively studied. In Griggs and Yeh (1992), it is shown that $\lambda_{2,1}(T)$ is either $\Delta + 1$ or $\Delta + 2$. But, Chang and Kuo (1996) disprove this conjecture by designing an efficient algorithm. An exhaustive survey on $L(h,k)$-labeling is available in (Calamoneri, 2014).

Different bounds for $\lambda_{0,1}(G)$ and $\lambda_{1,1}(G)$ were obtained for various type of graphs. The upper bound of $\lambda_{0,1}(G)$ of any graph G is $\Delta^2 - \Delta$ [12], where Δ is the degree of the graph. The upper bound of $\lambda_{1,1}(G)$ of any graph G is $\Delta^2 - 2$ (Chang & Kuo, 1996).

Calamoneri, Caminiti, and Petreschi (2009) proved that an interval graph G can be $L(h,k)$-labeled with span at most $\max(h, 2k)\Delta$, also they shown that for circular-arc graphs $\lambda_{h,k}(G) \le \max(h, 2k)\Delta + hw$.

The generalization of $L(h,k)$-labeling have been studied in Ghosh and Pal (2016, 2018), Ghosh, Paul, and Pal (2017), Amanathulla and Pal (2016, 2017a, 2017b, 2017c, 2017d, 2018), and Amanathulla, Sahoo, and Pal (n.d.).

DIFFERENCE BETWEEN NODE COLORING AND $L(h,k)$-LABELING

Vertex coloring is the starting point of the subject. The problem is: find a way of coloring the nodes of a graph such that no two adjacent nodes are colored using same color. This is called *node coloring problem*. In mathematical and computer representations, non-negative integers are used as a color.

$L(h,k)$-labeling problem is a variation of node coloring problem where the adjacent nodes and the nodes at distance two both are to be consider. For the case of $h = 1$ and $k = 0$, $L(h,k)$-labeling is same as node coloring problem. In this case, there is no restriction for distance two nodes, thus $L(1,0)$-labeling problem is nothing but node coloring problem.

The difference between optimal coloring and optimal $L(2,1)$-labeling is illustrated in Figure 1. In optimal coloring only three colors 0, 1, 2 are used while in $L(2,1)$-labeling five labels/colors 0, 2, 3, 4, 5 are used, i.e. the span is 5.

INTERSECTION GRAPH

The intersection graphs have lot of applications in real life problems, for this reason it has own classification number $05C62$. Any graph $G = (V, E)$ is called as an *intersection graph* for a finite family F of non-empty sets if there exists a bijective mapping between V and F so that two sets in F have non-empty intersection if and only if their corresponding nodes in V are adjoining to each other. F is called as an intersection model of G. For an intersection model F, we use $G(F)$ to denote the intersection graph.

Depends on the characteristic of the set F, one can get different intersection graphs. The most famous intersection graphs are given below.

- Interval graphs (IG) (F is the set of intervals on a real line)
- Circular-arc graphs (CAG) (F is the set of arcs around a circle)

Figure 1. Optimal coloring and optimal L(2,1)-labeling of graphs

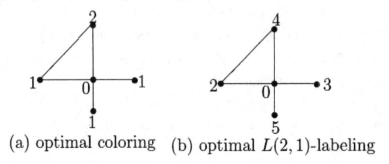

(a) optimal coloring   (b) optimal $L(2, 1)$-labeling

- Permutation graphs (PG) (F is the set of line segments between two parallel lines)
- Trapezoid graphs (TG) (F is the set of trapeziums between two parallel lines)

Different types of intersection graphs are described below:

Interval Graphs

Interval graph is a very important subclasses of intersection graph, it has many applications in real life situation. The definition of interval is given below.

Definition 2: (Interval graph) An undirected graph $G = (V, E)$ is an interval graph if there exists a bijection between node set V and a set of intervals I on real line $\mathbb{R}$ so that two nodes are adjacent in G if and only if their corresponding intervals have non-empty intersection.

Let $I = \{I_1, I_2, ..., I_n\}$, where $I_k = [a_k, b_k]$, $k = 1, 2, ... n$ be a set of intervals on real line, a_k and b_k are the left and right endpoints of I_k respectively. We draw a node v_k for the interval $I_k, k = 1, 2, ..., n$. Two nodes v_p and v_q are joined by an edge iff there corresponding intervals have non-empty intersection. The interval representation of G is denoted by I and the graph G is referred to as the IG for I. Also, it is to be noted that an interval I_k of I and a node v_k of V are one and same thing.

An interval representation and its corresponding IG are shown in Figure 2.

Circular-Arc Graphs

Circular-arc graph (CAG) is also an important subclasses of intersection graph. It is superset of interval graph. This graph is defined below.

Definition 3: An undirected graph $G = (V, E)$ is called a CAG if there exists a family of arcs A around a circle and a bijective mapping between nodes of G and arcs in A, so that two different nodes are adjoining in G iff the corresponding arcs intersects in A. Such a family of arcs called an arc representation for G.

Figure 2. An interval representation and the corresponding interval graph

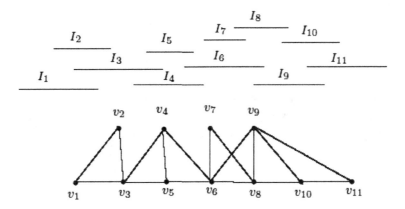

It is noted that every arcs must cover the circle, because otherwise it reduces to an IG.

Let $A = \{A_1, A_2, ..., A_n\}$ be a set of arcs around a circle. While going in dextral direction, the point at which we first encounter an arc will be called the *starting point* of the arc. Similarly, the point at which we leave an arc will be called the *finishing point* of that arc. An arc A_k is denoted by a closed interval $[s_k, f_k], k = 1, 2, ..., n$, where, s_k is the counterclockwise end i.e., starting point and f_k is the clockwise end i.e., finishing point and obviously $s_k < f_k$.

A CAG and its corresponding circular-arc representation have displayed in Figure 3.

Permutation Graph

The definition of permutation graph(PG) is displayed below.

Definition 4: A graph $G = (V, E)$ with nodes $V = \{1, 2, ..., n\}$ is called a PG if a permutation π over $\{1, 2, ..., n\}$ exists so that $(p, q) \in E$ iff

$$(p - q)(\pi^{-1}(p) - \pi^{-1}(q)) < 0.$$

A PG can be vizualized in a particular way, called matching diagram or permutation diagram of the PG.

Intuitively, for any permutation $\pi = (\pi_1, \pi_2, ..., \pi_n)$ on $S = \{1, 2, ..., n\}$, a permutation graph can be formed in the following visual way. Line up the numbers 1 to n horizontally on a line L_1 (called top line or upper channel). On the line beneath it, line up corresponding permutation such that π_p is below p on a line L_2 (called bottom line or lower channel). Then join each p and π_p^{-1} with a line segment which is corresponding to node v_p of the PG. The resulting diagram is known as a matching diagram or permutation diagram. In the permutation graph corresponding to π, two nodes v_p and v_q are adjoining iff the line segment $[p, \pi_p^{-1}]$ and $[q, \pi_q^{-1}]$ intersect. A permutation diagram and its corresponding permutation graph are displayed in Figure 4.

Figure 3. A circular-arc graph and its corresponding circular-arc representation

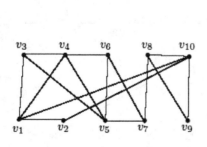 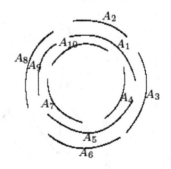

Figure 4. A permutation diagram and its corresponding permutation graph

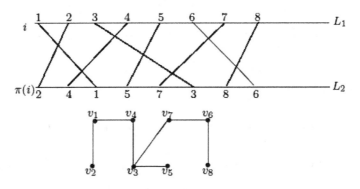

Trapezoid Graphs

The definition of this graph is given below.

Definition 5: (Trapezoid graph) A trapezoid graph consists of two horizontal lines L_1 (top line) and L_2 (bottom line) and a set of trapezoids $T = \{T_1, T_2, ..., T_n\}$ with corner points lying on these two lines. An undirected graph $G = (V, E)$ with node set $V = \{v_1, v_2, ..., v_n\}$ and edge set $E = \{e_1, e_2, ..., e_m\}$ is called a TG when a trapezoid representation exists with trapezoid set T, such that any node $v_k \in V$ corresponds to a trapezoid $T_k \in T$ and an edge $(v_k, v_l) \in E$ iff T_k and T_l intersect.

Any trapezoid T_k within these two lines is known by four corner points a_k, b_k, c_k and d_k which represent the upper left, upper right, lower left and lower right corner points respectively. Without loss of generality, we assume that no two trapezoids share a same endpoint. The term node or nodes and trapezoid are used interchangeably whenever the context is unambiguous. It is assumed that $T_p < T_q$ or $v_p < v_q$ iff $a_p < a_q$. Figure 5 shows an trapezoid representation and its corresponding trapezoid graph.

Figure 5. A trapezoid diagram and its corresponding trapezoid graph

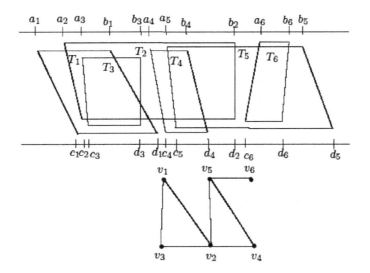

L(h,k)-LABELING OF INTERVAL GRAPH

In this section $L(0,1)$-, $L(1,1)$-, $L(2,1)$-labeling problems of interval graph are investigated and obtain a tighter upper bound for these problems.

L(0,1)-Labeling of Interval Graph

Here, some results regarding $L(0,1)$-labeling of interval graphs have been presented. A special type of set is defined below which we use later.

Definition 6: Let B_α be the set of 2-nbd nodes of v_α with index greater than α. That is, $B_\alpha = \{v_j \in V \mid d(v_j, v_\alpha) = 2$ and $v_j > v_\alpha\}$.

Definition 7: For an interval graph G let S_i, be a set of nodes such that

1.   All the nodes of S_i are adjacent to v_i
2.   No two nodes of S_i are adjacent
3.   Each S_i is maximal.

In the next lemma, a relation between $\lambda_{0,1}(G)$ and $\max |S_i|$, $i = 1, 2, ..., n$ is established.

Lemma 1: ([18]) Let $k = \max\limits_{v_i \in V} |S_i|$, $i = 1, 2, ..., n$. Then at least k labels are needed to label an interval graph by $L(0,1)$-labeling.

Algorithm for *L*(0,1)-Labeling of Interval Graphs

The strategy of our proposed algorithm is as follows. First of all we partition the node set V into a minimum number of disjoint sets (say p) in such a way that each set forms a clique. Let $V = \{U_1, U_2, ..., U_p\}$ be such a partition, where $U_i \cap U_j = \varphi$ for all, $i, j = 1, 2, ..., p$, $i \neq j$ and each U_i is a clique.

The formal algorithm to label the nodes of an interval graph by $L(0,1)$-labeling is now given below.

Algorithm L01 (Paul, Pal & Pal, 2013)

Input: Interval represent of the interval graph G.
Output: Optimal $L(0,1)$-labeling of G.
Step 1. Construct a set $L = \{0, 1, \cdots, k-1\}$.
Step 2. Construct all the partitions and label the partitions
$U_1, U_2, \cdots, U_k$ by $0, 1, \cdots, k-1$ respectively. That is, $f(U_i) = i - 1$,
$i = 1, 2, ..., k$.
Step 3. Set $f(U_{k+1}) = (f(U_k) + 1) \pmod{|L|}$.
That is, the value of $f(U_{k+1})$ must be equal to one of the values of
$f(U_1), ..., f(U_k)$.
Suppose $f(U_{k+1}) = f(U_s)$, $1 \leq s \leq k$.

Step 4. Find the P-set corresponding to U_{k+1}. That is, P_s^{k+1}.

Step 5. If $P_s^{k+1} = \varphi$, i.e. if there is no node $v_q \in U_s$ such that $d(v_q, v_r) \leq 2$ for all $v_r \in U_{k+1}$, then $f(U_{k+1})$ is the valid label for the nodes of U_{k+1}, then $f(U_{k+1})$ is valid for U_{k+1}.

Step 6. If $P_s^{k+1} = U_s$, i.e. if all the nodes of U_s are at a distance 2 from at least one node of U_{k+1}, then $f(U_{k+1})$ is not valid label for U_{k+1}.

Step 7. If some nodes of U_s are at a distance 2 from at least one node of U_{k+1}, then two cases may aries.

Step 7.1. If $U_{s+1} \cup P_s^{k+1}$ form a clique then shift the nodes of P_s^{k+1} from U_s to U_{s+1} and set $f(P_s^{k+1}) = f(U_{s+1})$.

Now check the validity of the label $f(U_{s+1})$ for the new partition U_{s+1} proceed same as above.

If $f(U_{s+1})$ is valid for U_{s+1} then $f(U_{k+1})$ is valid for U_{k+1}.

Otherwise, $f(U_{k+1})$ is not valid for U_{k+1}.

Step 7.2. If $U_{k+1} \cup P_s^{k+1}$ does not form a clique then either Situation 1 or Situation 2 occurs.

Subcase 7.2.1: If Situation 1 occurs then $f(U_{k+1})$ is not a valid label for U_{k+1}.

Subcase 7.2.2: If Situation 2 occurs then

1.   Find q_s^{k+1} and Q_s^{k+1}.

2.   Remove the labels of all the nodes of Q_s^{k+1} and label the nodes of Q_s^{k+1} in the following fashion.

Find the minimum index node (say v_y) from the set Q_s^{k+1}.

Find T_y and L_y.

Set $f(v_y) = \min\{L - L_y\}$.

Now find the next minimum index node (unlabelled) of the set Q_s^{k+1} and to label this node, follow the same procedure.

If all the nodes of Q_s^{k+1} are labelled by the above way, then $f(U_{k+1})$ is a valid label for U_{k+1}.

Otherwise, (i.e. if $L - L_y = \varphi$ for some $y \in Q_s^{k+1}$) $f(U_{k+1})$ is not a valid label for U_{k+1}.

Step 8. If the label $f(U_{k+1})$ for U_{k+1} is not valid, then reset $f(U_{k+1})$ as $(f(U_{k+1}) + i)(mod\,|L|)$, and repeat the above steps for $i = 2, 3, ..., k-1$.

If $f(U_{k+1})$ is not valid label then use a new label which does not

belongs to L. This new label is taken as the least integer which does not belongs to L, let it be j. Update L as $L \cup \{j\}$ and set $f(U_{k+1}) = j$.
  end L01

The upper bound of $L(0,1)$-labeling of an interval graph is provided in the next theorem.

    Theorem 1: (Paul, Pal & Pal, 2013) For an interval graph G, $\lambda_{0,1}(G) \leq \Delta$.

L(1,1)-Labeling of Interval Graph

In this section, $L(1,1)$-labeling problem on interval graphs are discussed. To solve the problem we use the concept of square of graphs. A new linear time algorithm is presented to compute G^2 from G when G is an interval graph. Also a linear time algorithm is designed to find all the maximal cliques of G^2 from G. Finally, it is shown that $L(1,1)$-labeling number (i.e. $\lambda_{1,1}$) of an interval graph can be computed in linear time.

Definition of square of an interval graph is stated below.

    Definition 8: Let $G = (V, E)$, where $V = \{v_1, v_2, \ldots, v_n\}$ be an interval graph. The square of G denoted by G^2 of G is a graph having the same node set as G and having an edge connecting v_i to v_j if and only if v_i to v_j are at distance at most two in G.

    Lemma 2: If G is an interval graph then G^k is also an interval graph.

Here we present some notations which are necessary for the rest of the chapter.

In this section, it is assumed that the intervals in I are indexed by increasing right endpoints, that is, $b_1 < b_2 < \cdots < b_n$. Let us define an array $e = \{e_1, e_2, \ldots, e_{2n}\}$ which contains the $2n$ endpoints on n intervals of an interval graph G on real line $\mathbb{R}$ in increasing order. That is, the array e is the collection of all a_j's and b_j's for all $j = 1, 2, \ldots, n$. For each element e_i of e, three fields, $e_i.val$, $e_i.int$, and $e_i.type$ are define as follows.

$e_i.val$ = value of the real line of the ith endpoint e_i,

$e_i.int = k$ if e_i is the endpoint of the interval I_k,

$$e_i.type = \begin{cases} a, \text{if the endpoint } e_i \text{ left endpoint} \\ b, \text{if the endpoint } e_i \text{ right endpoint.} \end{cases}$$

Computation of G²

We assume that the distance between any two consecutive $e_j, j = 1, 2, \ldots, n$ (for the input graph G) is one unit on $\mathbb{R}$. Our main aim is to compute the array E from e by adjusting the value of e_j on real line. Since there are uncountable number of points between two consecutive e_j's, therefore, we can increase the value of e_j (i.e. $e_j.val$) and set $e_j.val$ between two consecutive $e_j.val$'s with there proper position to compute the graph G^2. We call the new value of $e_j.val$ as $e_{j'}.val$. Finally, we set E_i, $i = 1, 2, ..., 2n$ for the i-th endpoints of the interval representation of G^2. The G^2 for an interval graph can be computed in $O(m + n)$ time.

The computation of maximum clique of G^2 is given by the following algorithm.

Algorithm MCG2

```
Input: Interval representation of an interval graph  G = (V,E)
Output: Maximal cliques of  G², i.e.  Cᵢ's of  G².
Step 1. Find  m(vᵢ), i = 1,2,...,n .
Step 2. Compute the set  M . Let it be  M = {v₁', v₂',...vₖ'}
Step 3. for  j = 1 to  k
 Step 3.1  Cⱼ = N₁[vⱼ']
 Step 3.2 mark the nodes of  Cⱼ in  V
if all the nodes of  V are marked then
     stop
else
     set  j ← j+1 and goto Step 3.1
end if;
end for;
end MCG2
```

From the above algorithm one can correctly computes all maximal cliques (C's) in $O(m + n)$ time. An important result is presented by the following theorem.

Theorem 2: $L(1,1)$-labeling problem of an interval graph G is equivalent to the node coloring problem of G^2.

Proof. In $L(1,1)$-labeling, the label difference of two nodes is at least 1 if the nodes are adjacent and the label difference of two nodes is also at least one if the nodes are at distance two. Now in G^2, $d(v_i, v_j) = 1$ if and only if v_i and v_j are at distance at most two in G. Thus, for the graph G^2 there is no restriction for distance two nodes as $d(v_i, v_j) = 2$ in G converted to $d(v_i, v_j) = 1$ in G^2. Therefore, $L(1,1)$-labeling problem of G converted to $L(1,0)$-labeling problem of G^2. In $L(1,0)$-labeling problem there is no restriction for distance two nodes and if two nodes are adjacent then there label difference is at least one. Again, $L(1,0)$-labeling problem is nothing but a simple node colouring problem. Hence the result.

L(2,1)-Labeling of Interval Graph

In this subsection, it is assumed that the intervals in I are indexed by increasing left endpoints, that is, $a_1 < a_2 < \cdots < a_n$ ($v_1 < v_2 < ... < v_n$). By using this ordering, the following results hold, which are the foundation to $L(2,1)$-labeling for interval graphs.

Lemma 3: ([17]) A graph G is an interval graph if and only if there exists an ordering of its nodes $v_1 < v_2 < \cdots < v_n$ such that if $v_i < v_j < v_l$ and $(v_i, v_l) \in E$ then $(v_i, v_j) \in E$.

Lemma 4: ([8]) The maximal cliques of an interval graph G can be linearly ordered in such a way that for every node $v \in G$, the maximal cliques containing v occurs consecutively.

Algorithm for *L*(2,1)-Labeling of Interval Graphs

Now, we present a greedy algorithm to $L(2,1)$-label an interval graph. Based on this algorithm a new upper bound of $L(2,1)$-labeling of an interval graph in terms of Δ and ω is given. We prove that $L(2,1)$-labeling number of an interval graph does not exceed $\Delta + \omega$.

Consider the node set $V = \{v_1, v_2, \ldots, v_n\}$, where $v_1 < v_2 < v_3 < \ldots < v_n$ and $1, 2, \ldots, n$ are the indices of corresponding nodes. We denote $f(v_i)$, the label of the node v_i, for all $i = 1, 2, \ldots, n$. We define two sets A_i and B_i below.

$$A_i = \{v_j : d(v_j, v_i) = 1, j < i\} \text{ and } B_i = \{v_j : d(v_j, v_i) = 2, j < i\}.$$

Clearly, the indices of the nodes of the sets A_i and B_i are strictly less then i. Also we denote a node v_i^* as the minimum index node of the set A_i and $I_{i^*} = [a_{i^*}, b_{i^*}]$ is the interval corresponding to the node v_i^*. Some useful results related to the sets A_i and B_i for $i = 1, 2, \ldots, n$ are given below.

Lemma 5 (Paul, Pal & Pal, 2015): For each node $v_k \in V$, $A_k \cup \{v_k\}$ forms a clique, and $\max_k |A_k| = \omega - 1$, where ω is the size of the maximum clique.

Lemma 6: For each node $v_k \in V$, $|A_k| + |B_k| \leq \Delta$.

Proof. Let v_k^* be the minimum index node of the set A_k. So, all the nodes of A_k are adjacent to v_k^* as by Lemma 5, $A_k \cup \{v_k\}$ forms a clique. Since v_k^* is minimum among all the nodes of A_k, so the left endpoint of the interval I_{k^*} (interval corresponding to the node v_k^*) is minimum among all the left endpoints of the intervals corresponding to the nodes of the set A_k (see Figure 6).

So, all the intervals corresponding to the nodes of B_k must have non empty intersection with I_{k^*}. Thus, all the nodes of B_k are adjacent to v_k^*. That is, $|A_k| + |B_k| \leq deg(v_k^*)$. Again, $deg(v_k^*) \leq \Delta$. Therefore, $|A_k| + |B_k| \leq \Delta$.

Now we design a greedy algorithm to $L(2,1)$-label all the nodes of an interval graph.

Figure 6. Relation between A_k and B_k

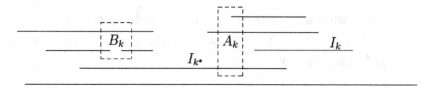

Algorithm L21

```
Input: Set of ordered intervals represents the interval graph.
//assume that the intervals are ordered with respect to the left endpoints.
That is,
```
$a_i < a_j$ for $i < j$ and hence $v_1 < v_2 < ... < v_n$ .//
```
Output:
```
$f(v_i)$, the $L(2,1)$-label of v_i, $i = 1, 2, ..., n$.
```
for each i = 1 to n do
  Compute
```
A_i and B_i.
```
  Let j be the smallest non-negative integer such that
```
$$j \notin \{f(v_k) - 1, f(v_k), f(v_k) + 1 \mid v_k \in A_i\} \cup \{f(v_l) \mid v_l \in B_i\}$$
$$f(v_i) = j ;$$
```
end for;
 end L21
```

Some useful results of this algorithm are discussed below. The algorithm $L21$ can correctly label an interval graph by $L(2,1)$-labeling method using $O(m+n)$ time. The new upper bound of $\lambda_{2,1}$ for an interval graph is stated in the following theorem.

Theorem 3: ([21]) For any interval graph $G = (V, E)$, $\lambda_{2,1}(G) \leq \Delta + \omega$.

Proof. Let $v_k \in V$. The label $f(v_k)$ of v_k depends on all labeled nodes of A_k and B_k, by Algorithm L21. Again, each non-zero labeled nodes of A_k forbid at most 3 integers, zero labeled node of A_k forbid 2 integers (0,1 only) and each labeled node of B_k forbid at most 1 integer.

Now two cases will arise.

Case 1. If $D_k = A_k - \{v_k^*\}$, i.e. there is no node $v_l \in A_k - \{v_k^*\}$ such that $f(v_l) = 0$.

Thus in this case, $f(v_k)$ forbid at most $3 \mid A_k \mid + \mid B_k \mid$ integers. Again, we found that each labeled node of the set $A_k - \{v_k^*\}$ creates at least one common forbidden integer. Thus, for all the labeled nodes of $A_k - \{v_k^*\}$ there is at least $\mid D_k \mid = \mid A_k \mid -1$ number of common forbidden integers. Among $3 \mid A_k \mid + \mid B_k \mid$ integers, $\mid A_k \mid -1$ integers are common, i.e. they are repeated. Thus,

$$f(v_k) \leq 3 \mid A_k \mid + \mid B_k \mid -(\mid A_k \mid -1)$$

or

$$f(v_k) \leq 2 \mid A_k \mid +\Delta - \mid A_k \mid +1$$

or

$$f(v_k) \leq \mid A_k \mid +\Delta +1.$$

Case 2. If $D_k \subset A_k - \{v_k^*\}$. Then there exists at least one node $v_l \in A_k - \{v_k^*\}$ such that $f(v_l) = 0$. But, no two nodes of $A_k - \{v_k^*\}$ have label 0 as it violates the condition of $L(2,1)$-labeling. So, there exist only one node $v_l \in A_k - \{v_k^*\}$ such that $f(v_l) = 0$. Thus, in A_k, $(|A_k| - 1)$ nodes get non-zero label and one node gets zero label. Therefore, in this case $f(v_k)$ forbid at most $3(|A_k| - 1) + 2 + |B_k|$ integers. Here 2 comes from the fact that a node of the set A_k with label 0 forbid only two integers. Again, there exists at least $|A_k| - 2$ common forbidden integers among $3(|A_k| - 1) + 2 + |B_k|$ as $|D_k| = |A_k| - 2$. Hence,

$$f(v_k) \leq \{3(|A_k| - 1) + 2 + |B_k|)\} - (|A_k| - 2)$$

or

$$f(v_k) \leq 2|A_k| + (\Delta - |A_k|) + 1$$

or

$$f(v_k) \leq |A_k| + \Delta + 1.$$

Hence, from the both cases, $f(v_k) \leq |A_k| + \Delta + 1$.
Therefore,

$$\lambda_{2,1}(G) \leq max|A_k| + \Delta + 1 = (\omega - 1) + \Delta + 1 = \Delta + \omega .$$

Hence, $\lambda_{2,1}(G) \leq \Delta + \omega$.

To illustrate the above algorithm we consider an interval graph of 13 nodes (see Figure 6) and label this graph by algorithm $L21$. First we consider the node v_1. For v_1, $A_1 = \varphi$ and $B_1 = \varphi$. So the smallest non-negative integer for v_1 is 0. Thus, we set $f(v_1) = 0$. Next, for the node v_2, $A_2 = \{v_1\}, B_2 = \varphi$. Therefore, the smallest non-negative integers for v_2 does not belong to

$$\{f(v_1) - 1, f(v_1), f(v_1) + 1\},$$

i.e. $\{0,1\}$. Thus, we set $f(v_2) = 2$. Proceed in the same way we get

$$f(v_3) = 4, f(v_4) = 6, f(v_5) = 8, f(v_6) = 1.$$

Similarly, for the node v_7, $A_7 = \{v_5, v_6\}$ and $B_7 = \{v_1, v_3, v_4\}$. Thus the set of smallest non-negative integers do not belong to

Figure 7. An interval graph G labelled by L(2,1)-labeling

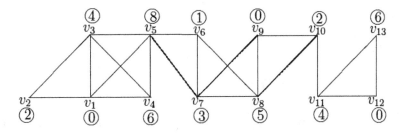

$\{f(v_6) - 1, f(v_6), f(v_6) + 1, f(v_5) - 1, f(v_5), f(v_5) + 1\} \cup \{f(v_1), f(v_3), f(v_4)\}$, is

$\{0, 1, 2, 7, 8, 9\} \cup \{0, 4, 6\}$,

i.e.

$\{0, 1, 2, 4, 6, 7, 8, 9\}$.

So, the minimum available label for the node v_7 is 3. Thus, $f(v_7) = 3$. Proceed same as above we label the other nodes of the graph of Fig. 6.

$L(h,k)$-LABELING OF CIRCULAR-ARC GRAPHS

In this section, three problems namely $L(0,1)$-, $L(1,1)$-, $L(2,1)$-labeling problems of circular-arc graph have been studied and obtain good results for these problems.

Let G be a circular-arc graph with arcs set I. Let us consider the following notations:

1. $L_0(I_k)$: the set of labels which are used before labeling the arc I_k for any arc $I_k \in I$.
2. $L_1(I_k)$: the set of labels which are used to label the nodes at distance one from the arc I_k before labeling the arc I_k, $I_k \in I$.
3. $L_2(I_k)$: the set of labels which are used to label the nodes at distance two from the arc I_k, before labeling the arc I_k, $I_k \in I$.
4. $L_{1\vee2}(I_k)$: the set of labels which are used to label the nodes at distance either one or two from the arc I_k, before labeling the arc I_k, for any arc $I_k \in I$.
5. f_j: the label of the arc I_j, $I_j \in I$
6. L: the label set, i.e. the set of labels used to label the circular-arc graph G completely.

It can be verified that $L_{1\vee2}(I_k) = L_1(I_k) \cup L_2(I_k)$ for any arc $I_k \in I$.

Definition 9: For a circular-arc graph G we define a set of arcs S_{I_j}, for each $I_j \in I$ such that

1. all arcs of S_{I_j} are adjacent to I_j.

2. no two arcs of S_{I_j} are adjacent.

3. each S_{I_j} is maximal.

L(0,1)-Labeling of Circular Arc Graphs

In this section, we present some lemmas related to the proposed algorithm. Also an algorithm is designed to solve $L(0,1)$-labeling problem on circular-arc graphs, along with time complexity. The following results are proved in [26].

1. If $L_1(I_j) - L_2(I_j) \neq \varnothing$ then $f_j = l$, where $l \in L_1(I_j) - L_2(I_j)$ for any arc $I_j \in I$.
2. If $L_0(I_j) - L_2(I_j) \neq \varnothing$ then $f_j = l$, where $l \in L_0(I_j) - L_2(I_j)$ for any arc $I_j \in I$.
3. If $L_0(I_j) - L_2(I_j) = \varnothing$ then $f_j \neq l$, where $l \in L_0(I_j) - L_2(I_j)$ and $f_j = m$ where $m = max\{L_0(I_j)\} + 1$ for any arc $I_j \in I$.

Now we discuss about the bounds of $\lambda_{0,1}(G)$ for a circular-arc graphs. The lower bound for $L(0,1)$-labeling is given by $\lambda_{0,1}(G) \geq k - 1$, where $k = \max_{I_j \in I} |S_{I_j}|$, $j = 1, 2, 3, ..., n$. For any arc $I_k \in I$, $|L_2(I_k)| \leq \Delta$ and using these results we can proved that $\lambda_{0,1}(G) \leq \Delta$, which gives the upper bound of $L(0,1)$-labeling of circular-arc graph.

Algorithm for L(0,1)-labeling of circular-arc graph

Now we design an algorithm to $L(0,1)$-label a circular-arc graph.

Algorithm L01

Input: A set of ordered arcs I of a circular-arc graph.
//assume that the arcs are ordered with respect to left end points (i.e. in clockwise
direction namely $I_1, I_2, I_3, ..., I_n$) where $I = \{I_1, I_2, I_3, ..., I_n\}$ //
Output: f_j, the L(0,1)-label of I_j, $j = 1, 2, 3, ..., n$.
Initialization: $f_1 = 0$;
  $L_0(I_2) = \{0\}$;
  for each $j = 2$ to $n - 1$ compute $L_1(I_j)$ and $L_2(I_j)$
  if $L_1(I_j) - L_2(I_j) \neq \varnothing$ then $f_j = l$, and set $L_0(I_{j+1}) = L_0(I_j)$,
  for any $l \in L_1(I_j) - L_2(I_j)$;
  else if $L_0(I_j) - L_2(I_j) \neq \varnothing$ then $f_j = m$, and set $L_0(I_{j+1}) = L_0(I_j)$,
      where $m \in L_0(I_j) - L_2(I_j)$;

```
else  f_j = p ,  where  p = max{L_0(I_j)}+1  and set
        L_0(I_{j+1}) = L_0(I_j) ∪ {p} ;
end for;
if  L_1(I_n) − L_2(I_n) ≠ ∅  then  f_n = q ,  where  q ∈ L_1(I_n) − L_2(I_n)
else if  L_0(I_n) − L_2(I_n) ≠ ∅  then  f_n = r ,  where  r ∈ L_0(I_n) − L_2(I_n) ,
else  f_n = s ,  where  s = max{L_0(I_n)}+1 ;
end  L01
```

Theorem 4: ([26]) The Algorithm $L01$ correctly labels the nodes of a circular-arc graph using $L(0,1)$-labeling condition. The maximum label used by this algorithm is Δ .

Proof. Let $I = \{I_1, I_2, I_3, ..., I_n\}$, also let $f_1 = 0$, $L_0(I_2) = \{0\}$.

Case 1: If the set $L_0(I_2)$ is sufficient to label the whole graph then the result is obviously true and $\lambda_{0,1}(G) = 0$.

Case 2: If we use extra label then we have to show that the set $L_0(I_2)$ is not sufficient to label the graph.

Suppose we label the arc I_k. This arc can not be labeled by a label from the set $L_2(I_k)$. In this case, obviously, $L_1(I_k) - L_2(I_k) = \varnothing$, otherwise, $f_k = l$ where

$$l \in L_1(I_k) - L_2(I_k) \subseteq L_0(I_k).$$

In this case, $L_0(I_k) - L_2(I_k) = \varnothing$, otherwise, $f_k = m$, where

$$m \in L_0(I_k) - L_2(I_k) \subseteq L_0(I_k).$$

So, all the labels in $L_0(I_k)$ are already use to label the arcs which are at distance two from the arc I_k before labeling I_k. So, there is no scope to label the arc I_k by a label from the set $L_0(I_k)$. So, we must label the arc I_k by an extra label m, i.e. $f_k = m$, where $m = L_0(I_k) + 1$; otherwise the condition of $L(0,1)$-labeling is violated.

If

$$L_0(I_k) = \{0, 1, 2, ..., \Delta\},$$

then,

$$L_0(I_k) - L_2(I_k) \neq \varnothing.$$

According to our proposed algorithm, we need additional label if

$$L_0(I_k) - L_2(I_k) = \varnothing.$$

But

$$L_0(I_k) - L_2(I_k) \neq \varnothing,$$

so additional label is not required to label the arc I_k. This is true for any arc I_k. Hence, $\lambda_{0,1}(G) \leq \Delta$.

Theorem 5: ([26]) Any circular-arc graph can be $L(0,1)$-labeled using $O(n\Delta^2)$ time, where n and Δ represents the number of nodes and the degree of the graph G.

Proof. Let L be the label set and $|L|$ be its cardinality. According to the algorithm $L01$,

$$|L_1(I_k)| \leq |L| \text{ and } |L_2(I_k)| \leq |L|$$

for any $I_k \in I$. So $L_1(I_k) - L_2(I_k)$ can be computed using at most $|L| . |L| = |L|^2$ time. Here $L_2(I_k) \subseteq L_0(I_k)$ and both $L_0(I_k)$ and $L_2(I_k)$ are subsets of

$$\{0, 1, 2, \ldots, |L| - 1\},$$

so $L_0(I_k) - L_2(I_k)$ can be computed using $|L|$ time. Again, the union of the set $L_0(I_k)$ and a singleton set can be done in unit time, since the sets are disjoint. This process is repeated for $n-1$ times. So, the total time complexity for the algorithm $L01$ is

$$O((n-1)|L|^2) = O(n|L|^2).$$

Since, $|L| \leq \Delta$, therefore the running time for the algorithm $L01$ is $O(n\Delta^2)$.

Let us consider the circular-arc graph of Fig. 8 to illustrate the algorithm $L01$.

Now $I = \{I_1, I_2, I_3, \ldots, I_{10}\}$ and also $\Delta = 4$.

$f_j = $ the label of the arc I_j, for $j = 1, 2, 3, \ldots, 10$.

$$f_1 = 0, L_0(I_2) = \{0\}.$$

$$L_1(I_2) - L_2(I_2) = \{0\} - \varnothing = \{0\} \neq \varnothing.$$

So,

$$f_2 = 0, L_0(I_3) = L_0(I_2) = \{0\}.$$

$$L_1(I_3) - L_2(I_3) = \{0\} - \{0\} = \varnothing.$$

Also,

$$L_0(I_3) - L_2(I_3) = \{0\} - \{0\} = \varnothing.$$

Therefore,

$$f_3 = \max\{L_0(I_3)\} + 1 = 0 + 1 = 1,\ L_0(I_4) = L_0(I_3) \cup \{1\} = \{0\} \cup \{1\} = \{0,1\}.$$

$$L_1(I_4) - L_2(I_4) = \{0,1\} - \{0\} = \{1\} \neq \varnothing.$$

So,

$$f_4 = 1,\ L_0(I_5) = L_0(I_4) = \{0,1\}.$$

$$L_1(I_5) - L_2(I_5) = \{1\} - \{0\} = \{1\} \neq \varnothing.$$

So,

$$f_5 = 1,\ L_0(I_6) = L_0(I_5) = \{0,1\}.$$

Similarly,

$$f_6 = 2,\ f_7 = 2,\ f_8 = 3,\ f_9 = 3 \text{ and } f_{10} = 3.$$

Therefore,

$$\lambda_{0,1}(G) = max\{L\} = 3.$$

L(1,1): Labeling of Circular-Arc Graphs

By extending the idea of $L(0,1)$-labeling, we design an algorithm for $L(1,1)$-labeling of circular-arc graphs. In this subsection we present some lemmas related to our work, upper bound of $L(1,1)$-labeling, an algorithm $L11$ and time complexity of the proposed algorithm $L11$. The following results are very useful to develop this section, all results are proved in Amanathulla and Pal (2016).

1. If $L_0(I_j) - L_{1\vee 2}(I_j) \neq \varphi$, then $f_j = l$, where $l \in L_0(I_j) - L_{1\vee 2}(I_j)$ for any $I_j \in I$.
2. If $L_0(I_j) - L_{1\vee 2}(I_j) = \varnothing$, then $f_j \neq l$, for any $l \in L_0(I_j) - L_{1\vee 2}(I_j)$ but $f_j = m$, where $m = \max\{L_0(I_j)\} + 1$.

Figure 8. Illustration of Algorithm L01

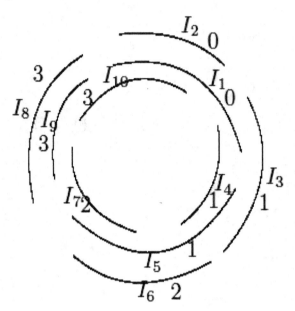

3. For any circular-arc graph G, $L_{1\vee2}(I_k) \subseteq L_0(I_k)$ for any arc I_k of G.

4. For any circular-arc graph G, $L_0(I_k) - L_{1\vee2}(I_k) \neq \varnothing$, for any arc I_k of G, where $L_0(I_k) = \{0,1,2,...,2\Delta\}$.

The following theorem present the upper bound of $L(1,1)$-labeling of circular-arc graph.

Theorem 6: (Amanathulla & Pal, 2016) For any circular-arc graph G, $\lambda_{1,1}(G) \leq 2\Delta$, where Δ is the degree of the graph G.

Proof. Let $I = \{I_1, I_2, I_3, ..., I_n\}$, also let $f_1 = 0$, $L_0(I_2) = \{0\}$. If the graph has only one node then obviously, $\lambda_{1,1}(G) = 0$. If the graph contains more than one node then the set $L_0(I_2)$ is insufficient to label the whole graph. Now we are going to label the arc I_k. We can not label the arc I_k by the label in the set $L_{1\vee2}(I_k)$. In this case obviously, $L_0(I_k) - L_{1\vee2}(I_k) = \varphi$, otherwise, $f_k = l$, where

$$l \in L_0(I_k) - L_{1\vee2}(I_k) \subseteq L_0(I_k).$$

So, all the labels in $L_0(I_k)$ are already used to label the arcs which are in distance one or two from the arc I_k before labeling the arc I_k. So, there is no scope to label the arc I_k by the label in the set $L_0(I_k)$.

Hence, we label the arc I_k by an additional label m, i.e. $f_k = m$, where $m = \{L_0(I_k)\} + 1$; otherwise the condition of $L(1,1)$-labeling is violated.

If $L_0(I_k) = \{0,1,2,...,2\Delta\}$, then, $L_0(I_k) - L_{1\vee2}(I_k) \neq \varnothing$. According to our proposed algorithm, we need additional label if $L_0(I_k) - L_{1\vee2}(I_k) = \varnothing$. But $L_0(I_k) - L_{1\vee2}(I_k) \neq \varphi$, so additional label is not required to label the arc I_k. This is true for any arc I_k. Hence, $\lambda_{1,1}(G) \leq 2\Delta$.

Algorithm for L(1,1)-Labeling of Circular-Arc Graph

In this subsection we present an algorithm to $L(1,1)$-label a circular-arc graphs.

Algorithm L11 (Amanathulla & Pal, 2016)

Input: A set of ordered arcs I of a circular-arc graph.
//assume that the arcs are ordered with respect to left end points (namely
$I_1, I_2, I_3, ..., I_n$), where $I = \{I_1, I_2, I_3, ..., I_n\}$ //
Output: f_j, the L(1,1) label of I_j, $j = 1, 2, 3, ..., n$.
Initialization: $f_1 = 0$;
$L_0(I_2) = \{0\}$;
for each $j = 2$ to $n-1$ compute $L_{1\lor 2}(I_j)$
if $L_0(I_j) - L_{1\lor 2}(I_j) \neq \varnothing$ then $f_j = l$, and set $L_0(I_{j+1}) = L_0(I_j)$
where $l \in L_0(I_j) - L_{1\lor 2}(I_j)$;
else $f_j = m$, where $m = max\{L_0(I_j)\} + 1$ and set
$L_0(I_{j+1}) = L_0(I_j) \cup \{m\}$;
end for;
if $L_0(I_n) - L_{1\lor 2}(I_n) \neq \varnothing$ then $f_n = p$, where p is any integer of the
set $L_0(I_n) - L_{1\lor 2}(I_n)$;
else $f_n = q$, where $q = max\{L_0(I_n)\} + 1$;
end L11.

The Algorithm $L11$ correctly labels the nodes of a circular-arc graph using $L(1,1)$-labeling condition and the maximum label used by this algorithm is 2Δ.

To illustrate the algorithm we consider a circular-arc graph of Figure 9.

Figure 9. Illustration of Algorithm L11

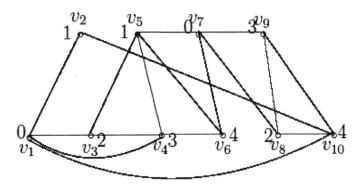

Let $I = \{I_1, I_2, I_3, \ldots, I_{10}\}$ and also $\Delta = 4$.

f_j = The label of the arc I_j, for $j = 1, 2, 3, \ldots, 10$.

$f_1 = 0$, $L_0(I_2) = \{0\}$.

$L_0(I_2) - L_{1\vee 2}(I_2) = \{0\} - \{0\} = \varnothing$.

Therefore

$f_2 = \max\{L_0(I_2)\} + 1 = 0 + 1 = 1$, $L_0(I_3) = L_0(I_2) \cup \{f_2\} = \{0\} \cup \{1\} = \{0,1\}$.

$L_0(I_3) - L_{1\vee 2}(I_3) = \{0,1\} - \{0,1\} = \varnothing$.

Therefore

$f_3 = \max\{L_0(I_3)\} + 1 = 1 + 1 = 2$, $L_0(I_4) = L_0(I_3) \cup \{f_3\} = \{0,1\} \cup \{2\} = \{0,1,2\}$.

$L_0(I_4) - L_{1\vee 2}(I_4) = \{0,1,2\} - \{0,2,1\} = \varnothing$.

Therefore

$f_4 = \max\{L_0(I_4)\} + 1 = 2 + 1 = 3$,

$L_0(I_5) = L_0(I_4) \cup \{f_4\} = \{0,1,2\} \cup \{3\} = \{0,1,2,3\}$.

$L_0(I_5) - L_{1\vee 2}(I_5) = \{0,1,2,3\} - \{3,2,0\} = \{1\} \neq \varnothing$.

So

$f_5 = 1$, $L_0(I_6) = L_0(I_5) = \{0,1,2,3\}$.

$L_0(I_6) - L_{1\vee 2}(I_6) = \{0,1,2,3\} - \{1,3,2,0\} = \varnothing$.

Therefore,

$f_6 = \max\{L_0(I_6)\} + 1 = 3 + 1 = 4$, $L_0(I_7) = L_0(I_6) \cup \{f_6\} = \{0,1,2,3\} \cup \{4\} = \{0,1,2,3,4\}$.

Similarly,

$$f_7 = 0, \; f_8 = 2, \; f_9 = 3, \; f_{10} = 4.$$

Hence, $\lambda_{1,1}(G) = \max L = 4$.

L(2,1)-Labeling of Circular-Arc Graphs

The main idea to $L(2,1)$-label a circular-arc graph is to convert the given graph to an interval graph by deleting some suitable arcs. To convert a circular-arc graph to an interval graph draw a line L from the center of the circle perpendicular to the arcs (see Figure 9). Let C be the set of arcs which intersect the line L. Remove the arcs of C from the set of arcs I. If we consider the set of arcs $I - C$ as line segments, then they form a set of intervals on a real line (Figure 10) and these intervals form an interval graph.

Thus, one can conclude the following that the arcs of $I - C$ induced an interval representation on a real line (see Figure 9).

Lemma 7: The nodes corresponding to the arcs of C form a clique.

Proof. The arcs of C must intersect the line L which is drawn from the center of circle. That is, any two arcs intersect each other. Thus, the nodes corresponding to the arcs of C form a clique.

The following algorithm labels a CAG by $L(2,1)$-labeling.

Algorithm L21CA (Paul, Pal & Pal, 2014)

Input: Circular-arc representation of the given circular-arc graph G.
Output: $L(2,1)$-labeling of $G = (V, E)$.

Figure 10. A circular-arc representation and the corresponding circular-arc graph G

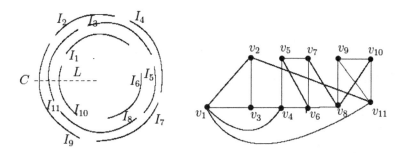

Figure 11. The set of intervals obtained from Figure 10 by deleting the arcs of C

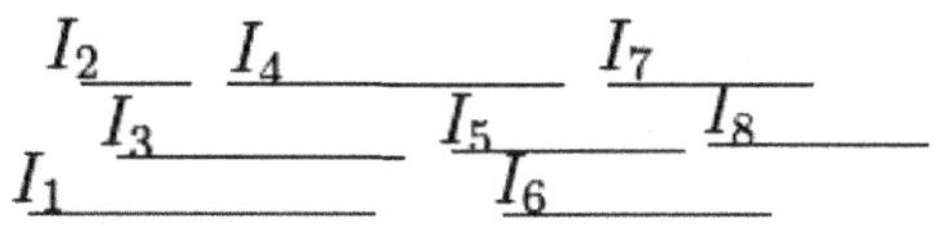

Step 1: Draw a straight line L and let C be the set of arcs intersecting L.

Step 2: Remove C from the circular-arc representation. Let this reduced graph be

$G' = (V', E')$, where V' is the set of nodes corresponding to the arcs $I - C$.

Step 3: Compute the first maximal clique C_1 of G'.

Step 4: Arrange the nodes of the clique C_1 according to the decreasing order of the

degree of the nodes. If the degree of two nodes $v_i, v_j \in C_1$ are equal then we

put v_j before v_i if $i > j$, in the order and let this new clique be C_1'.

Step 5: // Label all the nodes of C_1' //

for each node $v_k \in C_1'$

set $f(v_k) = 2(i - 1)$ // i represents the position of the node v_k in C_1' //

Step 6: Compute the set $V' - C_1$.

If $V' - C_1 = \varphi$ then

stop;

       otherwise go to Step 5;

For $i = m$ to n do steps 7 to 11// v_m is the first node of $V' - C_1$ //

Step 7:

Compute $A_i = \{v_j : d(v_j, v_i) = 1 \text{ and } j < i\}$;

and $B_i = \{v_j : d(v_j, v_i) = 2 \text{ and } j < i\}$.

Step 8: Compute for each node $v_j \in A_i$

$$r_{A_i}(f(v_j)) = \{f(v_j) - 1, f(v_j), f(v_j) + 1\} \cap (Z^+ \cup \{0\})$$

and for each node $v_j \in B_i$

$$r_{B_i}'(f(v_j)) = \{f(v_j)\}.$$

Step 9: $S_i^1 = \bigcup_{v_j \in A_i} (r_{A_i}(f(v_j)))$, //set of positive integers forbid by all the nodes

of A_i //

$S_i^2 = \bigcup_{v_j \in B_i} (r_{B_i}'(f(v_j)))$. //set of positive integers forbid by all the nodes of B_i //

Step 10: Compute $R_i = S_i^1 \bigcup S_i^2$ // R_i is the set of positive integers unavailable

to label v_i //

Step 11: $f(v_i) = \min\{(Z^+ \cup \{0\}) - R_i\}$

Let F be the highest label, i.e. maximum of $f(v_i)$.

Step 12: Now label the nodes corresponding to the arcs of C. Without loss of

generality let $C = \{I_1, I_2, ..., I_p\}$.

for $j = 1$ to p

$f(v_j) = F + 2 + 2(j - 1)$; // $f(v_j)$ is the label of the node corresponding to the

arc $I_j \in C$ .//

```
end for;
 end L21CA
```

Theorem 7: (Paul, Pal & Pal, 2014) For any circular-arc graph $G = (V, E)$, $\lambda_{2,1}(G) \leq \Delta + 3\omega$, where Δ and ω represent the maximum degree of the nodes and size of maximum clique of G respectively.

Proof. Let G' be an interval graph corresponding to the intervals representation of $I - C$ (by Lemma 7). Δ_1 and ω_1 represent the degree of the nodes and size of maximum clique of G' respectively. Clearly, $\Delta_1 \leq \Delta$ and $\omega_1 \leq \omega$ as $G' \subseteq G$. Thus, by Theorem 3, $\lambda_{2,1}(G') \leq \Delta_1 + \omega_1$. Again, the nodes of C forms a clique, and it can be labelled by the integers $0, 2, 4, \ldots, 2(|C|-1)$. Thus, to label all the nodes corresponding to the arcs of C, at most $2(\omega - 1) + 1$ additional labels are required as max $|C| = \omega$. These additional labels must be started from $\Delta_1 + \omega_1 + 2$ as from Theorem 3, $\lambda_{2,1}(G') \leq \Delta_1 + \omega_1$.

Therefore,

$$\lambda_{2,1}(G) \leq \Delta_1 + \omega_1 + 2 + 2(\omega - 1).$$

Thus,

$$\lambda_{2,1}(G) \leq \Delta + 3\omega(\because \Delta_1 \leq \Delta \, and \, \omega_1 \leq \omega).$$

Hence the result.

It is proved that the time complexity of algorithm L21CA is $O(n^2)$, where n is the number of nodes of the graph (Paul, Pal & Pal, 2014).

L(h,k)-LABELING OF PERMUTATION GRAPHS

In this section, we studied two problems namely, $L(0,1)$- and $L(2,1)$-labeling of permutation graphs.

L(0,1)-Labeling of Permutation Graphs

Now, some particular types of sets and terms are defined which are used to describe the $L(0,1)$-labeling.

$$S_i = \{j \in V(G) : d(j, i) = 2 \, and \, j < i\}.$$

$$a_i = \min\{j \in V(G) : d(i, j) = 1 \, and \, j < i\},$$

but if no such a_i exists, then let $a_i = 0$.

$$b_i = \{j \in V(G) : d(i, j) = 1, j > i \, and \, \pi^{-1}(j) \, is \, minimum \, in \, the \, bottom \, line \, L_2\},$$

if there is no such b_i exists, then let $b_i = 0$.

$$P_i = \{x \in V(G) : a_i < x < i \, and \, \pi^{-1}(x) < \pi^{-1}(b_i)\}$$

$$Q_i = \{y \in V(G) : \pi^{-1}(b_i) < \pi^{-1}(y) < \pi^{-1}(i) \, and \, y < a_i\}$$

$$P_{i'} = \{x \in V(G) : a_i < x < i \, and \, \pi^{-1}(x) < \pi^{-1}(i)\}$$

$$Q_{i'} = \{y \in V(G) : \pi^{-1}(b_i) < \pi^{-1}(y) < \pi^{-1}(i) \, and \, y < i\}$$

$$S_{i'} = S_i \cup \{i\}$$

S_i^* is the set of unlabelled nodes of the set $S_{i'}$

S_i^{**} is the set of nodes in ascending order of the set S_i^*

$f(i)$ denotes the label of the node i.

k denotes the number of labelled nodes.

Z' denote the set of all non negative integers.

From definition of $P_i, Q_i, P_{i'}$ and $Q_{i'}$, clearly, $P_i \subseteq P_{i'}$ and $Q_i \subseteq Q_{i'}$.

It can be proved that, for any node i, $S_i = P_i' \cup Q_i'$.

The algorithm to $L(0,1)$-label a permutation graph is given below:

Algorithm L01 (Paul, Pal & Pal, 2015)

```
Input: Matching diagram i, π(i), i = 1,2,...,n .
Output: L(0,1)-labeling of permutation graph.
Step 1. Initialization
  i = n
  k = 0 //k represents the number of labelled nodes.
  Initially all the nodes are unlabelled.//
Step 2. Choose the node i.
Step 3. Find Sᵢ∪{i} and let it be Sᵢ′ .
Step 4. Compute the unlabelled nodes of Sᵢ′. Let the set of these nodes be Sᵢ* .
Step 5. If Sᵢ* = φ then set i ← i − 1 and go to Step 1.
  else
Step 5.1. Find Sᵢ** . Let Sᵢ** = {i₁,i₂,...,iₘ};
Step 5.2. Set j = i₁ . //j indicates the nodes of the set Sᵢ** .//
Step 5.3. Compute f(N₂(j));
```

Step 5.4. $f(j) = \min\{Z' - f(N_2(j))\}$. // minimum available label.//

Step 5.5. Record the current value of k.

Step 5.6. Set $j \leftarrow j+1$ and go to Step 5.3. This process will continue until $j = i_m$.

Step 6. If $k < n$ then set $i \leftarrow i-1$ and go to Step 1. Process will continue until

$k = n$.

end L01

Lemma 8: If x, y be two nodes such that $x \in P_i$ and $y \in Q_i$, then $d(x,y) = 1$.

Proof. For a node i, let P_i and Q_i be non empty. Let $x \in P_i$ and $y \in Q_i$. From the definition of P_i and Q_i, $y < a_i$ and $a_i < x$, that is, $y < a_i < x$. Therefore, $y < x$.

Again,

$$\pi^{-1}(x) < \pi^{-1}(b_i) \text{ and } \pi^{-1}(b_i) < \pi^{-1}(y).$$

Therefore,

$$\pi^{-1}(x) < \pi^{-1}(b_i) < \pi^{-1}(y).$$

That is,

$$\pi^{-1}(x) < \pi^{-1}(y).$$

Thus,

$$(x - y)(\pi^{-1}(x) - \pi^{-1}(y)) < 0.$$

Therefore, there is an edge between x and y. That is, $d(x,y) = 1$.

Theorem 8: For a permutation graph G, $\lambda_{0,1}(G) \leq \Delta - 1$.

Proof. Since the graph is connected, for each node i, either $a_i \neq 0$ or $b_i \neq 0$ or both are non-zero. For example, for the node n, $a_n \neq 0$, but $b_n = 0$.

Suppose we are in the midst of the algorithm and choose a node i. Now, the aim is to label some nodes of S_i^{**}. Assumed that $S_i^{**} \neq \varphi$. To label the nodes of S_i^{**}, different cases may arise.

Case 1. When $a_i \neq 0$, and $b_i = 0$.

In this case, $Q_{i'} = \varphi$ and therefore $S_i = P_{i'}$. In Figure 11, the nodes corresponding to the dotted lines are the elements of the set $P_{i'}$. Let $x \in P_{i'}$. So,

$$a_i < x < i \text{ and } \pi^{-1}(x) < \pi^{-1}(i) < \pi^{-1}(a_i).$$

Figure 12. When $a_i \neq 0$ and $b_i = 0$

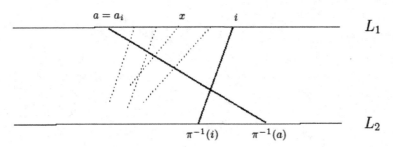

Therefore,

$$(a_i - x)(\pi^{-1}(a_i) - \pi^{-1}(x)) < 0.$$

So, $d(a_i, x) = 1$, i.e. all the nodes of $P_{i'}(= S_i)$ are adjacent to a_i. Again, i is adjacent to a_i. Hence, all the nodes of $S_{i'}$ are adjacent to a_i. Therefore, $|S_{i'}| \leq deg(a_i) \leq \Delta$. Thus, to label the nodes of $S_{i'}$, at most Δ labels are needed (including 0) as the nodes are labelled by $L(0,1)$-labeling. Obviously, S_i^{**} can be labelled by using labels from 0 to $\Delta - 1$ as $S_i^{**} \subseteq S_{i'}$.

Case 2. When $a_i = 0$, and $b_i \neq 0$.

In this case, $P_{i'} = \varphi$, so $S_i = Q_{i'}$. In Figure 12, the nodes corresponding to the dotted lines are the elements of the set $Q_{i'}$. Now, let $y \in Q_{i'}$. So,

$$\pi^{-1}(b_i) < \pi^{-1}(y) < \pi^{-1}(i) \text{ and } y < i < b_i\}.$$

That is,

$$(y - b_i)(\pi^{-1}(y) - \pi^{-1}(b_i)) < 0.$$

Figure 13. When $a_i = 0$ and $b_i \neq 0$

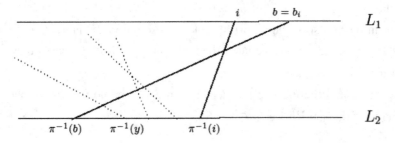

Therefore, $d(y, b_i) = 1$. So, all the nodes of the set $Q_i (= S_i)$ are adjacent to b_i. Again, i is adjacent to b_i. So, all the nodes of $S_{i'}$ are adjacent to b_i. Therefore, $\mid S_{i'} \mid \leq deg(b_i) \leq \Delta$. Thus S_i^{**} can be labelled by using colours from 0 to $\Delta - 1$ as $S_i^{**} \subseteq S_{i'}$.

Case 3. When $a_i \neq 0$ and $b_i \neq 0$.

In this case, either $P_{i'}$ or $Q_{i'}$ are already labelled in previous steps.

Subcase I: When $P_{i'} \neq \varphi$ and $Q_{i'} = \varphi$ then this case is similar as in Case 1. So the conclusion is same.

Subcase II: When $P_{i'} = \varphi$ and $Q_{i'} \neq \varphi$ then this case is similar as in Case 2. So the conclusion is same.

Subcase III: When $P_{i'} \neq \varphi$, $Q_{i'} \neq \varphi$ and $P_{i'}$ are already labelled.

In this case, the unlabelled nodes are the nodes of $Q_i (\subseteq Q_{i'})$. Let y be any arbitrary node of the set Q_i. Now, from Lemma 8, $d(x, y) = 1$ for some $x \in P_i$. Thus, the available labels for y are $f(x)$ or $f(l)$, where $d(y, l) = 1$ and $l \notin P_i$. The node l is already labelled because when we choose the node b_i (in algorithm $i = b_i$) then $l \in P'_{b_i}$. Thus, l is already labelled. Again, from Cases 1 and 2, $f(x)$ or $f(l) \in [0, \Delta - 1]$ and so, $f(y) \in [0, \Delta - 1]$. Therefore, all the nodes of Q_i can be labelled by using labels from 0 to $\Delta - 1$.

Subcase IV: When $P_{i'} \neq \varphi$, $Q_{i'} \neq \varphi$ and $Q_{i'}$ are already labelled.

Conclusion of this subcase is same as Subcase III. That is, at most $\Delta - 1$ labels are required to label all the nodes of the set P_i.

Hence the result.

L(2,1)-Labeling of Permutation Graphs

Here we have discuss about *L*(2,1)-labeling of permutation graphs and obtain a tighter upper bor for this problem.

An Algorithm for *L*(2,1)-Labeling of Permutation Graph

A subset S of $V(G)$ is called an i-stable set (or i-independent set) if the distance between any two nodes in S is strictly greater than i. A 1-stable set is usually known as independent set. A maximal

Figure 14. When $a_i \neq 0$ and $b_i \neq 0$

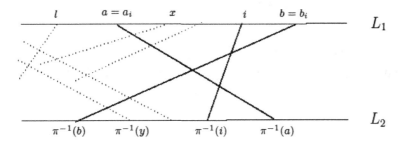

i-stable set S of the set $F \subseteq V$ of nodes is an i-stable subset of F such that S is not a proper subset of any other i-stable subset of G contained in F.

```
Input: A permutation graph G = (V, E).
Output: Value of maximum label k.
Initialization: S_{-1} = φ, V = V(G), i = 0.
Step 1: If S_{i-1} ≠ φ then
set F_i = {x ∈ V : x is unlabeled and d(x, y) ≥ 2 for all y ∈ S_{i-1}}
else F_i = V.
If F_i ≠ φ then compute S_i (maximal 2-stable subset of F_i)
else set S_i = φ.
Step 2: Label all the nodes of S_i by i.
Step 3: V ← V - S_i.
Step 4: If V ≠ φ then set i ← i+1 and go to step 1.
Step 5: This process is continued until V = φ. Set k = i (number of itera-
tions).
Stop.
```

The value of k computed by the above algorithm is an upper bound of $\lambda_{2,1}(G)$. Let $C \subset V$ be the set of nodes which are labeled by the largest label k by algorithm CK. Let $x \in C$.

Let us define three subsets $I_1(x)$, $I_2(x)$ and $I_3(x)$ as follows.

$$
\begin{aligned}
I_1(x) &= \{i : 0 \le i \le k-1 \text{ and } d(x, y) = 1 \text{ for some } y \in S_i\} \\
I_2(x) &= \{i : 0 \le i \le k-1 \text{ and } d(x, y) \le 2 \text{ for some } y \in S_i\} \\
I_3(x) &= \{i : 0 \le i \le k-1 \text{ and } d(x, y) \ge 3 \text{ for all } y \in S_i\}
\end{aligned}
$$

Therefore, the sum of the cardinalities of the sets $I_2(x)$ and $I_3(x)$ is k. That is, $|I_2(x)| + |I_3(x)| = k$. Again for any $i \in I_3(x)$, $x \notin F_i$ since otherwise $S_i \cup \{x\}$ would be a 2-stable subset of F_i, which contradicts the choice of S_i. That is, $d(x, y) = 1$ for some $y \in S_{i-1}$, i.e. $i - 1 \in I_1(x)$. Since for every $i \in I_3(x), i - 1 \in I_1(x)$. Thus $|I_3(x)| \le |I_1(x)|$.

Hence,

$$\lambda_{2,1}(G) \le k = |I_2(x)| + |I_3(x)| \le |I_2(x)| + |I_1(x)|. \tag{1}$$

Let G be a permutation graph and it is labeled by the algorithm CK. If $x \in C$, $f(x) = k$. The sets $I_1(x), I_2(x)$ and $I_3(x)$ are defined for each $x \in C$. We now analysis the cardinalities of the sets $I_1(x)$ and $I_2(x)$ as

$$\lambda_{2,1}(G) \leq \mid I_2(x) \mid + \mid I_1(x) \mid .$$

For a permutation graph G,

$$\mid I_2(x) \mid \leq max\{3\Delta - 2, 4\Delta - 8\},$$

for all $x \in C$ (the prove of this result is found in [20]).

Theorem 9: (Paul, Pal & Pal, 2015) For a permutation graph G,

$$\lambda_{2,1}(G) \leq max\{4\Delta - 2, 5\Delta - 8\}.$$

Proof. In a permutation graph G, for all $x \in C$, $\mid I_1(x) \mid$ is less than Δ. Thus from equation 1,

$$\begin{aligned}
\lambda_{2,1}(G) &\leq &\mid I_2(x) \mid + \mid I_1(x) \mid \\
&\leq & max\{3\Delta - 2, 4\Delta - 8\} + \Delta \\
&\leq & max\{4\Delta - 2, 5\Delta - 8\}.
\end{aligned}$$

Hence the theorem.

L(h,k)-LABELING OF TRAPEZOID GRAPHS

In this section, $L(0,1)$- and $L(2,1)$-labeling of trapezoid graphs are discussed and a good upper bound is obtained.

L(0,1)-Labeling of Trapezoid Graphs

Here, $L(0,1)$-labeling of trapezoid graph is investigated. Let us define some usefull terms:

$$S_i = \{v_j \in V(G) : d(v_i, v_j) = 2 \ and \ v_j < v_i\}.$$

$$S_{i'} = S_i \cup \{i\}.$$

Again, for each trapezoid $T_i, i = 1, 2, \ldots, n$, let us define two sets M_i and N_i as follows.

$$M_i = \{v_j \in V(G) : a^i < b_j < a_i \ and \ d_j < c_i\}. \ \text{If} \ a^i = 0 \ \text{then} \ M_i = \varphi .$$

$N_i = \{v_j \in V(G) : c^i < d_j < c_i \text{ and } b_j < a_i\}$. If $c^i = 0$ then $N_i = \varphi$.

It is easy to verify that, for any node $v_i, S_i = M_i \cup N_i$.

An Algorithm for *L*(0,1)-Labeling of Trapezoid Graph

In this section, an efficient algorithm is presented to $L(0,1)$-label of a trapezoid graph and it is shown that the upper bound of $\lambda_{0,1}(G)$ of a trapezoid graph is $\Delta - 1$.

Algorithm TL01 (Paul, Pal & Pal, n.d.)

```
Input: Trapezoid diagram  T_i, i = 1, 2, ..., n  of a given trapezoid graph  G = (V, E).
Output:  f(v_i),  the  L(0,1)-label of the node  v_i ∈ V.
Step 1. For  i = 1  to  n
  find  S_i  and  S_{i'}.
Step 2. Initialization
         i = n
         k = 0  // k  represents the number of nodes which are labelled at any
stage. Initially all the nodes are taken as unlabelled.//
Step 3. Choose the node  v_i  and compute  S''_i.
Step 4. If  S''_i = φ  then
      set  i ← i - 1  and go to Step 3.
  else
        for each  v_j ∈ S''_i
Step 4.1. Compute  f(N_2(v_j));
Step 4.2.  f(v_j) = min{Z' - f(N_2(v_j))} . // minimum available label.//
end for;
end if;
Step 5. Record the current value of  k .
Step 6. If  k < n  then set  i ← i - 1  and go to Step 3. Process will continue
until
         k = n .
       end L01
```

Theorem 10: (Paul, Pal & Pal, n.d.) The algorithm TL01 take $O(mn\Delta)$ time to $L(0,1)$-label all the nodes of a trapezoid graph satisfying $\lambda_{0,1}(G) \leq \Delta - 1$.

L(2,1)-Labeling of Trapezoid Graphs

In this section, we present a greedy algorithm to $L(2,1)$-label a trapezoid graph. Based on this algorithm an upper bound of $L(2,1)$-labeling of a trapezoid graph is obtained. Here, we prove that $\lambda_{2,1}(G)$ of a trapezoid graph does not exceed $5\Delta - 4$.

Consider the node set $V = \{v_1, v_2, \ldots, v_n\}$, where $v_1 < v_2 < \ldots < v_n$ and $1, 2, \ldots, n$ are the indices of the corresponding nodes. We denote $f(v_i)$, the label of the node v_i, for all $i = 1, 2, \ldots, n$. We define two sets A_i and B_i below.

$$A_i = \{v_j : d(v_j, v_i) = 1, j < i\} \text{ and } B_i = \{v_j : d(v_j, v_i) = 2, j < i\},$$

It can be proved that a node $v_i \in V$, $|B_i| \leq 2\Delta - 4$.

Now we design a greedy algorithm to label all the nodes of a trapezoid graph.

Algorithm TGL21

```
Input: Set of ordered nodes of a trapezoid graph.
//assume that the trapezoid are ordered with respect to the upper left corner
points.
```
That is, $a_i < a_j$ for $v_i < v_j$ and hence $v_1 < v_2 < \ldots < v_n$. //
```
Output:
```
$f(v_i)$, the L(2,1)-label of v_i, $i = 1, 2, \ldots, n$.
```
for each
```
$i = 1$ to n do
```
  Let
```
j be the smallest non-negative integer such that
$$j \notin \{f(v_k) - 1, f(v_k), f(v_k) + 1 \mid v_k \in A_i\} \cup \{f(v_l) \mid v_l \in B_i\}$$
$$f(v_i) = j;$$
```
end for;
end TGL21
```

Theorem 11: The time complexity of algorithm TGL21 is $O(m + n\Delta)$, where Δ, m, n represent maximum degree, number of edges and number of nodes respectively.

Proof. Label of a node v_i, i.e. $f(v_i)$ depends only on the labels of the sets A_i and B_i. Now, A_i can be computed by computing the adjacency of the node v_i. That is, computation of all A_i's takes $O(\sum deg(v_i))$, i.e. $O(m)$ time. Again, let T_{i_1} and T_{i_2} be two trapezoids such that $a_{i_1} = a^i$ and $c_{i_2} = c^i$. Thus, all the members of the set B_i are adjacent to either v_{i_1} or v_{i_2} or both. Therefore, all the B_i's can be computed in

$$O(\sum deg(v_{i_1}) + \sum deg(v_{i_2}))$$

time, which is at most $O(2\sum\Delta)$, i.e. $O(n\Delta)$. Thus, the overall time complexity of algorithm TGL21 is $O(m + n\Delta)$.

The upper bound of $L(2,1)$-labeling of trapezoid graph is stated in the next theorem.

Theorem 12: For a trapezoid graph G, $\lambda_{2,1}(G) \leq 5\Delta - 4$.

Proof. Suppose we are in the midst of the algorithm and going to colour the ith node, i.e. v_i. Clearly, label of the node v_i depends only on the set A_i and B_i as we label the nodes of the graph by $L(2,1)$-labeling. Thus the total number of forbidden colours for the node v_i is at most $3\mid A_i\mid + \mid B_i \mid$ as each 1-nbd node of v_i forbibd at most 3-integers and each 2-nbd node forbid at most 1-integer. Again A_i is the set of 1-nbd nodes of v_i with index less than i. So, obviously $\mid A_i \mid\leq \Delta$, for all i. Again, $\mid B_i \mid\leq 2\Delta - 4$. Thus,

$$f(v_i) \leq 3\mid A_i\mid + \mid B_i \mid$$

or,

$$f(v_i) \leq 3\Delta + 2\Delta - 4$$

or,

$$f(v_i) \leq 5\Delta - 4$$

Hence, $\lambda_{2,1}(G) \leq 5\Delta - 4$.

REFERENCES

Amanathulla, & Pal. (2016a). - and -labeling problems on circular-arc graphs. *International Journal of Soft Computing*, *11*(6), 343-350.

Amanathulla, & Pal. (2016b). - and -labeling problems on circular-arc graphs. *International Journal of Control Theory and Applications*, *9*(34), 869-884.

Amanathulla & Pal. (2017c). $L(3,2,1)$- and $L(4,3,2,1)$-labeling problems on interval graphs. *AKCE International Journal of Graphs and Combinatorics*, *14*, 205-215.

Amanathulla, S., & Pal, M. (2017a). -labeling problems on interval graphs. *International Journal of Control Theory and Applications*, *10*(1), 467–479.

Amanathulla, S., & Pal, M. (2017b). -labeling problems on permutation graphs. *Transylvanian Review*, *25*(14), 3939–3953.

Amanathulla, S., & Pal, M. (2017v). Labeling problems on circular-arc graphs. *Far East Journal of Mathematical Sciences*, *102*(6), 1279–1300. doi:10.17654/MS102061279

Amanathulla, S., & Pal, M. (2018). Surjective $L(2,1)$. *Journal of Intelligent & Fuzzy Systems*, *35*(1), 739–748. doi:10.3233/JIFS-171176

Amanathulla, S., Sahoo, S., & Pal, M. (n.d.). Labeling numbers of square of paths, complete graphs and complete bipartite graphs. *Journal of Intelligent & Fuzzy Systems*. doi:10.3233/JIFS-172195

Bertossi, A. A., & Bonuccelli, M. A. (1995). Code assignment for hidden terminal interference avoidance in multi-hop packet radio networks. *IEEE/ACM Transactions on Networking*, *3*(4), 441–449. doi:10.1109/90.413218

Calamoneri, T. (2014). The $L(h,k)$-labeling problem, an updated survey and annotated bibliography. *The Computer Journal*, *54*(8), 1–54.

Calamoneri, T., Caminiti, S., Petreschi, R., & Olariu, S. (2009). On the $L(h,k)$. *Networks*, *53*(1), 27–34. doi:10.1002/net.20257

Chang, G. J., & Kuo, D. (1996). The $L(2,1)$. *SIAM Journal on Discrete Mathematics*, *9*(2), 309–316. doi:10.1137/S0895480193245339

Ghosh, S., & Pal, A. (2016). $L(3,1)$-labeling of some simple graphs. *Advanced Modeling and Optimization*, *18*(2), 243–248.

Ghosh, S., & Pal, A. (2018). Exact algorithm for -labeling of cartesian product between complete bipartite graph and path. *ACES Conference*.

Ghosh, S., Paul, S., & Pal, A. (2017). -Labeling of cartesian product of complete bipartite graph and path. *Journal of Informatics and Mathematical Sciences*, *9*(3), 685–698.

Golumbic. (2004). *Algorithmic graph theory and perfect graphs* (2nd ed.). Elsevier.

Goncalves, D. (2008). On the $L(d,1)$-labelinng of graphs. *Discrete Mathematics*, *308*, 1405–1414.

Griggs, J. R., & Yeh, R. K. (1992). Labeling graphs with a condition at distance two. *SIAM Journal on Discrete Mathematics*, *5*(4), 586–595. doi:10.1137/0405048

Havet, F., Reed, B., & Sereni, J. S. (2008). L(2, 1)-labeling of graphs. *Proceedings 19th Annual ACM-SIAM Symposium on Discrete Algorithms, SODA 2008*, 621-630.

Jin, X. T., & Yeh, R. K. (2004). Graph distance-dependent labeling related to code assignment in compute networks. *Naval Research Logistics*, *51*, 159–164.

Jonas, K. (1993). *Graph coloring analogues with a condition at distance two: $L(2,1)$-labelings and list λ-labelings* (Ph.D. Thesis). University of South Carolina, Columbia, SC.

Khan, N., Pal, M., & Pal, A. (2012a). Labeling of cactus graphs. *Mapana Journal of Science*, *11*(4), 15–42. doi:10.12723/mjs.23.2

Khan, N., Pal, M., & Pal, A. (2012b). -Abeling of cactus graphs. *Communications and Network*, *4*(01), 18–29. doi:10.4236/cn.2012.41003

Kral, D., & Skrekovski, R. (2003). A theory about channel assignment problem. *SIAM Journal on Discrete Mathematics*, *16*(3), 426–437. doi:10.1137/S0895480101399449

Olariu, S. (1991). An optimal greedy heuristic to color interval graphs. *Information Processing Letters*, *37*(1), 21–25. doi:10.1016/0020-0190(91)90245-D

Paul, Pal, & Pal. (n.d.a). $L(0,1)$-labeling of trapezoid graphs. *International Journal of Applied Computational Mathematics*. DOI doi:10.100740819-017-0372-y

Paul, S., Pal, M., & Pal, A. (2013). An efficient algorithm to solve $L(0,1)$-labeling problem on interval graphs. *Advanced Modelling and Optimization*, *15*(1), 31–43.

Paul, S., Pal, M., & Pal, A. (2014). Labeling of circular-arc graph. *Annals of Pure and Applied Mathematics*, *5*(2), 208–219.

Paul, S., Pal, M., & Pal, A. (2015). Labeling of permutation and bipartite permutation graphs. *Mathematical in Computer Science*, *9*(1), 113–123. doi:10.100711786-014-0180-2

Paul, S., Pal, M., & Pal, A. (2015). Labeling of interval graphs. *Journal of Applied Mathematics and Computing*, *49*(1), 419–432. doi:10.100712190-014-0846-6

Paul, S., Pal, M., & Pal, A. (2015). Labeling of permutation graphs. *Journal of Mathematical Modelling and Algorithms*, *14*(4), 469–479. doi:10.100710852-015-9280-5

Paul, Pal, & Pal. (n.d.b). $L(2,1)$-labeling of trapezoid graphs. *Communicated*.

Saha, A., Pal, M., & Pal, T. K. (2007). Selection of programme slots of television channels for giving advertisement: A graph theoretic approach. *Information Science*, *177*(12), 2480–2492. doi:10.1016/j.ins.2007.01.015

Yeh, R. K. (1990). *Labeling graphs with a condition at distance two* (Ph.D Thesis). University of South Carolina, Columbia, SC.

Chapter 8
Set–Valuations of Graphs and Their Applications

Germina K. Augusthy

Central University of Kerala, India

ABSTRACT

A set-valuation of a graph G=(V,E) assigns to the vertices or edges of G elements of the power set of a given nonempty set X subject to certain conditions. A set-indexer of G is an injective set-valuation $f:V(G)\rightarrow 2^X$ such that the induced set-valuation $f^\oplus:E(G)\rightarrow 2^X$ on the edges of G defined by $f^\oplus(uv)=f(u)\oplus f(v)$ $\forall uv \in E(G)$ is also injective, where $\oplus$ denotes the symmetric difference of the subsets of X. Set-valued graphs such as set-graceful graphs, topological set-graceful graphs, set-sequential graphs, set-magic graphs are discussed. Set-valuations with a metric, associated with each pair of vertices is defined as distance pattern distinguishing (DPD) set of a graph (open-distance pattern distinguishing set of a graph (ODPU)) is $\emptyset \neq M \subseteq V(G)$ and for each $u\in V(G)$, $f_{M(u)}=\{d(u,v): v \in M\}$ be the distance-pattern of u with respect to the marker set M. If f_M is injective (uniform) then the set M is a DPD (ODPU) set of G and G is a DPD (ODPU)-graph. This chapter briefly reports the existing results, new challenges, open problems, and conjectures that are abound in this topic.

INTRODUCTION

Labeling is a term used in technical sense for naming objects using symbolic format drawn from any universe of discourse such as the set of numbers, algebraic groups or the power set 2^X of a 'ground set' X. The objects requiring labeling could come from a variety of fields of human interest such as chemical elements, radio antennae, spectral bands and plant/animal species. Further, categorization of objects based on certain clustering rules might lead to derived labels from the labels of objects in each cluster; for instance labels *a* and *b* of two individual elements in a dyad {A,B} could be used to derive a labeling for the dyad in a way that could reflect a relational combination of the labels *a* and *b*. To be specific, A and B are assigned labels *a,b* from an algebraic group, whence the dyad {A,B} is assigned the label *a*b* where * is the group operation. Such assignments are generally motivated by a need to optimize on the number of symbols used to label the entire discrete structure so that the structure could be effectively

DOI: 10.4018/978-1-5225-9380-5.ch008

encoded for handling its computerized analysis. In general, graph labelings, where the *basic elements* (i.e., vertices and/or edges) of a graph are assigned elements of a given set or subsets of a nonempty 'ground set' subject to given conditions, have often been motivated by practical considerations such as the one mentioned above. They are also of theoretical interest on their own right. 'Graph labeling' as an independent notion using numbers was first introduced in the mid sixties. Most graph labeling methods trace their origin to one introduced by Rosa (1967).

Even though the study of graceful graphs and graceful labeling methods was introduced by Rosa (1965) the term *graceful graph* was used first by Golomb (1972). Rosa (1965) defined a β-*valuation* of a (p,q)-graph G as an injection f from the vertices of G to the set $\{0,1,\ldots,q-1\}$ such that, when each edge xy is assigned the label $|f(x)-f(y)|$, the resulting edge labels are all distinct. In a graceful labeling of a graph G the resulting edge labels must be distinct and take values $1,2,\ldots,q$. The study of graceful labelings of a graph is a prolific area of research in graph theory. The graceful labeling problem is to determine which graphs are graceful. Proving a graph G is or is not graceful involves either producing a graceful labeling of G or showing that G does not admit a graceful labeling. While the graceful labeling of graphs is perceived to be a primarily theoretical topic in the field of graph theory, gracefully labelled graphs often serve as models in a wide range of applications. Such applications include coding theory and communication network addressing. Bloom and Golomb (1977) give a detailed account of some of the important applications of gracefully labelled graphs. That 'all trees are graceful' is a long-standing conjecture known as the "Ringel–Kotzig Conjecture" (Acharya, Rao & Arumugam, 2008).

A seminal departure from assigning numbers to the basic elements of a given graphs G was made in Acharya (2001) by suggesting to consider set-valued functions instead, motivated by certain considerations in social psychology. Interpersonal relationships depend on personal attitudes of the individuals in any social group. When opinions are expressed by the individuals to others in the group, the types of interpersonal interactions get affirmed and/or modified. On the other hand, such affirmations and/or modifications in various types of interpersonal interaction in the group could induce change in the attitudes of the persons in the group. In fact, it was this revisory socio-psychological phenomenon that motivated a study of *total set-valuations* $h : V(G) \cup E(G) \to 2^X$, viz. assignment of subsets of a given set to the basic elements of a given graph with a variety of constraints motivated either by theoretical or by practical considerations (Acharya, 1983). In this chapter, we give a brief report of the existing results, new challenges, open problems and conjectures that are abound in this area, of set-valuations of a finite simple graph.

For standard terminology and notation in graph theory, hypergraph theory and signed graph theory not given here, the reader may refer respectively to Berge (1973), Harary (1972), and Harary and Norman (1953). In this paper, by a *graph* we shall mean a finite undirected graph without loops or multiple edges.

SET-VALUATIONS

A *set-valuation* of a graph $G=(V, E)$ is simply an assignment of elements of the *power set* 2^X of a given nonempty 'ground set' X to the basic elements of G; set-valuations have a variety of origins (Acharya, 2001). In particular, a *set-indexer* of G is defined to be an injective 'vertex set-valuation' $f : V(G) \to 2^X$ such that the induced 'edge set-valuation' $f^{\oplus} : E(G) \to 2^X$ on the edges of G defined by

$$f^{\oplus}(uv) = f(u) \oplus f(v) \forall uv \in E(G)$$

is also injective, where '$\oplus$' denotes the operation of taking the symmetric difference of the subsets of X. Further, G is said to be *set-graceful* if there exists a set-indexer $f : V(G) \to 2^X$ such that

$$f^{\oplus}(E(G)) := \{f^{\oplus}(e) : e \in E(G)\} = 2^X - \{\phi\},$$

such a set-indexer being called a *set-graceful labeling* of G. In Acharya (2001), it is proved that for every graph G there exists a *topological set-indexer* (or, a *T-set-indexer*), which is a set-indexer $f : V(G) \to 2^X$ such that the family

$$f(V(G)) = \{f(u) : u \in V(G)\}$$

is a topology on X, thereby establishing a link between graph theory and point-set topology.

In Acharya (1983), a set-indexer f of a given graph $G=(V,E)$ is called a *segregation* of X on G if

$$f(V(G)) \cap f^{\oplus}(E((G)) = \phi$$

and if, in addition,

$$f(V(G)) \cup f^{\oplus}(E((G)) = 2^X$$

then f is called a *set-sequential labeling* of G. A graph is then called *set-sequential* if it admits a set-sequential labeling with respect to some set X. Recently, it has been proved that the path P_{2^n-1} is not set-sequential for $n \in \{2,3\}$ and is set-sequential for every value of $n \geq 4$. In general, the problem of determining set-sequential trees is open (Acharya, 1983) Also, it has been shown that complete graph K_5 of order 5 is the only set-sequential Eulerian graph (Acharya, 2010).

Since, by their very definitions, set-graceful, topologically set-graceful and set-sequential graphs have exponential orders or sizes it is not hard to see that most graphs do not fall under any of these classes of graphs. *Even within the classes of graphs satisfying the order or size conditions for a graph to be in any of these classes, we surmise that similar conclusion holds.* Hence, it becomes important to have many infinite families of such graphs towards gaining deeper insight into the properties of these very special graphs. Special investigations have been initiated in this area (Acharya et al., 2008, 2009; Acharya & Germina, 2010, Acharya et al., 2012, 2008); in one of these, given an arbitrary graph G, a method has been described to generate infinite ascending chains of set-graceful graphs, topologically set-graceful graphs and set-sequential graphs with G as an *initializing graph* and such that at each stage of construction the preceding graph is an induced subgraph of the succeeding 'host' graph.

Given any set-valuation $f : V \to 2^X$ of a graph $G=(V,E)$, the hypergraphs

$$H^V = (X, f(V)), f(V) = \{f(v) : v \in V\}$$

and

$$H^E = (X, f^{\oplus}(E)), f^{\oplus}(E) = \{f^{\oplus}(e) : e \in E\}$$

are called respectively the *vertex set-valuation* (V_f-)hypergraph of G and *edge set-valuation* (E_f-) hypergraph of G; recently these hypergraphs are being studied (e.g., see Acharya Germina & Paul, 2010; Paul & Germina, 2011, 2012).

Next, for any set $F \subseteq E$ let $\displaystyle\sum_{x \in F} h(x)$ stand for any expression

$$(h(u_x) \oplus h(v_x)) \oplus (h(u_y) \oplus h(v_y)) \oplus \ldots$$

when F is expressed as $\{u_x v_x \in E : x \in F\}$. The following result is an analogue of a well known property of arbitrary networks, called *Kirchhoff's Voltage Law* (KVL); the analogy could be seen by treating 2^X as an additive 'voltage group' where the 'addition' is the binary operation of taking symmetric difference between any two subsets of X.

Proposition 1. *Let G=(V,E) be any graph (Acharya, 1983). Then, for any total set-valuation* $h : V \cup E \to 2^X$ *and for any cycle C,* $\displaystyle\sum_{x \in E(C)} h(x) = \phi.$

Corollary 2. *(Acharya, 2010) If G=(V,E) is any finite Eulerian graph, then for any set-valuation* $f : V \to 2^X$, $\displaystyle\sum_{x \in E} f^{\oplus}(x) = \phi.$

Proposition 3. *(Acharya, 1983) For any graph G=(V,E) any total set-valuation* $h : V \cup E \to 2^X$ *and any path P of length at least three, joining vertices u and v,* $\displaystyle\sum_{x \in E(P)} h(x) = h(u) \oplus h(v).$

Corollary 4. *(Acharya, 2010) Let G=(V,E) be any graph,* $f : V \to 2^X$ *be any injective vertex set-valuation of G and u, v be any two arbitrarily given distinct vertices of G. Then, for no u−v path P of length at least three, one has*

$$\sum_{x \in E(P)} f^{\oplus}(x) = \phi.$$

Example 5. *Let G=(V,E) be isomorphic to a path*

$$P = (u = u_0, u_1, u_2, \ldots, u_k = v)$$

where k≥1. Suppose G admits an injective set-valuation $f : V \to 2^X$ *such that* $\displaystyle\sum_{e \in E(P)} f^{\oplus}(e) = \phi$. *Then, by Corollary 4, we get k≤2, whence we have* $G \in \{P_1, P_2, P_3\}$ *where* P_n denotes the path with n vertices (and hence of length n−1).

Example 6. *From the way we are able to give an injective set-valuation of* $P_n, n = 3$, a set-valuation f of the star $K_{1,n}$ is suggested, which is such that $f^{\oplus}$ is also injective; however, note that $h = f \cup f^{\oplus}$ is not injective.

Observe in the set-valuations of graphs explained above that the induced set-valuation are also injective on their own right which need not be so in general. This motivated the following definition.

Definition 7. (Acharya, 1983) *Let G=(V,E) be a graph, X be a nonempty set and* 2^X *denote the set of all subsets of X. A set-indexer of G is an injective set-valued function* $f : V(G) \rightarrow 2^X$ *such that the function* $f^{\oplus} : E(G) \rightarrow 2^X - \{\phi\}$, defined by $f \oplus (uv) = f(u) \oplus f(v)$ for every uv∈E(G), is also injective.

Definition 7 can be thought of as applicable to infinite graphs as well.

Theorem 8. (Acharya, 1983) *Every graph has a set-indexer.*

The infimum of the cardinalities of the sets Y with respect to which G has a set-indexer is hence defined to be the *set-indexing number* of G, denoted $\sigma(G)$. Hence, the problem is actually interesting to find $\sigma(G)$ for any given graph G, especially when G is finite. For example, $\sigma(K_4) = 3$.

Theorem 9. (Acharya, 1983) *Let G be any graph, X be a non-empty set and* $f : V(G) \rightarrow 2^X$ *be any assignment to the vertices of G. Then, the mapping* $\bar{f} : V(G) \rightarrow 2^X$ *defined by* $\bar{f}(u) \equiv f(u)(= X - f(u))$ $\forall u \in V(G)$ *is a set assignment of the vertices of G.*

Theorem 10. (Acharya, 1983) *For any (p,q)-graph G,*

$$\left\lceil \log_2(q + 1) \right\rceil \leq \sigma(G) \leq p - 1,$$

where $\lceil r \rceil$ denote the least integer not less than the real number r.

Remark 11. (Acharya, 1983) $| E(G) | \leq 2^{|X|} - 1$ *and hence*

$$V(G) - 1 \geq \sigma(G) \geq \left\lceil \log_2(E(G) + 1) \right\rceil.$$

Clearly, $\sigma(K_n) = n - 1$ for 1≤n≤5.

Theorem 12. (Acharya & Hegde, 1985) *If G is a (p,q)-graph with p≥6 then σ(G)≤p−2 and this bound is attained by* K_6 *and* K_7.

Corollary 13. (Acharya & Hegde, 1985) *If G is a graph with σ(G)=|V(G)|−1 then |V(G)|≤5.*

Lemma 14. (Acharya & Hegde, 1985) *If f is a set-assignment to the vertices of* K_n *for n≥3 such that* $f^{\oplus}$ *is injective then, f is also injective.*

Another motivation from social psychology to study assignment problems on graphs comes from voltage graphs (or gain networks) (Acharya, 1982). A *voltage graph* (*gain graph*) is an ordered triple $(G,M,\underline{s})$ where G is an undirected graph, M is an arbitrary algebraic group, called the voltage group, and $\underline{s}$ is a function assigning elements of M to the edges of G such that $\underline{s}(u,v)\underline{s}(v,u)=e$, the identity element of M, and each edge uv of G is regarded as a symmetric pair of arcs (u,v) and (v,u); $\underline{s}$ is called a voltage assignment of G. If in particular, when M is taken to be the group M_n of all n-dimensional vectors of +1's and −1's under ordinary componentwise multiplication; then the voltage assignment s_n is called an n-signing of G. A particular case is the signed graph introduced by Harary, Norman, and Cartwright (1965). We can interpret an n-signing s_n of a graph G=(V,E) as an assignment of subsets of a set

$Z = \{z_1, z_2, \ldots, z_n\}$ to the elements as follows: There is a natural one-to-one correspondence $\psi : 2^Z \to M_n$ obtained by setting $\psi(A) = (a_1, a_2, \ldots, a_n)$ for $A \in 2^Z$ such that $a_i = +1$, if $z_i \in A$ and $z_i = -1$ if $z_i / \in A$, for each $i \in \{1,2,\ldots,n\}$. The interested reader may refer to [4] for more results.

SET-GRACEFUL GRAPHS

Recall that a graph $G=(V,E)$ is said to be *set-graceful* (Acharya, 1983) if there exist a set X and a set-indexer $f : V(G) \to 2^X$ such that $f^{\oplus}(E(G)) = 2^X - \{\phi\}$; and Acharya (1983) called the minimum size of the set X with respect to which G is set-graceful the set-graceful index $\gamma(G)$.

Following are some fundamental results on set-graceful graphs.

Theorem 15. (Acharya, 1983) *If G is a set-graceful (p,q)-graph with $\gamma(G)=m$ then, $q = 2^m - 1$ and p≤q+1.*

Theorem 16. (Acharya, 1983) *Every set-graceful graph G with q edges and r vertices can be embedded in a set-graceful graph, with q edges and q+1 vertices.*

Theorem 17. (Acharya, 1983) *Every connected set-graceful graph with q edges and q+1 vertices is a tree of order $p = 2^m$ and for every natural number m such a tree exists.*

Theorem 18. (Vijayakumar, 2007) *For any integer m≥2, the path P_{2^m} with 2^m vertices is not set-graceful.*

Theorem 19. (Acharya, 1983) *A necessary condition for a graph G=(V,E) to have a set-graceful labeling with respect to a set X of cardinality n is that it be possible to partition V(G) into two subsets V_e and V_o such that the number of edges joining the vertices of V_o with those of V_e is exactly 2^{n-1}.*

If a (p,q)-graph is set-graceful then $q = 2^m - 1$ for some positive integer m. This implies *almost all graphs* of order p, and hence almost all graphs are not set-graceful. Further, for every positive integer m, there exists a set-graceful graph of size $q = 2^m - 1$. However, not all (p,q)-graphs with $q = 2^m - 1$ are set-graceful as, for instance, it is not difficult to verify that the complete graph K_5 of order 5 is not set-graceful. More generally the following more results are well known.

Theorem 20. (Acharya, 1983) *If K_n is set-graceful and $\gamma(K_n) = m$ then*

$$n = \frac{1}{2}(1 + \sqrt{2^{m+3} - 7}).$$

Theorem 21. (Mollard, Payan & Shixin, 1987) *The complete graph K_n is set-graceful if and only if* n∈{2,3,6}.

Theorem 22. (Mollard, Payan & Shixin, 1987) *A necessary condition for a complete graph K_n to be set-graceful with respect to a set X is that (n−2) is a perfect square.*

The condition is not sufficient as K_{11} is not set graceful.

Theorem 23. (Mollard, Payan & Shixin, 1987) *The cycle C_n is set-graceful if and only if $n = 2^m - 1$ for some integer m≥2.*

Acharya (1983) considered special eulerian circuits which yield a set-graceful labeling of the cycle of length $2^{n+1} - 1$ and called such eulerian circuits of D_n^* successful ones and others as unsuccessful. Acharya (1983) raised the following problem.

Problem 24. (Acharya, 1983) *Determine (at least one) successful eulerian circuits in D_n^* if they exist.*

Lemma 25. (Acharya, 1983) *There is a unique cycle of length 2 in D_n.*

Lemma 26. (Acharya, 1983) *In any successful eulerian circuit in D_n^*, (x,A),(A,Ā), and ⌈(x,Ā),(Ā,A) cannot occur in that order for any x≠A.*

Lemma 27. (Acharya, 1983) *There exists an eulerian circuit α in D_n^* such that for every x≠A both the arc pairs (x,A), (A,Ā) and ⌈(x,Ā), (Ā,A) do not occur simultaneously.*

The following Conjecture of Acharya (1983) is yet to be settled.

Conjecture 28. (Acharya, 1983) *In D_n^*, all Eulerian circuits are successful if and only if n=2.*

Theorem 29. (Acharya, 1983) *Let G be any graph and u be any vertex of G. Then for any set-assignment $f : V(G) \rightarrow 2^X$ to the vertices of G there exists a set-assignment $h : V(G) \rightarrow 2^X$ to the vertices of G such that h(u)=∅ and $f^{\oplus} = g^{\oplus}$.*

Corollary 30. (Acharya, 1983) *Let G be any graph and $O_X(G)$ denote the set of all optimal set-indexers f of G with respect to a set X such that f(u)=∅ for some u∈V(G). Then, $O_X(G)$ is non-empty.*

Theorem 31. (Acharya, 1983) *If $f : V(G) \rightarrow 2^X$ is an optimal set-indexer of a graph G then $\bigcup_{u \in V(G)} f(u) = \phi$.*

Theorem 32. (Acharya, 1983; Acharya et al., 2008) *Every connected set-graceful graph with q edges and q+1 vertices is a tree of order $p = 2^m$ and for every natural number m such a tree exists.*

In fact the star $K_{1,2^n-1}$ was the graph used in the proof of Theorem 32. One takes set X with |X|=n, assigns ∅ to the center of the star and all the nonempty remaining subsets of X are then assigned to the remaining vertices of the star in a one-to-one manner. In fact, one may not limit n to be finite in the labeling whence described work show that the star whose center has order higher infinite degree than the order of the set X with respect to which one obtains the graceful set-valuation.

Theorem 33. (Acharya et al., 2008) *If a tree is set-graceful with respect to a set X of cardinality m, then its order is 2^m.*

It is important to note here that not every tree of order 2^m need be set-graceful as, for instance, it is not difficult to verify that the path P_4 of length 3 is not set-graceful.

Theorem 34. *(Acharya et al., 2008) For any integer n≥2 the path P_{2^n} is not set-graceful.*

Following conjecture appeared in (Acharya et al., 2008).

Conjecture 35. *For every natural number n, there exists a set-graceful tree of diameter n−1*

The conjecture is proved for *n*=1,2,3 and 4 in Acharya et al. (2008)

Theorem 36. (Mollard, Payan & Shixin, 1987) *For any natural number n, C_{2^n-1} is set-graceful.*

If δ_i denotes the diameter of the cycle C_{2^i-1} then $d_i = d_{i-1} + 2^{i-2}$ whence,

$$diam(C_{2^n-1}) = \left\lfloor \frac{2^n-1}{2} \right\rfloor$$

for any integer $n \geq 2$. Thus we have, for any integer m of the form $\left\lceil \dfrac{2^n - 1}{2} \right\rceil$ for some natural number n, there exists a set-graceful graph of diameter m.

It is interesting to note that every set-graceful (p,q)-graph $G=(V,E)$ with respect to a set X of cardinality n can be embedded in a set-graceful $(q+1,q)$-graph H. This may be achieved as follows. Let f be a set-indexer of G. Then $2^X - f(V)$ has $m(G) = 2^n - p$ elements each of which does not appear as a set assigned to a vertex in (G,f). Then adjoin $m(G)$ isolated points to G and assign to them the sets from $2^X - f(V)$.

Mentioned below are some of the important results on set-graceful graphs appeared in (Acharya et al., 2009; Kumar, 2009; Princy, 2007).

1. If H is a set-graceful graph with n edges $(n \geq 1)$ and $n+1$ vertices then the join of H and $\overline{K}_m$ is set-graceful if and only if $m = 2^{n_1} - 1, n_1 \in N$.

2. If S_n denote the star with $2^n - 1$ spokes and $m = 2^{n_1} - 1$ for $n_1 \in N$, then the join $S_n + \overline{K}_m$ is set-graceful.

3. $P_n + \overline{K}_m$ is set-graceful if $n \leq 2$ and $m = 2^{n_1} - 1$ where $n_1 \in N$.

4. $P_n + \overline{K}_m$ is not set-graceful for all $n \neq 2^{x_1}$, and for all $m \neq 2^{x_2} - 1$ for $x_1 > 2$, $x_1, x_2 \in N$.

5. $P_n + \overline{K}_m$ is not set-graceful for $n = 2^2$ and $m = 2^{n_1} - 1$ for $n_1 \in N$.

6. Let T be a caterpillar with the path P_m having $V(P_m)=\{v_1,v_2,...,v_m\}$. Then T is set-graceful with respect to a set of cardinality m if $d(v_i) = 2^i + 1$, $1 \leq i \leq m-1$.

7. Let T be a caterpillar with the path K_2 and let $V(K_2) = \{v_1, v_2\}$. Then T is set-graceful with respect to a set of cardinality m if $d(v_1) = 2^{m-1} + 1$ and $d(v_2) = 2^{m-1} - 1$. Let X be a set with $|X|=m$. A uniform binary tree with one pendant edge added at the root vertex having $2^m - 1$ edges is set-graceful.

8. The splitting graph $S'(G)$ of a set-graceful graph G is not necessarily set-graceful.

9. $S'(P_n)$ is not set-graceful for all n.

10. Let G be a (p, q)-graph. $S'(G)$ is not set-graceful for $q \equiv 0,2,3 \pmod 4$. Further the only possible values for q so that $S'(G)$ could be set-graceful are $21,85,341,\dots$.

11. $K_{3,5}$ is not set-graceful.

12. Let G be a set-graceful graph. Then the *corona* of G_f and K_1, that is $G_f \odot K_1$ is set-graceful if G_f is the full augmentation of G.

Acharya and Hegde (1983) proved that any graph G can be embedded in a set-graceful graph; further, he showed that every set-graceful (p,q)-graph $G=(V,E)$ with respect to a set of cardinality n can be embedded in a set-graceful $(q+1,q)$-graph H, where by an *embedding* of G one means identifying an induced subgraph in H that is isomorphic to G. Such a 'host' graph H of G together with its set-graceful labeling is considered as *fully augmented*. Fully augmented set-graceful graph of a set-graceful (p,q)-graph G can be obtained by adding $2^n - p$ isolated vertices with labels as those subsets of X that are not present as vertex labels in the set-graceful labeling given on G.

If G is a set-graceful (p,q)-graph, then G_f will denote the *full augmentation* of a set-graceful labeling f of G in the sense that G_f contains G as an induced subgraph and f taken in its extended form as the set-graceful labeling of $H = G + \overline{K}_m$ defined therein (*i.e.* f restricted to the vertices of G is the original set-graceful labeling of G). Given any set-graceful $(q+1,q)$-graph H, there exists an infinite ascending chain $H =: H_0 \subset H_1 \subset H_2 \subset \dots$ of set-graceful graphs $H_1, H_2, \dots$, such that H_i is an induced subgraph of H_{i+1} for every nonnegative integer i and H_i is a fully augmented connected graph of order

$$| V(H_i) |= q - i + 1 + \sum_{r=1}^{i} 2^{n_r}$$

for every $i \geq 1$. Acharya *et al.* [15] proved that every graph can be embedded into a connected set-graceful graph and the problems of determining the clique number, the independence number and the chromatic number of a set graceful graph are NP-complete. For more results on set-graceful graphs, the interested reader may refer to Acharya (1983, 2001, 2010), Acharya and Hegde (1985), Acharya et al. (2008, 2012, 2009, 2008), Acharya and Germina (2010) Hegde (1991), Kumar (2009), and Princy (2007).

Motivated from the following theorem, Acharya defined the concept of Topologically set-graceful graphs.

Theorem 37. (Acharya, 1983) *For every graph G, there exists a set-indexer* $f : V(G) \to 2^X$ *such that the family* f(G)={f(u):u∈V(G)} *is a topology on X.*

TOPOLOGICALLY SET-GRACEFUL GRAPHS

Acharya proved in Acharya (1983) that for every graph G, there exists a set-indexer $f : V(G) \to 2^X$ such that the family $f(V):=\{f(u):u \in V(G)\}$ is a topology on X, called a *topological set-indexer* (or, a *top-set-indexer* in short) of G with respect to X, thereby establishing another interesting link between the theory of graphs and point-set topology, the earlier known such link being a one-to-one correspondence between the set of all transitive digraphs on a given set V and the set of all topologies on V, pointed out by E. Sampathkumar & K.H. Kulkarni (1973); hence, it would be of much independent interest to investigate the inter-linkage between the notions of top-set-indexers of a graph and of transitive digraphs. Further, for a finite graph G, the *top-set-indexing number(Acharya, 1983)* of G, denoted $t(G)$, is the smallest cardinality of a set X with respect to which G admits a top-set-indexer and any such top-set-indexer of G is said to be *optimal*. It is obvious that, in general, $\sigma(G) \leq t(G) \leq \sigma_g(G)$. Further, Acharya (2001) called G a *topologically set-graceful* (or, *top-set-graceful*) *graph* if G satisfies $\sigma_g(G) = t(G)$ and any optimal top-set-indexer of such a graph a *top-set-graceful labeling* of G. In general, if G is a graph and f is a top-set-indexer of G then the graph G together with f, denoted G^f, is said to be *topologised* by f (or, f *topologises* G). Also, if f is a top-set-indexer of G then the members of $\{\emptyset\} \cup f(V)$ are said to be *f-open* and those subsets of X that are not in $f(V)$ are said to be *f-closed* in the topological sense.

Given a topology τ on a nonempty set X, let G_τ denote the class of all graphs $G=(V,E)$ that admit a set-indexer $f : V \to 2^X$ of G such that $f(V)=\tau$; G is then a *realization* of τ. Construct a star whose vertices represent the members of τ in a one-to-one manner, with the center labelled by the empty set $\emptyset$

and all the other (pendant) vertices labelled by the nonempty τ-open sets; thus, $K_{1,|\tau|-1} \in G_\tau$. We shall call τ *graceful* if a realization G of τ is set-graceful (Acharya et al., 2008, 2012; Kumar, 2009; Princy, 2007).

By definition, for a top-set-graceful graph $G=(V,E)$ together with a top-set-graceful labeling $f : V \rightarrow 2^X, \{\phi\} \cup f(V)$ forms a topology on X. Hence, it is of interest to see precisely for which class of graphs these two definitions coincide. The following result answers one of the conjectures raised by Acharya (1983, 2001).

Theorem 38. (Acharya et al., 2009) *The complete graph K_n is set-graceful if and only if* $n \in \{1,2,3,6\}$.

The following answers the same question for T-set-graceful complete graphs.

Theorem 39. (Acharya et al., 2009) *The complete graph K_n is topologically set-graceful if and only if* $n \leq 3$.

Theorem 40. (Acharya et al., 2009) *Every graph can be embedded in a connected topologically set-graceful graph.*

Theorem 41. (Acharya et al., 2009) *Let G be a graph. Then there exists an infinite sequence* $(H = G_1, G_2, ...)$ *of connected topologically set-graceful graphs G_i where H contains G as an induced subgraph and G_i contains G_{i-1} as an induced subgraph for all integers i≥2.*

Since almost all labelled graphs are not topologically set-graceful it might be fruitful to find some classes of graphs which are not topologically set-graceful. Following are some of the classes of graphs which come under this category.

Theorem 42. (Acharya et al., 2012) *The cycle C_n is topologically set-graceful if and only if n=3.*

It is not possible to have a graceful topology with 2^n open sets on a set X of cardinality $n+m$, for $n,m \in N$, where N denotes the set of natural numbers. A topology τ with 2^n open sets is a graceful topology of a graph G if and only if the size of G is $2^n - 1$.

Lemma 43. (Acharya et al., 2009; Acharya et al., 2012) *The n-dimensional cube $Q_n, n > 1$ is not T-set-graceful.*

For, a T-set-graceful (p,q)-graph G, we have $p = 2^l + 2^{k-l} - 1$ and $q = 2^k - 1$, for some k and l. Hence, the grid $P_m \times P_n$ for $m,n \in N$ and $C_m \times C_n$ for $m,n \in N$ are not T-set-graceful. Similarly, if G_1 is a T-set-graceful (p_1, q_1)-graph and G_2 is a T-set-graceful (p_2, q_2)-graph then $G_1 + G_2$ and $G_1 \cup G_2$ are not T-set-graceful and if $G_1 \times G_2$ is T-set-graceful then p_1 and p_2 are both odd.

Lemma 44. (Acharya et al., 2009; Acharya & Germina, 2010) *An r-regular graph is not T-set-graceful for $r \equiv 0,1,3 \pmod 4$.*

Hence, if G is an r-regular T-set-graceful graph then $r \equiv 2 \pmod 4$. The converse of this result is not true. For instance, if $r=2$ then, by Theorem 42, no cycle $C_n, n \geq 4$, is T-set-graceful. Next, for any integer r, $6 \leq r \equiv 2 \pmod 4$ we have K_{r+1} which is not T-set-graceful.

Problem 45. (Acharya et al., 2009; Acharya & Germina, 2010) *For $6 \leq r \equiv 2 \pmod 4$, determine the class of r-regular T-set-graceful graphs.*

Further, we have the following conjecture.

Conjecture 46. (Acharya et al., 2008; Acharya et al., 2009) *Every graph G can be embedded as an induced subgraph in an r-regular T-set-graceful graph for any integer r that satisfies $r - \Delta \equiv (k+2) \pmod 4$ for some nonnegative integer k.*

Theorem 47. (Acharya et al., 2012, Princy, 2007) *For trees, the notion of set-gracefulness and the notion of top-set-gracefulness are equivalent.*

The following most fundamental result mimics a well known result of P. Erdös quoted by Golomb (1972). Towards this end, we need to recall the result due to Harary (1972) that the number of labelled trees with p vertices is p^{p-2}.

Theorem 48. (Acharya et al., 2012; Princy, 2007) *Almost all finite labelled trees are not top-set-graceful.*

By virtue of Theorem 47 and the fact that all the trees are covered by the class of labeled trees, we have the following interesting result which is somewhat surprising in view of quite a contrasting analogue of it in the theory of *graceful graphs*.

Corollary 49. (Acharya et al., 2012) *Almost all finite trees are not top-set-graceful.*

In fact, there do exist exponential order trees that are not even set-graceful. For example, it is known that the path P_{2^n} for $n \geq 2$ is not set-graceful (Acharya & Hedge, 1985; Hegde, 1991) and hence cannot be top-set-graceful. On the other hand, a number of exponential order trees are known to be set-graceful (e.g., see Acharya, 2001; Acharya & Hegde, 1985; Acharya et al., 2012; Hegde, 2009).

Thus, it becomes quite interesting to determine the class of set-graceful trees. Further, which of them are top-set-graceful?

Conjecture 50. (Acharya et al., 2012) *There exists a 'good' characterization of (top) set-graceful trees.*

Given a topology τ on a nonempty set X, let G_τ denote the class of all graphs $G=(V,E)$ that admit a set-indexer $f : V \rightarrow 2^X$ of G such that $f(V)=\tau$; G is then a *realization* of τ. Further, the topology τ is called *graceful topology* if a realization G of τ is set-graceful.

Let τ be a graceful topology on the set X with $|X|=n$. Take $i \in X$. Then, the realization of τ has an edge of label $X \setminus \{i\}$. Correspondingly, there exist two sets A_i and B_i in τ so that $A_i \Delta B_i = X \setminus \{i\}$. Then, it is necessary that if $A_i \cap B_i \neq \phi$, then $A_i \cap B_i = \{i\} \in \tau$ and if $A_i \cap B_i = \phi$, then

$$A_i \Delta B_i = A_i \cup B_i = X \setminus \{i\} \in \tau.$$

Accordingly, let I_0 and I_1 be defined as follows:

$$I_0 = \{i \in X : A_i \cap B_i \neq \phi\},$$

$$I_1 = \{i \in X : A_i \cap B_i = \phi\}.$$

Then for every $i \in I_0$, $\{i\} \in \tau$ and for every $j \in I_1$, $X \setminus \{j\} \in \tau$.

For any graceful topology τ with respect to a set X, I_0 is nonempty, since otherwise, if I_0 is empty then for every $i \in X$,

$$i \in I_1 = \{i \in X : A_i \cap B_i = \phi\},$$

whence $X \setminus \{i\} \in \tau$. But since τ is a topology,

$$\bigcup_{i=1}^{n-1} X \setminus \{i\} = \{n\} \in \tau,$$

a contradiction. Hence, every graceful topology on an arbitrarily given nonempty set X contains at least one singleton set.

Hence, it is natural to ask whether we can determine the number of singleton sets that are necessarily to be in τ so that τ is a graceful topology. The following theorem gives a complete answer to this question.

Theorem 51. (Acharya et al., 2012) *A topology τ_m in which there are exactly m singleton sets $A_1, A_2, \ldots, A_m$ is a graceful topology if and only if it contains all supersets and all subsets of $A_1 \cup A_2 \cup \ldots \cup A_m$.*

Corollary 52. (Acharya et al., 2012) *The minimal cardinality of a graceful topology, containing k singleton sets, on a set of cardinality n is $2^k + 2^{n-k} - 1$.*

Theorem 53. (Acharya et al., 2012) *If $l \geq 1$ and $2^l + 2^{n-l} - 1 = S$, then there exist graceful topologies with cardinalities*

$$S, S + 2^1 - 1, S + 2^1, S + 2^2 - 1, S + 2^2, \ldots, S + 2^{n-l} - 1.$$

Theorem 54. (Acharya et al., 2012) *Let X be a set of cardinality n and τ_n be the discrete topology on X. Then, at most $2^{(n-1)(2^n-1)}$ labelled set-graceful realizations of τ_n can be constructed.*

Corollary 55. (Acharya et al., 2012) *There exist graceful topologies with cardinality n where $1 \leq n \leq 27$*

Theorem 56. (Acharya et al., 2012) *Let X be a set of cardinality n and τ_n be the discrete topology on X. Then, at most $2^{(n-1)(2^n-1)}$ labelled set-graceful realizations of τ_n can be constructed.*

Corollary 57. (Acharya et al., 2012) *Let τ_i be a graceful topology on X. Then $\tau_i \times \tau_i$ can be partitioned into $2^n - 1$ equivalence classes.*

Further, in view of Corollary 57, Acharya *et al.* pursued their study in finding newer classes of top-set-graceful graphs and, in that spirit, we shall find some such classes. Of course, still there could be special classes of top-set-graceful graphs whose characterizations turn out to be NP-complete.

Theorem 58. (Acharya et al., 2012) *The number of distinct graceful topologies on a set X of cardinality n is $2^n - 1$.*

As well known, two topologies τ_1 and τ_2 on X are isomorphic if there exists a bijection $f: X \to X$ such that $A \in \tau_1$ if and only if $f(A) \in \tau_2$.

Theorem 59. (Acharya et al., 2012) *Almost all labelled graphs are not top-set-graceful.*

The next result gives an infinite class of finite graphs, which are not trees, that are top-set-graceful.

Theorem 60. (Acharya et al., 2012) *$K_2 + \overline{K_t}$ is top-set-graceful with respect to X of cardinality m if and only if $t = 2^{m-1} - 1$.*

By a *topologically full augmentation* of a set-graceful labeling f of a graph G is the number of isolated vertices that need to be added to G such that by assigning distinct nonempty subsets from $2^X \setminus f(V(G))$ to the isolated vertices the resulting extension F of f to the so *augmented graph G^** is its top-set-graceful labeling. More results may be found in (*Acharya, 1983;* Acharya & Germina, 2010; Acharya et al., 2009; Acharya et al., 2012; Kumar, 2009; Princy, 2007).

TOPOGENIC GRAPHS

Let X be any nonempty set, $T(X)$ denote the set of all topologies on X and let $\tau \in T(X)$. We shall say that τ is *'graphical'* (Acharya, Germina & Joy, 2011) if there exist a graph $G=(V,E)$ and a set-labeling $f : V \to 2^X$ of G such that $f(V) \cup f^\oplus(E) =: \tau$, where

$$f^\oplus(E) := \{ f^\oplus(e) : e \in E \}.$$

Construct a graph $G=(V,E)$ with vertex set V such that $f:V \longrightarrow \tau$ is a bijection and edge set

$E=\{\{A,B\}:A,B \in \tau$ and $A \cap B = \varnothing\}$.

Then, by the very definition of a topology on X,

$$uv \in E \Leftrightarrow f(u) \cap f(v) = \phi \Leftrightarrow f^\oplus(uv) = f(u) \oplus f(v)$$
$$= (f(u) \cup f(v)) - (f(u) \cap f(v)) = f(u) \cup f(v) \in \tau$$

Since f is a bijection, it is easy to see that $f(V) \cup f^\oplus(E) = \tau$. Thus, we have

Proposition 61. (Acharya, Germina & Joy, 2011) *Every topology on a nonempty set X is graphical.* This motivates to define *topogenic set-indexer* as

Definition 62. (Acharya, Germina & Joy, 2011) *A graph G=(V,E) is topogenic with respect to a nonempty 'ground set' X if it admits a topogenic set-indexer, which is a set-indexer* $f : V \to 2^X$ *such that* $f(V) \cup f^\oplus(E)$ *is a topology on X.*

Theorem 63. (Acharya, Germina & Joy, 2011) *For every positive integer n, there exists a connected topogenic graph of order n.*

Consider a topogenic set-indexer $f : V(G) \to 2^X$ of a (p,q)-graph $G=(V,E)$ and let $\tau_f = f(V) \cup f^\oplus(E)$. The number of distinct f-open sets, viz., $|\tau_f|$ is called the *topogenic strength* [19] of f over G and if G is finite, the minimum (respectively maximum) of $|\tau_f|$ taken over all possible topogenic set-indexers f of G is denoted $\wp^0(G)$ (respectively $\wp^1(G)$). Because of the injectivity of f and $f^\oplus$, we must have

$$\wp^0(G) \leq | f(V) \cup f^\oplus(E) | \leq \wp^1(G) \leq p + q - k,$$

where k is the number of vertices of G that are adjacent to the vertex w for which $f(w) = \varnothing$ (such a vertex w exists since τ_f is a topology on X). Moreover, $p \leq \wp^0(G)$. Further, since $\phi \notin f^\oplus(E(G))$, $q \leq \wp^0(G) - 1$ or, equivalently, $q + 1 \leq \wp^0(G)$. Thus, for a topogenic (p,q)-graph G, $p \leq \wp^0(G)$ and $q + 1 \leq \wp^0(G)$. (See [19, 57, 58].)

Since a non-set-graceful graph may still be topogenic, we need to examine whether K_4, which is not set-graceful as such (Acharya & Hegde, 1985) is topogenic. We shall indeed see that K_4 is not topogenic. In Germina and Joy (2009) it is proved that K_4, K_5, K_6 are not topogenic. In fact Germina and Joy (2009) proved the following theorem.

Theorem 64. (Germina & Joy, 2009) $K_p, p \in \{1, 2, 3, 6\}$ *are the only set-graceful complete graphs that are (gracefully) topogenic.*

However, as observed above, non-set-graceful graphs could be topogenic, even gracefully topogenic. Therefore, it would be of potential interest to determine such complete graphs.

Conjecture 65. (Germina & Joy, 2009) *For every integer* $p \geq 7, K_p$ *is not topogenic.*

Finiding the total number of labelled topologies $T(n)$ one can define on a set X of cardinality n, is still an open question. Also, there is no known simple formula giving $T(n)$ for at least some specific values of n. For small values of n, $T(n)$ may be found; for example, $T(1)=1$, $T(2)=4$, and $T(3)=29$. For $n \geq 4$, the calculations are complicated.

The *topogenic index (Acharya, Germina & Joy, 2011)* of a graph G is defined as the least cardinality of a ground set X such that there is a topology τ on X which acts as a topogenic set-indexer of a graph H having the least order and containing G as an induced subgraph; this number is denoted as $\Upsilon(G)$. If G is a topogenic graph then $\Upsilon(G)$ is just the least cardinality of a ground set X such that there is a topology τ on X which acts as a topogenic set-indexer of G. Topogenic graphs are studied in Germina and Joy (2009, 2010), Joy (2011), Acharya, Germina, and Joy (2011). We list some of the results.

1. The star $K_{1,2^n-1}$ is gracefully topogenic for any positive integer n.
2. Every graph can be embedded as an induced subgraph of a gracefully topogenic graph.
3. The totally disconnected graph, which is characterized by the relations, $V \neq \varnothing$ and $E = \varnothing$, is topogenic.
4. The complete bipartite graph $K_{m,n}$ is topogenic for all positive integers m and n.
5. The complete tripartite graph $K_{1,m,n}$ is topogenic for all positive integers m and n.

BI-SET-GRACEFUL GRAPHS

Definition 66. (Princy, 2007) *A graph G is said to be bi set-graceful if both G and its line graph are set-graceful. The Line graph of G, denoted by L(G) has E(G) as its vertex set with two vertices of L(G) are adjacent whenever the corresponding edges of G are adjacent..*

Proposition 67. (Princy, 2007) *A uniform Binary tree* T_n *with one pendant edge added at the root vertex having* $2^m - 1$ *edges is bi set graceful if and only if* $n \leq 2$.

Proposition 68. (Princy, 2007) *Star* S_n *with* $2^n - 1$ *spokes is bi set-graceful if and only if* $n \leq 2$.

Proposition 69. (Princy, 2007) *The complete graph* K_n *on n vertices is bi set-graceful if and only if* $n \leq 3$.

Theorem 70. (Princy, 2007) *An r-regular connected (p,q)-graph G is bi set-graceful if and only if* $r=2$ *and* $q = 2^n - 1$ *for some* $n \in N$.

SET-SEQUENTIAL GRAPHS

A graph G is said to be set-sequential (Acharya, 1983) if there exist a nonempty set X and a bijective set-valued function

$$f : V(G) \cup E(G) \rightarrow 2^X - \{\phi\}$$

such that

$$f(uv)=f(u)\oplus f(v)$$

for every $uv \in E(G)$. We quote here some of the interesting results.

Theorem 71. (Acharya, 1983) *If $G=(V,E)$ is a connected set-sequential (p,q)-graph, then $G + K_1$ is set-graceful.*

Thus, one has the following straight forward result, giving a necessary condition for a (p,q)-graph to be set-sequential.

Theorem 72. (Acharya, 1983) *If a (p,q)-graph is set-sequential, then $p+q = 2^m - 1$, for some positive integer m.*

Corollary 73. (Acharya, 1983) *No (p,q)-graph with $p+q \cong 0 (mod\ 2)$ is set-sequential.*

Theorem 74. (Acharya et al., 2012) *If G is set-graceful then $G \cup \overline{K_t}$ is set-sequential for some positive integer t.*

For every positive integer m, there exists a set-sequential (p,q)-graph with $p+q = 2^m - 1$. For instance, take the star $G = K_{1,2^{m-1}-1}$ and assign the non-empty subsets of the set $X=\{1,2,3,\ldots,m\}$ as follows: Assign X to the central vertex and assign the first $2^{m-1} - 1$ nonempty subsets of X in their natural lexicographic order to the pendant vertices of G in a one -to-one manner. It is easy to verify that this assignment results into a set-sequential labeling of G. The converse is not true as, for instance, the path P_4 of length 3 shows. Using the fact that by the adjunction of one new vertex w with $\varnothing$ as its label to a set-sequentially labelled graph H and then making w adjacent to all the vertices of H yields a set-graceful graph as also a necessary condition for a graph to be set-graceful (Acharya, 1983), the following result that gives a necessary condition for a graph to be set-sequential.

Theorem 75. *(Hegde, 1989) If a (p,q)-graph G has a set-sequential labeling with respect to a set X of cardinality $m \geq 2$ then there exists a partition of the vertex set $V(G)$ of G into two non-empty sets A and B such that $|A| \geq |B|$ and the number of edges joining vertices of A with those of B is exactly $2^{m-1} - |B|$.*

Acharya and Hegde (1985) have given the following conjecture, which was later on disproved by Hegde (1991).

Conjecture 76. (Acharya & Hegde, 1985) *For every integer $m \geq 2$ such that $m' = 2^{m+3} - 7$ is a perfect square, the complete graph K_n of order $n = \frac{n}{2}[\sqrt{m'} - 1]$ is set-sequential*

Theorem 77. *(Hegde, 1989) The complete graph K_n is set sequential with respect to a set X of cardinality $m \geq 2$, if and only if n=2 and 5.*

Acharya and Hegde (1985) conjectured that 'No path P_m is set-sequential for any integer $m>2$', which was later on disproved by Mehta and G. R Vijayakumar (2008)

Theorem 78. (Acharya et al, 2012) *A star* $K_{1,p}$ is set-sequential if and only if $p = 2^{n-1} - 1$ for some integer $n\geq2$.

Theorem 79. (Germina, Abhishek & Princy, 2008) *Binary trees are not set-sequential.*

The following problems are open.

Problem 80. (Acharya, 1983) *Characterize set-sequential graphs, in particular set-sequential trees.*

Problem 81. (Acharya, 1983) *Given a set-graceful graph G, does there exist a connected set-sequential graph H such that G is an induced subgraph of H?*

An immediate observation from the very definition is that a necessary condition for a (p,q)-graph $G=(V,E)$ to admit a set-sequential labeling is $p + q + 1 = 2^m$ for some positive integer m (Acharya & Hegde, 1985). But the following embedding regarding set sequential graph gives the NP-completeness of determining the clique number and the chromatic number of a connected set sequential graph.

Theorem 82. (Acharya et al, 2009) *Every (p,q)-graph of order $p\geq5$ can be embedded into a connected set-sequential graph.*

Note that the induced subgraph of H, on vertices other than those of G and v_1, is a bipartite graph since the edges in this induced subgraph are only between S_4 and S_5 If the chromatic number and the clique number of G are ≥3, then the chromatic number $\chi(H)=\chi(G)+1$ and the clique number $\omega(H)=\omega(G)+1$. Therefore, the problems of determining the chromatic number and the clique number of a connected set sequential graph are NP-complete (Acharya et al, 2009).

Following result describes a method of constructing an ascending chain of set-sequential graphs for an arbitrarily given graph G containing it as an induced subgraph.

Theorem 83. (Acharya et al, 2009) *Let G be any graph. Then, there exists an infinite sequence* $H := (H = G_1, G_2, \ldots)$ *of set-sequential graphs where H contains G as an induced subgraph and* G_i *contains* G_{i-1} *as an induced subgraph for all integers* $i\geq2$.

We suggest the reader refer to Acharya (1983, 2010), Acharya et al. (2012), Acharya and Hegde (1985), Acharya et al. (2008), Acharya and Germina (2010), Acharya et al. (2008), Hegde (1989), Kumar (2009), and Princy (2007).

SET-SEQUENTIAL TOPOGENIC GRAPHS

Let $G=(V,E)$ be any graph and X be any set. A set-indexer f of G is called a *segregation* of X on G if the sets $\{f(u):u\in V(G)\}$ and $\{f^\oplus(e) : e \in E(G)\}$ are disjoint and if, in addition their union is the set $Y(X)=\tau-\varnothing$ for some topology τ on X, then f is called a *sequential topogenic labeling* of G (Germina, Joy & Thomas, 2010). A graph is called *sequential topogenic* if it admits a sequential topogenic labeling with respect to some set X. The sequential topogenic index $\gamma(G)$ of a graph G is the least cardinality of a set X with respect to which G has a sequential topogenic labeling. Further, if $f : V \cup E \to 2^X$ is a sequential topogenic labeling of G with $|X|=\gamma(G)$ we call f an optimal sequential topogenic labeling of G. Germina *et al.* [59] proved many classes of topogenic graphs such as the complete bipartite graph $K_{m,n}$ is sequential topogenic for every non-negative integers m,n.

They (Germina, Joy & Thomas, 2010) proved the any arbitrary graph G can be embedded as an induced subgraph of a set-graceful (set-sequential, topologically set-graceful, topogenic, sequentially topogenic) graph which is set-graceful (set-sequential, topologically set-graceful, topogenic, sequentially topogenic) and studied the complexity in determining the various parameters of graph like chromatic number, clique number, independence number, domination number etc. of set-graceful (set-sequential, topologically set-graceful, topogenic, sequentially topogenic, bitopological) graphs.

SET-MAGIC GRAPHS

A graph G is said to be *set-magic* if it edges can be assigned distinct subsets of a set X such that for every vertex u of G union of the subsets assigned to the edges incident at u is X, such a set assignment to the edges of G being called a s*et-magic labeling* of G *(Acharya, 1983)*.

Following are some interesting results on set-magic graphs.

Theorem 84. *(Acharya, 2001) For any finite graph G having a set-magic labeling* $f : E(G) \to 2^X$ we must have $| E(G) | \leq 2^X$ which gives

$$\left\lceil \log_2 | E(G) | \right\rceil \leq | X |.$$

Hence, if m=m(G) denotes the least cardinality of a set with respect to which G has a set-magic labeling then,

$$\left\lceil \log_2 | E(G) | \right\rceil \leq m(G).$$

Lemma 85. (Sedlacek, 1976) *Consider any integer m≥2 and let $\underline{m}$={1,2,...,m}. Order the set* $2^{\underline{m}}$ *by putting A<B for distinct* $A, B \in 2^{\underline{m}}$ *if and only if either*

|A|<|B| *or* |A|=|B| *and* min(A−B)<min(B−A).

Let $A_1, A_2, ..., A_{2^{\underline{m}}}$ *be the increasing sequence obtained in accordance with the ordering relating $<$ defined above. Then for each* $i, 1 \leq i \leq 2^{m-1}$ *one has* $A_i = \underline{m} - A_p$ *where* $p = 2^m - i + 1$. *That is,* A_i *and* A_p *are complements of each other in* $\underline{m}$.

Problem 86. (Acharya & Hegde, 1985) *Determine the graphs, finite and infinite which admit set-magic labeling f such that |f(e)=|f(e')| for any two edgese, e' in the component of G.*

Theorem 87. (Sedlacek, 1976) *For every integer m≥3,* $m(W_n) = 1 + \left\lceil \log_2 n \right\rceil$

Theorem 88. (Sedlacek, 1976) *There exists a connected infinite graph G=(V,E) with a set-magic labeling g with |g(e)|<∞ for each e∈E(G).*

Theorem 89. *(Vijayakumar, 2007) An infinite graph G has a set-magic labeling f such that |f(e)|<∞ for every e∈E(G) if and only if d(u)=|V(G)| ∀e∈V(G).*

Another interesting class of set-magic labeling (Acharya & Hegde, 1985), of a (finite or infinite) graph G are those $f : E(G) \to 2^X$ with the property that $\{f(e) : e \in E_u\}$ is a partition of X for each

$u \in V(G)$. Such a set-magic labeling may be called *partitioning set-magic labeling* of G, and a graph G which admits such a set-magic labeling may be called *partition set-magic*.

The class of partition set-magic graphs is a subclass of multicolorable graphs. A graph $G=(V,E)$ is said to be *multicolorable* if it admits a multicoloring, which is essentially a set-assignment $f : E(G) \rightarrow 2^X$ such that $\{f(e) : e \in E_u\}$ is a partition of X for each $v \in V(G)$. Thus, an injective multicoloring is same as a partitioning set-magic labeling and vice versa.

Alternatively, we may regard an *n-multicoloring* of a graph G as assignment of one or more colors from the color set $X = \{X_1, X_2, \ldots, X_n\}$ to the edges of G so that at each vertex u of G every color appears on exactly one edge incident at u. Equivalently, an *n*-multicoloring of a graph $G=(V,E)$ may be thought of as assignment λ of *n*-tuples from the involutory group M_n to the edges of G and for each $i \in n$, exactly one of the *n*-tuples as assigned to the edges in E_u has a -1 entry at the *i*th coordinate. Thus, satisfy,

$$\Pi_{x \in E_u} \lambda(x) = \overline{I} \forall u \in V(G) (\overline{I} = -1).$$

In general, an *n*-assigning $\lambda : E(G) \rightarrow M_n$ is called an *odd n-signing* if it satisfies

$$\Pi_{x \in E_u} \lambda(x) = \overline{I} \forall u \in V(G) (\overline{I} = -1),$$

(Acharya, 1983) that is an odd number of *n*-tuples assigned by λ to the edges in E_u have a -1 entry in their *i*-th coordinate for each $i \in n$ and for each $u \in V(G)$. In terms of the *n*-tuple representation ψ of set of assignments

$$\Pi_{x \in E_u} \lambda(x) = \overline{I} \forall u \in V(G) (\overline{I} = -1)$$

is equivalent to the set theoretic condition

$$\Sigma_{x \in E_u} \psi^{-1}(\lambda(x)) = X \forall u \in V(G).$$

Theorem 90. *(Berge, 1973) Let G be a simple graph. Then a regular multigraph H can be obtained from G by edge multiplication (i.e., replacement of some edges by several parallel edges) if and only if for every independent set S of vertices in G we have,*

(i)|N(S)|≥|S|

(ii)|N(S)|=|S|⇒N(N(S))=S.

Theorem 91. *(Berge, 1973) A simple graph G=(V,E) has a multicoloring if and only if some regular multigraph H=(V,F) obtained from G by edge multiplication satisfies:*

$$m_H(S, V - S) \geq \vartriangle(H)$$

for every S⊆V(G) with |S| odd where for a graph K, and for any two sets A,B of vertices of $Km_k(A, B)$ denotes the number of edges in A and the other in B.

Definition 92. (Acharya, 1983) *Let H be multigraph and* δ(H) *denote the minimum of the vertex degree* $d_H(u)$ *in H. Acharya called H a uniform degree parity (or u.d.p) multigraph if the degrees of the vertices of H are of the same parity. H is said to be odd (or even) if the number of vertices in H is odd(even). A spanning subgraph of H in which the degrees of the vertices are all odd is called an* odd degree factor of *H.*

Lemma 93. (Acharya & Hegde, 1985) *If H is multigraph having an odd degree factor then, H is an even graph.*

Acharya (1983) defined *multigraph H is odd degree factorable* if it can be written as the edge disjoint union degree factors $H_1, H_2, \ldots$ and the collection $\{H_i\}$ is then called an odd degree factorization. Acharya (1983) also defined the *multiplication index of G* denoted by κ(G) as the least integer n for which G has an n-multicoloring. If G is not multicolorable then κ(G)=∞. He proved a necessary condition for a graph G to have a generalized multicoloring (see Acharya (1983) and conjectured the following.

Conjecture 94. (Acharya & Hegde, 1985) *The conditions (i) All the vertex degree in* H_g *have the parity of n, and (ii)* H_g *has an odd degree factorization, are sufficient for a graph G to have a generalized multicoloring*

κ(G)≥∆*(G)≥∆(G), where (G) denote the least integer *t* for which a *t*-regular multigraph can be obtained from G by edge multiplication. If G is a graph having a generalized multicoloring then an odd degree factorable multigraph H may be obtained by the edge multiplication. Hence, let δ*(G) denote the least of the minimum vertex degrees δ(H) of the degree factorable multigraphs H obtainable from G by edge multiplication, and let $\kappa_g(G)$ denote the least integer n for which G has a generalized n-multicoloring ; and he called $\kappa_g(G)$ the generalized multicoloring index of G and $\kappa_g(G) = \infty$ if G has no generalized multicoloring. $\kappa_g(G) \geq \delta^*(G) \geq \delta(G)$, where $\delta^*(G)$ denote the least of the minimum vertex degrees δ(H) of the degree factorable multigraphs H obtainable from G by edge multiplication.

Definition 95. (Acharya & Hegde, 1985) *Given a set-assignment* $h : V(G) \cup E(G) \to 2^X$ *to the elements of a graph G=(V,E) its norm |h| is defined as the number*

$$| h | = \min_{x \in V \cup E} | h(x) |.$$

Remark 96. (Acharya & Hegde, 1985) *Put*

$$| h |_V = \min_{u \in V} | (u) |, | h |_E = \min_{e \in E} | h(e) |,$$

then h is a set-magic labeling of G whenever

1.   The restriction map h/E is injective, and

2. $\bigcap_{e \in E_u} h(e) = h(u) = X$ for each u∈V(G).

If $OSM_X(G)$ denotes the set of all optimal set-magic labelings with respect to the set X of a set-magic graph G then the set-magic number $\left.|G|\right._m$ defined by

$$\left.|G|\right._m = \max_{h \in OSM_X(G)} \left.|h|\right._E.$$

Also, an optimal set-magic labeling h with $\left.|h|\right._E = \left.|G|\right._m$ is said to be extremal (Acharya, 1983) . Any two $g_1, g_2 \in OSM_X(G)$ are said to be equivalent [89], written $g_1 \sim g_2$ if there is a permutation π of X, and an automorphism ψ of G such that

$$g_1(uv) = \pi g_2(\psi(u)\psi(v)) \forall uv \in E(G).$$

A set-magic graph G is said to be uniquely set-magic [10] if $g_1 \sim g_2$ for each pair $g_1, g_2 \in OSM_X(G)$.

Theorem 97. (Acharya & Hegde, 1985) *For every integer n≥2, there are only a finite number of graphs G for which m(G)=n.*

Sedlacek (1976) claimed that a graph G is set-magic if and only if G has at most one vertex of degree one, which was easily disproved by the argument that K_2 has two vertices of degree one; but it is set-magic. Vijayakumar [105] proved the following result.

Theorem 98. *(Vijayakumar, 2007) A graph G is set-magic if and only if it has at most one pendant vertex.*

Problem 99. *(Vijayakumar, 2007) For a set-magic graph G, find the best possible upper bound m(G).*

Theorem 100. *(Vijayakumar, 2007) For a infinite graph G the following are equivalent.*

1. *G has set-magic labeling f such that |f(e)|<∞.*
2. *G has set-magic labeling f such that |f(e)|<∞ for all e∈E and |f(e)|=|f(e')| whenever and e' are edges in the same connected component of G.*
3. *For k∈N, G has a k-magic labeling.*
4. *G has 2-magic labeling.*
5. *For all v∈V(G), deg v=|V|.*

Theorem 101. *(Vijayakumar, 2007) If an infinite graph G has a set-magic labeling f which satisfies the condition |f(e)|<∞ for all e∈E then degv=|V| for all v∈V(G).*

Theorem 102. *(Vijayakumar, 2007) Any infinite graph G has a set-magic labeling f which satisfies the condition deg v=|V| for all v∈V(G), has a set-magic labeling satisfying |f(e)|=2 for all e∈E(G).*

Theorem 103. *(Vijayakumar, 2007) For an infinite graph G=(V,E) the following are equivalent*

1. *G has a set-magic labeling f such that |f(e)|<∞ for all e∈E.*
2. *G has a set-magic labeling f such that |f(e)|<∞ for all e∈E and |f(e)|=|f(e')| whenever, e, e' are edges in the same connected component of G.*
3. *G has a set-magic labeling satisfying |f(e)|=η for all e∈E(G), where η is a positive integer.*

4.   *G has a set-magic labeling satisfying |f(e)|=2 for all e∈E(G).*

5.   *deg v=|V| for all v∈V(G).*

DISTANCE-PATTERNS OF VERTICES IN A GRAPH

Let $G=(V,E)$ be a given connected simple (p,q)-graph, $\varnothing \neq M \subseteq V(G)$ and $u \in V(G)$. Then, the *M-distance-pattern* of u is the set

$$f_M(u) = \{d(u,v) : v \in M\}.$$

Acharya [48, 6] while defining this new concept, following were the problems identified.

Problem 104. (Germina, 2011) *For what structural properties of the graph G, the function f_M is injective (or respectively uniform)?*

Problem 105. (Germina, 2011) *Characterize DPD-graphs having the given DPD-number.*

Problem 106. (Germina, 2011) *Which graphs G have the property that every k-subset of V(G) is a DPD-set of G. Solve this problem in particular when $k=\varrho(G)$?*

Problem 107. (Germina, 2011) *Which graphs G have exactly one $\varrho(G)$-set?*

Problem 108. (Germina, 2011) *For which values of n it is possible to extract a proper n-distance coloring of a given graph G using a distance-pattern function as a listing of colors for the vertices?*

Problem 109. (Germina, 2011) *Given any positive integer k, does there exist a graph G with $\varrho(G)=k$?*

Acharya (2006), while sharing his many incisive thoughts during our discussion in June 2008, introduced a new approach, namely, distance neighborhood pattern matrices (dnp-matrices), to study dpd-graphs, as follows.

For an arbitrarily fixed vertex u in G and for any nonnegative integer j, we let

$$N_j[u] = \{v \in V(G) : d(u,v) = j\}.$$

Clearly, $N_0[u]=\{u\}$ for all $u \in V(G)$ and $N_j[u] = V(G) - V(C_u)$ whenever j exceeds the eccentricity $\varepsilon(u)$ of u in the component C_u to which u belongs. Thus, if G is connected then, $N_j[u] = \phi$ if and only if $j > \varepsilon(u)$. If G is a connected graph then the vectors

$$\bar{u} = \left(\left|N_0[u]\right|, \left|N_1[u]\right|, \left|N_2[u]\right|, \ldots, \left|N_{\bar{a}(u)}[u]\right| \right)$$

associated with $u \in V(G)$ can be arranged as a $p \times (d_G + 1)$ nonnegative integer matrix D_G given by

$$
\begin{array}{ccccccccc}
1 & |N_1[v_1]| & |N_2[v_1]| & \ldots & |N_{\varepsilon(v_1)}[v_1]| & 0 & 0 & 0 \\
1 & |N_1[v_2]| & |N_2[v_2]| & \ldots & \ldots & |N_{\varepsilon(v_2)}[v_2]| & 0 & 0 \\
\ldots & \ldots & \ldots & \ldots & \ldots & \ldots & \ldots & \ldots \\
\ldots & \ldots & \ldots & \ldots & \ldots & \ldots & \ldots & \ldots \\
1 & |N_1[v_p]| & |N_2[v_p]| & \ldots & \ldots & \ldots & \ldots & |N_{\varepsilon(v_p)}[v_p]|
\end{array}
$$

where d_G denotes the diameter of G: we call D_G the *distance neighborhood pattern* (or, *dnp-*) *matrix* of G.

In general, for an arbitrarily given nonempty subset M of vertices in G, the *M-dnp* matrix D_G^M of G is defined by replacing $N_j[v_i]$ by $N_j^M[v_i]$ for all indices j and for all vertices $v_i \in M$ in the above dnp-matrix. It is important to note here that all the parameters, like the eccentricities $\varepsilon(v_i)$ and the diameter, are then to be with respect to the *marker set M*. Thus, D_G^M would be a nonnegative $p \times (d_G + 1)$ matrix. We will denote by D_G^{*M} the $(0,1)$-matrix obtained from D_G^M by replacing all its nonzero entries by 1.

In an attempt to solve these problems many researchers (Ananthakumar & Germina, 2011; Koshey, 2012; Koshy & Germina, 2010, 2010; Thomas, 2010; Thomas & Germina, 2010, 2010, Germina, 2011, 2010; Germina & Koshy, 2009; Germina, Joseph & Jose, 2010; Germina & Joseph, 2011; Germina & Kurian, 2011, 2012; Germina & Jose, 2011; Germina & Marykitty, 2012) studied different concepts of distance pattern of vertices in a graph.

DISTANCE PATTERN DISTINGUISHING (DPD-SET OF A GRAPH)

Let $G=(V,E)$ be a given connected simple (p,q)-graph with diameter d_G, $\varnothing \neq M \subseteq V(G)$ and for each $u \in V(G)$, let

$$
f_M(u) = \{d(u,v) : v \in M\}
$$

be the distance-pattern of u with respect to the marker set M. If f_M is injective then the set M is a distance pattern distinguishing set of G and G is a DPD-graph (Germina, 2011; Acharya, 2006). For a given connected simple (p,q)-graph, $G=(V,E)$ and an arbitrary nonempty subset $M \subseteq V(G)$ of G and for each $v \in V(G)$, define

$$
N_j^M[u] = \{v \in M : d(u,v) = j\}.
$$

The $p \times (d_G + 1)$ nonnegative integer matrix $D_G^M = (|N_j^M[v_i]|)$ is called the *M*-distance neighborhood pattern (or, M-dnp) matrix of G.

Many interesting results are established in Germina (2011, 2010), Germina, Joseph and Jose (2010), Germina and Joseph (2011), Germina and Jose (2011), and Ananthakumar and Germina (2011). Some of them are listed below.

1. For any (p,q)-graph G, $V(G)$ is a DPD-set if and only if G isomorphic to K_1.

2. Let G be any graph having a DPD-set M. Then, any vertex of G is adjacent to at most two pendent vertices. Further, if G has a vertex with exactly two pendent vertices adjacent to it then, exactly one of them belongs to M.

3. If a block G of order $p \geq 3$ has a DPD-set M then, G is not complete and $3 \leq |M| \leq p-1$.

4. There is no DPD-graph of diameter two, except P_3.

5. Let G be a (p,q)-graph.

Then for any positive integer k, $1 \leq k \leq p-1$, G is k-DPD-set uniform if and only if $G \cong K_1$ or $G \cong K_2$.

6. For any graph G, $\varrho(G)=1$ if and only if G is a path.

7. A tree T of order $p \geq 2$ has a DPD-set of cardinality $p-1 (\neq 2)$ if and only if T is isomorphic to a path or to the tree consisting of the path

$$P_6 := (v_1, v_2, v_3, v_4, v_5, v_6)$$

with one other vertex w that is adjacent to v_3 or v_4.

8. Every path

$$P_n = (v_1, v_2, v_3, \ldots, v_n), n \geq 4$$

has a DPD-set $M = \{v_i, v_{i+1}, v_{i+3}\}$ for every fixed i, $1 \leq i \leq n-3$.

9. Let T be any caterpillar with distance between any two pendent vertices greater than two. Then T has a DPD-set.

10. In any graph G, a nonempty $M \subseteq V(G)$ is a DPD-set if and only if no two rows of D_G^{*M} are identical.

11. For any DPD-graph G possessing a nontrivial DPD-set, all the nonzero entries in the first column of D_G^M are unity and their number is less than the number of rows.

12. Let G be a graph with DPD-set M and the M-DNP matrix D_G^M is such that the rows of D_G^{*M} are the elements of a basis of the Euclidean space R^n. Then $G \cong P_n$, a path on n vertices.

OPEN DISTANCE PATTERN UNIFORM (ODPU)-SETS OF GRAPH

We can also associate with each vertex u of a graph $G=(V,E)$ its *open A-distance pattern* (or, 'ODP' in short)

$$f_A^0(u) = \{d(u,v) : v \in A, u \neq v\}$$

and intend to study graphs in which every vertex has the same open distance pattern; we call such graphs *ODP-uniform graphs* (or, simply, 'ODPU-graphs'), where the *set-valued function* (or, *set-valuation*) f_A^o is called the *open distance pattern uniform* (or, a *ODPU-*) *function* and A is called an *ODPU-set* of G. *ODPU-number* of a graph G, denoted $\varsigma(G)$, is the minimum cardinality of an ODPU-set in G; if G does not possess an ODPU-set then we postulate that $\varsigma(G)=0$. Following are some interesting results we could establish on ODPU graphs (see Germina, 2011).

1. A tree T has an ODPU-set M if and only if $T \cong P_2$.
2. In any graph G, if there exists an ODPU-set M, then $M \subseteq C(G)$.
3. If G has an ODPU-set M then

$$\max\{f_M^o(v)\} = \mid f_M^o(v) \mid = r(G) \forall v \in V(G).$$

4. Let $G=(V,E)$ be any graph and $M \subseteq V$. Then, M is an ODPU-set if and only if

$$\max\{f_M^o(v)\} = \mid f_M^o(v) \mid = r(G),$$

for all $v \in V(G)$.

5. There is no graph with ODPU-number three.
6. A connected graph G is an ODPU-graph if and only if the center $C(G)$ of G is an ODPU-set.
7. Every ODPU-graph G satisfies, $r(G) \leq d(G) \leq r(G)+1$.
8. A graph with radius 1 and diameter 2 is an ODPU-graph if and only if there exists an $M \subset V(G)$ with $|M| \geq 2$ such that the induced subgraph $\langle M \rangle$ is complete and any vertex in $\langle G-M \rangle$ is adjacent to all the vertices of M.
9. For any ODPU-graph G, every ODPU-set in G is a total dominating-set of G.
10. For every integer $n \geq 3$ there is a graph G with ODPU-number $n+2$. We have proved that 3 cannot be the ODPU number of any graph. Hence, for an ODPU-graph, the number three is forbidden as the ODPU-number. Thus, 1 and 3 are the only two numbers forbidden as ODPU-numbers of any graph. Any graph G (may or may not be connected) with every vertex having positive degree and no vertex has full-degree can be embedded into an ODPU-graph H with G as an induced subgraph of H of order $|V(G)|+2$ such that $V(G)$ is an ODPU-set of the graph H.
11. A bipartite ODPU-graph $G=(X \cup Y,E)$ with the bipartition $\{X,Y\}$ of its vertex set V has ODPU-number 4 if and only if the set X has at least two vertices of degree $|Y|$ and the set Y has at least two vertices of degree $|X|$.
12. Let H be a connected graph with radius $r \geq 2$. Then the new graph $K = H[G_1, G_2, \ldots, G_n]$ has the same radius and diameter as that of H.
13. Given a finite integer $n \neq 1,3$, any graph G can be embedded in an ODPU-graph H with ODPU-number n and G as an induced subgraph of H.

Some of these results and open problems may be found in Jose (2009) and Germina (2011). Following problems are open.

Problem 110. *In an ODPU-graph G, what is the maximum order of a collection of pairwise disjoint ODPU-sets?*

Problem 111. *Characterize a graph that possess an ODPU-set M such that ⟨M⟩ is a block.*

Problem 112. *Characterize graphs in which every total dominating set is an ODPU set.*

DISTANCE-PATTERN SEGREGATED (DPS) GRAPHS

Let $G=(V,E)$ be a (p,q) graph. Given an arbitrary nonempty subset M of vertices in G, each vertex u in G is associated with the set

$$f_M{}'(u) = \{d(u,v) : v \in M\}$$

where $d(x,y)$ denotes the usual distance between the vertices x and y in G, called the M-distance pattern of u. G is called a *distance-pattern segregated* (or, in short, dps) graph if $f_M{}'$ is independent of the choice of $u \in M$ and injective set-valued function when restricted to the set $V-M$. The set M is called *distance-pattern segregating* (dps) *set* for G. The graph G itself is a *dps-graph* if G admits a dps-set. The least cardinality of dps-set in G is called *dps-number* denoted by $\sigma(G)$. We have proved many results on dps-graphs and dps-number of a graph (See Germina, 2011; Thomas & Germina, 2010; Koshy & Germina, 2010; Koshy, 2012). Some of them are listed below: (1) Complete graph K_n is a dps-graph having a dps-set of cardinality $n-1$. (2) If T is isomorphic to either star $K_{1,k}$ or $K_{1,k+1}$ or a graph obtained by subdividing at most two legs of $K_{1,k+1}$ or one leg of $K_{1,k}$, then $\sigma(T) \leq k$.

COMPLEMENTARY DISTANCE PATTERN UNIFORM (CDPU) GRAPH

Consider M be any non-empty subset of $V(G)$. For each vertex u in G if the distance pattern $f_M(u)$ is independent of the choice of $u \in V-M$, then G is called a Complementary Distance Pattern Uniform (CDPU) Graph, the set M is called the CDPU set. The least cardinality of CDPU set in G is called the *CDPU number* of G, denoted $\sigma(G)$. Listed below are some results under CDPU-graphs (See [48, 28, 32, 33, 29, 26].

1. Every self centered graph of order p has a CDPU set M with $|M| \leq p-2$.
2. If G is a self-median graph of order $n(2n-13)$, $n \geq 8$, then $\sigma(G) \leq 2n(n-7)$.
3. Let G be a graph with n vertices. If G is a self centered graph, then $1 \leq \sigma(G) \leq n-2$. If G is not a self centered graph, then $1 \leq \sigma(G) \leq n-r$, where r is the number of vertices with maximum eccentricity.
4. A graph G has $\sigma(G)=1$ if and only if G has atleast one vertex of full degree.
5. For all integers
6. $a_1 \geq a_2 \geq \ldots \geq a_n \geq 2$, $\sigma(K_{a_1,a_2,\ldots,a_n}) = n$.
7. $\sigma(C_n) = n-2$, if n is odd and $\sigma(C_n) = n/2$, if $n \geq 8$ is even.
8. $\sigma(G + \overline{K_m}) \leq m$ if G has no vertex of full degree. Following problems are open.

Problem 113. *Characterize graphs G in which every minimal CDPU-set is independent.*

Problem 114. *What is the maximum cardinality of a minimal CDPU set in G?*

Problem 115. *Determine whether every graph has an independent CDPU-set.*

Problem 116. *Characterize minimal CDPU-sets.*

Problem 117. *For any graph G find good bounds for σ(G).*

In Germina (2011) and Germina and Koshy (2009), it is also studied Independent CDPU graph (if there exists an independent CDPU set) for *G*. We classified many classes of independent CDPU graphs and calculated the ICDPU number of various classes of graphs.

DISTANCE-COMPATIBLE SET-LABELING (DCSL) GRAPHS

A *distance-compatible set-labeling (dcsl)* is an injective set-assignment $f: V(G) \to 2^X$, X a nonempty ground set, such that the corresponding induced function

$$f^{\oplus} : V(G) \times V(G) \to 2^X - \{\phi\},$$

defined by

$$f^{\oplus}(u, v) = f(u) \oplus f(v)$$

satisfies

$$| f^{\oplus}(u, v) | = k_{(u,v)} d(u, v)$$

for all distinct $u, v \in V(G)$, where $d(u,v)$ is the distance between u and v and $k_{(u,v)}$ is a constant; G is a *dcsl-graph* if it admits a dcsl. Further, G is *integrally dcsl* if all the proportionality constants $k_{(x,y)}$ are integers and such a dcsl of G is referred to as an *integral dcsl* of G. A dcsl f of a graph G is *k-uniform dcsl* if the constants of proportionality

$$k_{(x,y)}, (x, y) \in V(G) \times V(G)$$

are all equal to k; G itself is a *k-uniform dcsl graph* if it admits a *k-uniform dcsl*. The minimum cardinality of a ground set X such that G admits a 1-uniform dcsl graph is called the 1-uniform dcsl index δ_d of graph G. We have identified many classes of *k*-uniform dcsl graphs for $k \geq 1$ and also studied (k,r)-arithmetic dcsl graphs and characterized (k,r)-arithmetic complete dcsl graphs and also proved that all trees admit 1-uniform dcsl. We also established the relationship between *k*-uniform graphs and l_1 graphs and found that *k*-uniform graphs are generalization of l_1 graphs. The topic is of special interest for further investigation because of its applications in cryptography and signalling problems. We also studied the hypergraph connection of dcsl graphs. (See Koshy & Germina, 2010; Thomas & Germina, 2010; Thoams, 2010; Germina, 2010, 2011; Germina & Koshy, 2009).

Many conjectures and open problems have been identified for further investigation.

Conjecture 118. *If a graph contains an odd cycle as an induced subgraph then it is not 1-uniform dcsl.*

Problem 119. *Prove or disprove: For n≥4, no graph C_n has a 1-uniform dcsl f such that f(u)=∅ for some $u \in C_n$.*

Problem 120. *Prove or disprove: There exists a unicyclic graph G with its unique cycle having odd length such that f(u)=∅ for some u∈V(G), where f is a 1-uniform dcsl.*

Problem 121. *Prove or disprove: If a graph G has an odd cycle as an induced subgraph then G does not admit a 1-uniform dcsl.*

Problem 122. *For any even integer n≥4, consider any 1-uniform dcsl $f : V(G) \rightarrow 2^X$. It defines a hypergraph $M_f = (X, E_n)$ where*

$$E_n = \{f(v_i) : 1 \leq i \leq n\}.$$

What properties of hypergraphs can be identified in H_f with the characteristics of the cycle C_n? What is its cyclomatic number?

BI-DISTANCE PATTERN UNIFORM GRAPHS

A graph $G=(V,E)$ is *Bi-Distance Pattern Uniform* (Bi-DPU) if there exists $M{\subseteq}V(G)$ such that the *M*-distance pattern $f_M(u)=\{d(u,v): v\epsilon M\}$ is identical for all *u* in *M* and $f_M(v)$ is identical for all *v* in *V\M*. The set *M* is called a *Bi-DPU set* (Germina & Kurian, 2012, 2013). Some of the interesting results are listed below (Germina & Kurian, 2012, 2013). A non self-centered graph *G* has exactly two eccentricities and *Cen(G)* is self-centered then *G* is a Bi-DPU graph, where *Cen(G)* is the subgraph induced by the central vertices of *G*.

1. A tree *T* is a Bi-DPU graph if and only if $T \cong K_{1,n}$ or $B_{m,n}$ where $B_{m,n}$ is a bistar.
2. A graph *G* is a Bi-DPU graph with $\varsigma_B(G) = 1$ if and only if *G* has at least one vertex of full degree.

OPEN DISTANCE PATTERN COLORING OF A GRAPH

Given a connected (*p,q*)-graph $G=(V,E)$ of diameter $d(G)$, $∅{\neq}M{\subseteq}V(G)$ and a nonempty set $X=\{1,2,3,\ldots,d(G)\}$ of colors of cardinality $d(G)$, let f_M^o be an assignment of subsets of *X* to the vertices of *G*, such that $f^o{}_M(u)=\{d(u,v): u\epsilon M, u{\neq}v\}$ where $d(u,v)$ is the usual distance between *u* and *v*. Given such a function f_M^o for all vertices in *G*, an induced edge function $f_M^{\oplus}$ of an edge

$$uv \in E(G), f_M^{\oplus}(uv) = f_M^o(u) \oplus f_M^o(v).$$

We call f_M^o an *M*-open distance pattern coloring of G, if no two adjacent vertices have same f_M^o and if such an *M* exists for a graph *G*, then *G* is called an *M*-open distance pattern colorable graph. The *M*-open distance pattern edge coloring number of a graph *G* is the cardinality of $f_M^{\oplus}(G)$, taken over all $M \subseteq V(G)$, denoted by $\eta(G)$ (see Germina & Marykutty, 2012).

SET-VALUATIONS OF DIGRAPHS

Given a simple digraph *D*=(*V,A*) and a set-valuation $f : V \to 2^X$, to each arc (*u,v*) in *D* we assign the set *f*(*u*)−*f*(*v*). A set-valuation *f* of a given digraph *D*=(*V,A*) is a *set-indexer* of *D* if both *f* and its 'arc-induced function' g_f, defined by letting $g_f(u,v) = f(u) - f(v)$ for each arc (*u,v*) of *D*, are injective.

Further, *f* is *arc-bounded* (Acharya et al., 2009) if

$$\left| g_f(u,v) \right| < \left| f(v) \right|, \text{for each } (u,v) \in A.$$

Lemma 123. (Acharya et al., 2009) *The out-degree od(v) of any vertex v in a set-indexed digraph (D,f) is at most* $2^{|f(v)|}$.

Theorem 124. (Acharya et al., 2009) *Every digraph admits a set-indexer.*

The following problem is open (Acharya et al., 2009).

Problem 125. *What is the least cardinality of a ground set X with respect to which a given digraph D admits a set-indexer?*

If ω'(*D*) denotes the least cardinality of a ground set *X* with respect to which the given digraph *D* admits a set-indexer then from the proof of Theorem 124 one can infer that

$$\omega'(D) \leq |V(D) \cup A(D)|. \tag{2}$$

It would be interesting to determine digraphs *D* for which equality is attained in (2). The question whether the bound in (2) could be improved in general is open.

Note that any set-indexer *f* of a digraph *D*=(*V,A*) has the property that

$$| g_f(u,v) | = | f(u) - f(v) | \leq | f(u) |$$

for any arc (*u,v*), but $| g_f(u,v) |$ could be equal to, or larger than, |*f*(*v*)|. Thus, for any given digraph *D*=(*V,A*) and for any arc-binding set-indexer *f* we have

$$| g_f(u,v) | \leq \min\{| f(u) |, | f(v) | -1\} \text{ for every } (u,v) \in A \tag{3}$$

The following problems arise (Acharya et al., 2009).

Problem 126. *Characterize digraphs D=(V,A) that admit arc-binding set-indexers f such that*

$$| g_f(u,v) |= \min\{| f(u) |,| f(v) | -1\} \text{ for every (u,v)} \in \text{A} \qquad (4)$$

Problem 127. *What is the least cardinality of a ground set X with respect to which a given digraph D admits an arc-binding set-indexer?*

Definition 128. *A set-indexer f of a given digraph D=(V,A) is arc-binding if* $| g_f(u,v) |<| f(v) |$, *for each (u,v)*$\in$A.

Theorem 129. (Acharya et al., 2009) *Every digraph admits an arc-binding set-indexer.*

If $\omega^{ab}(D)$ denotes the least cardinality of a ground set X with respect to which the given digraph D admits an arc-binding set-indexer then from the proof of Theorem 129 one can infer that

$$w^{ab}(D) \leq 2n \text{ , where n=|V(D)|} \qquad (5)$$

It would be interesting to determine digraphs D for which equality is attained in (5). Again, the question whether the bound in (5) could be improved in general is open. In any case, we have for any digraph D,

$$w'(D) \leq w^{ab}(D) \qquad (6)$$

Problem 130. (Acharya et al., 2009) *Characterize digraphs D for which equality holds in (6).* For digraphs D of order n, the inequalities (5) and (6) can be put together as

$$w'(D) \leq w^{ab}(D) \leq 2n \qquad (7)$$

Another problem that naturally arises is the following.

Problem 131. *Find a 'good' lower bound for $\omega'(D)$.*

SET-VALUATIONS OF SIDIGRAPHS

Given a simple signed digraph $S=(V,A,\sigma)$ and a set-valued function, in general called a *set-valuation*, assigns a subset of a nonempty 'ground set' X to each element (i.e., a vertex and/or an arc) of S. In particular, a set-valuation $f : V \rightarrow 2^X$ is called a *vertex set-valuation* of S, a set-valuation $g : A \rightarrow 2^X$ is called an *arc set-valuation* of S and a set-valuation $h : V \cup A \rightarrow 2^X$ is called a *total set-valuation* of S. Further, an injective vertex set-valuation $f : V \rightarrow 2^X$ is called a *vertex set-labeling* of S if the induced arc set-valuation $g_f : A \rightarrow 2^X$ defined by letting for each $(u,v) \in A$

$$g_f(u,v) = f(u) - f(v)$$

satisfies, for each $(u; v) \in A; \sigma(u; v) = (-1)^{|f(u)-f(v)|}$

Furthermore, a vertex set-labeling f of S is called a *vertex set-indexer* if g_f is also injective. This note attempts to answer the question whether every signed digraph admits a vertex set-labeling (set-indexer).

Several open problems are posed and new directions of study of the notion and its applications are suggested in Acharya (2012). We list here some of the open problems cited in Acharya (2012).

Problem 132. (Acharya, 2012) *Characterize graphs G that satisfy equality in (3).*

Problem 133. (Acharya, 2012) *Characterize canonical signing of the vertex set-labelings (VSVC) of signed digraphs.*

Problem 134. (Acharya, 2012) *Characterize finite sequences of integers that are degree sequences of signed digraphs.*

Marking of a signed digraph $S=(V,E,\sigma)$ is simply a function $\mu:V\rightarrow\{-1,+1\}$. It is *degree-compatible* if it satisfies

$$\mu(v)=-1 \Leftrightarrow d(v)<0, v\in V \tag{10}$$

and *canonical* if

$$\mu(v) = -1 \Leftrightarrow \partial^-(v) \equiv 1(\mathrm{mod}\, 2), v \in V \tag{11}$$

Definitions (10) and (11) motivate one to regard S as *degree-canonical* if

$$d(v) < 0 \Leftrightarrow \partial^-(v) \equiv 1(\mathrm{mod}\, 2), v \in V \tag{12}$$

Problem 135. (Acharya, 2012) *Does every degree-canonical signed digraph admit a vertex set-labeling?*

Acharya in his paper (Acharya, 2012)comments that if a signed digraph does not admit a vertex set-labeling then it must be an unbalanced signed digraph. It also implies that no connected unbalanced signed graph admits a vertex set-labeling; note, however, that it does not preclude the possibility of having weakly connected signed digraphs, balanced as well unbalanced ones, that admit vertex set-labelings. Thus, he [8] raise the following problem.

Problem 136. (Acharya, 2012) *Characterize signed digraphs that admit vertex set-labelings.*

The following conjecture, if true, will pose serious obstacles towards solving this Problem.

Conjecture 137. (Acharya, 2012) *Every signed digraph can be embedded as an induced subgraph in a signed digraph that admits a vertex set-labeling.*

LINEAR HYPERGRAPH SET-INDEXER (LHSI)

For a graph $G=(V,E)$ and a non-empty set X, a *linear hypergraph set-indexer* (LHSI) is a function $f : V(G) \rightarrow 2^X$ satisfying the following conditions: (i)f is injective (ii) the ordered pair $H_f(G)=(X, f(V))$, where $f(V)=\{f(v):v\in V(G)\}$, is a linear hypergraph, (iii) the induced set-valued function $f^\oplus : E \rightarrow 2^X$, defined by

$$f^\oplus(uv) = f(u) \oplus f(v), \forall uv \in E$$

is injective, and (iv)

$$H_{f^{\oplus}}(G) = (X, f^{\oplus}(E))$$

where

$$f^{\oplus}(E) = \{f^{\oplus}(e) : e \in E\},$$

is a linear hypergraph. Recently these hypergraphs are being studied (see Acharya, Paul & Germina, 2010; *Paul, 2012;* Paul & Germina, 2011; Paul & Germina, 2012).

Theorem 138. (Acharya, Paul & Germina, 2010) *Let G=(V,E) be a (p,q)-graph and let the* $f : V \rightarrow 2^X$ be an LHSI of G. Let u be any vertex of G with its vertex degree d(u)≥2. Then, |f(u)|≤3.

Theorem 139. (Paul & Germina, 2011) *Let G=(V,E) be a graph and let* $f : V(G) \rightarrow 2^X$ *be an* LHSI of G. Let u be any vertex of G with d(u)≥4. Then, |f(u)|≤2.

Theorem 140. (Acharya, Paul & Germina, 2010) *For a simple graph G admitting an LHSI* $f : V(G) \rightarrow 2^X$, |X| can be any arbitrary positive integer greater than $I_L(G)$ if and only if G contains a pendent vertex.

Proposition 141. (Paul & Germina, 2012) *If G=(V,E) is a (p,q)-graph without pendent vertices and isolated vertices, then* $I^{UL}(G) \leq 2p$.

Theorem 142. (Paul & Germina, 2012) *For a (p,q)-graph G with δ(G)≥3,* $I^{UL}(G) \leq \dfrac{3p}{2}$.

Theorem 143. (Paul & Germina, 2011) *If a graph G admits a 3-uniform LHSI, then G contains no cycles of length ≤4.*

Theorem 144. (Paul & Germina, 2012) *If G is a (p,q)-graph with 2≤δ(G)≤Δ(G)≤3, then* $I^{UL}(G) \leq 3p - q$.

Theorem 145. (Acharya, Paul & Germina, 2010) *A graph G without isolated points admits a 3-uniform LHSI if and only if (1) Δ(G)≤3 and (2) girth g(G)≥5.*

Theorem 146. (Paul & Germina, 2011) *If G is a (p,q)-graph with 2≤δ(G)≤Δ(G)≤3 and girth(G)≥5, then* $I^{UL}(G) = 3p - q$.

Theorem 147. (Paul & Germina, 2012) *If G is a conn(p, q)-graph with 2≤δ(G)≤Δ(G)≤3 and girth g(G)≥5, there exists a 3-uniform LHSI f of G satisfying the following.*

1. $\mu(H_f(G) = \mu(G)$
2. $\mu(H_{f^{\oplus}}(G)) = \mu(L(G))+q$, where L(G) represents the line graph of G.

CONCLUSION AND SCOPE

This chapter on Set-valuations of graphs which includes a various type of set indexers, we hope that it will pave the way for any researcher for studying the topic. The conjectures and open problems identified in various sections appears would be quite interesting, for further investigation. In this perspective, the authors wish a general study on set-valuations of graphs would be a long term goal. In 2009, S. M. Hegde, in his paper Set colorings of graphs, European J. Combin. 30, (2009), 986-995, introduced the notion of set colorings of a graph *G* as an assignment of distinct subsets of a finite set *X* of *n* colors to the vertices of G such that all the colors of the edges which are obtained as the symmetric differences of the subsets assigned to their end-vertices are distinct. Additionally, if all the sets on the vertices and edges of G are the set of all nonempty subsets of X, then the coloring is said to be a strong set-coloring and G is said to be strongly set-colorable. However, the concept of set colorings of graphs and strong set-coloring defined in 2009 paper of Hegde are respectively, nothing but, the set-valuation of graphs and set indexer of a graphs defined by Acharya in 1983.

REFERENCES

Abhishek, K. (2009). *New Directions in the Theory of Set-Valuations of Graphs* (Ph.D. Thesis). Kannur University, Kannur, Kerala, India.

Abhishek & Germina. (2012). On Set valued graphs. *J. Fuzzy Set Valued Analysis*. doi: .doi:10.5899/2012/jfsva-00127

Acharya. (1982). *Two structural criteria for voltage graphs satisfying Kirchoff's voltage law*. Res. Rep. No. HCS/DST/409/7/82-83.

Acharya, Abhishek, & Germina. (2012). Hypergraphs of minimal arc bases in a digraph. *J. Combin. Inform. Syst. Sci, 37*(2-4), 329–341.

Acharya. (1983). On d-sequential graphs. *J. Math. Phys. Sci., 17*(1) 21–35.

Acharya, Germina, & Paul. (2010). Linear hypergraph set-indexres of graph. *Int. Math. Forum, 5*(68), 3359–3370.

Acharya, & Germina,, & Joy. (2011). Set-valuation of digraphs. *Global J. Pure Appl. Math., 7*(3), 237–243.

Acharya, & Germina, & Joy. (2011). Topogenic graphs. *Advanced Stud. Contemp. Math., 21*(2), 139–159.

Acharya, & Germina, & Koshy. (2012). A creative survey on complementary distance pattern uniform sets of vertices in a graph. *J. Combin. Inform. Syst. Sci, 37*(1), 291–308.

Acharya, & Germina, Abhishek, & Slater. (2012). Some new results on set-graceful and set-sequential graphs. *J. Combin. Inform. Syst. Sci, 37*(2-4), 239–249.

Acharya, & Germina, Abhishek, & Slater. (2012). Some new results on set-graceful and set-sequential graphs. *J. Combin. Inform. Syst. Sci., 37*(2-4), 39–249.

Acharya, Germina, & Ahbishek, Rao, & Zaslavsky. (2009). Point- and arc-reaching sets of vertices in a digraph. *Indian J. Math., 51*(3), 597–609.

Acharya, B. D. (1983). Set-Valuations and their Applications. MRI Lecture Notes in Applied Math., 2, 1-17.

Acharya, B. D. (2001). Set-indexers of a graph and set-graceful graphs. *Bull. Allahabad Math. Soc., 16*, 1–23.

Acharya, B. D. (2010). K_5 is the only Eulerian set-sequential graph. *Bulletin of the Calcutta Mathematical Society*, *102*(5), 465–470.

Acharya, B. D. (2012). Set-valuations of a signed digraph. *J. Combin. Inform. Syst. Sci.*, *37*(2-4), 145–167.

Acharya, B. D., & Germina, K. A. (2010). Unigeodesic Graphs and Related Recent Notions, International Journal of Algorithms. *Computing and Mathematics*, *3*(1), 89–92.

Acharya, B. D., & Germina, K. A. (2011). Distance compatible set-labeling of graphs. *Indian J. Math. Comp. Sci. Jhs.*, *1*, 49–54.

Acharya, B. D., Germina, K. A., Princy, K. L., & Rao, S. B. (2008). On set-valuations of graphs. In *Proc. II Group Discussion, Labeling of Discrete Structures and Their Applications*. Narosa Publishing House.

Acharya, B. D., Germina, K. A., Princy, K. L., & Rao, S. B. (2008). Graph labellings, embedding and NP-completeness theorems. *J. Combin. Math. Combin. Computing*, *67*, 163–180.

Acharya, B. D., Germina, K. A., Princy, K. L., & Rao, S. B. (2009). On set-valuations of graphs: Embedding and NP-completeness theorems for set-graceful, topologically set-graceful and set-sequential graphs. *J. Discrete Math. Sci. Cryptog.*, *12*(4), 481–487. doi:10.1080/09720529.2009.10698249

Acharya, B. D., Germina, K. A., Princy, K. L., & Rao, S. B. (2012). Topologically set-graceful graphs. *J. Combin. Inform. Syst. Sci.*, *37*(2-4), 309–328.

Acharya, B. D., & Hegde, S. M. (1985). Set-sequential graphs. *National Academy Science Letters*, *8*(12), 387–390.

Acharya, B. D., Rao, S. B., & Arumugam, S. (2008). Embeddings and NP-Complete Problems for Graceful Graphs. In B. D. Acharya, S. Arumugam, & A. Rosa (Eds.), *Labeling of Discrete Structures and Applications* (pp. 57–62). New Delhi: Narosa Publishing House.

Ananthakumar, R., & Germina, K. A. (2011). Distance pattern distinguishing sets in a graph. *Advanced Stud. Contemp. Math*, *21*(1), 107–114.

Andrasfai. (1991). *Flows, Matrices*. Adam Hilger.

Benoumhani, M. (2006). The Number of topologies on a finite set. *J. Integer Sequences*, *9*(2), 9–10.

Berge, C. (1973). *Graphs and Hypergraphs*. Amsterdam: North-Holland.

Berge, C. (1989). *Hypergraphs*. Amsterdam: North-Holland.

Bhargav, T. N., & Ahlborn, T. J. (1968). On topological spaces associated with digraphs. *Acta Math. Acad. Scient. Hungar*, *19*(1-2), 47–52. doi:10.1007/BF01894678

Bindhu, K. (2010). *Advanced Studies on Labeling of graphs and Hypergraphs and Related Topics* (Ph.D. Thesis). Kannur University, Kannur, Kerala, India.

Bindhu,, K., & Germina, K.A. (2010). Distance compatible set-labeling index of graphs. *Int. J. Contemp. Math. Sci.*, *5*(19), 911–919.

Bindhu, K., & Germina, K.A. (2010). (k,r)-Arithmetic dcsl labeling of graphs. *Int. Math. Forum, 5*(45), 2237–2247.

Bindhu,, K., Germina, K.A., & Joy. (2010). On Sequential topogenic graphs. *Int. J. Contemp. Math. Sci., 5*(36), 1799–1805.

Bloom, G. S., & Golomb, S. W. (1977). Applications of numbered undirected graphs. *Proceedings of the IEEE, 65*(4), 562–570. doi:10.1109/PROC.1977.10517

Buckley, F., & Harary, F. (1990). Distances in Graphs. Addison–Wesley.

Cosyn, E. (2002). Coarsening a knowledge structure. *Journal of Mathematical Psychology, 46*(2), 123–139. doi:10.1006/jmps.2001.1376

deBruijn, N. G. (1959). Generalization of Polya's fundamental theorem in enumerative combinatorial analysis. *Indagationes Mathematicae, 21*, 59–69. doi:10.1016/S1385-7258(59)50008-6

Deza & Sikirić. (2008). Geometry of Chemical Graphs: Polycycles and Two-Faced Maps. Encyc. Math. Appl., 119.

Evans, J. W., Harary, F., & Lynn, M. S. (1967). On Computer enumeration of finite topologies. *Communications of the ACM, 10*(5), 295–298. doi:10.1145/363282.363311

Fiksel, J. (1980). Dynamic evolution of societal networks. *The Journal of Mathematical Sociology, 7*(1), 27–46. doi:10.1080/0022250X.1980.9989897

Gallian. (2011). A Dynamic survey of graph labelling. *Electronic J. Combin., Dynamic Surveys.*

Germina & Koshy. (2009). Independent complementary distance pattern uniform graphs. *J. Math. Combin., 4*, 63–74.

Germina & Joy. (2009). Topogenic graphs: II. Embeddings. *Indian J. Math., 51*(3), 645–661.

Germina & Koshy. (2010). Complementary distance pattern uniform graphs. *Int. J. Contemp. Math. Sci., 5*(55), 2745–2751.

Germina, Joseph, & Jose. (2010). Distance neighbourhood pattern matrices. *European J. Pure Appl. Math., 3*(4), 748–764.

Germina & Joy. (2010). Enumeration of graphical realization. *Int. J. Algorithms Comput. Math., 3*(1), 31–46.

Germina, Joy, & Thomas. (2010). On Sequential topogenic graphs. *Int. J. Contemp. Math. Sci., 5*(36), 1799–1805.

Germina, Joy, & Thomas. (2010). On Sequential topogenic graphs. *Proc. Int. Conf. on Mathematics and Computer Sci., (5-6), 131–134.

Germina & Koshy. (2011). New perspectives on CDPU graphs. *General Math. Notes, 4*(1), 90–98.

Germina & Joseph. (2011). Some general results on distance pattern distinguishable graphs. *Int. J. Contemp. Math. Sci., 6*, 713–720.

Germina & Kurian. (2011). Bi-distance pattern uniform number. *Int. Math. Forum, 7*(27), 1303–1308.

Germina & Jose. (2012). A creative review on distance pattern distinguishing sets in a graph. *J. Combin. Inform. Syst. Sci, 37*(2-4), 267–278.

Germina & Kurian. (2012). Bi-distance pattern uniform number. *Int. Math. Forum, 7*(27), 1303–1308.

Germina, & Kurian. (2013). Bi-distance pattern uniform graphs. *Proc. Jangjeon Math. Soc., 16*(1), 87–90.

Germina, & Bindhu, K. (2010). On bitopological graphs. *Int. J. Algorithms Comput. Math., 4*(1), 71–79.

Germina, K. A. (2010). Distance pattern distinguishing sets in a graph. *Int. Math. Forum, 5*(34), 1697–1704.

Germina, K. A. (2012). Uniform Distance-compatible set-labelings of graphs. *J. Combin. Inform. Syst. Sci, 37*(2-4), 179–188.

Germina, K. A. (2013). Out set-magic digraphs. *J. Discrete Math. Sci. Cryptog., 16*(1), 45–59. doi:10.1080/09720529.2013.778458

Germina, K. A. (2011). *Set-valuations of Graphs and Applications*. Technical Report, DST Grant-In-Aid Project No. SR/S4/277/05. Department of Science and Technology (DST), Govt. of India.

Germina, K. A., Abhishek, K., & Princy, K. L. (2008). Further results on set-valued graphs. *J. Discrete Math. Sci. Cryptog, 11*(5), 559–566. doi:10.1080/09720529.2008.10698208

Germina & Jose. (2011). Distance pattern distinguishable trees. *Int. Math. Forum, 12*, 591–604.

Germina & Marykutty. (2012). Open distance pattern coloring of a graph. *J. Fuzzy Set Valued Analysis*. doi:10.5899/2012/jfsva-00144

Golomb, S. W. (1972). How to Number a Graph. In R. C. Read (Ed.), *Graph Theory and Computing* (pp. 23–37). New York: Academic Press. doi:10.1016/B978-1-4832-3187-7.50008-8

Golomb, S. W. (1974). The Largest graceful subgraph of the complete graph. *The American Mathematical Monthly, 81*(5), 499–501. doi:10.1080/00029890.1974.11993597

Golumbic, M. C. (1980). *Algorithmic Graph Theory and Perfect Graphs*. San Diego, CA: Academic Press.

Harary, F. (1967). *Graph Theory and Theoretical Physics*. London: Academic Press.

Harary, F. (1972). *Graph Theory*. Reading, MA: Addison–Wesley.

Harary, F., Norman, R. Z., & Cartwright, D. W. (1965). *Structural Models: An Introduction to the Theory of Directed Graphs*. New York: Wiley.

Harary & Norman. (1953). *Graph Theory as a Mathematical Model in Social Science*. University of Michigan Institute of Social Research.

Hegde, S. M. (1989). *Numbered Graphs and their Applications* (Ph. D thesis). University of Delhi, Delhi.

Hegde, S. M. (1991). On set-valuations of graphs. *National Academy Science Letters, 14*(4), 181–182.

Hegde, S. M. (1993). On k-sequential graphs. *National Academy Science Letters, 16*(11-12), 299–301.

Hegde, S. M. (2008). On set-labelings of graphs. In B. D. Acharya, S. Arumugam, & A. Rosa (Eds.), *Labelings of Discrete Structures and Applications* (pp. 97–107). New Delhi, India: Narosa Publishing House.

Hegde, S. M. (2009). Set colorings of graphs. *European Journal of Combinatorics*, *30*(4), 986–995. doi:10.1016/j.ejc.2008.06.005

Jose, B. K. (2009). Open distance pattern uniform graphs. *Int. J. Math. Combin.*, *3*, 103–115.

Jose, B. K. (2009). Open distance-pattern uniform graphs. *Int. J. Math. Combin.*, *3*, 103–115.

Joy, Germina, & Thomas. (2009). On Sequential topogenic graphs. *Int. J. Contemp. Math. Sci.*, *5*, 597–609.

Joy. (2011). *A Study on Topologies Arising from Graphs and Digraphs* (Ph.D. Thesis). Kannur University, Kannur, Kerala, India.

Joy & Germina. (2010). Topogenic set-indexers: Extended abstract. *Proc. National Workshop on Graph Theory Applied to Chemistry*, 112–114.

Joy & Germina. (2012). On gracefully topogenic graphs. *J. Combin. Inform. Syst. Sci.*, *37*, 279–289.

Koshy, B. (2012). *Labelings of Graphs and Hypergraphs* (Ph.D. Thesis). Kannur University, Kannur, Kerala, India.

Koshy & Germina. (2009). Independent complementary distance pattern uniform graphs. *Int. J. Math. Combin.*, (4), 63–74.

Koshy & Germina. (2010). M-Complementary distance uniform matrix of a graph. *Int. Math. Forum*, *5*(45), 2225–2235.

Koshy & Germina. (2010). Distance pattern segregated graphs. *Proc. Int. Conf. on Mathematics and Computer Science*, 5-6, 135-138.

Mehta, A. R., & Vijayakumar, G. R. (2008). A note on ternary sequences of strings of 0 and 1. *AKCE Int. J. Graphs Combin.*, *5*(2), 175–179.

Mollard, M., & Payan, C. (1989). On two conjectures about set-graceful graphs. *European Journal of Combinatorics*, *10*(2), 185–187. doi:10.1016/S0195-6698(89)80047-2

Mollard, M., Payan, C., & Shixin, S. (1987). *Graceful problems*. In Seventh Hungarian Colloquium on Finite and Infinite Combinatorics, Budapest, Hungary.

Ore, O. (1962). *Theory of Graphs, Amer. Math. Soc. Colloq. Publ* (Vol. 38). Providence, RI: Amer. Math. Soc.

Patil, G. H., & Chaudhary, M. S. (1995). A recursive determination of topologies on finite sets. *Indian Journal of Pure and Applied Mathematics*, *26*(2), 143–148.

Paul, V. (2012). *Labeling of Graphs and Set-Indexing Hypergraphs and Related Topics* (Ph.D. Thesis). Kannur University, Kannur, Kerala, India.

Paul, V., & Germina, K. A. (2011). On 3-uniform linear hypergraph set indexers of a graph. *Int. J. Contemp. Math. Sci., 6*(18), 861–868.

Paul, V., & Germina, K. A. (2012). On Structural properties of 3-uniform linear hypergraph set indexers of a graph. *Advances Theor. Appl. Math., 7*(1), 95–104.

Paul & Germina. (2012). On linear hypergraph set indexers of a graph: Cyclomatic number, cyclicity and 2-colorability. *J. Combin. Inform. Syst. Sci., 37*(2 4), 227–237.

Peay, E. R. (1982). Structural models with qualitative values. *The Journal of Mathematical Sociology, 8*(2), 161–192. doi:10.1080/0022250X.1982.9989921

Princy, K. L. (2007). *Some Studies on Set-Valuations of Graphs–Embedding and NP-Completeness* (Ph.D. Thesis). Kannur University, Kannur, Kerala, India.

Rosa, A. (1967). *On certain valuations of the vertices of a graph, in Theory of Graphs (Int. Symposium, Rome, July 1966)*. Gordon and Breach.

Sampathkumar, E. (2006). Generalized graph structures. *Bull. Kerala Math. Assoc., 3*(2), 67–123.

Sampathkumar, E., & Germina, K. A. (2010). k-Transitive digraphs and topologies. *Int. Math. Forum, 15*(63), 111–3119.

Sampathkumar, E., & Kulkarni, K. H. (1973). Transitive digraphs and topologies on a set. *J. Karnatak Univ. Sci., 18*, 266–273.

Sampathkumar, E., Neelagiri, P. S., & Venkatachalam, C. V. (1988). Odd and even colorings of a graph. *J. Karnatak Univ. Sci., 33*, 128–133.

Sedlacek, J. (1976). Some properties of magic graphs, in Graphs, Hypergraphs and Block Systems. *Proc. Conf.,* 247–253.

Sedlacek, J. (1976). On magic graphs. *Mathematica Slovaca, 26*(4), 329–335.

Sharp, H. Jr. (1968). Cardinality of finite topologies. *J. Combin. Theory, 5*(1), 82–86. doi:10.1016/S0021-9800(68)80031-6

Stephen, D. (1968). Topology on finite sets. *The American Mathematical Monthly, 75,* 739–741.

Teichert, H.-M., & Sonntag, M. (2004). Competition hypergraphs. *Discrete Applied Mathematics, 143*(1-3), 324–329. doi:10.1016/j.dam.2004.02.010

Vijayakumar, G. R. (2007). A note on set-magic graphs. In *Proc. II Group Discussion, Labelling of Discrete Structures and their Applications*. Narosa Publishing House.

Vijayakumar, G. R. (2007). Set-magic labellings of infinite graphs. In *Proc. II Group Discussion, Labelling of Discrete Structures and their Applications*. Narosa Publishing House.

Vijayakumar, G. R. (2011). A *note on set-graceful graphs.* arXiv:1101.2729

Chapter 9
Sumset Valuations of Graphs and Their Applications

Sudev Naduvath

https://orcid.org/0000-0001-9692-4053

Christ University, India

Germina K. Augusthy

Central University of Kerala, India

Johan Kok

Tshwane Metropolitan Police Department, South Africa

ABSTRACT

Graph labelling is an assignment of labels to the vertices and/or edges of a graph with respect to certain restrictions and in accordance with certain predefined rules. The sumset of two non-empty sets A and B, denoted by A+B, is defined by $A+B=\{a=b: a\in A, b\in B\}$. Let X be a non-empty subset of the set $\mathbb{Z}$ and $\mathscr{P}(X)$ be its power set. An textit{sumset labelling} of a given graph G is an injective set-valued function $f: V(G)\to\mathscr{P}_0(X)$, which induces a function $f^*: E(G)\to\mathscr{P}_0(X)$ defined by $f^*(uv)=f(u)+f(v)$, where $f(u)+f(v)$ is the sumset of the set-labels of the vertices u and v. This chapter discusses different types of sumset labeling of graphs and their structural characterizations. The properties and characterizations of certain hypergraphs and signed graphs, which are induced by the sumset-labeling of given graphs, are also done in this chapter.

INTRODUCTION

In graph theory, graph labelling is an assignment of labels to the vertices and/or edges of a graph with respect to certain restrictions and in accordance with certain predefined rules. For the past few decades, graph labelling has become a fruitful research area in graph theory. Different graph labellings have resulted from practical considerations. They are not only of theoretical interests but have many practical implications also. For many applications, the edges or vertices of a given graph are given labels that are meaningful in the context.

DOI: 10.4018/978-1-5225-9380-5.ch009

Graph Labelling

Valuation of a graph G is a one to one mapping of the vertex set of G on to the set of all integers $\mathbb{Z}$. Graph labeling has become a fertile research area in graph theory after the introduction of the notion of certain types number valuations, called β-valuations, of graphs in Rosa (1967). The most popular one among the number valuations of graphs is β-*valuation* or *graceful labelling* of a graph (see Golomb, 1972; Rosa, 1967) which is defined as an injective function

$$f : V\big(G\big) \to \big\{1, 2, 3, ..., |E|\big\}$$

such that, when each edge xy is assigned the label $\big|f\big(x\big) - f\big(y\big)\big|$, the resulting edge labels are distinct. A graph $G\big(V, E\big)$ is said to be *edge-graceful* if there exists a bijection

$$f : E \to \big\{1, 2, 3, ..., |E|\big\}$$

such that the induced mapping

$$f^{+} : V \to \big\{0, 1, \cdots, |V| - 1\big\}$$

given by

$$f^{+}\big(u\big) = f\big(uv\big)\big(\bmod |V|\big)$$

taken over all edges uv is a bijection.

Motivated by various problems in social networks, a set analogue of number valuations called set-valuation of graphs, has been introduced in Acharya (1983). In the number valuations, the elements of a graph are assigned to numbers while in the set-valuations, the elements of a graph are assigned to sets subject to certain conditions. The mathematical definition of a set-valuation of a graph concerned is as given below (see Acharya, 1983):

For a graph G and a non-empty set X, a *set-labeling* or a set-valuation of G, with respect to the set X, is an injective set-valued function

$$f : V\big(G\big) \to \mathcal{P}\big(X\big)$$

such that the function

$$f^{\oplus} : E\big(G\big) \to \mathcal{P}\big(X\big) - \big\{\varnothing\big\}$$

is defined by

$$f^{\oplus}\left(uv\right) = f\left(u\right) \oplus f\left(v\right)$$

for every $uv \in E\left(G\right)$, where $\mathcal{P}\left(X\right)$ is the set of all subsets of X and $\oplus$ is the symmetric difference of sets. The definition can be generalised by allowing any binary operation $*$ of two sets in place of the symmetric difference $\oplus$ in the above definitions.

Many significant and interesting studies on set-labelling of graphs have been done in the last two decades (see [1-10, 32, 33, 91]).

Sumsets of Two Sets of Numbers

The *sumset* of two non-empty sets A and B, denoted by $A + B$, is defined by

$$A + B = \left\{a + b : a \in A, b \in B\right\}$$

(c.f. Nathanson, 1996). Note that unlike other operations of sets, sumsets can be defined only for number sets. This area is very much interesting and significant not only because not much articles on sumsets are available in the literature, but also the studies on the structural characteristics and cardinality of sumsets in context of graph labelling are also promising. If either A or B is uncountable, then $A + B$ is also uncountable. Also, $A + \varnothing = \varnothing$. in view these, we consider the finite non-empty subsets of $\mathbb{Z}$, the set of integers for our current discussions.

If A, B and C are non-empty sets of integers such that $C = A + B$, where $A \neq \left\{0\right\}$ and $B \neq \left\{0\right\}$, then A and B are said to be the *non-trivial summands* of the set C and C is said to be the *non-trivial sumset* of A and B. It can also be noted that for a non-empty set A containing the element 0, $A + \left\{0\right\} = A$. In this case $\left\{0\right\}$ and A are said to be the *trivial summands* of A. The following result provides the bounds for the cardinality of the sumset of two non-empty sets of integers.

Theorem 1.1 (Nathanson, 1996) If A and B are two non-empty sets of integers, then

$$\left|A\right| + \left|B\right| - 1 \leq \left|A + B\right| \leq \left|A\right|\left|B\right|.$$

Two ordered pairs $\left(a, b\right)$ and $\left(c, d\right)$ in $A \times B$ may be *compatible* if $a + b = c + d$. If $\left(a, b\right)$ and $\left(c, d\right)$ are compatible, then we write $\left(a, b\right) \sim \left(c, d\right)$. Note that the compatibility relation $\sim$ is an equivalence relation (see [59, 60]). A compatibility class of an ordered pair $\left(a, b\right)$ in $A \times B$ with respect to the integer $k = a + b$ is the subset of $A \times B$ defined by

$$\left\{\left(c, d\right) \in A \times B : \left(a, b\right) \sim \left(c, d\right)\right\}$$

and is denoted by $[(a,b)]_k$ or C_k. Note that $(a,b) \in [(a,b)]_k$, and hence each compatibility class $[(a,b)]_k$ is non-empty.

If a compatibility class contains exactly one element, then it may be called a trivial compatibility class. The *saturation number* of a compatibility class, denoted by $\eta(C_k)$, is the maximum number of elements possible in a compatibility class and a compatibility class with the maximum number of elements is called a saturated compatibility class. It is proved in Sudev and Germina (2014) that the cardinality of a saturated class in $A \times B$ is $\min(|A|,|B|)$. For example, consider the sets $A = \{2,4,6,8\}$ and $B = \{1,3,5\}$. Then,

$$A + B = \{3,5,7,9,11,13\}.$$

The compatibility classes in $A \times B$ are

$$C_3 = \{(2,1)\}$$

$$C_5 = \{(2,3),(4,1)\}$$

$$C_7 = \{(2,5),(4,3),(6,1)\}$$

$$C_9 = \{(4,5),(6,3),(8,1)\}$$

$$C_{11} = \{(6,5),(8,3)\}$$

$$C_{13} = \{(8,5)\}.$$

Here, the compatibility classes C_7 and C_9 are saturated classes for $A \times B$.

The set $A \times B$ need not always contain a saturated class. In other words, there is a possibility that the cardinality of every compatibility class of $A \times B$ is fewer than $\eta(C_k)$. In such cases, a compatibility class of $A \times B$ that contains a maximum number of elements (more than all other classes of $A \times B$), but fewer than $\min(|A|,|B|)$, is called a *maximal compatibility class*. Note that all saturated compatibility classes are maximal compatibility classes, but a maximal compatibility class need not be a saturated class, which can be verified from the following example.

Consider the set $A = \{1,2,3\}$ and $B = \{2,3,5\}$. Then,

$$A + B = \{3,4,5,6,7,8\}.$$

The compatibility classes are

$$C_3 = \left\{ (1,2) \right\}$$

$$C_4 = \left\{ (1,3),(2,2) \right\}$$

$$C_5 = \left\{ (2,3),(3,2) \right\}$$

$$C_6 = \left\{ (1,5),(3,3) \right\}$$

$$C_7 = \left\{ (2,5) \right\}$$

$$C_8 = \left\{ (3,5) \right\}.$$

In the above example, C_4, C_5 and C_6 are maximal compatible classes but none of the compatibility classes is a saturated class in $A \times B$.

Note that only one representative element of each compatibility class C_k of $A \times B$ contributes an element to their sumset $A + B$ and all other $r_k - 1$ elements are avoided.

SUMSET LABELLING OF GRAPHS

Using the concepts of the sumset of two non-empty finite sets of integers, the notion of a sumset labelling of a graph was introduced as follows.

Definition 2.1 (Naduvath & Augustine, 2018) Let X be a non-empty subset of the set $\mathbb{Z}$ and $\mathcal{P}(X)$ be its power set. A *sumset labelling* or an *integer additive set-labelling* of a given graph G is an injective set-valued function

$$f : V(G) \to \mathcal{P}_0(X),$$

which induces a function

$$f^+ : E(G) \to \mathcal{P}_0(X)$$

defined by

$$f^+\left(uv\right) = f\left(u\right) + f\left(v\right),$$

where $f\left(u\right) + f\left(v\right)$ is the sumset of the set-labels of the vertices u and v.

It is proved that a sumset labelling can be defined for every graph (see Sudav & Germina, 2014). The graphs with sumset labellings defined on them are called sumset graphs. A sumset indexer or an integer additive set-indexer of a graph G is a sumset labelling

$$f : V\left(G\right) \to \mathcal{P}_0\left(X\right)$$

of G such that the induced function

$$f^+ : E\left(G\right) \to \mathcal{P}_0\left(X\right)$$

defined by

$$f^+\left(uv\right) = f\left(u\right) + f\left(v\right)$$

is also injective (see Germina & Anandavally, 2012).

Figure 1 illustrates a graph with a sumset labelling.

As mentioned earlier, unlike other binary operations on sets, the sumset has some unique properties. Hence, the graphs which admit graphs will have many interesting properties including many exciting structural characteristics. This makes the study on sumset graphs significant and promising.

Figure 1. An illustration to sumset graphs

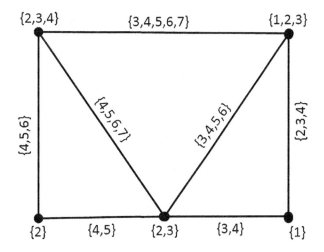

A Broad Classification of Sumset Graphs

The sumset labellings of graphs (and hence the sumset graphs) can be broadly classified into two categories.

1. **Intrinsic Sumset Graphs**: This type includes the sumset graphs which are characterised with respect to the nature of the (individual) set-labels and their elements. The Type-I sumset graphs are classified in terms of the nature of elements in the set-labels (such as arithmetic progressions, sequences etc.) and the cardinality of the set-labels.
2. **Extrinsic Sumset Graphs**: This type includes the sumset graphs which are characterised in terms of the nature of the collection of the set-labels of the graphs. That is, the graphs are classified according to whether the collection of vertex set-labels or the collection of the edge set-labels or both together follow certain properties. For example, the characterisation of sumset graphs is done in accordance with whether the collection of set-labels forms a topology of the ground set, the power set of the ground set, a filter of the ground set etc.

In the following discussion, we will investigate the properties and structural characterisations of the graphs which admit the above-mentioned types of sumset graphs.

SOME INTRINSIC SUMSET GRAPHS

Some basic definitions of sumset graphs based on the cardinality of set-labels are as given below:

1. The cardinality of the set-label of an element (vertex or edge) of a graph G is called the *set-indexing number* of that element.
2. A sumset labelling is said to be k-*uniform* if

$$\left|f^+\left(e\right)\right| = k, \forall e \in E\left(G\right).$$

In other words, a connected graph G is said to have a k-uniform sumset labelling if all its edges have the same set-indexing number k.

3. The vertex set $V\left(G\right)$ is said to be l-uniformly set-indexed, if

$$\left|f\left(v\right)\right| = l, \forall v \in V\left(G\right).$$

A graph is said to be $\left(k, l\right)$-*uniform* if

$$\left|f\left(v\right)\right| = l, \forall v \in V\left(G\right)$$

and

$$\left| f^+ \left(e \right) \right| = k, \forall e \in E \left(G \right).$$

Let X be a finite set of integers and let $f : V \left(G \right) \to \mathcal{P}_0 \left(X \right)$ be a sumset labelling of a graph G on n vertices. Then, it can be observed that X contains at least $\left\lceil \log_2 \left(n + 1 \right) \right\rceil$ elements (see [46, 60]. It can be observed that if $V \left(G \right)$ is l-uniformly set-indexed, then the cardinality of X is given by $\binom{|X|}{l} \geq n$ (cf. [46, 60]).

Three types of sumset labellings of graphs, based on the cardinalities of the set-labels assigned to the elements of graphs, have broadly been studied in the recent literature. They are strong, weak and arithmetic sumset-labellings. If f is a sumset labelling of G, then by Theorem 1.1, it can be noted that

$$\left| f \left(u \right) \right| + \left| f \left(v \right) \right| - 1 \leq \left| f^+ \left(uv \right) \right| = \left| f \left(u \right) + f \left(v \right) \right| \leq \left| f \left(u \right) \right| \left| f \left(v \right) \right|. \tag{1}$$

The notion of weak sumset labelling originated from the question of whether the set-indexing numbers of edges could be equal to those of their end vertices. By the definition of the sumsets, we know that $|A|, |B| \leq |A + B|$. What are the conditions required for the existence of equality in this condition? By Theorem 1.1, one can observe that $|A + B| = |A|$ if and only if $|B| = 1$. Hence, the cardinality of the sumset of two sets is equal to the cardinality of one of them if and only if the other set is a singleton set. Moreover, $|A + B| = |A| = |B|$ if and only if the sets A and B and the sumset $A + B$ are all singleton sets. The elements (vertices and edges) with singleton set-labels (that is, with set-indexing number 1) may be called mono-indexed elements.

The investigations on the lower bound in the above expression, lead us to the notions of different types of arithmetic sumset labellings and the investigations on the upper bound provided the concepts and results of strong sumset labelling of graphs. We shall now go through all these types of sumset labellings one by one.

Weak Sumset Labelling of Graphs

Definition 3.1 (Sudev & Germina, 2014) Let $f : V \left(G \right) \to \mathcal{P}_0 \left(X \right)$ be a sumset labelling defined on a graph G. Then, f is said to be a *weak sumset labelling* if the set-indexing number of every edge of G is equal to the set-indexing number of one or both of its end vertices.

A graph with a weak sumset labelling defined on it may be called a weak sumset graph. Figure 2 depicts a weak sumset graph.

Invoking the properties of the cardinalities of non-empty sets and their sumsets mentioned above, a necessary and sufficient condition for a graph G to have a weak sumset labelling has been established in Sudev and Germina (2014) as follows.

Lemma 3.1 (Sudev & Germina, 2014) *A sumset labelling f of a graph G is a weak sumset labelling of G if and only if at least one end vertex of every edge in G has a singleton set-label.*

Figure 2. An illustration to weak sumset graphs

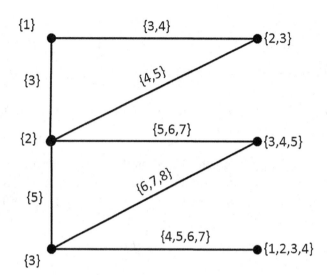

In other words, a graph G is a weak sumset graph if and only if it is bipartite or it has mono-indexed edges. The minimum number of mono-indexed edges required for a graph G to admit a weak sumset labelling is said to be the sparing number of G. The sparing number of a graph G is denoted by $\varphi(G)$. For any subgraph H of a sumset labelling-graph G, it can be noted that the induced mapping $f\mid_H$ is a weak sumset labelling for H. Therefore, for any subgraph H of a sumset labelling-graph G, $\varphi(H) \le \varphi(G)$.

A set of vertices V' of a graph G is said to have the maximal incidence in G, if the number of edges in G having at least one end vertex in V' is maximum over all other set of vertices in G. The following results discuss the sparing number of arbitrary graphs.

Theorem 3.2 (Sudev & Germina, 2014) *Let G be a given weak sumset graph and I be an independent set in G which has the maximal incidence in G. Then, the sparing number of G is the $\left| E(G-I) \right|$.*

Theorem 3.3 (Sudev & Germina, 2014) *Let $G(V,E)$ be a connected weak sumset labelling-graph and let E_I be a maximal set of edges of G whose one end vertex is in an independent set of G. Then, the sparing number of G is $\left| E \right| - \left| E_I \right|$.*

Theorem 3.4 (Sudev & Germina, 2014) If $G(V,E)$ be a non-empty graph and I^* be an independent set having maximum incidence in G. Then

$$\varphi(G) = \left| E \right| - \sum_{v_i \in I^*} d(v_i).$$

The sparing number of a r-regular graph is determined in the following result.

Corollary 3.5 (Sudev & Germina, 2014) If G is an r-regular graph, then $\varphi(G) = \left| E \right| - r\alpha$, where α is the independence number of G.

216

The spring number of different graph classes, derived graph classes, graph operations and graph products have been discussed in recent literature (ref Chithra, Sudev, Germina, 204, 2015; Sudev & Germina, 2014a, 2014b, 2014c, 2014d, 2014e, 2014g, 2014h, 2015a, 2014i, 2015b).

The following result discussed the sparing number of the union of two graphs.

Theorem 3.6 (Sudev & Germina, 2014) Let G_1 and G_2 be two subgraphs of a graph G such that $G = G_1 \cup G2$. Then,

$$\varphi(G) = \varphi(G_1) + \varphi(G_2) - \varphi(G_1 \cap G_2).$$

Moreover, if G_1 and G_2 are edge-disjoint subgraphs of G, then

$$\varphi(G_1 \cup G_2) = \varphi(G_1) + \varphi(G_2).$$

The following results discussed the sparing number of the join of two graphs.

Theorem 3.7 (Sudev & Germina, 2014) The sparing number of the join of two non-empty graphs $G_1(V_1, E_1)$ and $G_2(V_2, E_2)$ is

$$\min\left\{\varphi(G_1) + |E_2|(\beta_1 + 1), \varphi(G_2) + |E_1|(\beta_2 + 1)\right\},$$

where β_1, β_2 are the covering numbers of G_1 and G_2 respectively.

Theorem 3.8 (Sudev & Germina, 2014) Let G be the join of two graphs $G_1(V_1, E_1)$ and $G_2(V_2, E_2)$. Then,

$$\varphi(G_1 + G_2) = |E_1| + |E_2| + |V_1||V_2| - \max\left\{\left(|V_2|(|I_1| - 1) + |E_{I_1}|\right), \left(|V_1|(|I_2| - 1) + |E_{I_2}|\right)\right\},$$

where, for $j = 1, 2$ and I_j is an independent set of the graph G_j with the maximal incidence in G_j and E_{I_j} is the set of edges in $G_1 + G_2$ with their one end vertex in I_j.

The sparing number of the Cartesian product two weak sumset graphs is determined in the following result (see Naduvath & Augustine, 2018; Sudev & Germina, 2014).

Theorem 3.9 Let $G_1(V_1, E_1)$ and $G_2(V_2, E_2)$ be two non-empty graphs. Then, the sparing number of their Cartesian product $G_1 \square G_2$ is

$$\varphi(G_1 \square G_2) = |V_1|\varphi(G_2) + |V_2|\varphi(G_1).$$

The sparing number of the Corona product two weak sumset graphs is determined in the following result (see [46, 69]).

Theorem 3.10 Let $G_1\left(V_1, E_1\right)$ and $G_2\left(V_2, E_2\right)$ be two non-empty graphs. The sparing number of their corona is given by

$$\varphi\left(G_1 \odot G_2\right) = \varphi\left(G_1\right) + \left|V_1\right|\left|E_2\right|.$$

The set-labelling number of a given graph G is the minimum cardinality of the ground set X so that the function

$$f : V\left(G\right) \to \mathcal{P}_0\left(X\right)$$

is a sumset labelling of G. Then, set-labelling number of a graph G is denoted by $\varsigma\left(G\right)$. Then, the weak set-labelling number of a graph G is the minimum cardinality of the ground set X, such that f is a weak sumset labelling of G and is denoted by $\varsigma^*\left(G\right)$. The weak set-labelling number of some fundamental graphs have been determined in Sudev, Germina & Chithra (2014).

For $k > 1$, graph G admits a weakly k-uniform sumset labelling, if and only if one end vertex of every edge of G is mono-indexed and the other end vertex must be set-labelled by distinct non-singleton sets of cardinality k. That is, a graph G is a weakly uniform sumset graph if and only if it is bipartite. Moreover, note that the sparing number of bipartite graphs is 0.

Strong Sumset Labelling of Graphs

Definition 3.2 (Sudev & Germina, 2015) Let f is a sumset labelling defined on a given graph G. Then f is said to be a *strong sumset labelling* of G if

$$\left|f^+\left(uv\right)\right| = \left|f\left(u\right) + f\left(v\right)\right| = \left|f\left(u\right)\right|\left|f\left(v\right)\right|.$$

A graph which admits a strong sumset labelling or a strong sumset indexer is called a strong sumset graph or a strong sumset graph. Figure 3 gives an example of strong sumset graphs:

If f is a strong sumset labelling of a graph G, then it can be noted that the compatibility classes in the Cartesian product of the set-labels of any two adjacent vertices of G are all trivial classes.

The difference-set of a set A, denoted by D_A is the set defined by

$$D_A = \left\{\left|a - b\right| : a, b \in A\right\}.$$

In view of this notion, the following theorem established a necessary and sufficient condition for the sumset labelling f of a graph G to be a strong sumset labelling of G.

Theorem 3.11 (Naduvath & Augustine, 2018; Sudev & Augustine, 2015) *A graph G admits a strong sumset labelling if and only if the difference-sets of the set-labels of any two adjacent vertices in G are disjoint.*

Figure 3. An example of strong sumset graphs

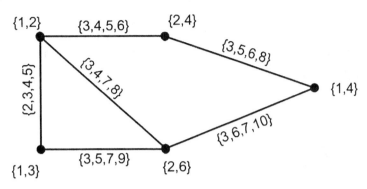

The minimum number of pairwise-disjoint difference-sets of the set-labels of vertices required in a graph G so that G admits a strong sumset labelling is called the nourishing number of G and is denoted by $\varkappa(G)$. Any two vertices in a complete graph K_n are adjacent to each other and hence for K_n to admit a strong sumset labelling, the difference-sets of the set-labels of all its vertices must be pairwise disjoint. Therefore, the nourishing number of K_n is n. Therefore, for any graph on n vertices, we have $\varkappa(G) \leq n$. Furthermore, if G contains a complete subgraph H of order $r \leq n$, then the set-labels of all vertices of H must have pairwise disjoint difference-sets. The nourishing numbers of certain fundamental graph classes and graph powers have been determined in Sudev and Germina (2015). (see Table).

If a graph G admits a strongly k-uniform sumset labelling, then it can be noted that the set-indexing number of every edge of G is the product of the set-indexing numbers of its end vertices. Hence, if G is a strongly k-uniform sumset graph, then each vertex of G must have some set-indexing number d_i, which is a divisor of k. Hence, $V(G)$ can be partitioned into at most n sets, say $(X_1, X_2, \ldots, X_l)$ such that each X_i consists of the vertices of G having the set-indexing number d_i, where l is the number of divisors of the integer k. In Sudev and Germina (2014) it is established that all bipartite graphs admit a strongly k-uniform sumset labelling, for any positive integer k.

The following theorem established a condition required for a complete graph to admit a strongly k-uniform sumset labelling.

Theorem 3.12 (Sudev & Germina, 2015) *A strongly k- uniform sumset labelling of a complete graph K_n is a (k,l)-uniform sumset labelling, where $l = \sqrt{k}$.*

In view of the fact that the number of divisors of a non-square integer is even and the number of divisors of a perfect square integer is odd, we have interesting results as stated below:

Theorem 3.13 (Sudev & Germina, 2015) *A Let k be a non-square integer. Then, a graph G admits a strongly k-uniform sumset labelling if and only if G is bipartite or a union of disjoint bipartite components.*

Theorem 3.14 (Sudev & Germina, 2015) *A Let the graph G has a strongly k-uniform sumset labelling. Then*

1. G contains at most one component that is a clique.

2. If G has a component which is a clique, then k is a perfect square.

As a result of the above theorems, one can infer that a connected non-bipartite graph G admits a strongly k-uniform sumset labelling if and only if k is a perfect square and this sumset labelling is (k, l)-uniform, where $l = \sqrt{k}$.

Arithmetic Sumset Labelling of Graphs

In Nathanson (1996), it is proved that

$$|A + B| = |A| + |B| - 1$$

if and only if the elements of both A and B are in arithmetic progressions with the same common difference. Hence, in this context, the graphs whose vertices and edges having the set-labels are arithmetic progressions arise much interest. If the context is clear, the common difference of the set-label of an element of a graph may be called the common difference of that element. The deterministic ratio of an edge of G is the ratio, $k \geq 1$ between the deterministic indices of its end vertices. It is also assumed that every set-label of the elements of G should have at least three elements.

Definition 3.3 (Sudev & Germina, 2018) An *arithmetic sumset labelling* of a graph G is a sumset labelling f of G, with respect to which the set-labels of all vertices and edges of G are arithmetic progressions. Similarly, an arithmetic sumset indexer of G is a sumset labelling of G, such that f^+ is also injective.

A graph with an arithmetic sumset labelling is called an arithmetic sumset graph. The following result provides a necessary and sufficient condition for a graph to admit an arithmetic sumset labelling.

Theorem 3.15 (Sudev & Germina, 2018) A graph G admits an arithmetic sumset labelling f if and only if for any two adjacent vertices in G, the deterministic ratio of every edge of G is a positive integer, which is less than or equal to the set-indexing number of its end vertex having a smaller common difference.

The set-indexing number of an edge of an arithmetic sumset graph G, in terms of the set-indexing numbers of its end vertices, is determined in the following theorem.

Theorem 3.16 (Sudev & Germina, 2018) Let G be an arithmetic sumset graph and let d_i and d_j be the common differences of two adjacent vertices v_i and v_j in G. If $\left|f\left(v_i\right)\right| \leq \left|f\left(v_j\right)\right|$, then for some positive integer $1 \leq k \leq \left|f\left(v_i\right)\right|$, the edge $v_i v_j$ has the set-indexing number

$$\left|f\left(v_i\right)\right| + k\left(\left|f\left(v_j\right)\right| - 1\right),$$

where f is an arithmetic sumset labelling of G.

We note that if the set-labels of two adjacent vertices are arithmetic progressions with the same common difference, say d, then the set-label of the corresponding edge is also an arithmetic progression with same common difference d. Hence, we have

Definition 3.4 (Sudev & Germina, 2018; Sudev & Germina, 2015) Let f be an arithmetic sumset labelling defined on a given graph G. If all the elements of G have the same common difference under f, then f is said to be an *isoarithmetic sumset labelling*.

A graph which has an isoarithmetic sumset labelling is called an isoarithmetic sumset graph. It is proved that every graph G can have an isoarithmetic sumset labelling (cf. Sudev & Germina, 2015). Invoking Theorem 1.1, the set-indexing number of edges of an isoarithmetic sumset graph can be found out in Sudev and Germina (2015) as stated in the following result.

Theorem 3.17 (Sudev & Germina, 2018) Let G be a graph with an arithmetic sumset labelling f defined on it. Then, f is an isoarithmetic sumset labelling on G if and only if the set-indexing number of every edge of G is one less than the sum of the set-indexing numbers of its end vertices.

The following theorems answer the questions whether an isoarithmetic sumset labelling f defined on a graph G be a uniform sumset labelling.

Theorem 3.18 *An isoarithmetic sumset labelling of a graph G is a uniform sumset labelling if and only if $V(G)$ is uniformly set-indexed or G is bipartite.*

Theorem 3.19 *An isoarithmetic sumset labelling f of a graph G is an r-uniform sumset labelling if and only if every component of G is either bipartite or its vertex set is l-uniformly set-indexed, where*

$$l = \frac{1}{2}(r+1).$$

Let f be an arithmetic sumset labelling defined on a given graph G such that $V(G)$ is l-uniformly set-indexed. Then, in view of the above theorem, one can observe that f is an isoarithmetic sumset labelling of G if and only if G is a $(2l-1)$-uniform sumset graph (see Sudev & Germina, 2015). It is also proved in Sudev and Germina (2015) that no isoarithmetic sumset labelling defined on a given graph G can be a strong sumset labelling of G. This result immediate from the fact that $m+n-1=mn$ if and only if either $m=1$ or $n=1$.

For an isoarithmetic sumset graph G, the number of saturated classes in the Cartesian product of the set-labels of any two adjacent vertices in G is one greater than the difference between the cardinality of the set-labels of these vertices. Moreover, exactly two compatibility classes, other than the saturated classes, have the same cardinality in the Cartesian product of the set-labels of these vertices (see Sudev & Germina, 2015). Also, if $V(G)$ is l-uniformly set-indexed, then there exists exactly one saturated class in the Cartesian product of the set-labels of any two adjacent vertices in G (see Sudev & Germina, 2015).

Definition 3.5 (Sudev & Germina, 2018) An arithmetic sumset labelling f of a graph G, under which the deterministic ratio of each edge e of G is a positive integer greater than 1 and less than or equal to the set-indexing number of the end vertex of e having smaller common is called a biarithmetic sumset labelling of G. If the deterministic ratio of every edge of G is the same, then f is called an identical biarithmetic sumset labelling.

In other words, a biarithmetic sumset labelling of a graph G is an arithmetic sumset labelling f of G, for which the common differences of any two adjacent vertices v_i and v_j in G, denoted by d_i and d_j respectively such that $d_i < d_j$, holds the condition $d_j = kd_i$ where k is a positive integer such that $1 \langle k \leq |f(v_i)|$. A graph G with a biarithmetic sumset labelling may be called a biarithmetic sumset graph.

The following theorem states a necessary and sufficient condition for a graph to have an identical biarithmetic sumset labelling.

Theorem 3.20 (Sudev & Germina, 2018) *A graph G admits an identical biarithmetic sumset labelling if and only if it is bipartite.*

The following theorem describes the set-indexing number of the edges of a biarithmetic sumset graph.

Theorem 3.21 (Sudev & Germina, 2018) *Let G be a graph which admits an arithmetic sumset indexer, say f. Let v_i and v_j be two adjacent vertices in G with the deterministic indices d_i and d_j, such that $d_j = kd_i$, where k is a positive integer such that $1 \langle k \leq |f(v_i)|$. Then, the set-indexing number of the edge $v_i v_j$ is*

$$\left| f(v_i) \right| + k \left(\left| f(v_j) \right| - 1 \right).$$

A necessary and sufficient condition for a biarithmetic sumset labelling of G to be a strong sumset labelling is described in the result given below.

Theorem 3.22 (Sudev & Germina, 2018) *A biarithmetic sumset labelling f of a graph G is a strong sumset labelling of G if and only if the deterministic ratio of every edge of G is equal to the set-indexing number of its end vertex having a smaller deterministic index.*

Invoking Theorem 3.22, it is proved that an identical biarithmetic sumset labelling of a graph G is a strong sumset labelling of G if and only if one partition of $V(G)$ is k-uniformly set-indexed, where k is the deterministic ratio of the edges of G.

Let G be a biarithmetic sumset graph. Let v_i and v_j be two adjacent vertices in G, where v_i has the smaller common difference. Let k be the deterministic ratio of the edge $v_i v_j$. Then, a compatible class in $f(v_i) \times f(v_j)$ is a saturated class if and only if

$$\left| f(v_i) \right| = \left(\left| f(v_j) \right| - 1 \right) k + r, r > 0.$$

Also, the number of saturated classes in $f(v_i) \times f(v_j)$ is

$$\left| f(v_i) \right| - \left(\left| f(v_j) \right| - 1 \right) k$$

(see [78]). Moreover, for $1 \leq p \leq n - 1$, there are exactly $2k$ compatibility classes containing p elements. It is clear that if $f(v_i) < f(v_j)$, then there is no saturated class in $f(v_i) \times f(v_j)$. Moreover, if $\left| f(v_i) \right| = pk + q$, where p, q are integers such that $p \leq \left(\left| f(v_j) \right| - 1 \right)$ and $q < k$, then

1. if $q = 0$, then $\left(\left| f(v_j) \right| - p + 1 \right) k$ compatibility classes are maximal compatibility classes and contain p elements.

2. if $q > 0$, then $\left(\left\| f\left(v_j\right) \right\| - p - 1 \right)k + q$ compatibility classes are maximal compatibility classes and contain $\left(p + 1\right)$ elements.

Definition 3.6 (Sudev & Germina, 2017) A *prime arithmetic sumset labelling* of a graph G is an arithmetic sumset labelling $f : V\left(G\right) \to \mathcal{P}_0\left(X\right)$, where for any two adjacent vertices in G the deterministic index of one vertex is a prime multiple of the common difference of the other, where this prime integer is less than or equal to the set-indexing number of the second vertex.

In other words, an arithmetic sumset labelling f is a prime arithmetic sumset labelling of G if for any two adjacent vertices v_i and v_j of G with the common differences d_i and d_j respectively, such that $d_i \le d_j$, $d_j = p_i d_i$ where p_i is a prime number such that $1 \langle p_i \le \left| f\left(v_i\right) \right| $ (refer to Sudev & Germina, 2017). The following theorem states a necessary and sufficient condition for a graph to admit a prime arithmetic sumset labelling.

Theorem 3.23 (Sudev & Germina, 2017) *A graph G admits a prime arithmetic sumset labelling if and only if it is bipartite.*

A vertex arithmetic sumset labelling (or edge arithmetic sumset labelling) of a graph is a sumset labelling of G in which all vertex set-labels (or edge set-labels) are arithmetic progressions. If all the vertex set-labels of a graph G are arithmetic progressions and the edge set-labels are not arithmetic progressions with respect to a sumset-labelling, then such a sumset labelling is called semi-arithmetic sumset labelling (cf. Sudev & Germina, 2014).

A vertex-arithmetic sumset labelling f of a graph G, under which the differences d_i and d_j of the set-labels $f\left(v_i\right)$ and $f\left(v_j\right)$ respectively for two adjacent vertices v_i and v_j of G, holds the conditions $d_j = k d_i$ and k is a non-negative integer greater than $\left| f\left(v_i\right) \right|$ is called the semi-arithmetic sumset labelling of the first kind (cf. Sudev & Germina, 2014). A vertex-arithmetic sumset labelling f of a graph G, under which the differences d_i and d_j of the set-labels $f\left(v_i\right)$ and $f\left(v_j\right)$ respectively for two adjacent vertices v_i and v_j of G are not multiples of each other, is called the semi-arithmetic sumset labelling of the second kind (cf. Sudev & Germina, 2014).

It is proved in Sudev and Germina (2014) that every first kind semi-arithmetic sumset labelling of a graph G is a strong sumset labelling of G. The following result established a necessary and sufficient condition for a semi-arithmetic sumset labelling of G to be a uniform sumset labelling.

Theorem 3.24 (Sudev & Germina, 2014) *A semi-arithmetic sumset labelling of the first kind of a graph G is a uniform sumset labelling if and only if either G is bipartite or $V\left(G\right)$ is uniformly set-indexed.*

The following theorem discusses the conditions required for a semi-arithmetic sumset labelling of the second kind to be a strong sumset labelling.

Theorem 3.25 (Sudev & Germina, 2014) *Let f be a semi-arithmetic sumset labelling defined on G. Also, let*

$$\left| f\left(v_j\right) \right| = q.\left| f\left(v_i\right) \right| + r, 0 < r < \left| f\left(v_i\right) \right|.$$

Then, f is a strong sumset labelling if and only if $q > \left| f\left(v_i\right)\right|$ or the differences d_i and d_j of two set labels $f\left(v_i\right)$ and $f\left(v_j\right)$ respectively, are relatively prime.

One can note that an arithmetic sumset labelling with arbitrary differences need not have any saturated compatibility classes. Let f be an arithmetic sumset labelling with arbitrary common differences on a graph G and

$$\left| f\left(v_j\right)\right| = q.\left| f\left(v_i\right)\right| + r \,.$$

Also, let q_1 and q_2 be the positive integers such that $q_1 \cdot \mid f\left(v_j\right) = q_2.r$. Then, the number of elements in a maximal compatible class of

$$f\left(v_i\right) \times f\left(v_j\right) \text{ is } \left\lceil \frac{\left|\left| f\left(v_j\right)\right|\right|}{q_1} \right\rceil .$$

Let f be a semi-arithmetic sumset labelling of the first kind of a graph G and let v_i and v_j be two adjacent vertices in G . Then, we can observe that all the compatibility classes in $f\left(v_i\right) \times f\left(v_j\right)$ are trivial classes.

SOME EXTRINSIC SUMSET GRAPHS

Another classification of sumset graphs is based on the nature and structure of the collection of set-labels of the vertices and edges of graphs. In this section, we examine different types of sumset labellings in this category.

Among these types of sumset graphs, the most interesting graphs are those the collections of whose set-labels forms the topologies of the ground set X . In the following section, we discuss and characterise the graphs whose set-labels form topologies of X .

Topological and Topogenic Sumset Graphs

Definition 4.1 (Sudev & Germina, 2015) Let G be a graph and let X be a non-empty set of integers. A sumset labelling $f : V\left(G\right) \to \mathcal{P}_0\left(X\right)$ is called a *topological sumset labelling* of G if $f\left(V\left(G\right)\right) \cup \left\{\varnothing\right\}$ is a topology on X . A graph G which admits a topological sumset labelling is called a topological sumset graph.

For a finite set X of integers, let the given function $f : V\left(G\right) \to \mathcal{P}_0\left(X\right)$ be a sumset labelling on a graph G . Since the set-label of every edge uv is the sumset of the sets $f\left(u\right)$ and $f\left(v\right)$, it can be observed that $\left\{0\right\}$ cannot be the set-label of any edge of G . Moreover, since f is a topological sumset

Figure 4. A topological sumset graph

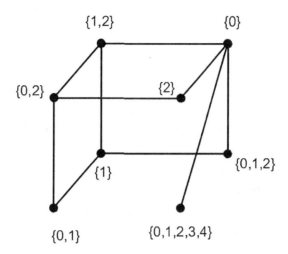

labelling defined on G, X must be the set-label of some vertex, say u, of G and hence the set $\{0\}$ will be the set-label of a vertex, say v, and the vertices u and v are adjacent in G.

Figure 4 provides an illustration to a topological sumset graph:

Let f be a topological sumset labelling of a given graph G with respect to a non-empty finite ground set X. Then, $\mathcal{T} = f\big(V(G)\big) \cup \{\varnothing\}$ is a topology on X. Then, the graph G is said to be a f-graphical realisation (or simply f-realisation) of $\mathcal{T}$. The elements of the sets $f(V)$ are called f-open sets in G. Existence of graphical realisations for certain topologies of a given set X is established in the following theorem.

Theorem 4.1 (Sudev & Germina, 2015) *Let X be a non-empty finite set of integers. A topology $\mathcal{T}$ of X, consisting of the set $\{0\}$ is graphically realisable.*

Let X be the ground set and $\mathcal{T}$ be a topology on X and let $f : V(G) \to \mathcal{P}_0(X)$ is a topological sumset labelling of a graph G. One can observe the following structural properties of a topological sumset graph.

1. An element x_r in X can be an element of the set-label $f(v)$ of a vertex v of G if and only if $x_r + x_s \leq l$, where x_s is any element of the set-label of an adjacent vertex u of v in G and l is the maximal element in X.
2. The vertices whose set-labels containing the maximal element of the ground set X are pendant vertices which are adjacent to the vertex having the set-label $\{0\}$.
3. If G has only one pendant vertex and if G admits a topological sumset labelling, then X is the only vertex set-label in G, which contains the maximal element of X.
4. the minimum number of pendant edges incident on a particular vertex of a topological sumset graph is equal to the number of f-open sets in $f\big(V(G)\big)$ containing the maximal element of the ground set X,

5.   the minimum number of pendant vertices of a topological sumset graph G is the number of f-open sets in $\mathcal{T}$, each of which is the non-trivial summand of at most one f-open set in $\mathcal{T}$.

A necessary and sufficient condition for a given graph to admit a topological sumset labelling is provided in the following theorem.

Theorem 4.2 (Sudev & Germina, 2015) *A graph G admits a topological sumset labelling if and only if G has at least one pendant vertex.*

In view of Theorem 4.2, it is to be noted that all path graphs and trees admit topological sumset labellings. For $n \geq 3$, neither a cycle C_n nor a complete graph K_n admits a topological sumset labelling. For $k \geq 2$, no k-connected graph admits a topological sumset labelling. Also, for $m, n \geq 2$, no complete bipartite graph $K_{m,n}$ admits a topological sumset labelling.

Let G be a graph with a pendant vertex v which admits a topological sumset labelling, say f, with respect to a ground set X. Let f_1 be the restriction of f to the graph $G - v$. Then, there exists a collection $\mathcal{B}$ of proper subsets of X which, together with $\{\varnothing\}$, form a topology on the union of all elements of $\mathcal{B}$ (see Sudev & Germina, 2015).

If $\mathcal{T}$ is the indiscrete topology of the ground set X, the only graph G which admits a topological sumset labelling with respect $\mathcal{T}$ is the trivial graph K_1. If X is a two point set, say $X = \{0, 1\}$, then the topology $\mathcal{T}_1 = \{\varnothing, \{0\}, X\}$ and $\mathcal{T}_2 = \{\varnothing, \{1\}, X\}$ are the Sierpinski's topologies. Then, the only graph G which admits a topological sumset labelling with respect to a Sierpinski's topology is the graph K_2. A graph G, on n vertices, admits a topological sumset labelling with respect to the discrete topology on the ground set X if and only if G has at least $2^{|X|-1}$ pendant vertices which are adjacent to a single vertex of G (see Sudev & Germina, 2015). In view of this fact, we observe that a graph on even number of vertices does not admit a topological sumset labelling with respect to the discrete topology on the ground set X. A star graph $K_{1,r}$ admits a topological sumset labelling with respect to the discrete topology on the ground set X, if and only if $r = 2^{|X|} - 2$.

Definition 4.2 (Sudev & Germina, 2015) Let X be a finite non-empty set of integers. A sumset labelling f of a given graph G, defined by $f : V(G) \to \mathcal{P}_0(X)$, is said to be a *topogenic sumset labelling* of G, with respect to the ground set X, if

$$\mathcal{T} = f(V(G)) \cup f^+(E(G)) \cup \{\varnothing\}$$

is a topology on X.

Figure 5 depicts a topogenic sumset graph.

A graph G that admits a topogenic sumset labelling is said to be a topogenic sumset graph. Let $\mathcal{T}$ be a topology on X, which is neither the indiscrete topology nor the discrete topology on X. Then, $\mathcal{T}_1 = \{\varnothing, \{a\}, X\}$ where $a \in X$. Then, the graph K_2 admits a topogenic sumset labelling with respect to the set X if and only if $a = 0$.

Figure 5. A topogenic sumset graph

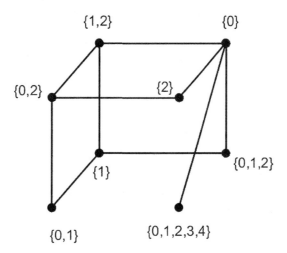

If $f : V(G) \to \mathcal{P}_0(X)$ is a topogenic sumset labelling of G, then it is possible for a vertex $v \in V(G)$ and an edge $e \in E(G)$, that $f(v) = f^+(e)$. In the following lemma, we discuss the minimum number of elements in a topogenic sumset graph G having the same set-labels.

Let X be a non-empty finite ground set and let $\mathcal{T}$ be a topology on X. Then, a graph G is said to be a topogenic graphical realisation of $\mathcal{T}$ if there exists a topogenic sumset labelling f on G such that

$$f\big(V(G)\big) \cup f^+\big(E(G)\big) \cup \{\varnothing\} = \mathcal{T}.$$

The result below described a necessary and sufficient condition for the existence of a topogenic graphical realisation for a given topology on the ground set X.

Theorem 4.3 (Sudev & Germina, 2015) *Let X be a non-empty finite ground set and let $\mathcal{T}$ be a topology on X. Then, a graph G is a topogenic graphical realisation of $\mathcal{T}$ if and only if every non-empty set of $\mathcal{T}$ is either a summand of some other elements of $\mathcal{T}$ or the sumset of two elements of $\mathcal{T}$.*

The following theorem determined the order and size of a topogenic sumset graph.

Theorem 4.4 (Sudev & Germina, 2015) *Let G be a topogenic sumset graph with respect to a topology $\mathcal{T}$ of a given ground set X and let $\mathcal{T}$ contains ρ sets which are not the sumsets of any other sets in $\mathcal{T}$. Then, $|V(G)| \geq \rho$ and $|E(G)| \geq \tau - \rho$, where $\tau = |\mathcal{T}|$.*

Sumset Filter Graphs

Given a set X, a partial ordering $\subseteq$ can be defined on the power set $\mathcal{P}(X)$ by subset inclusion, turning $\big(\mathcal{P}(X), \subseteq\big)$ into a lattice. A *filter* on X, denoted by $\mathcal{F}$, is a non-empty subset of the power set $\mathcal{P}(X)$ of X which has the following properties.

1. $\varnothing \notin \mathcal{F}, X \in \mathcal{F}$.

2. $A, B \in \mathcal{F} A \cap B \in \mathcal{F}$.
3. $A \in \mathcal{F}, A \subset B, B \in \mathcal{F}$ where B is a non-empty subset of X.

In view of the definition of filters mentioned above, the notion of a sumset filter labelling of a graph is defined as follows.

Definition 4.3 (Sudev, Chithra & Germina, 2015) Let X be a finite set of integers. Then, a sumset labelling $f : V(G) \to \mathcal{P}_0(X)$ is said to be a *sumset filter labelling* of G if $\mathcal{F} = f(V)$ is a proper filter on X. A graph G which admits a sumset filter labelling is called a sumset filter graph.

The following is a necessary and sufficient condition for a sumset labelling f of a given graph G to be a sumset filter labelling of G.

Theorem 4.5 (Sudev, Chithra & Germina, 2015) *A sumset labelling f defined on a given graph G with respect to a non-empty ground set X is a sumset filter labelling of G if and only if the following conditions hold.*

1. $0 \in X$.
2. *every subset of X containing 0 is the set-label of some vertex in G.*
3. *0 is an element of the set-label of every vertex in G.*

If a graph G admits a sumset filter labelling, then it can be observed that G has $2^{|X|-1}$ vertices. Moreover, for a sumset filter graph G, only one vertex of G can have a singleton set-label as the only possible singleton set-label in $\mathcal{F}$ is $\{0\}$. Another interesting property for sumset filter graphs is that

$$f^+(E(G)) \subseteq f(V(G)).$$

Figure 6 is an illustration of a sumset filter graph.

Another interesting structural property of a sumset filter graph has been proved in Sudev, Chithra, and Germina (2015) as follows:

Theorem 4.6 (Sudev, Chithra & Germina, 2015) *If a graph G admits a sumset filter labelling, with respect to a non-empty ground set X, then G must have at least $2^{|X|-2}$ pendant vertices that are incident on a single vertex of G.*

Figure 6. A sumset filter graph

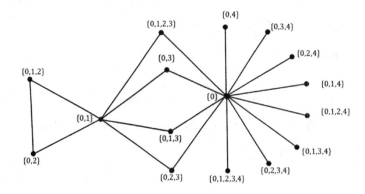

If a given graph G admits a sumset filter labelling f, then it can be seen that every element of the collection $\mathcal{F} = f\left(V\left(G\right)\right)$ belongs to some finite chain of sets in $\mathcal{F}$ of the form

$$\left\{0\right\} = f\left(v_1\right) \subset f\left(v_2\right) \subset f\left(v_3\right) \subset \ldots f\left(v_r\right) = X .$$

The following are some of the immediate observations on the structural characteristics of sumset filter graphs, which are proved in Sudev and Gerfmina (2015).

1. The existence of a sumset filter labelling is not a hereditary property. That is, a sumset filter labelling of a graph need not induce a sumset filter labelling for all of its subgraphs.
2. For $n \geq 3$, no paths P_n admits a sumset filter labelling. No cycles admit sumset filter labellings and as a result, neither Eulerian graphs nor Hamiltonian graphs admit sumset filter labellings. Neither complete graphs nor complete bipartite graphs admit sumset filter labellings. For $r > 2$, complete r-partite graphs also do not admit sumset filter labellings.
3. Graphs having an odd number of vertices never admits a sumset filter labelling.
4. Removing any non-leaf edge of a sumset filter graph preserves the sumset filter labelling of that graph. This property is known as the monotone property.

Graceful and Sequential Sumset Graphs

Another classification of sumset graphs is according to whether all of the nonempty elements of $\mathcal{P}\left(X\right)$, the power set of X, appears as the set-labels of of the elements of the graphs concerned. Hence, the notion of graceful sumset labelling of graphs have been introduced in [86] as follows:

Definition 4.4 (Sudev & Germina, 2019) Let G be a graph and let X be a non-empty set of integers. A sumset labelling $f : V\left(G\right) \to \mathcal{P}_0\left(X\right)$ is said to be a *sumset graceful labelling* or a *graceful sumset labelling* of G if

$$f^+\left(E\left(G\right)\right) = \mathcal{P}\left(X\right) \setminus \left\{\varnothing, \left\{0\right\}\right\}.$$

A graph with a graceful sumset labelling may be called a graceful sumset graph. Figure 7 illustrates a sumset graceful graph.

The structural properties of a graceful sumset graph are very much interesting a study in this direction has been done in [86]. Some important properties of them are as follows:

If $f : V\left(G\right) \to \mathcal{P}_0\left(X\right)$ is a sumset graceful labelling of a given graph G, then $\left\{0\right\}$ must be the set-label of a vertex in G. The vertices of a graceful sumset graph G, whose set-labels are the trivial sumsets of any two subsets of X, must be adjacent to the vertex v that has the set-label $\left\{0\right\}$.

If x_1 and x_2 are the minimal and second minimal non-zero elements of the ground set X of a graceful sumset graph G, then it can be noted that the vertices of G that have the set-labels $\left\{x_1\right\}$ and $\left\{x_2\right\}$, must be adjacent to the vertex v that has the set-label $\left\{0\right\}$. If A_i and A_j be two distinct subsets of the

Figure 7. A sumset graceful graph

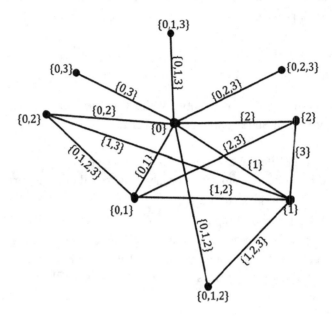

ground set X and let x_i and x_j be the maximal elements of A_i and A_j respectively, then A_i and A_j can be the set-labels of two adjacent vertices of a sumset graceful graph G only if $x_i + x_j \leq x_n$, the maximal element of X.

Let $f : V(G) \to \mathcal{P}_0(X)$ be a sumset graceful labelling of a given graph G and let x_n be the maximal element of X. If A_i and A_j are set-labels of two adjacent vertices, say v_i and v_j, then $A_i + A_j$ is the set-label of the corresponding edge v_iv_j. Hence, for any $x_i \in A_i, x_j \in A_j$, $x_i + x_j \leq x_n$ and hence, if one of these two sets consists of the maximal element of X, then the other set cannot have a non-zero element. Hence, x_n is an element of the set-label of a vertex v of G if v is a pendant vertex that is adjacent to the vertex labelled by $\{0\}$. It can also be noted that if G is a graph without pendant vertices, then no vertex of G can have a set-label consisting of the maximal element of the ground set X.

Let f be a sumset graceful labelling defined on G. Then,

$$f^+(E) = \mathcal{P}(X) \setminus \{\varnothing, \{0\}\}.$$

Therefore,

$$|E(G)| = |\mathcal{P}(X)| - 2 = 2^{|X|} - 2 = 2\left(2^{|X|-1} - 1\right).$$

That is, G has even number of edges. Therefore, the cardinality of the ground set X is $|X| = \log_2\left[|E(G)| + 2\right]$.

Other important structural properties of a graceful sumset graph are explained in the following theorems.

Theorem 4.7 (Sudev & Germina, 2019) *If a graph* G *admits a sumset graceful labelling* f *with respect to a finite ground set* X, *then the vertices of* G, *having trivial summands of any subset of* X *as their set-labels, are the pendant vertices of* G.

Theorem 4.8 (Sudev & Germina, 2019) *Let* G *be a sumset graceful graph which admits a sumset graceful labelling* f *with respect to a finite non-empty set* X. *Then,* G *must have at least* $|X| - 1$ *pendant vertices.*

A graph G which admits a graceful sumset indexer with respect to the set X is said to be a graceful graph-realisation of the set X with respect to the sumset labelling f. It is proved in [86] that there exists a non-bipartite graceful graph-realisation G corresponding to any non-empty finite set of integers containing the element 0.

Let X be a non-empty finite set of integers. Then, a graph G admits a graceful sumset labelling if and only if the following conditions hold (see Sudev & Germina, 2019).

1. $0 \in X$ and $\{0\}$ be a set-label of some vertex, say v, of G.

2. The minimum number of pendant vertices in G is the number of subsets of X which are the trivial summands of any subsets of X.

3. The minimum degree of the vertex v is equal to the number of subsets of X which are not the sumsets of any two subsets of X and not the non-trivial summands of any other subsets of X.

4. The minimum number of pendant vertices that are adjacent to a given vertex of G is the number of subsets of X which are neither the non-trivial sumsets of any two subsets of X nor the non-trivial summands of any subsets of X.

Another type of sumset graph is the one with the vertex set-labels and the edge set-labels together form the set $\mathcal{P}_0(X)$. Hence, the notion of a sumset sequential labelling of a given graph has been introduced in (Sudev & Germina, 2015) as follows:

Definition 4.5 (Sudev & Germina, 2019) A sumset labelling f of G is said to be a *sumset sequential labelling* or a *sequential sumset labelling* if the induced function

$$f^*\left(G\right) = f\left(V\left(G\right)\right) \cup f^+\left(E\left(G\right)\right) = \mathcal{P}_0\left(X\right),$$

where f^* is defined by

$$f^*\left(x\right) = \begin{cases} f\left(x\right) & \text{if } x \in V\left(G\right), \\ f^+\left(x\right) & \text{if } x \in E\left(G\right). \end{cases}$$

A graph G which admits a sumset sequential labelling may be called a sequential sumset graph (see Figure 8 for illustration).

It can be noted that every sumset graceful labelling of a graph G is also a sumset sequential labelling of G.

Figure 8. A sumset sequential graph

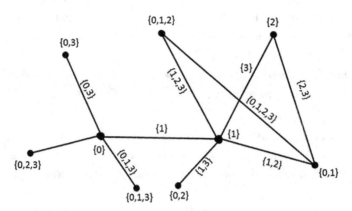

If G is a sumset sequential graph without isolated vertices and the function f^* is an injective function, then one can note that no vertex of G can have a set-label $\{0\}$. Moreover, if one vertex v of G has the set label $\{0\}$, then v should be an isolated vertex of G and no edge of G can also have the set label $\{0\}$.

The following result provides a relation connecting the size and order of a given sumset sequential graph G and the cardinality of its ground set X.

Proposition 4.9 (Sudev & Germina, 2015) *Let G be a graph on n vertices and m edges. If f is a sumset sequential labelling of a graph G with respect to a ground set X, then $m + n = 2^{|X|} - \left(1 + \kappa\right)$, where κ is the number of subsets of X which are the set-labels of both vertices and edges.*

Two sets A and B are said to be of same parity if their cardinalities are simultaneously odd or simultaneously even. Then, the following theorem is on the parity of the vertex set and edge set of G.

Proposition 4.10 (Sudev & Germina, 2015) *Let f be a sumset sequential labelling of a given graph G, with respect to a ground set X. Then, if $V\left(G\right)$ and $E\left(G\right)$ are of same the parity, then κ is an odd integer and if $V\left(G\right)$ and $E\left(G\right)$ are of different parity, then κ is an even integer, where κ is the number of subsets of X which are the set-labels of both vertices and edges.*

It is shown in Sudev and Germina (2015) that a graph G that admits a sumset sequential labelling with respect to a non empty ground set X should have at least ρ vertices, where ρ is the number of elements in $\mathcal{P}_0\left(X\right)$ which are not the sumsets of any two elements of $\mathcal{P}_0\left(X\right)$. Let $\mathcal{B}$ be the collection of subsets of X which are neither the sumsets of any two subsets of X nor their sumsets are subsets of X. If $\mathcal{B}$ is non-empty, then for a sequential sumset graph G, it is established in Sudev and Germina (2015) that

1. $\{0\}$ is the set-label of a vertex in G,

2. the minimum number pendant vertices in G is equal to the cardinality of $\mathcal{B}$. and as a result, a sequential sumset graph G should have at least one pendant vertex.

HYPERGRAPHS ASSOCIATED WITH SUMSET GRAPHS

In terms of the set-labels of different sumset graphs, some new types of hypergraphs graphs have been introduced and studied the properties of these discrete structures in Sudev and Germina (2015). Three new types of hypergraphs associated with the set-labels of the elements of given sumset graphs have been introduced and their structural properties have been studied in Sudev and Germina (2015).

Definition 5.1 (Sudev & Germina, 2015) Let $X \subset \mathbb{Z}$ and a set-valued function $f : V\left(G\right) \to \mathcal{P}_0\left(X\right)$ be a sumset labelling of a given graph G. Then, the hypergraph $\mathcal{H}(X, \mathcal{L}\left(V\right)$ is called the *vertex set-label hypergraph* (VSL-hypergraph) of G with respect to the sumset labelling f where

$$\mathcal{L}\left(V\right) = \left\{ f\left(v\right) : v \in V\left(G\right) \right\}.$$

The VSL-hypergraph of a sumset graph G is denoted by $\mathcal{H}_G^f$.

In a similar way, the notion of edge set-label hypergraphs of a given sumset graph G has been defined in Sudev and Germina (2015) as follows.

Definition 5.2 (Sudev & Germina, 2015) Let $X \subset \mathbb{Z}$ and $f : V\left(G\right) \to \mathcal{P}_0\left(X\right)$ be a sumset labelling of a given graph G. Then, a hypergraph $\left(X, \mathcal{L}\left(E\right)\right)$ is called the *edge set-label hypergraph* (ESL-hypergraph) of G with respect to the sumset labelling f, where

$$\mathcal{L}\left(E\right) = \left\{ f^+\left(e\right) : e \in E\left(G\right) \right\}.$$

The ESL-hypergraph of a sumset graph G is denoted by $\mathcal{H}_G^{f^+}$.

Another type of hypergraph associated with a given sumset labelling graph G, namely total set-label hypergraph, is defined as follows.

Definition 5.3 (Sudev & Germina, 2015) Let $X \subset \mathbb{Z}$ and $f : V\left(G\right) \to \mathcal{P}_0\left(X\right)$ be a sumset labelling of a given graph G. Then, a hypergraph $(X, \mathcal{L}\left(G\right)$ is called *the total set-label hypergraph* (TSL-hypergraph) of G with respect to the sumset labelling f, where

$$\mathcal{L}\left(G\right) = \begin{cases} f\left(x\right) & \text{if } x \in V\left(G\right) \\ f^+\left(x\right) & \text{if } x \in E\left(G\right). \end{cases}$$

The TSL-hypergraph of a sumset graph G is denoted by $\mathcal{H}_G^*$.

In the first observation, it is clear that if a sumset graph G has no pendant vertex, then its vertex set-label hypergraph $\mathcal{H}_G^f$ has some isolated vertices. A hypergraph is said to be a connected hypergraph if its intersection graph is connected. In view of the above facts, a necessary condition for a vertex set-label hypergraph of a sumset graph to be connected is established in [82] as follows:

Proposition 5.1 (Sudev & Germina, 2015) *If the vertex set-label hypergraph of a sumset graph G is a connected hypergraph, then G must have some pendant vertices.*

The converse of Proposition 5.1 is also not true in general. The existence of a connected VSL-hypergraph for a given sumset graph G is established in the following proposition.

Proposition 5.2 (Sudev & Germina, 2015) *Let G be a graph with at least one pendant vertex. Then, there exists a sumset labelling f defined on G such that the corresponding VSL-hypergraph $\mathcal{H}_G^f$ is connected.*

Invoking Proposition 5.1 and Proposition 5.2, a necessary and sufficient condition for the VSL-hypergraph $\mathcal{H}_G^f$ of a given sumset graph G to be a connected hypergraph has been proposed in [82] as given below.

Theorem 5.3 (Sudev & Germina, 2015) *Let $f : V\left(G\right) \to \mathcal{P}_0\left(X\right)$ be a sumset labelling defined on a given graph G, where X is a non-empty set of integers. Then, the VSL-hypergraph $\mathcal{H}_G^f$ of G is a connected hypergraph if and only if the following conditions hold.*

1. $0 \in X$ *and* $\left\{0\right\}$ *is the set-label of a vertex, say v_1 of G*

2. *There exists at least one pendant vertex, say v_i in G which is adjacent to v_1 and the set-label of which contains the elements of X which are the trivial summands of any element of X.*

A hypergraph $\mathcal{H}$ is said to be a uniform hypergraph (or a k-uniform hypergraph) if all the hyperedges of G are of the same size (or the size k). The following results are obvious conditions for the VSL-hypergraphs to be uniform hypergraphs.

Proposition 5.4 (Sudev & Germina, 2015) *Let f be a sumset labelling defined on a graph G. Then, the VSL-hypergraph $\mathcal{H}_G^f$ of Gf is a k-uniform hypergraph if and only if $V\left(G\right)$ is k-uniformly set-indexed.*

A hypergraph is said to be a *linear hypergraph* if any two of its hyperedges have at most one element (vertex) in common. Hence, the conditions required for the hypergraphs associated to a given sumset graph to be linear hypergraphs is provided in the following result.

Theorem 5.5 (Sudev & Germina, 2015) *Let G be a graph with some pendant vertices, which admits a sumset labelling f. Then, the VSL-hypergraph $\mathcal{H}_G^f$ of G is a connected linear hypergraph if and only if the set labels of any two vertices of G contain at most one common element and for every vertex in G, there exists at least one vertex in G such that their set-labels have exactly one common element.*

It can be noticed that the VSL-hypergraphs associated with the topological sumset graphs, topogenic sumset graphs and certain types of graceful and sequential sumset graphs are obviously connected.

The similar results for edge set-label hypergraphs have also been discussed in Sudev and Germina (2015). A necessary condition for the ESL-hypergraph $\mathcal{H}_G^{f^+}$ of G to have isolated vertices is as follows.

Proposition 5.6 (Sudev & Germina, 2015) *Let $f : V\left(G\right) \to \mathcal{P}_0\left(X\right)$ be a sumset labelling of a given graph G. If $0 \notin X$, then its ESL-hypergraph $\mathcal{H}_G^{f^+}$ has some isolated vertices.*

The following proposition established a condition for the ESL-hypergraph $\mathcal{H}_G^{f^+}$ of a sumset graph G to be a connected hypergraph.

Proposition 5.7 (Sudev & Germina, 2015) *Let G be a graph with at least one pendant vertex. Then, there exists a sumset labelling f defined on G such that the ESL-hypergraph $\mathcal{H}_G^{f^+}$ is connected.*

In this context also, we can see that the ESL-hypergraphs associated with the topological sumset graphs and topogenic sumset graphs are connected hypergraphs.

In view of the above propositions, a necessary and sufficient condition for an ESL-hypergraph to be a connected hypergraph has been proved in Sudev and Germina (2015) as follows.

Theorem 5.8 (Sudev & Germina, 2015) *Let G be a graph which admits a sumset labelling $f : V(G) \to \mathcal{P}_0(X)$, where X is a non-empty set of integers. Then, the ESL-hypergraph $\mathcal{H}_G^{f^+}$ of G is a connected hypergraph if and only if the following conditions hold.*

1.   *$0 \in X$ and $\{0\}$ is the set-label of a vertex, say v_1 of G,*

2.   *there exists at least one pendant vertex, say v_j, in G which is adjacent to v_1 and the set-label of which contains the elements of X that are the trivial sumsets of any elements of X.*

Two hypergraphs are said to isomorphic if their intersection graphs are isomorphic. A necessary condition for the isomorphism between the VSL-hypergraph and the ESL-hypergraph of a sumset labelling graph is provided in Sudev and Germina (2015) as follows.

Theorem 5.9 (Sudev & Germina, 2015) *If the VSL-hypergraph and the ESL-hypergraph of a sumset graph G are isomorphic, then G is a unicyclic graph.*

A necessary and sufficient condition for an ESL-hypergraph of a sumset graph G to be a uniform hypergraph has been discussed (Sudev & Germina, 2015) as given below.

Proposition 5.10 [82] *Let f be a sumset labelling defined on a graph G. Then, the ESL-hypergraph $\mathcal{H}_G^{f^+}$ of G is a k-uniform hypergraph if and only if f is a k-uniform sumset labelling defined on G.*

A necessary and sufficient condition for the ESL-hypergraph $\mathcal{H}_G^{f^+}$ of a sumset labelling graph G to be a connected linear hypergraph is as follows.

Theorem 5.11 (Sudev & Germina, 2015) *Let G be a graph with pendant vertices, which admits a sumset labelling f. Then, the ESL-hypergraph $\mathcal{H}_G^{f^+}$ of G is a connected linear hypergraph if and only if*

1.   *For any two adjacent pairs of vertices (v_i, v_j) and (v_r, v_s) there exists at most one element in $f(v_i) \times f(v_j)$ that is compatible with an element in $f(v_r) \times f(v_s)$ and vice versa.*

2.   *For every pair of adjacent vertices (v_i, v_j) of G, there exist at least one more pair of adjacent vertices (v_r, v_s) such that we can find a unique element $(a, b) \in f(v_i) \times f(v_j)$ and a unique element $(c, d) \in f(v_r) \times f(v_s)$ such that (a, b) and (c, d) are compatible.*

The existence of a connected total set-label hypergraph for a given sumset graph is established in the following result.

Proposition 5.12 (Sudev & Germina, 2015) *For any given graph G, there exists a sumset labelling f such that the corresponding TSL-hypergraph is connected.*

The following theorem established a necessary and sufficient condition for the TSL-hypergraph of a sumset graph G to be a uniform hypergraph.

Theorem 5.13 (Sudev & Germina, 2015) *The TSL-hypergraph of a sumset graph G be a uniform hypergraph if and only if G is 1-uniform.*

A necessary condition for a TSL-hypergraph $\mathcal{H}_G^*$ of a sumset graph G to be a uniform hypergraph is proved in Sudev and Germina (2015) as follows.

Theorem 5.14 *If the TSL-hypergraph $\mathcal{H}_G^*$ of a given sumset graph G is a uniform hypergraph, then $\mathcal{H}_G^*$ is a disconnected hypergraph.*

SUMSET LABELLED SIGNED GRAPHS

A signed graph (see Zaslavsky, 1982, 2012), denoted by $S(G,\sigma)$, is a graph $G(V,E)$ together with a function $\sigma : E(G) \to \{+,-\}$ that assigns a sign, either $+$ or $-$, to each ordinary edge in G. The function σ is called the signature or sign function of S, which is defined on all edges except half edges and is required to be positive on free loops. An edge e of a signed graph S is said to be a positive edge if $\sigma(e) = +$ and an edge $\sigma(e)$ of a signed graph S is said to be a negative edge if $\sigma(e) = -$.

A simple cycle (or path) of a signed graph S is said to be balanced if the product of signs of its edges is $+$. A signed graph is said to be a balanced signed graph if it contains no half edges and all of its simple cycles are balanced. It is to be noted that the number of all negative signed graph is balanced if and only if it is bipartite. Two sets A and B are said to be same parity vertices if the cardinality of both are simultaneously odd or simultaneously even. The sets which do not have the same parity are usually called opposite parity sets. In a set-valued graph G, two vertices u and v are said to be same parity vertices if their set-labels are same parity sets. Otherwise, they are said to be opposite parity vertices.

Sumset Signed Graphs

The structural characteristics of the signed graphs associated with different types of sumset graphs have been studied in Sudev and Germina (2015), Sudev, Ashraf and Germina (2019), and Sudev and Germina, 2015). The notion of a sumset labelling of a signed graph is introduced in Sudev and Germina (2015) as follows.

Definition 6.1 (Sudev, Germina, 2015) Let $X \subseteq \mathbb{Z}$ and let S be a signed graph, with corresponding underlying graph G and the signature σ. An injective function $f : V(S) \to \mathcal{P}_0(X)$ is said to be a *sumset labelling* of S if f is a sumset labelling of the underlying graph G and the signature of S is defined by $\sigma(uv)) = (-1)^{|f(u)+f(v)|}$. A signed graph which admits a sumset labelling is called an sumset signed graph (sumset signed graph) and is denoted by S_f.

Definition 6.2 (Sudev, Germina, 2015) An sumset labelling f of a signed graph S is called *a weak sumset labelling* or a *strong sumset labelling* or an *arithmetic sumset labelling* of S, in accordance with the sumset labelling f of the underlying graph G is a weak sumset labelling or a strong sumset labelling or an arithmetic sumset labelling of the corresponding underlying graph G.

The following result provides a necessary and sufficient condition for the existence of a balanced signed graph corresponding to a strongly uniform sumset graph.

Theorem 6.1 (Sudev, Germina, 2015) *A strongly k-uniform sumset signed graph S is balanced if and only if the underlying graph G is a bipartite graph or $\sqrt{k}$ is an even integer.*

The following result provides a solution to this problem of finding the conditions required for a strongly uniform sumset signed graph to be clusterable.

Proposition 6.2 (Sudev, Germina, 2015) *A strongly k-uniform sumset signed graph S is clusterable if and only if the underlying graph G is bipartite and k is an odd integer.*

If the underlying graph G of a strong sumset signed graph S is bipartite, then S is balanced if and only if the number of negative edges in S in every cycle of G is even. This is possible only when the number of distinct pairs of adjacent vertices, having odd parity set-labels, in every cycle of S is even. Therefore, we have.

Proposition 6.3 (Sudev, Germina, 2015) *Let S be a strong sumset signed graph with the underlying graph G, bipartite. Then, S is clusterable if and only if the number of distinct pairs of adjacent vertices having odd parity set-labels is even.*

The following result describes the conditions required for the clusterability of (non-uniform) strong sumset signed graphs whose underlying graph G is a bipartite graph.

Proposition 6.4 (Sudev, Germina, 2015) *The strong sumset signed graph, whose underlying graph G is a bipartite graph, is clusterable if and only if there exist at least two adjacent vertices in S_f with odd parity set-labels.*

A necessary and sufficient condition for a strong sumset signed graph to be clusterable is described in the following theorem.

Theorem 6.5 (Sudev, Germina, 2015) *A strong sumset signed graph S is clusterable if and only if every odd cycle of the underlying graph G has at least two adjacent vertices with even parity set-label and at least two adjacent vertices with odd parity set-label.*

Balance and clusterability of the induced signed graphs of weak sumset graphs are described in the following theorems.

Proposition 6.6 (Sudev, Germina, 2015) *A weakly k-uniform sumset signed graph is always balanced.*

The following theorem discussed the clusterability of weakly uniform sumset signed graphs.

Theorem 6.7 (Sudev, Germina, 2015) *A weakly k-uniform sumset signed graph S is clusterable if and only if k is a positive odd integer.*

Balance of a weak sumset signed graph whose underlying graph is a bipartite graph is discussed in the following result.

Proposition 6.8 (Sudev, Germina, 2015) *A weak sumset signed graph S, whose underlying graph G is a bipartite graph, is balanced if and only if the number of odd parity non-singleton set-labels in every cycle of S is even.*

The following theorem establishes a necessary and sufficient condition for a weak sumset signed graph whose underlying graph is a bipartite graph.

Theorem 6.9 (Sudev, Germina, 2015) *The weak sumset signed graph S, whose underlying graph G is a bipartite graph, is clusterable if and only if some non-singleton vertex set-labels of S are of the odd parity.*

The balance property of weak sumset signed graphs whose underlying graphs are non-bipartite have been discussed the following theorem.

Theorem 6.10 (Sudev, Germina, 2015) *A weak sumset signed graph* S *, whose underlying graph* G *is a non-bipartite graph, is not balanced.*

Corollary 6.11 (Sudev, Germina, 2015) *A weak sumset signed graph* S *, whose underlying graph* G *is non-bipartite, is not clusterable.*

The following theorem for an edge of an arithmetic sumset graph to be a positive edge.

Theorem 6.12 (Sudev, Ashraf & Germina, 2019) *Let* $\S$ *be a signed graph which admits an arithmetic sumset labelling* f *of a graph* G *. Then, an edge* uv *is a positive edge of a sumset signed graph if and only if*

1. *The set-labels* $f\left(u\right)$ *and* $f\left(v\right)$ *are of different parity, provided the deterministic ratio of the edge* uv *is odd.*
2. *The set-label of the end vertex, with a minimum common difference, is of even parity, provided the deterministic ratio of the edge* uv *is even.*

The following theorem establishes a necessary and sufficient condition for an arithmetic sumset signed graph to a balanced signed graph.

Theorem 6.13 (Sudev, Ashraf & Germina, 2019) *A sumset signed graph* $\S$ *is balanced if and only if its underlying graph* G *is bipartite.*

The following theorem described a necessary and sufficient condition for an isoarithmetic sumset signed graph to be balanced.

Theorem 6.14 (Sudev, Germina, 2015) *An isoarithmetic sumset signed graph* S *is balanced if and only if every cycle in* S *has even number of distinct pairs of adjacent vertices having the same parity set-labels.*

The following result provides the required conditions required for an isoarithmetic sumset signed graph to be clusterable.

Proposition 6.15 (Sudev, Germina, 2015) *An isoarithmetic sumset signed graph* S *is clusterable if and only if* S *contains some disjoint pairs of adjacent vertices having the same parity set-labels.*

All set-labels mentioned in this section are arithmetic progressions so that the given signed graph S admits an arithmetic sumset labelling. Invoking the above results, we establish the following theorem for an edges of an arithmetic sumset graph to be a positive edge.

Theorem 6.16 (Sudev, Germina, 2015) *Let* S *be a signed graph which admits an arithmetic sumset labelling* f *. Then, an edge* uv *is a positive edge of a sumset labelled signed graph if and only if*

1. *The set-labels* $f\left(u\right)$ *and* $f\left(v\right)$ *are of different parity, provided the deterministic ratio of the edge* uv *is odd.*
2. *The set-label of the end vertex, with a minimum common difference, is of even parity, provided the deterministic ratio of the edge* uv *is even.*

The following theorem established a necessary and sufficient condition for an arithmetic sumset signed graph to a balanced signed graph.

Theorem 6.17 (Sudev, Germina, 2015) *An arithmetic sumset signed graph* S *is balanced if and only if its underlying graph* G *is bipartite.*

A spanned signed subgraph S' of a signed graph S is a signed graph which preserves signature and whose underlying graph H is a spanning subgraph of the underlying graph G of S. The following result is an obvious and immediate from the corresponding definition of the balanced signed graphs.

Theorem 6.18 (Sudev, Germina, 2015) *Let S' be a spanned signed subgraph of a balanced arithmetic sumset signed graph S. Then, S' is balanced with respect to induced labelling and signature if and only if the following conditions are hold.*

1. *The set $E\left(S \setminus S'\right)$ contains even number of negative edges in S, if the signed graph S is edge disjoint.*
2. *The set $E\left(S \setminus S'\right)$ contains an odd number of negative edges in S if some of the negative edges are common to two more cycles in S.*

Switching of Sumset Signed Graphs

The *switching* of a signed graph is the process of changing the sign of the edges in S whose end vertices have da ifferent sign. We denote the switched signed graph of a signed graph S by S'. The following theorem discusses the nature of the switched signed graph of a weak sumset signed graph.

Theorem 6.19 (Sudev, Germina, 2015) *The switched signed graph of a weak sumset signed graph is always an all-negative signed graph.*

Invoking the above theorem, a necessary and sufficient condition for the switched signed graph S' of a weak sumset signed graph S is balanced is established in the following result.

Theorem 6.20 (Sudev, Germina, 2015) *The switched signed graph S' of a weak sumset signed graph S is balanced if and only if the underlying graph of S (and S') is a bipartite graph.*

The following result discusses the conditions required for the switched signed graph of a strong sumset signed graph to be a homogeneous signed graph.

Theorem 6.21 (Sudev, Germina, 2015) *The switched signed graph S' of a strong sumset signed graph S is an all-negative signed graph if and only if at least one end vertex of every edge of S is an odd parity vertex.*

From the last paragraph of the above proof, we can observe an immediate result as stated below.

Theorem 6.22 (Sudev, Germina, 2015) *The switched signed graph S' of a strong sumset signed graph S is an all-positive signed graph if and only if all vertices in S are even parity vertices.*

A necessary and sufficient condition for a signed graph S to admit an arithmetic sumset labelling is found out in [90] as given below.

Theorem 6.23 (Sudev, Germina, 2015) *A signed graph S admits an arithmetic sumset labelling f if and only if for every edge of S, the set-labels of its end vertices are arithmetic progressions with common differences d_u and d_v such that $d_u \leq d_v$ and its deterministic ratio $\dfrac{d_v}{d_u}$ is a positive integer less than or equal to $\left| f\left(u\right) \right|$.*

The following theorem discussed the nature of the switched signed graphs of isoarithmetic sumset signed graphs.

Theorem 6.24 (Sudev, Germina, 2015) *The switched signed graphs of an isoarithmetic sumset signed graph S is an all-negative homogeneous signed graph.*

The following theorem discussed a necessary and sufficient condition for the switched signed graphs of isoarithmetic sumset signed graphs to be balanced.

Theorem 6.25 (Sudev, Germina, 2015) *The switched signed graph S' of an isoarithmetic sumset signed graph S is balanced if and only if the underlying graph G of S is bipartite.*

Next, assume that the deterministic ratio k of the edges in an arithmetic sumset signed graph S is greater than 1. Then, we have the following theorem.

Theorem 6.26 (Sudev, Germina, 2015) *The switched signed graph S' of an arithmetic sumset signed graph S is an all-negative homogeneous signed graph if and only if S contains no edge with even deterministic ratio, whose end vertices with a greater common difference is of even parity.*

In view of Theorem 6.26, we can establish a necessary and sufficient condition for the switched signed graph of a given arithmetic sumset signed graph to be an all-positive homogeneous signed graph as follows.

Theorem 6.27 (Sudev, Germina, 2015) *The switched signed graph S' of an arithmetic sumset signed graph S is an all-positive signed graph if and only if the following conditions hold.*

1. *The deterministic ratio of every edge of S is even; and*
2. *The end vertex with greater common difference of every edge of S is an even parity vertex.*

Theorem 6.28 (Sudev, Germina, 2015) *If S is a prime arithmetic sumset signed graph with deterministic ratio $p > 2$, then its switched signed graph S' is an all-negative signed graph.*

Theorem 6.29 (Sudev, Germina, 2015) *The switched signed graph S' of a prime arithmetic sumset signed graph S, with deterministic ratio $p > 2$, is balanced if and only if its underlying graph is a bipartite graph.*

MODULAR SUMSET LABELLING OF GRAPHS

Let n be a positive integer. We denote the set of all non-negative integers modulo n by $\mathbb{Z}_n$ and its power set by $\mathcal{P}(\mathbb{Z}_n)$. The modular sumset of the two subsets A and B of $\mathbb{Z}_n$, denoted by $A + B$, is the set defined by

$$A + B = \left\{ x : a + b \equiv x \ (mod \ n), a \in A, b \in B \right\}.$$

Through out our discussion, $A + B$ is the sumset of A and B. It can also be noted that

$$A, B \subseteq \mathbb{Z}_n A + B \subseteq \mathbb{Z}_n.$$

Then, using the concepts of modular sumsets of sets and analogous to the definition of sumset labellings of graphs, the notion of modular sumset labelling of a graph has been introduced in Naduvath (2017) as follows:

Definition 7.1 (Naduvath, 2017) A function $f : V(G) \to \mathcal{P}(\mathbb{Z}_n)$, whose induced function

$$f^+\left(uv\right): E\left(G\right) \to \mathcal{P}\left(\mathbb{Z}_n\right)$$

is defined by

$$f^+\left(uv\right) = f\left(u\right) + f\left(v\right),$$

is said to be a *modular sumset labelling* if f is injective.

It is proved in Naduvath (2017) that every graph admits a modular sumset labelling. A graph G which admits a modular sumset labelling is called an modular sumset graph. In View of the corresponding results on the bounds for the cardinality of sumsets of two sets, the cardinality of the set-label of a modular sumset graph G is given in the following theorem.

Theorem 7.1 (Naduvath, 2017) *Let* $f : V\left(G\right) \to \mathcal{P}\left(\mathbb{Z}_n\right)$ *be a modular sumset labelling of a given graph* G *. Then, for any edge* $uv \in E\left(G\right)$, *we have*

$$\left|f\left(u\right)\right| + \left|f\left(v\right)\right| - 1 \le \left|f^+\left(uv\right)\right| = \left|f\left(u\right) + f\left(v\right)\right| \le \left|f\left(u\right)\right|\left|f\left(v\right)\right| \le n.$$

Analogous to the notions of sumset labelled graphs, the following terms and definitions have been defined for modular sumset graphs.

1. The cardinality of the set-label of an element of G is said to be the set-labelling number of that element.
2. If all the vertices of a graph G have the same set-labelling number, say l, then we say that $V\left(G\right)$ is l-uniformly set-labelled.
3. A modular sumset labelling of G is said to be a k-uniform modular sumset labelling if the set-labelling number of all edges of G have the same set-labelling number k.
4. The elements of G having the set-labelling number 1 are called the mono-indexed elements of G.
5. The smallest value of n such that $f : V\left(G\right) \to \mathcal{P}\left(\mathbb{Z}_n\right)$ is a modular sumset labelling of a given graph G is called the modular sumset number of G. The modular sumset number of a graph G is denoted by $\eta\left(G\right)$.

An edge can be mono-indexed if and only if its end vertices are mono-indexed. The set-labelling number of an edge of a given graph G is equal to the set-labelling number of both of its end vertices if and only if the edge and its end vertices are mono-index.

One may note that the modular sumset number of a graph G is at least $1 + \log_2 m$ (see Naduvath, 2017).

The following result establishes the condition for a sumset to have the same cardinality of one or both of its summands.

Proposition 7.2 (Naduvath, 2017) *Let A and B be two non-empty subsets of $\mathbb{Z}_n$. Then, $|A + B| = |A|$ (or $|A + B| = |B|$) if and only if either $|A| = |B| = \mathbb{Z}_n$ or $|B| = 1$ (or $|A| = 1$). More over, $|A + B| = |A| = |B|$ if and only if $|A| = |B| = \mathbb{Z}_n$ or $|A| = |B| = \mathbb{Z}_n$.*

Hence, analogous to the corresponding notion of sumset labelled graphs (see [34, 60]), we have the following definition.

Definition 7.2 (Naduvath, 2017) A modular sumset labelling f of a graph G is said to be a *weak modular sumset labelling* of G if the set-labelling number of every edge of G is equal to the set-labelling number of at least one of its end vertices.

A graph G which admits a weak modular sumset labelling is called a weak modular sumset graph. It is to be noted that for a weak modular sumset graph, no two adjacent vertices can have non-singleton set-labels. As in the case of

Hence, analogous to the corresponding of weak sumset labelled graphs the following results are valid for weak modular sumset graphs.

Theorem 7.3 (Naduvath, 2017) *A graph G admits a weak modular sumset labelling if and only if G is bipartite or contains mono-indexed edges.*

Theorem 7.4 (Naduvath, 2017) *A graph G admits a weakly uniform modular sumset labelling if and only if G is bipartite.*

The notion of the weak modular sumset number of a given graph is introduced in [47] as follows.

Definition 7.3 (Naduvath, 2017) The *weak modular sumset number* of a graph G is defined to be the minimum value of n such that a modular sumset labelling $f : V(G) \to \mathcal{P}(\mathbb{Z}_n)$ is a weak modular sumset labelling of G.

The following theorem determines the weak sumset number of an arbitrary graph G in terms of its covering and independence numbers.

Theorem 7.5 (Naduvath, 2017) *Let G be a modular sumset graph and α and β be its covering number and independence number respectively. Then, the weak modular sumset number of G is $max\{\alpha, r\}$, where r is the smallest positive integer such that $2^r - r - 1 \geq \beta$.*

The weak modular sumset number of many fundamental graph classes has been determined in the paper (Sudev & Germina, 2015).

The minimum cardinality of the ground set X when the given graph G admits a weakly uniform modular sumset labelling the following result.

Theorem 7.6 (Naduvath, 2017) *Let G be a graph with covering number α and independence number β and let G admits a weakly k-uniform modular sumset labelling, where $k < \alpha$ being a positive integer. Then, the minimum cardinality of the ground set $\mathbb{Z}_n$ is $max\{\alpha, r\}$, where r is the smallest positive integer such that $\binom{r}{k} \geq \beta$.*

Analogous to the corresponding notions on strong sumset labelled graphs, the following notion has been introduced in (Naduvath, 2017).

Definition 7.4 (Naduvath, 2017) *Let $f : V(G) \to \mathcal{P}(\mathbb{Z}_n)$ be a modular sumset labelling defined on a given graph G. Then, f is said to be a strong modular sumset labelling if for the associated function*

$$f^+ : E\left(G\right) \to \mathcal{P}\left(\mathbb{Z}_n\right), \left|f^+\left(uv\right)\right| = \left|f\left(u\right)\right|\left|f\left(v\right)\right| \; \forall uv \in E\left(G\right).$$

A graph which admits a strong modular sumset labelling is called a strong modular sumset graph.

Then, a necessary and sufficient condition for a graph to admit a strong modular sumset labelling has been proved in Naduvath, 2017) as given below.

Theorem 7.7 (Naduvath, 2017) *A modular sumset labelling* $f : V\left(G\right) \to \mathcal{P}\left(\mathbb{Z}_n\right)$ *of a given graph* G *is a strong modular sumset labelling of* G *if and only if*

$$D_{f\left(u\right)} \cap D_{f\left(v\right)} = \varnothing, \forall uv \in E\left(G\right),$$

where $\left|f\left(u\right)\right|\left|f\left(v\right)\right| \leq n$.

Analogous to the weak modular sumset number of a graph G, we can define the strong modular sumset number of G as the minimum cardinality required for the ground set $\mathbb{Z}_n$ so that G admits a strong modular sumset labelling. The choice of ground set $\mathbb{Z}_n$ is very important in this context because n should be sufficiently large so that the vertices of the given graph can be labelled in such a way that the difference sets of these set-labels of all adjacent vertices must be pairwise disjoint.

Analogous to the corresponding results on strong sumset labelled graphs and strongly uniform sumset labelled graphs proved in Sudev and Germina (2014), the following results are valid for strong modular sumset graphs also.

Theorem 7.8 (Naduvath, 2017) *For a positive integer* $k \leq n$, *a modular sumset labelling* $f : V\left(G\right) \to \mathcal{P}\left(\mathbb{Z}_n\right)$ *of a given connected graph* G *is a a strongly* k*-uniform modular sumset labelling of* G *if and only if either* k *is a perfect square or* G *is bipartite.*

Theorem 7.9 (Naduvath, 2017) *For a positive non-square integer* $k \leq n$, *a modular sumset labelling* $f : V\left(G\right) \to \mathcal{P}\left(\mathbb{Z}_n\right)$ *of an arbitrary graph* G *is a a strongly* k*-uniform modular sumset labelling of* G *if and only if either* G *is bipartite or a disjoint union of bipartite components.*

For a positive integer $k \leq n$, the maximum number of components in a strongly k-uniform modular sumset graph is as follows.

Proposition 7.10 (Naduvath, 2017) *Let* f *be a strongly* k*-uniform modular sumset labelling of a graph* G *with respect to the ground set* $\mathbb{Z}_n$. *Then, the maximum number of components in* G *is the number of distinct pairs of divisors* r *and* s *of* k *such that* $rs = k$.

It can be observed that a strongly k-uniform modular sumset graph can have a non-bipartite component if and only if k is a perfect square. More over, a strongly k-uniform modular sumset graph G can have at most one non-bipartite component.

A modular sumset labelling $f : V\left(G\right) \to \mathcal{P}\left(\mathbb{Z}_n\right)$ of a given graph G is a *maximal modular sumset labelling* of G if the set-label of every edge of G is the ground set $\mathbb{Z}_n$ itself. The conditions required for a graph to admit a maximal modular sumset labelling has been found out in [47].

Proposition 7.11 (Naduvath, 2017) *The modular sumset labelling* $f : V\left(G\right) \to \mathcal{P}\left(\mathbb{Z}_n\right)$ *of a given graph* G *is a maximal modular sumset labelling of* G *if and only if for every pair of adjacent vertices* u *and* v *of* G *some or all of the following conditions hold.*

1. $\left| f(u) \right| + \left| f(v) \right| \geq n$ *if* $D_{f(u)} \cap D_{f(v)} \neq \varnothing$. *The strict inequality holds when* $D_{f(u)}$ *and* $D_{f(v)}$ *are arithmetic progressions containing the same elements.*

2. $\left| f(u) \right| \left| f(v) \right| \geq n$ *if* $D_{f(u)} \cap D_{f(v)} = \varnothing$.

A necessary and sufficient condition for a strong modular sumset labelling of a graph G to be a maximal modular sumset labelling of G .

Theorem 7.12 (Naduvath, 2017) *Let* f *be a strong sumset-labelling of a given graph* G . *Then,* f *is a maximal sumset-labelling of* G *if and only if* n *is a perfect square or* G *is bipartite or a disjoint union of bipartite components.*

SCOPE FOR FURTHER STUDIES

More properties and characteristics of different types of sumset labellings, both uniform and non-uniform, are yet to be investigated. Many problems regarding the admissibility of sumset labelling by various graph classes, graph operations and graph products are still open. The studies on sumset graphs whose set-labels are well-known sequences are also promising for further studies. Some of the problems in this area identified are the following.

Problem 1. Characterise the sumset graphs, whose vertex set-labels are arithmetic progressions, but the edge set-labels are not arithmetic progressions.

Problem 2. Characterise the arithmetic sumset graphs, the deterministic ratios of whose edges are prime numbers.

A sumset labelling f of a graph G is said to be a bitopological sumset labelling of G if both the collections $f(V) \cup \{\varnothing\}$ and $f^+(E) \cup \{\varnothing, \{0\}\}$ are topologies of the ground set X .

Problem 3. Verify the existence of graphical realisation of topological sumset labellings corresponding to different standard topologies.

Problem 4. Verify the existence of certain types of sumset labellings for a given graph G such that the collections of its vertex set-labels and edge set-labels form different algebraic structures such as semigroups, σ -algebra etc. of the ground set X and determine the structural properties of the graphs which admit these types of sumset labellings.

Analogous to the different types of sumset graphs, the investigation can be extended to the field of modular sumset graphs also. All these facts highlight a wide scope for further investigations in this area.

REFERENCES

Abhishek, K. (2007). *New directions in the theory of set-valuations of graphs* (Ph. D Thesis). Kannur University, Kannur, India.

Abhishek, K. (2013). Set-valued graphs-II. *J. Fuzzy Set Valued Anal., 2013*, 1–16. doi:10.5899/2013/jfsva-00149

Abhishek, K. (2015). A note on set-indexed graphs. *J. Discrete Math. Sci. Cryptography*, *18*(1-2), 31–40. doi:10.1080/09720529.2013.867637

Abhishek, K. (2015). Set-valued graphs: A survey. *J. Discrete Math. Sci. Cryptography*, *18*(1-2), 55–80. doi:10.1080/09720529.2014.894306

Acharya, B. D. (1983). *Set-valuations and their applications. MRI Lecture Notes in Applied Mathematics, No.2*. Allahabad: The Mehta Research Institute of Mathematics and Mathematical Physics.

Acharya, B. D. (2001). Set-indexers of a graph and set-graceful graphs. *Bull. Allahabad Math. Soc.*, *16*, 1–23.

Acharya, B. D. (2012). Set-valuations of signed digraphs. *J. Combin. Inform. System Sci.*, *37*(2-4), 145–167.

Acharya, B. D., & Germina, K. A. (2013). Set-valuations of graphs and their applications: A survey. *Ann. Pure Appl. Math.*, *4*(1), 8–42.

Acharya, B. D., Germina, K. A., & Paul, V. (2010). Linear hypergraph set-indexers of a graph. *Int. Math. Forum*, *5*(68), 3359-3370.

Acharya, B. D., Germina, K. A., Princy, K. L., & Rao, S. B. (2008). On set-valuations of graphs. In Labelling of Discrete Structures and Applications. Narosa Pub. House.

Akiyama, J., Avis, D., Chavtal, V., & Era, H. (1981). Balancing signed graphs. *Discrete Applied Mathematics*, *3*(4), 227–233. doi:10.1016/0166-218X(81)90001-9

Anandavally, T. M. K. (2013). A characterisation of 2 -uniform IASI graphs. *Int. J. of Contemp. Math. Sci.*, *8*(10), 459–462.

Apostol, T. M. (1989). *Introduction to analytic number theory*. New York: Springer-Verlag.

Berge, C. (1979). *Graphs and hypergraphs*. Amsterdam: North-Holland.

Berge, C. (1989). *Hypergraphs: Combinatorics of finite sets*. Amsterdam: North-Holland.

Berge, C. (2001). *Theory of graphs*. Dover Pub.

Bollabás, B. (1998). *Modern graph theory* (International Edition). Springer.

Bondy, J. A., & Murty, U. S. R. (2008). *Graph theory*. Springer.

Brandstädt, A., Le, V. B., & Spinrad, J. P. (1999). *Graph classes: A survey*. Philadelphia: SIAM.

Burton, D. M. (2007). *Elementary number theory*. New Delhi: Tata McGraw-Hill Inc.

Chartrand, G., & Lesniak, L. (1996). *Graphs and digraphs*. CRC Press.

Chartrand, G., & Zhang, P. (2005). *Introduction to graph theory*. McGraw-Hill Inc.

Chithra, K. P., Sudev, N. K., & Germina, K. A. (2014). The sparing number of the Cartesian product of certain graphs. *Commun. Math. Appl.*, *5*(1), 23–30.

Chithra, K. P., Sudev, N. K., & Germina, K. A. (2014). A study on the sparing number of the corona of certain graphs, *Res. Review. Discrete Math. Structure (London, England)*, *1*(2), 5–15.

Chithra, K. P., Sudev, N. K., & Germina, K. A. (2015). On the sparing number of the edge-corona of graphs. *International Journal of Computers and Applications*, *118*(1), 1–5. doi:10.5120/20706-3025

Clark, W. E. (2002). *Elementary number theory*. Department of Mathematics, University of South Florida.

Clarke, J., & Holton, D. A. (1991). *A first look at graph theory*. Singapore: World Scientific Pub.

Deo, N. (1974). *Graph theory with application to engineering and computer science*. Delhi: Prentice Hall of India Pvt. Ltd.

Diestel, R. (2010). *Graph theory*. New York: Springer-Verlag.

Dugundji, J. (1966). *Topology*. Boston: Allyn and Bacon.

Gallian, J. A. (2018). A dynamic survey of graph labelling. *The Journal of Combinatorics*, DS-6.

Germina K.A., (2011). *Set-valuations of a graph and applications*. Final Technical Report, DST Grant-In-Aid Project No.SR/S4/277/05, The Dept. of Science and Technology (DST), Govt. of India.

Germina, K. A., & Anandavally, T. M. K. (2012). Integer additive set-indexers of a graph: Sum square graphs. *J. Combin. Inform. System Sci.*, *37*(2-4), 345–358.

Germina, K. A., & Sudev, N. K. (2013). On weakly uniform integer additive set-indexers of graphs. *Int. Math. Forum*, *8*(37), 1827-1834. DOI: 10.12988/imf.2013.310188

Golomb, S. W. (1972). How to number a graph. In R. C. Read (Ed.), *Graph Theory and Computing* (pp. 13–22). Academic Press.

Gross, J., & Yellen, J. (1999). *Graph theory and its applications*. CRC Press.

Hammack, R., Imrich, W., & Clavzar, S. (2011). *Handbook of product graphs*. CRC Press.

Harary, F. (1953). On the notion of balance of a signed graph. *The Michigan Mathematical Journal*, *2*(2), 143–146.

Harary, F. (1969). *Graph theory*. New Delhi: Narosa Pub. House.

Harary, F., & Palmer, E. M. (1973). *Graphical enumeration*. Academic Press Inc.

Joshi, K. D. (1983). *Introduction to General Topology*. New Delhi: New Age International.

Joshi, K. D. (2003). *Applied discrete structures*. New Delhi: New Age International.

Krishnamoorthy, V. (1966). On the number of topologies of finite sets. *The American Mathematical Monthly*, *73*(2), 154–157.

McKee, T. A., & McMorris, F. R. (1999). *Topics in intersection graph theory*. Philadelphia: SIAM.

Munkers, J. R. (2000). *Topology*. Delhi: Prentice Hall of India.

Naduvath, S. (2017). A study on modular sumset labelling of graphs. *Discrete Mathematics, Algorithms, and Applications*, *9*(1), 1–16. doi:10.1142/S1793830917500392

Naduvath, S., & Augustine, G. (2018). *An introduction of sumset valued graphs*. Beau Bassin, Mauritius: Lambert Academic Publ.

Nathanson, M. B. (1996). *Additive number theory: Inverse problems and geometry of sumsets*. New York: Springer.

Ore, O. (1962). *Theory of Graphs*. American Math. Soc. Colloquium Pub.

Paul, V. (2012). *Labeling and set-indexing hypergraphs of a graph and related areas* (PhD Thesis). Kannur Univ., Kannur, India.

Paul, V., & Germina, K. A. (2011). On uniform linear hypergraph set-indexers of graphs. *Int. J. Contemp. Math. Sci.*, *6*(18), 861–868.

Princy, K. L. (2007). *Some studies on set-valuations of graph-embedding and NP-completeness* (Ph. D Thesis). Kannur Univ., Kannur, India.

Rao, S. B., & Germina, K. A. (2011). Graph labelings and complexity problems: A review. In P. Panigrahi & S. B. Rao (Eds.), *Graph Theory Research Directions*. New Delhi: Narosa Pub. House.

Roberts, F. S. (1978). *Graph theory and its practical applications to problems of society*. Philadelphia: SIAM.

Rosa, A. (1967). On certain valuation of the vertices of a graph. In *Theory of Graphs*. New York: Gordon and Breach.

Rousa, I. K. (2008). *Sumsets and structures*. Budapest: Alfréd Rényi Institute of Mathematics.

Singh, G. S. (1998). A note on labelings of graphs. *Graphs and Combinatorics*, *14*, 201–207.

Stoll R.R., (1979). *Set theory and Logic*. Dover pub.

Sudev, N. K. (2015). *Set-valuations of discrete structures and their applications* (PhD Thesis). Kannur University, Kannur, India.

Sudev, N. K., Ashraf, P. K., & Germina, K. A. (2019).Integer additive set-valuations of signed graphs. *TWMS J. Appl. Engg. Math.*

Sudev, N. K., Chithra, K. P., & Germina, K. A. (2015). On integer additive set-filter graphs. *J. Abst. Comput. Math.*, *3*, 8–15.

Sudev, N. K., & Germina, K. A. (2014). On integer additive set-indexers of graphs. *Int. J. Math. Sci. Eng. Appl.*, *8*(II), 11–22.

Sudev, N. K., & Germina, K. A. (2014). A characterization of weak integer additive set-indexers of graphs. *J. Fuzzy Set Valued Anal.*, *2014*, 1–7. doi:10.5899/2014/jfsva-00189

Sudev, N. K., & Germina, K. A. (2014). *Weak integer additive set-indexers of graph operations*, Global J. Math. Sci. Theory. *Practical*, *6*(1), 25–36.

Sudev, N. K., & Germina, K. A. (2014). A study of semi-arithmetic integer additive set-indexers of graphs. *Int. J. Math. Sci. Eng. Appl.*, *8*(III), 157–165.

Sudev, N. K., & Germina, K. A. (2014). On the sparing number of certain graph structures. *Annals Pure Appl. Math.*, *6*(2), 140–149.

Sudev, N. K., & Germina, K. A. (2014). Associated graphs of certain arithmetic IASI graphs. *Int. J. Math. Soft Comput.*, *4*(2), 71–80.

Sudev, N. K., & Germina, K. A. (2014). A note on the sparing number of graphs. *Adv. Appl. Discrete Math.*, *14*(1), 51–65.

Sudev, N. K., & Germina, K. A. (2014). Further studies on the sparing number of graphs, *TechS Vidya e-J. Res.*, *2*, 25–36.

Sudev, N. K., & Germina, K. A. (2014). Weak integer additive set-indexers of certain graph products. *J. Inform. Math. Sci.*, *6*(1), 35–43.

Sudev, N. K., & Germina, K. A. (2014). A Characterisation of strong integer additive set indexers of graphs. *Commun. Math. Appl.*, *5*(3), 101–110.

Sudev, N. K., & Germina, K. A. (2014). A study on semi-arithmetic set-indexers of graphs. *Int. J. Math. Sci. Eng. Appl.*, *8*(III), 157–165.

Sudev, N. K., & Germina, K. A. (2015). Weak integer additive set indexes of certain graph classes. *J. Discrete Math. Sci. Cryptography*, *18*(2-3), 117–128. doi:10.1080/09720529.2014.962866

Sudev, N. K., & Germina, K. A. (2015). Some new results on weak integer additive set-labelling of graphs. *International Journal of Computers and Applications*, *128*(5), 1–5. doi:10.5120/ijca2015906514

Sudev, N. K., & Germina, K. A. (2015). Some new results on strong integer additive set indexers of graphs. *Discrete Mathematics, Algorithms, and Applications*, *7*(1), 1–11. doi:10.1142/S1793830914500657

Sudev, N. K., & Germina, K. A. (2015). A note on the sparing number on the sieve graphs of certain graphs, *Appl. Mathematical Notes*, *15*, 29–37.

Sudev, N. K., & Germina, K. A. (2015). A study on the nourishing number of graphs and graph power. *Math.*, *3*, 29–39. doi:10.3390/math3010029

Sudev, N. K., & Germina, K. A. (2015). A study on topological integer additive set-labelling of graphs. *Electron. J. Graph Theory Appl.*, *3*(1), 70–84. doi:10.5614/ejgta.2015.3.1.8

Sudev, N. K., & Germina, K. A. (2015). On certain arithmetic integer additive set-indexers of graphs. *Discrete Mathematics, Algorithms, and Applications*, *7*(3), 1–15. doi:10.1142/S1793830915500251

Sudev, N. K., & Germina, K. A. (2015). Some new results on weak integer additive set-labelled graphs. *International Journal of Computers and Applications*, *128*(1), 1–5. doi:10.5120/ijca2015906514

Sudev, N. K., & Germina, K. A. (2015). On the hypergraphs associated with certain integer additive set-labelled graphs. *J. Adv. Res. Appl. Math.*, *7*(4), 23–33. doi:10.5373/jaram.2287.021015

Sudev, N.K., & Germina, K.A. (2015). On integer additive set-sequential graphs. *Int. J. Math. Combin.*, *2015*(3), 125-133.

Sudev, N. K., & Germina, K. A. (2015). A study on topogenic integer additive set-labelled graphs. *J. Adv. Res. Pure Math.*, *7*(3), 15–22. doi:10.5373/jarpm.2230.121314

Sudev, N. K., & Germina, K. A. (2015). Integer additive set-valuations of signed graphs. *Carpathian Math. Publ.*, *7*(2), 236–246.

Sudev, N. K., & Germina, K. A. (2015). Switched signed graphs of integer additive set-valued signed graphs. *Discrete Mathematics, Algorithms, and Applications*, *9*(4), 1–10. doi:10.1142/S1793830917500434

Sudev, N. K., & Germina, K. A. (2017). A study on prime arithmetic integer additive set-indexers of graphs. *Proyecciones J. Math.*, *36*(2), 195–208. doi:10.4067/S0716-09172017000200195

Sudev, N. K., & Germina, K. A. (2018). Arithmetic integer additive set-indexers of certain graph operations. *J. Inform. Math. Sci.*, *10*(1-2), 321–332. doi:10.26713/jims.v10i1+&+2.617

Sudev, N. K., & Germina, K. A. (2019). A study on integer additive set-graceful graphs. *Southeast Asian Bulletin of Mathematics*.

Sudev, N. K., Germina, K. A., & Chithra, K. P. (2014). Weak set-labelling number of certain integer additive set-labelled graphs. *International Journal of Computers and Applications*, *114*(2), 1–6. doi:10.5120/19947-1772

Thomas, B. K. (2009). *Advanced studies on labelling of graphs and hypergraphs and related areas* (PhD Thesis). Kannur University, Kannur, India.

Voloshin, V. I. (2009). *Introduction to graph and hypergraph theory*. New York: Nova Science Pub.

West, D. B. (2001). *Introduction to graph theory*. Pearson Education Inc.

Zaslavsky, T. (1981). Characterizations of signed graphs. *Journal of Graph Theory*, *5*(4), 401–406. doi:10.1002/jgt.3190050409

Zaslavsky, T. (1982). Signed graphs. *Discrete Applied Mathematics*, *4*(1), 47–74. doi:10.1016/0166-218X(82)90033-6

Zaslavsky, T. (2010). *Balance and clustering in signed graphs*. Hyderabad, India: C R Rao Adv. Instt. of Mathematics Statistics and Computer Science.

Zaslavsky, T. (2012). Signed graphs and geometry. *J. Combin. Inform. System Sci.*, *37*(2-4), 95–143.

KEY TERMS AND DEFINITIONS

Arithmetic Sumset Labeling: A sumset labeling in which the set-labels of all vertices and edges of the graph are arithmetic progressions.

Graceful Sumset Labeling: A sumset labeling with the collection of all edge set-set-labels together with the null set and {0} form the power set of the ground set.

Sequential Sumset Labeling: A sumset labeling with the collection of all vertex set-set-labels and edge-set-labels together with the nullset form the powerset of the ground set.

Set-Indexing Number of an Element: The cardinality of the set-label of that element.

Set-Label Hypergraphs: The hypergraphs whose elements are the set-labels of the elements of a sumset graph.

Strong Sumset Labeling: A sumset labeling in which the set-indexing number of every edge is the product of the set-indexing numbers of its end vertices.

Sumset Labeling: An assignment of number sets to the elements of a graph such that the set-label of each edge is the sumset of the set-labels of its end vertices.

Sumset Signed Graph: A signed graph with a sumset labeling and whose signs are induced by the set-labels of its elements.

Topogenic Sumset Labeling: A sumset labeling with the collection of all vertex set-set-labels and edge-set-labels together with the nullset form a topology of the ground set.

Weak Sumset Labeling: A sumset labeling in which the set-indexing number of each edge is equal to that of at least one of its end vertices.

Chapter 10
Tripartite and Quadpartite Size Ramsey Numbers for All Pairs of Connected Graphs on Four Vertices

Chula J. Jayawardene
University of Colombo, Sri Lanka

ABSTRACT

A popular area of graph theory is based on a paper written in 1930 by F. P. Ramsey titled "On a Problem on Formal Logic." A theorem which was proved in his paper triggered the study of modern Ramsey theory. However, his premature death at the young age of 26 hindered the development of this area of study at the initial stages. The balanced size multipartite Ramsey number mj (H,G) is defined as the smallest positive number s such that Kj×s→ (H,G). There are 36 pairs of (H, G), when H, G represent connected graphs on four vertices (as there are only 6 non-isomorphic connected graphs on four vertices). In this chapter, the authors find mj (H, G) exhaustively for all such pairs in the tripartite case j=3, and in the quadpartite case j=4, excluding the case $m_4 (K_4,K_4)$. In this case, the only known result is that $m_4 (K_4,K_4)$ is greater than or equal to 4, since no upper bound has been found as yet.

HISTORY

After the publication of the original paper by F. P. Ramsey, the resurrection of Ramsey Theory for graphs emerged in the paper by Paul Erdös and George Szekeres, published around 1935.

Theorem (by Paul Erdös and George Szekeres)

For any $m \geq 2$ and $n \geq 2, r(m,n)$ exists and it satisfies

$$r(m,n) \leq r(m-1,n) + r(m,n-1)$$

DOI: 10.4018/978-1-5225-9380-5.ch010

and

$$r\left(m,n\right) \leq \binom{m+n-2}{m-1}.$$

Following the publication of the paper by Paul Erdös and George Szekeres, the exact determination of these numbers for small graphs has been attempted by many mathematicians. Unfortunately, their progress was hindered by the stubbornly resistant ($r\left(5,5\right)$), which is presently known to lie between 43 and 48 (see Vigleik Angeltveit and Brendan D. McKay (Angelteveit, 2017) for the best upper bound of 48 and Exoo (Exoo, 1989) for the best lower bound of 43). Finding sharp bounds for ($r\left(6,6\right)$) appears to be an even more arduous task. This has been eloquently expressed by the great mathematician Paul Erdös using the following quotation (see 1990 Scientific American article by Ronald Graham and Joel Spencer).*Suppose aliens invade the earth and threaten to obliterate it in a year's time unless human beings can find the Ramsey number for red five and blue five ($r\left(5,5\right)$). We could marshal the world's best minds and fastest computers, and within a year we could probably calculate the value. If the aliens demanded the Ramsey number for red six and blue six ($r\left(6,6\right)$), however, we would have no choice but to launch a pre-emptive attack.* – Paul Erdös (1990 Scientific American article by Ronald Graham & Joel Spencer)

Researchers are now trying to approach this problem by using new techniques to investigate whether the lower bound 43 (Exoo, 1989) and the upper bound 48 (Angelteveit (2017)) could be improved. Off-shoots of classical Ramsey numbers, introduced by Burger and Vuuren (2004), Syfrizal et al., (2005, 2012), are the multipartite Ramsey numbers.

Let K_N denote the complete graph on N vertices. For any two graphs, say H, G (without loops and parallel edges), we say that $K_N \rightarrow (H,G)$, if for any red/blue colouring of K_N, given by $K_N = H_R \oplus H_B$, there exists a red copy H in H_R or a blue copy G in H_B. In Ramsey theory one important part deals with the exact determination of Ramsey numbers, $r(H,G)$, defined as the smallest positive number N such that $K_N \rightarrow (H,G)$. Formally, the balance multipartite graph $K_{j\times s}$ can be viewed as a graph consisting of the vertex set

$$V\left(K_{j\times s}\right) = \left\{v_{m,n} : m \in \left\{1,2,...j\right\} and\, n \in \left\{1,2,...s\right\}\right\}$$

and the edge set

$$E\left(K_{j\times s}\right) = \left\{\left(v_{m,n}, v_{m',n'}\right) : m,m' \in \left\{1,2,...,j\right\}, m \neq m'\, and\, n,n' \in \left\{1,2,...,s\right\}\right\}.$$

In such a graph, the set of vertices in the mth partite set is denoted by $\left\{v_{mn} \mid n \in \left\{1,2,...,s\right\}\right\}$. The Ramsey number $m_j(H,G)$ is defined as the smallest positive number s such that that $K_{j\times s} \rightarrow (H,G)$.

Until now, the exact value $m_j(H,G)$, has been found only for a few pairs of graphs when $\left|V\left(G\right)\right| \le 5$, $\left|V\left(H\right)\right| \le 5$ and $j \ge 3$. In this chapter, we restrict our attention to the tripartite case $j = 3$, and the quadpartite case $j = 4$ and find the exact value of $m_j(H,G)$ exhaustively for all 36 pairs (H,G) in both cases (excluding the exact value of $m_4\left(K_4, K_4\right)$).

SOME RECENT OFFSHOTS OF CLASSICAL RAMSEY NUMBERS

Some other offshoots of classical Ramsey numbers are

- Convex Ramsey numbers (Jayawardene et al., 2017b),
- The Bondy-Erdos conjecture (Bollobas, Jayawardene et al., 2000).

Convex Ramsey Numbers

Convex Ramsey numbers, originated in a talk given by H. Harborth at an A. M. S. meeting held in Memphis (in 1997), are related to a Geometric variation of Classical Ramsey numbers. In this meeting, Harborth considered a regular graph K_n drawn in the 2 dimensional Euclidian plane with its vertices placed in a regular n- gon, where the n vertices of the complete graph are labelled by the vertices v_1, v_2, ..., v_n where $v_1 \le v_2 \le ... \le v_n$.

In such a situation, Harborth defined a convex cycle of length l, which is defined as a cycle of K_n consisting of the l vertices $w_1, w_2, ... w_n$ satisfying $v_1 \le w_1 \le w_2 \le ... \le w_l \le v_n$. Such a cycle was denoted by $C_l{}^c$ and the length l path generated by removing the (w_1, w_l) from the $C_l{}^c$ was denoted by $P_l{}^c$. As expected, we say that $K_N \to (P_k{}^c, P_l{}^c)$, if for any red/blue colouring of K_N, given by $K_N = H_R \oplus H_B$, there exists a red copy of $P_k{}^c$ in H_R or a blue copy of $P_l{}^c$ in H_B. Consequently he defined $r_c(P_k{}^c, P_l{}^c)$ as the smallest positive integer N such that $K_N \to (P_k{}^c, P_l{}^c)$.

This problem remained unsolved for almost two decades, and was eventually solved by Jayawardene, Rousseau and Harborth in 2017. In this paper the following theorem is proved.

Theorem: $r_c\left(P_k^c, P_l^c\right) = 1 + \left(k-1\right)\left(l-2\right)$ if $l \ge k \ge 3$.

This theorem was proved by using the following four lemmas:

Lemma 1: *A two-coloring of a convex n-gon with $n > \left(m-1\right)\left(l-2\right)$ that has no blue P_m^c and no red P_l^c contains a red $\left(l-1\right)$-Block.*

Lemma 2: *The two-coloring of the convex r-gon (where $r = 1 + \left(k-1\right)\left(l-2\right)$) that has no blue P_k^c, no red P_l^c contains a red $\left(l-1\right)$-Block S_1. Let $v_{1,1}$ denote the first vertex appearing after the rightmost vertex of S_1. Then the remaining $v_{1,1}$ oriented convex $\left(r-l+1\right)$-gon does not contain a blue $v_{1,1}$ oriented P_{k-2}^c.*

Lemma 3: *The two-coloring of the convex r-gon (where $r = 1 + \left(k-1\right)\left(l-2\right)$) that has no blue P_k^c, no red P_l^c contains a red $\left(l-1\right)$-Block S_1. Let $v_{1,1}$ denote the first vertex appearing after the* *rightmost vertex of* S_1. *Then the remaining* $v_{1,1}$ *oriented convex* $(r-l+1)$-*gon does not contain a blue* $v_{1,1}$ *oriented* P_{k-2}^c.

Lemma 4: *The two-coloring of a convex* $r-(k-2)(l-1)$-*gon (where* $r = 1+(k-1)(l-2)$) *obtained by removing the vertices of* S_1 *and* $v_{j-1,1}$ *critical red* $(l-1)$-*Blocks* S_j *for* $2 \leq j \leq k-2$ *successively from the initial* r-*gon will contain a blue* P_k^c.

In addition to this Jayawardene et al (2017b), were also able to show that this result could be generalized to

Theorem 2 *For* t *colors and for* $3 \leq l_1 \leq l_2 \leq ... \leq l_t$ *we have*

$$r_c(P_{l_1}^c, P_{l_2}^c, ..., P_{l_t}^c) = 1 + (l_t - 2)\prod_{i=1}^{t-1}(l_i - 1).$$

The Bondy - Erdos Conjecture

In 1973, Bondy & Erdös proved the following Theorem (see Bondy et al., 1973)

Theorem *For* $n \geq m^2 - 2$,

$$r(C_n, K_m) = (m-1)(n-1) + 1$$

Later on, it was conjectured in Schelp et al., (1978) that

$$r(C_n, K_m) = (m-1)(n-1) + 1$$

is in fact true for $n \geq m$, where $(n,m) \neq (3,3)$. The case $m = 3$ and $m = 4$ was proved by Faudree et al., in 1974. Consequently, Bollobas, Jayawardene et al., (2000) proved the result for $m = 5$. The best upper bound found up to now is by Nikiforov (2005), where he proved that the result is true for $m \geq 34$. However, the conjecture remains partially unsolved even today, as the conjecture has not been proven for $10 \leq m \leq 33$.

NOTATION

Let G and H be finite graphs without loops or multiple edges. In the arguments which follow, a two-colouring of the edges of K_N will mean a mapping from $E(K_N)$ into {red, blue}. The resulting set of red edges will be denoted by H_R, as will the corresponding graph induced by these edges. Similarly, the set of blue edges (or the blue subgraph) will be denoted by H_B. In such a case, we denote $K_N = H_R \oplus H_B$. Moreover, we say that $K_N \rightarrow (H,G)$, if for any red/blue colouring of K_N given by $K_N = H_R \oplus H_B$, there exists a red copy of H in H_R or a blue copy of G in H_B. It should be noted that $r(H,G)$ is defined as the smallest integer N such that $K_N \to (H, G)$. Likewise, Ramsey number $m_j(H,G)$ is defined as the smallest positive number s such that $K_{j \times s} \to (H, G)$. Standard notation will be used for graph theoretic functions of red and blue subgraphs; for example, for any red/blue colouring of $K_{j \times s}$, given by $K_{j \times s} = H_R \oplus H_B$, δ_R refers to the minimum degree in the red graph H_R. Standard symbols will be used for cycles, complete graphs, and complete multipartite graphs.

SUMMARY

In the main section, we will prove how these exact values can be found using various Graph Theory techniques. The most important results pertaining to this chapter can be summarized using the following two tables.

Tripartite Case: $m_3(H,G)$ for All Pairs of Connected Graphs on 4 Vertices

See Table 1.

Quadpartite Case: $m_4(H,G)$ for All Pairs of Connected Graphs on 4 Vertices

See Table 2.

Main Section

As addressed in Jayawardene (2018), we have the following useful theorem.

Theorem *If $j \geq 3$, then*

$$m_j\left(K_{1,3} + e, K_{1,3} + e\right) = \begin{cases} 1 & j \geq 7 \\ 2 & j = 6 \\ \infty & j \in \{3, 4, 5\} \end{cases}$$

Table 1. All values of $m_3(H,G)$

$m_3(H,G)$ $j = 3$ Case	Graph H	Graph G					
		P_4	$K_{1,3}$	C_4	$K_{1,3} + e$	B_2	K_4
Row 1	P_4	2	3	3	3	3	
Row 2	$K_{1,3}$	3	3	3	3	4	
Row 3	C_4	3	3	3	3	4	
Row 4	$K_{1,3} + e$	3	3	3			
Row 5	B_2	4	4	4			
Row 6	K_4						

Table 2. All values of $m_4(H,G)$

$m_4(H,G)$ $j = 4$ Case	Graph H	Graph G					
		P_4	$K_{1,3}$	C_4	$K_{1,3} + e$	B_2	K_4
Row 1	P_4	2	2	2	2	2	
Row 2	$K_{1,3}$	2	2	2	3	3	
Row 3	C_4	2	2	2	2	3	
Row 4	$K_{1,3} + e$	2	3	2			
Row 5	B_2	2	3	3			
Row 6	K_4						∞

Proof. If $j \geq 7$, since $r\left(K_{1,3}+e, K_{1,3}+e\right) = 7$ (Jayawardene, 2018), we get $m_j(K_{1,3}+e, K_{1,3}+e) = 1$.

Next color the graph $K_{6\times 1} = H_R \oplus H_B$, such that $H_R = 2K_3$. Then the graph has no red $K_{1,3}+e$ and has no blue $K_{1,3}+e$. Therefore, $m_6(K_{1,3}, K_{1,3}+e) \geq 2$. Next to show, $m_6(K_{1,3}+e, K_{1,3}+e) \leq 2$, consider any red/blue coloring given by $K_{6\times 2} = H_R \oplus H_B$, such that H_R contains no red $K_{1,3}+e$ and H_B contains no blue $K_{1,3}+e$. As $m_6\left(C_3, K_{1,3}+e\right) = 2$ from (Jayawardene, 2018), there is a red C_3, in H_R. Without loss of generality assume that the red C_3, is induced by say v_{11}, v_{21}, v_{31}. But then if we consider the vertex v_{11} it must be adjacent in blue to all of the vertices of v_{41}, v_{42}, v_{52}, v_{62} as otherwise would result in a red $K_{1,3}+e$. But then all the edges (v_{41}, v_{52}), (v_{41}, v_{62}), (v_{42}, v_{52}), (v_{42}, v_{62}) and (v_{52}, v_{62}) will be forced to be red as otherwise it will result in a blue $K_{1,3}+e$.

But then the vertex set $S = \{v_{41}, v_{42}, v_{52}, v_{62}\}$ will contain a red $K_{1,3}+e$, a contradiction. Thus, $m_6(K_{1,3}+e, K_{1,3}+e) \leq 2$. Therefore, we get $m_6\left(K_{1,3}+e, K_{1,3}+e\right) = 2$.

When

$$j \in \{3,4,5\}, \ m_3\left(C_3, K_{1,3}+e\right) = \infty$$

follows from (Jayawardene, 2018), Therefore, as C_3, is a subgraph $K_{1,3}+e$, it follows that,

$$m_j\left(K_{1,3}+e, K_{1,3}+e\right) = \infty \ for \ j = \{3,4,5\},$$

as required.

Corollary

As a direct consequence of the above theorem, we see that

$$m_j\left(G, H\right) = \infty \ when \ j \in \{3,4\},$$

for any $G, H \subseteq K_{1,3}+e$. This gives the required values of the two tables corresponding to the cells of rows 4, 5 and 6 when $H = K_{1,3} + e, B_2$ and K_4. Moreover, the exact values of the two tables correspond-

Figure 1. Diagram related to the proof of $m_6(K_{1,3}+e, K_{1,3}+e) \leq 2$

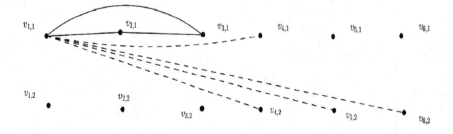

ing to the cells of row 4 when $H = P_4, K_{1,3}$ and C_4, also follow from Jayawardene (2018). As $m_j(G, H) = m_j(H, G)$, we will also get all values of the two tables corresponding to the cells of column 4 ($H = K_{1,3} + e$) when $G = P_4, K_{1,3}$ and C_4.

Next, by Jayawardene et al., (2017d) and Jayawardene et al., (2017a), we have the following theorems related to the 5th row of the two tables.

Theorem *If* $j \geq 3$, *then*

$$m_j(B_2, P_4) = \begin{cases} 4 & if \ j = 3 \\ 2 & if \ j \in \{4, 5\} \\ 1 & if \ j \geq 6 \end{cases}$$

Proof. Consider the coloring $K_{3 \times 3} = H_R \oplus H_B$ where H_B consist only of the three 3-cycles in

$$\left\{ v_{1i} v_{2i} v_{3i} v_{1i} : i \in \{1, 2, 3\} \right\}.$$

Then $K_{3 \times 3}$ has neither a red B_2 nor a blue P_4. Therefore $m_3(B_2, P_4) \geq 4$.

Now consider any red-blue colouring of $K_{3 \times 4}$. Assume it has no red B_2 and no blue P_4. Then each of the subgraphs H_1 and H_2 where

$$V(H_1) = \cup_{j=1}^{3} \left\{ v_{ij} : i \in \{1, 2, 3\} \right\}$$

and

$$V(H_2) = \cup_{j=2}^{4} \left\{ v_{ij} : i \in \{1, 2, 3\} \right\}$$

has a blue P_3. By relabeling we can assume them to be $v_{11} v_{21} v_{31}$ and $v_{14} v_{24} v_{33}$. As $K_{3 \times 4}$ is blue P_4 free, v_{14} and v_{31} are not incident to any blue edges other than $v_{14} v_{24}$ and $v_{31} v_{21}$ respectively. Then the red edges

$$\left\{ v_{14} v_{2i} : i \in \{2, 3\} \right\} \cup \left\{ v_{31} v_{2i} : i \in \{2, 3\} \right\} \cup \left\{ v_{14} v_{31} \right\}$$

forces a red B_2, a contradiction. Therefore, $m_3(B_2, P_4) = 4$.

Next consider the coloring $K_{5 \times 1} = H_R \oplus H_B$. where H_B consist only of the edge $v_{21} v_{31}$ and the 3-cycle $v_{41} v_{51} v_{11} v_{41}$. Then $K_{5 \times 1}$ has neither a blue P_4 nor a red B_2. Hence

$$m_4(B_2, P_4) \geq m_5(B_2, P_4) \geq 2.$$

Now consider any red-blue coloring of $K_{4 \times 2}$ which is blue P_4 free. Then it has a red C_3, say $v_{11} v_{21} v_{31} v_{11}$ (as $m_4 \left(C_3, P_4 \right) = 2$ by Jayawardene (2017c)), Suppose each of v_{41} and v_{42} are incident to two vertices of $V \left(C_3 \right)$ in blue. Then $K_{4 \times 2}$ has a blue P_4, a contradiction. Therefore at least one of v_{41} or v_{42} (say v_{41}) is incident to two vertices (say v_{11} and v_{21}) of $V \left(C_3 \right)$ in red. Then the red edges $v_{41} v_{11}, v_{41} v_{21}$ together with the red C_3 forms a red B_2. Therefore, $2 \geq m_4 \left(B_2, P_4 \right)$. Thus, we get

$$m_4 \left(B_2, P_4 \right) = m_5 \left(B_2, P_4 \right) = 2 .$$

Consider any red-blue coloring of $K_{6 \times 1}$ with no blue P_4. Then it has a red C_3, say $v_{11} v_{21} v_{31} v_{11}$ (as $m_6 \left(C_3, P_4 \right) = 1$ by Jayawardene (2017c)). Now considering v_{41} and v_{51} and arguing as above, it can be proved that $m_j \left(B_2, P_4 \right) = 1$ when $j \geq 6$.

Theorem *If* $j \geq 3$, *then*

$$m_j \left(B_2, K_{1,3} \right) = \begin{cases} 4 & if \ j = 3 \\ 3 & if \ j = 4 \\ 2 & if \ j \in \left\{ 5, 6 \right\} \\ 1 & if \ j \geq 7 \end{cases}$$

Proof. Consider a colouring of $K_{3 \times 3} = H_R \oplus H_B$ where H_B consists of blue C_9 as indicated in the following diagram.

Figure 2. When H_B consists of blue a C_9

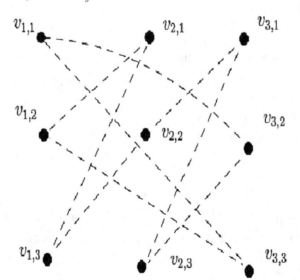

Then $K_{3\times3}$ has neither a red B_2 nor a blue $K_{1,3}$. Therefore, $m_4\left(B_2, K_{1,3}\right) \geq 4$.

Now consider any colouring of $K_{3\times4} = H_R \oplus H_B$. Assume the graph has no blue $K_{1,3}$. Then $K_{4\times3}$ has a red C_3, say $v_{11}v_{21}v_{31}v_{11}$ (as $m_3\left(C_3, K_{1,3}\right) = 3$ by Jayawardene (2017c). Let $W = \left\{v_{11}, v_{21}, v_{31}\right\}$. In order to avoid a red B_2 each vertex of $V\left(K_{4\times3}\right) \setminus W$ is adjacent in blue to one vertex in W.

By the pigeon hole principle this results in a blue $K_{1,3}$, a contradiction. Therefore, $m_3\left(B_2, K_{1,3}\right) \leq 4$. Hence $m_3\left(B_2, K_{1,3}\right) = 4$.

Next as $m_4\left(C_3, K_{1,3}\right) = 3$ (see Jayawardene, 2017c), and C_3 is a subgraph of B_2, we get $m_4\left(B_2, K_{1,3}\right) \geq 3$.

Now consider any colouring of $K_{4\times3} = H_R \oplus H_B$. Assume the graph has no blue $K_{1,3}$. Then $K_{4\times3}$ has a red C_3, say $v_{11}v_{21}v_{31}v_{11}$ (as $m_4\left(C_3, K_{1,3}\right) = 3$ by Jayawardene (2017c). In order to avoid a red B_2 each vertex in $\left\{v_{41}, v_{42}, v_{43}\right\}$ is adjacent in blue to two vertices in $\left\{v_{i1} : i \in \{1,2,3\}\right\}$ and further v_{12} is adjacent in blue to one vertex in $\left\{v_{i1} : i \in \{2,3\}\right\}$. By pigeon hole principle this results in a blue $K_{1,3}$, a contradiction. Therefore, $m_4\left(B_2, K_{1,3}\right) \leq 3$. Hence $m_4\left(B_2, K_{1,3}\right) = 3$.

Next consider any colouring of $K_{6\times1} = H_R \oplus H_B$ where H_B consist only of the two blue cycles, $v_{11}v_{21}v_{31}v_{11}$ and $v_{41}v_{51}v_{61}v_{41}$. Then $K_{6\times1}$ has niether a red B_2 nor a blue $K_{1,3}$. Therefore,

$$m_5\left(B_2, K_{1,3}\right) \geq m_6\left(B_2, K_{1,3}\right) \geq 2.$$

Now consider any red-blue coloring of $K_{5\times2}$. Assume the graph has no blue $K_{1,3}$. Then there is a red C_3 say $v_{11}v_{21}v_{31}v_{11}$ (since $m_5\left(C_3, K_{1,3}\right) = 2$ by Jayawardene (2017c). If every vertex in $\left\{v_{41}, v_{42}, v_{51}, v_{52}\right\}$ is adjacent in blue to at least two vertices of $V\left(C_3\right)$ then the graph has a blue $K_{1,3}$, a contradiction. Therefore there is a vertex in $\left\{v_{41}, v_{42}, v_{51}, v_{52}\right\}$ which is adjacent in red to two vertices of $V\left(C_3\right)$. This forces a red B_2. Hence $m_5\left(B_2, K_{1,3}\right) \leq 2$. Therefore, $m_j\left(B_2, K_{1,3}\right) = 2$ when $j \in \{5,6\}$.

Finally, we have $m_j\left(B_2, K_{1,3}\right) = 1$ for $j \geq 7$ as $r\left(B_2, K_{1,3}\right) = 7$ (see Jayawardene, 2017d).

Theorem 2:

$$m_j\left(C_4, B_2\right) = \begin{cases} 1 & \text{if } j \geq 7 \\ 2 & \text{if } j = \{5,6\} \\ 3 & \text{if } j = 4 \\ 4 & \text{if } j = 3 \end{cases}$$

Proof. **Case 1:** Clearly $m_j(C_4, B_2) = 1$ if $j \geq 7$, since $r(C_4, B_2) = 7$ (see Jayawardene, 2017d).

Case 2: Let $j \in \{5,6\}$. Consider the coloring of $K_{j\times1} = H_R \oplus H_B$, generated by $H_R = C_5$ and $H_R = 2K_3$ when $j = 5$ and $j = 6$ respectively. Then, $K_{j\times1}$ has no red C_4 or a blue B_2. Therefore, we obtain that $m_5(C_4, B_2)$

≥ 2 and $m_6(C_4, B_2) \geq 2$. Next we have to show $m_5(C_4, B_2) \leq 2$. For this consider any coloring consisting of (red, blue) given by $K_{5 \times 2} = H_R \oplus H_B$, such that H_R contains no red C_4 and H_B contains no red B_2. Then since $m_5(C_4, C_3) = 2$, without loss of generality we may assume that $(v_{1,1}, v_{2,1}, v_{3,1})$ is a blue cycle. Define $T = \{v_{4,1}, v_{4,2}, v_{5,1}, v_{5,2}\}$ and $S = \{v_{1,1}, v_{2,1}, v_{3,1}\}$. Then, if any vertex of T is adjacent to two vertices of S in blue, it will result blue B_2, contrary to our assumption. Therefore, we will be left with the option every vertex of T is adjacent to at least two vertices of S in red. But then as $|S| = 3$ there will be two vertices of T adjacent in red to the same pair of vertices in S. This will result in a red C_4, a contradiction. From this we can conclude that $m_j(C_4, B_2) \leq 2$ if $j \in \{5,6\}$. That is, $m_j(C_4, B_2) = 2$ if $j \in \{5,6\}$. We are left with the following two cases, namely $j = 4$(case 3) and $j = 3$(case 4).

Case 3: $j = 4$

Consider the coloring of $K_{4 \times 2} = H_R \oplus H_B$, generated by H_R illustrated in the following figure.

Then H_R will contain three disjoint triangles except for two triangles containing a common vertex. Thus the red-blue coloring generated by the figure will be such that H_R contains no red C_4 and H_B contains no blue B_2. Therefore, we will get $m_4(C_4, B_2) \geq 3$. To show that $m_3(C_4, B_2) \leq 3$ consider any coloring consisting of (red, blue) given by $K_{3 \times 4} = H_R \oplus H_B$, such that H_R contains no red C_4 and H_B contains no blue B_2. Since $m_4(C_4, C_3) = 2$ without loss of generality, we get that, $v_{1,1} v_{2,1} v_{3,1} v_{1,1}$ is a blue cycle. Define $S = \{v_{1,1}, v_{2,1}, v_{3,1}\}$ and $T = \{v_{4,1}, v_{4,2}, v_{4,3}\}$. If any vertex of T has adjacent in blue to 2 vertices of S we will get a blue B_2 and if any two vertex of T has adjacent in red to the same 2 vertices of S we will get a red C_4. Therefore, we may assume that $v_{4,1}$ is adjacent in blue to $v_{1,1}$ and in red to $v_{2,1}$ and $v_{3,1}$; $v_{4,2}$ is adjacent in blue to $v_{2,1}$ and in red to $v_{1,1}$ and $v_{3,1}$; $v_{4,3}$ is adjacent in blue to $v_{3,1}$ and in red to $v_{1,1}$ and $v_{2,1}$.

Figure 3. Case 3

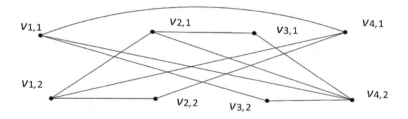

Figure 4. The first scenario

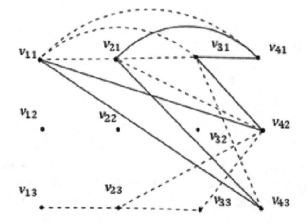

Also in the remaining 6 vertices (in $S^c \cap T^c$) must contain a blue P_3 as $m_3(C_4, P_3) = 2$. Thus, without loss of generality we may assume that $v_{1,3} v_{2,3} v_{3,3}$ is a blue P_3. In order to avoid a red C_4, all three vertices of $\{v_{1,3}, v_{2,3}, v_{3,3}\}$ must be adjacent in blue to at least two vertices of T. Thus, without loss of generality this gives rise to two possible scenarios illustrated in Figure 5 and Figure 6 respectively.

In the first scenario, in order to avoid a blue B_2 both $(v_{1,3}, v_{4,2})$ and $(v_{1,3}, v_{3,3})$ must be a red edges. Then in order to avoid a red C_4, $(v_{1,1}, v_{3,3})$ must be a blue edge; in order to avoid a blue B_2, $(v_{3,3}, v_{2,1})$ and $(v_{1,1}, v_{2,3})$ must be red edges; in order to avoid a red C_4, $(v_{3,1}, v_{2,3})$, $(v_{1,3}, v_{4,3})$ and $(v_{1,3}, v_{4,1})$ must be a blue edges. But then in order to avoid a blue B_2, $(v_{2,3}, v_{4,3})$ must be red. In order to avoid a red C_4, $(v_{2,3}, v_{4,1})$ must be a blue edge.

In order to avoid a blue B_2, $(v_{3,3}, v_{4,1})$ must be a red edge and in order to avoid a red C_4, $(v_{1,3}, v_{2,1})$ must be a blue edge. The red/blue graph generated is illustrated in the above figure.

But then if $(v_{1,3}, v_{3,1})$ is red we get a red C_4 and if is blue we get a blue we get a B_2, a contradiction.

In the second scenario, we may assume that $v_{4,1}$ is adjacent in blue to $v_{2,3}$ and $v_{3,3}$; $v_{4,2}$ is adjacent in blue to $v_{1,3}$ and $v_{3,3}$; $v_{4,3}$ is adjacent in blue to $v_{1,3}$ and $v_{3,3}$. Next in order to avoid a blue B_2, $(v_{3,3}, v_{4,3})$, $(v_{2,3}, v_{4,2})$, $(v_{1,3}, v_{4,1})$ and $(v_{2,3}, v_{3,1})$ must be a red edges. Then in order to avoid a red C_4, $(v_{1,1}, v_{2,3})$ must be a blue edge; in order to avoid a blue B_2, $(v_{3,3}, v_{1,1})$ must be a red edge. In order to avoid a red C_4, $(v_{2,1}, v_{3,3})$ must be a blue edge; in order to avoid a blue B_2, $(v_{1,3}, v_{2,1})$ must be a red edge; and in order to avoid a red C_4, $(v_{1,3}, v_{3,1})$ must be a blue edge. But then the vertices in $\{v_{1,3}, v_{3,1}, v_{4,3}, v_{2,3}\}$ will induce a blue B_2, a contradiction.

Case 4: $j = 3$

Consider the coloring of $K_{3 \times 4} = H_R \oplus H_B$, generated by H_R and H_B illustrated in the following figure. The red blue coloring generated by the following figure will be such that H_R contains no red C_4 and H_B contains no red B_2. Therefore, we will get $m_3(C_4, B_2) \geq 4$.

To show that $m_3(C_4, B_2) \leq 4$, consider any coloring consisting of (red,blue) given by $K_{3 \times 4} = H_R \oplus H_B$ such that there is no red C_4 in H_R or a red B_2 in H_B. Subject to these conditions will first show the following three claims.

Figure 5. The final graph generated by the first scenario

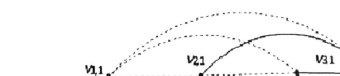

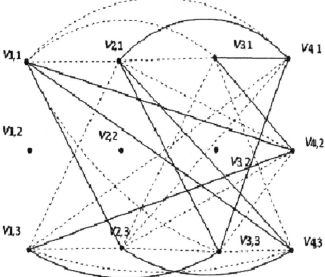

Figure 6. The second scenario

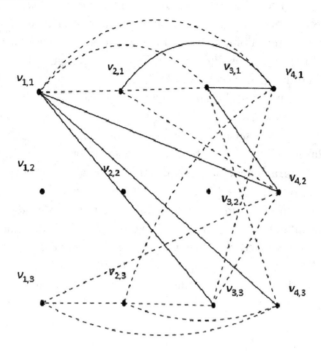

Figure 7. Case 4

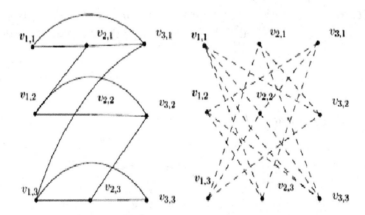

Notation: Let $1 \leq i, j \leq 4$. A vertex $v \in K_{3 \times 4}$ having blue degree $i + j$ is said to consist of a blue (i, j) split if v is adjacent in blue to i vertices of one partite set and j vertices of the other partite set.

Claim 1: All vertices of $K_{3 \times 4}$ have blue degree at most five.

Proof of Claim 1: Suppose $v_{1,4}$ has blue degree at least 6. Without loss of generality, there are two possibilities. Then one of the following two scenarios must be true. The first $v_{1,4}$ is adjacent in blue to all vertices of $S = \{v_{2,1}, v_{2,2}, v_{2,3}, v_{2,4}, v_{3,3}, v_{3,4}\}$ and the second $v_{1,1}$ is adjacent in blue to all vertices of $S = \{v_{2,1}, v_{2,2}, v_{2,3}, v_{3,1}, v_{3,2}, v_{3,3}\}$.

In the first scenario to avoid a blue B_2, both $v_{3,3}$ and $v_{3,4}$ will have to be adjacent in red to at least three vertices of the second partite set. This will result in a red C_4 containing $v_{3,3}$ and $v_{3,4}$, a contradiction. Hence the claim follows.

In the second scenario, In order to avoid a blue B_2 all edges between $\{v_{2,1}, v_{2,2}, v_{2,3}\}$ and $\{v_{3,1}, v_{3,2}, v_{3,3}\}$. Next, applying $m_3(C_4, C_3) = 3$ to $K_{3\times3}$ consisting of the first three elements of the three partite sets, we obtain a blue B_2 containing $v_{1,4}$. Hence the claim follows.

Claim 2: $K_{3\times4}$ has at least one vertex of blue degree five.

Proof of Claim 2: Applying $m_3(C_4, C_3) = 3$ to $K_{3\times3}$ consisting of the first three elements of the three partite sets, without loss of generality we obtain a blue C_3 containing $S = \{v_{1,1}, v_{2,1}, v_{3,1}\}$. Next, as there is no blue B_2, each vertex outside S will have to be adjacent in red to at least one vertex of S. Thus by pigeon-hole principle at least one vertex must have degree greater than 5. Thus by Claim 1, we can conclude that S has at least one vertex of blue degree five as required.

Claim 3: $K_{3\times4}$ has at least one vertex of blue (3,2) split.

Proof of Claim 3: Suppose that the claim is false. Let $v_{1,4}$ be a vertex having a blue (4,1) split. In particular, suppose that $v_{1,4}$ is adjacent in blue to all vertices of
$S = \{v_{2,1}, v_{2,2}, v_{2,3}, v_{2,4}, v_{3,4}\}$ and adjacent in red to all vertices of $T = \{v_{3,1}, v_{3,2}, v_{3,3}\}$. Since there is no blue B_2 without loss of generality we may assume that $v_{3,4}$ adjacent in red to all vertices of $T = \{v_{2,1}, v_{2,2}, v_{2,3}\}$. Next we deal with 2 possible options.

Option a: $v_{3,4}$ is adjacent in blue to at least one vertex of $\{v_{1,1}, v_{1,2}, v_{1,3}\}$ (say $v_{1,3}$). Then, since there is no red C_4 and by Claim 1, we would get that $v_{1,3}$ will be a blue (3,2) split.

Option b: $v_{3,4}$ is adjacent in red to all three vertices of $\{v_{1,1}, v_{1,2}, v_{1,3}\}$. Then, since there is no red C_4 and by Claim 1, we would get that $v_{2,4}$ will be a blue (3,2) split.

Now let us try to complete the proof of $j = 3$ case. According to lemma 3, $v_{1,4}$ be a vertex having a blue (3,2) split.

In particular, suppose that $v_{1,4}$ is adjacent in blue to all vertices of $S = \{v_{2,2}, v_{2,3}, v_{2,4}, v_{3,3}, v_{3,4}\}$ and adjacent in red to all vertices of $T = \{v_{2,1}, v_{3,1}, v_{3,2}\}$.

Figure 8. The generated graph for both options a and b, when j = 3

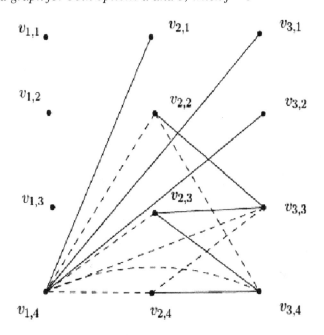

Since there is no blue B_2 or a red C_4, without loss of generality we may assume that $v_{3,4}$ adjacent in red to $v_{2,3}, v_{2,4}$ and that $v_{3,3}$ adjacent in red to $v_{2,2}, v_{2,3}$. Next as there is no red C_4, we would get that $(v_{2,4}, v_{3,3})$. and $(v_{2,2}, v_{3,4})$ are blue edges. This is illustrated in the following figure.

Next we get that as in claim 3, there are two possible subcases to consider.

Subcase 4.1: $(v_{2,3}, v_{1,3})$ is red. First in order to avoid a red C_4 both $(v_{1,3}, v_{2,2})$ and $(v_{1,3}, v_{2,4})$ will have to be blue. Next in order to avoid a blue B_2 both $(v_{1,3}, v_{3,3})$ and $(v_{1,3}, v_{3,4})$ will have to be red. But this would give us $v_{1,3} v_{3,4} v_{2,3} v_{3,3} v_{1,3}$ is a red C_4.

Subcase 4.2: $v_{2,3}$ is adjacent in blue to all three vertices of $\{v_{1,1}, v_{1,2}, v_{1,3}\}$. Next, as there is no red

Figure 9. The final graph of case 4, when j = 3

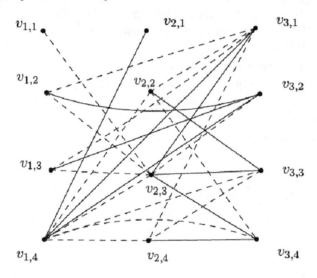

Figure 10. The H_B needed to show $m_6(C_4, K_4) \geq 4$

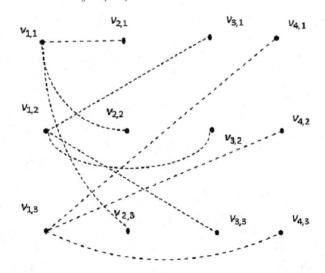

C_4, $v_{2,3}$ will be adjacent to at least one vertex of $\{v_{3,1}, v_{3,2}\}$ in blue. Without loss of generality assume $(v_{2,3}, v_{3,2})$ is blue. Then, by Claim 1, $(v_{2,3}, v_{3,1})$ will be red. Next, as there is no red C_4, $(v_{2,2}, v_{3,1})$ and $(v_{2,4}, v_{3,1})$ will have to be blue.

In order to avoid a blue B_2 the vertex $v_{3,2}$ must be adjacent to two vertices of $\{v_{1,1}, v_{1,2}, v_{1,3}\}$ in red. Without loss of generality assume that $(v_{3,2}, v_{1,2})$ and $(v_{3,2}, v_{1,3})$ are red. But then in order to avoid a red C_4, $(v_{3,1}, v_{1,2})$ and $(v_{3,1}, v_{1,3})$ are blue. Consider four vertices, $v_{3,1}$ is adjacent to in blue, given by $W = \{v_{1,2}, v_{1,3}, v_{2,2}, v_{2,4}\}$. In order to avoid a blue B_2 there can be at most one blue edge among them. That is there are three red edges in the subgraph induced by W. Exhaustive search will show that in each of the possibilities either $v_{1,2}\, v_{3,2}\, v_{1,3}\, v_{2,2}\, v_{1,2}$ or $v_{1,2}\, v_{3,2}\, v_{1,3}\, v_{2,4}\, v_{1,2}$ will be a red C_4, a contradiction.

Next, by Syfrizal et al., (2005), and Jayawardene (2018), we obtain all values related to the 1st and 2nd rows of the two tables. This, along with $m_j\big(G, H\big) = m_j\big(H, G\big)$, gives us all the required values of the two tables, excluding the entries $m_4(C_4, K_4)$ and $m_4(K_4, C_4)$.

The Unsolved Case m$_4$(C$_4$,K$_4$) (as Clearly m$_3$(C$_4$,K$_4$) = ∞)

Color the graph $K_{4 \times 3} = H_R \oplus H_B$ such that H_R consists of the following graph. Then, we can see that the graph has no red C_4 and no blue K_4. Therefore, $m_6\,(C_4,\, K_4) \geq 4$. However, as mentioned before, we are only able to find a lower bound, while the upper bound remains an unsolved problem.

REFERENCES

Angeltveit, V., & Mckay, B. D. (2017). *R(5, 5) ≤ 48.* arXiv:1703.08768

Bollobas, B., Jayawardene, C. J., Sheng, Y. J., Ru, H. Y., Rousseau, C. C., & Min, Z. K. (2000). On a Conjecture Involving Cycle-Complete Graph Ramsey Numbers. *The Australasian Journal of Combinatorics*, *22*, 63–71.

Burger, A. P., & Vuuren, V. J. H. (2004). Ramsey numbers in Complete Balanced Multipartite Graphs. Part II: Size Numbers. *Discrete Mathematics*, *283*(1-3), 45–49. doi:10.1016/j.disc.2004.02.003

Exoo, G. (1998). A lower bound for r(5, 5). *Journal of Graph Theory*, *13*(1), 97–98. doi:10.1002/jgt.3190130113

Jayawardene, C. J. (2018). On a Ramsey problem involving the 3-Pan Graph. *Annals of Pure and Applied Mathematics*, *16*(2), 437–441.

Jayawardene, C. J., & Hewage, T. (2017a). On a Ramsey problem involving quadrilaterals. *Annals of Pure and Applied Mathematics*, *13*(2), 297–304. doi:10.22457/apam.v13n2a16

Jayawardene, C. J., Rousseau, C. C., & Harboth, H. (2017b). On Path Convex Ramsey Numbers. *Journal of Graph Theory*, *86*(3), 286–294. doi:10.1002/jgt.22126

Jayawardene, C. J., & Samerasekara, L. (2017c). Size Ramsey numbers for C_3 versus all graphs G up to 4 vertices. *National Science Foundation*, *45*(1), 67–72. doi:10.4038/jnsfsr.v45i1.8039

Jayawardene, C. J., & Samerasekara, L. (2017d). Size Multipartite Ramsey numbers for K_4-e versus all graphs G up to 4 vertices. *Annals of Pure and Applied Mathematics, 13*(1), 9–26. doi:10.22457/apam.v13n1a2

Syafrizal, S., & Baskoro, E. T. (2012). Lower bounds of the size multipartite Ramsey numbers. *The 5th Mathematics, AIP Conf. Proc., 1450,* 259-261.

Syafrizal, S., Baskoro, E. T., & Uttunggadewa, S. (2005). The size multipartite Ramsey number for paths. *Journal Combin. Math. Combin. Comput., 55,* 103–107.

Chapter 11
Energy of Graphs

Harishchandra S. Ramane
Karnatak University, India

ABSTRACT

The energy of a graph G is defined as the sum of the absolute values of the eigenvalues of its adjacency matrix. The graph energy has close correlation with the total pi-electron energy of molecules calculated with Huckel molecular orbital method in chemistry. A graph whose energy is greater than the energy of complete graph of same order is called hyperenergetic graph. A non-complete graph having energy equal to the energy of complete graph is called borderenergetic graph. Two non-cospectral graphs are said to be equienergetic graphs if they have same energy. In this chapter, the results on graph energy are reported. Various bounds for graph energy and its characterization are summarized. Construction of hyperenergetic, borderenergetic, and equienergetic graphs are reported.

INTRODUCTION

The energy of a graph is the sum of the absolute values of the eigenvalues of its adjacency matrix. It has a correlation with the total π-electron energy of a molecule in the quantum chemistry as calculated with the Huckel molecular orbital method (Gutman & Polansky, 1986).

Let G be a finite, simple, undirected graph with vertex set $V(G)$ and edge set $F(G)$ The number of vertices of G is denoted by n and the number of edges of G is denoted by m. If $V(G) = \{v_1, v_2, ..., v_n\}$ then the adjacency matrix of G is a square matrix $A(G) = [a_{ij}]$ of order n in which $a_{ij} = 1$, if the vertex v_i is adjacent to the vertex v_j and $a_{ij} = 0$, otherwise. The characteristic polynomial of $A(G)$ denoted by $\Phi(G: \lambda) = \det(\lambda I - A(G))$, where I is an identity matrix of order n. The roots of the equation $\Phi(G: \lambda) = 0$ are called the eigenvalues of G and they are labeled as $\lambda_1, \lambda_2, ..., \lambda_n$. Their collection is called the spectrum of G denoted by $Spec(G)$ (Cvetkovic, Doob & Sachs, 1980).

If $\lambda_1, \lambda_2, ..., \lambda_k$ are the distinct eigenvalues with respective multiplicities $m_1, m_2, ..., m_k$ then we write

$$Spec(G) = \begin{pmatrix} \lambda_1 & \lambda_2 & \cdots & \lambda_k \\ m_1 & m_2 & \cdots & m_k \end{pmatrix}.$$ Since $A(G)$ is a real symmetric matrix, its eigenvalues are real and can

be ordered as $\lambda_1 \geq \lambda_2 \geq ... \geq \lambda_n$.

DOI: 10.4018/978-1-5225-9380-5.ch011

Two non-isomorphic graphs are said to be cospectral if they have same spectra. Details about the graph spectra can be found in the book (Cvetkovic, Doob & Sachs, 1980) and for graph theoretic terminology one can refer the book (Harary, 1999).

One of the chemical applications of spectral graph theory is based on the correspondence between the graph eigenvalues and the molecular orbital energy level of π-electron in conjugated hydrocarbons (Gutman & Polansky, 1986).

The molecular graph of a hydrocarbon is obtained as follows: the carbon atoms are represented by the vertices and two vertices are adjacent if and only if there is a carbon-carbon bond. Hydrogen atoms are ignored.

Within the Huckel molecular orbital (HMO) method (Huckel & Quantentheoretische Beitrage zum Benzolproblem, 1931), the energy level of π-electron in molecules of conjugated hydrocarbons are related to the eigenvalues of a molecular graph as $\varepsilon_i = \alpha + \beta_i$ where α and β are empirical constants of the HMO model. The total energy of π-electrons denoted by E_π is

$$E_\pi = \sum_{i=1}^{n} g_i \varepsilon_i \,,$$

where g_i is the occupation number with energy ε_i and $g_1 + g_2 + \ldots + g_n = n$. This yields

$$E_\pi = n\alpha + \beta \sum_{i=1}^{n} g_i \lambda_i \tag{1}$$

For majority of conjugated hydrocarbons $g_i = 2$ if $\lambda_i > 0$ and $g_i = 0$ if $\lambda_i < 0$. Therefore Eq. (1) can be written as

$$E_\pi = n\alpha + 2\beta \sum_{+} \lambda_i \,,$$

where $\sum_{+}$ indicates the summation over positive eigenvalues of the molecular graph.

Figure 1. Molecule and its molecular graph

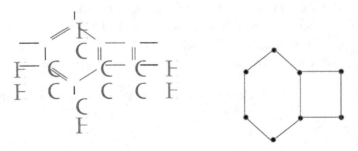

Since the sum of all graph eigenvalues is equal to zero, we arrive at

$$E_\pi = n\alpha + \beta\sum_{i=1}^{n} |\lambda_i|. \tag{2}$$

Because n, α and β are constants, the only non trivial term on the right hand side of Eq. (2) is the sum of the absolute values of the eigenvalues of the molecular graph. Thus in 1978, Ivan Gutman (Gutman, 1978) put forward the mathematical definition of graph energy as the sum of the absolute values of the eigenvalues of a graph G and is denoted by $E(G)$. That is

$$E(G) = \sum_{i=1}^{n} |\lambda_i|.$$

Characteristic polynomial of a graph given in Fig. 2 is $\lambda^4 - 5\lambda^2 - 4\lambda$ and eigevalues are 2.5616, 0, – 1 and – 1 .5616. Hence its energy is E(G) $\approx$ 5.1232.

A comprehensive results on the graph energy are given in the book (Li, Shi, & Gutman, 2012).

The complete graph K_n on n vertices has eigenvalues $n-1$ and -1 ($n-1$ times). Hence $E(K_n) = 2(n-1)$.

The complete bipartite graph $K_{p,q}$ has eigenvalues $\pm\sqrt{pq}$ and 0 ($p+q-2$ times). Thus $E(K_{p,q}) = 2\sqrt{pq}$.

Further the energy of a star $S_n = K_{1,\,n-1}$ is $E(K_{1,n-1}) = 2\sqrt{n-1}$.

Let C_n and P_n denotes the cycle and path on n vertices respectively.

$$E(C_n) = \begin{cases} 4\cot(\pi/n) \text{ if } n \equiv 0(\mathrm{mod}\,4) \\ 4\cos ec(\pi/n) \text{ if } n \equiv 2(\mathrm{mod}\,4) \\ 2\cos ec(\pi/2n) \text{ if } n \equiv 1(\mathrm{mod}\,2) \end{cases}$$

$$E(P_n) = \begin{cases} 2\cos ec(\pi/2(n+1)) - 2 \text{ if } n \equiv 0(\mathrm{mod}\,2) \\ 2\cot(\pi/2(n+1)) - 2 \text{ if } n \equiv 1(\mathrm{mod}\,2) \end{cases}.$$

Coulson (Coulson, 1940) gives the energy of a graph in terms of integration.

Figure 2. Graph and its adjacency matrix

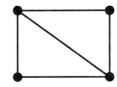

$$\begin{bmatrix} 0 & 1 & 1 & 1 \\ 1 & 0 & 1 & 0 \\ 1 & 1 & 0 & 1 \\ 1 & 0 & 1 & 0 \end{bmatrix}$$

Theorem 1.1 (Coulson, 1940): Let G be a graph with n vertices and characteristic polynomial $\Phi(G:\lambda)$. Then

$$E(G) = \frac{1}{\pi} \int_{-\infty}^{\infty} \left[n - \frac{i\lambda\Phi'(G:i\lambda)}{\Phi(G:i\lambda)} \right] d\lambda ,$$

where $i = \sqrt{-1}$ and $\Phi'(G:\lambda)$ is the derivative of $\Phi(G:\lambda)$.

Proof: Let $\Phi(G:z)$ be the polynomial in z of degree n and let $\lambda_1, \lambda_2, ..., \lambda_n$ be its roots. Then

$$\Phi(G:z) = \prod_{j=1}^{n} (z - \lambda_j) \text{ and consequently}$$

$$\frac{\Phi'(G:z)}{\Phi(G:z)} = \sum_{j=1}^{n} \frac{1}{z - \lambda_j} .$$

Therefore

$$\frac{z\Phi'(G:z)}{\Phi(G:z)} = \sum_{j=1}^{n} \frac{z}{z - \lambda_j} = n + \sum_{j=1}^{n} \frac{\lambda_j}{z - \lambda_j} .$$

Therefore

$$\frac{z\Phi'(G:z)}{\Phi(G:z)} - n = \sum_{j=1}^{n} \frac{\lambda_j}{z - \lambda_j} \to 0 \text{ as } |z| \to \infty .$$

Consider the contour Γ shown in Fig. 3.
By Cauchy's formula

$$\frac{1}{2\pi i} \oint_{\Gamma} \frac{dz}{z - z_0} = \begin{cases} 1 \, if \, z_0 \in \Gamma \\ 0 \, if \, z_0 \notin \Gamma \end{cases}$$

Therefore

$$\frac{1}{2\pi i} \oint_{\Gamma} \left[\frac{z\Phi'(G:z)}{\Phi(G:z)} - n \right] dz = \frac{1}{2\pi i} \oint_{\Gamma} \sum_{j=1}^{n} \frac{\lambda_j}{z - \lambda_j} dz$$

Figure 3. Contour Γ

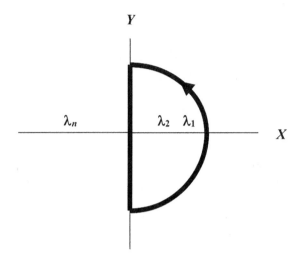

$$= \sum_{j=1}^{n} \frac{\lambda_j}{2\pi i} \oint_{\Gamma} \frac{dz}{z - \lambda_j} = \sum_{\lambda_j \geq 0} \lambda_j = \frac{E(G)}{2}.$$

In the limiting case when Γ becomes infinitely large, the only non vanishing contribution to the above integral comes from the integration along the y – axis. Thus

$$E(G) = \frac{1}{\pi i} \oint_{\Gamma} \left[\frac{z\Phi'(G:z)}{\Phi(G:z)} - n \right] dz$$

$$= \frac{1}{\pi i} \int_{-\infty}^{\infty} \left[\frac{z\Phi'(G:z)}{\Phi(G:z)} - n \right] dz + \frac{1}{\pi i} \int_{\infty}^{-\infty} \left[\frac{z\Phi'(G:z)}{\Phi(G:z)} - n \right] dz$$

$$= 0 + \frac{1}{\pi i} \int_{\infty}^{-\infty} \left[\frac{iy\Phi'(G:iy)}{\Phi(G:iy)} - n \right] d(iy)$$

$$= \frac{1}{\pi} \int_{-\infty}^{\infty} \left[n - \frac{iy\Phi'(G:iy)}{\Phi(G:iy)} \right] dy = \frac{1}{\pi} \int_{-\infty}^{\infty} \left[n - \frac{i\lambda\Phi'(G:i\lambda)}{\Phi(G:i\lambda)} \right] d\lambda.$$

2. Bounds for Energy

If $\lambda_1, \lambda_2, \ldots, \lambda_n$ are the eigenvalues of G. Then,

$$\sum_{i=1}^{n} \lambda_i = 0 \,,\ \sum_{i=1}^{n} \lambda_i^2 = 2m \text{ and } \sum_{i<j} \lambda_i \lambda_j = -m \,.$$

Following bounds are due to McClelland (McClelland, 1971).

Theorem 2.1 (McClelland, 1971): Let G be a graph with n vertices and m edges. Then,

$$\sqrt{2m + n(n-1) \mid \det(A(G)) \mid^{2/n}} \leq E(G) \leq \sqrt{2mn} \,.$$

Proof: Lower bound: Since the geometric mean of positive numbers is not greater than their arithmetic mean,

$$\frac{1}{n(n-1)} \sum_{i \neq j} \mid \lambda_i \mid \mid \lambda_j \mid \geq \prod_{i \neq j} \left(\mid \lambda_i \mid \mid \lambda_j \mid \right)^{1/n(n-1)}$$

$$= \prod_{i=1}^{n} \mid \lambda_i \mid^{2/n} = \mid \det(A(G)) \mid^{2/n} \,.$$

Therefore

$$(E(G))^2 = \sum_{i=1}^{n} \mid \lambda_i \mid^2 + 2\sum_{i<j} \mid \lambda_i \mid \mid \lambda_j \mid$$

$$= 2m + \sum_{i \neq j} \mid \lambda_i \mid \mid \lambda_j \mid$$

$$\geq 2m + n(n-1) \mid \det(A(G)) \mid^{2/n} \,.$$

Therefore

$$E(G) \geq \sqrt{2m + n(n-1) \mid \det(A(G)) \mid^{2/n}} \,.$$

Upper Bound: By Cauchy-Schwartz inequality,

$$\left(\sum_{i=1}^{n} \mid \lambda_i \mid \right)^2 \leq n \sum_{i=1}^{n} \mid \lambda_i \mid^2 \,.$$

Therefore

$$E(G) \leq \sqrt{2mn} \,.$$

Theorem 2.2 (Gutman, 2001): For a graph G with m edges,

$$2\sqrt{m} \leq E(G) \leq 2m \,.$$

Equality on left side holds if and only if G consists of a complete bipartite graph $K_{p,q}$ such that $pq = m$ and arbitrary many isolated vertices. Equality on right hand side holds if and only G consists of m copies of K_2 and arbitrary many isolated vertices.

Definition (Adiga & Rakshith, 2016): The regular graph G which is neither complete nor empty is said to be strongly regular with parameters (n, r, a, c) if it is n-vertex, r-regular graph and every pair of adjacent vertices has a common neighbours and every pair of nonadjacent vertices has c common neighbours.

Following bound is due to Koolen and Moulten (Koolen & Moulton, 2001):

Theorem 2.3 (Koolen & Moulton, 2001): Let G be a graph with n vertices and m edges. If $2m \geq n$, then,

$$E(G) \leq \frac{2m}{n} + \sqrt{(n-1)\left[2m - \left(\frac{2m}{n}\right)^2\right]} \,.$$

Equality holds if and only if G consists of $n/2$ copies of K_2 or $G \cong K_n$ or G is non-complete strongly regular graph with two nontrivial eigenvalues both having absolute values equal to

$$\sqrt{(2m - (2m/n)^2)/(n-1)} \,.$$

Proof: Suppose $\lambda_1, \lambda_2, \ldots, \lambda_n$ are the eigenvalues of G. Cauchy-Schwarz inequality is,

$$\left(\sum_{i=1}^{n} a_i b_i\right)^2 \leq \left(\sum_{i=1}^{n} a_i^2\right)\left(\sum_{i=1}^{n} b_i^2\right)$$

Let $a_i = 1$, $b_i = |\lambda_i|$, $i = 2, 3, \ldots, n$.
Therefore

$$\left(\sum_{i=2}^{n} |\lambda_i|\right)^2 \leq (n-1)\sum_{i=2}^{n} |\lambda_i|^2 \,.$$

$$(E(G) - \lambda_1)^2 \leq (n-1)(2m - \lambda_1^2) \,.$$

Therefore

$$E(G) \leq \lambda_1 + \sqrt{(n-1)(2m - \lambda_1^2)} \qquad (3)$$

Consider the function,

$$f(x) \leq x + \sqrt{(n-1)(2m - x^2)}$$

It is decreasing function in the interval $(\sqrt{2m/n}, \sqrt{2m})$ and attains maximum for $x = 2m/n$. Since $\lambda_1 \geq 2m/n$, by Eq. (3), $f(\lambda_1) \leq f(2m/n)$.
Therefore

$$E(G) \leq \frac{2m}{n} + \sqrt{(n-1)\left[2m - \left(\frac{2m}{n}\right)^2\right]}.$$

For equality:
Since the eigenvalues of $(n/2)K_2$ are ± 1 (with multiplicity $n/2$) and the eigenvalues of K_n are $n - 1$ and -1 (with multiplicity $n - 1$), it is easy to check that if G is one of the graph specified in the theorem, then the equality holds.

Conversely, if equality holds then $\lambda_1 = 2m/n$. It follows that G is regular with degree $2m/n$. Since equality must also holds in Cauchy-Schwarz inequality, we have

$$|\lambda_i| = \sqrt{(2m - (2m/n)^2)/(n-1)}$$

for $2 \leq i \leq n$. Hence either G has two eigenvalues with equal absolute values, in which case $G \cong (n/2)K_2$ or G has two eigenvalues with distinct absolute values in which case $G \cong K_n$ or G has three eigenvalues with distinct absolute values equal to $2m/n$ or

$$\sqrt{(2m - (2m/n)^2)/(n-1)},$$

in which case G must be non-complete strongly regular graph as required.

Corollary 2.4 (Walikar, Ramane, & Hampiholi, 1999): If G is an r-regular graph on n vertices, then,

$$E(G) \leq r + \sqrt{r(n-1)(n-r)}.$$

A 2-(v,k,λ)-design is a collection of k-subsets or blocks of a set of v points, such that each 2-set of points lies in exactly λ blocks. The design is called symmetric in case the number of blocks b equal v (Hall, 1986). The incidence matrix B of a 2-(v,k,λ)-design is the $v \times b$ matrix defined so that for each point x and block S, $B_{x,S} = 0$ if $x \notin S$ and $B_{x,S} = 1$, otherwise. The incidence graph of a design is defined to be the graph with adjacency matrix

$$\begin{bmatrix} O & B \\ B^T & O \end{bmatrix}.$$

Following result is obtained by Koolen and Moulton (Koolen & Moulton, 2003) for bipartite graphs

Theorem 2.5 (Koolen & Moulton, 2003): Let G be a bipartite graph with $n \geq 3$ vertices and m edges. If $2m \geq n$, then

$$E(G) \leq \frac{4m}{n} + \sqrt{(n-2)\left[2m - 2\left(\frac{2m}{n}\right)^2\right]}.$$

Equality holds if and only if at least one of the following statements holds

1. $n = 2m$ and $G \cong mK_2$.
2. $n = 2t$, $m = t^2$ and $G \cong K_{t,t}$ and
3. $n = 2\nu$, $2\sqrt{m} < n < 2m$ and G is the incident graph of a symmetric 2-(ν,k,λ)-design with $k = 2m/n$ and $\lambda = k(k-1)/(\nu-1)$.

A graph is said to be semi-regular bipartite with parameters (s_1, s_2) if it is bipartite, and vertices in one vertex class have equal degree s_1 and vertices in other vertex class have equal degree s_2.

If $d_1, d_2, \ldots, d_n$ are the degrees of the vertices of G, then the first Zagreb index (Gutman, N. Trinajstic, 1972) is defined as

$$M_1(G) = \sum_{i=1}^{n} d_i^2.$$

Following results is obtained in (Zhou, 2004).

Theorem 2.6 (Zhou, 2004): If G is a graph with n vertices and m edges and Zagreb index $M_1(G)$, then

$$E(G) \leq \sqrt{\frac{M_1(G)}{n}} + \sqrt{(n-1)\left[2m - \frac{M_1(G)}{n}\right]}.$$

Equality holds if and only if G is either $(n/2)K_2$ or K_n, or a non-complete connected strongly regular graph with two nontrivial eigenvalues both with absolute value

$$\sqrt{(2m - (2m/n)^2)/(n-1)}$$

or nK_1.

In the following theorem an upper bound for the energy of bipartite graphs of a given partite is given.

Theorem 2.7 (Zhou, 2008): Let G be a bipartite graph with $n \geq 2$ vertices, $m \geq 1$ edges, the first Zagreb index $M_1(G)$ and an (n_1, n_2)-bipartition where $n_1 \leq n_2$. If $M_1(G) \geq nm/n_1$, then

$$E(G) \leq 2\sqrt{\frac{M_1(G)}{n}} + 2\sqrt{(n_1 - 1)\left[m - \frac{M_1(G)}{n}\right]}.$$

Equality holds if and only if G is K_{n_1,n_2} or $n_1 K_{1,n_2/n_1}$ or the incidence graph of a symmetric 2-(n_1,k,λ)-design with $n_1 > k = m/n_1$ and $\lambda = k(k-1)/(n_1 - 1)$ or a connected semi-regular bipartite graph with exactly five distinct eigenvalues.

Let a_u denotes the average of the degrees of the neighbours of a vertex u in G. Let $T_{p,q}$ be the tree obtained by joining the centers of p-copies of star $S_q = K_{1,q-1}$ to a new vertex, where $p, q \geq 1$. Obviously $T_{p,q}$ has $pq+1$ vertices and $T_{1,q} = S_{q+1}$ and $T_{p,1} = S_{p+1}$.

Theorem 2.8 (Zhou, 2008): Let G be an acyclic graph with n vertices and $m \geq 1$ edges. Let

$$s = \max\{d_u + a_u - 1 \mid u \in V(G)\}$$

and

$$a = a(n) = \begin{cases} 2 \ \text{if } n \text{ is even} \\ 3 \ \text{if } n \text{ is odd} \end{cases}.$$

Then

$$E(G) \leq 2\sqrt{s} + 2\sqrt{(n - a)(2m - 2s)}.$$

Equality holds if and only if

$$G = (n/2)K_{1,1} \text{ or } G = K_{1,m} \cup (n - 1 - m)K_1$$

if n is even and

$$G = ((n - 1)/2)K_{1,1} \text{ or } G = K_{1,m} \cup (n - 1 - m)K_1 \text{ or } G = T_{p,2} \cup (n - 2p - 1)K_{1,1}$$

Figure 4. $T_{3,4}$

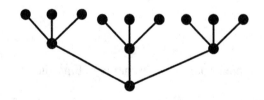

with $1 \leq p \leq (n-1)/2$ if n is odd.

Theorem 2.9 (Koolen, 2001): Let G be a graph on n vertices. Then

$$E(G) \leq \frac{n}{2}\left(\sqrt{n}+1\right),$$

with equality if and only if G is strongly regular graph with parameters

$$\left(n, \frac{n+\sqrt{n}}{2}, \frac{n+2\sqrt{n}}{4}, \frac{n+2\sqrt{n}}{4}\right).$$

Proof: Suppose G has n vertices and m edges. If $2m \geq n$, then by Theorem 2.3,

$$E(G) \leq \frac{2m}{n} + \sqrt{(n-1)\left[2m - \left(\frac{2m}{n}\right)^2\right]} = f(m). \tag{4}$$

The function $f(m)$ is maximum when $m = \dfrac{n^2 + n\sqrt{n}}{4}$. Substituting this in Eq. (4), we get

$$E(G) \leq \frac{n}{2}\left(\sqrt{n}+1\right).$$

Moreover the equality holds if and only if G is strongly regular graph with parameters

$$\left(n, \frac{n+\sqrt{n}}{2}, \frac{n+2\sqrt{n}}{4}, \frac{n+2\sqrt{n}}{4}\right).$$

If $2m \leq n$, then by Theorem 2.2, $E(G) \leq 2m \leq n$. Hence the proof follows immediately.
In (Haemers, 2008) it is showed that there are strongly regular graph with parameter

$$\left(n, \frac{n+\sqrt{n}}{2}, \frac{n+2\sqrt{n}}{4}, \frac{n+2\sqrt{n}}{4}\right)$$

for (i) $n = 4^p$, $p \geq 1$; (ii) $n = 4^p q^4$, $p, q \geq 1$ and (iii) $n = 4^{p+1}q^2$, $p \geq 1$ and $4q - 1$ is a prime power or $2q - 1$ is a prime number or q is a square or $q < 167$.

There exist the maximal energy graphs of order $n = 4k^4$ for k which are strongly regular graphs with parameters $(4k^4, 2k^4 + k^2, k^4 + k^2, k^4 + k^2)$ (Haemers & Xiang, 2010).

Following result gives the upper bound for the energy of bipartite graph in terms of the number of vertices and characterizes the graphs for which this bound is sharp.

Theorem 2.10 (Koolen & Moulton, 2003): Let G be a bipartite graph on $n \geq 3$ vertices. Then

$$E(G) \leq \frac{n}{\sqrt{8}}(\sqrt{n} + \sqrt{2}).$$

With equality if and only if $n = 2\nu$ and if G is the incidence graph of a

$2\text{-}(\nu, (v + \sqrt{\nu})/2, (\nu + 2\sqrt{\nu})/4)$-design.

By Corollary 2.4,

$$E(G) \leq r + \sqrt{r(n-1)(n-r)} = B_0$$

for an r-regular graph. Balakrishnan (Balakrishnan, 2004) showed that for $\varepsilon > 0$ there exists infinitely many n-vertex regular graphs of degree r such that $\dfrac{E(G)}{B_0} < \varepsilon$. In the same paper he raised the following problem.

Problem 2.11 (Balakrishnan, 2004): Given a positive integer $n \geq 3$, does there exist an r-regular graph G for some r, $0 < r < n$ such that $E(G)/B_0 > 1 - \varepsilon$, where $B_0 = r + \sqrt{r(n-1)(n-r)}$?

An affirmative answer to this question is given in (Li, Li, & Shi, 2010)(Walikar, Ramane, & Jog, 2008), not for general, but when $n \equiv 1 \pmod 4$.

Theorem 2.12 (Li, Li, & Shi, 2010)(Walikar, Ramane, & Jog, 2008): Given any positive integer $n \geq 5$, $n \equiv 1 \pmod 4$, and $\varepsilon > 0$, there exists an r-regular graph G on n vertices for some r, $0 < r < n$ such that $\dfrac{E(G)}{B_0} > 1 - \varepsilon$.

Proof: Let G be the Paley graph on n vertices (Godsil & Royle, 2001), which is strongly regular graph with parameters, $(n, (n-1)/2, (n-5)/4, (n-1)/4)$. It is regular graph of degree $(n-1)/2$ and its eigenvalues are $(n-1)/2$, $(-1 + \sqrt{n})/2$ and $(-1 - \sqrt{n})/2$ with multiplicities 1, $(n-1)/2$ and $(n-1)/2$ respectively.

Therefore,

$$E(G) = \left|\frac{n-1}{2}\right| + \left|\frac{-1+\sqrt{n}}{2}\right|\left(\frac{n-1}{2}\right) + \left|\frac{-1-\sqrt{n}}{2}\right|\left(\frac{n-1}{2}\right) = \frac{(n-1)(\sqrt{n}+1)}{2}.$$

The bound B_0 of $E(G)$ is

$$B_0 = \frac{n-1}{2} + \sqrt{\left(\frac{n-1}{2}\right)(n-1)\left(n - \frac{n-1}{2}\right)} = \frac{(n-1)(1 + \sqrt{n+1})}{2}.$$

Hence,

$$\frac{E(G)}{B_0} = \frac{\sqrt{n}+1}{(1+\sqrt{n+1})} \text{ tends to 1 as } n \to \infty.$$

This shows that there exists an *r*-regular graph *G* on *n* vertices, $n \equiv 1(\mathrm{mod}\ 4)$ for some *r*, $0 < r < n$ such that $E(G)/B_0 > 1 - \varepsilon$.

Let $S_n = K_{1,\,n-1}$ be the star and P_n be the path on *n* vertices.

Theorem 2.13 (Gutman, 1977): For any tree *T* on *n* vertices, $E(S_n) \le E(T) \le E(P_n)$.

Among all trees with *n* vertices, star has minimum energy and path has maximum energy.

Let $T_1(n)$ be obtained by joining a vertex to a terminal vertex of S_{n-1}. Let $T_2(n)$ be the tree obtained by joining two vertices to a terminal vertex of S_{n-2}. Let $T_3(n)$ be the tree obtained by joining a vertex of P_2 to a terminal vertex of S_{n-2}. Let $T_4(n)$ be the tree obtained by joining a middle vertex of P_5 to the terminal vertex of P_{n-5}.

Theorem 2.15 (Gutman, 1977): If *T* is any tree on *n* vertices different from S_n, $T_1(n)$, $T_2(n)$, $T_3(n)$, $T_4(n)$ and P_n, then

$$E(S_n) < E(T_1(n)) < E(T_2(n)) < E(T_3(n)) < E(T) < E(T_4(n)) < E(P_n).$$

Let $A_n(k)$ and $B_n(k)$ be the trees as shown in Fig. 6.

Theorem 2.16 (Walikar & Ramane, 2005): Let $A_n(k)$ be the tree as in Fig. 6, then for any two integers *n* and *k*,

$$E(A_n(1)) < E(A_n(2)) < \ldots < E(A_n(\lfloor (n/2) - 1 \rfloor)).$$

Theorem 2.17 (Walikar & Ramane, 2005): Let $B_n(k)$ be the tree as in Fig. 6, then for any two integers *n* and *k*,

Figure 5. S_9, $T_1(9)$, $T_2(9)$, $T_3(9)$, $T_4(9)$

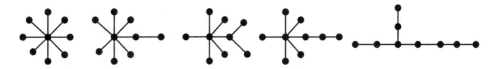

Figure 6. $A_n(k)$ and $B_n(k)$

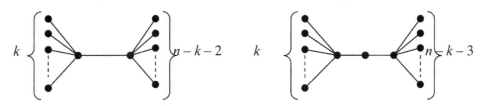

$$E(B_n(1)) < E(B_n(2)) < \ldots < E(B_n(\lfloor (n-3)/2 \rfloor)).$$

Theorem 2.18 (Walikar & Ramane, 2005)**:** For any two integers n and k, $E(A_n(k+1)) < E(B_n(k))$. Energy of some more trees can be found in (Walikar, Ramane, & Hampiholi, 2001).

3. Hyperenergetic, Non-Hperenergetic and Borderenergetic Graphs

McClelland (1971) showed that for molecular graphs of conjugated hydrocarbons (which possess relatively few edges and for which

$$n/2 \leq m \leq 3n/2),\ E(G) \approx a\sqrt{2mn}$$

where $a \approx 0.9$. According to this $E(G)$ is monotonically increasing function of m and n. In view of this observation, in 1978 Ivan Gutman (1978) conjectured that among all graphs with n vertices the complete graph K_n (which posses maximum number of edges) has maximum energy. That is, for any graph G of order n,

$$E(G) \leq E(K_n) \leq 2(n-1).$$

This conjecture is not true (Cvetkovic & Gutman, 1986). There are graphs whose energy exceeds the energy of K_n. This laid to the concept of hyperenergetic graphs (Gutman, 1999). A graph G is said to be hyperenergetic if $E(G) > 2(n-1)$ and it is said to be non-hyperenergetic if $E(G) < 2(n-1)$. A non-complete graph whose energy is equal to $2(n-1)$ is called borderenergetic (Gong, et al., 2015).

The first systematic construction of hyperenegetic graphs was given by Walikar, Ramane and Hampiholi (Walikar, Ramane, & Hampiholi, 1999).

Definition: A line graph G, denoted by $L(G)$ is a graph whose vertex has one-to-one correspondence with the edges of G and two vertices in $L(G)$ are adjacent if and only if the corresponding edges are adjacent in G.

Lemma 3.1 (Sachs, et al., 1967): Let G be an r-regular graph with n vertices and m edges. Then

$$\Phi(L(G);\lambda) = (\lambda + 2)^{m-n}\ \Phi(G;\lambda + 2 - r)$$

Theorem 3.2 (Walikar, Ramane, & Hampiholi, 1999): Let $G \cong L(K_p)$. Then G is hyperenergetic for all $p \geq 5$.

Proof: $\Phi(K_p;\lambda) = (\lambda - p + 1)(\lambda + 1)^{p-1}$.

By Lemma 3.1,

$$\Phi(L(K_p);\lambda) = (\lambda - 2p + 4)(\lambda - p + 4)^{p-1}(\lambda + 2)^{p(p-3)/2}.$$

Therefore

$$E(L(K_p)) = |2p - 4| + |p - 4|(p-1) + |-2|(p(p-3)/2)$$

$= 2p^2 - 6p.$

Which is greater than $E(K_{p(p-1)/2}) = p^2 - p - 2$ for $p \geq 5$.

Hence $G \cong L(K_p)$ is hyperenergetic for $p \geq 5$.

The line graph of complete bipartite graph $K_{p,p}$, $p \geq 3$ and the line graph of the cocktail party graph (a regular graph on $2k$ vertices with regularity $2k - 2$) and their complements are hyperenergetic (Walikar, Ramane, & Hampiholi, 1999).

Definition: The vertices and edges of G are referred as its elements. Let G be a graph with vertex set $V(G)$ and edge set $F(G)$. The total graph of G, denoted by $T(G)$ is a graph with vertex set $V(G) \cup F(G)$ and two vertices in $T(G)$ are adjacent if and only if the corresponding elements of G are adjacent or incident in G.

Theorem 3.3 (Acharya, Rao, & Walikar, 2003): For any r-regular graph G of order n,

(i) $L(G)$ is hyperenergetic if $r \geq 4$;

(ii) $T(G)$ is hyperenergetic if $r \geq 6$;

where $L(G)$ is the line graph and $T(G)$ is the total graph of G.

Proof: (i) If $\lambda_1, \lambda_2, \ldots, \lambda_n$ are the eigenvalues of a regular graph G, then by Lemma 3.1, the eigenvalues of $L(G)$ are $\lambda_i + r - 2$, $i = 1, 2, \ldots, n$ and -2 ($m - n$ times).

$$E(L(G)) = \sum_{i=1}^{n} |\lambda_i + r - 2| + |-2| (m - n)$$

$$\geq \left| \sum_{i=1}^{n} (\lambda_i + r - 2) \right| + 2(m - n)$$

$$= n(r - 2) + 2(m - n) = 2m + n(r - 4)$$

The graph $L(G)$ is hyperenergetic if $E(L(G)) > 2(m - 1)$.

That is if $2m + n(r - 4) > 2m - 2$. It holds if $r \geq 4$.

(ii) If $\lambda_1, \lambda_2, \ldots, \lambda_n$ are the eigenvalues of a regular graph G, then the eigenvalues of T(G) are (Cvetkovic, Doob, & Sachs, 1980)

$$\frac{1}{2} \left[2\lambda_i + r - 2 \pm \sqrt{4\lambda_i + r^2 + 4} \right], i = 1, 2, \ldots, i \text{ and } -2 \ (m - n \text{ times}).$$

Therefore

$$E(T(G)) = \sum_{i=1}^{n} \left| \frac{1}{2} \left[2\lambda_i + r - 2 \pm \sqrt{4\lambda_i + r^2 + 4} \right] \right| + |-2|(m-n)$$

$$= \left| \sum_{i=1}^{n} \frac{1}{2} \left[2\lambda_i + r - 2 \pm \sqrt{4\lambda_i + r^2 + 4} \right] \right| + 2(m-n)$$

$$= n(r-2) + 2(m-n) = 2m + n(r-4).$$

The order of $T(G)$ is $m + n$. Therefore $T(G)$ is hyperenergetic if $E(T(G)) > 2(m + n - 1)$.
That is

$$2m + n(r-4) > 2(m + n - 1).$$

It holds as $r \geq 6$.

By the Theorem 3.2 we see that there are hyperenergetic graphs with $n = p(p-1)/2$ vertices for $p \geq 5$. In particular for $n = 10, 15, 21, \ldots$.

Following theorem gives the constructed hyperenergetic graphs on n vertices, for all $n \geq 9$.

Theorem 3.4 (Gutman, 1999): There are hyperenergetic graphs on n vertices for all $n \geq 9$.

Proof: Let v be a vertex of the complete graph K_n, $n \geq 3$ and let e_i, $i = 1, 2, \ldots, k$, $1 \leq k \leq n - 1$ be its distinct edges, all being incident to v. The graph $Ka_n(k)$ is obtained by deleting e_i, $i = 1, 2, \ldots, k$ from K_n. In addition $Ka_n(0) \cong K_n$.

For $n \geq 3$ and $0 \leq k \leq n - 1$, the eigenvalues of $Ka_n(k)$ are -1 ($n - 3$ times) and three roots $\lambda_1, \lambda_2, \lambda_3$ of the equation

$$\lambda^3 - (n-3)\lambda^2 - (2n - k - 3)\lambda + (k-1)(n-1-k) = 0,$$

of which two (say λ_1 and λ_2) are positive and one (say λ_3) is negative.
Therefore

$$E(Ka_n(k)) = n - 3 + |\lambda_1| + |\lambda_2| + |\lambda_3|$$

$$= n - 3 + \lambda_1 + \lambda_2 - \lambda_3.$$

Thus

$$E(Ka_n(k)) > E(K_n) = 2(n-1) \text{ if } \lambda_1 + \lambda_2 - \lambda_3 > n + 1.$$

This is true for

$$k = 2, n \geq 10; k = 3, n \geq 9; k = 4, n \geq 9; k = 5, n \geq 10; k \geq 6 \text{ and } n \geq k + 4.$$

Hence the proof.

Hou et al. (2001) showed that the line graph of any graph with n vertices and m edges, $n \geq 5$ and $m \geq 2n$ is hyperenergetic. Also, the line graph of any graph of any bipartite graph with n vertices and m edges, $n \geq 7$, $m \geq 2(n-1)$ is hyperenergetic.

Let $S \subseteq \{1, 2, ..., n\}$ with the property that if $i \in S$ then $n - i \in S$. The graph G with vertex set $V = \{v_0, v_1, ..., v_{n-1}\}$ in which v_i is adjacent to v_j if and only if $i - j \pmod{n} \in S$ is called a circulant graph. Some classes of circulant graphs are hyperenergetic (Stevanovic & Stankovic, 2005).

A graph G on n vertices is said to be non-hyperenergetic if $E(G) < 2(n-1)$.

Theorem 3.5 (Gutman, et. al, 2000): A graph G with n vertices and m edges such that $m < 2n - 2$ cannot be hyperenergetic.

Proof: By Theorem 2.3 we have

$$E(G) \leq \frac{2m}{n} + \sqrt{(n-1)\left[2m - \left(\frac{2m}{n}\right)^2\right]}$$

If

$$\frac{2m}{n} + \sqrt{(n-1)\left[2m - \left(\frac{2m}{n}\right)^2\right]} < 2(n-1)$$

then G cannot be hyperenergetic. (5)

Eq. (5) gives that

$$\left[m - 2(n-1)\right]\left[m - \frac{n(n-1)}{2}\right] > 0$$

Above equation holds true for $m > n(n-1)/2$ and for $m < 2(n-1)$.

The condition $m > n(n-1)/2$ is not possible. Therefore there remains $m < 2(n-1)$. Hence the proof.

Huckel graph is a graph in which every vertex has degree at most 3 (Gutman & Polansky, 1986). Hence Huckel graph has at most $3n/2$ edges. Thus in Huckel graph $m < 3n/2 < 2(n-1)$ for all $n > 4$. Hence no Huckel graph is hyperenergetic (Gutman, et. al, 2000).

Definition: Let G_1 and G_2 are graphs with disjoint vertex sets.

(i) H_a is union of G_1 and G_2.
(ii) H_b is obtained by identifying one vertex of G_1 and one vertex of G_2.
(iii) H_c is obtained by connecting one vertex of G_1 and one vertex of G_2 by an edge.
(iv) H_d is obtained by connecting two vertices of G_1 with two vertices of G_2.
(v) H_e is obtained by connecting one vertex of G_1 with two vertices of G_2.

Theorem 3.6 (Walikar, 2001): Let G_i be the graph with n_i vertices and m_i edges such that $m_i \leq 2n_i - 2$, $i = 1, 2$. Then H_a, H_b, H_c, H_d and H_e defined in above definition are non-hyperenergetic.

Figure 7.

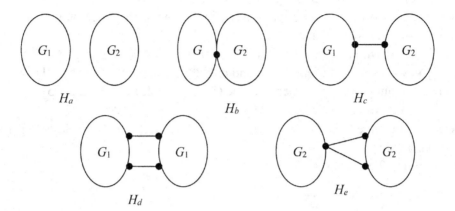

Proof: Let G_1 be the graph with n_1 vertices and m_1 edges such that $m_1 \leq 2n_1 - 2$ and G_2 be the graph with n_2 vertices and m_2 edges such that $m_2 \leq 2n_2 - 2$.

(i) $H_a = G_1 \cup G_2$. Hence $n = n_1 + n_2$ and

$$m = m_1 + m_2 \leq 2n_1 - 2 + 2n_2 - 2 = 2(n_1 + n_2) - 4 < 2(n_1 + n_2) - 2 = 2n - 2.$$

Therefore H_a is non-hyperenergetic.

(ii) $n = n_1 + n_2 - 1$ and

$$m = m_1 + m_2 \leq 2n_1 - 2 + 2n_2 - 2 = 2(n_1 + n_2 - 1) - 2 = 2n - 2.$$

Hence H_b is non-hyperenergetic.

Cases (iii), (iv) and (v) are analogous.

Corollary 3.7 (Walikar, 2001): All graphs whose blocks has average degree less than 3.5 are non-hyperenergetic. In particular all trees, all graphs in which every edge belongs to at most one cycle (cactii) are non-hyperenergetic.

A non-complete graph G is said to be borderenergetic if $E(G) = 2n - 2$ (Gong, et al., 2015). The smallest non-complete borderenergetic graph has seven vertices and it is unique depicted in Fig. 8. It is an integral graph with eigenvalues $5, 1, -1, -1, -1, -1, -2$. Borderenergetic graphs on 7, 8, 9 and 10 vertices are reported in (Gong, et al., 2015)(Shao & Deng, 2016).

Lemma 3.8 (Sachs, 1962): If $\lambda_1, \lambda_2, \ldots, \lambda_n$ are the eigenvalues of a regular graph G of order n and of degree r, then the eigenvalues of $\bar{G}$, the complement of G, are $n - r - 1$ and $-\lambda_i - 1$, $i = 2, 3, \ldots, n$.

Let C_n denotes the cycle on n vertices.

Theorem 3.9 (Gong et al., 2015): Let p, q and t be non-negative integers and let $p + q = 2$. Then $pC_4 \cup qC_6 \cup tC_3$ is borderenergetic.

Figure 8.

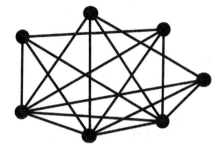

Proof: The eigenvalues of C_4 are 2, 0, 0, – 2; eigenvalues of C_6 are 2, 1, 2, – 1, – 1, – 2 and the eigenvalues of C_3 are 2, – 1, – 1. The order of $pC_4 \cup qC_6 \cup tC_3$ is $4p + 6q + 3t$ and is regular of degree 2. By Lemma 3.8, the eigenvalues of $\overline{pC_4 \cup qC_6 \cup tC_3}$ are $4p + 6q + 3t – 3$, – 3 ($p+q+t – 1$ times), 1 ($p+q$ times), 0 ($2q+2t$ times), – 1 ($2p$ times) and – 2 ($2q$ times). Therefore

$$E\left(\overline{pC_4 \cup qC_6 \cup tC_3}\right) = 10p + 14q + 6t - 6.$$

Recall that $p + q = 2$. Then

$$E\left(\overline{pC_4 \cup qC_6 \cup tC_3}\right) = 2(4p + 6q + 3t - 1),$$

which shows that $\overline{pC_4 \cup qC_6 \cup tC_3}$ is borderenergetic.

Corollary 3.10 (Gong et al., 2015): For each integer $n \geq 7$, there exists a connected non-complete borderenergetic graph of order n.

4. Equienergetic Graphs

Two graphs G_1 and G_2 are said to be equienergetic if $E(G_1) = E(G_2)$. The concept of equienergetic graphs was introduced by Balakrishnan (2004) and Brankov et al. (2004) independently at same time. Co-spectral graphs, for obvious reason, are equienergetic. If O_k is the k-vertex graph without edges and G is any graph, then G and $G \cup O_k$ are equienergetic. The simplest non-trivial example is formed by the triangle and the quadrangle (G_a and G_b in Fig. 9); their eigenvalues are 2, – 1, – 1 and 2, 0, 0, – 2, respectively and therefore $E(G_a) = E(G_b) = 4$. The simplest equienergetic graphs with non-integer energies are the 4-vertex path (G_c in Fig. 9) and 6-vertex star (G_d), with eigenvalues

$$\sqrt{(3+\sqrt{5})/2}, \ \sqrt{(3-\sqrt{5})/2}, \ -\sqrt{(3-\sqrt{5})/2}, \ -\sqrt{(3+\sqrt{5})/2} \text{ and } \sqrt{5}, 0, 0, 0, -\sqrt{5},$$

respectively, both having energy equal to $2\sqrt{5}$. In addition to the quadrangle G_b, there are two more non-cospectral 4-vertex graphs, having energy equal to 4: G_e (with eigenvalues 1, 1, – 1, –1) and G_f (with eigenvalues 2, 0, – 1, – 1). These graphs are not connected. The smallest pair of non-cospectral, connected equienergetic graphs with the same number of vertices are the pentagon (G_g) and the tetragonal pyramid (G_h), whose eigenvlaues are

$$2,\ (\sqrt{5}-1)/2,\ (\sqrt{5}-1)/2,\ -(\sqrt{5}+1)/2,\ -(\sqrt{5}+1)/2 \text{ and } \sqrt{5}+1,0,0,\ -\sqrt{5}+1,-2,$$

respectively, and for which $E(G_g) = E(G_h) = 2\sqrt{5}+2$.

Borderenergetic graphs, given in Section 3 are also equienergetic with the complete graphs. It is interesting to find non-cospectral equienergetic graphs having same number of vertices and having same number of edges.

Consider the graphs H_1 and H_2 given in Fig. 10. The characteristic polynomials of H_1 and H_2 are

$$\Phi(H_1: \lambda) = (\lambda - 4)(\lambda - 1)^4(\lambda + 2)^4.$$

and

$$\Phi(H_2: \lambda) = (\lambda - 4)(\lambda - 2)(\lambda - 1)^2(\lambda + 1)^2(\lambda + 2)^3.$$

Graphs H_1 and H_2 are non-cospectral and $E(H_1) = E(H_2) = 16$.

Figure 9. Non-cospectral equienergetic graphs $E(G_a) = E(G_b) = E(G_e) = E(G_f) = 4$, $E(G_c) = E(G_d) = 2\sqrt{5}$, $E(G_g) = E(G_h) = 2\sqrt{5} + 2$

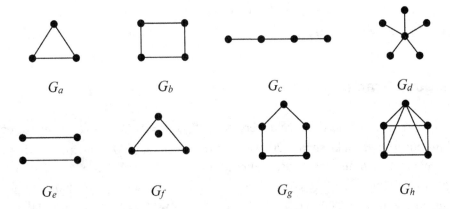

Figure 10. Equienergtic graphs

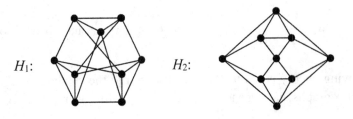

Let G be a graph and $L^1(G) = L(G)$ be its line graph. Further, let $L^k(G) = L(L^{k-1}(G))$, $k \geq 2$, be the iterated line graphs of G, where $L^0(G) = G$. If G is a regular graph on $n = n_0$ vertices and of degree $r = r_0$, then $L(G)$ is a regular graph on $n_1 = nr/2$ vertices and of degree $r_1 = 2r - 2$. Consequently all iterated line graphs $L^k(G)$ of a regular graph G are regular. The order of $L^k(G)$ is (Buckley, 1981) (Buckley, 1993)

$$n_k = \frac{1}{2} n_{k-1} r_{k-1} = \frac{n}{2^k} \prod_{i=0}^{k-1} r_i = \frac{n}{2^k} \prod_{i=0}^{k-1} (2^i r - 2^{i+1} + 2)$$

and degree is

$$r_k = 2^k r - 2^{k+1} + 2.$$

Theorem 4.1 (Ramane, 2004): Let G be a regular graph of on n vertices and of degree $r \geq 3$, then

$$E(L^2(G)) = 2nr(r-2). \tag{6}$$

Proof. Let $\lambda_1, \lambda_2, \ldots, \lambda_n$ be the eigenvalues of G. Then by Lemma 3.1, the eigenvalues of $L(G)$ are

$$\left. \begin{array}{ll} \lambda_i + r - 2, & i = 1,2,\ldots,n \\ and -2, & nr(r-2)/2 \; times \end{array} \right\} \tag{7}$$

Again applying Lemma 3.1 to Eq. (7), the eigenvalues of $L^2(G)$ are

$$\left. \begin{array}{ll} \lambda_i + 3r - 6, & i = 1,2,\ldots,n \\ 2r - 6, & n(r-2)/2 \; times \\ and -2, & nr(r-2)/2 \; times \end{array} \right\} \tag{8}$$

The eigenvalues of a regular graph of degree r, satisfy the condition $-r \leq \lambda_i \leq r$, $i = 1,2,\ldots,n$ (Cvetkovic & Doob, 1980).

If $r \geq 3$ then $\lambda_i + 3r - 6 \geq 4r - 6 \geq 0$ and $2r - 6 \geq 0$. Therefore by Eq. (8)

$$E(L^2(G)) = \sum_{i=1}^{n} |\lambda_i + 3r - 6| + |2r - 6| \frac{n(r-2)}{2} + |-2| \frac{nr(r-2)}{2}$$

$$= 2nr(r-2),$$

since $\sum_{i=1}^{n} \lambda_i = 0$ (Cvetkovic & Doob, 1980).

Corollary 4.2 (Ramane et al., 2004)**:** Let G_1 and G_2 be two regular graphs on n vertices and of degree $r \geq 3$. Then $L^2(G_1)$ and $L^2(G_2)$ are equienergetic.

Proof. Follows directly from Eq. (6). $\square$

Corollary 4.3 (Ramane et al., 2005)**:** Let n_k and r_k be the number of vertices and degree of $L^k(G)$ respectively, then for any $k \geq 1$,

(i) $E(L^{k+1}(G)) = 2n_k(r_k - 2)$.

(ii) $E(L^{k+1}(G)) = 2n(r-2) \prod_{i=0}^{k-1} (2^i r - 2^{i+1} + 2)$

(iii) $E(L^k(G)) = 4(n_k - n_{k-1}) = 4n_k \left(\dfrac{r_k - 2}{r_k + 2} \right). \square$

By above Corollary 4.3 (ii) we see that the energy of second and higher order line graphs of a regular graph G of degree greater than 2 is solely depend on the order n and degree r of G.

Corollary 4.4 (Ramane et al., 2005)**:** Let G_1 and G_2 be two non-cospectral regular graphs on n vertices and of degree $r \geq 3$. Then for any $k \geq 2$, both $L^k(G_1)$ and $L^k(G_2)$ are regular, non-cospectral, possessing same number of vertices, same number of edges and equienergetic. In addition if G_1 and G_2 are connected then $L^k(G_1)$ and $L^k(G_2)$ are also connected.

An analogous results for the complement of $L^k(G)$ are obtained in (Ramane, 2004) (Walikar et al., 2007). Let $\overline{G}$ be the complement of G.

Theorem 4.5 (Ramane, 2004)**:** If G is a regular graph of order n and of degree $r \geq 3$, then

$$E\left(\overline{L^2(G)}\right) = (nr - 4)(2r - 3) - 2.$$

Corollary 4.6 (Walikar et al., 2007)**:** Let n_k and r_k be the number of vertices and degree of $L^k(G)$ respectively, then for any $k \geq 1$,

$$E\left(\overline{L^k(G)}\right) = \left[\frac{n}{2^{k-2}} \prod_{i=0}^{k-2} (2^i r - 2^{i+1} + 2) \right] (2^{k-1} r - 2^k + 1) - 2 \,.$$

Corollary 4.7 (Walikar et al., 2007)**:** Let G_1 and G_2 be two non-cospectral regular graphs on n vertices and of degree $r \geq 3$. Then for any $k \geq 2$, both $\overline{L^k(G_1)}$ and $\overline{L^k(G_2)}$ are regular, non-cospectral, possessing same number of vertices, same number of edges and equienergetic.

Let G be a simple graph with vertex set $V = \{v_1, v_2, \ldots, v_n\}$. The extended double cover of G denoted by G^* is the bipartite graph with bipartition (X, Y) where $X = \{x_1, x_2, \ldots x_n\}$ and $Y = \{y_1, y_2, \ldots, y_n\}$ in which x_i and y_j are adjacent if and only if $i = j$ or v_i and v_j are adjacent in G.

Theorem 4.8 (Xu & Hou, 2007)**:** Let G_1 and G_2 be r-regular graphs of order n with $r \geq 3$. Then

(i) $(L^2(G_1))^*$ and $(L^2(G_2))^*$ are equienergetic bipartite graphs and

$$E((L^2(G_1))^*) = E((L^2(G_2))^*) = nr(3r-5).$$

(ii) $\left(\overline{L^2(G_1)}\right)^*$ and $\left(\overline{L^2(G_2)}\right)^*$ are equienergetic bipartite graphs and

$$E\left[\left(\overline{L^2(G_1)}\right)^*\right] = E\left[\left(\overline{L^2(G_2)}\right)^*\right] = (5nr-16)(r-2)+nr-8.$$

(iii) $\overline{\left(L^2(G_1)\right)^*}$ and $\overline{\left(L^2(G_2)\right)^*}$ are equienergetic bipartite graphs and

$$E\left[\overline{\left(L^2(G_1)\right)^*}\right] = E\left[\overline{\left(L^2(G_2)\right)^*}\right] = (2nr-4)(2r-3)-2.$$

R. Balakrishnan (2004) has given another method for construction of equienergetic graphs.

Definition: The tensor product of two graphs G_1 and G_2 is the graph $G_1 \otimes G_2$ with vertex set $V(G_1) \times V(G_2)$ and in which the vertices (u_1, u_2) and (v_1, v_2) are adjacent if and only if u_1 is adjacent to v_1 in G_1 and u_2 is adjacent v_2 in G_2.

Theorem 4.9 (Balakrishnan, 2004): If G_1 and G_2 are any two graphs then $E(G_1 \otimes G_2) = E(G_1)E(G_2)$.

Proof: Let the eigenvalues of G_1 and G_2 be $\lambda_1, \lambda_2, ..., \lambda_n$ and $\mu_1, \mu_2, ..., \mu_n$ respectively then the eigenvalues $G_1 \otimes G_2$ are $\lambda_i \mu_j$, $i = 1, 2, ..., n_1$ and $j = 1, 2, ..., n_2$ (Cvetkovic, Doob, & Sachs, 1980). Therefore

$$E(G_1 \otimes G_2) = \sum_{i=1}^{n_1}\sum_{j=1}^{n_2}\left|\lambda_i\mu_j\right| = \sum_{i=1}^{n_1}\left|\lambda_i\right|\sum_{j=1}^{n_2}\left|\mu_j\right| = E(G_1)E(G_2).$$

Using Theorem 4.9, Balakrishnan (2004) showed that

$$E(K_2 \otimes K_2 \otimes G) = E(C_4 \otimes G) = 4E(G).$$

Thus he constructed pairs of non-cospectral equienergetic graphs for $n \equiv 0(\mod 4)$.

Figure 11. Pentagon G, Four sided pyramid H

 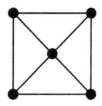

Similar construction is done by D. Stevanovic (2005) for $n \equiv 0 \pmod 5$. Let G be the pentagon and H be a tetragonal pyramid as shown in Fig. 11 and their energy is $E(G) = E(H) = 2\sqrt{5} + 2$.

Let

$$G_{ki} = \underbrace{G \otimes \cdots \otimes G}_{i} \underbrace{\otimes H \otimes \ldots \otimes H}_{k-i}.$$

Then by Theorem 4.9, the energy of G_{ki} is $(2\sqrt{5} + 2)^n$. The largest eigenvalue of G_{ki} is equal to $2^i (\sqrt{5} + 1)^{k-i}$ and so no two of these graphs are cospectral. Thus the graphs G_{ki}, $i = 1, 2, \ldots, k$ form a family of k mutually non-cospectral connected equienergetic graphs having the same number of vertices.

Another result of this kind was obtained by Indulal and Vijayakumar (2007).

Definition: The Cartesian product of two graphs G and H is the graph $G{\times}H$ whose vertex set is $V(G){\times}V(H)$ and two vertices (u_1, u_2) and (v_1, v_2) are adjacent in $G{\times}H$ if either $u_1 = v_1$ and u_2 is adjacent to v_2 in H or u_1 is adjacent to v_1 in G and $u_2 = v_2$.

Let l and k be positive integers such that $l \geq 2k$. Let K_l be the complete graph on l vertices. Let G be an n-vertex graph whose spectrum lies in the interval $[-k, +k]$. Then

$$E((K_l)^k \times G) = E\left(\underbrace{K_l \times K_l \times \cdots \times K_l}_{k} \times G \right) = 2nk(l-1)^k.$$

Thus for any equienergetic graphs G_1 and G_2 and for any k and l satisfying above stated requirements, the pairs $(K_l)^k{\times}G_1$ and $(K_l)^k{\times}G_2$ are equienergetic (Indulal & Vijayakumar, 2007).

Definition: Let G be a graph with vertex set $V(G) = \{v_1, v_2, \ldots, v_n\}$. Take another set of vertices $U = \{u_1, u_2, \ldots, u_n\}$. Define a graph DG whose vertex set is $V(G) \cup U$ and whose edge set consists only of the edges joining u_i to the neighbours of v_i in G for $i = 1, 2, \ldots, n$. The resulting graph DG is called the identity duplication graph.

Definition: Let G be a graph with vertex set $V(G) = \{v_1, v_2, \ldots, v_n\}$. Take another copy of G with vertices labeled as $\{u_1, u_2, \ldots, u_n\}$. Make u_i adjacent to the neighbours of v_i in G for $i = 1, 2, \ldots, n$. The resulting graph is called double cover and is denoted by D_2G.

If $\lambda_1, \lambda_2, \ldots, \lambda_n$ are the eigenvalues of G then the eigenvalues of DG are λ_i and $-\lambda_i$, $i = 1, 2, \ldots, n$ and the eigenvalues of D_2G are $2\lambda_i$, $i = 1, 2, \ldots, n$ and 0 (n times). Therefore DG and D_2G form a pair of noncospectral equienergetic graphs (Indulal & Vijayakumar, 2006). Also $G{\otimes}K_2$ and D_2G form a pair of noncospectral equenergetic graphs.

Definition: Let G be an r-regular graph with vertex set $V(G) = \{v_1, v_2, \ldots, v_n\}$. Introduce a set of n isolated vertices $\{u_1, u_2, \ldots, u_n\}$ and make each u_i adjacent to the neighbours of v_i in G for $i = 1, 2, \ldots, n$. Then introduce a set of $k \geq 0$ isolated vertices and make all of them adjacent to all vertices of G. The resulting graph is denoted by $H_k(G)$.

Indulal and Vijayakumar (Indulal & Vijayakumar, 2006) proved that

$$E(H_k(G)) = \sqrt{5}\left[E(G) - r + \sqrt{r^2 + \frac{4nk}{5}}\right].$$

Using this they constructed equienergetic graphs for $n \geq 20$.

Definition: The join $G_1 \vee G_2$ of two graphs G_1 and G_2 is the graph obtained by joining every vertex of G_1 with every vertex of G_2.

Lemma 4.10 (Ramane et al., 2007): Let G_i is an r_i-regular graph of with n_i vertices, $i = 1, 2$. Then

$$E(G_1 \vee G_2) = E(G_1) + E(G_2) + \sqrt{(r_1 + r_2)^2 + 4(n_1 n_2 - r_1 r_2)} - (r_1 + r_2).$$

Proof: If G_i is a regular graph of degree r_i with n_i vertices, $i = 1, 2$, then (Cvetkovic, Doob, & Sachs, 1980) (Finck et al., 1965)

$$\Phi(G_1 \vee G_2 : \lambda) = \frac{[(\lambda - r_1)(\lambda - r_2) - n_1 n_2]}{(\lambda - r_1)(\lambda - r_2)} \Phi(G_1 : \lambda)\Phi(G_2 : \lambda),$$

which gives

$$(\lambda - r_1)(\lambda - r_2)\Phi(G_1 \vee G_2 : \lambda) = [(\lambda - r_1)(\lambda - r_2) - n_1 n_2]\Phi(G_1 : \lambda)\Phi(G_2 : \lambda).$$

Let

$$P_1(\lambda) = (\lambda - r_1)(\lambda - r_2)\Phi(G_1 \vee G_2 : \lambda)$$

and

$$P_2(\lambda) = [(\lambda - r_1)(\lambda - r_2) - n_1 n_2]\Phi(G_1 : \lambda)\Phi(G_2 : \lambda).$$

The sum of the absolute values of the roots of $P_1(\lambda) = 0$ is

$$E(G_1 \vee G_2) + r_1 + r_2. \tag{9}$$

And the sum of the absolute values of the roots of $P_2(\lambda) = 0$ is

$$E(G_1) + E(G_2) + \left|\frac{r_1 + r_2 + \sqrt{(r_1 + r_2)^2 + 4(n_1 n_2 - r_1 r_2)}}{2}\right|$$

$$+ \left|\frac{r_1 + r_2 - \sqrt{(r_1 + r_2)^2 + 4(n_1 n_2 - r_1 r_2)}}{2}\right|$$

$$= E(G_1) + E(G_2) + \sqrt{(r_1 + r_2)^2 + 4(n_1 n_2 - r_1 r_2)} \; . \tag{10}$$

Since $P_1(\lambda) = P_2(\lambda)$, equating (9) and (10) we get

$$E(G_1 \vee G_2) = E(G_1) + E(G_2) + \sqrt{(r_1 + r_2)^2 + 4(n_1 n_2 - r_1 r_2)} \; - (r_1 + r_2).$$

Indulal and Vijayakumar (Indulal & Vijayakumar, 2006) constructed the pairs of equienergetic graphs on n vertices for $n = 6, 14, 18$ and $n \geq 20$. Liu and Liu (Liu & Liu, 2008) constructed pairs of equienergetic graphs on n vertices for $n \geq 10$. Ramane and Walikar (Ramane et al., 2007) proved the existence of noncospectral equienergtci graphs on n vertices for all $n \geq 9$.

Theorem 4.11 (Ramane et al., 2007): There exists a pair of connected non-cospectral, equienergetic graphs on n vertices for all $n \geq 9$.

Proof: Consider the graphs H_1 and H_2 as shown in Fig. 10.

Both H_1 and H_2 are connected regular graphs on nine vertices and of degree four and $E(H_1) = E(H_2)$ $= 16$.

A complete graph K_p is regular graph on p vertices and of degree $p - 1$.

Knowing $E(K_p) = 2(p - 1)$, by Lemma 4.10, we have

$$E(H_1 \vee K_p) = E(H_2 \vee K_p) = p + 11 + \sqrt{(p + 3)^2 + 4(5p + 4)} \; .$$

Figure 12. Equienergetic trees

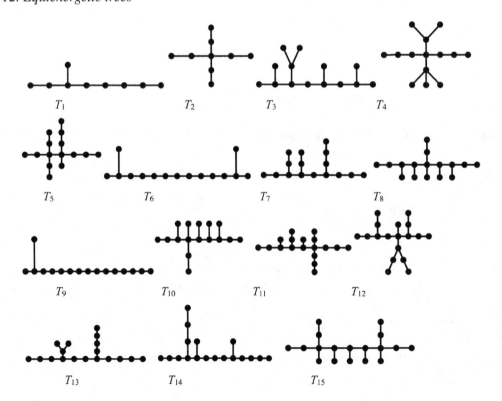

Thus $H_1 \vee K_p$ and $H_2 \vee K_p$ are equienergetic. Moreover $H_1 \vee K_p$ and $H_2 \vee K_p$ are non-cospectral, since H_1 and H_2 are non-cospectral. Further $H_1 \vee K_p$ and $H_2 \vee K_p$ are connected and possess equal number of vertices $n = 9 + p$, $p = 0, 1, 2, \ldots$ $\square$

In fact, in place of K_p, if any r-regular graph H of order p is taken in above theorem then (Ramane et al., 2009)

$$E(H_1 \vee H) = E(H_2 \vee H) = 12 + E(H) - r + \sqrt{36p + (r-4)^2} \ .$$

Adiga and Rakshith (2016) gave another construction of equienergetic graphs on n vertices for $n \geq 11$.

A chemical tree is a tree in which no vertex has a degree greater than four (Gutman & Polansky, 1986). V. Brankov, D. Stevanovic and I. Gutman (2004) obtained pairs of equienergetic chemical trees for $n \leq 18$ vertices. For this they downloaded from the site (McKay, n.d.), all trees with n vertices, $n = 4, 5, \ldots, 18$, computed their eigenvalues and energies and recorded those having equal energies. There are 205001 trees with $4 \leq n \leq 18$ of which 103444 are chemical trees. After ignoring the cospectral trees, there are 27 pairs and 5 triplets of noncospectral equienergetic trees. Of them only six pairs (T_1, T_2), (T_3, T_4), (T_5, T_6), (T_7, T_8), (T_9, T_{10}), (T_{11}, T_{12}), and one triplet (T_{13}, T_{14}, T_{15}), consisted of equienergetic chemical trees, which are depicted in Fig. 12. The trees T_{13} and T_{14} are cospectral.

It should be noted that until now no systematic method is known for construction of pairs of equienergetic trees (Gutman, 2015).

REFERENCES

Acharya, Rao, & Walika. (2003). *Energy of signed digraphs*. Lecture Notes, Group Discussion on Energy of a Graph. Karnatak University.

Adiga, C., & Rakshith, B. R. (2016). On spectra of variants of the corona of two graphs and some new equienergetic graphs. *Discussiones Mathematicae. Graph Theory, 36*(1), 127–140. doi:10.7151/dmgt.1850

Balakrishnan, R. (2004). The energy of a graph. *Linear Algebra and Its Applications, 387*, 287–295. doi:10.1016/j.laa.2004.02.038

Brankov, V., Stevanovic, D., & Gutman, I. (2004). Equienergetic chemical trees. *Journal of the Serbian Chemical Society, 69*(7), 549–553. doi:10.2298/JSC0407549B

Buckley, F. (1981). Iterated line graphs. *Congr. Numer., 33*, 390–394.

Buckley, F. (1993). The size of iterated line graphs. *Graph Theory Notes, New York, 25*, 33–36.

Coulson, C. A. (1940). On the calculation of the energy in unsaturated hydrocarbon molecules. *Proceedings of the Cambridge Philosophical Society, 36*(2), 201–203. doi:10.1017/S0305004100017175

Cvetkovic, D. M., Doob, M., & Sachs, H. (1980). *Spectra of Graphs*. New York: Academic Press.

Cvetkovic, D. M., & Gutman, I. (1986). The computer system GRAPH: A useful tool in chemical graph theory. *Journal of Computational Chemistry, 7*, 640–644. doi:10.1002/jcc.540070505

Finck, H. J., & Grohmann, G. (1965). Vollstandiges Produkt, chromatische Zahl und charakteristisches Polynom regular Graphen I. *Wiss. Z. TH Ilmenau, 11*, 1–3.

Godsil, C., & Royle, G. (2001). *Algebraic Graph Theory*. New York: Springer. doi:10.1007/978-1-4613-0163-9

Gong, S., Li, X., Xu, G., Gutman, I., & Furtula, B. (2015). Borderenergetic graphs. *MATCH Commun. Math. Comput. Chem., 74*, 321–332.

Gutman, I. (1977). Acyclic systems with extremal Huckel π-electron energy. *Theoretica Chimica Acta, 45*(2), 79–87. doi:10.1007/BF00552542

Gutman, I. (1978). The energy of a graph. *Ber. Math. Stat. Sekt. Forschungszentrum Graz, 103*, 1–22.

Gutman, I. (1999). Hyperenergetic molecular graphs. *Journal of the Serbian Chemical Society, 64*, 199–205.

Gutman, I. (2001). The energy of a graph: old and new results. In A. Betten, A. Kohnert, R. Laue, & A. Wassermann (Eds.), *Algebraic Combinatorics and Applications* (pp. 196–211). Berlin: Springer. doi:10.1007/978-3-642-59448-9_13

Gutman, I. (2015). Open problems for equienergetic graphs. *Iranain J. Math. Chem., 6*, 185–187.

Gutman, I., Hou, Y., Walikar, H. B., Ramane, H. S., & Hampiholi, P. R. (2000). No Huckel graph is hyperenergetic. *Journal of the Serbian Chemical Society, 65*(11), 799–801. doi:10.2298/JSC0011799G

Gutman, I., & Polansky, O. E. (1986). *Mathematical Concepts in Organic Chemistry*. Berlin: Springer-Verlag. doi:10.1007/978-3-642-70982-1

Gutman, I., & Trinajstic, N. (1972). Graph theory and molecular orbitals, Total π-electron energy of alternants hydrocarbons. *Chemical Physics Letters, 17*(4), 535–538. doi:10.1016/0009-2614(72)85099-1

Haemers, W. H. (2008). Strongly regular graphs with maximal energy. *Linear Algebra and Its Applications, 429*(11-12), 2719–2723. doi:10.1016/j.laa.2008.03.024

Haemers, W. H., & Xiang, Q. (2010). Strongly regular graphs with parameters $(4m^4, 2m^4 + m^2, m^4 + m^2, m^4 + m^2)$ exist for all $m > 1$. *European Journal of Combinatorics, 31*, 1553–1559. doi:10.1016/j.ejc.2009.07.009

Hall, M. (1986). *Combinatorial Theory*. New York: Wiley.

Harary, F. (1999). *Graph Theory*. New Delhi: Narosa Publishing House.

Hou, Y., & Gutman, I. (2001). Hyperenergetic line graphs. *MATCH Commun. Math. Comput. Chem., 43*, 29–39.

Huckel, E. (1931). Quantentheoretische Beitrage zum Benzolproblem. *Zeitschrift fur Physik, 70*, 204–286. doi:10.1007/BF01339530

Indulal, G., & Vijayakumar, A. (2006). On a pair of equienergetic graphs. *MATCH Commun. Math. Comput. Chem., 55*, 83–90.

Indulal, G., & Vijayakumar, A. (2007). Energies of some non-regular graphs. *Journal of Mathematical Chemistry*, *42*(3), 377–386. doi:10.100710910-006-9108-7

Koolen, J. H., & Moulton, V. (2001). Maximal energy graphs. *Advances in Applied Mathematics*, *26*(1), 47–52. doi:10.1006/aama.2000.0705

Koolen, J. H., & Moulton, V. (2003). Maximal energy bipartite graphs. *Graphs and Combinatorics*, *19*(1), 131–135. doi:10.100700373-002-0487-7

Li, X., Li, Y., & Shi, Y. (2010). Note on the energy of regular graphs. *Linear Algebra and Its Applications*, *432*(5), 1144–1146. doi:10.1016/j.laa.2009.10.023

Li, X., Shi, Y., & Gutman, I. (2012). *Graph Energy*. New York: Springer. doi:10.1007/978-1-4614-4220-2

Liu, J., & Liu, B. (2008). Note on a pair of equienergetic graphs. *MATCH Commun. Math. Comput. Chem.*, *59*, 275–278.

McClelland, B. J. (1971). Properties of latent roots of a matrix: The estimation of π-electron energies. *The Journal of Chemical Physics*, *54*(2), 640–643. doi:10.1063/1.1674889

Ramane, H. S., Gutman, I., Walikar, H. B., & Halkarni, S. B. (2004). Another class of equienergetic graphs. *Kragujevac J. Math.*, *26*, 15–17.

Ramane, H. S., & Walikar, H. B. (2007). Construction of equienergetic graphs. *MATCH Commun. Math. Comput. Chem.*, *57*, 203–210.

Ramane, H. S., Walikar, H. B., & Gutman, I. (2009). Equienergetic graphs. *J. Comb. Math. Comb. Comput.*, *69*, 165–173.

Ramane, H. S., Walikar, H. B., Rao, S. B., Acharya, B. D., Hampiholi, P. R., Jog, S. R., & Gutman, I. (2004). Equienergetic graphs. *Kragujevac J. Math.*, *26*, 5–13.

Ramane, H. S., Walikar, H. B., Rao, S. B., Acharya, B. D., Hampiholi, P. R., Jog, S. R., & Gutman, I. (2005). Spectra and energies of iterated line graphs of regular graphs. *Applied Mathematics Letters*, *18*(6), 679–682. doi:10.1016/j.aml.2004.04.012

Sachs, H. (1962). Uber selbstkomplementare Graphen. *Publicationes Mathematicae (Debrecen)*, *9*, 270–288.

Sachs, H. (1967). Uber Teiler, Faktoren und charakteristiche Polynome Von Graphen, Teil II. *Wiss. Z. TH Ilmenau*, *13*, 405–412.

Shao, Z., & Deng, F. (2016). Correcting the number of borderenergetic graphs of order 10. *MATCH Commun. Math. Comput. Chem.*, *75*, 263–266.

Stevanovic, D. (2005). Energy and NEPS of graphs. *Linear and Multilinear Algebra*, *53*(1), 67–74. doi:10.1080/03081080410001714705

Stevanovic, D., & Stankovic, I. (2005). Remarks on hyperenergetic circulant graphs. *Linear Algebra and Its Applications*, *400*, 345–348. doi:10.1016/j.laa.2005.01.001

Walikar & Ramane. (2005). Energy of trees with edge independence number two. *Proc. Nat. Acad. Sci., India, 75*(A), 137 – 140.

Walikar, H. B., Gutman, I., Hampiholi, P. R., & Ramane, H. S. (2001). Nonhyperenergetic graphs. *Graph Theory Notes, New York, 51*, 14–16.

Walikar, H. B., Ramane, H. S., Gutman, I., & Halkarni, S. B. (2007). On equienergetic graphs and molecular graphs. *Kragujevac J. Sci., 29*, 73–84.

Walikar, H. B., Ramane, H. S., & Hampiholi, P. R. (1999). On the energy of a graph. In R. Balakrishnan, H. M. Mulder, & A. Vijaykumar (Eds.), *Graph Connections* (pp. 120–123). New Delhi: Allied Publishers.

Walikar, H. B., Ramane, H. S., & Hampiholi, P. R. Energy of trees with edge independence number three. In *Proc. Nat. Conf. Math. Comput. Models*. Allied Publishers.

Walikar, H. B., Ramane, H. S., & Jog, S. R. (2008). On an open problem of R. Balakrishnan and the energy of products of graphs. *Graph Theory Notes, New York, 55*, 41–44.

Xu, L., & Hou, Y. (2007). Equienergetic bipartite graphs. *MATCH Commun. Math. Comput. Chem., 57*, 363–370.

Zhou, B. (2004). Energy of graphs. *MATCH Commun. Math. Comput. Chem., 51*, 111–118.

Zhou, B., & Ramane, H. S. (2008). On upper bounds for energy of bipartite graphs. *Indian Journal of Pure and Applied Mathematics, 39*, 483–490.

Chapter 12
Applying Graph Theory to Detect Cases of Money Laundering and Terrorism Financing

Natalia G. Miloslavskaya

(iD) https://orcid.org/0000-0002-1231-1805

National Research Nuclear University MEPhI (Moscow Engineering Physics Institute), Russia

Andrey Nikiforov

National Research Nuclear University MEPhI (Moscow Engineering Physics Institute), Russia

Kirill Plaksiy

National Research Nuclear University MEPhI (Moscow Engineering Physics Institute), Russia

Alexander Tolstoy

National Research Nuclear University MEPhI (Moscow Engineering Physics Institute), Russia

ABSTRACT

A technique to automate the generation of criminal cases for money laundering and financing of terrorism (ML/FT) based on typologies is proposed. That will help an automated system from making a decision about the exact coincidence when comparing the case objects and their links with those in the typologies. Several types of subgraph changes (mutations) are examined. The main goal to apply these mutations is to consider other possible ML/FT variants that do not correspond explicitly to the typologies but have a similar scenario. Visualization methods like the graph theory are used to order perception of data and to reduce its volumes. This work also uses the foundations of information and financial security. The research demonstrates possibilities of applying the graph theory and big data tools in investigating information security incidents. A program has been written to verify the technique proposed. It was tested on case graphs built on the typologies under consideration.

DOI: 10.4018/978-1-5225-9380-5.ch012

INTRODUCTION

The project is motivated by the fact that its research area lies at the intersection of such important challenges of modern society as the transition to advanced digital intelligent technologies and big data processing, as well as a counteraction to terrorism and ideological extremism, technogenic, biogenic, sociocultural, and cyber threats and other sources of danger. In particular, this arises a problem of necessary formation of typologies (EAG-2, n.d.) and automation of searching and revealing schemes of suspicious activity in the field of combating ML/FT (AML/CFT) (EAG-1, n.d.). While solving the problem, the unique characteristics of these schemes demand new methodology and techniques in investigating them. Volumes of heterogeneous data to be analyzed in a short time have the explicit characteristics of big data. Since information on typologies should not become known in a criminal environment, its protection must be provided.

The significance of the project is determined by the technique proposed and its software implementation to automate the generation of new criminal schemes (for example, "Peso" and "commission scheme") based on the typologies but are not their exact copies. This feature has hampered the existing automated systems in the comparative analysis of the analyzed objects and links between them with the objects and their links in the typologies as to conclude whether they are exactly identical or not. The advantages of graphs and big data are also investigated and applied in the analysis and processing of financial investigations that should be protected. The project results may be of interest for further scientific research in the AML/CFT area (EAG-1, n.d.), and their practical significance may be proved in the applications of the proposed technique.

The scientific novelty of the research is that using the suggested technique based on the graph approaches, the schemes of criminal cases can be multiplied, bringing millions of all possible variants of one case to be processed later and analyzed for searching similar schemes with big data tools. Thus new rules for identifying ML/FT schemes could be created, and the automation of AML/CFT crime detection could be simplified.

Big Data Techniques in Investigating Money Laundering and Financing of Terrorism

In solving problems of information and financial security big data technologies can be used to receive and process huge data sets in a short time. In our case big data tools assume a huge gain in term of in time not only the collection but in sorting data, filling different databases, etc.

The main methods while working with big data (Barsegyan at al., 2008), without which the study of huge amounts of information will be problematic, can be listed as the following:

- Associative rules which reflect frequent dependencies (or associations) between objects or events. The dependencies found in the information array are represented in the form of rules and used to analyze the nature of data and predict an occurrence of events;
- Decision trees (classification and regression). The task of classification is to define an object class by its characteristics. The problem of regression, as well as the classification problem, can be determined from the studied property of the object; Unlike functions: the classical set of classes, and also the set of real numbers;

- Artificial neural network (ANN), a mathematical model, as well as its software or hardware implementation, is built on an organization and functioning principles of biological neural networks, a living organism nerve cell networks. INS can work with an instructor or without it. In both types, the ANN studies the interactions of the stimulus and responses to the reactions. Getting a large sample of such "questions" allows anyone to perform forecast of various events;
- Cluster analysis. The task of clustering is to search for independent groups (clusters) and their characteristics in the entire set of analyzed data which provides a better understanding of the data. Moreover, the grouping of homogeneous objects reduces their number and consequently facilitates the analysis, then each kind of information is processed by the ANN.

It is necessary to consider the following priority tasks in order to provide prompt access to the most complete information to respond to crimes immediately:

- Informing of non-standard activity when working with these data and a assessing primary a suspicious activity and potential damage from it;
- Determining the extent of potential prevalent vulnerabilities, which can be used to implement criminal schemes;
- Analysis of current events in order to predict a possible implementation of a particular criminal activity;
- Planning corrective measures and assessing their effectiveness;
- Assessing an organization's procedures for ensuring confidential information's security, etc.

Special technical solutions are required to perform these tasks. Big data technologies (Cielen, Meysman, & Ali, 2016) ensure a more detailed and comprehensive picture of security vulnerabilities and threats, detection, and prevention in the future fraud cases or technical failures and so on, based on the system data from the equipment's state up to the analysis of network traffic. The following tasks could be distinguished among those that can be solved with the help of big data technologies:

- Analysis of data on transactions in financial institutions systems in order to identify fraudulence;
- Analysis of data from video surveillance systems and access control systems simultaneously;
- Analysis of hardware and software event logs depending on which it is possible to identify connections and predict a potentially dangerous behavior of users and programs;
- Automatic detection of incidents among all the events occurred and so on.

Since the data on transactions and various transactions exceeded long ago the limits of conventional data, big data tools are used for anti-ML/CFT (AML/CFT), for example, SAS Anti-Money Laundering System (SAS AML), which was specially created for banking systems with the view of monitoring bank operations, consolidating information, detecting of suspicious activities (SAS AML, n.d.). These tools allow to get rid of unnecessary data leaving only the necessary information (Dreżewski et al., 2015).

The project offers to use big data technologies considered not only in terms of information security (IS) but also in the context of tasks of financial investigations (FI). The issues, which may arise in these areas, are listed below, where the tasks of ensuring IS for which big data technologies can be used may include the following:

- Issues about confidential data protection of FI objects, in particular, ensuring their integrity;
- Assessment of risk of unauthorized access and leakage of confidential information, for example, in the investigation in progress and the evidence found;
- The problem of improper storage of confidential information;
- Information flows analysis on all the used protocols of transmitted data;
- Risk of personal data misuse by third parties, etc.

The interests of FI demand to study the following:

- Correspondence analysis between FR evidence and case objects;
- Financial activities analysis and suspicious activity identification;
- Improper storage of information constituting bank secrecy;
- Accompanying documentation check in order to search for inconsistencies and possible clues;
- Funds misuse or other assets analysis, etc.

Big data technologies can also be used for forensics as the information and analytical expertise on the incident. With these technologies, one can automate a search and gather evidence which will greatly facilitate professionals work.

In IT management or other projects, big data technologies allow obtaining useful statistics on the users' behavior and their preferences that helps to perform the tasks above.

Here the nine most common ML/FT typologies are considered: goods and services cost overstatement in invoices; goods and services cost underestimation in invoices; goods and services multiple invoicing; goods or services in short supply; goods or services excess supply; incorrect description of goods and services; foreign black market exchange operations for pesos exchange; carousel scheme (FATF, 2007) of value-added tax (VAT) evasion (FATF, 2006); and commission scheme (Bulancev et al., 2008). Typology schemes were used to test the technique obtained. Such schemes as carousel, commission and issuing several invoices for a multi-party deal were considered (FATF, 2008).

ML/FT Schemes

The carousel scheme differs from conventional ones such as smuggling when there is a limited market for sale. Criminal groups participating in the carousel scheme are not competitors for one another. Such a scheme assumes circulation of goods as many times as possible without paying the VAT. The goods are sold only within the "carousel" at the highest price at the time of sale. The higher the price, the higher is the amount of VAT. Such a complex system requires close coordination in order to ensure goods present in the proper place at the appropriate time. There is a reason to suspect that the groups exchange their information and resources to carry out the fraudulent scheme. In the UK, an analysis of the current carousel shows that criminal organizations hand over goods and companies to each other for a further use in fraudulent schemes.

Another method involves issuance of several invoices for one international trade transaction. People are able to justify multiple payments for the same goods batch or the same service by invoicing several times ML/TF. The involvement of several credit and financial institutions in such additional payment transactions increases the complexity of transactions. Moreover, even if there is a fact of multiple payments for goods or services, there are numerous explanations which do not contradict the legal terms since payment terms change, previous payment instructions have been specified and the latest payment penalty has been paid, etc. In contrast to the method of overstating and understating the goods price in the invoice, in this case, it is necessary to take into account that the exporter or importer does not need to indicate the actual price of goods or services, which do not correspond to reality in the documentation.

In addition to the practice of manipulating export and import prices, those involved in ML/FT often indicate in documents services and goods amount, quality or type, which do not correspond to reality. For example, an exporter can ship a relatively inexpensive product and include it in an invoice under the guise of a more expensive or completely different product. As a result, there are discrepancies between the shipping and customs documents and the actually shipped goods. Invalid descriptions are also used when selling services, such as providing financial advice, consulting services and market research services.

A commission scheme also used in this work to test the efficiency of the proposed method has been considered. Foreign (as a rule, offshore) company purchases goods from a foreign seller, sends it at Russian price of purchase and specifies this price as contractual in a contract. The contract provides Russian company obligations to conduct customs clearance of the goods in the mode of import to the Russian customs territory for a free sale. The Russian company must accept the goods for safekeeping. The goods owner has not changed yet. When importing customs duties are paid, they are based on the contract goods price. Then the foreign and Russian companies conclude a commission agreement while the foreign one acts as a commitment and the Russian one as a commission agent. It is determined by the contract that commission agent acts in transactions with goods buyers on his own behalf but in the interests of the foreign partner. The amount of compensation is agreed in advance: it can be a goods value percentage declared at the border, as well as some specific amount with a premium in the form of price percentage when it is sold. The goods are sold at market prices. The received income consists of the initial goods cost, commissions, and foreign enterprise income. The commission agent pays taxes on commissions. Taxes on turnover are minimal since the principal amounts are posted on off-balance accounts.

The "Currency operations on the black market for exchanging pesos" ("Peso") crime scheme is also of interest. Its scenario of the simplest currency transaction includes the following:

1. The Colombian drug smuggler smuggles drugs into the US and sells them there for cash;
2. The drug dealer agrees to sell US dollars below the nominal exchange rate to a broker working with pesos for Colombian pesos;
3. The broker transfers funds to the drug dealer in pesos from his bank account in Colombia, which excludes the further involvement of the drug syndicate in the operation;
4. The broker working with pesos structures "sprays" the dollar amount in the US banking system in order to avoid meeting the reporting requirements and pools these funds on its US bank account;
5. The broker identifies a Colombian importer who needs $$ to purchase goods from an American exporter;
6. The broker who works with pesos agrees to pay the American exporter (on behalf of the Colombian importer) money from his US bank account;
7. The American exporter ships goods to Colombia;
8. The Colombian importer sells the goods (often they are expensive, for example, personal computers, household electronics) for pesos and returns the money to the broker. Thus the broker replenishes the stock of pesos.

At the moment each financial intelligence unit (FIU) uses its own methods for automated crime search by typologies, but they do not include all possible variants of these typologies. The idea of this research is to search for financial crimes not only by typologies but rather by their variations (Mehmet, Wijesekera, & Buchholtz, 2014).

ML/FT Information in the Graph Form

Visualization methods like the graph theory are proposed to be used to order perception of data and to reduce its volumes. With the expansion of IT for various operations on data, graphs have become one of the most frequently used in programming. The data visualization algorithm with the help of graphs was considered in "Graphs in programming: processing, visualization and application" (Kasyanov & Evstigneev, 2003). The authors apply their method for presenting information.

Visualization methods have been used to simplify data management. For example, the graph theory (Harary, 1969) simplifies the perception of information and reduces its volumes using varied approaches and tuning to the specifics of the problem being solved. With the computer technology's wide-spreading for various operations on data since the beginning of the 21st century, graphs have begun to be used in programming.

The graph theory is well applicable to the AML/CFT purposes to give a holistic and consolidated view of the various entities involved in financial crime and the relationships existing between them. In this research, the ML/FT case graph construction consists of two main stages shown below.

1. *Selection of case objects (figurants) and their properties.* The objects involved are individuals who are physical and legal entities participating directly or indirectly in the case. When the graph has been formed, the objects are represented by graph nodes (links between these objects are edges). Their names, roles as well as properties are signed next to each of these nodes. The object's properties play an important role in formalization - they are indicators, which help in searching of a particular physical or legal person involved in the ML/FT crime. The properties are determined based on the identity of a person involved in a particular activity. This method is applicable for specifying the investigation object because it allows describing entirely its main features. The number of properties depends primarily on the complexity of the "picture" under consideration: the more complex is, the more it is necessary not to lose the correct perception of the system under study. The examples of properties may be a territorial feature (whether involved person is a resident or non-resident of the country, in which main investigation actions take place), a natural or legal person, as well as the organizational and legal form of involved person, signs of "suspicion," which indicate that a person may be implicated in a particular type of criminal activity and possible multiplicity of object (for example, person's account who, in cases of importance, is allocated to a separate object), etc.
2. *Links between the figurant's qualification and their species.* Links reflect various participants' interactions and their relationship with each other. When forming the graph, the links are represented as its edges. Each link has a name written next to it.

The link types and their periodicity must be distinguished. In international trade, the cases of multiple criminal in nature operations are rather common. For example, communication types can be a one-off transaction, multiple transactions with a half-year break, etc. Such operations are performed between several criminal persons involved. The link types for individuals can be friendly, business, etc. For communications between individuals and legal entities, they are called as a head, founder, employee and client. The legal entities they are the founder's, partner's or client's relations. For ML/FT cases, such a distinction is quite sufficient since all possible link types between FI objects are distinguished. Special attention should be paid to the national relations.

In addition to specifying the link's name and type, other characteristics should be specified for this connection. For example, the type of currency in which the funds are transferred, the transfer amount from which the payment is made and other characteristics of this link.

In accordance with the actions described above, the sequence for Peso typology the first stage of working with the plot is the identification of involved case persons. The case plot is the essence of the case (dispute) outlined in a concise and accessible language. These are a broker, non-resident importer, exporter-resident, unknown individual (possible multiplicity of such persons = n, suspicion of drug trafficking), broker's bank account in country A (resident account), broker's bank account in country B (non-resident account). The broker's accounts in this case are represented by separate objects to emphasize their role and significance in the case. In other cases it is recommended to take the bank as an object in which an account is entered.

Here is a link consideration between case objects: shipment of goods (between the participants: exporter-resident, importer-non-resident), payment under the contract on behalf of the importer (between the exponents: an exporter-resident, broker's resident account), guarantee agreement (between the persons: a broker, (Non-resident importer)), funds transfer (between involved persons: broker, non-resident importer), relations (between the brokers, broker's broker and non-resident broker's account), contract's approvement (between the participants: broker, unknown individuals (possible multiplicity of such persons = n)), currency exchange operations with cash (between the persons involved: broker, unknown individual (possible multiplicity of such persons = n)), currency receipt under the contract (between persons: an unknown person (possible multiplicity Such persons = n), non-resident bank account of the broker), funds transfer (broker's resident account, broker's resident account). The case graph is shown in Fig. 1.

Each graph describing a particular case consists of a set of subgraphs. The subgraph is a graph containing a certain subset of vertices and edges linked to them. The original graph is called a supergraph in relation to a subgraph. It should be noted that a subgraph consisting of a single object is not a subgraph.

A sugraph (partial graph) is a graph containing all the vertices (nodes) and a subset of the original graph edges.

The following algorithm of actions in order to select a subgraph from the main graph is proposed:

- Selection of all possible subsets of vertices and a subset of edges linked to them which represent a separate complete logical action, event, operation between these vertices (objects);
- Representation of a subgraph of the studied case graph in the Visio program and signing all nodes and edges correspond to the objects and connections participating;
- Calculation of invariants for a selected subgraph using the built-in functions of the IBM I2 program.

Invariants are calculated by the real case graph subgraph and then compared with the invariants by the graphs subgraphs built according to the typologies. If the real case subgraphs' and the typologies' invariants coincide, then it is assumed that these subgraphs are isomorphic. That means that the case under consideration is suspicious, and it requires further analysis and processing. A few cases based on

Figure 1. Case "Peso" graph

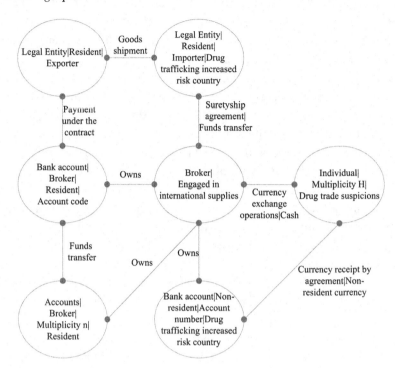

the ML/FT typologies in the field of trade were examined, and the technique operation was demonstrated in practice.

The "foreign exchange transactions on the black market for the exchange of pesos" case supergraph is shown in Figure 1. This graph has seven nodes and nine edges. It is a graph that contains a small number of input objects (nodes) and links (edges) but even from such a graph, it is possible to select certain subgraphs. Subgraph 1 is good for consideration with the following nodes: a broker, individual | multiplicity n, non-resident broker bank account | high-risk country; and with edges: a contract | currency exchange operations | cash, currency receipts under the contract, ownership of an account. Figure 2 shows the highlighted subgraph.

Case steps 2 and 3 can be seen here: The drug dealer (an individual | frequency n | suspicion of drug trafficking) agrees to sell US dollars below the nominal exchange rate to a broker (a broker |engaged in international supply) working with pesos for Colombian pesos; The broker transfers funds to the drug dealer in pesos from his bank account in Colombia (a bank account | non-resident | in a country with an increased risk of drug trafficking | account code), which excludes further involvement of the drug syndicate in the operation.

At its core subgraph 1 describes the currency exchange that drug traders need much. After selling drugs in the United States and receiving the currency in dollars, dealers need to exchange them for the Colombian pesos for further use in their country. This exchange is carried out by a broker who receives cash in dollars and then transfers an amount approximately equivalent to received one but in pesos from his bank account in Colombia (with a commission for providing the service).

Using the IBM I2 program, the general statistics (invariants) for subgraph 1 presented in Table 1 were calculated.

Figure 2. Subgraph 1

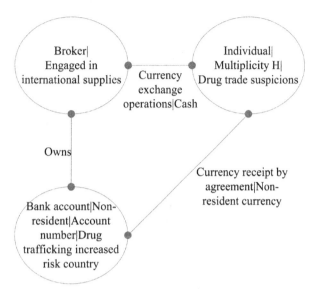

A graph invariant in the graph theory is usually a certain numerical value or a regulated set of values (a hash function), which characterizes the graph structure. It is independent of the way the vertices or graphical image of the graph are designated.

Then the subgraph 2 of the peso black-currency exchange case's supergraph is of interest.

Objects (nodes) of subgraph No. 2 are: a broker, legal entity | non-resident | importer | country with increased risk of drug trafficking, legal entity | resident | Exporter, Broker Bank Account | resident | account code. The subgraph 2 edges are: a guarantee agreement, funds transfer, shipment of goods, payment under the agreement on behalf of the importer, ownership.

Figure 3 shows the selected subgraph 2.

Table 1. Invariants for Subgraph 1

Subgraph	1
Number of nodes	3
Number of edges	3
Average degree	2
Weighted average degree	2
Diameter	1
Density	1
Modularity	0
Connected components	1
Average clustering coefficient	1
Average path length	1

Figure 3. Subgraph 2

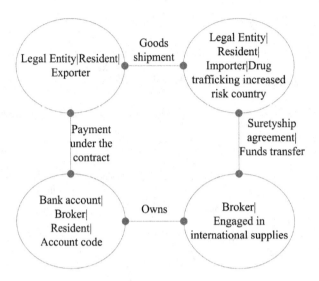

Case steps 5 to 8 are carried out in subgraph 2: a broker identifies a Colombian importer who needs US dollars to purchase goods from an American exporter; the broker who works with pesos agrees to pay the American exporter (on behalf of the Colombian importer) money from his US bank account; the American exporter ships goods to Colombia; the Colombian importer sells the goods (often they are expensive, for example, personal computers, household electronics) for pesos and returns the money to the broker. Thus the broker replenishes the stock of pesos.

Using the IBM I2 program, the general statistics (invariants) for subgraph 2 presented in Table 2 were calculated.

A complex subgraph is specified by two parameters m and n (Kasyanov & Kasyanova, 2013). The parameter m is responsible for the number of objects united by a serial link in the operations chain (Fig.

Table 2. Subgraph 2 invariants

Subgraph	2
Number of nodes	4
Number of edges	4
Average degree	2
Weighted average degree	2
Diameter	2
Density	0,667
Modularity	0
Connected components	1
Average clustering coefficient	0
Average path length	1,333

Figure 4. Objects' communication: a) sequential, b) parallel

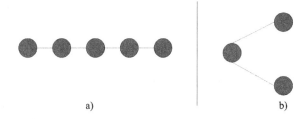

a)   b)

4a). The parameter n is responsible for the number of parallel links between the objects or a chain of objects (Fig. 4b).

The $m^\times n$ subgraph mutation process consists of the following steps:

- To form the case graph subgraph that is a subject to mutation;
- To select a random number R in the range from 1 to ($m^\times n$-1);
- To duplicate a selected subgraph R times;
- To create necessary number links.

The two examples illustrating the process of mutation are considered below.

Example 1. Take the graph "A" with $m = 2$ and $n = 1$ (Fig. 5a). Perform a subgraph mutation with nodes 1-2. Duplicate the selected subgraph 1 time and get the mutated graph "A*" (Fig. 5b).

Example 2. Consider graph "B" with $m = 4$ and $n = 1$ (Fig. 6a). Perform a subgraph mutation with nodes 2-3-4-5. Duplicate the selected subgraph twice and obtain the mutated graph "B*" (Fig. 6b).

The mutation of graph "B" can be applied to the ML/FT commission scheme described since dozens of one-day firms with various connections between them can participate in this scheme. They create several parallel object chains by their interaction, which are grouped further in one independent object in the final result.

The concept of a graph object's synonym was also used in the research. The synonyms are case objects which are mutually replaceable. For example, it is possible to change one object for another in the graph/subgraph without changing the general meaning and structure under consideration (Mehmet & Wijesekera, 2013).

Figure 5. Graph "A" (a) and its mutation (b)

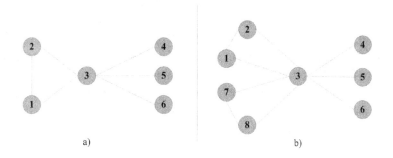

a)   b)

Figure 6. Graph "B" (a) and its mutation (b)

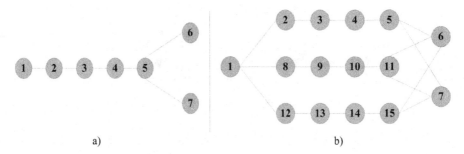

a)   b)

The synonyms are used here for a foreign trade operation generating process in the schemes based on the ML/FT typologies. The synonyms may be pawnshops, companies with the signs of fictitiousness, offshore organizations, insurance companies, shell companies, and others since these objects are used for the same purpose, to move money, imitate the real trade and procurement and other operations and remove them from the source of the origin. In fact, these facilities are only intermediaries through which the money flows.

Numerous substitutions of some objects presented in the typologies for others are manifested mainly at the stage of ML/FT process stratification (multiplication) since this stage implies actions related to concealing the origin of criminal funds and mislead the law enforcement agencies and FI services.

The nine most common ML/FT typologies were considered: goods and services cost overstatement in invoices; goods and services cost underestimation in invoices; goods and services multiple invoicing; goods or services in short supply; goods or services excess supply; incorrect description of goods and services; foreign black market exchange operations for pesos exchange; carousel scheme of VAT evasion and commission scheme.

An example of commission scheme graph is considered in Fig. 7 with the designations: (1) |Supply commission agreement on raw materials, goods |Supply payment|, (2) Supply contract for raw materials and goods, (3) Cash, (4) |Cash| Supply payments for goods, equipment, machinery|, (5) | Supply contract for raw materials, goods| Delivery payment|.

Subgraph 1 is allocated from this graph and will be responsible for the final steps of ML/FT (Fig. 8a). With the help of synonyms some objects are replaced with others, namely the pawnshop is replaced with an insurance company, which will accept premiums in the form of "dirty" money and also pay money to individuals ("cashing out"), and insurance company with a one-day company, which will buy goods and services from an offshore company. Then new subgraph appears, and it does not change the essence of this scheme. Subgraph 1 with objects replaced by their synonyms is shown in Fig. 8b.

In generating new schemes based on the ML/FT typologies, the "synonymization" process was got directly after the subgraph structure's mutation process. The new objects' properties, which were obtained by duplicating a subgraph, were replaced by their synonyms' properties with a probability of 0.5 since this coefficient is optimal for the increasing number of objects' and properties' mutations. If one uses, for example, 0.01, then there will be slight changes (properties and objects will change too rarely), which is not informative. The same is also the case for 0.99 (the new structure will have little in common with the original graph). This approach is quite logical since it increases the variety of new graphs (Von Landesberger et al., 2011).

Figure 7. Commission scheme's graph

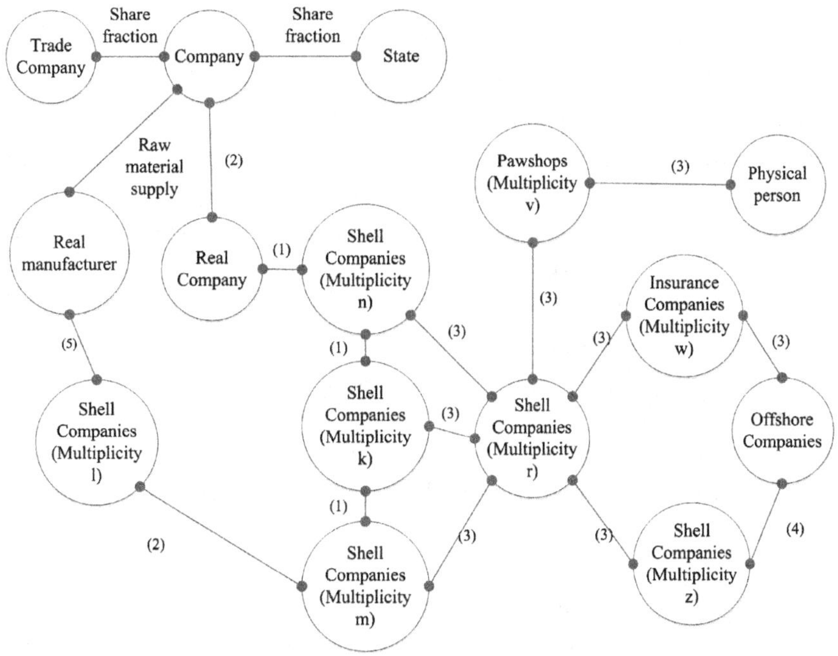

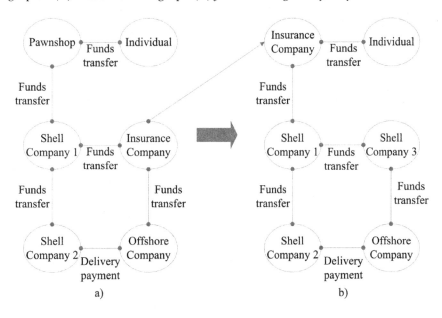

Figure 8. Subgraph 1 (a) and a new subgraph (b) formed using the synonyms

In the graph theory, the edge graph L (G) of a non-oriented graph G is the graph L (G) representing the neighborhood of the edges of G.

One of the earliest and most important theorems on edge graphs belongs to Whitney who proved that the graph G structure is completely determined by the edge graph. In other words, the entire graph can be recovered from the edge graph.

Figure 9 show the formation of the edge graph L (G) from the original graph G. In the edge graph L (G) any node of L (G) is an edge of the original graph G. Two vertices of L (G) are adjacent if and only if their corresponding edges have a common node in G. Each node of the edge graph L (G) is labeled with a pair of vertex numbers of the corresponding edge in the original graph G. For example, a vertex with label 1,3 corresponds to the edge of the original graph between vertices 1 and 3. Node 1.3 is adjacent to three other nodes: 1.4, 1.2 (corresponding to edges with common vertex 1 in the initial graph) and 4.3 (corresponding to edges with common vertex 3 in the initial graph).

The graph edge theory applied to graphs constructed according to the ML/FT typologies goes like this. It is supposed that there is the following form of graph N (Fig. 10).

The sequence when forming the model graph W is as follows according to the general type of the ML/FT commission scheme:

- The edge of the original graph N constructed by the typology is replaced by the edge graph's node;
- The so-called edge graph allocated from the original one is performed whereas the nodes remain unchanged at this stage;
- The contiguity matrix for the edge graph (Table 3) is constructed as consisting of zeros and ones and according to the following rule: if the original graph N edge is connected through a common node with another original graph N edge, then at the intersection of their coordinates (ordinal numbers) there will be "1", otherwise appears "0".

Figure 9. Obtaining the edge graph from the original graph: a) original graph G; b) L (G) nodes are created from arcs of G; c) Addition edges to L (G); d) edge graph L (G)

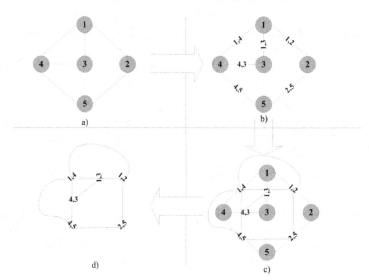

Figure 10. Graph N

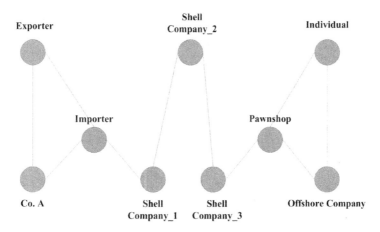

Table 3. The contiguity matrix for edge graph

		1	2	3	4	5	6	7	8	9	10
1		0	1	1	1	0	0	0	0	0	0
2		1	0	1	0	0	0	0	0	0	0
3		1	1	0	1	0	0	0	0	0	0
4		1	0	1	0	1	0	0	0	0	0
5		0	0	0	1	0	1	0	0	0	0
6		0	0	0	0	1	0	1	0	0	0
7		0	0	0	0	0	1	0	1	1	0
8		0	0	0	0	0	0	1	0	1	1
9		0	0	0	0	0	0	1	1	0	1
10		0	0	0	0	0	0	0	1	1	0

The search for the edge graph's nodes, which have a power of 2 (i.e. the sum of ones in a row or column of contiguity matrix for the edge graph), is performed.

The original graph's nodes and edges, which correspond to the edge graph's nodes with power 2 (i.e. the edge graph node must be between the original graph nodes), are replaced by a common node in the model graph. The process of forming a model graph from the original graph N is shown in Fig. 11.

This sequence allows one to take into account any addition or reduction in the case under consideration, which is in the ML/FT straight chain, i.e. they have a consistent connection with each other. Such objects serve as structural elements of the operations chain, through which criminal funds pass (Battista et all, 1998).

The connection between the nodes in graphs can be linear or parallel. The two nodes connected through an edge have a linear connection (denoted as a circle in a rectangle). The nodes connected through an edge with the same ancestor node form a parallel relationship (denoted as a circle in the circle). The number of combined objects is indicated above each designation.

Figure 11. Obtaining a model graph from the original graph N

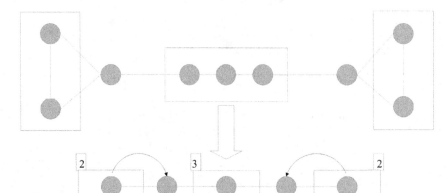

An example illustrating the sequence of actions is considered below. According to the "Peso" typology described earlier ML/FT relies on graph G (Fig. 12), by which the model graph is further defined (Fig. 13).

The model graph is isomorphic to graph G. Isomorphism is a mathematical and logic concept, which means the relationship between any two objects of an identical structure (Silyutin, 2004). Two graphs are said to be isomorphic if there exists a vertices permutation, for which they coincide. In other words, two graphs are said to be isomorphic if there exists a one-to-one correspondence between their vertices and edges that preserves contiguity and incidence (the graphs differ only in the names of their vertices) (Fig. 14).

Figure 12. Graph G for the ML/FT typology

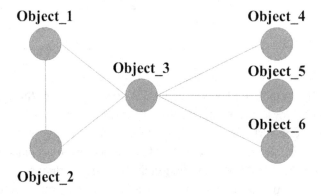

Figure 13. Model graph

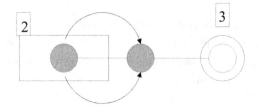

Figure 14. Graph isomorphism

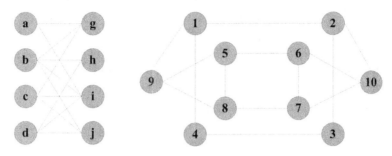

There is a one-to-one correspondence between the isomorphic elements: each element or the relation between them of one object corresponds exactly to one element (and relation) of another object and vice versa.

Two graphs F (V, E) and F* (V*, E*) are isomorphic if there is a one-to-one correspondence between the sets of vertices V and V* so that any two vertices in graph F are adjacent if and only if the adjacent vertices in graph F* are adjacent. Adjacency is a concept used for only two edges or only two vertices: two edges incident to one vertex are said to be contiguous; two vertices incident to one edge are also called adjacent.

Here is a real case K, which should be checked for the ML/FT signs. Case K graph is shown in Fig. 15.

Since case K graph belongs to a graph's subset of model graphs and the model graph is isomorphic to the graph N constructed from the typologies (Cormen at al., 2009), a conclusion can be made about the relationship between case K graph and graph G constructed from the typologies and therefore the criminal character of a case under investigation. The model graph is the one, in which vertices correspond to letters, and each vertex is related to a set of word identifiers with a corresponding letter (Golberg, 1989). Direct comparison of case K graph and the graph constructed from the typologies is not possible since these graphs are not isomorphic (Malinin & Malinina, 2009). The model graph extracted from the graph constructed from the typology is carried out primarily in order to increase the processing speed and to compare the graphs constructed for real cases with the graphs constructed from the typologies. In the model graph, only important case attributes remain (figurants, their properties, links, and their properties).

Figure 15. Case K graph

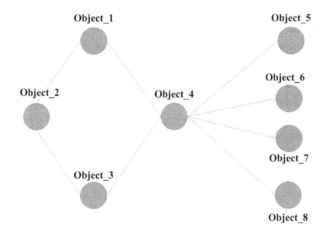

The described edge graph theory can be applied to a variety of currently known ML/FT typologies, in particular, those discussed earlier. As a result of actions performed with the graph, an array of initial typologies with all their generated synonyms is obtained and written into the typology database. After that, by using the big data tools (for example, OpenRefine, DataCleaner) all typologies and synonyms are filled with information gathered from the case under investigation. When using the big data tools, there is a search performed among the completed variants, in which the combination of all information used gives the same implementation as in the typology in the compiled database. Finding the correct implementation of a typology or a synonym will mean finding a possible criminal scheme, which requires subsequent verification and appropriate confirmation.

Structure of the Database Created

Certain bases are needed for the storage of collected and derived data. The so-called graph database management systems (DBMS), for example, the Neo4j platform (Neo4j, n.d.), the DataStax Enterprise Graph (DataStax, n.d.) and others, have been appeared for this purpose. Unfortunately, such tools are appropriate for the graph applications and are not efficient at processing large amounts of accumulated data, for which the "big data" term was introduced. On the other hand, they are not optimized to handle queries that span the entire database as it is needed in the case of the ML/FT typologies. Therefore, in the framework of this research graph DBMSs cannot be used. The reason is in an urgent need for specialized tools to work with the ever-growing data. Numerous platforms such as Hadoop (Apache, n.d.) have been developed, representing systems with a set of various tools for analyzing and processing big data.

The database created to test the technique proposed includes information about the names of the cases based on the ML/FT typologies using foreign trade operations (CASES), objects involved in these cases (OBJECTS) and their characteristics (FEATURES), subgraphs of the main cases graphs (SUBGRAPHS), synonyms of objects (SYNONYMS), and operations that case objects make (OPERATIONS) (Fig. 16). On the figure, (FK) is a foreign key. In the context of relational databases, an FK is a field (or collection of fields) in one table that uniquely identifies a row of another table or the same table. Putting simpler, the FK is defined in a second table, but it refers to the primary key or a unique key in the first table.

This database was also used for creating new cases. A PL/SQL procedure was written that fully implements all the steps of generating new schemes based on typology-based cases.

Summary

There is a method proposed for generating schemes based on the cases constructed according to typologies which includes several stages of copying the selected case, selecting the subgraphs of this case, mutating the subgraphs and replacing the properties of the new generated objects with the properties of their synonyms. All stages of this methodology were programmed and successfully passed both debugging and checking on case graphs built according to typologies. In order to implement a specific example of developed methodology approbation and its program some common typologies of ML/FT were considered including: overstating and understating goods cost and services in invoices, multiple invoicing for goods and services, not full delivery of goods or services, supply of excess volume of goods or services, inaccurate description of goods and services, foreign exchange operations on the black market by exchanging pesos, the carousel scheme of VAT evasion and commission scheme.

Figure 16. Structure of the database created

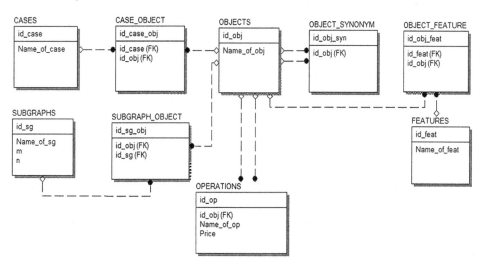

A technique for generating new schemes based on the cases constructed using the ML/FT typologies was worked out. It includes the following stages:

1. Choice of all schema elements with their properties and all objects' replication;
2. Case creation;
3. Objects' creation with the corresponding serial numbers;
4. Defining all elements of the intermediate tables "Object Properties", "Object in the Case", "Subgraph of the Case" and table modification with OPERATIONS using the new keys;
5. Creation of subgraphs with new objects; object identification numbers in newly created subgraph should differ from object identification numbers in the original subgraph;
6. New objects appropriation to synonyms;
7. Structure mutation without changing properties of constituent elements;
8. Assignment of the case subgraph created;
9. Defining new objects necessary for number formation for the selected subgraph;
10. Defining the necessary number of links between the objects;
11. Synonyms mutation;
12. All case objects bust and synonyms presence check;
13. Replacing those objects that have synonyms' object properties with found synonyms' properties.

This technique was implemented as a separate software module (~ 200 lines of code). The procedural block-structured language PL/SQL was chosen (PL/SQL Tutorial, n.d.) for this task. It is an extension of the SQL language designed to work with the Oracle DBMS, and it provides application developer and interactive user with the following key features:

- Subprogram implementation as a separate block including the use of nested blocks;
- Packages, procedures, and functions stored in the database;
- Provision of an interface for calling external procedures;

- Support for both SQL data types and types introduced in the PL/SQL;
- Application of an explicit and implicit cursor, as well as FOR loop operator for the cursor;
- PL/SQL variables and attribute cursor initiation that allow referring to data type or element structure;
- Collection and object types' introduction;
- Support a set of control and loop operators;
- Mechanism implementation for handling exceptions.

The Oracle SQL Developer was chosen to implement the technique proposed. This is a tool for writing SQL queries, developing PL/SQL packages, procedures, functions, and triggers. The Oracle SQL Developer supports the development and administration of applications and allows to export and import data and structures. In fact, it is a powerful tool for the PL/SQL language to develop version control for migrating from other databases and administering communication between different databases (Kulikov, 2016).

As a result of the program run, the new data was gained for the newly created case on its basis. The new case graph (Peso graph) was constructed that clearly shows the changes obtained (namely, the number of resident bank accounts of the broker was increased). Mutations occurred in both subgraphs (in the second subgraph one of objects "individual" was deleted).

The crime schemes in the form of graphs were prepared using capabilities of Visio.

In Fig. 17 there is a case graph before changing. In Fig. 18 there is a graph after the corresponding change.

The following designations were used here: (1) currency exchange operations, (2) currency receipt by agreement, (3) owns, (4) funds transfer, (5) payment under the contract, and (6) goods shipment.

The results of the program test are presented in table 4.

CONCLUSION

For filling the database with data, it is clear that for this type of research it is impossible to get realistic financial data to evaluate the proposed technique. Financial institutions do not share their financial information as it is considered confidential. Therefore, to evaluate the software created synthetic financial data was generated and used as test data.

After multiple tests conducted, it was concluded that the created database scheme describes the subject area adequately. The prepared script met all the requirements.

Analyzing the graphs obtained and taking into account all the information, it can be concluded that with graphs mutation and apparent change in the structure the essence of the criminal scheme remains

Table 4. Test results

N°	Action	Result	Time, s
1	Creating a database scheme	successfully	10
2	Filling the database with data	successfully	16
3	Executing generation script	successfully	21

Figure 17. Schematic representation before mutation

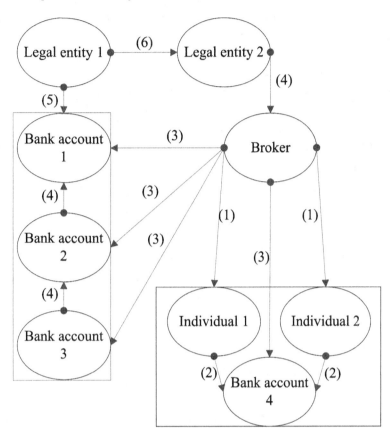

Figure 18. Schematic representation after mutation

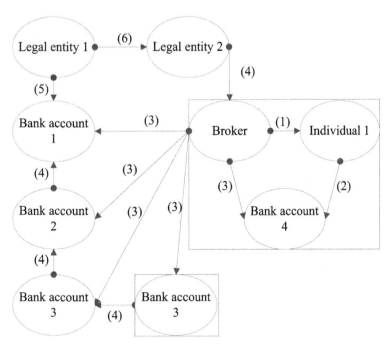

unchanged. If there are differences from the strictly stated typologies, the criminal schemes will be found with a relatively small probability. This approach presupposes a different way of treating typologies, viz. the study of all the possible options set for criminal schemes implementation. The technique for generating typologies' synonyms created for this task does not depend on the direct correspondence of a crime typology and allows the investigator to detect criminal schemes with various kinds of changes.

The novelty of the research is that using the suggested technique based on graph approaches, the schemes of criminal cases can be multiplied, obtaining millions of all possible variants of one case. Later these variants will be processed (analyzed for searching similar schemes) with big data tools. On this basis, it is possible to create new rules for identifying ML/FT schemes, which will simplify the automation of ML/FT crime detection.

ACKNOWLEDGMENT

This work was supported by the MEPhI Academic Excellence Project (agreement with the Ministry of Education and Science of the Russian Federation of August 27, 2013, project no. 02.a03.21.0005) and by the Russian Foundation for Basic Research (project no. 18-07-00088/18).

REFERENCES

Apache Hadoop. (n.d.). Retrieved October 28, 2018, from http://hadoop.apache.org/

Barsegyan, A. A., Kupriyanov, M. S., Stepanenko, V. V., & Holod, I. I. (2008). *Technologies for data analysis: data mining, visual mining, text mining, OLAP* (p. 384). Saint Petersburg: BHV-Petersburg.

Battista, G. D., Eades, P., Tamassia, R., & Tollis, I. G. (1998). *Graph drawing: algorithms for the visualization of graphs*. Prentice Hall PTR.

Bulancev, V.Y., Zaripov, I.A., Kuren'kov, D.B., Lukashenko, S.N., & Miroshkin A.A. (2008). "Shine and poverty" of tax planning for imports. *International Banking*.

Cielen, D., Meysman, A., & Ali, M. (2016). Introducing data science: big data, machine learning, and more, using Python tools. Manning Publications Co.

Cormen, T. H., Leiserson, C. E., Rivest, R. L., & Stein, C. (2009). Introduction to algorithms. MIT Press.

DataStax Enterprise Graph Super-Powering Your Data Relationships. (n.d.). Retrieved October 28, 2018, from https://www.datastax.com/products/datastax-enterprise-graph/

Dreżewski, R., Sepielak, J., & Filipkowski, W. (2015). The application of social network analysis algorithms in a system supporting money laundering detection. *Information Sciences*, *295*, 18–32. doi:10.1016/j.ins.2014.10.015

Golberg, D. E. (1989). *Genetic algorithms in search, optimization, and machine learning*. Reading, MA: Addison Wesley.

Kasyanov, V. N., & Evstigneev, V. A. (2003). *Graphs in programming: processing, visualization and application.* SPb.: BHV-Petersburg.

Kasyanov, V. N., & Kasyanova, E. V. (2013). Information visualisation based on graph models. *Enterprise Information Systems, 7*(2), 187–197. doi:10.1080/17517575.2012.743188

Kulikov, S. S. (2016). Working with MySQL, MS SQL Server and Oracle in the examples of EPAM Systems. Minsk: BOFF.

Malinin, L., & Malinina, N. (2009). Graph isomorphism in theorems and algorithms. Moscow: Librocom.

Mehmet, M., & Wijesekera, D. (2013). *Detecting the evolution of money laundering schemes.* Paper presented at Ninth Annual IFIP WG 11.9 Digital Forensics, Orlando, FL.

Mehmet, M., Wijesekera, D., & Buchholtz, M. F. (2014). Money laundering detection framework to link the disparate and evolving schemes. *Journal of Digital Forensics, Security and Law, 8*(3), 40–70.

Neo4j – Platform for connected data. (n.d.). Retrieved October 28, 2018, from https://neo4j.com/ PL/SQL

SAS Anti-Money Laundering (SAS AML). (n.d.). Retrieved October 28, 2018, from http://www. tadviser.ru/index.php/%D0%9F%D1%80%D0%BE%D0%B4%D1%83%D0%BA%D1%82:SAS_Anti-Money_Laundering_(SAS_AML)

Silyutin, D. S. (2004). *Genetic search of graph isomorphism controlled by multi-agent system* (Doctoral dissertation). Taganrog State University of Radio Engineering.

The Eurasian Group on Combating Money Laundering and Financing of Terrorism (EAG-1). (n.d.). *Glossary of definitions of the Eurasian Group on AML/CFT.* Retrieved October 28, 2018, from https://eurasiangroup.org/en/glossary

The Eurasian Group on Combating Money Laundering and Financing of Terrorism (EAG-2). (n.d.). *Typologies research topics.* Retrieved October 28, 2018, from https://eurasiangroup.org/en/typologies-research-topics

The Financial Action Task Force (FATF). (2006). *Trade-based money laundering.* Retrieved October 28, 2018, from http://www.fatf-gafi.org/publications/methodsandtrends/documents/trade-basedmoney-laundering.html

The Financial Action Task Force (FATF). (2007). *Laundering the proceeds of VAT carousel fraud.* Retrieved October 28, 2018, from http://www.fatf-gafi.org/documents/documents/launderingtheproceedsofvat carouselfraudreport.html

The Financial Action Task Force (FATF). (2008). *Best practices on trade based money laundering.* Retrieved October 28, 2018, from http://www.fatf-gafi.org/media/fatf/documents/recommendations/ BPP%20Trade%20Based%20Money%20Laundering%202012%20COVER.pdf

Tutorial. (n.d.). Retrieved October 28, 2018, from https://www.tutorialspoint.com/plsql/

Von Landesberger, T., Kuijper, A., Schreck, T., Kohlhammer, J., van Wijk, J. J., Fekete, J. D., & Fellner, D. W. (2011). Visual analysis of large graphs: state-of-the-art and future research challenges. Computer Graphics Forum, 30(6), 1719-1749.

Chapter 13
Formalization and Discrete Modelling of Communication in the Digital Age by Using Graph Theory

Radi Romansky
Technical University of Sofia, Bulgaria

ABSTRACT

Globalization is an important characteristic of the digital age which is based on the informatization of the society as a social-economical and science-technical process for changing the information environment while keeping the rights of citizens and organizations. The key features of the digital age are knowledge orientation, digital representation, virtual and innovative nature, integration and inter-network interactions, remote access to the information resources, economic and social cohesion, dynamic development, etc. The graph theory is a suitable apparatus for discrete presentation, formalization, and model investigation of the processes in the modern society because each state of a process could be presented as a node in a discrete graph with connections to other states. The chapter discusses application of the graph theory for a discrete formalization of the communication infrastructure and processes for remote access to information and network resources. An extension of the graph theory like apparatus of Petri nets is discussed and some examples for objects investigation are presented.

INTRODUCTION

The Information Society is an organization of distributed information resources in the global network with remote access and information sharing. Each user could access different resources via Internet and use them in the virtual space in a centralized or decentralized manner. Different approaches for access to distributed objects via Internet are determined. For example, the research subject presented in (Kravets, 2018) is a maintenance service distributed organization possessing a network of affiliates and representatives. This paper considers some problems and peculiarities of modelling and efficient

DOI: 10.4018/978-1-5225-9380-5.ch013

development in social and economic monitoring and the obtained results could be applied both in static and dynamic configuration of distributed companies. Another proposal is a visible light communication as an alternative optical-based wireless communication is discussed in (Adiono, 2018) and the paper presents the design, implementation, and demonstration of the TCP/IP data-exchange over visible light.

The modern society has many opportunities of the contemporary digital world which can be used at the stage of preliminary description, conceptual modelling, initial formulation of the process or object, development and implementation. It should be known that each new digital technology has challenges for the security and privacy of the user's data (Romansky, 2017). On the other hand, the modern society proposes services as e-access, e-society, e-policy, e-democracy, e-voting, e-governance/e-government, e-learning, e-health, e-business/e-commerce, e-consultation, e-inclusion, etc., which could improve and extend the communications between citizens and institutions in the digital age (Kerr, 2013). This is possible on the base of the key features of the digital age – knowledge orientation, digital representation of the objects, virtual nature, molecular structure, integration and inter-network interactions, remote access to the information resources, innovative nature, economic and social cohesion, dynamic development, global scale, etc.

The modern society is built on the base of different information technologies (IT), including grid technology, social communications, cloud and mobile cloud computing, Internet of Things (IoT), Big Data Analysis (Kharchenko, 2018). The combination of some new technologies permits to create different new features. For example, a combination of smart homes with mobile communication is discussed in (Olawumi, 2017) and the authors (Yu, 2016) present several challenges that smart grids could be made for cyber-physical systems. Another problem – cloud speech recognition, can be applied in different areas. In this reason, the combination of cloud services, speech control and cyber-physical systems is the object of discussion in (Škraba, 2018). The article describes development of an algorithm for efficient harvesting of speech recognition cloud services with application to the cyber physical systems.

In the digital world all accessed and used resources should be regarded as an integration of information with technological and technical units. This point of view permits to describe the digital environments as formalized structures with separate objects and relations (connections) between them. For example, (Bolnokina, 2019) discusses multistage systems, at the input of which comes a stream of requests that require executing a set of serial-parallel jobs for their support. The purpose of the article is to make formalization (by nonlinear objective functions and recursive constrains) of the problem of assigning specialists to jobs and finding suitable algorithm to solve it.

The purpose of this chapter is to make a brief survey of the information basis of the modern society with realization of a global informatization and increasing effectiveness of the using information resources based on the components of the digital world. The purpose is to summarize the main key features and components of the modern society and to present the possibilities to apply the graph theory for discrete formalization of information processes and communications in the e-space. An investigation of a component of the digital space could be conducted by using designed model and preliminary phase must be formalization of the investigated objects. The graph theory proposes an extension in the sphere of discrete investigation by using the Petri Net (PN) apparatus which is discussed in end of the chapter.

INFORMATIZATION AS A BASE FOR BUILDING THE MODERN SOCIETY

Informatization – Terminological Aspects, Definitions and Development

"Informatization" should be regarded as a fundamental term for building the Information Society related to the role of the term "Industrialization" for the Industrial Society. UNESCO determines this term as *"development and application of methods and tools for collecting, processing, storing and dissemination of information that permits forming new knowledge and their using for management of the processes in the society"*. Informatization of the society is a social-economical and science-technical process for changing the social information environment, based on creation of optimal conditions for realization of the information necessities by keeping the rights of citizens and organizations. This important process for the modern society has some special features listed below.

- Using information resources and the information culture of citizens;
- Creating different possibilities for using knowledge of human activity;
- The information is basic resource for all activities in the society;
- Using contemporary Information & Communication Technologies (ICT);
- Developing common information space.

First introducing the term "informatization" has been made independently by authors of two publications – Marc Porat (in 1977) and S. Nora & A. Minc (in 1978). The goal of these publications was to present the role of the information for the transformation the Industrial Age to the Information Age. In this reason, the term sets the foundations of the modern society by determining the important role of the information. This definition was used for the first time on government level in the report "Informatization of the society" requested by Valeri Jiscar d'Esten, President of the Republic of France.

In the next decades of the 20th century the term "informatization" is object of discussions and different definitions related to the development of the modern society. For example, acad. A. E. Ershov (Russia) determines the informatization as *"a complex of measures secured full using of correct and true knowledge in all social activities"* (Afinogenov, 2013). The development of contemporary information technologies (IT) and extension of the digital communications via the global network permits G. Wang (1994) to determine the "informatization" as a process of using information and Information Technologies (IT) to force the economic, political, social and cultural status of the society and to grow the speed, quantity, and popularity of information production and distribution. Some years later Randy Kluver (Kluver, 2004) determined this Wang's definition as a "phenomenon". In the book "Information society: A retrospective view" (Dordick, 1993) the "informatization" is defined by a complex of three primary dimensions – infrastructure, economic, and social. The authors determine the end of the industrial age and the transition to the new information age.

The new 21st century adds new nuances to the definition of the term and determines the "informatization" as a process through which new communication technologies are used as a mean for further development as a nation becomes more and more an Information Society (formulated by Everett Rogers in 2000). Some years later Kim (2004) proposed to measure the informatization in a country using a composite measure made up of the following extraneous variables: Education, R&D Expenditure, Agricultural Sector and Intellectual Property. Kim also suggests three approaches to conceptualizing informatization based on: (1) Economic (measured with economic data); (2) Technologic (measured by

ICT data - number of computers per population); (3) Amount of information (e.g. number of published technological journals).

The contemporary technologies permit to update the informatization in the context of the modern society. For example, (Wang, 2011) focuses on the research of the application of cloud computing in education informatization. The main idea is to extend the educational technologies as virtualization, network storage technology distributed and parallel computing technologies, networking, etc., by using the cloud computing technology. In (Guo, 2014) a new advanced service-oriented networked manufacturing model based on cloud computing is proposed in which the informatization of manufacturing resources is determined as a core part for realization. With the increasing role of the informatization over the last years creates new challenges and could be a risk for confidential calculations in key objects. This problem is discussed in (Vorobiev, 2017) and corresponding recommendations and architectural concepts for protection of confidential calculation and information processing on the base of internal electronic base are proposed. Another problem is discussed in (Li, 2018) – the paper presents an approach to evaluation accounting informatization as an important part of enterprise informatization which *"affects the accounting and finance operational efficiency"*. This problem is evaluated from multiple aspects with a goal to find the strengths of corporative accounting informatization. The paper proposes constructing the evaluation index system from the aspects of strategic position, infrastructure construction, implementation of the informatization, and application efficiency. An entropy & cross entropy method based on intuitionistic fuzzy environment is proposed for evaluation results obtaining and analysing.

The importance of the informatization in the modern society is proven by the need for serious and effective teaching in the era of globalization, discussed in (Zhang, 2018). The article deals with a research on evaluation model of the effective teaching in the era of informatization with main goal improving the traditional classroom efficiency and the author proposes a new frame work to evaluate the college teaching in the context of information age. An additional effect of informatization can be sought in the social sphere of the society. The need for socially competent specialists in the modern society, who are able to quickly adapt to the changing life is discussed in (Sergeeva, 2018). The paper deals with the socialization of the personality under the conditions of the globalization and informatization of the society and the authors determine that *"the problem of personal social competency development is of an interdisciplinary character"*. A comparative analysis must be made for establishing social competency in the modern society to investigate different components – personal, social cognitive, axiological, etc.

Approaches, Components and Models of Informatization

The researches in the filed of information management are important part of the society in the 21st century and for the development of modern civilization with globalization in the economic, political and cultural spheres. The article (Melnyk, 2018) determines the informatization of the modern society as *"one of the most important attributes and resources of human activity, which grows into knowledge"*. In this reason, the paper presents a research for conceptualization of information management as a science based on the ICT and additional mechanisms for informatization that will help to identify the information and innovation potential of the society. The author formulates a concept of information management as one of the most modern trends in the development of the 21st century society and finally in the conclusion he determines that the formulated concept of information management *"should be introduced in all areas of the development of society, including high school, which requires the formation of a creative personality as a subject of an informative and innovative society"*. Another publication discusses the social and economic structure of society (Goncharov, 2018) and determines that this structure which is based on information economy by nature avoids most social and economic problems. It could be determined *two basic theoretical-methodical approaches* for informatization of the society:

- **Technical Approach:** The goal of the informatization is to develop basic technical and technological means and tools for the sphere of manufacturing and management and Information and Communication technologies (ICT) are the means for increasing work performance. In this aspect the paper (Aiqun, 2018) proposes a new methodology of the informatization construction of the higher education institutions for increasing the ICT productivity and to improve the governance framework of the IT education capability.
- **Sociological Approach:** It regards the informatization as a process uniting all spheres of the human activity with reflection to the all persons in the society – their knowledge, moral, economic interests, personal developments, etc. In this reason, the informatization of the society is a set of connected and related factors – technical, economic, social, political, cultural, etc. The relationship between the new economy and civil society in the digital age is discussed in (Zhyvko, 2018) and the conclusion is that a specific methodology for studying this interaction should be developed. The author points out that the most important role of governments should be the creation effective conditions for financial support of the social and economic development and the transfer of knowledge.

The summarization permits to determine the informatization as a technological, social and cultural process which change the life of the individuals in the modern society and needs high activities of the government and all ICT users for forming adequate information culture. In this reason, some basic technic and technological components of the informatization could be determined:

- **Electronization:** Using electronic technologies in different spheres of the human activity: manufacturing, management, education, research, social and cultural spheres, etc.
- **Mediatization:** Process for improvement of the means for the information collecting, keeping and dissemination by using electronic (digital) tools and devices for quick transfer between users.
- **Computerization:** Process for improvement methods for searching and processing of the information by using computer tools.
- **Intellectualization:** Process of development of the human knowledge and possibilities for creation new information for increasing intellectual potential of the society, including with using artificial intellect.

A new term *"Internetalization"* is proposed in the last decade of 20th century to describe building and development of the global network and world information system (www) – the fundament of processes for *globalization*.

As a result, in the beginning of 21st century a new term *"internetalization"* has been proposed to describe the increasing the role of the contemporary network world and the world information system (www) as a fundament of processes for globalization and development of the modern society. This new point of view is based not only on the Internet communications in the education (Bell, 2001), information sharing and changing (Aggarwal, 2003), international marketing (Alrawi, 2007), etc., but on the new technology IoT (Rosemann, 2013), for example.

Table 1. Point of view of the society for the Informatization

Positive Sides	Negative Sides
Free development of the individuals;	"Automatization" of the man;
Information Society;	Dehumanization of the life;
Socialization of the society;	Cultural level decreasing;
Communication development;	Very much information;
Increasing freedoms;	Isolation of the man;
Decentralization;	Centralization;
Extended participation in the public life;	Increasing the state administration;
Rationalizations;	Manipulation of the men;
Economic resources;	Concentration;
Decentralization of the manufacturing;	Standardization;
New products;	Increasing of the life complexity;
Diversification.	Dehumanization of the labour.

The public's opinion on informatization is not unambiguous. A summarization of some important key estimates is presented in the table 1.

There are some separate models for informatization of the society and a brief summary is presented below.

1. **"West Model"**: Approach accepted by the industrial developed countries in Europe and U.S.
   a. **"European Path"**: Searching of the balance between the total control by the state and the "seclusion" of the market. The countries form the European Union have own programs for informatization with social direction and state regulation.
   b. **"American-English Path"**: Model of social-economic development in which the state functions are minimized and the main role is of the private business. Full liberalization of the IT-market and limited management of the state in the field of telecommunications.
2. **"Asian Model"**: It is based on the effective cooperation between the state administration and the market. The main principle is "coexistence and co-development". It is the model of Japan and "Asian tigers" (South Korea, Taiwan, Singapore, and Hong Kong)
   a. **"Japan Path"**: It is based on the idea that each know its status with regard to others and has behavior in line with the objectives of its group. Each management unit takes care of their subordinates, and they are bound to implement the directives unconditionally. So state and business have a joint responsibility for the emerging problems and their solution. The main focus is on the development of ICT in order to improve the quality of life of the people.
   b. **"Asian Tigers"**: The fundament is the "Japan model" but with taking into account different starting conditions. Their economic development has been supported by Japanese capital investments, but they have adopted the philosophy of "business-state". The "country-market" relationship is sociological and political rather than purely economic.
3. **"Model of Developing Countries"**: Two ideologies for societal informatization (building of the IS) based on telecommunication reform - Latin American and Asian. Possible strategies are fundamentally different - privatization (transfer of state-owned enterprises and activities for full or partial private control and control) and liberalization (alleviation of the conditions for entry of third parties into the market, including monopoly producers of goods).

a. **The "Latin American Model":** Was launched with privatization to rapidly improve the quality of telecommunication services at the cost of abandoning market.

b. **The "Asian Model":** Is characterized by the introduction of competition, leaving any serious privatization for later.

c. **The "Indian Model":** The chosen path is based on careful changes, gradual and retention of national cultural roots by accepting human capital as core capital. The model is defined as an intermediate variant - it is closer to the Latin American range and depth, but due to the slower rate of change - close to the Asian one.

KEY FEATURES AND COMPONENTS OF THE MODERN SOCIETY

As presented in the previous section, at the end of the 20th century (1998-2000) the term "informatization" accept the contents of application of the contemporary tools for information processing, ICT and WWW (World Wide Web) in the development of the society. The goal is to build a developed global communication infrastructure and increasing the effectiveness of using determined information resources based on system computerization of the society. In this reason the information could be regarded as a product of the intellectual activity of the society. The main paradigm is *"If over 50% of the people in the society is busy in the sphere of information-intellectual services, the society is changed as information"*. It can be accepted that the Information Society (IS) is created at the beginning of 21st century and the basic features are presented below.

- **Essence**: A society in which information is a major product, and the aim is to ensure the prosperity of the economic, social and cultural spheres, with the realization of processes and interactions in society and the economy being realized through the global network and the contemporary ICT.
- **Objective:** Effective implementation of modern ICTs to improve the social, economic and cultural status of society through the rapid and effective exchange of data between different organizations, administrative structures and businesses, as well as providing citizens with various e-services and remote access capabilities and use of information resources.
- **Main Task of the IS:** By combining different components to provide conditions for efficient and modern management by creating information environments, systems and platforms for remote access to information resources and their use.
- **Components:** Present a diverse range of possible IS activities as e-access, e-servicing, e-society, e-policy, e-democracy, e-voting, e-governance/e-government, e-learning, e-health, e-business/e-commerce, e-consultation, e-banking, e-inclusion, etc.

The global goal of the modern society can be divided into three areas:

- **Public Sphere:** To create citizens' conditions for effective e-servicing in ensuring their rights to Knowledge, information and protection;
- **Manufacturing Sphere:** Providing complex automation for all branches of material production;
- **Scientific Sphere:** Ensuring the continuous development of science in order to guarantee the scientific justification of the emerging problems in the informatization of society.

Figure 1. Interaction between IT and the business in the modern society

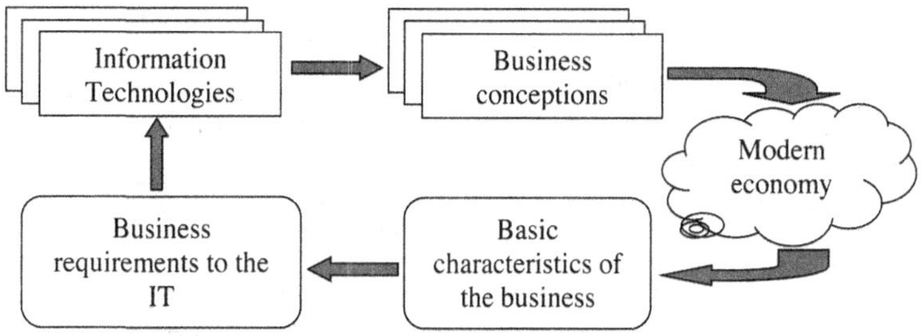

These parts of the global goal determine the characteristics of the contemporary business and create a new economy based on the contemporary IT (Figure 1): • Transition from purely economic to complex solving the problems; • Main goal of management is cost-saving and resource-efficiency; • More attention is paid to consumer value than the economic value; • Active cooperation between different organizations.

The main elements of the contemporary society can be determined as: *Information and Communication Technologies* (a set of computer equipment, software, computer networks for data transmission and created local and distributed information systems with databases); *Information resources* (information and associated technical equipment, program tools, description data and information objects that are designed, built, developed and managed to maintain the information service); *Information security* (measures and procedures to ensure an acceptable level of protection of information resources from accidental or deliberate acts such as disclosure, alteration or destruction).

Based on the classification of Don Tapscott (Tapscott, 2014) the basic key features of the modern society are summarized below which permit transition from traditional (concentrate) model of business organization to the new distributed model based on the contemporary ICT (Figure 2).

- Knowledge orientation. Knowledge is the main production and knowledge-based technical tools are being disseminated. Control systems evolve to knowledge-processing systems. Knowledge and education are a top priority.
- Digital representation of objects. Digitalization of all communications between people and institutions, including documents and services.
- Virtual nature. Global virtualization of the real physical and network environment by creating cloud resources, virtual stores (data centers), mobile cloud applications, intelligent agent technology, IoT and big data analysis (see Figure 2 – "Virtual model of organization").
- Molecular structure. Transition from centralized (administrative-command hierarchy) system to the distributed services and creating individual workers and multifunctional components and services.
- Integration and inter-network interactions. Changing the enterprise structure with moving from the centralized to the decentralized architecture by using distributed in the global network resources and building private virtual networks (see Figure 2 – "Public networks").
- Convergence. Convergence of key sectors of the economy and organizational structures.
- Economic and social cohesion. Building a "society without frontiers", creating social communities with information sharing and dissemination, including between institutions.

Figure 2. Models for organization and management of the company activity

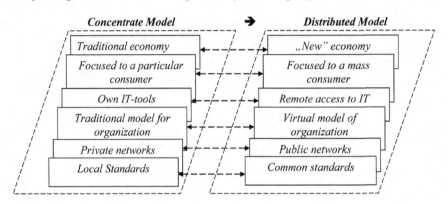

- Innovative nature. Economy based on innovation in all spheres and using new research results in all activities, including in management, marketing (see Figure 2 – "New economy").
- Transformation of the "producer-user" relationship. Removing the boundary between manufacturer and user with extending the focus to the many consumers by developing information systems and software products. Users can participate in testing, tuning, and even when creating a product with orientation to specific application (see Figure 2 – "Focused to a mass consumer").
- Global Scale. Knowledge has no limits and the economy becomes a global activity with cooperation in the e-space and using common standards for manufacturing and management (see Figure 2 – "Common standards"). This permits the execution of individual tasks to be done in different places, including at home, increasing the interdependence of individual countries by using accessible technologies in the global environment (see Figure 2 – "Remote access to IT").
- Existence of contradictions. There are massive social contradictions between workers and dismissed workers with unnecessary knowledge, between knowledgeable and unknowing, between those who have access to information flows and who do not have such access.
- Dynamics. IS operates in real-time mode, trading is electronic, communications are instantaneous, up-to-date information on a manageable process is available with all the necessary parameters. All this reduces the life cycle of production.

These key features reflect the design of business information systems and 3-level architecture is proposed (Figure 3):

Level [1] – Processing sub-system for basic processing of the input data, its economic analysis, formulation of hypothesises, reports preparation, etc.;

Level [2] – Information sub-system which includes databases, archives, and other information resources for business activity support;

Level [3] – Control sub-system for generation logical solutions about information processing.

The modern society has many components presented as services or as activities based on the ICT and globalization of the information resources. A brief summary is presented below.

- **e-Access:** Technological capabilities to connect and interact in an electronic (network) environment.

Figure 3. Organization of a business information system

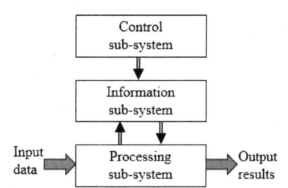

- **e-Society:** Covers the availability and usability of computers and the Internet by citizens. The development is associated with the rapid penetration of computer technology and network communications into the home, with the number of computer-engaging people constantly growing.
- **e-Policy:** This is an expression of the political will to build the IS that covers the actions and intentions of the administration (legislative, executive, judiciary and municipalities) to "electronize" the economy and democratic processes in society.
- **e-Democracy:** Reflects the use of ICT in various forms of democratic life, providing access to political processes and all government information, the ability to easily connect with state institutions, and people's involvement in policy determination.
- **e-Voting:** An opportunity to vote through the Internet, which is a so-called internet democracy (i-democracy), seen as a particular case of e-democracy.
- **e-Learning:** Includes computer provision, Internet access in schools and universities, their presence in the web, the use of ICT by students, and the implementation of new information technologies in lifelong learning. Option is e-education/e-training.
- **e-Governance:** Reflects the democratic principles of strengthening individual rights in public life - the right to seek, receive and disseminate information and ideas, the protection of personal data and the integrity of correspondence, the integration of people with disabilities, access to public information, the improvement of the quality of public services.
- **e-Government:** A narrower concept of e-governance, which is defined as "a system combining the use of ICT by the state administration with organizational changes and newly acquired skills leading to the improvement of public services and the democratic process." It is linked to services provided to citizens and businesses electronically by government entities.
- **e-Health:** Application and development of modern ICT principles and capabilities in the field of national healthcare, providing the opportunity for on-line (direct) access to information resources of the health and health insurance institutions, observing the principles of protection of personal data and the right to privacy.
- **e-Business:** Includes potential opportunities and the real use of ICT and the Internet for business and cooperation with partners, internal and external communication and business process integration in the enterprise. The main prerequisites for development are the provision of IT equipment to the global network and the enhancement of IT staff education.

- **e-Commerce:** An element of e-business centered on the "electronization" of supply and sales, which is associated with web-based shops (e-commerce), e-pay via the Internet, banking (e-banking), invoicing and subsequent delivery.
- **e-Consultation:** Way Of gathering opinions of certain persons or of society as a whole on a specific policy issue without the obligation to take a decision in accordance with the outcome.
- **e-Inclusion:** Political and technological empowerment of citizens by electronic means regardless of their age, gender, education, socio-economic status, language, special needs and place of residence.

and other as *e-campaigning, e-justice, e-legislation, e-parliament, e-referendum, e-petitioning, e-polling, e-mediation,* etc.

The globalization permits different activities in the network space based on using and sharing information. All components presented above use distributed information resources stored in computer systems or global storages and should be accessed via communication environment. There are two basic manners for using information resources:

1. Centralization of all resources in a fixed computer for remote access by access to the server;
2. Decentralization of the resources and access based on message passing communications between distributed systems.

The second method can be realized by two models – centralized access (all information resources are located in one central station and users must make access to this system) and decentralized access (each node in the network has information resources stored in own computer memory and could provide them on the base of request received by other station) – Figure 4. Such a communication infrastructure could be described as a finite set of nodes with communication lines (finite set of arcs) between them – directed or undirected graph structure.

GRAPH THEORY AS A BASIS FOR DISCRETE FORMALIZATION OF INFORMATION PROCESSES IN THE E-SPACE

Any investigation of a given object (system or process) requires the formulation of an initial concept (conceptual model) and presentation (abstract description) through appropriate formalization. In case of complex objects, a preliminary decomposition is recommended for determining the final number of sub-objects with relative autonomy regarding the input-output connections. . The meaning is defining a finite number of sub-functions *f, g, h*, realizing the joint global function $\varphi(x) = f \circ g \circ h = f(g(h(x)))$ – Figure 5.

The Graph Theory is very suitable for representing discrete system and processes in the modern society. It is known that each information process could be described as a sequence of states with relation (transitions) between them. Many publications discuss the using of graphs for solving concrete problems. For example, the scheme of the investigated object (process of developing software in import substitution projects) is represented in the form of directed graph (Kravets, 2017) when the elements (nodes) are connected by arcs as a tree structure.

When developing interactive and distributed environments in the e-space (digital space of the global network), it is necessary to carry out a preliminary analysis of possible relationships and interactions

Figure 4. Two basic models for organization the remote access to resources

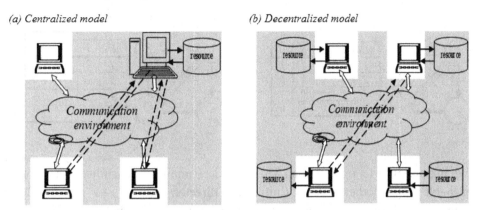

(a) Centralized model

(b) Decentralized model

Figure 5. Decomposition of a complex system

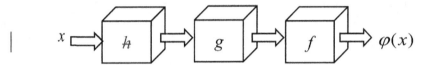

between the different components of the environment. An important stage is the study of ongoing processes in providing remote access to information resources, servicing user requests, providing information content, etc. In the network space, the information processes for accessing a remote resource can be viewed as a conversion of a sequence of events (process states) with transitions between them, which permits to apply the Graph Theory.

It is known that the graph is a discrete structure $G(V, A)$ with final non-empty set of nodes $V=\{V_1,...,V_n\}$ and a set of arcs $A=\{a_{ij}\}$ between couples of nodes $a_{ij}:V_i \rightarrow V_j$ (also final and non-empty set). The type of arcs determines the type of graphs:

- Oriented graph (all arcs have a fixed direction) or non-oriented graph (two-way connection);
- Binary graph ($\forall a_{ij} \in \{0, 1\}$) or weight graph ($\forall a_{ij} = k \geq 0$).

Each graph is defined by a matrix of neighbourhood (MN) with elements $MN[i,j]$ presenting connections between nodes $a_{ij}:V_i \rightarrow V_j$. Definitions of a binary (a) and weight (b) graphs are given below and the graphical presentation is shown in Figure 6.

(a) $\quad MN\left[i, j\right] = \begin{cases} 0; & if \, \nexists a_{ij}; for \, \forall i, j \\ 1; & if \, \exists a_{ij}; for \, \forall i, j \end{cases}$

(b) $\quad MN\left[i, j\right] = \begin{cases} 0; & if \, \exists a_{ij} \, and \, i = j \\ k; & if \, \exists a_{ij} \, and \, i \neq j; k \in N^+ \\ \infty; & if \, \nexists a_{ij}; for \, \forall i, j \end{cases}$

Figure 6. Binary and weight graphs

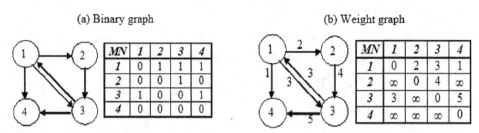

(a) Binary graph

MN	1	2	3	4
1	0	1	1	1
2	0	0	1	0
3	1	0	0	1
4	0	0	0	0

(b) Weight graph

MN	1	2	3	4
1	0	2	3	1
2	∞	0	4	∞
3	3	∞	0	5
4	∞	∞	∞	0

One of the important tasks in the Graph Theory is determining reachability in a graph. A node V_K (final) is reachable from another node V_N (initial) if a directed path $<V_N \rightarrow V_K>$ exists. If an opposite path exists, the node V_K is opposite-reachable from the node V_N. The reachability is presented by Matrix of reachability ($MR[i,j]$) which consist of elements (values) which determine the existence of a path for each couple of nodes. For a binary graph the elements are 1 (for existing path) or 0 (if a path is missing) and for a weight graph the elements must present the shortest path between each couple of nodes. Figure 7 shows a graphical interpretation of the Dijkstra's algorithm for determining the minimum path in graph from the initial node "1" to the final node "5". The minimum rank 2 determines the shortest path between these two nodes

The main e-space objects can be considered as independent, territorially remote units with their own internal structure and functionality. This allows a digital space to be described as a discreet set of nodes representing the individual geographical points entering into relationships with one another, i.e. for the formalization to use a directed weighting graph with discrete and finite sets: V – set of nodes presented the components of the space; A – set of arcs for presentation physical line for connection with weights d_{ij}, corresponding to the communication parameters between the individual locations of the objects. 3 groups are determined in the abstract model (Figure 8):

- *Users,* which realize multiple remote access to one or more units in the e-space (digital / network environments);
- *Communication Medium* for transportation of the requests for access to an information content (traditionally the global network is used for realization this component);

Figure 7. Determining the shortest path in a non-ordered graph

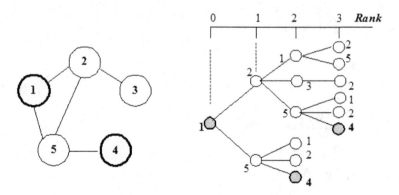

Figure 8. Abstract formalization of the digital space

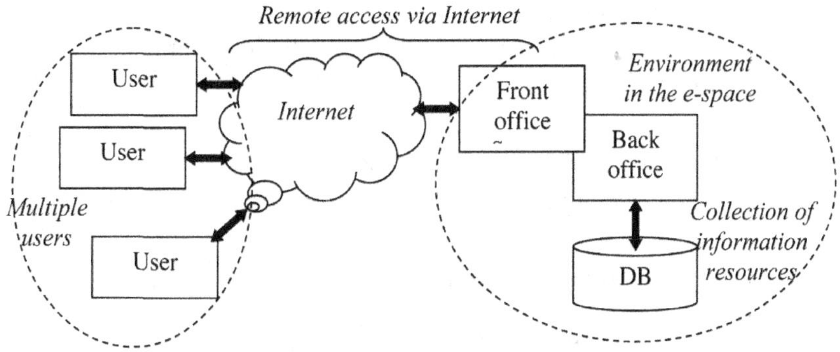

- *Resources* which are the objects for remote access by users.

The formalization describes the distributed e-space as a discrete set of nodes (remote components) with relation between them – this permits to define directed weight graph on the base of the following two discrete sets (Romansky, 2010):

$V = \{V_1, ..., V_n\}$ – set of nodes presenting all distributed components of the e-space (sources of requests for remote access, distributed information resources, tools for communication, etc.);

$A = \{a_{ij} / i,j = 1,2,..., n\}$ – set of the connected arcs between couples independent nodes presented by weight coefficients d_{ij}, which correspond to the communication parameters between the objects' localizations.

COMMUNICATION INFRASTUCTURE AND FORMAL DESCRIPTION OF DIGITAL SPACE

Communication infrastructure has some basic parameters, for example:

Network Infrastructure - provides the necessary communication environment for access to distributed information resources.

Provided services - a key feature of the capabilities of the network space represented by various distributed systems, digital spaces, sites, etc., such as remote access to resources, text, speech, video and other information, access to distributed database, and etc.).

The time for requesting a customized service is defined as the average statistic time from the time the message is transmitted to the moment it is received, the purpose is to successfully satisfy the user request rather than the performance (client / server model).

The cost of information processing for network services is formed by taking into account the value of the means of I/O operations, transmission, storage and processing of the data.

For the purpose of the formal description of digital space, the initial set V may be divided into separate subsets which allow to define the specific functions of the components of the information service:

- $U = \{U_i / i=1,..., N\}$, $U \neq \emptyset$ – set of users which make remote access to distributed information and other type resources by generating request *req: Ui → Rj* ; for $\forall U_i \in U$ and $\forall R_j \in R$;

- $R = \{R_j \, / \, j=1,\ldots, M\}$, $R \neq \varnothing$ – set of resources providing information content *Inf: Rj* → $U_i^`$ (for $\forall U_i \in U$ and $\forall R_j \in R$) or services for common use;
- $T = \{T_q \, / \, q=1,\ldots,K\}$, $T \neq \varnothing$ – set of technical and technological components (transmitters) of the network space used for transmission of information objects in the two directions (requests and content);
- $D = \{D_{IN}, D_{OUT}\}$ – distributors of information objects collecting algorithm, protocols and other tools for distributing and directing the requests and information blocks in e-space.

Based on this sets the formalization of the digital space could be determined as an ordered structure $eSPACE=\{U, R, T, D\}$ with two types of relations between components: $req : U_i \xrightarrow{T(q)} R_j$ – transfer of requests; $Inf : R_j \xrightarrow{T(q)} U_i$ – delivery of service and/or information (see Figure 9).

As a summary $V = U \cap R \cap T \cap D$ is the model of e-space and the distance between nodes is defined by d_{ij} ($i \neq j$; $i,j \in \{1 \div n\}$). This permits to form a *matrix of the distances DM* with dimension '$n \times n$' and elements $d_{ii}=0$, by which could be determined the minimum paths between independent nodes in the structure.

Let's define two binary parameters $u_{ik} \in \{0,1\}$ and $r_{jk} \in \{0,1\}$, that correspond to the real locations of the users $U_i \in U$ and/or of the resource $R_j \in R$ in a node $V_k \in V$. They could be described by the expressions:

$$u_{ik} = \begin{cases} 1, & \text{if user } U_i \in U \text{ is located in the node } V_k \in V \\ 0, & \text{otherwise} \end{cases}$$

Figure 9. Formalization of the communication in the digital space

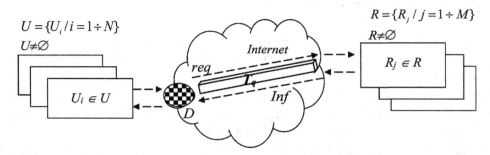

Table 2.

DM	V_1	...	V_j	...	V_n
V_1					
...					
V_i			d_{ij}		
...					
V_n					

$$r_{jk} = \begin{cases} 1, & \text{if } R_j \in R \text{ is located in the node } V_k \in V \\ 0, & \text{otherwise} \end{cases}$$

The actual localization of the components in the physical nodes of *V* is important for the concretization of the information servicing processes. For the general situation is assumed that $U \cap R \neq \emptyset$ and that one or more nodes can contain objects from the both groups (*U* and *R sets*). This allows the formation of two types of relations in eSPACE (Figure 10):

$$\exists V_k \in V \rightarrow \left(V_k \in U \right) \& \left(V_k \in R \right) \tag{1}$$

$$\exists V_k \in V \rightarrow \left(V_k \in U \right) \& \left(V_k \neg \in R \right) or \left(V_k \neg \in U \right) \& \left(V_k \in R \right) \tag{2}$$

Based on the actual localizations of the active objects in the eSPACE it could be constructed two matrixes

$$UL = \{UL\left[i,k\right] = u_{ik}; i = 1 \div N; k = 1 \div n \ ;$$

$$RL = \{RL\left[j,k\right] = r_{jk}; j = 1 \div M; k = 1 \div n \ ;$$

by which is defined a new vector using the procedure:

FOR *k=1,2,…,n* DO

Figure 10. Two types of relations in eSPACE

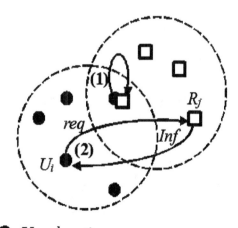

● *User location*
☐ *Information resource location*

$$VL[k] = \begin{cases} 1, & if \left\{ \sum_{i=1}^{N} UL[i,k] \geq 1 \right\} \& \left\{ \sum_{i=1}^{M} RL[j,k] \geq 1 \right\} \\ 0, & otherwise \end{cases}$$

According the procedure the elements $VL[k] = 1$ determine relation type (1) and the elements $VL[k] = 0$ determine the relations type (2) – Figure 11.

Finally, the formalization permits to provide a deterministic investigation of the processes in e-space by establishing binary relations between active nodes in graph structure. Each transaction could be described by ordered couple (u, r), for $u \in U$ and $r \in R$ and the following relation scheme is realized:

$$BR \subset V \times V = \{(u,r) \, / \, u,r \notin U \cap R\}; \forall u \in U \subset V \, \& \, \forall r \in R \subset V\}$$

It is assumed that a real transfer via network medium exists only if the relation between two nodes is of type (2). An example of any possible situation is shown in Figure 12.

EXTENSION OF GRAPH THEORY FOR DISCRETE MODELLING OF PROCESSES IN MODER SOCIETY

Analytical modelling is based on mathematical methods of describing the behaviour and physical properties of the system under study as mathematical objects and relationships between them. Generally, the mathematical model is an expression or system of equations representing the object as an analytical dependence on a type $Y=F(X, S)$ or $\Phi(X, S, Y) = 0$.

E-space processes are stochastic (probable) but are developed in a discrete environment by computer and communication tools. This determines the heterogeneous character of the investigated objects in the space. This permits to apply *deterministic modelling* by means of the discrete mathematics and discrete

Figure 11. Constructing matrix VL on the base of the matrixes UL and RL

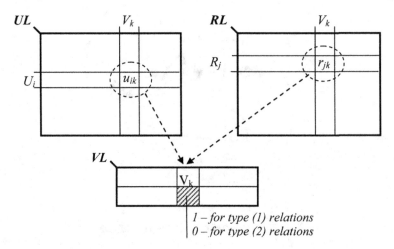

Figure 12. Example for relation scheme defining

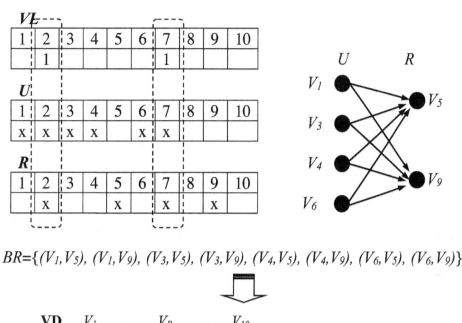

$$BR=\{(V_1,V_5),\ (V_1,V_9),\ (V_3,V_5),\ (V_3,V_9),\ (V_4,V_5),\ (V_4,V_9),\ (V_6,V_5),\ (V_6,V_9)\}$$

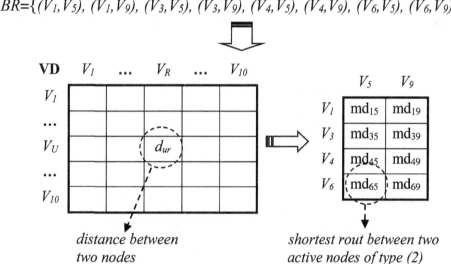

distance between two nodes

shortest rout between two active nodes of type (2)

structures (automats, graphs, Petri net, etc.). An opportunity for describing and investigation of objects in the digital spice is Theory of Petri Nets which could be regarded as an extension of the Graph Theory.

The modelling by using Petri nets (PNs) relates to formalization of the investigated process or object as a sequence of events, named transactions that could be activated if connected conditions (presented by marked places) are realized. Each PN-model could be defined as an ordered triple $PN = (P, T, F)$, where: $P = \{p_1, p_2, ..., p_m\}$ – set of places; $T = \{t_1, t_2, ..., t_n\}$ – set of transactions; $F \subseteq (P{\times}T){\cup}(T{\times}P)$ – set of relations (set of arcs). The sets P and T are final sets ($m{\geq}0$, $n{\geq}0$, $P{\cap}T{=}\varnothing$). The set F includes ordered couples (p_i, t_j) which define relations $P{\rightarrow}T$ and $T{\rightarrow}P$, and two functions are constructed – $I, O \in \mathbf{N}^{|P|{\times}|T|}$, named input and output functions.

The extended PN definition $PN = (P, T, F, W, \mu_0)$ includes two additional elements: $W: F{\rightarrow}\{1,2,3,...\}$ – weight function for each arc and $\mu_0: P \rightarrow \{0,1,2,3,...\}$ – initial marking for PN execution starting.

The matrix approach permits to define the PN by using two matrix: $\mathbf{D}^-$ – input matrix (present the input places for each transaction) and $\mathbf{D}^+$ – output matrix (present the output places for each transaction):

$\mathbf{D}^-[j, i] = \eta(p_i, I(t_j)); \forall p_i \in P; j=1 \div m$

and

$\mathbf{D}^+[j, i] = \eta(p_i, O(t_j)); \forall p_i \in P; j=1 \div n.$

The definition $PN = (P, T, \mathbf{D}^-, \mathbf{D}^+)$ permits to present each transaction t_j as a vector $e[j]$ with n elements that all of them are 0, but only element "j" is 1.

The defined PN could be described as a directed multi-graph where functions I and O are presented by arcs between places (rings) and transactions (rectangle or segment). The modelling by using PN is based on execution of marked net that each position has an integer count of marks $k \geq 0$ that could be changed during the PN evolution $\mu_0 \rightarrow \mu_1 \rightarrow \mu_2 \rightarrow \dots$.

An abstract model based on the formalization made in the previous section is shown in Figure 13a. Three basic primitives could be determined in the model as abstract segments – users, transmitters and resources. The functionality of distributers could be included in the functions of the transmitters. In this reason, the general scheme of a PN-model can be determined as it is shown in Figure 13b.

Figure 13. Developing a Petri Net model (PN-model) of processes in e-Space

a) Abstract model of a process for access to a resource in the e-space

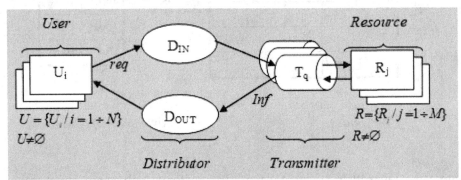

b) General scheme of the PN-model

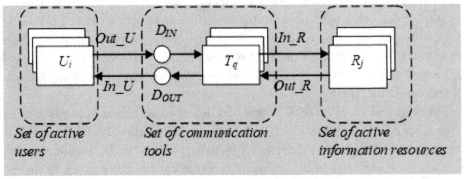

Table 3. Input and output matrixes

D^-	p_1	p_2	p_3	D_{IN}	D_{OUT}	D^+	p_1	p_2	p_3	D_{IN}	D_{OUT}
t_1	2	0	0	0	0	t_1	1	1	0	0	0
t_2	0	0	1	0	0	t_2	1	0	0	0	0
t_3	0	1	0	0	0	t_3	0	0	0	1	0
t_4	0	0	0	0	1	t_4	0	0	1	0	0

Each of the tree basic components of the general scheme is presented below as a separate formal description by the PN apparatus. Verification of the functionality of each analytic model is made by execution started by initial marking. The PN-models of the components are presented by analytical definition and graphical presentation as a directed multi-graph with nodes (positions and transactions) with directed arcs between them.

Definition of the Component "User"

Events (nodes type transaction): t_1 – generation of a request for access to distributed information resource; t_2 – received information block processing; t_3 – sending a request to the distributed medium for routing; t_4 – an information block enters into the input buffer from the distributed medium.

Conditions (nodes type position): p_1 – availability of the user to work with the distributed resource; p_2 – presence of a request in the input buffer; p_3 – presence of information block in the output buffer.

Analytical PN-Model

$$P = \{p_1, p_2, p_3\} \Rightarrow |P| = 3$$

$$I(t_1) = \{p_1, p_1\}$$

$$I(t_2) = \{p_3\}$$

$$I(t_3) = \{p_2\}$$

$$I(t_4) = \{In\_U \equiv D_{OUT}\}$$

$$T = \{t_1, t_2, t_3, t_4\} \Rightarrow |T| = 4$$

$$O(t_1) = \{p_1, p_2\}$$

$$O(t_2) = \{p_1\}$$

$$O(t_3) = \{Out\_U \equiv D_{IN}\}$$

$$O(t_4) = \{p_3\}$$

Definition of the Component "Resource"

Events (nodes type transaction): t_R – processing of a request for remote access to the information resource; t_S – giving the access to the information resource (an information block sending).

Conditions (nodes type position): p_A – presence of entered in the input buffer request for access and using of distributed resource; p_B – readiness to give of a distributed resource; p_C – presence of processed request; p_D – the access to the distributed resource is given and the information block is directed to the network medium.

Analytical PN-Model

$$P = \{p_A, p_B, p_C, p_D\} \Rightarrow |P| = 4$$

$$I(t_R) = \{p_A, p_B\}$$

$$I(t_S) = \{p_C, p_C\}$$

$$T = \{t_R, t_S\} \Rightarrow |T| = 2$$

$$O(t_R) = \{p_C, p_C\}$$

$$O(t_S) = \{p_B, p_D\}$$

Definition of the Component "Transmitter"

Events (nodes type transaction): t_{T1} – request transmission through the distributed (network) medium based on the routing algorithm; t_{T2} – information block (contents) transmission through the distributed medium to the user (respond returning).

Table 4. Input and output matrixes

D⁻	p_A	p_B	p_C	p_D	**D⁺**	p_A	p_B	p_C	p_D
t_R	1	1	0	0	t_R	0	0	2	0
t_S	0	0	2	0	t_S	0	1	0	1

Table 5. Input and output matrixes

D⁻	D_{IN}	D_{OUT}	p_{T1}	p_{T2}	In_R	Out_R	**D⁺**	D_{IN}	D_{OUT}	p_{T1}	p_{T2}	In_R	Out_R
t_{T1}	1	0	1	0	0	0	t_{T1}	0	0	0	1	1	0
t_{T2}	0	0	0	1	0	1	t_{T2}	0	1	1	0	0	0

Conditions (nodes type position): p_{T1} – readiness to transmit a request to node (distributed resource) in the communication medium (availability of rout and communication resource); p_{T2} – readiness to transmit information block to the user (availability of free communication resource on the rout).

Analytical PN-Model

$$P = \{p_{T1}, p_{T2}\} \Rightarrow |P| = 2$$

$$I(t_{T1}) = \{D_{IN}, p_{T1}\}$$

$$I(t_{T2}) = \{Out\_R, p_{T2}\}$$

$$T = \{t_{T1}, t_{T2}\} \Rightarrow |T| = 2$$

$$O(t_{T1}) = \{In\_R, p_{T2}\}$$

$$O(t_{T2}) = \{D_{OUT}, p_{T1}\}$$

Figure 14 shows generalization of defined primitives and presents one-access process in e-Space with a set of positions

$$P = \{p_1, p_2, p_3, p_4, D_{IN}, D_{OUT}, p_{T1}, p_{T2}, p_A, p_B, p_C, p_D\}$$

and a set of transition

$$T = \{t_1, t_2, t_3, t_4, t_{T1}, t_{T2}, t_R, t_S\}.$$

To investigate a multiple-access in e-Space an extension of the PN-model definition should be made, for example, the analytical definition for the communication medium only will be as presented below:

$$P = \{D_{IN}, D_{OUT}, \{p_{T1j}, p_{T2j} \ / \ j = 1 \div K\}\};$$

Figure 14. Graphical interpretation of the PN-model of e-Space (base segment)

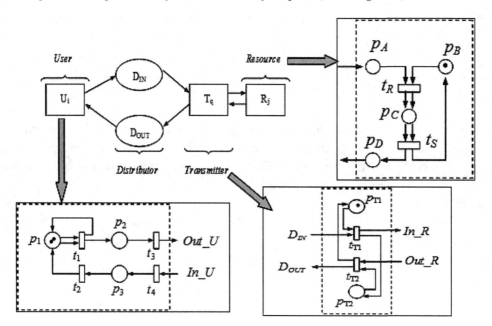

$$I(t_{T1j}) = \{D_{IN}, p_{T1j}\}$$

$$I(t_{T2j}) = \{Out\_R_j, p_{T2j}\}$$

$$T = \{t_{T1j}, t_{T2j} / j = 1 \div K\}$$

$$O(t_{T1j}) = \{In\_R_j, p_{T2j}\}$$

$$O(t_{T2j}) = \{D_{OUT}, p_{T1j}\}$$

for $j = 1, 2, ..., K$

The united PN-model formed on the base of Figure 14 which permits to investigate one-access process in the digital medium is presented in Figure 15 with an initial marking $\mu_0 = (2, 0, 0, 0, 0, 1, 0, 0, 1, 0, 0)$ which permits activation of the transaction t_1 only.

The evaluation of the PN is $\mu_0 \rightarrow \mu_1 \rightarrow \mu_2 \rightarrow \mu_3 \rightarrow \mu_4 \rightarrow \mu_5 \rightarrow \mu_6 \rightarrow \mu_7 \rightarrow (\mu_8 = \mu_0)$ with markings below and tree of reachability shown in Figure 16:

$\mu_0 = (2, 0, 0, 0, 0, 1, 0, 0, 1, 0, 0) \rightarrow$ permitted transaction t_1;

$\mu_1 = (1, 1, 0, 0, 0, 1, 0, 0, 1, 0, 0) \rightarrow$ permitted transaction t_3;

Figure 15. Multigraph for defining united PN-model

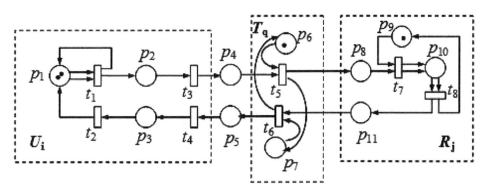

$\mu_2=(1, 0, 0, 1, 0, 1, 0, 0, 1, 0, 0) \rightarrow$ permitted transaction t_5;

$\mu_3=(1, 0, 0, 0, 0, 0, 1, 1, 1, 0, 0) \rightarrow$ permitted transaction t_7;

$\mu_4=(1, 0, 0, 0, 0, 0, 1, 0, 0, 2, 0) \rightarrow$ permitted transaction t_8;

$\mu_5=(1, 0, 0, 0, 0, 0, 1, 0, 1, 0, 1) \rightarrow$ permitted transaction t_6;

$\mu_6=(1, 0, 0, 0, 1, 1, 0, 0, 1, 0, 0) \rightarrow$ permitted transaction t_4;

$\mu_7=(1, 0, 1, 0, 0, 1, 0, 0, 1, 0, 0) \rightarrow$ permitted transaction t_2;

$\mu_8 \equiv \mu_0$

Another example for using the graph theory in the modern society is an investigation of a secure access to information resources in a corporative information system realised on the base of access to distributed resources. There are many information resources in the digital space located in the network nodes, in the cloud and/or data centres, private networks, etc. The access of these resources must be regulated according to their type.

Figure 17 presents an initial graph formalization of example structure of a corporative system which supports and processes two types of resources – public (regulation is not needed) and private (stored in internal or external storages and access must be strong regulated). The correct architecture of such system consists of two administrative sub-systems (Front Office and Back Office) and one information sub-system. The activity of the access regulation is distributed between the administrative sub-systems. The functionality of the first sub-system includes collection of audit information for each access (permitted or not-permitted) which must be stored in supported audit files system (time of the access, IP address, requested resource is possible, other attributes of the access). This information could be used for a statistical report preparation.

The Front Office is input/output portal which communicates with users accessing the corporative system. Two procedures must be realized in this level – initial registration of each new user and identification of already registered user which sends a request for access system resources. A personal profile

Figure 16. Tree of reachability of the united PN-model execution

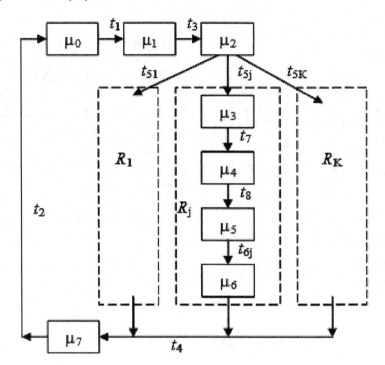

Figure 17. Preliminary formalization of a system structure

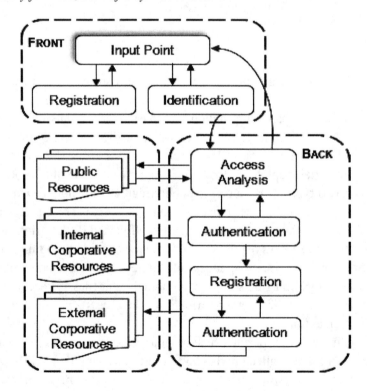

(file with personal data) is created for each new user during the registration and this information should be stored in a protected archive which must be accessed by authorised person only. The role of the procedure for identification should guarantee legitimate access to the system resources of the back office.

The Back Office sub-system organizes the internal administration of the resources and must organize their protection based on preliminary analysis of the each request and determining the level of accessed resource. This permit to select the level of protection and permits to activate procedures for user's authentication and user's authorization. These procedures work with using the information completed and stored in an administrative database which consists of personal profiles for authorized persons of the staff and it is a part of Digital Rights Management System (DRMS). Many of the external information resources traditionally can be stored in a cloud which provides own security mechanisms, but it is good if the corporative system also provides additional security measures for access.

The functionality of the proposed structure is supported by some important procedures for information security and personal data protection and can be described by the directed graph shown in Figure 18. The choice of the graph apparatus is made based on the following conditions:

- Number of investigated events at the structure is finite and this permit defining a finite set of states presented by nodes in a directed graph.
- Each observed event in the architecture has constant behaviour presented by fixed procedure (algorithm), but realization is made under a stochastic flow.
- Graph nodes present all observed events (sub-processes) and transitions between nodes describe possible realization of processes.

The nodes present the main procedures, for example: (1) – input point which organizes the communications with the users; (2) – registration of a new user; (3) – procedure for preliminary identification; (4) – input in the back office and procedure for access analysing; (5) – procedure for authentication of the access and legitimate request; (6) – determining of the requested resource and select the level of protection and security means using with a registration of an audit information ; (7) – procedure for authorization for the requested resource with two results (permitted or not-permitted access).

The users of the corporative system could be two types – internal (for example staff of the organization) and external (clients, corporative partners, distributors, etc.). The internal users can work directly with the back office sub-system, but additional means and tools for identification and access regulation can be used (badges, magnetic cards, limited access to some rooms, different passwords, biometric identification, etc.). The external users access the components of the corporative system (including internal

Figure 18. Graph description of the system functionality

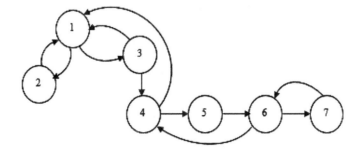

local network, computers, memories, file systems with information, etc.) by remote access and serious measures to provide secure access must be implemented on different levels. For example, the procedure "authentication" could be realized in two steps. The first step is "identification" and could be realized in the Front Office. The second step is "verification" and must be realized in the Back Office. On the other hand, the procedure "authorization" must be must be consistent with the rules of the internal DRMS (Digital Rights Management System) system. This is very important because the authorization ensures access to personal profiles (files with personal data) and to administrative documents (corporative archive). These information resources must be reliable protected by organizational and technical measures according to the rules of the data protection law and the regulation GDPR and the ePrivacy Directive. An additional discussion on the problem is made in (Romansky, 2017c) where a stochastic investigation of secure access to corporative resources is presented. Figure 19 presents the proposed algorithmic scheme for functional organization of a system for information security and data protection which is based on procedures for access control and digital rights management discussed above.

Deterministic description and investigation can be made by using the classic PN apparatus where the main procedures are presented as transitions, and the condition are presented as positions. The theoretical-set definition is presented below (Romansky, 2016).

Formal definition of Petri Net:

$$PN = \{T, P, I, O\}, T \cap P = \varnothing$$

Set of transactions $T = \{t_i \,/\, i=1 \div 9\}$: t_1 – remote user's access to the input point; t_2 – registration of a new user; t_3 – activate the identification procedure; t_4 – access to public resources; t_5 – authentication procedure for access to corporative (private) resources (T - correct; F – incorrect); t_6 – finishing the work; t_7 – authorization procedure (rights checking) by using DRMS tools (T – correct; F – incorrect); t_8 – access to external corporative resources; t_9 – access to internal corporative resources.

Set of positions $P = \{p_j \,/\, j=1 \div 5\}$: p_1 – user's access is activated; p_2 – access of a registered user; p_3 – legitimated user's access; p_4 – successful authentication; p_5 – successful authorization.

The graph scheme of the model based on the presented theoretical-set definition is shown in the Figure 20.

Table 6.

Input Functions	Output Functions
$I(t_1) = \varnothing$ $I(t_2) = \{p_1, p_1\}$ $I(t_3) = \{p_1, p_2\}$ $I(t_4) = I(t_5) = I(t_6) = \{p_3\}$ $I(t_7) = \{p_4\}$ $I(t_8) = I(t_9) = \{p_5\}$	$O(t_1) = \{p_1, p_1\}$ $O(t_2) = \{p_1, p_2\}$ $O(t_3) = O(t_4) = O(F.t_5) = O(F.t_7) = \{p_3\}$ $O(T.t_5) = \{p_4\}$ $O(t_6) = O(t_8) = \varnothing$ $O(T.t_7) = \{p_5\}$ $O(t_9) = \{p_4\}$

Figure 19. Algorithmic scheme for processes in system for information security

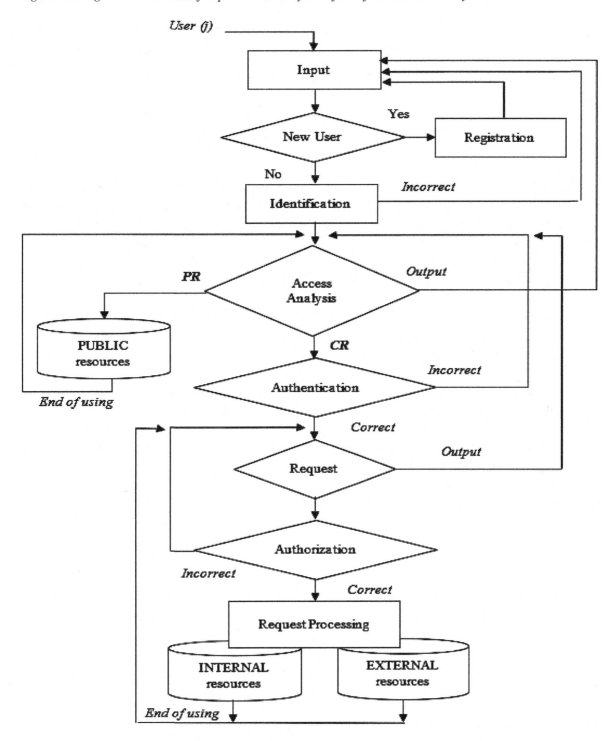

Figure 20. Graph presentation of the defined PN-model

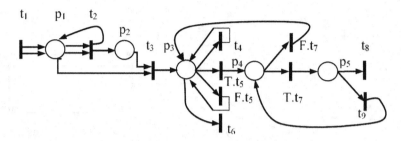

The evolution of the defined PN-model is presented by the tree of possible transactions given in the Figure 21. Marking μ_j (j=0, 1, 2, 3, 4, 5) for each new state is presented in the figure. The investigation of the PN-model could be made based on the analysis of the evolution tree and the characteristics are determined below.

- **Reachability:** The model permits cyclic recurrence with reiteration of some phases based on activation of the selected transition.
- **Liveness:** This characteristic deals with initial marking μ_0, bearing in mind that minimum one permitted transaction for each step of the evolution exists.
- **Blocking:** The PN-model has not any blocked marking and the model is active, so there are not conflict situations.
- **Boundless:** The model is 2-limited because the number of marks at each position during the evaluation $<\mu_0\rightarrow\mu_1\rightarrow\mu_2\rightarrow \ldots \rightarrow\mu_0>$ is no more than 2, i.e. $\Sigma\mu(p_i) \leq 2$.

Another research by using the Markov Chain apparatus is presented in (Romansky, 2017c) where the proposed main procedures are formalized as a discrete states with stochastic transitions between them. Selected parameters (a, b and c) are determined as a controlled for the analytical model designing and experiments carried out – the graph of the states of this model is shown in Figure 22.

Figure 21. Evaluation tree construction of the defined PN-model.

$$\mu_0 = (0,0,0,0,0) \xrightarrow{t1} \mu_1 = (2,0,0,0,0)$$

$$\mu_1 \xrightarrow{t2} \mu_2 = (1,1,0,0,0) \xrightarrow{t3} \mu_3 = (0,0,1,0,0)$$

$$\mu_3 \xrightarrow{t4} \mu_3$$

$$\mu_3 \xrightarrow{t5F} \mu_3$$

$$\mu_3 \xrightarrow{t6} \mu_0$$

$$\mu_3 \xrightarrow{t5T} \mu_4 = (0,0,0,1,0) \xrightarrow{t7F} \mu_3$$

$$\mu_4 \xrightarrow{t7T} \mu_5 = (0,0,0,0,1) \xrightarrow{t9} \mu_4$$

$$\mu_5 \xrightarrow{t8} \mu_0$$

$$\mu_0 \rightarrow \mu_1 \rightarrow \mu_2 \rightarrow \mu_3 \begin{cases} \rightarrow \mu_3 \\ \rightarrow \mu_3 \\ \rightarrow \mu_4 \begin{cases} \rightarrow \mu_5 \begin{cases} \rightarrow \mu_0 \\ \rightarrow \mu_4 \end{cases} \\ \rightarrow \mu_3 \end{cases} \\ \rightarrow \mu_0 \end{cases}$$

Figure 22. Graph of the state of the designed Markov model

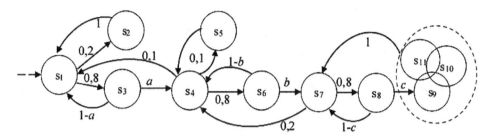

The model is based on the discrete set $S = \{s_j/\mathrm{j}=1\div 11\}$ whit the states listed below.

s_1 – remote access to the input point;

s_2 – procedure for registration of a new user;

s_3 – procedure for identification of a user access to the corporative system;

s_4 – procedure for analysis the type of the required access to the corporative resources after input in the back office of the e-servicing corporative system;

s_5 – realization of a process for using a public resource which is unlimited without restriction;

s_6 – procedure for authentication of the user based on information from the personal profile created by the DRMS;

s_7 – procedure for analysing user's request for access to the private information and system resources;

s_8 – procedure of authorization based on information about user's rights defined by DRMS;

s_9 – process for realization of the user's request after successful authorization; the access could be performed into one of the following two directions;

s_{10} – access and using internal corporative information resources;

s_{11} – access and using external corporative information resources.

The analytical solution of the model permits to investigate the usability of the procedures and additional evaluation based on the Statistical Software Develve has been made.

CONCLUSION

The modern society has many opportunities which could be investigated by using graph theory. Not only in the areas of digital communications, access to information and using the new technologies based of ICT, but it could be suitable and useful in different technical research. For example, the presentation of a designed artificial neural network is made by directed graph. In this reason, principle scheme of a controller implemented with three-layered neural network based on the backpropagation procedure is discussed in (Rafajlovski, 2018). This solution permits to identify the unknown dynamics of the designed system and can be combined with a reference model in order to reach the required control accuracy. Other application of the Graph Theory is made in (Mohammadi, 2019) for analysing the specific of the IT-based business processes based on Service-Oriented Architecture (SOA). Graph strictures are used for presented shared capabilities by shared services, service encapsulation and communication, relations between SAO layers, etc.

On the other hand, PN-apparatus is well used tool for modelling and investigation systems and processes. A production line is modelled based on intelligent agents and Petri nets in (Christoforidis, 2018). Especially, PNs are used for discrete presentation of linear production, linear paths to same result, dedicated machine model and multifunctional machine model.

The preliminary formalization of processes in the digital world by using different graph structures permits to design analytical models for investigation important characteristics of processes and systems. The determined analytical modelling, including graphs and Petri nets, is one possible direction, but other theories can be applied. In this reason, theory of Markov processes, and particularly Markov Chain apparatus is very useful for modelling and investigation of processes with stochastic nature. This possibility is discussed in the book (Romansky, 2017b). On the other hand, preliminary formalization and stochastic analytical modelling by using Markov Chain for the purpose of architectural design of a web-application is presented in (Romansky, 2018). Each analytical investigation could be extended by realization of a full multi-factor experimental plan with using statistical software, for example, or using suitable apparatus for simulation.

REFERENCES

Adiono, T., Fuada, S., Saputo, R. A., & Luthfi, M. (2018). Internet access over visible light. *Proceedings of the 2018 IEEE International Conference on Consumers Electronics*. Retrieved from https://ieeexplore. ieee.org/abstract/document/8552123

Afinogenov, Gr. (2013). Andrei Ershov and the Soviet Information Age. *Kritika: Explorations in Russian and Eurasian History*, *14*(3), 561–584. doi:10.1353/kri.2013.0046

Aggarwal, A. A. (2003). Internetalization of end-users. *Journal of Organizational and End User Computing*, *15*(1), 54–56. doi:10.4018/joeuc.2003010104

Aiqun, Z. (2018). An IT capability approach to informatization construction of higher education Institutions. *Procedia Computer Science*, *131*, 683–690. doi:10.1016/j.procs.2018.04.312

Alrawi, K. (2017). The internet and international marketing. *Competitiveness Review*, *17*(4), 222–233. doi:10.1108/10595420710844316

Bell, J., Deans, K., Ibbotson, P., & Sinkovics, R. (2001). Towards the "internetalization" of international marketing education. *Marketing Education Review*, *22*(9), 69–79. doi:10.1080/10528008.2001.11488758

Bolnokina, E. V., Oleinikova, S. A., & Kravets, O. J. (2019). Determination of optimal composition of team of executors for multistage service system. *International Journal on Information Technologies and Security*, *11*(1), 51-58.

Christoforidis, G., Stykas, V. A., & Kasso, T. (2018). Simulated comparison of push/pull production with committed and non-committed automated guided vehicles. *Proceedings of the 32ⁿᵈ International conference on Information Technologies - InfoTech-2018*.

Dordick, H. S., & Wang, G. (1993). *Information society: A retrospective view*. Sage Publications, Inc.

Goncharov, V. N. (2018). Informatization of society: social and economic aspects of development. *Proceedings of the 8th International scientific Conference Informatization of society: Socio-Economic, Socio-Cultural and International Aspects*.

Guo, L., Wang, S., Kang, L., Li, Q., Chen, G., & Li, C. (2014). A method of manufacture resource informatization in cloud manufacturing. *Journal of Software Engineering*, *8*(1), 32-40. Doi:10.3923/jse.2014.32.40

Kerr, A., & Waddington, J. (2013). E-Communications: An aspect of union renewal or merely doing thing electronically? *British Journal of Industrial Relations*, *52*(4), 658–681. doi:10.1111/bjir.12010

Kharchenko, V. (2018). Big Data and Internet of Things for safety critical applications: Challenges, methodology and industry cases. *International Journal on Information Technologies and Security*, *10*(4), 3-16.

Kluver, R. (2004). Globalization, informatization, and intercultural communication. In Fr. E. Jandt (Ed.), Intercultural Communication: A Global Reader (pp. 425-437). SAGE Publications.

Kravets, O. J., Kosorukov, O., & Utyusheva, L. (2017). Specific features of graphical-analytical representation of a process model of managing innovation activity in the framework of software development for import substitution projects. *Journal of Theoretical and Applied Information Technology*, *95*(14), 3337-3351.

Kravets, O. J., Choporov, O. N., & Bolnokin, V. E. (2018). Mathematical models and algorithmization of monitoring control an affiliated network in maintenance service distributed organization. *Quality – Access to Success*, *19*(167), 68-72.

Li, M., Wei, W., Wang, L., & Qi, X. (2018). Approach to evaluating accounting informatization based on entropy in intuitionistic fuzzy environment. *Entropy (Basel, Switzerland)*, *20*(6), 476. doi:10.3390/e20060476

Melnyk, V. V. (2018). Information management as a factor of innovative society development. *Humanities Bulletin of Zaporizhzhe State Engineering Academy*, *74*(74), 39–47. doi:10.30839/2072-7941.2018.149651

Mohammadi, M., & Mukhtar, M. (2018). Service-Oriented Architecture and process modelling. *Proceedings of the 32nd International Conference on Information Technologies – InfoTech-2018*.

Olawumi, O., Väänänen, A., Haataja, U., & Toivanen, P. (2017). Security issues in smart homes and mobile health system: Threat analysis, possible countermeasures and lessons learned. *International Journal on Information Technologies and Security*, *9*(1), 31–52.

Rafajlovski, G., & Digalovski, M. (2018). Controlling induction motors. *Proceedings of the 32nd International conference on Information Technologies – InfoTech-2018*.

Romansky, R. (2017a). A survey of digital world opportunities and challenges for user's privacy. *International Journal on Information Technologies and Security*, *9*(4), 97-112.

Romansky, R. (2017b). *Information servicing in distributed learning environments. Formalization and model investigation*. Saarbrüken, Germany: LAP LAMBERT Academic Publishing.

Romansky, R., & Kirilov, K. (2018). Architectural design and modelling of a web based application for GDPR clarification. *AIP Conference Proceedings (American Institute of Physics), Proceedings of the 44th International Conference on Applications of Mathematics in Engineering and Economics (AMEE'18).* 10.1063/1.5082121

Romansky, R., & Noninska, I. (2016). Discrete formalization and investigation of secure access to corporative resources. *International Journal of Engineering Research and Management, 3*(5), 97–101.

Romansky, R., & Noninska, I. (2017c). Stochastic investigation of secure access to the resources of a corporative system. *International Journal of Scientific & Engineering Research, 8*(1), 578-584.

Romansky, R., & Parvanova, E. (2010). Formalization and discrete modelling of the information servicing in distributed learning environment. *Communication & Cognition, 43*(1&2), 1-15.

Rosemann, M. (2013) The Internet of Things: new digital capital in the hands of customers. *Business Transformation Journal, 9,* 6-15. Retrieved from http://eprints.qut.edu.au/66451/

Sergeeva, M. G., Karavanova, L. Z., Bereznatskaya, M. A., Klychkov, K. E., Loktionova, T. E., & Chauzova, V. A. (2018). Socialization of a personality under the conditions of globalization and informatization of the society. *Revista Espacios, 39*(21), 28. Retrieved from http://www.revistaespacios.com/a18v39n21/a18v39n21p28.pdf

Škraba, A., Stanovov, V., Semenkin, E., Koložvari, A., & Kofjač, D. (2018). Development of algorithm for combination of cloud services for speech control of cyber-physical systems. *International Journal on Information Technologies and Security, 10*(1), 73-82.

Tapscott, D. (2014). *The digital economy anniversary edition: Rethinking promise and peril in the age of networked intelligence* (2nd ed.). McGraw-Hill Education.

Vorobiev, E. G., Petrenko, S. A., Kovaleva, I. V., & Abrosimov, I. K. (2017). Organization of the entrusted calculations in crucial objects of informatization under uncertainty. *Proceedings of the 2017 XX IEEE International Conference on Soft Computing and Measurements.* 10.1109/SCM.2017.7970566

Wang, B., & Xing, H. Y. (2011). The application of cloud computing in education informatization. *Proceedings of the 2011 International Conference on Computer Science and Service System.* 10.1109/CSSS.2011.5973921

Yu, X., & Xue, Y. (2016). Smart grids: A cyber-physical systems perspective. *Proceedings of the IEEE, 104*(5), 1058–1070. doi:10.1109/JPROC.2015.2503119

Zhang, J. (2018). Research on evaluation model of the effective EFL teaching in the era of informatization. *Journal of Language Teaching and Research, 9*(4), 738-745.

Zhivko, M. (2018). New economy: synergy of informatization and global civil society. *Journal of European Economy, 17*(1), 34-55.

KEY TERMS AND DEFINITIONS

DRMS: Digital rights management system.
GDPR: General data protection regulation.
I/O: Input/output.
ICT: Information and communication technologies.
IoT: Internet of things.
IS: Information society.
IT: Information technologies.
PN: Petri net.
WWW: World wide web.

Chapter 14
Graph Theory:
Novel Multiple-Attribute Decision-Making Effect

Seethalakshmi R.
SASTRA University (Deemed), India

ABSTRACT

Mathematics acts an important and essential need in different fields. One of the significant roles in mathematics is played by graph theory that is used in structural models and innovative methods, models in various disciplines for better strategic decisions. In mathematics, graph theory is the study through graphs by which the structural relationship studied with a pair wise relationship between different objects. The different types of network theory or models or model of the network are called graphs. These graphs do not form a part of analytical geometry, but they are called graph theory, which is points connected by lines. The various concepts of graph theory have varied applications in diverse fields. The chapter will deal with graph theory and its application in various financial market decisions. The topological properties of the network of stocks will provide a deeper understanding and a good conclusion to the market structure and connectivity. The chapter is very useful for academicians, market researchers, financial analysts, and economists.

INTRODUCTION

Graph Theory – Relevant to Mathematics

Mathematics acts a critical and essential need in different fields. One of the significant roles in Mathematics is played by Graph theory of structural models. The structural arrangement paves the way for innovative methods models in various disciplines for better strategic decisions

In mathematics, graph theory is the study through graphs by which the structural relationship studied with a pairwise relationship between different objects. A graph consists of lines or arcs connected by nodal points or vertices which are also connected edges. The different types of network theory or models

DOI: 10.4018/978-1-5225-9380-5.ch014

or model of the network are called as graphs. These graphs do not form a part of analytical geometry, but they called as graph theory which is ' points connected by lines.'

Elements of Graph Theory

A pair that has set of nodes V(G) and a set of edges E(G) is called a graph, represented by G = (V, E). The adjacency relation of a graph G is a symmetric binary relation induced by the edge set E(G) on a vertex set V(G). Dual nodes i and j are said to be adjacent when there exists an edge between i and j.

Degree

The number of edges incident on a node 'i' is representing the amount of that node and is named by 'd_i' There are two types of graph namely 'Direct graph' and 'Indirect graph.' The 'in degree, di-' and 'out degree, di+,' of node i are to be distinguished in the direct graph.

Directed and Undirected Graph

The nodes in an undirected graph are connected by bidirectional edges. On the other hand, the nodes in a direct graph are connected by edges that point in one direction.

Regular Graph

A regular graph consists of nodes all of which are of the same degree. For example, all nodes in of 'k-regular' graph are of degree 'k.'

Adjacency Matrix

The *adjacency matrix*, $\mathbf{A}(G)$ in the graph G, is a square matrix of order n in which the entry a_{ij} is 1 if the edge $e_{ij} \in E(G)$, otherwise a_j is 0. The matrix $\mathbf{A}$ is symmetric, i.e. $a_{ij} = a_{ji} \forall i, j \in V(G)$ for an undirected graph.

Path

A sequence of edges and nodes which are connected in such a way that all nodes are distinct from each other. For example, P_k denotes a path with 'k' nodes. The number of edges in a path is said to be the path length and is represented by d(u, v).

Diameter

The utmost or maximum of d(u,v) over all nodes u and v is said to be the diameter of a graph.

Walk

The list $\{v_0, e_{01}, v_1, ..., v_{k-1}, e_{k-1,k}, v_k\}$, of alternating edges and nodes is a walk.

Trail

The no repeated edge with the walk is known as a *trail and with no* repetitive node is a path.

Geodesic Distance

The shortest path distance between the two nodes are called as Geodesic distance.

Circuit

A closed trail is called as a circuit. In a circuit, the endpoints of a trail are the same.

Cycle

If a circuit has no repeated node, then it is called a *cycle. A cycle* on *n* nodes is denoted by C_n. A cycle can be a circuit whereas a cycle need not be a circuit.

Tree

A graph which is connected without any cycles is known as Tree.
A tree refers to a graph connected with no cycles.

Spanning Tree

A subset of graph G is called as a spanning tree. In a spanning tree, all the vertices have the smallest number of edges connected to them. It has no cycles and said to be connected.
Some simple graphs are given in Figure 1.

Sub Graph

A subset of nodes and its connecting edges of a graph G, when drawn as a graph, is called as sub-graph G_1. In other words, the set of vertices and edges of a subgraph G_1 are all subsets of G.

Figure 1.

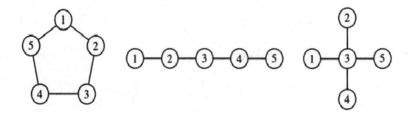

Connectedness

The each pair of nodes which is connected by a path in a graph, then it is said to be connected. A graph that is not connected is disconnected. The maximally connected subgraphs form the components of a graph G.

Weighted Graph

If a nonnegative real number w(v) called weight of v, is assigned to each vertex v in a graph G, then the graph is said to be a weighted graph.

Complete Graph

A graph with each node connected by every other node is called as a total or complete graph. A complete graph of 'n' nodes is denoted by K_n.

Bipartite Graphs

If the vertices of a graph G can be divided into two sets V_1(with n_1 nodes) and V_2(with n_2 nodes) such that every vertex in one set is connected to a vertex in another set, then it is called as a bipartite graph and denoted as $K_{n1,n2}$

Random Graphs

A graph with randomly but similarly distributed edges is called a random graph. They possess no distinct and unique models.

Network

A graph in which the nodes are connected through edges is called as a network, and they adhere to the principles of graph theory. Discrete mathematics and networks find extensive application in modeling road traffic regulation systems, fluid flow in pipes, the flow of currents in an electrical circuit

Centrality

The position of a node in a network determines how important it is. A measure of this is centrality (Borgatti, 2005; Borgatti, and Everett, 2006; Freeman, 1978). The amount of links that is incident to a node is termed as Degree centrality. The number of steps taken to reach a node from any other node in a network is closeness centrality. The amount or number of of paths passing through a node is called as betweenness centrality. So, centrality tells the significance of a node and helps as a basis for choosing the correct measure as the application demands.

Market Graph

Data of stock market when represented in the form of network is called as a market graph. The graph shows the correlation between the opening prices of a pair of stocks over a specific period.

The Visibility Graph

L. Lacasa et al. was the first man to propose the visibility graph algorithm in 2008 in which the number or values of the time series is schemed or plotted using the perpendicular bars. This was done to characterize time series. In other words, the time series is charted or mapped into a network. The vertical bars are linked at the top.

Definition

Two random or arbitrary data values (t_1, y_1) and (t_2, y_2) are visible, and as a result they turn into two linked or associated by means of nodes of the related or linked graph (Bebarta, &Venkatesh, 2016), if any other data (t_3, y_3) located between them fills the prerequisite:

$$y_3 < y_2 + (y_1 - y_2) \frac{t_{2-t_3}}{t_{2-t_1}} .$$

The visibility graph of a time series with 10 data values as mapped from a histogram is shown in Fig2. As can be seen, the bars in the histogram are linked at the top and the same is reflected in the visibility graph also.

Figure 2.

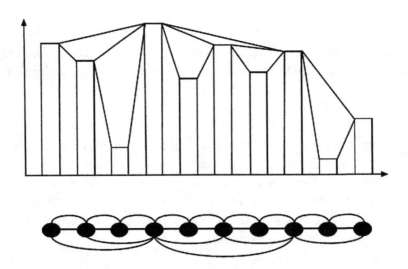

Cliques

A clique of a graph G is a total or absolute sub graph of G with all possible edges. The largest possible clique that can be formed from a graph is the maximum clique. The subsequent definition is simplified and generalized concept of the clique. That is, in its place of cliques one can think about dense sub graphs, or *quasi-cliques*. A γ -*clique* C_γ, also called a *quasi-clique*, V is the subset such that $G(C_\gamma)$ has no less than or atleast $q(q - 1)/2$ edges, where q is the cardinality (i.e., vertices number) of C_γ.

Significance

Generally, in order to understand and to represent a variety of application program in different areas, graph theoretical fundamentals are employed. It includes the analysis of molecules, bonds construction and study of atoms in chemistry. In biology, the habitat of a species forms the vertices and the movement way forms the edges of a ecosystem graph. It seems to be helpful in the study of rearing of gander, spread of diseases and the effect of movement that influences different species, to predict the changes in the population size of a species using predator-prey relationships, carry capacity, and population density. Graphical theory is of great use in Operations Research too. The salesman problem who involved in travelling, the smallest or shortest spanning tree in the graph with weights, acquiring an ideal match of employment for men and finding the shortest way between two vertices in a graph are few examples. Transport networks, activity networks, and game theory are also demonstrated using graphs. Large, complicated projects such as PERT (Project Evaluation Review Technique) and CPM (Critical Path Method) are planned and scheduled successfully using graph theories (Pranav & Chirag,2013). An efficient way of performing a competitive task in engineering, economics, or war science can easily be explored using game theory. A digraph is used to represent the method of a finite game. Here, positions are vertices and the game moves are represented by edges. The internal structure of the stock market is clearly visible and easily understood when the stock market data are represented in graphs.

Chemistry

Graph theory plays a vital role in chemistry, where the structure of the chemical is both Quantitative and Qualitative Structure Property and Structure Activity relations etc., are characterised by topological indices. From a chemical graph, it is possible to analyse the consequences of connectivity and this pertains to chemical graph theory. The model of any chemical system comprising of diffent clusters, reactions, molecules, polymers, crystals etc can be symbolized by a chemical graph (Turro Angew,1986) Atoms(Trinajstic, 1988), molecules, fragments of molecules, atom groups, orbitals, intermediaries, electrons, the sites of a chemical system and rescheduling or rearrangements, steps of elementary reaction interactions of bonds and nonds, Vander Waals force that exist between them are generally shown in chemical graph. The chemical graph of a molecule can be indicated in the form of a matrix. To say in detail, the atoms and their interconnectivity through various bonds can be indicated as an adjacency matrix or a distance matrix. The polynomials of these matrices and their eigen values describe the molecular structure which is of great use in study of quantitative structure property/activity relationships (Mishra, 2003).

Figure 3.

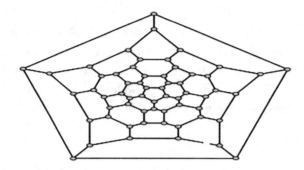

Fullerene

Molecules made of only Carbon or Boron atoms are called as Fullerenes. Their structure cannot be easily observed in nature. Graph theory eases the understanding of the structure of these molecules. Levis Dot structure is a representation of a molecule in planar form. This structure shows the chemical symbol of the atoms of molecules and a line joining them if they are bonded. The Levis dot structure can easily be transformed to a graph. The atoms form the vertices and the bonds form the edges of the graph. Figure 3 shows the graphical representation of Buckminsterfullerene C_{60}. The connection of vertices is such that each vertex is connected to 3 other vertices. The faces are pentagonal or hexagonal which is the unique feature of a fullerene graph. They can also be defined as the planar projection of the Platonic and Archimedean Solids (Johnson, 2010).

Biology

A biological system can also be graphically shown for proteins, genes or additional organic /biological components (interacting species) form the nodes and their interconnections form the edges of a biological graph. They show a better understanding of the cellular functions (Albert and Barabási, 2002; Newman, 2003; Schadt, 2005; et al, 2007; François, 2007).Graph theory enables the easy understanding of the otherwise hard to interpret protein interactions (within the cell) in computational biology. Cell functions (protein interactions) and the cause of diseases can henceforth be studied with ease. The protein-protein interaction (PPI) network has proteins as nodes and the interactions forming the edges as shown in the example figure 4.

Likewise the protein-molecule interactions as that of nucleic acids or metabolites can also be graphically represented with their molecules at nodes and the interactions forming the edges. Graphical representation is an organized way of depicting biological information for sure. It is a major breakthrough in the research on human diseases.

The Relationship Between Various E-commerce Domain and Digital Communication

E-commerce is another domain where graph theory has set foot providing world class solutions for major problems. A recommender system has to face 2 major issues

Figure 4.

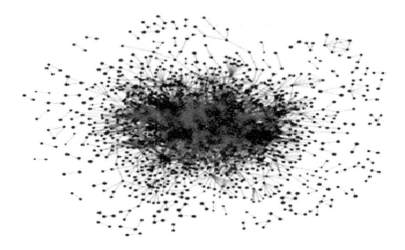

- with only sparse transaction data available, recommendations have to be made and
- bringing together the varying input data to arrive at a high-quality recommendation

Graph theory extends its support to solve this issue. Convert the available data into a graph with consumers and products at nodes and sales transactions as edges (line joining the nodes). This consumer-product graph helps to sort out the best recommendation. This approach of conversion or transformation of attributes of products, similarities of products and consumers, and the Probabilistic Relational Model(PRM)) is formed in the literature of relational learning. Sales transaction data is generated by studying the features and topology of consumer product graphs and relationship mechanism.

As there are two types of nodes - 1. consumers and 2. products interconnected by edges - sales transactions or other types of consumer-product interactions, the consumer-product graph is a bipartite graph. Thus, the triplet - a set of consumer nodes (C), a set of product nodes (P), and a set of links between consumer and product nodes (E) together constitute the consumer-product graph G_{cp}. Figure 5 (Zan Huang, 2005) is an example for consumer-product graph plotted with data in sales transaction table in Figure 6. A consumer-product graph clearly outstands a random graph where recommendations are based on the sales transactions.

Small -World Networks

Networks that are sparse, highly clustered and have small average length are called Small-world networks. 80% of second neighbors of a node X are also neighbors of X. Even then, the normal or average distance (between any two nodes) is as low trait or characteristics path extent of a random graph. Generation of graph for small-world networks has been modeled by Watts-Strogatz in 1998 after wide (Watts and Strogatz, 1998). Yet, this model fails to explain the expertise in finding short paths through such networks. The simulation of 'wildfire' like spread of an infectious disease by Watts and Strogatz is an example for small-world network. It has been stated by (Adamic,1999) that the world wide web is also a small-world network. Hyperlink modeling, crawling, and finding authoritative sources are some of the ways for effective functioning of search engines as stated by Adamic.

Figure 5.

Transaction		
Consumer	**Product**	**Attribute**
c1	p2	...
c1	p4	...
c2	p2	...
c2	p3	...
c2	p4	...
c3	p1	...
c3	p3	...

Figure 6.

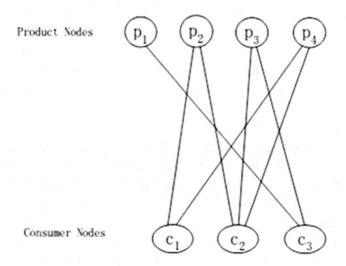

Consumer-product graph

Software Engineering

Software engineering has far wide benefitted by graph theory. Data flow diagrams are drawn with transformations at vertices and data flow as edges to pictorially represent requirement specification. It becomes easy to show relations among modules in the design stage while amid testing, the control flow of a program related with McCabe's multifaceted nature measure which utilizes directed graphs for tending to the sequence of executed instructions and etc. The graph theory and network modeling find equal application in Software Process Management too. Abstract syntax tree (Srinuvasu and Prasad, 2016) forms the basis of creating graphs in a class-oriented model. In such graphs, vertices are the classes and edges are nothing but relations between the classes. The type of relationship represents the weight of the graph. Once this is done, graph mining algorithms help in identifying the subgraphs, frequent subgraphs and isomorphic graphs

Marketing

Marketing also makes use of network theories. Network modeling seems to be a better choice than traditional ones. Yet, there are as such no standards for modeling and network data are difficult to collect. The structural position of their network members and relations are clearly depicted through graphs (Freeman 1984; Hage and Harary, 1983; Moreno, 1953). In a typical marketing graph, nodes represent the actors and edges refer to the relations between actors. The edges vary with type of relation. A line with double headed arrow indicates mutual relation and a line with arrow on one end indicates one side relation.

In Operations Research

Graphs are used to pictorially express and find suitable solutions for Operation Research problems though they have more constraints. For example, the transport of commodity from one destination to the other is represented by transport network from which ways of either increasing the flow or reducing the cost within the suggested flow can be found out.

To make it more clear, let us consider a set of n projects to be performed. Not all projects are independent; some can be done only after the completion of the other and so on. A graph showing this type of network is a *directed acyclic graph which has no directed cycles*. As an example, let us consider a directed edge $i \rightarrow j$ *in a graph such that* project i depends on project j. let p_i *be the* profit of each project assigned at the end of the project. However, positive costs are given to some projects with negative profit. Let X be a subset of all interdependent projects. We should find a subset of largest profit earning projects. But, in case of all projects resulting in positive costs, nothing can be done. The projects are partitioned into two subsets S(selected projects) and T *(rejected projects)*. Our problem is reduced to a minimum cut problem, and the best way to maximize profit can be easily found out. Prim's Algorithm and Kruskal's algorithm are used to find minimum distance spanning tree.

Prim's Algorithm

1. Starting from a node,
2. Finding the minimum path connected to the node,
3. Finding the minimum path connected to the network created,
4. Continuing the algorithm to visit all nodes.

Kruskal's Algorithm

1. Ordering edges based on their weight,
2. Starting from the edge with minimum weight and adding two nodes of it to the tree,
3. Choosing the next edge and repeating the previous step,
4. Continuing the algorithm to visit all the nodes.

Financial Flow Networks

Assets

Resources or properties owned by an individual or a business unit which is expected to be of economic advantage in future are termed as assets. The sum of money that is owed or duedis referred to as debt.

The management of funds between providers and users based on the supply and demand is a challenging task for the financial systems. Generally, the government, banks, manufacturing companies and households' mortgages are providers of fund which is used by borrowers like students, home buyers and businesses. This system can be represented by flow networks. Funds providers form the source nodes, borrowers the sink nodes, the financial intermediaries form the nodes that are neither source nor sink nodes and the liquidity and maturity transformations between them form the edges of the graph.

The financial system which comprises by a set of the operators $\Omega = \{\omega_i \,|\, i = 1, ..., n\}$, such as insurance companies, commercial bank and other mediators that are indirectly or directly linked towards on another by some or other financial obligation. Let d_{ij} be the value of the accountability concerned by agent i and by the j. Then statements, denoting the statements of assets and liabilities or the of the associate value includes Ω is $a_i + c_i = e_i + d_i + h_i$ where:

(i) $a_i \in R^+$ is the external value of assets held by agent i
(ii) $c_i = \sum_j d_{ji}$ is the sum of the *internal assets* held by agent i
(iii) $d_i = \sum_j d_{ij}$ is the sum of the value of the internal debt of agent i
(iv) h_i is the exterior debt of agent i
(v) e_i value of the equity of agent i

From this, we can say that the assets are equal to the debt plus the equity of an agent i, or from the budget identity, equity is equal to your total amount of debt ducted from assets of an agent i. A financial flow network is a multi-source stream network N= {L, A, T, H,Ω) such that N is associated, directed and associated graph with two sources and some sink nodes, where:

1. $\Omega = \{\omega_i\}$ is a set of n nodes that stand for the financial mediators.
2. $A = \{a^k\}$ is the sink node that represents the exterior possessions detained by affiliates in Ω.
3. H is a source node that represents debts of the households claims (bonds, deposits etc)aligned with the agents or against in Ω.
4. T is another source node that represents the equity of the shareholders (Ω).
5. L is a the set of directed edges from the source nodes to the mediator nodes, and from the intermediary nodes to the sink nodes.
6.   The capacity function is denoted as T:L$\rightarrow$R + which is a map acquaintances each edge in the network with its appropriate value.

Exclusive Application in Financial Market

Clustering

The extent to which nodes in a graph cluster together is an important aspect of socio-economic networks. A node, say 'i' is linked to another node 'j' which in turn is linked to node 'k' and node 'k' is linked to node 'i' is called as transitivity and the frequency of transitivity is clustering. For any given node, the average of the fraction of all pairs of nodes that are both linked to that node and those that are linked to each other is a measure of the clustering of that particular node. Average of this measure for all the nodes indicates the magnitude of transitivity of that network.

Def

Any institution like commercial bank or investment bank, mutual funds, pension funds that facilitates channelling of funds between two parties is called a 'financial intermediary'.

Risk management in financial market involves detecting the factors affecting financial markets, analyse them and arrive at a decision that does not lead to high degree risks. This is the core job of the investors, managers and policy-makers. Financial analysts seek the help of social sciences and mathematics professionals for a better analysis of financial market. Graph theory and network modeling offer a helping hand to the financial people too in market analysis.

The shares of a company and derivatives are traded in a stock market where buyers and sellers exchange their holdings. It is also the best place where a company can raise its capital. The stock market is woven around a lot of parameters such as prices, volume of trade, valuation etc. Nevertheless, the percentage change is a major player. There are many internal and external factors that influence the value of a particular stock in the market. In fact, the value of a stock goes up or down depending on the value of any other stock resulting in stock market crashes sometimes. The stock market hierarchy can be best understood from the plot of minimum spanning trees with stock price returns (Rehan, 2013).

The index returns of worldwide stock exchanges can be explored to arrive at a correlation based network (Bonanno et al, 2000). The analysis of stock exchange indices of the world wide stock market is complicated when compared to single stock market as the opening and closing time of the stock market is not uniform throughout and each country uses a different currency.

(Kyungsik et al., 2007) identified the degree of distribution of nodes, edge density and size of communities when examining all joint-stock companies in Karachi Stock Exchange in 2003.Suppose that $r_i(t)$ is the stock price return, which is defined as follows:

$r_i(t) = \ln [p_i(t+\Delta t) / p_i(t)]$

where $p_i(t)$ is the stock price at time t. The association coefficient between the stocks i and j is defined as

$$C_{ij} = \frac{\overline{r_i r_j} - \overline{r_i}\,\overline{r_j}}{\sqrt{\left(\overline{r_i^2} - \overline{r_i}^2\right)\left(\overline{r_j^2} - \overline{r_j}^2\right)}}$$

It is clearly evident from his findings that power law forms the basis of the stock network. All the edges with weight less than the threshold level θ were not included. In numerical terms, all edges (that connect the stocks) with $Cij \geq \theta$ are only plotted in the market graph. In other words, the values of such stocks change in a similar pattern in accordance with θ. Thus the internal structure of the stock market can very clearly be understood by studying the market graph. A similar study of Chinese stock by Wei Jiang et al. (2009) explored the possibility of clustering of similar stock communities. The stock market hierarchy so arrived reveals the truth that stocks within the same sector were similar. Also, the master or key stock that determines the movement of other stocks in a sector can be easily circled out from this correlation network. High degree stocks (those that are connected to many other stocks) determine the market movement, capital stocks in particular. The smaller the correlation network, sharper will be the further market movement (up or down) (Mandere, 2009).

In a market graph consisting of various cliques, there is a high correlation between all the stocks in a clique. Stocks that behave similarly are always grouped together. The clique with maximum number of mutually correlated stocks is the maximum clique which is found using the standard "best-in" greedy algorithm in a standard integer programming formulation (Bomzeet al, 1999).

Edge Density

The numbers of edges in a graph divided by the greatest possible number of edges, is called as the edge density. It changes with the change in the correlation threshold level. Market graphs for various degrees of correlation can be plotted. The size of the cliques and independent sets vary accordingly. The change in edge density over successive time periods was studied (Boginski et al, 2006). For a correlation threshold of 0.5, the edges that corresponded to similar stocks were grouped. It was found that the edge density in the last period of time was 8.5 times that of the first. This is indicative of the increase in globalization of the stock market.

The Stock Network Method Based on the Visibility Graph

It is important to extract stock network from the stock market. Many researchers describe the stock market as time series (Amihud, 2002; Bebarta and Venkatesh, 2016; Shumway and Stoffer, 2011) considering the stock income over a defined period or on a day. However, they do not give a picture of the foreign stock companies. The stock of the various organizations of other countries are replicated as modeling of stock
Stock modeling involves the following stepladder:

Step 1: Collection of closing price values of all companies stock using stock software.
Step 2: Representing the above closing value of each stock as a perpendicular bar. The height of the vertical bar is the value of closing price. A taller bar means a higher closing price.
Step 3: There is random arrangement of data for each year.
Step 4: A straight line connects 2 vertical bars directly if there is no bar on the way between them.
Step 5: Plotting this into a graph with vertical bars as nodes and line joining the bars as edges of network

Researchers repeatedly arrange the yearly data of vertical bars randomly. Applying visibility graph method, complex networks can be constructed from the range of data which is an essential part of studying the stock market. Though the network construction methods are available in plenty, the visibility graph method is the opted one as it is efficient in studying the stock market.

Crash

When there is a sudden but significant fall of stock prices, the stock market is said to have 'crashed'. This is dreadful as it affects a nation's economy greatly. If we consider such crashes in US market in the past for example, the worst was the 'Great Wall Street Crash of 1929'. The causes were many. Yet, continuous panic selling declined the entire market. Pictorial representation of the stock market pattern enables quick and easy analysis than the textual representation. Minimum spanning tree in graphs helps in showing changes in the stock market. One can get a concise and clear figure of price values, stock clusters and relationship between stocks. Stocks in the same business sectors formed clusters independent of the period when they appeared. The cause for market crash is varied; may be due to various reasons like change of CEO, terrorist attacks, or the enhanced growth or change of technology and its popping. Whatever it is, the overall market stock is affected to a greater extent. Market crash results in the variations of all stocks irrespective of whether they are highly interrelated or have no apparent or little associationship. As predicted by (Yinghua Zhang, 2009), during market crash, the total minimum distance of a Minimum Spanning Tree is small but it increases when the stock market is not affected by the effects of crash.

Volatility

Stock markets are volatile. Volatility is a measure of risk and uncertainty. Selling on peak values and buying at low times produces more money and great returns, the volatility is said to be high. The compounded returns are identified by the the standard deviations of various financial instruments for a particular time which determine the volatality. A 50% volatile stock is high risk as it can decrease or increase up to half its value. Yinghua Zhang chose three time windows with the return dynamics to study the daily volatility pattern. The behavior of the real stock market can correctly be predicted by the network technique. When volatility is higher, price fluctuation would be wider.

Practical Applications of Net Work Models in Finance:

A collection of edge disjoint paths in a financial flow network is equivalent to a number of different venues from a designated entity to another, such that no other entities are used more than once.

- The capacity of nodes in a financial flow network shows the handling capacity of a financial body.
- The most substantial possible ways of allocating money excluding money sharing with other financial entities is represented by the maximum matching in a bipartite graph.
- Couples or individuals getting a loan from banks, maximizing the loan amount by banks are some of the assignment problems represented and solved in a financial flow network

CONCLUSION

To get meaningful information from correlation coefficient matrix and study market dynamics, stock market topology study is a powerful tool. Minimum spanning tree concept is helpful in monitoring the dynamics of stock markets in an easier way. It also helps in understanding connectivity between markets of countries and reveals certain anomalies in the market

We studied the basics of graph theory and their applications. Network modeling is useful for stock market analysis. Network topology explains the market structure and connectivity clearly to the academicians, market researcher, financial analyst and economist benefit by studying this chapter.

REFERENCES

Adamic, L. (1999). *The Small World Web*. Retrieved from http: //www. parc. xerox. com/istl / groups / i a /www / small world paper. html

Albert, R., & Barabási, A. L. (2002). The statistical mechanics of complex networks. *Reviews of Modern Physics*, *74*(1), 47–97. doi:10.1103/RevModPhys.74.47

Amihud, Y. (2002). Illiquidity and stock returns: Cross-section and time-series effects *. *Journal of Financial Markets*, *5*(1), 31–56. doi:10.1016/S1386-4181(01)00024-6

Bebarta & Venkatesh. (2016). A *Low Complexity FLANN Architecture for Forecasting Stock Time Series Data Training with MetaHeuristic Firefly Algorithm*. Springer.

Boginski, V., Sergiy, B., & Pardalos, P. M. (2006). Mining market data: A network approach. *Computers & Operations Research*, *33*(11), 3171–3184. doi:10.1016/j.cor.2005.01.027

Bomze, I. M., Budinich, M., Pardalos, P. M., & Pelillo, M. (1999). The maximum clique problem. In D.-Z. Du & P. M. Pardalos (Eds.), *Handbook of combinatorial optimization* (pp. 1–74). Dordrecht: Kluwer Academic Publishers. doi:10.1007/978-1-4757-3023-4_1

Bonanno, Vandewalle, & Mantegna. (2000). Taxonomy of stock market indices. *Physical Review E*, *62*, R7615.

Borgatti, S. (2005). Centrality and network flow. *Social Networks*, *27*(1), 55–71. doi:10.1016/j.socnet.2004.11.008

Borgatti, S., & Everett, R. (2006). A graph-theoretic perspective on centrality. *Social Networks*, *28*(4), 466–484. doi:10.1016/j.socnet.2005.11.005

Chuang, H. Y., Lee, E., Liu, Y. T., Lee, D., & Ideker, T. (2007). Network-based classification of breast cancer metastasis. *Molecular Systems Biology*, *3*, 140–150. doi:10.1038/msb4100180 PMID:17940530

François, K. (2007). *Biological Network. Complex Systems and Interdisciplinary Science – Volume 3*. World Scientific Publishing Company.

Freeman, L. C. (1978). Centrality in social networks conceptual clarification. *Social Networks*, *1*(3), 215–239. doi:10.1016/0378-8733(78)90021-7

Freeman, L. C. (1984). Turning a profit from mathematics: The case of social networks. *The Journal of Mathematical Sociology, 10*(3-4), 343–360. doi:10.1080/0022250X.1984.9989975

Hage, P., & Harary, F. (1983). *Structural models in anthropology*. Cambridge, UK: Cambridge University Press.

Huang. (2005). *Graph-based Analysis for E-commerce Recommendation* (PhD dissertation). The University of Arizona.

Johnson. (2010). *Molecular Graph Theory, A Masters Project*. Worcester Polytechnic Institute.

Kyungsik, K., Soo Yong, K., & Deock-Ho, H. (2007). Characteristic of Networks in Financial Markets. *Computer Physics Communications, 177*(1-2), 184–185. doi:10.1016/j.cpc.2007.02.037

Lacasa, L., Luque, B., Ballesteros, F., Luque, J., & Nuño, J. C. (2008). From time series to Complex networks: The visibility graph. *Proceedings of the National Academy of Sciences of the United States of America, 105*(13), 4972–4975. doi:10.1073/pnas.0709247105 PMID:18362361

Mandere, E. (2009). *Financial networks and their applications to the stock market* (Thesis). Graduate College of Bowling Green State University.

Mishra, B. K. (n.d.). *Molecular (graph) characteristics of some hydrocarbons through graph theory*. Retrieved from http://citeseerx.ist.psu.edu/viewdoc/download?doi=10.1.1.604.8277&rep=rep1&type=pdf

Moreno, J. L. (1953). Who Shall Survive? New York: Beacon House Inc. Beacon, N,Y. Library of Congress Catalog Card No. 53-7284

Newman, M. (2003). The structure and function of complex networks. *SIAM Review, 45*(2), 167–256. doi:10.1137/S003614450342480

Pranav, P., & Chirag, P. (2013). Various Graphs and Their Applications in Real World. International Journal of Engineering Research and Technology, 2(12).

Rehan, Haran, Chauhan, & Grover. (2013). Visualizing the Indian stock market: A complexnetworks approach. *International Journal of Advances in Engineering and Technology, 6*(3), 1348–1354.

Schadt, E. E., Lamb, J., Yang, X., Zhu, J., Edwards, S., Guhathakurta, D., ... Lusis, A. J. (2005). An integrative genomics approach to infer causal associations between gene expression and disease. *Nature Genetics, 37*(7), 710–717. doi:10.1038/ng1589 PMID:15965475

Shumway & Stoffer. (n.d.). *Time Series Analysis and Its Applications*. Springer-Verlag.

Srinuvasu, M. A., & Prasad, D. D. (2016). Class-Oriented Model Graph Design Based on Abstract Syntax Tree. International Journal of Computer Science and Communication, 7(2).

Trinajstic, N. (1988). *MATH/CHEM/COMP 1987* (R. C. Lacher, Ed.; Vol. 83). Amsterdam: Elsevier.

Turro & Angew. (1986). Geometric and Topological thinking in Organic Chemistry. *Chem. Int. Ed. Engl, 25*, 882.

Watts, D., & Strogatz, S. (1998). Collective Dynamics of 'Small-World' Networks. *Nature, 393*(6), 440–442. doi:10.1038/30918 PMID:9623998

Wei-Qiang, H., Xin-Tian, Z., & Shuang, Y. (2009). A Network Analysis of the Chinese Stock Market. *Physica, 388*(14), 2956–2964. doi:10.1016/j.physa.2009.03.028

Zhang. (2009). *Stock market network topology analysis based on a minimum spanning Tree approach* (Thesis). Graduate College of Bowling Green State University.

Chapter 15
Influential Nodes in Social Networks:
Centrality Measures

Kousik Das
Vidyasagar University, India

Rupkumar Mahapatra
Vidyasagar University, India

Sovan Samanta
https://orcid.org/0000-0003-3200-8990
Tamralipta Mahavidyalaya, India

Anita Pal
National Institute of Technology Durgapur, India

ABSTRACT

Social network is the perfect place for connecting people. The social network is a social structure formed by a set of nodes (persons, organizations, etc.) and a set of links (connection between nodes). People feel very comfortable to share news and information through a social network. This chapter measures the influential persons in different types of online and offline social networks.

INTRODUCTION

Humans are social person and best decision maker. They cannot stay alone, they want to make friends, share news, information etc. Social network (Fig.1) is the perfect place for connecting people. Social network is a social structure formed by a set of nodes (persons, organizations etc.) and a set of links (connection between nodes). People feel very comfort to share news, information through social network. Recently online social networks like Twitter, Facebook and LinkedIn etc. have grown extremely popular in human life. A statistic (Tab.1) on social network in 2018 is given below.

DOI: 10.4018/978-1-5225-9380-5.ch015

Table 1. A statistic of population and network users.

Sl. No.	Particulars	Numbers (Billion)
01	World population	7.7
02	Mobile users	5.135
03	Internet users	4.021
04	Social networks users	3.196
05	Facebook users	2.27\

Figure 1. An example of social network.

There are some advantages of social networks in our daily life. In the field of education, it has big applications. Students, teachers, researchers and scientists have been spending lots of time in the social networks to talk about education works. Schools and colleges also use social network to keep in touch with students and teachers. Many companies also use social network to reach at customers. So it is the place for marketing products. But there are also some disadvantages of social network. Cybercrime is a big problem for people. Sexual exploitations and addictions to social network is also disadvantages for human. A fundamental problem in social network is to find influential node or central node (for example, head of cybercrime in crime network) within the network. Centrality indicates most important node or central node or influential node within a network. Hence measure of centrality is a very essential task in social network.

Many centrality measures (Das, Samanta and Pal, 2018) have been introduced over the years. The very simple centrality measure is degree centrality (Shaw, 1954). It is measured by the number of direct connections. This measure considers only local importance. Closeness centrality (Sabidussi, 1996) of a node is defined by the inverse sum of geodesic distances from that node to all others nodes. Betweenness centrality (Shaw, 1954; Freeman, 1977; Freeman et al., 1991) is defined by the number of the shortest paths pass through a node. This measure gives also global information in the network. Influential node is closely related to high degree, high betweenness and high closeness. Beside these standard centrality measures, there are many centrality measures i.e. eccentricity centrality (Hage and Harary, 1995), information centrality (Stephenson and Zellen, 1989), stress centrality (Shimbel, 1953), eigenvector centrality (Bonacich, 1972; Bonacich, 2001), katz centrality (Katz, 1953), subgraph centrality (Estrada and Rodriguez, 2005), functional centrality (Rodriguez, Estrada and Gutierrez, 2006), leverage centrality (Joyce, 2010), k-shell centrality (Das, et al, 2018; Kitsak et al., 2010), neighbourhood coreness centrality (Bae and Kim, 2014), neighbourhood centrality (Lin, 2015).

Thus influential node in social network is a very important topic. In this chapter, our objective is to discuss some useful centrality measures and some of their applications (Wassermann and Faust, 1994) in real life. This chapter will inspire all to creative and joyful.

LITERATURE REVIEW

Due to importance of influential node in real life, many centrality measures were introduced and developed day to day. The first measure of centrality was introduced by A. Bavelas (1950) for connected graph and applied in communication network (Bavelas, 1948; Bavelas, 1950). Based on shortest path, stress centrality was proposed by A. Shibel (1953) to measure amount of communication (Shimbel, 1954). To measure the relative degree of influence of a node, Katz (1953) centrality were introduced by Katz(1953). Beauchamp(1965) proposed an improvement of centrality index (Beauchamp, 1965) containing points and also graphs. Hence Sabbidussi(1966) claimed about improvement of centrality index (Sabidussi, 1966). Nieminen (1974) modified some axioms of centrality index (Nieminen, 1974).

Freeman (1978) developed three type measures of centrality (Freeman, 1978) i.e absolute type, comparative type and centralization of whole network and examine on small network. Based on information transmission among vertices, information centrality (Stephenson and Zelen, 1989) introduced by Stephenson and Zelen (1989). Borgatti and White (1994) generalized geodesic centrality measures (White and Borgatti, 1994) and Everett and Borgatti (1999) extended the standard centrality measures (Evereett and Borgatti, 1999) to apply for groups and classes and also as individuals.

A faster algorithm for measure of betweenness centrality (Brandes, 2001) proposed by Brandes(2001) which shorten time and space. Estrada and Rodriguez (2005) introduced subgraph centrality14] to measure the number of times of a vertex involve in the different connected subgraph of the network. Rodríguez et al. (2006) generalized the subgraph centrality as functional centrality (Rodriguez et al., 2006). Bonacich (2007) discussed some unique properties for eigenvector centrality (Bonacich, 2007). The measures of centrality on weighted network (Opsahl, Agneessens, and Skvoretz, 2010) improved by Opsahl et al. (2010). Joyce et al. (2010) introduced leverage centrality measure (Joyce, 2010) and applied in human brain.

k-shell decomposition (Das, et al, 2018; Kitsak et al., 2010) first introduced by Kitsak (2010) and proposed that the influential nodes are staying within core of network. Zeng et al. (2013), Liu (2014) modified and improved methods to find more acceptable in the ranking list (Lin, Ren, and Guo, 2014; Zeng and Zhang, 2013). Bae and Kim (2014) introduced neighborhood coreness centrality (Bae and Kim, 2014) that ranks all the nodes of a network. Liu et al. (2015) proposed nth step neighborhood centrality (Lin, 2015) to find influential node. Hence Wang et al. (2017) proposed weighted neighborhood centrality (Wang et al., 2017) to find more accurate ranking list of influential nodes. The authors with their contributions (Tab.2) are listed below.

List of notations used in this chapter are given in table(Tab.3)

Table 2. A gist of contribution

Authors	Contributions
Bavelas (1948,1950)	First defined centrality measure to apply in communication network (Bavelas, 1948; Bavelas, 1950).
Beauchamp (1965)	Improved index of centrality in graph (Beauchamp, 1965).
Linton C. Freeman (1979)	Developed three mathematical model on centrality (degree, closeness, betweenness) (Freeman, 1978)
Philip Bonacich (1972)	Suggested that the eigenvector could make a good network centrality measure (Bonacich, 1972).
Freeman et al. (1991)	Introduced a new maesure of centrality based on the concept of network flows. It was conceptually similar to Freeman's measure but different from the original (Freeman, Borgatti, and White, 1991).
White and Borgatti (1994)	Generalized Freeman's geodesic centrality measures for betweenness on undirected graph to the more directed graphs (White and Borgatti, 1994)
Everett and Borgatti (1999)	Extended the standard network centrality measures of degree, closeness and betweenness to apply to groups and classes as well as individuals (Evereett and Borgatti, 1999).
U. Brandes (2001)	Introduced a faster algorithm for betweenness centrality which reduced both time and space of comparative analyses (Brandes, 2001).
Shaw (1954)	Introduced degree centrality as an index of important vertex (Shaw, 1954).
Bonacich (2007)	Discussed some unique properties of eigen vector centrality (Bonacich, 2007).
T. Opsal (2010)	Improved the measure of centrality in weighted network (Opsahl et al., 2010).
Kitsak (2010)	Found that the most efficient spreaders are located within the core of a network by k-shell decomposition (Kitsak et al., 2010).
Zeng (2013) & Liu (2015)	Improved the k-shell decomposition method and improved ranking method respectively (Lin, 2015; Zeng and Zhang, 2013).

GRAPH AND NETWORK

Every network can be represented by a graph $G(V, E)$ where V is the set of vertices and E is the set of edges.

Let $V = \{1, 2, 3, \ldots\ldots, n\}$ be a finite set of nodes. A relationship $g_{ij} \in \{0, 1\}$ between nodes i and j is given by following

$$g_{ij} = \begin{cases} 1, & \text{if there is a connection between } i \text{ and } j \\ 0, & \text{otherwise} \end{cases}$$

A *network* G is defined by a set containing nodes V and also links among these nodes.

A *social network* is a network where nodes are persons or groups or organizations and there are links or connections between nodes.

In Fig. 2, there are eight nodes and eleven links form a social network.

A *directed network* is a network where links between nodes having direction associated to each link.

A *weighted network* is a network where edges among the nodes have weights.

The *neighbor* of a node i in a network G is the collection of direct contacts of the node and is defined by

$$N_i(G) = \{j \in N : g_{ij} = 1\}$$

and the number of nodes adjacent to node i in G i.e. $d_i(G) = |N_i(G)|$, is called degree $d_i(G)$ of the node i.

A *walk* is formed by succession of nodes where two nodes are connected by a link and link may occur more than once. A trial is a walk in which all links are distinct. A path is a trial in which all nodes are distinct.

The *geodesic distance* $d(i,j)$ of the node i to the node j is number of links in a shortest path from node i to node j and is denoted by $d(i,j)$.

A *connected network* is a network having path among every two nodes within a network.

CENTRALITY MEASURES

Centrality measures can be remarked as a mathematical tool applied in the social network analysis to identify important element in the network. Mathematical definition of some centrality measures are described following.

Degree Centrality

Degree centrality (Shaw, 1954) of a node is a number that indicates the number of nodes directly linked to this node. Mathematically, degree centrality $C_D(x)$ is defined by $C_D(x) = d_x$ where d_x is the degree of a node x.

For normalization, degree centrality

$$C_D'(x) = \frac{d_x}{n-1}$$

where n is the size of the network.

Figure 2. Structure of a network.

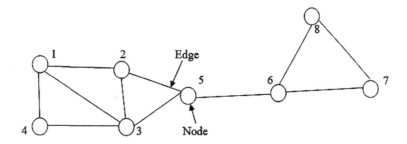

Table 3. Notations.

Notations	Meaning
V	Set of nodes
E	Set of edges
n	Cardinality of V
m	Cardinality of E
d_i	Degree of a node i
$d(i,j)$	Geodesic distance from node i to node j
σ_{st}	Number of shortest paths from node s to node t
$\sigma_{st}(v)$	Number of shortest paths from node s to node t through node v
C_D	Degree centrality
C_C	Closeness centrality
C_{EC}	Eccentricity centrality
C_I	Information centrality
C_S	Stress centrality
C_B	Betweenness centrality
C_K	Katz centrality
C_{SG}	Subgraph centrality
C_F	Functional centrality
C_L	Leverage centrality
C_{NC}	Neighborhood coreness centrality
C_N	Neighborhood centrality

Closeness Centrality

The centrality $C_C(x)$ according to closeness (Sabidussi, 1966) of a vertex x is given by

$$C_C(x) = \frac{1}{\sum_{y \in N} d(x,y)}, \; N = \text{set of vertices in the network.}$$

$d(x,y)$ is geodesic distance between the vertices x and y.

For normalization, closeness centrality

$$C_C'(x) = \frac{n-1}{\sum_{y \in N} d(x,y)}$$

where n is the size of the network.

Eccentricity Centrality

The eccentricity is the inverse of maximum geodesic distance between a vertex and any vertex of the network. Mathematically, the eccentricity (Hage and Harary, 1995) $C_{EC}(x)$ for a vertex x is given by

$$C_{EC}(x) = \frac{1}{\max_{y \in N} d(x,y)}$$

Information Centrality

It is related to closeness centrality and based on information which is containing in all possible paths among pairs of nodes.

Let A be the adjacency matrix of a network. Then information centrality (Stephenson and Zelen, 1989) $C_I(k)$ of a node k is given by

$$C_I(k) = \frac{1}{b_{kk} + (T - 2R)/n},$$

Where *for any fixed,*

D = diagonal matrix with degree values,
U = matrix with unitary elements

Stress Centrality

It is the first centrality index based on shortest path. Mathematically, stress centrality (Shimbel, 1954) $C_s(v)$ of a node v is given by

$$C_S(v) = \sum_{s \notin N} \sum_{t \notin N} \sigma_{st}(v)$$

where $\sigma_{st}(v)$ denotes the number of shortest paths between s and t containing v.

The same definition can be applied for edges given by

$$C_S(e) = \sum_{s \in N} \sum_{t \in N} \sigma_{st}(e)$$

where $\sigma_{st}(e)$ denotes the number of shortest paths between s and t containing e.

Betweenness Centrality

The distance from one vertex to other vertices is not only important, more important is that which vertices staying on which shortest paths among pairs of others vertices in communication networks. Betweenness centrality (Shaw, 1954; Freeman, 1978; Freeman, 1977) $C_B(x)$ of a vertex x in the network is given by

$$C_B(x) = \sum_{y \neq z \in N} \frac{\sigma_{st}(x)}{\sigma_{st}}$$

where $\sigma_{st}(x)$ denotes the number of shortest paths between s and t containing x, σ_{st} denotes the number of all shortest paths between s and t in the network.

Eigenvector Centrality

Let A be the adjacency matrix such that $a_{ij} = 1$ if node i is connected to node j and $a_{ij} = 0$ if not. Then eigenvector centrality (Bonacich, 1972; Bonacich, 2001) for node i is given by

$$Ax = \lambda x, \; \lambda x_i = \sum_{j=1}^{n} a_{ij} x_j \; i = 1, 2 \ldots n,$$

λ is a constant.

Katz Centrality

It measures influence by calculating the total number of walks between a pair of nodes within a network.

Let A be the adjacency matrix of a network. Elements (a_{ij}) of A are given by

$$a_{ij} = \begin{cases} 1, & \text{if there is a link between } i \text{ and } j \\ 0, & \text{otherwise} \end{cases}$$

Mathematically, Katz (1953) centrality $C_K(i)$ of a node i is given by

$$C_K(i) = \sum_{k=1}^{\infty} \sum_{j=1}^{n} \alpha^k \left(A^k \right)_{ji}$$

Where α is smaller than the reciprocal of the absolute value of the largest eigenvalue of A. The powers of A mean the presence of links among two nodes through neighbors.

Subgraph Centrality

The number of closed walks of length k starting and ending on the node i is given by the local spectral moments $\mu_k(i) = \left(A^k \right)_{ii}$, ith diagonal element of kth power of the adjacency matrix A of a network.

Mathematically, the subgraph centrality (Estrada and Rodriguez, 2005) $C_{SG}(i)$ of a node i is given by

$$C_{SG}(i) = \sum_{k=0}^{\infty} \frac{\mu_k(i)}{k!}$$

Functional Centrality

Mathematically, the functional centrality (Rodrigeuz et al., 2006) $C_F(i)$ of a node i is given by

$$C_F(i) = \sum_{j=0}^{\infty} a_j \mu_j(i)$$

The number of closed walks of length l is weighted by a_l. $\mu_k(i)$ are the local spectral moments obtained by the number of closed walks of length k starting and ending on the node i.

Leverage Centrality

Mathematically, leverage centrality (Joyce, 2010) $C_L(i)$ of a node i is given by

$$C_L(i) = \frac{1}{k_i} \sum_{N_i} \frac{k_i - k_j}{k_i + k_j}$$

This measure combines the degree k_i of a node and the degree k_j of each of its neighbour's averaged over all neighbours N_i.

K-Shell Centrality

K-shell centrality (Das, et al, 2018; Kitsak et al., 2010) of a node is equal to the k-shell value of the node. High k-shell value of nodes indicates the more central node. K-shell value is obtained by k-shell decomposition of the network.

Neighborhood Coreness Centrality

The neighborhood coreness (Bae and Kim, 2014) C_{NC} of a node v is defined as follows.

$$C_{NC}(v) = \sum_{w \in N(v)} ks(w)$$

Where $N(v)$ is the of the neighbors adjacent to node v and $ks(w)$ is the k-shell value of its neighbor node w. The extended neighborhood coreness of node v defined as:

$$C_{NC+}(v) = \sum_{w \in N(v)} C_{NC}(w)$$

$C_{nc}(w)$ is the neighborhood coreness of neighbor w of node v.

Neighborhood Centrality

Neighborhood centrality (Lin, 2015) $C_{NC}(i)$ of a node i is given by

$$C_N(i) = d_i + p \sum_{j \in N(i)} d_j + p^2 \sum_{k \in N(j) \backslash i} d_k + p^3 \sum_{l \in N(k) \backslash j} d_l + \ldots\ldots + p^n \sum_{s \in N(j) \backslash x} d_s$$

Where d_i is degree or k-shell value of node i, n is the step of neighbors and $p \in [0,1]$ is an adjustable parameter.

A gist (Tab.4) of centrality measures with time complexity, limitations and applications are given in the following table.

APPLICATIONS IN REAL LIFE

There are huge numbers of applications of centrality measures in many areas like sociology, psychology, economics, anthropology, biology, terrorist, traffic, neural, business etc.

Table 4. A comparison between centrality measures.

Centrality Measures	Time Complexity	Limitations	Applications Area
Degree centrality	O(m)	This measure counts only direct connections but does not take indirect connections.	Finding a person who influences directly others.
Closeness centrality	O(n³)	This measure counts the distance from one node to others but it does not depend on the path among nodes.	Determining a location that can spread information fast.
Eccentricity centrality	O(mn)	This measure take only maximum distance between nodes.	Determining a location that can spread information fast.
Information centrality	O(n³)	This measure does not consider shortest path but consider all paths	Many situations where flow of information along network is consider important.
Stress centrality	O(mn)	This measure considers all shortest path between any two vertices.	Determining shortest path in transportation problem.
Betweenness centrality	O(n³) or O(mn)	This measure counts only the distance between nodes but it does not differentiate among the nodes of the network.	Determining the node who controls the information among other nodes via connecting paths.
Eigenvector centrality	O(n²)	This measure counts only important links.	Determining the location for emergency facility.
K-shell centrality	O(m)	It differentiates among core nodes and non-core nodes but it takes all the nodes as same priority.	Finding super spreader node in a network.

- **Biological Networks:** In this network, nodes are gene, protein, metabolites and edges are interaction between them. Jeong (2001) used degree centrality to correlate the degree of a protein in the network (Jeong et al., 2001). Wuchty and Stadler (2003) applied closeness centrality to many biological networks (Wuchty and Stadler, 2003). Koschutzki and Schreiber (2005) applied five centrality measures (Koschutzki, 2005) (degree, closeness, eccentricity, betweenness, eigenvector) to PPI network and gene regulatory network (Koschutzki et al., 2008; Koschutzki et al., 2004).
- **Co-Authorship Network:** In this network, nodes are researcher or scientist and edges are co-author between them. Liu et al. (2005) applied centrality measures to co-authorship network (Liu et al., 2005) and found that betweenness centrality gives best result among them. Also Betweenness centrality was used by Chen (2006) in scientific literature (Chen, 2006) and applied to weighted co-authorship network (Liu et al., 2005; Liu et al., 2007) by Liu (2007).
- **Criminal Network:** In this network, nodes are criminals and links are connection between them. Cybercrimes are using internet to do financial theft. In criminology, centrality measures detect the leaders of the criminal network. Degree, closeness, betweenness was applied by Sparrow (1991) in criminal intelligence (Sparrow, 1991). Coles (2001) also analyzed crime groups (Cole, 2001) as social network.
- **Road Network:** Roadways and junctions between roadways form a road network. Everyone wants low cost shortest path when travelling. Centrality measures (degree, closeness, betweenness, and eigenvector) are useful to identify the traffic congestion (in the road network (Jayaweera et al. 2017).
- **Transportation Network:** Road, railways, urban street, airline, water path are different category of this network. Centrality measures (degree, closeness, betweenness) are used to analyze urban

transport network Betweenness centrality applied to analyze global structure of worldwide air transportation network (Guimera et al., 2005).

- **Student Network:** In school or college interactions or relationships between students of a class form a student network. Students improve their character, knowledge etc. The social network analysis explains the behavior of relationship among students. The degree centrality, betweenness centrality measures analyze relationship among students (Grunspan et al., 2014). Position of student in communication and interaction networks (Bruun and Brewe, 2013) is related with student's performance (Bruun and Brewe, 2013).

CONCLUSION

Users of social network are rapidly increasing day to day. Teenagers, college students, actors, politicians, sportsman, businessman all are engaging in social network to do their respective activities. But who is important in the network? To find the answer, in this chapter, we described briefly some standard centrality measure that can find most influential node in the social network. These measures have different time complexity, limitation and area of application. The area of application is not limited, we describe some few areas.

However centrality measures are very important topic in network analysis. But their proper application is also important and it is also an open issue always. This chapter will motivate the readers to apply centrality measures in some new areas and to introduce new centrality measures to solve human related problems.

REFERENCES

al Liu, L. G. (2007). Weighted network properties of Chinese nature science basic research. *Physica A*, *377*(1), 302–314. doi:10.1016/j.physa.2006.11.011

al Liu, X. (2005). Co-authorship networks in the digital library research community. *Information Processing & Management*, *41*(6), 1462–1480. doi:10.1016/j.ipm.2005.03.012

Bae, J., & Kim, S. (2014). Identifying and ranking influential spreaders in complex networks by neighborhood coreness. *Physica A*, *395*, 549–559. doi:10.1016/j.physa.2013.10.047

Bavelas, A. (1948). A mathematical model for group structures. *Applied Anthropology*, *7*, 16–30.

Bavelas, A. (1950). Communication patterns in task oriented groups. *The Journal of the Acoustical Society of America*, *22*(6), 725–730. doi:10.1121/1.1906679

Beauchamp, M. A. (1965). An improved index of centrality. *Behavioral Science*, *10*(2), 161–163. doi:10.1002/bs.3830100205 PMID:14284290

Bonacich, P. (1972). Factoring and weighing approaches to status scores and clique identification. *The Journal of Mathematical Sociology*, *2*(1), 113–120. doi:10.1080/0022250X.1972.9989806

Bonacich, P. (1987). Power and centrality: A family of measures. *American Journal of Sociology*, *92*(5), 1170–1182. doi:10.1086/228631

Bonacich, P. (2007). Some unique properties of eigenvector centrality. *Social Networks*, *29*(4), 555–564. doi:10.1016/j.socnet.2007.04.002

Bonacich, P., & Lloyd, P. (2001). Eigenvector-like measures of centrality for asymmetric relations. *Social Networks*, *23*(3), 191–201. doi:10.1016/S0378-8733(01)00038-7

Brandes, U. (2001). A faster algorithm for betweenness centrality. *Journal of Mathematical Sociology*, *25*(2), 163-177.

Bruun, J., & Brewe, E. (2013). Talking and learning physics: Predicting future grades from network measures and Force Concept Inventory pretest scores. *Physical Review Physics Education Research*, *9*(2), 020109. doi:10.1103/PhysRevSTPER.9.020109

Chen, C. M. (2006). CiteSpace II: Detecting and visualizing emerging trends and transient patterns in scientific literature. *Journal of the American Society for Information Science and Technology*, *57*(3), 359377. doi:10.1002/asi.20317

Coles, N. (2001). Analyzing serious crime groups as social network. *British Journal of Criminology*, *41*, 580–594. doi:10.1093/bjc/41.4.580

Das, K., Samanta, S., & Pal, M. (2018). Study on centrality measures in social networks: A survey. *Social Network Analysis and Mining*, 1–11.

Estrada, E., & Rodriguez-Velazquez, J. A. (2005). Subgraph centrality in complex networks. *Physical Review*, *71*, 056103. PMID:16089598

Everett, M. G., & Borgatti, S. P. (1999). The centrality of groups and classes. *The Journal of Mathematical Sociology*, *23*(3), 181–201. doi:10.1080/0022250X.1999.9990219

Freeman, L. C. (1977). A Set of Measures of Centrality Based on Betweenness. *Sociometry*, *40*(1), 35–41. doi:10.2307/3033543

Freeman, L. C. (1978). Centrality in Social networks conceptual clarification. *Social Networks*, *I*(3), 215–239. doi:10.1016/0378-8733(78)90021-7

Freeman, L. C., Borgatti, S. P., & White, D. R. (1991). Centrality in valued graphs: A measure of betweenness based on network flow. *Social Networks*, *13*(2), 141–154. doi:10.1016/0378-8733(91)90017-N

Grunspan, D. Z., Wiggins, B. L., & Goodreau, S. M. (2014). Understanding Classrooms through Social Network Analysis: A Primer for Social Network Analysis in Education Research. *CBE Life Sciences Education*, *13*(2), 167–178. doi:10.1187/cbe.13-08-0162 PMID:26086650

Guimera, R. (2005) The worldwide air transportation network: anomalous centrality, community structure, and cities global roles. *Proceedings of the National Academy of Sciences*. 10.1073/pnas.0407994102

Hage, P., & Harary, F. (1995). Ecentricity and centrality in networks. *Social Networks*, *17*(1), 57–63. doi:10.1016/0378-8733(94)00248-9

Jayaweera, I. M. L. N., Perera, K. K. K. R., & Munasinghe, J. (2017). Centrality Measures to Identify Traffic Congestion on Road Networks: A Case Study of Sri Lanka. *IOSR Journal of Mathematics*, *13*(2), 13–19. doi:10.9790/5728-1302011319

Jeong, H., Mason, S. P., Barabási, A.-L., & Oltvai, Z. N. (2001). Lethality and centrality in protein networks. *Nature*, *411*(6833), 41–42. doi:10.1038/35075138 PMID:11333967

Joyce, K. E., Laurienti, P. J., Burdette, J. H., & Hayasaka, S. (2010). A New Measure of Centrality for Brain Networks. *PLoS One*, *5*(8), 12200. doi:10.1371/journal.pone.0012200 PMID:20808943

Katz, L. (1953). A new status index derived from sociometric analysis. *Psychometrika*, *18*(1), 39–43. doi:10.1007/BF02289026

Kitsak, M., Gallos, L. K., Havlin, S., Liljeros, F., Muchnik, L., Stanley, H. E., & Makse, H. A. (2010). Identification of influential spreaders in complex networks. *Nature Physics*, *6*(11), 888–893. doi:10.1038/nphys1746

Koschutzki, D. (2005) Centrality Indices. Analysis: Methodological Foundations, 3418, 16–61. doi:10.1007/978-3-540-31955-9_3

Koschützki, D., & Schreiber, F. (2004) Comparison of centralities for biological networks. *German Conference on Bioinformatics, 53*, 199-206.

Koschutzki, D., & Schreiber, F. (2008). Centrality analysis methods for biological networks and their application to gene regulatory networks. *Gene Regulation and Systems Biology*, *2*, 193–201. doi:10.4137/GRSB.S702 PMID:19787083

Liu, J. G., Ren, Z. M., & Guo, Q. (2014). Ranking the spreading influence in complex networks. *Physica A*, *392*(18), 4154–4159. doi:10.1016/j.physa.2013.04.037

Liu, Y. (2015). Identify influential spreaders in complex networks: The role of neighborhood. *Physica A, 452*, 289–298.

Nieminen, J. (1974). On the centrality in a graph. *Scandinavian Journal of Psychology*, *15*(1), 322–336. doi:10.1111/j.1467-9450.1974.tb00598.x PMID:4453827

Opsahl, T., Agneessens, F., & Skvoretz, J. (2010). Node centrality in weighted networks. Generalizing degree and shortest paths. *Social Networks*, *32*(3), 245–251. doi:10.1016/j.socnet.2010.03.006

Rodriguez, J. A., Estrada, E., & Gutierrez, A. (2006). Functional centrality in graphs. *Linear and Multilinear Algebra*, *55*(3), 293–302. doi:10.1080/03081080601002221

Sabidussi, G. (1966). The centrality index of a graph. *Psychometrika, 31*(4), 581-603.

Shaw, M. E. (1954). Group structure and the behavior of individuals in small groups. *The Journal of Psychology*, *38*(1), 139–149. doi:10.1080/00223980.1954.9712925

Shimbel, A. (1953). Structural parameters of communication networks. *The Bulletin of Mathematical Biophysics*, *15*(4), 501–507. doi:10.1007/BF02476438

Sparrow, M. K. (1991). The application of network analysis to criminal intelligence: An assessment of the prospects. *Social Networks*, *13*(3), 251–274. doi:10.1016/0378-8733(91)90008-H

Stephenson, K., & Zelen, M. (1989). Rethinking centrality: Methods and examples. *Social Networks*, *11*(1), 1–37. doi:10.1016/0378-8733(89)90016-6

Wang, J. (2017). A novel weight neighborhood centrality algorithm for identifying influential spreaders in complex networks. *Physica A*, (17), 30121-8.

Wassermann, S., & Faust, K. (1994). *Social Network Analysis: Methods and Applications*. Cambridge, UK: Cambridge University Press. doi:10.1017/CBO9780511815478

White, D. R., & Borgatti, S. P. (1994). Betweenness centrality measures for directed graphs. *Social Networks*, *16*(4), 335–346. doi:10.1016/0378-8733(94)90015-9

Wuchty, S., & Stadler, P. F. (2003). Centers of complex networks. *Journal of Theoretical Biology*, *223*(1), 45–53. doi:10.1016/S0022-5193(03)00071-7 PMID:12782116

Zeng, A., & Zhang, C. J. (2013). Ranking spreaders by decomposing complex networks. *Physics Letters. [Part A]*, *377*(14), 1031–1035. doi:10.1016/j.physleta.2013.02.039

Chapter 16
Integration of Multiple Cache Server Scheme for User–Based Fuzzy Logic in Content Delivery Networks

Manoj Kumar Srivastav
Champdani Adarsh Sharmik Vidyamandir, India

Robin Singh Bhadoria
ⓘ https://orcid.org/0000-0002-6314-4736
Indian Institute of Information Technology Bhopal, India

Tarasankar Pramanik
ⓘ https://orcid.org/0000-0001-7582-2525
Khanpur Gangche High School, India

ABSTRACT

The internet plays important role in the modern society. With the passage of time, internet consumers are increasing. Therefore, the traffic loads during communication between client and its associated server are getting complex. Various networking systems are available to send the information or to receive messages via the internet. Some networking systems are so expensive that they cannot be used for the regular purpose. A user always tries to use that networking system that works on expansion of optimizing the cost. A content delivery network (CDN) also called as content distribution network has been developed to manage better performance between client and list of available servers. This chapter presents the mathematical model to find optimization among client and cache server during delivery of content based on fuzzy logic.

DOI: 10.4018/978-1-5225-9380-5.ch016

INTRODUCTION

Due to the invention of the computer, network services become increasingly popular around the world. Network services comprise two basic tools, one is computer hardware, and another is computer software. Comprising two tools data are transmitted from one computer to another computer. In computer networks, computers are generally connected in two ways: one is peer to peer networks system, and another is server-centric network or client-server network system. In server-centric network, several client computers are connected to a server which provide certain services to them (Mangili et al., 2016). In this scenario, clients have opportunity to make request from the server. But in peer-to-peer network, client computers are connected to each other. They can request and share data or information among them. Client-server is a software architecture model with two parts namely, (i) Client system (ii) Server system. This network architecture is very much easy to use and reliable due to its structural setup. Every clients gets information from a reliable source. Server helps in administering the whole setup and facilities delivered by them.

There exists one server to hose the resource which is requested and used by clients (Bhadoria et al., 2017). But in client-server model, traffic congestion is an important issue and for this reason the total network system becomes slow and sometimes it becomes failed. However, in peer-to-peer network, although there is no server-side support but, due to its client-client connection, the network system is too much fast for information interchange. As the reliability of information is at high priority, a client-server network is too much acceptable besides these disadvantages. To improve and use most of its advantages, a new distributed architecture Content Delivery Networks (CDN) has developed. There is required to optimize the cost for both the clients and servers to provide best of the services. Internet content delivery service improves the performance of delivering of content. In computer network architecture, 'Content' is a collection of information. This can be nicely represented using the notion of set theory as follows:

$$Content = \left\{ i : i \, is \, an \, information \, available \, for \, retrieval \, by \, client \right\}$$

So, contents may be web pages, audio clippings, video clippings, images, software, etc. To get the better performance, server-side has to make multiple copies of contents and distribute to different Cache servers. One has to design a management process so that the distance from the origin server to a client is minimum. In this process, nearest cache server serves for the client who is requesting content. This forms a network and requires optimizing the cost of content delivery whose diagrammatical representation is shown in Figure 1.

GENERAL WORKING CONCEPT BETWEEN CLIENT AND SERVER IN COMPUTER NETWORK

The computer network architecture of server-to-client model is categorized into three stages, which are as follows:

Figure 1. Diagrammatical representation of nearest cache server

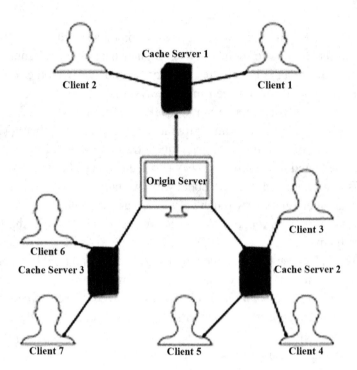

Resource/Information Requirement by Client

In computer network architecture, the tasks of each computer are to collect, store and deliver information to the clients who need it. To do this, computer has to process data to produce information. A set of instructions, called a program, manage the whole task. Each instruction of a program performs a specific task during execution done by a computer. Software is the collection of such programs, which is designed to perform whole task for the fulfillment of clients' need. Software are classified into two parts: (i) System Software and (ii) Application Software. All these flow of work is shown in Figure 2. System software is a software to run computer system whereas the Application software is designed to perform particular task when it is needed in a particular environment. For example, a 'Web browser' is a software application (Maddah-Ali & Niesen, 2015). A browser running on a computer behaves like a client and it request web server to get information. In computation of particular task, there are three stages and they are: input, processing and output.

Communication from Server to Client

Transferring information from one side to another side is called the Communication. A client makes a request to get information for particular content. A server gives services after getting request for particular content made by the client.

Figure 2. Layout of System and Application Software

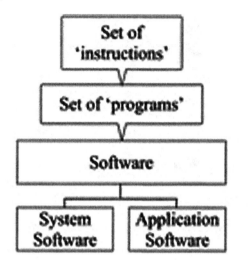

Figure 3. Relations among Server Cache and Server Client in CDNs.

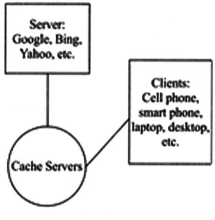

Delivery of Content (Information) to Client

Server is responsible to give its services to clients with respect to client's demand. Server is delivering the information to clients. Thus, delivery of content is the final stage where a server attends the requests from various clients.

CACHE SERVER PERFORMANCE IN CDNS

In computer network, information plays a key role. In client-server network model, server is responsible to deliver the informations to the client. Therefore, the performance is the time taken to transmit the information to client by the server (Srivastav & Nath, 2016). In general, there exists a server which serve different client at a time but, in CDN, there exists one central server and many cache servers. To improve the performance in CDN, the performance of cache servers must be increased. The CDNs is a combination of network of servers and it is hosted by a service provider in multiple locations. A CDN is based on distributed network system to handle large number of simultaneous request made by clients.

There are so many contents or services such as E-docs, Music/Audio (Broberg et al., 2009), Educational video, Movie, etc. which are required to deliver the end user by servers. Clients are persons or group of persons using cell phones, smart phones, laptop, desktop, etc. Generally, there exist two components in CDN (i) Origin Server (ii) Cache Server. Origin Server contains the original version of the content to distribute over the internet whereas cache server contains the duplicate version of the content.

An end user or client may request for particular content on the internet. When a client requests particular content on the internet, the role of the cache server (for particular contents) starts with the request with considering the geographical location of the consumer (Sarkar et al., 2016).

Following steps are essential workflow in CDNs.

Figure 4. Workflow of content requests and data processing in CDNs.

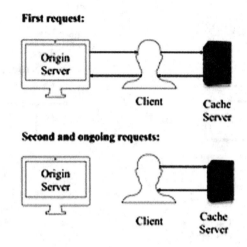

First request:

Origin Server — Client — Cache Server

Second and ongoing requests:

Origin Server — Client — Cache Server

Step 1: Client or end user sends a request for specific content
Step 2: Request redirected to the origin server
Step 3: Origin server returns resource requested to the cache server
Step 4: Cache server contains the duplicated copy of contents
Step 5: Cache server returns request to the cache server
Step 6: Client makes subsequent requests
Step 7: Cache server returned cached copy to the client.

Workflow of Content Information in CDNS

The workflow of content requests and data processing in CDNs is shown in Figure 4.

DNS Resolution for CDNs

A Domain Name System (DNS) is internetservice that translates a domain name into IP address. When a client requests a resource, the first step is to make a DNS request (Sahoo et al., 2016). The browser gives the domain name and expects to receive an IP address back. With the IP address, the browser can then contact the web server directly for a subsequent request (there are multiple layers of DNS caching). A domain name may have a single IP address or multiple IP addresses. If it is a simple blog or small commercial website, it may have a single IP address, and if it has a large web application, then it has multiple IP addresses.

Step 1: A web browser requests a resource (DNS request)
Step 2: After getting an IP address, the browser can direct request towards web server.
Step 3: A server handling the DNS request for a domain handled by CDN will assess an incoming appeal and control the best route to forward it to.
Step 4: At it is simplest, the DNS server does a geographic lookup based on the DNS resolver's IP address and then returns an IP address for an *edge server* that is physically closest to that area.

FUZZY LOGIC BASED MATHEMATICAL MODELLING FOR CACHE SERVER SCHEME IN CDNS

In recent days, a website provides different types of contents such as text, documents, scripts, media files, images, movies, audio clips, etc. (Ali et al., 2011). There are multiple cache servers,and any one of them fulfils user requirements. Here deciding about which cache server will serve the client is vague. In this case, one can use fuzzy logic to take better decision. Fuzzy logic is used to label fuzziness. Its membership function best characterizes this fuzziness. Membership function represents the degree of truth in fuzzy logic. Every cache servers are assigned a membership value between 0 and 1. Cache servers which will finally serve the client have membership value 1 (Ali et al., 2011).

During the final stage, content from a server to user may be either fully or partially delivered and it can be a case of undelivered. So, there exists scope to develop the concept of fuzzy set theory in CDNs.

Fuzzy set theory will help to develop policy-based techniques related to delivering of information. Fuzzy logic can deal with uncertainty and non-linearity that may be happening in CDNs.

CDN system exists on three pillars which are Origin Server, Cache Server and Client/End user. Generally, there exist one origin server and multiple cache servers in CDNs (Rodrigues et al., 2013). Numbers of a user may be single or multiple,and they may request separately or simultaneously. The whole working concept can be analysed in the following mathematical way:

Let

O = Set of origin Server = $\{\, o \,\}$,

C = Set of cached Server = $\left\{ c_1, c_2, c_3, \ldots, c_n \right\}$,

U = Set of end user the = $\left\{ u_1, u_2, u_3, \ldots, u_m \right\}$.

In general concept, client-server relation is worked as

Origin server $O \rightarrow$ End User U.

However, in CDN, it is working for better performance as follows

Origin server $O \rightarrow$ Cache Server $C \rightarrow$ End User U.

Role of a server depends on its geographic location. CDN uses multiple servers. Generally, there exist one origin server and many cache servers. For cache server $\left\{ c_1, c_2, c_3, \ldots, c_n \right\}$ to user u, distance D can be taken as

$$D = \min \left\{ dist\left(c_1, u\right), dist\left(c_2, u\right), dist\left(c_3, u\right), \ldots, dist\left(c_n, u\right) \right\}.$$

Content Based Selection for Multiple Cache Server

All cache servers process and execute requests in equal fashion. However, the distance between user and server plays an important role. So, here conditional approach in performing by the server exists.

1. Take the distance from a different cache server to the user.
2. Choose the minimum distance.
3. That server will serve to the user.

D = the event of measurement of minimum distance.

A comparative theory exists during measurement. C = the event of servicing done by cache server possessing minimum distance user requesting $\{c_1, c_2, ..., c_n\}$ as shown in Figure 4.1.

In the client-server model there are not any issues of distance so, here probabilistic approach will work independently concerning event D and C. But in the CDNs, the event D and C are not independent as Figure 5 discusses the comparison of number of users and event occurred for the said approach.

Optimization Approach for Handling Cache Server

To get an optimal solution in multi-dimension, we have to do the following thing (Venketesh & Venkatesan, 2009):

1. Distance took from (central node to different node at a different place) = minimum distance value (path 1, path 2,..., path n).

Figure 5. Comparison between Number of Users and Different Events Occurred.

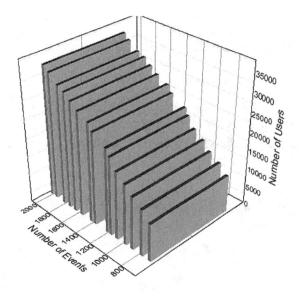

2. Time is taken from (central node to different nodes to different place) = time value concerning minimum distance (path 1, path 2,....., path n)
3. Cost expenditure (C) directly proportional to distance value i.e.

$C \propto D$, if time T is constant

Cost expenditure directly proportional to a time value i.e. $C \propto T$, if distance D is constant. So, $C \propto (D \cdot T)$, when time and distance vary. Therefore, $C = k(D \cdot T)$, where k is variation constant.

Keeping above concept in mind, performance from Cache Server to end user are represented by cost value C_{ij}, where C_{ij} represents cost values from cache a server C_i to user U_j, where $i = 1, 2, ..., m$ and $j = 1, 2, ..., n$. This ratio of distance versus available cache in the corresponding router, is demonstrated in Figure 6.

The clear cache available to distinct users from multiple domains would prevail if some users per router got recognised DNS for the mentioned session (Berger et al., 2017). This scenario is well depicted in Figure 7.

The Table 1 will work similar to Mathematical Transportation Problem. So, different types of optimisation methods which are used to solve a transportation problem can apply to the above matrix system (Stocker et al., 2017).

Note: (i) Here same cache server is working at the different end user, so it is treated as a distinct cache server during the process of optimisation (Bai et al., 2016). (ii) Availability of content or requirement of content is treated as in the form of numerical values.

The average download rate would subsequently depend on the available cache that is residing at a different router in the network. This chapter has experimented on two different domains like e-commerce and education. In Figure 8, there are two parallel row depicts the two different domain and its associated average download per cache are demonstrated accordingly.

Figure 6. Number of Cache Router per Distance (Incrementally).

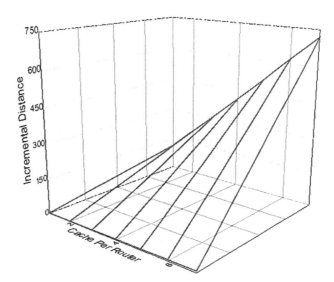

Figure 7. Number of Users (Domain Wise) Per Cache Routers.

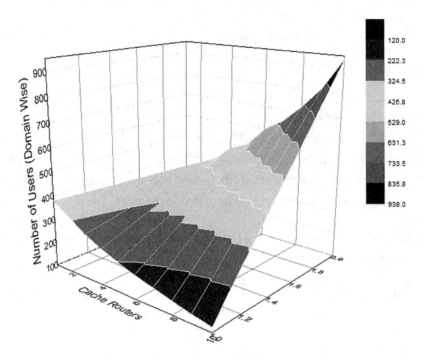

Figure 8. Average Download Rate (In GB) for Each Cache Server

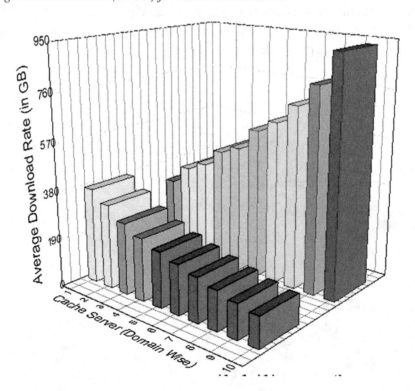

Table 1. Representation of Cost Values in Matrix Form

Cache Server	End User					Availability of Content
	U_1	U_2	U_3	..	U_m	
C_1	C_{11}	C_{12}	C_{13}	..	C_{1m}	A_1
C_2	C_{21}	C_{22}	C_{23}	..	C_{2m}	A_2
C_n	C_{n1}	C_{n2}	C_{n3}	..	C_{nm}	A_n
Requirement of Content	B_1	B_2	B_3	...	B_m	$\sum A_i - \sum B_j$

CHAPTER SUMMARY

Uses of the internet have been enormously increased in the past few decades. It has both pros and cons. However, due to its large facilities speed, efficient content management, availability of resources, etc. internet users are going to be increased. Now it is primary need the system to be secure and as well as fast regarding its utilization. Clients always try to get better performance from the web with an expenditure of minimum cost and on the other hand Service providers too wants to provide the content to the client with minimum cost. This is possible only when technology used to delivery of content is much advanced. Delivery of content depends on the location of Origin Server and Cache Server. This chapter describes a mathematical matrix using the concept of Transportation Theory to show the optimization among Cache Server and Client during delivery of content in CDNs. In future, there is a scope to develop this technology and to improve the performance of the system better.

REFERENCES

Ali, W., Shamsuddin, S. M., & Ismail, A. S. (2011). A survey of web caching and prefetching. *Int. J. Advance. Soft Comput. Appl*, *3*(1), 18–44.

Bai, B., Wang, L., Han, Z., Chen, W., & Svensson, T. (2016). Caching based socially-aware D2D communications in wireless content delivery networks: A hypergraph framework. *IEEE Wireless Communications*, *23*(4), 74–81. doi:10.1109/MWC.2016.7553029

Berger, D. S., Sitaraman, R. K., & Harchol-Balter, M. (2017). AdaptSize: Orchestrating the Hot Object Memory Cache in a Content Delivery Network. In NSDI (pp. 483-498). Academic Press.

Bhadoria, R. S., Chaudhari, N. S., & Vidanagama, V. T. N. (2017). Analyzing the role of interfaces in enterprise service bus: A middleware epitome for service-oriented systems. *Computer Standards & Interfaces*, *56*, 146–155.

Broberg, J., Buyya, R., & Tari, Z. (2009). MetaCDN: Harnessing 'Storage Clouds' for high performance content delivery. *Journal of Network and Computer Applications*, *32*(5), 1012–1022. doi:10.1016/j.jnca.2009.03.004

Maddah-Ali, M. A., & Niesen, U. (2015). Decentralized coded caching attains order-optimal memory-rate tradeoff. *IEEE/ACM Transactions on Networking*, *23*(4), 1029–1040. doi:10.1109/TNET.2014.2317316

Mangili, M., Martignon, F., & Capone, A. (2016). Performance analysis of Content-Centric and Content-Delivery networks with evolving object popularity. *Computer Networks*, *94*, 80–98. doi:10.1016/j.comnet.2015.11.019

Rodrigues, M., Moreira, A., Azevedo, E., Neves, M., Sadok, D., Callado, A., & Souza, V. (2013). On learning how to plan content delivery networks. In *Proceedings of the 46th Annual Simulation Symposium* (p. 13). Society for Computer Simulation International.

Sahoo, J., Salahuddin, M., Glitho, R., Elbiaze, H., & Ajib, W. (2016). A Survey on Replica Server Placement Algorithms for Content Delivery Networks. *IEEE Communications Surveys and Tutorials*.

Sarkar, D., Rakesh, N., & Mishra, K. K. (2016). Content delivery networks: Insights and recent advancement. In *Parallel, Distributed and Grid Computing (PDGC), 2016 Fourth International Conference on* (pp. 1-5). IEEE.

Shanmugam, K., Golrezaei, N., Dimakis, A. G., Molisch, A. F., & Caire, G. (2013). Femtocaching: Wireless content delivery through distributed caching helpers. *IEEE Transactions on Information Theory*, *59*(12), 8402–8413. doi:10.1109/TIT.2013.2281606

Srivastav, M.K., & Nath, A. (2016). Web Content Management System. *International Journal of Innovative Research in Advanced Engineering, 3*(3), 51-56.

Srivastav, M. K., & Nath, A. (2017). Content Delivery and Management System. *Proc. of IEEE, International Conference on Information, Communication, Instrumentation and Control (ICICIC 2017)*. 10.1109/ICOMICON.2017.8279058

Stocker, V., Smaragdakis, G., Lehr, W., & Bauer, S. (2017). The growing complexity of content delivery networks: Challenges and implications for the Internet ecosystem. *Telecommunications Policy*, *41*(10), 1003–1016. doi:10.1016/j.telpol.2017.02.004

Venketesh, P., & Venkatesan, R. (2009). A survey on applications of neural networks and evolutionarytechniques in web caching. *IETE Technical Review*, *26*(3), 171–180. doi:10.4103/0256-4602.50701

Chapter 17
Social Networks and Graph Theory in the Search for Distant Knowledge:
Studying the Field of Industrial Engineering

Alejandro Vega-Muñoz
https://orcid.org/0000-0002-9427-2044
Universidad San Sebastián, Chile

Juan Manuel Arjona-Fuentes
Universidad Loyola Andalucia, Spain

ABSTRACT

This chapter presents how the analysis of social networks supported in graph theory contributes to the search for "distant knowledge" in the field of industrial engineering, discipline of engineering that in its current form began in the early 20th century when the first engineers began to apply scientific theory to manufacturing. In particular, the case of Chilean documented scientific production in this area of engineering is analyzed as a category of the web of science distinguishing its degree of connection with the great knowledge, generating organizations worldwide, determining its high dissociation with the great contemporary theoretical referents, and recommending the way to reduce these problems in the future.

INTRODUCTION

This chapter presents how the analysis of social networks and graph theory contributes to the search for 'distant knowledge' in the field of industrial engineering, the discipline of engineering in its current form, began in the early 20th century when the first engineers began to apply scientific theory to manufacture (Daneshvar-Rouyendegh & Feryal-Can, 2012).

DOI: 10.4018/978-1-5225-9380-5.ch017

Industrial engineering creates engineering processes and systems that improve quality and productivity, besides is concerned with the development, improvement, and implementation of integrated systems of people, money, knowledge, information, equipment, energy, materials, analysis and synthesis, as well as the mathematical, physical and social sciences together with the principles and methods of engineering it design to specify, predict, and evaluate the results to be obtained from such systems or processes (Pinar, Günther & Fazleena, 2016). Frederick Taylor is been well known as the pioneer expert in management, engineer and the leader of the engineering movement by developing methodologies to improve efficiency in manufacturing without using the term, in those times, 'industrial engineering' as a concept (Maynard & Zandin, 2001).

Economic theory realizes that the efforts of industrialization and the increase of human capital with advanced knowledge, generate better conditions adding value to production and consequently greater progress to society (Solow, 1957). In this aspect, the industrialization technique generated in the field of Industrial Engineering should be a fundamental pillar, which justifies the importance of its study and academic progress. (Maynard & Zandin, 2001).

Particularly in the case of Chile, for a university system of sixty institutions of higher education, despite importance of this engineering field, there are only two doctoral programs in industrial engineering accredited by the National Accreditation Commission (2019) and knowledge that the country produces in the category of 'Engineering, Industrial' of WoS (2019) barely reaches 0.251% of the world knowledge production of the last forty years (1978-2017). And therefore, it needs to be incorporated into 'knowledge cluster' of this discipline, to at least access the knowledge spills studied by Gambardella and Giarratana (2010) and the collective generation of architectural knowledge, through network interaction routines (Tallman, Jenkins, Henry and Pinch, 2004: 265-266, Nilsson, 2019).

On the other hand, Giuliani (2005) warns that similar meso-characteristics, such as geographic and relational proximity, do not necessarily originate shared knowledge flows. Vega, Benítez & Yévenes (2005) also conclude that in order to take advantage of the multiple synergies of a cluster, the actors must consider and comply with certain norms and conditions they favor systemic action and strongly accentuate information flows. To which Huggins (2008) adds that many actors do not acquire their knowledge in their closest environment, because they promote their growth in innovation, and therefore require obtaining international border knowledge. In addition, actors with greater capacity for absorbing knowledge manage to remain more connected to global knowledge networks and not only depend on the local knowledge network.

Thus, as a result, it is possible to identify Chile's position in the 'cluster of knowledge' of Industrial Engineering and define the best approach strategy to contemporary actors of global reference, to take advantage of this.

SOCIAL NETWORKS AND GRAPH THEORY

According to Cárdenas-Tapia, Klingler-Kaufman and Rivas-Tovar (2012), graph theory is a discipline of discrete mathematics, with its own development; that contributes to the analysis of social networks by providing precise concepts to refer to properties of the social structure, quantification methods of those properties, through the actors representation (nodes) and their interactions (arcs) through a graph. Thus, already Lozares (1996), account for as from origin of his mathematical theory of graphs has given sustenance to the theory of social networks, which is exemplified in the seminal works of Bavelas (1948), Festinger (1949), Bavelas (1950), Cartwright and Zander (Eds) (1953), and Harary and Norman (1953).

Thus, adapting the language of graph theory, researchers are called 'nodes' and constitute individual points, and the connection between researchers is a line or 'edge', which is established if, for example, they have signed as co-authors in a joint publication (Cheng et al., 2019). Being able to graduate the nodes according to their number of edges. Given this arrangement, it is possible to use the analysis of social networks used to measure various forms of collaboration. (Ceballos et al., 2017).

For Prathap (2018, 2018b), an index of citations, that it accounts for a published scientific article usage, is represented as a complex and well-structured network of links, which counts the publications that have been cited and identifies the sources of those citations. Thus, when carrying out bibliometric and scientometric analysis, the established links can be reconciled through network analysis and graph theoretical techniques. Giving way to new lines of research within the 'science of metrics', on the network properties of the indexing of citations as a recursive improvement through iterative algorithm protocols that arise from graph theory, with more sophisticated approaches such as those observed in the social networks analysis (SNA). In general terms, it is important to note that there is a wide variety of indexes that have evolved from the impact of factor ideas was proposed by Eugene Garfield in 1955 and tend increasingly to measure results that individualize an author, a specific article or a combination from both. (Kim & Chung, 2018).

Delgado-Saab et al. (2016), already identified in a specific way as graph theory has expanded its possibilities of application in the field of sociology, during the last decades given the mathematical modeling of social networks and scientometrics. Determining for both of their: analysis factors, basic information of the graph and functionality, as shown in table 1.

According to Abbasi et al. (2018), analysis of social networks is one of the central themes of computer science and computational social sciences. It is therefore interesting to study the researchers structural networks influence and how their diversity impacts on the research communities results. That it allows accounting for the structural position of researchers (centrality) and the diversity in productivity (record count) and performance (h-index) of these, and therefore the power of academic collaborators in a research front.

The scientometrics and social network analysis, in combination, create a powerful tool in favor of research collaboration. While scientometrics provides mechanism measuring scientific performance, the social network analysis contributes to methods based on graph theory for measuring collaboration. And though exist theoretical synergies between scientometrics, social network analysis, and knowledge management, research management increasingly in scientometrics and social network analysis, but knowledge management consider sparingly to these techniques as antecedent. (Ceballos et al., 2017).

SEARCH FOR DISTANT KNOWLEDGE

Maskell (2014) from economic geography, distinguishes patterns of relevant behavior in the search for 'distant knowledge' and establishes the ways for their transfer and learning. Relieving the ideas previously highlighted by Gilsing et al (2008: 1719-1720), of technological distance, over geographical distance. Being the technological distance an operationalization of the cognitive distance.

Thus, it is possible to study the world knowledge cluster (knowledge clusters) in Industrial Engineering, as an agglomeration of actors around the production of epistemic knowledge (Evers, Gerke &

Table 1. Characteristics of mathematical modeling: social networks analysis and scientometrics analysis

Analysis	Analysis Factors	Basic Information (Graph)	Algorithms Types (Functionality)
Social Networks	1) Nodal Degree: can be interpreted by the opportunity to influence or be directly influenced. Predict leadership, knowledge, power. 2) Intermediation: Frequency in which a node appears in the shortest path that connects to two other nodes. 3) Popularity Indicator (Eigenvector): the node that has the Highest score is one that is connected to many well-connected nodes (node grade). 4) Structural holes: Nodes with joint sides, more sides of the node has more use. 5) Overlap: grouping of nodes, taking those nodes that (a) do not belong to a group or (b) belong to both, as nodes of overlap. 6) Egocentric Network: an 'ego' network is the sub-graph associated with a given node (the ego), with distance 1.	Nodes: Represent a person (actor) or an organization. Edge: Represent an instance of a social relationship, a type can be: Relationship, Role Based, Cognitive / Perceptual, Affective, Interaction, Affiliation.	Scroll adjacency matrix. Scroll Matrix away. Degree calculation of a node. Density Calculation of a node. Find the points of articulation. Go through the graph (Depth First Search, DFS).
Scientometrics	1) Authors by Area: is to locate the authors according to the specialty of their publications. 2) Publication Country: it is given by the country where the publication is registered, being indifferent to the nationality of the author. 3) Frequency of citations by author: are the references that are made about an author.	Nodes: Represent publications. Edges: There can be 3 types of directed edges: one represents the reference between the articles, another the self-citation and the third one to the articles referred to more than one article analyzed.	Adjacency matrix. Scroll Matrix away. Calculate the degree of a node. Density of a node. Points of articulation. Go through the graph (DFS)

Source: (Delgado-Saab et al., 2016)

Menkhoff, 2010), through relational scientometrics (Vega & Salinas, 2017), supporting the traditional scientometric study in the analysis of social networks and graph theory, being able to define a strategy of approach to 'distant knowledge' for Chile, having the position of Chile based on the two main techniques of identification of positions in social networks (Knoke and Laumann, 2012, p.460):

'Cohesion of the subgroup', centered on the cooperative interaction between two or more actors that maintain dense interactions with each other (cliques) and it is based on direct communication diffusion, examining the influence of interlocutors within the group and incorporating: opinions, behaviors, attitudes and policies. Thus, among actors into a subgroup, the elaboration of policies of one follows punctually those of another, both sharing costs and benefits of the interaction.

'Structural Equivalence', based on the competitive interaction between two or more actors who hold structurally similar positions, given their interactions with other network actors. Thus, among competing actors, measures that are applied by others in a position of similarity they are accepted and therefore the imitation behavior adopted by an actor is caused by other structurally equivalent within the network.

For a country with scarcely documented production of mainstream knowledge, such as Chile in Industrial Engineering, independently of the predefined option to improve its position in the global knowledge cluster, it is essential to identify references in the scientific production of this area of engineering.

DISTANT KNOWLEDGE IN THE FIELD OF INDUSTRIAL ENGINEERING

It is discussed below, as in the absence of a critical mass of researchers in a specific engineering area. The need to generate useful knowledge for the development of a country must be supplied externally, demanding the detection of relevant knowledge, it is distant in a geographically and cognitively sense, in order to generate new local knowledge, either autonomously or co-produced.

The Need to Develop More Chilean Research in Industrial Engineering

Chile is a country whose economy depends mainly on the services sector, which represents 64% of GDP, with the industry sector accounting for 34% of the economic composition of this country. Which despite having achieved in recent decades one of the lowest levels of poverty and the highest per capita income in Latin America, its inequality is the most extreme among the member countries of the Organization for Economic Cooperation and Development (Central Intelligence Agency, 2019). What it maintains the historical argument about the need to add more value to their products through industrialization technique, based on an evolution of a deterministic nature from science to technology, from technology to industry, and then industry to trade and as a final consequence to social progress (Escobar, 2005; Saavedra, 1985).

From the industrialization efforts of Chile in the 1960s, as an economic disaster measure that activates Chile's development model, following a cataclysm of proportions, an earthquake that affects two important cities in the south-central part of the country (in Concepción city on May 21, and in Valdivia city on May 22), to that it was added a tsunami that affected Chile between Concepción city and Chiloé island, and produced such effects into the entire Pacific Ocean area; that it was the impulse to create the Pacific Tsunami Warning Center (PTWC) in 1965. (National Historical Museum of Chile, 2009). The Production Development Corporation (CORFO[1]) was responsible for creating a set of companies and organizations to support the creation of companies. This policy could extend until 1973, with CORFO managing to control some 500 productive units that year (CORFO, 1962, Patrimonio Consultores, 2009).

These efforts of industrialization are diminished, and lack of advanced knowledge could be the answer, at least from the economic theory (Solow, 1957). As already mentioned, in this chapter the knowledge of 'main current' produced in Chile in the field of this study, the category 'Engineering, Industrial', in WoS (2019) barely reaches 0.251% of the world knowledge production in the last forty years (1978-2017). WoS Category 'Engineering, Industrial' in the Science Citation Index Expanded, includes resources that focus on engineering systems that integrate people, materials, capital, and equipment to provide products and services. Relevant topics covered in the category include operations research, process engineering, productivity engineering, manufacturing, computer-integrated manufacturing (CIM), industrial economics, and design engineering. (Clarivate, 2019).

And therefore, to face the problem of the production of scientific knowledge in Industrial Engineering, it is necessary to study this at the level of the system as a whole. And the theory of graphs is in it suitable as a support of the social networks analysis (SNA).

Considering the Chilean production in WoS for this field versus what happened worldwide, with the results extracted from WoS with the search vector {WC = (Engineering, Industrial) AND DOCUMENT TYPES: (Article) Timespan: 1978-2017. Indexes: SCI-EXPANDED}, we can see in Figure 1 how the production of knowledge in the WoS category under study in the world and Chile behaves in the time.

Figure 1. Documented scientific production in WoS of Chile and the world
Data Source: WoS, 2019

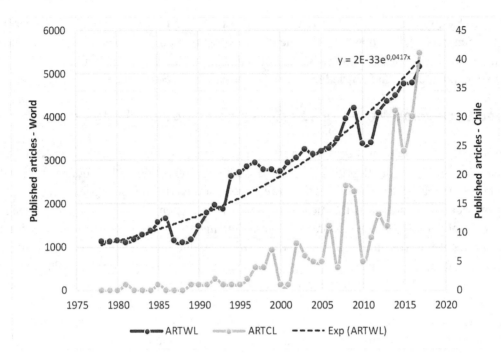

Table 2. Data fit with SPSS from Published articles - World to exponential regression

Coefficients					
	Unstandardized Coefficients		Standardized Coefficients	t	Sig.
	B	Std. Error	Beta		
PUB_YEARS	,042	,002	,959	20,875	,000
(Constant)	1,668E-33	,000			
The dependent variable is ln(ARTWL).					

Data Source: (WoS, 2019)

The worldwide data adjust to an exponential growth of the science for the Industrial Engineering and therefore they would give an account of the fulfillment of that bibliometric law (Price, 1976). On the contrary, the presence of zero values prevents the adjustment to an exponential growth of the 267 articles published from Chile and, therefore, there is no local evidence of critical mass investigating in the area of Industrial Engineering (for detail of the articles surveyed see the Appendix). a question that puts even more interest in the study of academic research developed in the small local group.

Figure 2 gives us evidence of the focus of research in industrial engineering in Chile, which differs in its emphasis from research developed in this area worldwide. Mainly because of the higher percentage of publications that occur in Chile at the intersection between the field under study and 'operations research and management science' and with 'computer science interdisciplinary applications'. As well as, the distancing of other categories that are related worldwide such as: ergonomics, psychology ap-

Figure 2. Percentage of joint contribution between industrial engineering and other WoS categories in the World and Chile
Data Source: WoS, 2019

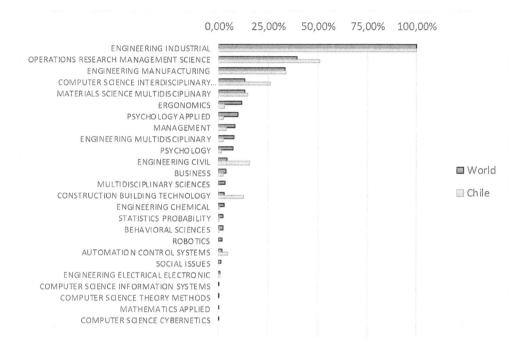

plied, management, engineering multidisciplinary and psychology, and in Chile they seem to be of small interdisciplinary interest for industrial engineering. Even more unusual is the strong emphasis is given to the production of interlocking articles with civil engineering and construction building technology, as well as to a lesser extent with automation control systems. These antecedents account for biases that seem to distance the research lines of Chilean industrial engineering of this specialty in the world.

Local Networks of Chilean Research in Industrial Engineering

As previously identified for the case of Chile, the scientific production in the category of Industrial Engineering between 1978 and 2017, generated a set of 267 research articles. In figure 3, a graph is observed for twenty two organizations that contribute in at least 5 articles, fifteen Chilean organizations and seven from other countries (despite contributing to 5 articles the Austral University of Chile (UACH) and the Technical University of Munich (TUM), were not incorporated because of their minor connection with the settings presented in the figure).

The graph of figure 3 shows the centrality of the Pontifical Catholic University of Chile (PUC), which contributes to 96 of the 267 studies done, in collaboration with thirteen of twenty-two institutions represented. Regarding foreign institutions, contributions to Chilean works are presented from universities in Spain (Polytechnic University of Catalonia (8), Polytechnic University of Valencia (7) and University of Seville (7)), New Zealand (University of Auckland (8)), Brazil (Federal University of Pernambuco (5)), United Kingdom (University of Kent (5)) and United States (University of Colorado - Boulder (5)).

Figure 3. Graph of Chilean scientific co-production in Industrial Engineering
Data Source: WoS, 2019

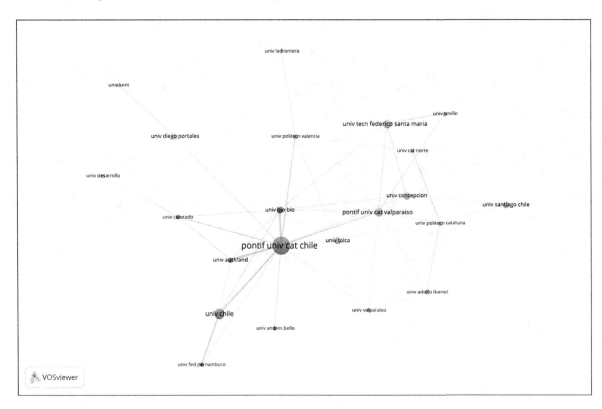

Being the largest international contribution to the Spanish, given the historical tie of Chile with that country, although none of these seven institutions of higher education is among the 100 organizations that produce more knowledge in the field of interest on which this paper investigates.

The scientific production of the Pontifical Catholic University of Chile in the WoS category of Industrial Engineering is mainly the product of academic work of its departments of: Engineering & Construction Management and Electrical Engineering. As we can see, in Chile, a scientific production abundant in Industrial Engineering, it is not necessarily the product of an Industrial Engineering Academic Department.

Chilean institutions total identified in the graph (figure 3), only five universities offer doctorates in industrial engineering. In these settings, both of accredited programs had had questions and objections to the scientific production of its academic faculty, as well as the non-accredited program, which among other observations, one of greater complexity to be overcome was just related to this topic.

Networks and Connections of the Chilean Research in Industrial Engineering

To the 106,475 articles indexed to the SCIE of WoS found worldwide, limitation restrictions were applied according to bibliometric laws commonly used in level I scientometrics or scientific production scientometrics (Vega & Salinas, 2017), in relation to the contemporary validity of writings, the main core of journals where academic discussion on industrial engineering is more intense or core of Bradford

Table 3. Chilean Ph.D. program in Industrial Engineering by Universities producing knowledge in that field

University	Acronym	PhD Program	Accredited
Pontifical Catholic University of Chile	PUC	Doctorate in Engineering Sciences, Industrial Engineering and Transport	Yes, 4 of 10 years.
Pontifical Catholic University of Valparaíso	PUCV	Doctorate in Industrial Engineering	No, not subjected
University of Santiago, Chile	USACH	Doctorate in Engineering Sciences, Industrial Engineering	No, rejected
University of Concepción	UDEC	Doctorate in Industrial Engineering	No, in process
Adolfo Ibáñez University	UAI	Doctorate in Industrial Engineering and Operations Research	Yes, 3 of 10 years.

Data Source: (National Accreditation Commission - Chile, 2014, 2016, 2018, 2018)

for that period, and the prolific authors publishing in those sources (journals) and for the most recent semi-temporal period.

To the level of detail, using the criteria proposed by Garfield (1987), Price (1963), Spinak (1998) and Vega & Salinas (2017), contemporaneity was restricted with median of more recent articles or semi-contemporary period of publications, delimiting thus articles to a total amount of 55,721 published between 2004-2017 and discarding the documents with obsolescence. The main core of five journals from a total amount of fifty-three journals was chosen over this subset, where 40.6% of articles in 'Engineering, Industrial' category were published between 2004-2017: Journal of Materials Processing Technology (7,437), International Journal of Production Research (5,084), International Journal of Production Economics (3,857), Computers Operations Research (3,145), and Computers Industrial Engineering (3,100). Those that correspond to the core of journals of the Journal Citation Report (JCR) of WoS where the mainstream discussion on this Engineering area is concentrated, bounded to 22,623 articles.

Finally, it was decided to restrict the 161 most prolific authors by publishing in those five journals for that period (sqrt(22,623)≈150). As a result, 4,813 contributions were obtained from these 161 authors with 20 or more published articles within the indicated restrictions, which by co-authorship are reduced to 3,978 articles that will compose the nodes of the global graph or 'knowledge cluster' to be studied, which are represented in the graph of figure 4.

The graph shows the presence of Chile in connection with 6 countries (in parentheses, the detail of organizations): Brazil (Pontifical Catholic University of Rio de Janeiro), Canada (HEC Montreal), Germany (Technical University of Munich), Italy (National Research Council), Japan (Hiroshima University) and Spain (University of Seville). But in fact, distant from the big world producers in the cluster of knowledge, like the People's Republic of China, United States of America, Taiwan, Canada and South Korea, whose contributions are shown in table 4.

Discounting repetitions, the set of 10 countries of table 4 concentrate 88% of the contributions with 3,500 articles and a high contribution to the co-authorship of these articles originated the People's Republic of China which reaches 44%. Table 5 lists the ten organizations in these countries that collaborate in the production of a third of this knowledge (1,185 articles).

By concentrating on the top ten authors of these organizations of high scientific production, 551 contributions are visualized, which discounted the co-authorship, resulting in 513 articles. Table 6 shows the indicators of the citation report of the WoS Core Collection, for the articles by these authors in the

Figure 4. Graph of contemporary world scientific co-production relevant in Industrial Engineering
Data Source: WoS, 2019

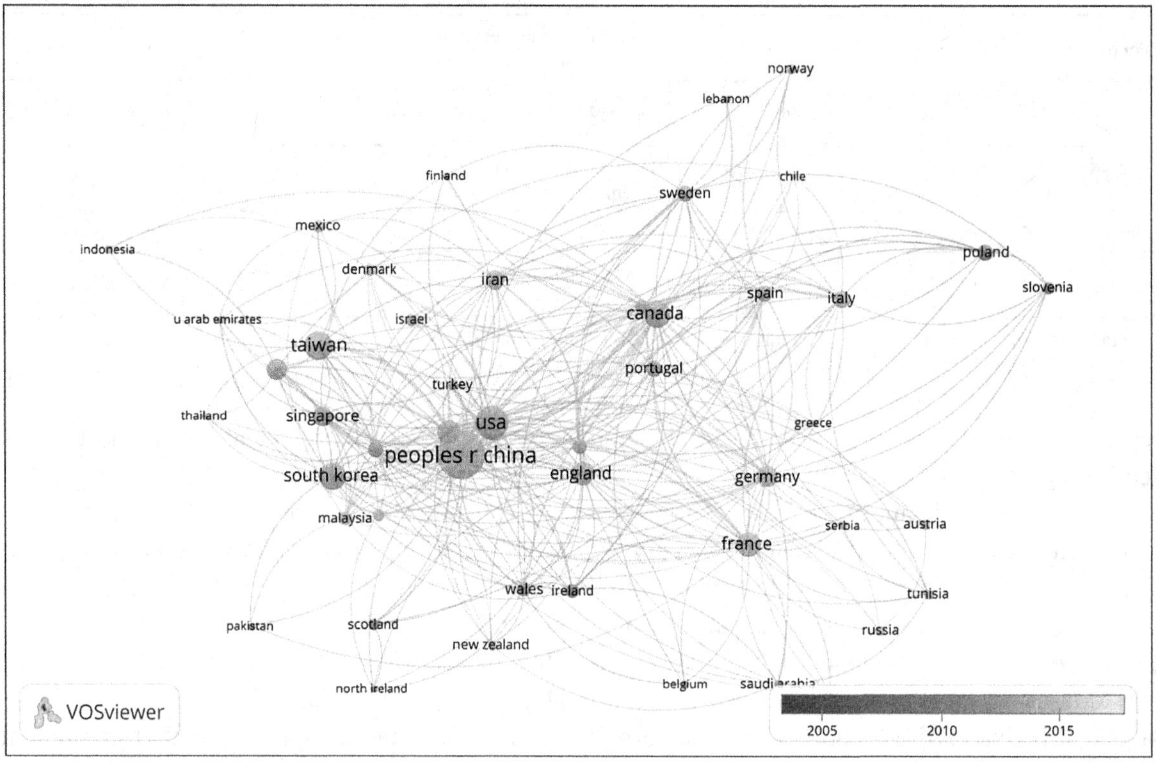

Table 4. Contribution to the world scientific production in industrial engineering by authors affiliation countries

Number	Field WoS: Countries/Regions	Record Count	% of 3,978
1	Peoples R China	1,752	44%
2	USA	624	16%
3	Taiwan	387	10%
4	Canada	335	8%
5	South Korea	307	8%
6	United Kingdom	291	7%
7	France	219	6%
8	India	208	5%
9	Japan	149	4%
10	Germany	147	4%

Data Source: (WoS, 2019)

Table 5. Organizations of high contribution to the production of world knowledge in industrial engineering

Number	Field WoS: Organizations-Enhanced	Record Count	% of 3,500
1	Hong Kong Polytechnic University	247	7%
2	Shanghai Jiao Tong University	131	4%
3	Tsinghua University	127	4%
4	Indian Institute of Technology System	123	4%
5	Chinese Academy of Sciences	114	3%
6	University of Montreal	107	3%
7	Xi'an Jiaotong University	99	3%
8	Northeastern University of China	94	3%
9	Harbin Institute of Technology	89	3%
10	Huazhong University of Science Technology	87	2%
10	University of Hong Kong	87	2%

Data Source: (WoS, 2019)

study period, which account for the citations received within the Core Collection in general and strictly from research articles (in both totals and without self-citations), in addition to the productivity (record count) and performance (h-index).

H-index allows us to correct the impact of an author's publications, regarding the author's ability to generate not only one, but a set of influential articles for the academic community. So this impact index is calculated by ordering the items from highest to lowest, respect to the number of citations received in WoS core collection and if h items are cited at least h times, the maximum number of h is the h index of That researcher, for example for the author Cheng TCE an h-index of 34 implies the author has, 34 articles cited at least 34 times each (Kim & Chung, 2018).

Table 6a. Citation report for articles results from Web of Science Core Collection between 1978 and 2017

Number	Field WoS: Authors	Record Count	% of 1,185	Sum of Times Cited	Without Self-Citations	h-Index	Average Citations per Item	Citing Articles	Without Self-Citations
1	Cheng TCE	137	12%	3,453	3,372	34	25,2	2.770	2.723
2	Tiwari MK	75	6%	1,223	1,197	19	16,3	1,129	1,112
3	Chan FTS	74	6%	1,901	1,862	26	25,7	1,669	1,647
4	Laporte G	55	5%	1,100	1,069	22	20,0	941	920
5	Huang GQ	42	4%	946	914	18	22,5	782	765
6	Jiang ZB	41	3%	338	315	10	8,2	302	288
7	Wang L	35	3%	1,346	1,319	20	38,5	1,047	1,032
8	Shankar R	34	3%	1,012	1,000	15	29,8	936	928
9	Liang L	30	3%	731	716	16	24,4	645	634
10	Wang ZR	28	2%	357	356	11	12,8	328	327

Data Source: (WoS, 2019)

Table 6b. Ranking of prolific authors based on the h-index of the Citation report WoS in table 6

Rank	h-Index	Field: Authors	Author's Affiliation Organization
1	34	Cheng TCE	Hong Kong Polytechnic University, Department of Logistics & Maritime Studies. SAR CN (https://lms.polyu.edu.hk/).
2	26	Chan FTS	Hong Kong Polytechnic University, Department of Industrial and Systems Engineering. SAR CN (https://www.polyu.edu.hk/ise/).
3	22	Laporte G	University of Montreal (HEC Montreal), Department of Decision Sciences. CA (https://www.hec.ca/mqg/).
4	20	Wang L	Tsinghua University, Department of Automation. (https://www.tsinghua.edu.cn/publish/auen/).
5	19	Tiwari MK	Indian Institute of Technology - Kharagpur, Department of Industrial & Systems Engineering. IN (http://www.iitkgp.ac.in/department/IM).
6	18	Huang GQ	University of Hong Kong, Department of Industrial and Manufacturing Systems Engineering. SAR CN (https://www.imse.hku.hk/).
7	16	Liang L	Chinese Academy of Sciences / University of Science & Technology of China, Department of Management Science. CN (http://en.business.ustc.edu.cn/audms/). / Hefei University of Technology, School of Management. CN (http://en.hfut.edu.cn/index.php).
8	15	Shankar R	Indian Institute of Technology - Delhi, Department of Management Studies. IN (http://dms.iitd.ac.in/).
9	11	Wang ZR	Harbin Institute of Technology, School of Material Science and Engineering. CN.
10	10	Jiang ZB	Shanghai Jiao Tong University, Department of Industrial Engineering and Management.

Data Source: (WoS, 2019)

Thus, the adjusted by the h-index, it is clear on table 6 the leadership of two researchers from Hong Kong Polytechnic University, professors T.C. Edwin Cheng (Faculty of Business Dean) and Felix T. S. Chan. Also, it is noteworthy the ascent of Professor Ling Wang from the Department of Automation at Tsinghua University, with 20 articles from at least 20 citations in WoS Core Collection. As well as the decreasing of Professor Zhibin Jiang who works in the Department of Industrial Engineering and Management at Shanghai Jiao Tong University, that despite having 41 articles published in WoS, only 10 of these have at least 10 citations. The thematic interaction of this group of top ten authors is reflected in Figure 5, standing out as the topics with the most links and occurrence: Supply Chain, Information Sharing and Demand.

In relation to this group of selected articles, only two articles with the participation of Chilean authors stand out: Gutiérrez-Jarpa, Obreque, Laporte and Marianov (2013) work that studies issues about location, coverage, transit systems, traffic capture and networks, and Gutiérrez-Jarpa, Laporte, Marianov and Moccia (2017), regarding rapid transit network design, modal competition and Multi-objective optimization. In which you can see the repeated work of professors Marianov (Department of Electrical Engineering, Pontifical Catholic University of Chile,) and Gutiérrez-Jarpa (School of Industrial Engineering, Pontifical Catholic University of Valparaiso,) in connection with Professor Gilbert Laporte, a prolific researcher at HEC Montreal (University of Montreal).

Figure 5. Thematic co-occurrence network in articles by top ten authors
Data Source: WoS, 2019

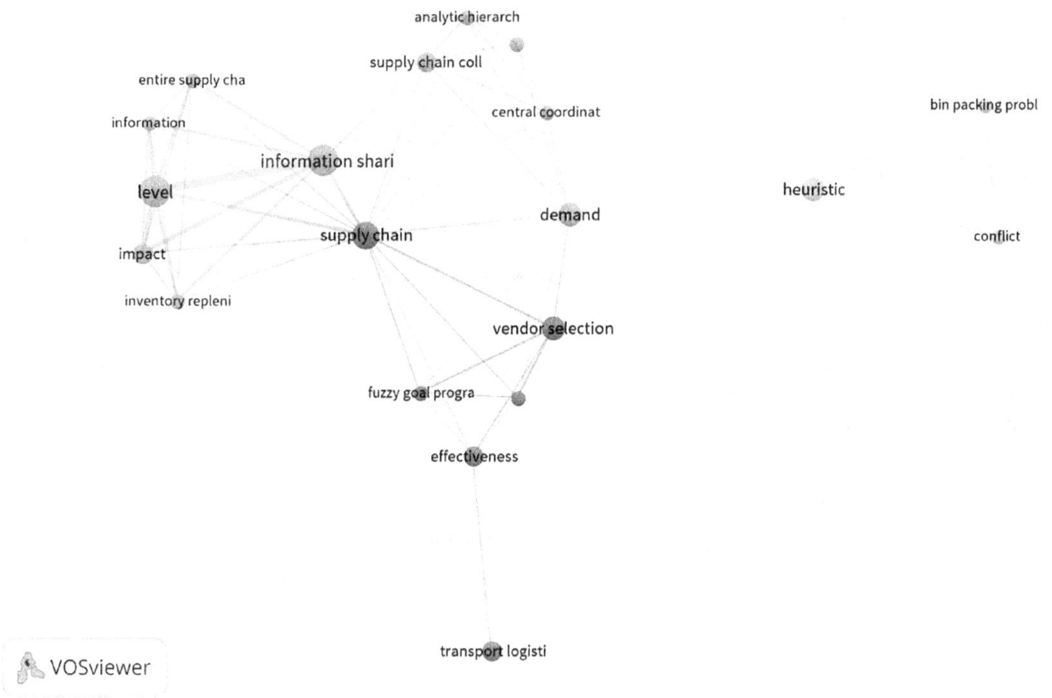

SOLUTIONS AND RECOMMENDATIONS

The low results of Chile in the context of the cluster of knowledge that investigates in industrial engineering need to be solved in favor of national socio-economic development. The scarce 267 articles with local contribution are not only strongly concentrated in a single higher education institution, but also are too disconnected from the large organizations of scientific production worldwide and relevant researchers.

The ways of reversing it can occur in the two fundamental alternative work lines, enunciated by Knoke and Laumann (2012):

1. In this scenery there is a high awareness of the specific area of knowledge to be developed and a high awareness of the appropriate sources that can help to co-produce this desired knowledge, the strategy that should prevail is the establishment of 'pipelines', which allow to joint formal work between Chilean researchers and those of world reference with the idea to establish a possible cohesive subgroup (Maskell, 2014, Henn & Bathelt, 2018), such as the one already established by Pontifical Catholic University of Chile and Pontifical Catholic University of Valparaiso with University of Montreal and which it is just beginning to take shape between Tsinghua University and some four Chilean universities through its Latin America Center. (Tsinghua University, 2018).

2. Another way to consider is an informal work that values the achievements of this group of prolific researchers studying their publications, incorporating them as sources of research in Chile, also incorporating the documents they cite as sources for local research and, above all, giving account of the directions of future research that these works show in order to achieve a structural equiva-

lence. Thus, indirectly, links are established and it manages to insert Chilean researchers into the work network of world-renowned researchers, for this reason they will not go unnoticed given the effect that new citations will have on their scientometric indexes, own effects of an invisible college construction strategy (Price, 1986).

FUTURE RESEARCH DIRECTIONS

It is of future interest of the authors to explore in terms of graph theory and the analysis of social networks the development of world-wide areas in the Industrial Engineering, but little developed in Chile, like interdisciplinarity in the categories: Ergonomics, Applied Psychology, Psychology, Management and Engineering Multidisciplinary.

Another area of interest to continue this research is, from the worldwide research, to reconceptualize the specific fields of interest in the research in industrial engineering, through the analysis of the keyword extraction methods, methods that have received high attention and among those that are network-based methods. (Yang et al., 2018).

In addition to research other areas of engineering worldwide and the possibilities of Chile's insertion in their global knowledge clusters, so that they can contribute to the economic and social development of the country.

CONCLUSION

Seventy years of early mathematical modeling efforts by Bavelas (1948) about groups structure, graph theory goes on giving support and sustenance to human behavioral study from the optic of the social sciences. Preponderance, abundance and information value in knowledge society, has brought to develop new study forms especially on the information of great socio-economic impact such as scientific information, giving way to relational scientometric that allows studying the knowledge production processes (what, how and for whom) and geoeconomics (where). But also establishing knowledge mobility needs in capital terms.

This work has revealed that Chilean scientific production in the industrial engineering field is a very small proportion of the global total, dominated by China, North America and Canada and at the level of organizations, by Chinese, Indian and Canadian universities. In addition, this Chilean production is strong0 concentrated in a few institutions and authors, where the dominant topics of study are not entirely coincident with the one generated by the cluster of knowledge in industrial engineering worldwide. All this makes it difficult to establish linkages and the generation of knowledge networks between Chilean institutions and those of the rest of the world.

In this sense, two complementary ways of development for Chilean research in Industrial Engineering are offered, on the one hand, it establishes institutional and personal contact lines between Chilean researchers and the main research teams in the world so that Chilean talent has a projection in the best positioned journals in Web of Science. This will require an effort of confluence between the research lines of Chilean authors and those of the most prolific authors worldwide. On the other hand, an effort to strengthen the internal research networks in Chile, head by leading Chilean researchers, and nurtured by world-renowned knowledge, it will allow an increase in the number of new researchers integrated into international networks.

All the above is not an immediate process, it will require time and economic and personal effort by institutions and researchers, but necessary for future socioeconomic development of the country.

ACKNOWLEDGMENT

The researchers of this study are both grateful to Data Science Lab of Engineering Faculty at Universidad de Playa Ancha (Chile) for the specialized software training and to Foundation 'School of International Affairs' (Chile) for institutional sponsorship provided in the field of Library and Information Sciences.

REFERENCES

Abbasi, A., Jalili, M., & Sadeghi-Niaraki, A. (2018). Influence of network-based structural and power diversity on research performance. *Scientometrics*, *117*(1), 579–590. doi:10.100711192-018-2879-3

Bavelas, A. (1948). A mathematical model for groups' structure. *Applied Anthropology*, *7*(3), 16-30. Available in: http://www.jstor.org/stable/44135428

Bavelas, A. (1950). Communication patterns in task-oriented groups. *The Journal of the Acoustical Society of America*, *22*(6), 725–730. doi:10.1121/1.1906679

Cárdenas-Tapia, M., Klingler-Kaufman, C., & Rivas-Tovar, L. A. (2012, October). *Structural analysis of a knowledge network through graph theory*. Paper presented at the XVII International Accounting and Administration Congress, School of Accounting and Administration at National Autonomous University of Mexico, Mexico City. Available in: http://congreso.investiga.fca.unam.mx/es/congreso_xvii.php

Cartwright, D., & Zander, A. (Eds.). (1953). *Group dynamics*. Londres: Tavistock.

Ceballos, H. G., Fangmeyer, J. J. Jr, Galeano, N., Juarez, E., & Cantu-Ortiz, F. J. (2017). Impelling research productivity and impact through collaboration: A scientometric case study of knowledge management. *Knowledge Management Research and Practice*, *15*(3), 346–355. doi:10.105741275-017-0064-8

Central Intelligence Agency. (2019). *The World Factbook*. Available in: https://www.cia.gov/library/publications/resources/the-world-factbook/

Cheng, F. F., Huang, Y. W., Tsaih, D. C., & Wu, C. S. (2019). Trend analysis of co-authorship network in Library Hi Tech. *Library Hi Tech*, *37*(1), 43–56. doi:10.1108/LHT-11-2017-0241

Clarivate. (2019). *Master Journal List. Scope Notes: Science Citation Index Expanded*. Available in: http://mjl.clarivate.com.uchile.idm.oclc.org/scope/scope_scie/

CORFO. (1962). *Twenty years of labor 1939 - 1959*. Santiago de Chile: Corporation for the Promotion of Production. Available in: http://www.memoriachilena.cl/602/w3-article-9037.html

Daneshvar-Rouyendegh, B., & Feryal-Can, G. (2012). Selection of working area for industrial engineering students. *Procedia: Social and Behavioral Sciences*, *31*, 15–19. doi:10.1016/j.sbspro.2011.12.008

Delgado-Saab, I., Carapaica, I., & Ortiz-Sosa, D. (2016). Software tool for problem analysis whose abstraction is based on graphs. *Tekhné, 1*(10). Available in: http://revistasenlinea.saber.ucab.edu.ve/temas/index.php/tekhne/article/view/3002

Escobar, A. (2005). Welcome to Cyberia. Notes for an anthropology of the cyberculture. *Revista de Estudios Sociales*, (22): 15–35. doi:10.7440/res22.2005.01

Evers, H. D., Gerke, S., & Menkoff, T. (2010). Knowledge clusters and knowledge hubs: Designing epistemic landscapes for development. *Journal of Knowledge Management, 14*(5), 678–689.

Festinger, L. (1949). The analysis of sociograms using matrix algebra. *Human Relations, 2*(2), 153–158. doi:10.1177/001872674900200205

Gambardella, A., & Giarratana, M. (2010). Organizational attributes and the distribution of rewards in a region: Managerial firms vs. knowledge clusters. *Organization Science, 21*(2), 573–586.

Garfield, E. (1987). Little Science, Big Science. . .and beyond gathers together the major works of Derek de Solla Price. *Essays of an Information Scientist, 10*(11), 72. Available in: http://www.garfield.library.upenn.edu/essays/v10p072y1987.pdf

Gilsing, V., Nooteboom, B., Vanhaverbeke, W., Duysters, G., & van der Oord, A. (2008). Network embeddedness and the exploration of novel technologies: Technological distance, betweenness centrality and density. *Research Policy, 37*(10), 1717–1731. doi:10.1016/j.respol.2008.08.010

Giuliani, E. (2005). *The Structure of Cluster Knowledge Networks Uneven, not Pervasive and Collective*. DRUID Working Papers 05-11, DRUID, Copenhagen Business School, Department of Industrial Economics and Strategy/Aalborg University, Department of Business Studies. Available in: https://ideas.repec.org/p/aal/abbswp/05-11.html

Gutiérrez-Jarpa, G., Laporte, G., Marianov, V., & Moccia, L. (2017). Multi-objective rapid transit network design with modal competition: The case of Concepción, Chile. *Computers & Operations Research, 78*, 27–43. doi:10.1016/j.cor.2016.08.009

Gutiérrez-Jarpa, G., Obreque, C., Laporte, G., & Marianov, V. (2013). Rapid transit network design for optimal cost and origin–destination demand capture. *Computers & Operations Research, 40*(12), 3000–3009. doi:10.1016/j.cor.2013.06.013

Harary, F., & Norman, R. Z. (1953). *Graph theory as a mathematical model in social science*. Ann Arbor, MI: Institute for Social Science.

Henn, S., & Bathelt, H. (2018). Cross-local knowledge fertilization, cluster emergence, and the generation of buzz. *Industrial and Corporate Change, 27*(3), 449–466. doi:10.1093/icc/dtx036

Huggins, R. (2008). The evolution of knowledge clusters. Progress and policy. *Economic Development Quarterly, 22*(4), 277–289. doi:10.1177/0891242408323196

Kim, K., & Chung, Y. (2018). Overview of journal metrics. *Science Editing, 5*(1), 16–20. doi:10.6087/kcse.112

Knoke, D., & Laumann, E. O. (2012). Social organization in national politics areas. Exploration of some structural hypothesis. In *Social network analysis. Origins, theories and applications*. Madrid: Centro de Investigaciones Sociológicas.

Lanktree, C., & Briere, J. (1991, January). *Early data on the trauma symptom checklist for children (TSC-C)*. Paper presented at the meeting of the American Professional Society on the Abuse of Children, San Diego, CA.

Lozares, C. (1996). La teoría de las redes sociales. *Papers*, *48*, 103–126. doi:10.5565/rev/papers/v48n0.1814

Maskell, P. (2014). Accessing remote knowledge - the roles of trade fairs, pipelines, crowdsourcing and listening posts. *Journal of Economic Geography*, *14*(5), 883–902. doi:10.1093/jeg/lbu002

Maynard, H. B., & Zandin, K. B. (Eds.). (2001). *Maynard's industrial engineering handbook* (5th ed.). New York: McGraw-Hill.

National Accreditation Commission - Chile. (2014). *Exempt Resolution N°639 of Postgraduate Accreditation: Doctorate in Engineering Sciences, Industrial Engineering given by the University of Santiago, Chile. Santiago de Chile*. Chile: National Accreditation Commission.

National Accreditation Commission - Chile. (2016). *Exempt Resolution N°852 of Postgraduate Accreditation: Doctorate in Engineering Sciences, Industrial Engineering and Transport given by the Pontifical Catholic University of Chile. Santiago de Chile*. National Accreditation Commission.

National Accreditation Commission - Chile. (2018). *Exempt Resolution N°1023 of Postgraduate Accreditation: Doctorate in Industrial Engineering and Operations Research given by the Adolfo Ibáñez University. Santiago de Chile*. Chile: National Accreditation Commission.

National Accreditation Commission - Chile. (2019). *Advanced Search for Accreditations*. Retrieved from: https://www.cnachile.cl/Paginas/buscador-avanzado.aspx

National Historical Museum of Chile. (2009). *Earthquakes in Chile*. Santiago de Chile: Origo Editions.

Nilsson, M. (2019). Proximity and the trust formation process. *European Planning Studies*, *27*(5), 841–861. doi:10.1080/09654313.2019.1575338

Patrimonio Consultores. (2009). *History of the Production Development Corporation 1939 - 2009*. Santiago de Chile: Patrimonio Consultores. Available in: http://repositoriodigital.corfo.cl/bitstream/handle/11373/7229/HISTORIA%20CORFO%20FINAL.pdf

Pinar, B. S., Günther, B., & Fazleena, J. (2016). A Novel Framework for Achieving Sustainable Value Creation through Industrial Engineering Principles. *Procedia CIRP*, *40*, 516–523. doi:10.1016/j.procir.2016.01.126

Prathap, G. (2018). Eugene Garfield: From the metrics of science to the science of metrics. *Scientometrics*, *114*(2), 637–650. doi:10.100711192-017-2525-5

Prathap, G. (2018b). A bibliometric tale of two cities: Hong Kong and Singapore. *Scientometrics*, *117*(3), 2169–2175. doi:10.100711192-018-2927-z

Price, D. (1976). A general theory of bibliometric and other cumulative advantage processes. *Journal of the Association for Information Science, 27*, 292–306. doi:10.1002/asi.4630270505

Price, D. (1986). *Little science, big science – and beyond*. Columbia University Press.

Price, D. J. S. (1963). *Little Science, Big Science*. Columbia University Press.

Saavedra, I. (1985). Philosophy, science, technology and development. *Ambiente y Desarrollo, 1*(3), 43-54. Available in: http://www.cipmachile.com/web/200.75.6.169/RAD/1984-85/3_Saavedra.pdf

Solow, R. M. (1957). Technical Change and the Aggregate Production Function. *The Review of Economics and Statistics, 39*(3), 312-320. Available in: http://www.jstor.org/stable/1926047

Spinak, E. (1998). Indicadores cienciométricos. *Ciência da Informação, 27*(2), 141-148. Available in: http://revista.ibict.br/ciinf/article/view/795

Tallman, S., Jenkins, M., Henry, N., & Pinch, S. (2004). Knowledge, clusters, and competitive advantage. *Academy of Management Review, 29*(2), 258–271. doi:10.5465/amr.2004.12736089

Tsinghua University. (2018). *Qiu Yong visits Chile to promote cooperation between China and Chile on Higher Education*. Available in: http://news.tsinghua.edu.cn/publish/thunews-en/9670/2018/20181219103403591578325/20181219103403591578325_.html

Vega, A., & Salinas, C. M. (2017). Scientific Production Analysis in Public Affairs of Chile and Peru. Challenges for a Better Public Management. *Lex, 15*(20), 463–478. doi:10.21503/lex.v15i20.1451

Vega, M. A., Benítez, G. J., & Yévenes, S. A. (2005). Generic strategies for regional development in the plastics industry supply chain. *Horizontes Empresariales, 4*(1), 31-45. Available in: http://revistas.ubiobio.cl/index.php/HHEE/article/view/2068

WoS. (2019). *Web of Science Core Collection. Advanced Search*. Available in: https://clarivate.com/products/web-of-science/web-science-form/web-science-core-collection/

Yang, L., Li, K., & Huang, H. (2018). A new network model for extracting text keywords. *Scientometrics, 116*(1), 339–361. doi:10.100711192-018-2743-5

ADDITIONAL READING

De Nooy, W., Mrvar, A., & Batagelj, V. (2018). *Exploratory Social Network Analysis with Pajek* (3rd ed.). New York, NY: Cambridge University Press; doi:10.1017/9781108565691

Jennex, M. E. (2019). *Effective Knowledge Management Systems in Modern Society*. Hershey, PA: IGI Global; doi:10.4018/978-1-5225-5427-1

Kisielnicki, J., & Sobolewska, O. (2019). *Knowledge Management and Innovation in Network Organizations: Emerging Research and Opportunities*. Hershey, PA: IGI Global; doi:10.4018/978-1-5225-5930-6

Meghanathan, N. (2018). *Centrality Metrics for Complex Network Analysis: Emerging Research and Opportunities*. Hershey, PA: IGI Global; doi:10.4018/978-1-5225-3802-8

Mesquita, L. F., Ragozzino, R., & Reuer, J. J. (2017). *Collaborative Strategy: Critical Issues for Alliances and Networks*. Cheltenham: Edward Elgar Publishing. doi:10.4337/9781783479580

Ortega, J. L. (2016). *Social Network Sites for Scientists: A Quantitative Survey*. Oxford: Chandos Publishing; doi:10.1016/C2015-0-00021-1

Özman, M. (2017). *Strategic Management of Innovation Networks*. Cambridge: Cambridge University Press; doi:10.1017/9781107775534

Pshenichny, C., Diviacco, P., & Mouromtsev, D. (2018). *Dynamic Knowledge Representation in Scientific Domains*. Hershey, PA: IGI Global; doi:10.4018/978-1-5225-5261-1

Wani, Z. A., & Zainab, T. (2019). *Scholarly Content and Its Evolution by Scientometric Indicators: Emerging Research and Opportunities*. Hershey, PA: IGI Global; doi:10.4018/978-1-5225-5945-0

Zweig, K. A. (2016). Graph Theory, Social Network Analysis, and Network Science. In A. Z. Katharina (Ed.), *Network Analysis Literacy. A Practical Approach to the Analysis of Networks*. Vienna: Springer; doi:10.1007/978-3-7091-0741-6_2

KEY TERMS AND DEFINITIONS

Bibliometrics: Fraction of the Library and Information Sciences that studies the scientific production edited in books and scientific papers through statistical methods.

Bradford Core: It comprises one or a small number of scientific journals in a specific area of knowledge, which concentrate a high number of articles (usually one third of these) and therefore concentrates the specialized discussion on that subject.

Centrality: It refers to the measurement on a vertex of a mentioned graph, which determines its relative importance within it. The centrality of a degree (the most basic and current), corresponds to the number of links that a node has with the others.

Impact Factor: General name where with the set of metrics on documented scientific activity is known, under the logic that this published knowledge should allow generating social and economic effects, as well as being the substrate for the generation of new knowledge.

Scientometrics: Instrument of sub-discipline of Sociology of Science, which deals with quantitatively evaluate the scientific activity documented in the scientific optical disciplinary sectoral economy or applied, sub-discipline devoted to the study of economics research and development sector (R&D).

ENDNOTE

[1] Organization created in April 1939 by President Mr. Pedro Aguirre-Cerda, within the reconstruction measures of the earthquake occurred 24th January in that year pulverizing the Chillán City and demolished the Concepción City, in the central zone of Continental Chile. (National Historical Museum of Chile, 2009).

APPENDIX

As a way to provide enough information, so that both reviewers and future researchers can replicate the development of this chapter. The WoS codes of the 267 articles studied for the case of Chile are listed below by commas:

WOS:000427766800010,WOS:000413878100032,WOS:000412962400008,WOS:000412962400009,WOS:000412962400010,WOS:000418207900046,WOS:000418207900048,WOS:000414814700038,WOS:000407402600005,WOS:000412361900053,WOS:000413126700008,WOS:000410468600018,WOS:000401878000003,WOS:000407657400038,WOS:000406983100010,WOS:000403723700037,WOS:000399511200002,WOS:000403120300014,WOS:000402929700024,WOS:000401381300003,WOS:000395602400003,WOS:000398563900013,WOS:000394079400015,WOS:000404699400001,WOS:000397462100002,WOS:000394196100016,WOS:000392889600012,WOS:000397371900011,WOS:000392897600026,WOS:000392044300004,WOS:000391078100010,WOS:000390071400003,WOS:000423134100004,WOS:000409171400006,WOS:000403781700001,WOS:000401523000014,WOS:000400511400017,WOS:000396676500006,WOS:000391899700004,WOS:000387195700017,WOS:000384776100002,WOS:000383008100008,WOS:000389112600020,WOS:000386913700011,WOS:000380600000012,WOS:000389219800017,WOS:000384309900006,WOS:000383131000008,WOS:000380594100014,WOS:000378662500022,WOS:000373553800014,WOS:000373632300009,WOS:000374365400003,WOS:000374365400006,WOS:000374774800009,WOS:000371189700019,WOS:000370908300009,WOS:000373740300007,WOS:000373740300014,WOS:000369464000005,WOS:000371844900011,WOS:000368865200007,WOS:000368589300008,WOS:000370079000007,WOS:000366779900013,WOS:000386426300004,WOS:000380161700014,WOS:000379604100006,WOS:000375208100002,WOS:000368589800005,WOS:000367773600006,WOS:000366141100001,WOS:000365125200006,WOS:000359869400007,WOS:000357224000025,WOS:000356208800002,WOS:000357505900009,WOS:000358885800012,WOS:000355709400001,WOS:000359936900010,WOS:000356110400016,WOS:000353403000009,WOS:000356180000011,WOS:000355929000003,WOS:000354550100005,WOS:000349380700001,WOS:000349591400027,WOS:000348893500013,WOS:000353113400001,WOS:000369434200002,WOS:000347736900024,WOS:000346342100010,WOS:000346342100004,WOS:000346341900009,WOS:000345469600025,WOS:000345729500023,WOS:000345507500030,WOS:000345544500007,WOS:000341466500010,WOS:000340300400007,WOS:000340428100018,WOS:000338622100010,WOS:000337013300035,WOS:000334982000011,WOS:000335880000006,WOS:000336669800024,WOS:000332747800003,WOS:000332747800006,WOS:000332747800010,WOS:000332747800009,WOS:000332747800027,WOS:000332747800028,WOS:000332747800015,WOS:000332747800030,WOS:000332747800016,WOS:000332747800005,WOS:000332747800008,WOS:000332747800023,WOS:000333790900004,WOS:000331498900010,WOS:000330088100021,WOS:000329383300010,WOS:000332446600009,WOS:000333986200005,WOS:000332416700010,WOS:000326614300012,WOS:000326610000018,WOS:000324224000023,WOS:000324844300009,WOS:000324467400003,WOS:000324962000013,WOS:000323569900015,WOS:000318156600004,WOS:000313706700009,WOS:000313706700044,WOS:000313706700061,WOS:000314116300008,WOS:000312007300004,WOS:000301216600028,WOS:000312148700003,WOS:000307002700006,WOS:000306032500008,WOS:000310337000006,WOS:000298532900014,WOS:000306200000005,WOS:000305530800006,WOS:000304816000011,WOS:000300741100017,WOS:000295302100014,WOS:000297402700047,WOS:000296390200002,WOS:000296507700009,W

OS:000294397900013,WOS:000288410200002,WOS:000294778500008,WOS:000292662300010,W
OS:000289642700009,WOS:000299476200001,WOS:000285199300012,WOS:000284747000021,W
OS:000281856200008,WOS:000281460600011,WOS:000275524100017,WOS:000271605000005,W
OS:000267320700006,WOS:000271116400025,WOS:000271116400042,WOS:000271116400046,W
OS:000265369200008,WOS:000269848000007,WOS:000268519100014,WOS:000267518000002,W
OS:000266892800008,WOS:000267063000003,WOS:000262882500006,WOS:000265167500015,W
OS:000265315000005,WOS:000264032200050,WOS:000271292100004,WOS:000272986700006,WO
S:000272986700009,WOS:000260690500055,WOS:000260690500059,WOS:000255671800003,WO
S:000259893100022,WOS:000253254500010,WOS:000257095200041,WOS:000255737600091,WO
S:000255583000025,WOS:000254982500036,WOS:000252564100013,WOS:000250369200014,WO
S:000250369200017,WOS:000252514600041,WOS:000250165500009,WOS:000250165500011,WO
S:000257068900003,WOS:000261495000009,WOS:000250637000011,WOS:000247805200009,WO
S:000245029700008,WOS:000244811800059,WOS:000245198600003,WOS:000235584900003,WO
S:000238892800031,WOS:000246634900005,WOS:000234645400012,WOS:000236781400015,WO
S:000236235100002,WOS:000236429900005,WOS:000236294600019,WOS:000236294600020,W
OS:000233711000014,WOS:000235728900003,WOS:000232937200010,WOS:000231082100009,W
OS:000229609100101,WOS:000226391000007,WOS:000225804800006,WOS:000224742300018,W
OS:000228245400005,WOS:000221346000011,WOS:000189018400013,WOS:00188968500014,W
OS:000187510100067,WOS:000187510100101,WOS:000220517000005,WOS:000185923400001,W
OS:000181096100003,WOS:000179418400010,WOS:000179163300033,WOS:000178711000031,W
OS:000178566900010,WOS:000176237300008,WOS:000175760900004,WOS:000175722300007,W
OS:000228244100002,WOS:000174334500021,WOS:000165349500001,WOS:000089576800010,WO
S:000083884000006,WOS:000083884000019,WOS:000083300900019,WOS:000083300900022,WO
S:000083170600006,WOS:00080065300011,WOS:000080065300034,WOS:000074574500003,WO
S:000074574500006,WOS:000072207900001,WOS:A1998YJ56900004,WOS:A1997YG83700004,
WOS:A1997YB79600018,WOS:A1997XH57700024,WOS:A1997WA73000131,WOS:A1996VD772
00009,WOS:A1996UQ66300006,WOS:A1995TW09100040,WOS:A1994NL36100007,WOS:A1993
KL48800052,WOS:A1992JT68400007,WOS:A1992HH22600004,WOS:A1991FE90000004,WOS:A
1990CV71400002,WOS:A1989U774500005,WOS:A1985AYH5200003, WOS:A1981MH07300021.
Regarding the 3,978 articles reviewed for the global analysis, they are related to the following search vector and the restrictions that are detailed:

(WC=(Engineering, Industrial)) AND DOCUMENT TYPES: (Article)
Refined by: PUBLICATION YEARS: (2017 OR 2016 OR 2015 OR 2014 OR 2013 OR 2012 OR 2011 OR 2010 OR 2009 OR 2008 OR 2007 OR 2006 OR 2005 OR 2004) AND SOURCE TITLES: (JOURNAL OF MATERIALS PROCESSING TECHNOLOGY OR INTERNATIONAL JOURNAL OF PRODUCTION RESEARCH OR INTERNATIONAL JOURNAL OF PRODUCTION ECONOMICS OR COMPUTERS OPERATIONS RESEARCH OR COMPUTERS INDUSTRIAL ENGINEERING) AND AUTHORS: (CHENG TCE OR TIWARI MK OR CHAN FTS OR DOBRZANSKI LA OR WANG L OR JABER MY OR DOLGUI A OR LI L OR LAPORTE G OR LIU Y OR CHEN J OR WANG Y OR KUMAR S OR YANG H OR LI Y OR GUNASEKARAN A OR GLOCK CH OR WANG J OR HUANG GQ OR KIM J OR ZHANG J OR GHARBI A OR WU CC OR JIANG ZB OR YANG J OR LI J OR LEE S OR LIANG L OR LEE JH OR LI XP OR SHANKAR R OR TAVAKKOLI-MOGHADDAM R OR ZANDIEH M OR ZHANG Y OR CHANG PC OR WANG SY OR WU J OR CHEN X OR LEE

WC OR WANG XY OR WU Z OR GAO L OR LIN J OR LIN SW OR MINNER S OR WANG JB OR GHOMI SMTF OR WANG Q OR WANG ZJ OR WANG ZR OR ZHANG H OR ZHANG L OR CHEN Y OR CHU CB OR DISNEY SM OR HASHMI MSJ OR TANG O OR WU Y OR CHIEN CF OR IM YT OR KANG CG OR KIM KH OR LEE J OR LI XY OR MOSHEIOV G OR STEVENSON M OR TEKKAYA AE OR WU CH OR YING KC OR ZANONI S OR ALTAN T OR CARDENAS-BARRON LE OR FRAMINAN JM OR LEE YH OR LI H OR LIAO CJ OR LIN YK OR SARKIS J OR WANG X OR XIE M OR ZHANG X OR GENDREAU M OR GOYAL SK OR KOPAC J OR LI JS OR LIU M OR MORI K OR NOF SY OR LAI KH OR LIU X OR PASTOR R OR PEARN WL OR PERSONA A OR SUN LY OR WAGNER SM OR ZHANG YF OR ZHENG L OR ZHOU J OR BATTINI D OR CHEN L OR CHUNG SH OR KHOO MBC OR LEE DH OR LI DC OR LI G OR LI X OR MAR- TINS PAF OR REZG N OR YANG Y OR YUAN SJ OR ZHANG M OR ZHANG Q OR ZHAO J OR CASTAGLIOLA P OR CORDEAU JF OR GEN M OR KIM YD OR LIU JY OR MORABITO R OR RAHMAN M OR SEGERSTEDT A OR TAKAHASHI K OR TANG JF OR TANG LX OR THURER M OR TU YL OR WANG F OR WANG H OR WANG YM OR YANAGIMOTO J OR YEUNG ACL OR ZHAO XD OR CHEN CC OR LI Z OR LIU XH OR QIN Y OR SARKER BR OR WANG CH OR WANG Z OR WEE HM OR WU T OR ZHANG JX OR BAYRAKTAR E OR CHEN H OR DAVIM JP OR GOVINDAN K OR GUPTA JND OR JAYARAM J OR KENNE JP OR LI B OR LI M OR LI YJ OR LIU Q OR MLADENOVIC N OR TIEU AK OR TORABI SA OR VAN TYNE CJ OR WU CW OR YANG M OR YU YG OR ZHANG C)

Timespan: 1978-2017. Indexes: SCI-EXPANDED.

Chapter 18
Recent Developments on the Basics of Fuzzy Graph Theory

Ganesh Ghorai
Vidyasagar University, India

Kavikumar Jacob
 https://orcid.org/0000-0002-2314-4600
Universiti Tun Hussein Onn Malaysia, Malaysia

ABSTRACT

In this chapter, the authors introduce some basic definitions related to fuzzy graphs like directed and undirected fuzzy graph, walk, path and circuit of a fuzzy graph, complete and strong fuzzy graph, bipartite fuzzy graph, degree of a vertex in fuzzy graphs, fuzzy subgraph, etc. These concepts are illustrated with some examples. The recently developed concepts like fuzzy planar graphs are discussed where the crossing of two edges are considered. Finally, the concepts of fuzzy threshold graphs and fuzzy competitions graphs are also given as a generalization of threshold and competition graphs.

INTRODUCTION

The problem of "Seven bridges of Koningsberg" gives birth to a new dimension of mathematics. In late 1730, Leohhard Euler solved the problem using a model named as "Graph". Since then, graphs are being used as a mathematical tool for modeling real world problems. Graphs can be used in the situation where the objects are connected by some rules or relations. For instance, tramway where tram depots are connected by roads, circuit design where nodes are connected by wires, etc. Nowadays, graphs are being used in games and recreational mathematics too. But, the demand of real world does not end here. There are a lot of uncertainty in the real world. One of them is vagueness or fuzziness. Thus researchers are now studying a new branch of mathematics called "Fuzzy graph". Rosenfeld (1975) first considered fuzzy relations on fuzzy sets and developed the theory of fuzzy graphs. Many real life applications such as time scheduling, networking, communication, image segmentation, data mining, etc. can be solved by fuzzy graphs.

DOI: 10.4018/978-1-5225-9380-5.ch018

The definition of a fuzzy graph is given below.

BACKGROUND

A *fuzzy graph* (Rosenfeld, 1975) $\xi = (V, \sigma, \mu)$ of the graph $G^* = (V, E)$ is a non-empty set V together with a pair of functions $\sigma : V \rightarrow [0,1]$ and $\mu : V \times V \rightarrow [0,1]$ such that for all $x, y \in V$,

$$\mu(x, y) \leq \min\{\sigma(x), \sigma(y)\},$$

where $\sigma(x)$ and $\mu(x, y)$ represent the membership values of the vertex x and of the edge (x, y) in ξ respectively. A loop at a vertex x in a fuzzy graph is represented by $\mu(x, x) \neq 0$. An edge is non-trivial if $\mu(x, y) \neq 0$.

An example of fuzzy graph is given Figure 1.

In Figure 1, u, v, w, s, t are the vertices with membership values 0.7, 0.6, 0.8 respectively. There are 6 edges $(u, v), (u, w), (v, w), (w, s), (w, t)$ and (s, t) with membership values 0.5, 0.7, 0.6, 0.7, 0.8 and 0.5 respectively.

A fuzzy graph can be drawn in many different ways by changing the positions of the vertices and drawing the edges by straight lines or curved lines. It is immaterial whether the edges are drawn straight line or curved, short or long. The important thing is the relationship among the vertices.

If the vertex set V and edge set E are finite then the graph is called a finite graph otherwise, it is called infinite graph. In general, a graph means a finite graph.

Directed and Undirected Fuzzy Graphs

Directed fuzzy graphs (or simply fuzzy digraph) are the fuzzy graphs in which the fuzzy relations between edges are not necessarily symmetric. The definition of directed fuzzy graph is as follows:

Definition 2.1 (Mordeson and Nair, 1996) *Directed fuzzy graph (fuzzy digraph)* $\overrightarrow{\xi} = (V, \sigma, \overrightarrow{\mu})$ *is a non-empty set* V *together with a pair of functions* $\sigma : V \rightarrow [0,1]$ *and* $\overrightarrow{\mu} : V \times V \rightarrow [0,1]$ *such that for all* $x, y \in V$, $\overrightarrow{\mu}(x, y) \leq \sigma(x) \wedge \sigma(y)$.

Figure 1. An example of fuzzy graph

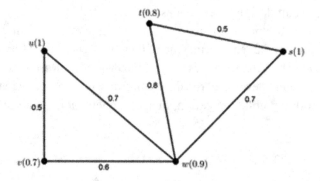

Since $\vec{\mu}$ is well defined, a fuzzy digraph has at most two directed edges (which must have opposite directions) between any two vertices. Here $\vec{\mu}(u,v)$ is denoted by the membership value of the edge $\overrightarrow{(u,v)}$. The loop at a vertex x is represented by $\vec{\mu}(x,x) \neq 0$. Here $\vec{\mu}$ need not be symmetric as $\vec{\mu}(x,y)$ and $\vec{\mu}(y,x)$ may have different values. The *underlying crisp graph of directed fuzzy graph* is the graph similarly obtained except the directed arcs are replaced by undirected edges.

Walks, Paths and Cycles

A **path** P in a fuzzy graph $\xi = (V, \sigma, \mu)$ is a sequence of distinct vertices

$$x = v_1, v_2, \ldots, y = v_n \, (n \geq 2)$$

such that

$$\mu(v_i, v_{i+1}) > 0, i = 1, 2, \ldots, (n-1).$$

Here, $n-1$ is called the **length** of the path P. A path P is a **cycle** if $v_1 = v_n$ and $n \geq 4$. That is $P = (V, \sigma, \mu)$ is a cycle in ξ if and only if (V, σ^*, μ^*) is a cycle in ξ^*. A cycle $P = (V, \sigma, \mu)$ is a **fuzzy cycle** if it contains more than one weak edge (i.e., there is no unique $(x,y) \in \nu^*$ such that $\mu(x,y) = \wedge\{\mu(u,v) : (u,v) \in \nu^*\}$).

$\min_{1 \leq i \leq n-1} \mu(v_i, v_{i+1})$ is called the strength of the path P. The maximum of the strengths of all paths in ξ from x to y is called the strength of connected between x and y and is denoted by $CONN_\xi(x,y)$.

We call ξ to be **connected** if $CONN_\xi(x,y) > 0$ for all $x, y \in V$. Otherwise, ξ is said to be disconnected.

Degree of a Vertex

Definition 2.2 *Let $\xi = (V, \sigma, \mu)$ be a fuzzy graph of the graph $G^* = (V, E)$. The degree of a vertex u is denoted by $d_\xi(u)$ where $d_\xi(u) = \sum_{(u,v) \in E} \mu(u,v)$.*

The minimum degree of ξ is

$$\delta(\xi) = \wedge\{d_\xi(u) : u \in V\}$$

and the minimum degree of ξ is

$$\Delta(\xi) = \vee\{d_\xi(u) : u \in V\}.$$

A fuzzy graph $\xi = (V, \sigma, \mu)$ is said to be *regular*(Nagoorgani and Radha, 2008) if $d(v) = k$, a positive real number, for all $v \in V$. If each vertex of ξ has same total degree k, then ξ is said to be a *totally regular* fuzzy graph. A fuzzy graph is said to be *irregular*(Nagoorgani and Latha, 2012), if there is a vertex which is adjacent to vertices with distinct degrees. A fuzzy graph is said to be *neighborly irregular*(Nagoorgani and Latha, 2012), if every two adjacent vertices of the graph have different degrees. A fuzzy graph is said to be *totally irregular*, if there is a vertex which is adjacent to vertices with distinct total degrees. If every two adjacent vertices have distinct total degrees of a fuzzy graph then it is called *neighborly total irregular*(Nagoorgani and Latha, 2012). A fuzzy graph is called *highly irregular*(Nagoorgani and Latha, 2012) if every vertex of G is adjacent to vertices with distinct degrees.

SOME SPECIAL FUZZY GRAPHS

In this section, we introduce some special types of fuzzy graphs like complete and strong fuzzy graph, bipartite fuzzy graph, fuzzy subgraph, complement of a fuzzy graph, homomorphism of fuzzy graphs etc.

Complete and Strong Fuzzy Graph

The definition of complete fuzzy graph is given by AL-Hawary.

Definition 3.1 (AL-Hawary, 2011) A fuzzy graph $\xi = (V, \sigma, \mu)$ is complete if

$$\mu(u, v) = \min\{\sigma(u), \sigma(v)\}$$

for all $u, v \in V$, where (u, v) denotes the edge between the vertices u and v.

ξ is said to be strong if

$$\mu(u, v) = \min\{\sigma(u), \sigma(v)\}$$

for all $(u, v) \in E$.

If an edge (x, y) of a fuzzy graph satisfies the condition

$$\mu(x, y) = \min\{\sigma(x), \sigma(y)\},$$

then this edge is called effective edge (Nagoorgani and Hussain, 2009). Two vertices are said to be effective adjacent if they are the end vertices of the same effective edge. Then the effective incident degree of a fuzzy graph is defined as number of effective incident edges on a vertex v. If all the edges of a fuzzy graph are effective, then the fuzzy graph becomes complete fuzzy graph. A pendent vertex in a fuzzy graph is defined as a vertex of an effective incident degree one. A fuzzy edge is called a fuzzy *pendant edge*(Samanta and Pal, 2013), if one end vertex is fuzzy pendant vertex. The membership value of the pendant edge is the minimum among the membership values of the end vertices.

Bipartite Fuzzy Graphs

Definition 3.2 (Samanta and Pal, 2015) *A fuzzy graph* $\xi = (V, \sigma, \mu)$ *is said to be bipartite if the vertex set* V *can be partitioned into two nonempty sets* V_1 *and* V_2 *such that* $\mu(v_1, v_2) = 0$ *if* $v_1, v_2 \in V_1$ *or* $v_1, v_2 \in V_2$. *Further, if*

$$\mu(v_1, v_2) = \min\{\sigma(v_1), \sigma(v_2)\}$$

for all $v_1 \in V_1$ *and* $v_2 \in V_2$, *then* ξ *is called a complete bipartite fuzzy graph and is denoted by* K_{σ_1, σ_2} *where* σ_1, σ_2 *are respectively the restrictions of* σ *to* V_1 *and* V_2.

Fuzzy Subgraph

The fuzzy graph $\xi' = (V, \tau, \nu)$ is called a *partial fuzzy subgraph* (Mordeson and Nair, 2000) of $\xi = (V, \sigma, \mu)$ if $\tau(x) \leq \sigma(x)$ for all $x \in V$ and $\nu(x, y) \leq \mu(x, y)$ for all $x, y \in V$.

Similarly, the fuzzy graph $\xi_1 = (V', \tau, \nu)$ is called a fuzzy subgraph of $\xi = (V, \sigma, \mu)$ induced by V' if $V' \subset V$, $\sigma(x) = \tau(x)$ for all $x \in V'$ and $\mu(x, y) = \nu(x, y)$ for all $x, y \in V'$. It is clear that fuzzy subgraph of a fuzzy graph is a special case of a partial fuzzy subgraph of the fuzzy graph.

Example 3.3 *Let* $\xi = (V, \sigma, \mu)$ *be a fuzzy graph (see Figure 2) where* $V = \{a, b, c\}$,

$$\sigma(a) = 0.5, \sigma(b) = 1, \sigma(c) = 0.9$$

and

$$\mu(a, b) = 0.4, \mu(b, c) = 0.8, \mu(a, c) = 0.5.$$

Let $\xi' = (V, \tau, \nu)$ *where*

$$\tau(a) = 0.4, \tau(b) = 0.9, \tau(c) = 0.7$$

Figure 2. Fuzzy graph with its partial fuzzy subgraph and fuzzy subgraph

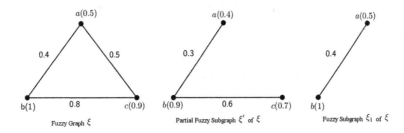

and

$\nu(a,b) = 0.3, \nu(b,c) = 0.6$.

Then clearly ξ' is a partial fuzzy subgraph of the fuzzy graph ξ. Also if $V' = \{a,b\}$ and $\xi_1 = \{V', \tau, \nu\}$ where $\tau(a) = 0.5, \tau(b) = 1$ and $\nu(ab) = 0.4$, then ξ_1 is the induced fuzzy subgraph of G induced by V'.

Complement of Fuzzy Graph

Mordeson and Nair (2000) defined the *complement* of a fuzzy graph $\xi = (V, \sigma, \mu)$ which is also a fuzzy graph $\xi' = (V, \sigma', \mu')$ where $\sigma'(u) = \sigma(u)$ for all $u \in V$ and

$$\mu'(u,v) = \begin{cases} 0, & if\ \mu(u,v) > 0, \\ \sigma(u) \wedge \sigma(v), & otherwise. \end{cases}$$

From this definition one can show that $(\xi')' = \xi$ if and only if ξ is strong. To avoid this situation, Sunitha and Vijaya Kumar (2002) modified this definition and proposed the following definition:

The complement of a fuzzy graph $\xi = (V, \sigma, \mu)$ is the fuzzy graph $\xi' = (V, \sigma', \mu')$ where $\sigma'(u) = \sigma(u)$ for all $u \in V$ and

$$\mu'(u,v) = \sigma(u) \wedge \sigma(v) - \mu(u,v)$$

for all $u, v \in V$.

In this case, one can check that $(\xi')' = \xi$.

Isomorphism of Fuzzy Graphs

A **homomorphism**(Bhutani, 1989) between two fuzzy graphs $\xi_1 = (V_1, \sigma_1, \mu_1)$ and $\xi_2 = (V_2, \sigma_2, \mu_2)$ is a map $f : V_1 \to V_2$ which satisfies $\sigma_1(x) \le \sigma_2(f(x))$ for all x\in V_1 and $\mu_1(x,y) \le \mu_2(f(x), f(y))$ for all $x, y \in V_1$, where V_1 is the set of vertices of G_1 and V_2 is that of G_2. A fuzzy graph G_1 is said to be **homomorphic** to G_2 if there exist an homomorphism between G_1 and G_2.

An **isomorphism**(Bhutani,1989) between two fuzzy graphs $\xi_1 = (V_1, \sigma_1, \mu_1)$ and $\xi_2 = (V_2, \sigma_2, \mu_2)$ is a bijective homomorphism $f : V_1 \to V_2$ which satisfies $\sigma_1(x) = \sigma_2(f(x))$ for all $x \in V_1$ and $\mu_1(x,y) = \mu_2(f(x), f(y))$ for all $x, y \in V_1$ where V_1 is the set of vertices of G_1 and V_2 is that of G_2. A fuzzy graph G_1 is said to be **isomorphic** to G_2 if there exist an isomorphism between G_1 and G_2.

Operations on Fuzzy Graphs

Many operations are defined on fuzzy graphs. Some of them are introduced here.

Definition 3.4 (AL-Hawary, 2011) *The semi-strong product of two fuzzy graphs* $G_1 = (V_1, \sigma_1, \mu_1)$ *and* $G_2 = (V_2, \sigma_2, \mu_2)$ *of the graphs* $G_1^* = (V_1, E_1)$ *and* $G_2^* = (V_2, E_2)$ *respectively, where it is assumed that* $V_1 \cap V_2 = \varnothing$ *, is defined to be the fuzzy graph*

$$G_1 \bullet G_2 = (\sigma_1 \bullet \sigma_2, \mu_1 \bullet \mu_2)$$

of the graph $G^* = (V_1 \times V_2, E)$ *such that*

$$E = \{(u, v_1)(u, v_2) \mid u \in V_1, v_1 v_2 \in E_2\} \cup \{(u_1, v_1)(u_2, v_2) \mid u_1 u_2 \in E_1, v_1 v_2 \in E_2\}$$

and

1. $(\sigma_1 \bullet \sigma_2)(u, v) = \sigma_1(u) \wedge \sigma_2(v)$ for all $(u, v) \in V_1 \times V_2$,
2. $(\mu_1 \bullet \mu_2)((u, v_1)(u, v_2)) = \sigma_1(u) \wedge \mu_2(v_1 v_2)$,
3. $(\mu_1 \bullet \mu_2)((u_1, v_1)(u_2, v_2)) = \mu_1(u_1 u_2) \wedge \mu_2(v_1 v_2)$.

Definition 3.5 (AL-Hawary, 2011) *The strong product of two fuzzy graphs* $G_1 = (V_1, \sigma_1, \mu_1)$ *and* $G_2 = (V_2, \sigma_2, \mu_2)$ *of the graphs* $G_1^* = (V_1, E_1)$ *and* $G_2^* = (V_2, E_2)$ *respectively, where it is assumed that* $V_1 \cap V_2 = \varnothing$ *, is defined to be the fuzzy graph*

$$G_1 \otimes G_2 = (\sigma_1 \otimes \sigma_2, \mu_1 \otimes \mu_2)$$

of the graph $G^* = (V_1 \times V_2, E)$ *such that*

$$E = \{(u, v_1)(u, v_2) \mid u \in V_1, v_1 v_2 \in E_2\} \cup \{(u_1, w)(u_2, w) \mid w \in V_2, u_1 u_2 \in E_1\} \cup \{(u_1, v_1)(u_2, v_2) \mid u_1 u_2 \in E_1, v_1 v_2 \in E_2\}$$

and

1. $(\sigma_1 \otimes \sigma_2)(u, v) = \sigma_1(u) \wedge \sigma_2(v)$ for all $(u, v) \in V_1 \times V_2$,
2. $(\mu_1 \otimes \mu_2)((u, v_1)(u, v_2)) = \sigma_1(u) \wedge \mu_2(v_1 v_2)$,
3. $(\mu_1 \otimes \mu_2)((u_1, w)(u_2, w)) = \sigma_2(w) \wedge \mu_1(u_1 u_2)$,
4. $(\mu_1 \otimes \mu_2)((u_1, v_1)(u_2, v_2)) = \mu_1(u_1 u_2) \wedge \mu_2(v_1 v_2)$.

Definition 3.6 (AL-Hawary, 2011) *The direct product of two fuzzy graphs* $G_1 = (V_1, \sigma_1, \mu_1)$ *and* $G_2 = (V_2, \sigma_2, \mu_2)$ *of the graphs* $G_1^* = (V_1, E_1)$ *and* $G_2^* = (V_2, E_2)$ *respectively such that* $V_1 \cap V_2 = \varnothing$, *is defined to be the fuzzy graph*

$$G_1 \sqcap G_2 = (\sigma_1 \sqcap \sigma_2, \mu_1 \sqcap \mu_2)$$

of the graph $G^* = (V_1 \times V_2, E)$ *such that*

$$E = \{(u_1, v_1)(u_2, v_2) \mid u_1 u_2 \in E_1, v_1 v_2 \in E_2\}$$

and

1. $(\sigma_1 \sqcap \sigma_2)(u, v) = \sigma_1(u) \wedge \sigma_2(v)$ for all $(u, v) \in V_1 \times V_2$,
2. $(\mu_1 \sqcap \mu_2)((u_1, v_1)(u_2, v_2)) = \mu_1(u_1 u_2) \wedge \mu_2(v_1 v_2)$.

Different Types of Arcs (Mathew and Sunitha, 2009)

An arc (u, v) in $\xi = (V, \sigma, \mu)$ is said to *strong* if $\mu(u, v) > 0$ and

$$\mu(u, v) \geq CONN_{\xi - (u,v)}(u, v)$$

where $\xi - (u, v)$ is the fuzzy subgraph obtained from ξ by deleting the arc (u, v).
   The arc (u, v) in ξ is said to be

1. α-strong if $\mu(u, v) > CONN_{\xi - (u,v)}(u, v)$,
2. β-strong if $\mu(u, v) = CONN_{\xi - (u,v)}(u, v)$,
3. δ-arc if $\mu(u, v) < CONN_{\xi - (u,v)}(u, v)$.

Example 3.7 *Let* $\xi = (V, \sigma, \mu)$ *be a fuzzy graph where* $V = \{u_1, u_2, u_3, u_4\}$, $\sigma(u_i) = 1$ *for* $i = 1, 2, 3, 4$ *and* $\mu(u_1, u_2) = 0.5$, $\mu(u_2, u_3) = 0.9$, $\mu(u_3, u_4) = 0.9$, $\mu(u_1, u_4) = 0.5$, $\mu(u_2, u_4) = 0.6$. *Here,*

$$\mu(u_2, u_3) = 0.9 > 0.6 = CONN_{\xi - (u_2, u_3)}$$

and

$$\mu(u_3, u_4) = 0.9 > 0.6 = CONN_{\xi - (u_3, u_4)}.$$

So (u_2, u_3) and (u_3, u_4) are α-strong arcs. Again,

$$\mu(u_1, u_2) = \mu(u_1, u_4) = 0.5 = CONNG_{\xi-(u_1,u_2)} = CONNG_{\xi-(u_1,u_4)}.$$

So, (u_1, u_2) and (u_1, u_4) are β-strong arcs. Also,

$$\mu(u_2, u_4) = 0.6 < 0.5 = CONNG_{\xi-(u_2,u_4)}$$

and so (u_2, u_4) is a δ-arc.

RECENT DEVELOPMENTS ON FUZZY GRAPHS

In this section, we give a very basic idea of fuzzy planar graph, fuzzy threshold graph and fuzzy competition graph with several examples. All these concepts have many important significance to model many real life problems.

Fuzzy Planar Graphs

Planarity is important in connecting the wire lines, gas lines, water lines, printed circuit design, etc. But, some times little crossing may be accepted to these design of such lines/ circuits. So fuzzy planar graph is an important topic for these connections.

A crisp graph is called non-planar graph if there is at least one crossing between the edges for all possible geometrical representations of the graph. Let a crisp graph G has a crossing for a certain geometrical representation between two edges (a, b) and (c, d). In fuzzy concept, we say that this two edges have membership values 1. If we remove the edge (c, d), the graph becomes planar. In fuzzy sense, we say that the edges (a, b) and (c, d) have membership values 1 and 0 respectively.

Figure 3. Example of a fuzzy graph where crossing has no significant

Let $\xi = (V, \sigma, \mu)$ be a fuzzy graph and for a certain geometric representation, the graph has only one crossing between two fuzzy edges $((w, x), \mu(w, x))$ and $((y, z), \mu(y, z))$. Then three cases may arise.

Case(i): If $\mu(w, x) = 1$ and $\mu(y, z) = 0$ or $\mu(w, x) = 0$ and $\mu(y, z) = 0$, then we say that the fuzzy graph has no crossing. For example, let $\mu(w, x)$ represents the crowdness of the road (w, x) (see Figure 3). In Figure 3, both $\mu(w, x)$ and $\mu(y, z)$ are tends to zero. Therefore, in this case, we can say that there is no crossing.

Case(ii): If $\mu(w, x)$ has value 1 or near to 1 and $\mu(y, z)$ has value near to 0, the crossing is not significant (see Figure 4).

Case(iii): If $\mu(w, x)$ has value near to 1 and $\mu(w, x)$ has value near to 1, then the crossing becomes highly significant for the planarity (see Figure 5).

Figure 4. Example of a fuzzy graph where crossing between edges is not significant

Figure 5. Example of a fuzzy graph where crossing between edges is highly significant

Before going to the definition fuzzy planar graph given by Samanta and Pal (2015), some co-related terms are discussed below.

In a fuzzy multigraph, when two edges intersect at a point, a value is assigned to that point. Suppose $\psi = (V, \sigma, E)$ be a fuzzy multigraph where E contains two edges $((a,b), \mu^k(a,b))$ and $((c,d), \mu^l(c,d))$ intersecting at a point P.

Strength of the fuzzy edge (a,b) can be measured by the value

$$I_{(a,b)} = \frac{\mu^k(a,b)}{\min\{\sigma(a), \sigma(b)\}}.$$

If $I_{(a,b)} \geq 0.5$, then the fuzzy edge is called strong, otherwise weak.

The intersecting value at the point P is given by

$$\mathcal{I}_P = \frac{I_{(a,b)} + I_{(c,d)}}{2}.$$

If the number of point of intersections in a fuzzy multigraph increases, planarity decreases. So for fuzzy multigraph, $\mathcal{I}_P$ is inversely proportional to the planarity.

Definition 4.1 (Samanta and Pal, 2015) *Let ψ be a fuzzy multigraph and for a certain geometrical representation $P_1, P_2, \ldots, P_z$ be the points of intersections between the edges. ψ is said to be fuzzy planar graph with degree of planarity f, where*

$$f = \frac{1}{1 + \{\mathcal{I}_{P_1} + \mathcal{I}_{P_2} + \ldots + \mathcal{I}_{P_z}\}}.$$

It is obvious that *f* is bounded and the range of *f* is $0 < f \leq 1$.

If there is no point of intersection for a certain geometrical representation of a fuzzy planar graph, then its degree of planarity is 1. In this case, the underlying crisp graph of this fuzzy graph is the crisp planar graph. If *f* decreases, then the number of points of intersection between the edges increases and obviously the nature of planarity decreases. From this analogy, one can say that every fuzzy graph is a fuzzy planar graph with certain degree of planarity.

Example 4.2 *Let us consider the fuzzy planar graph of Figure 6. Here $V = \{a, b, c, d, e\}$ is the non-empty vertex set with*

$$\sigma(a) = 0.8, \sigma(b) = 0.7, \sigma(c) = 0.6, \sigma(d) = 0.9, \sigma(e) = 0.9,$$

$$\mu^1(a,b) = 0.6, \mu^1(a,c) = 0.5, \mu^1(a,d) = 0.45, \mu^2(a,d) = 0.3, \mu^1(b,c) = 0.4, \mu^1(b,d) = 0.55,$$

$$\mu^1(c,d) = 0.6, \mu^1(a,e) = 0.8, \mu^1(c,e) = 0.5, \mu^1(e,d) = 0.8, \mu^1(e,b) = 0.7.$$

Figure 6. Example of fuzzy planar graph with degree of planarity 0.4683

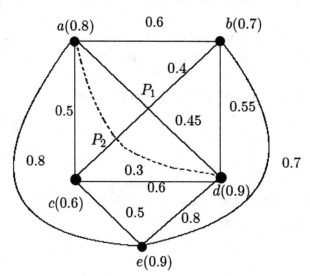

The fuzzy multigraph, (V, σ, E) has two point of intersections P_1 and P_2. P_1 is a point between the edges $((a,d), 0.45)$ and $((b,c), 0.4)$ and P_2 is between $((a,d), 0.3)$ and $((b,c), 0.4)$. For the edge $((a,d), 0.45)$,

$$I_{(a,d)} = \frac{0.45}{0.8} = 0.5625$$

and for the edge $((b,c), 0.4)$,

$$I_{(b,c)} = \frac{0.4}{0.6} = 0.6667.$$

For the first point of intersection P_1, intersecting value $\mathcal{I}_{P_1}$ is 0.6146 and that for the second point of intersection P_2, $\mathcal{I}_{P_2} = 0.5208$. Therefore, the degree of planarity for the fuzzy multigraph shown in Figure 6 is 0.4683.

Note 4.3 *The following important results are true for a fuzzy planar graph:*

1. If the degree of planarity of a fuzzy planar graph is greater than 0.5 then there is at most one intersection between two effective edges and if the degree of planarity is ≥0.67, then there is no intersection between two effective edges,

2. A fuzzy complete graph with five vertices K_5 or fuzzy complete bipartite graph $K_{3,3}$ has degree of planarity equal to 0.5.

Fuzzy Threshold Graphs

Threshold graphs play an important role in graph theory as well as in several applied areas such as psychology, computer science, scheduling, etc. These graphs can be used to control the flow of information between processors, much like the traffic lights used in controlling the flow of the traffic.

Chvatal and Hammer (1973) gives the name "threshold graphs" and studied the graphs for their application in set packing problems. Ordman [13] found the use of graphs in resource allocation problems. Chvatal and Hammer defined threshold graph as follows. A graph $G = (V, E)$ is a threshold graph when there exists non-negative reals $w_v, v \in V$ and t such that $W(U) \leq t$ if and only if $U \subseteq V$ is stable set where $W(U) = \sum_{v \in U} w_v$. So $G = (V, E)$ is a threshold graph whenever one can assign vertex weights such that a set of vertices is stable if and only if its total weight does not exceed a certain threshold. The threshold dimension, $t(G)$ of a graph G is the minimum number k of threshold subgraphs $T_1, T_2, \ldots, T_k$ of G that cover the edge set of G. Threshold partition number, denoted by $tp(G)$, is the minimum number of edge disjoint threshold subgraphs needed to cover $E(G)$.

Suppose that there are ATMs that are linked with bank branches. Each ATM has a threshold limit amount. This limit sets the maximum withdrawals per day. The goal is to determine a replenishment schedule for allocating cash inventory at bank branches to service a preassigned subset of ATMs. The problem can be modeled as a threshold graph since each ATM has a threshold of transactions. In reality, each ATM can have a different withdrawal limit. These withdrawal limits can be represented by a fuzzy set. The threshold limit can be set such that the branches can replenish the ATMs without ever hampering the flow of transactions in each ATM. Motivated by this example, we investigate use of the fuzzy threshold graph to model and solve this type of real world problems.

Definition of fuzzy threshold graphs are given by Samanta and Pal (2011) as below:

Definition 4.4 (Samanta and Pal, 2011) *A fuzzy graph* $\xi = (V, \sigma, \mu)$ *is called a fuzzy threshold graph if there exists non-negative real number T such that* $\sum_{u \in U} \sigma(u) \leq T$ *if and only if* $U \subset V$ *is stable set in* ξ.

Figure 7. Example of fuzzy threshold graph

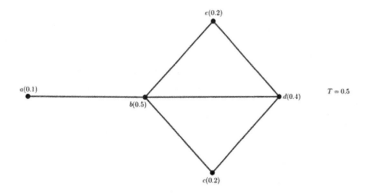

Example 4.5 *Let $\xi = (V, \sigma, \mu)$ be a fuzzy graph with vertex set $V = \{a, b, c, d, e\}$ such that*

$$\sigma(a) = 0.1, \sigma(b) = 0.5, \sigma(c) = 0.2, \sigma(d) = 0.2, \sigma(e) = 0.2.$$

Also, let T=0.5 and all cycles of the graph are fuzzy cycle. This graph is an example of fuzzy threshold graph (see Figure 7).

Fuzzy Competition Graphs

Cohen (1968) introduced the notion of competition graphs in connection with a problem in ecology. Let $\overrightarrow{D} = (V, \overrightarrow{E})$ be a digraph, which corresponds to a food web. A vertex $x \in V(\overrightarrow{D})$ represents a species in the food web and an arc $\overrightarrow{(x, s)} \in \overrightarrow{E}(D)$ means that x preys on the species s. If two species x and y have a common prey s, they will compete for the prey s. Based on this analogy, Cohen defined a graph which represents the relations of competition among the species in the food web. The competition graph $C(\overrightarrow{D})$ of a digraph $\overrightarrow{D} = (V, \overrightarrow{E})$ is an undirected graph $G = (V, E)$ which has the same vertex set V and has an edge between two distinct vertices $x, y \in V$ if there exists a vertex $s \in V$ and arcs $\overrightarrow{(x, s)}, \overrightarrow{(y, s)} \in \overrightarrow{E}(D)$. The competition graph is also applicable in channel assignment, coding, modeling of complex economic and energy systems, etc.

A lot of works have been done on competition graphs and its variations. In all these works, it is assumed that the vertices and edges of the graphs are precisely defined. But, in reality we observed that sometimes the vertices and edges of a graph can not be defined precisely. For example, in ecology, species may be of different types like vegetarian, non-vegetarian, strong, weak, etc. Similarly in ecology, preys may be tasty, digestive, harmful, etc. The terms tasty, digestive, harmful, etc. have no precise meanings. They are fuzzy in nature and hence, the species and preys may be assumed as fuzzy sets and inter-relationship between the species and preys can be designed by a fuzzy graph. This motivates the necessity of fuzzy competition graphs.

Figure 8. Example of fuzzy out-neighborhood and in-neighborhood of a vertex

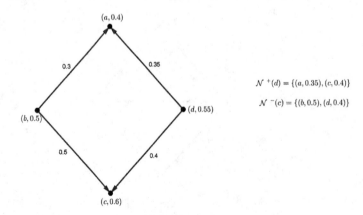

Like crisp graph, fuzzy out-neighborhood and fuzzy in-neighborhood of a vertex in directed fuzzy graph are defined below.

Definition 4.6 (Samanta and Pal, 2013) *Fuzzy out-neighborhood of a vertex v of a directed fuzzy graph* $\vec{\xi} = (V, \sigma, \vec{\mu})$ *is the fuzzy set* $\mathcal{N}^{+}(v) = (X_v^{+}, m_v^{+})$, *where* $X_v^{+} = \{u \mid \vec{\mu}(v, u) > 0\}$ *and* $m_v^{+} : X_v^{+} \to [0,1]$ *defined by* $m_v^{+}(u) = \vec{\mu}(v, u)$. *Similarly, fuzzy in-neighborhood of a vertex v of a directed fuzzy graph* $\vec{\xi} = (V, \sigma, \vec{\mu})$ *is the fuzzy set* $\mathcal{N}^{-}(v) = (X_v^{-}, m_v^{-})$, *where* $X_v^{-} = \{u \mid \vec{\mu}(u, v) > 0\}$ *and* $m_v^{-} : X_v^{-} \to [0,1]$ *defined by* $m_v^{-}(u) = \vec{\mu}(u, v)$.

Example 4.7 *Let* $\vec{\xi}$ *be a directed fuzzy graph. Let the vertex set be* $\{a, b, c, d\}$ *with membership values* $\sigma(a) = 0.4$, $\sigma(b) = 0.5$, $\sigma(c) = 0.6$, $\sigma(d) = 0.55$. *The membership values of arcs be* $\vec{\mu}(b, a) = 0.3$, $\vec{\mu}(d, a) = 0.35$, $\vec{\mu}(b, c) = 0.5$, *and* $\vec{\mu}(d, c) = 0.4$. *So*

$$\mathcal{N}^{+}(d) = \{(a, 0.35), (c, 0.4)\}.$$

$$\mathcal{N}^{-}(c) = \{(b, 0.5), (d, 0.4)\}.$$

(Note that $(a, \sigma(a))$ *represents the vertex a with membership value* $\sigma(a)$ *). It is shown in Figure 8.*

The definition of fuzzy competition graph is given by Samanta and Pal (2013) as follows.

Definition 4.8 (Samanta and Pal, 2013) *The fuzzy competition graph* $\mathcal{C}(\vec{\xi})$ *of a fuzzy digraph* $\vec{\xi} = (V, \sigma, \vec{\mu})$ *is an undirected fuzzy graph* $\xi = (V, \sigma, \mu)$ *which has the same fuzzy vertex set as in* $\vec{\xi}$ *and has a fuzzy edge between two vertices x,* $y \in V$ *in* $\mathcal{C}(\vec{\xi})$ *if and only if* $\mathcal{N}^{+}(x) \cap \mathcal{N}^{+}(y)$ *is non-empty fuzzy set in* $\vec{\xi}$ *and the edge membership value between x and y in* $\mathcal{C}(\vec{\xi})$ *is*

$$\mu(x, y) = (\sigma(x) \wedge \sigma(y)) h(\mathcal{N}^{+}(x) \cap \mathcal{N}^{+}(y)).$$

Example 4.9 *Let* $\vec{\xi}$ *be a directed fuzzy graph. Let the vertices with membership values of* $\vec{\xi}$ *be* $(a, 0.3)$, $(b, 0.6)$, $(c, 0.4)$, $(d, 0.5)$, $(e, 0.4)$ *with membership values of arcs be* $\vec{\mu}(b, a) = 0.2$, $\vec{\mu}(c, b) = 0.35$, $\vec{\mu}(d, c) = 0.2$, $\vec{\mu}(e, d) = 0.2$, $\vec{\mu}(e, a) = 0.25$, $\vec{\mu}(a, d) = 0.3$. *It is shown in Figure 9(a). The corresponding fuzzy competition graph is shown in Figure 9(b).*

Next, fuzzy k-competition graph is defined as a generalization of fuzzy competition graph.

Definition 4.10 *Let k be a non-negative number. The fuzzy k-competition graph* $\mathcal{C}_k(\vec{\xi})$ *of a fuzzy digraph* $\vec{\xi} = (V, \sigma, \mu)$ *is an undirected fuzzy graph* $\xi = (V, \sigma, \nu)$ *which has the same fuzzy vertex set as* $\vec{\xi}$ *and has a fuzzy edge between two vertices x,* $y \in V$ *in* $\mathcal{C}_k(\vec{\xi})$ *if and only if*

$$|\mathcal{N}^{+}(x) \cap \mathcal{N}^{+}(y)| > k.$$

Figure 9. Example of fuzzy competition graph

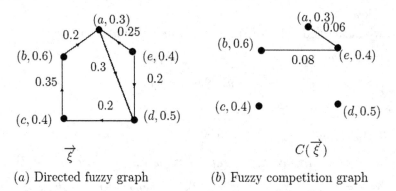

(a) Directed fuzzy graph   (b) Fuzzy competition graph

The edge membership value between x and y in $C_k(\vec{\xi})$ is

$$\nu(x,y) = \frac{(k'-k)}{k'}[\sigma(x) \wedge \sigma(y)]h(\mathcal{N}^+(x) \cap \mathcal{N}^+(y)),$$

where

$$k' = |\mathcal{N}^+(x) \cap \mathcal{N}^+(y)|.$$

When *k*=0, the fuzzy *k*-competition graph becomes fuzzy competition graph. An example of fuzzy 0.2-competition graph is given below.

Example 4.11 *Let $\vec{\xi}$ be a directed fuzzy graph with membership values of vertices of $\vec{\xi}$ are $(x, 0.4)$, $(y, 0.6)$, $(a, 0.6)$, $(b, 0.7)$, $(c, 0.8)$, $(d, 0.65)$ and the membership values of arcs be $\overrightarrow{\mu(x,a)} = 0.3$, $\overrightarrow{\mu(x,b)} = 0.35$, $\overrightarrow{\mu(x,c)} = 0.36$, $\overrightarrow{\mu(x,d)} = 0.4$, $\overrightarrow{\mu(y,a)} = 0.6$, $\overrightarrow{\mu(y,b)} = 0.5$, $\overrightarrow{\mu(y,c)} = 0.45$,*

Figure 10. Example of fuzzy 0.2-competition graph

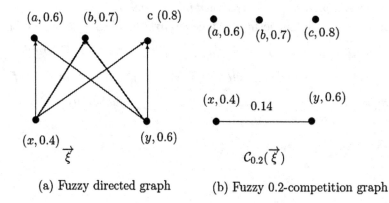

(a) Fuzzy directed graph   (b) Fuzzy 0.2-competition graph

$\overrightarrow{\mu(y,d)} = 0.35$ *(see Figure 10(a)). The corresponding fuzzy 0.2-competition graph is shown in the Figure 10(b).*

FUTURE RESEARCH DIRECTIONS

Due to the importance of the graphs in real life situations under different environments, one can go through the detailed study on the recent developments of bipolar fuzzy graphs, interval-valued fuzzy graphs, intuitionistic fuzzy graphs, m-polar fuzzy graphs, etc. as a natural extension of this work.

CONCLUSION

The theoretical concepts of fuzzy graphs are extremely useful tool in solving the combinatorial problems in different areas including algebra, number theory, geometry, topology, operation research, optimization, computer science, etc. The fuzzy models give more precision, flexibility, and comparability to the system as compared to the classical and fuzzy models. So, in this chapter some basic definitions related to fuzzy graph are given with several numerical examples. Finally, the concepts of fuzzy planar graph, fuzzy threshold graph, fuzzy competition graph and fuzzy k-competition graph are introduced.

REFERENCES

Al-Hawary, T. (2011). Complete fuzzy graphs. *International J Math Combin, 4*, 26–34.

Bhutani, K. R. (1989). On automorphism of fuzzy graphs. *Pattern Recognition Letters, 9*(3), 159–162. doi:10.1016/0167-8655(89)90049-4

Bhutani, K. R., Moderson, J., & Rosenfeld, A. (2004). On degrees of end nodes and cut nodes in fuzzy graphs. *Iranian Journal of Fuzzy Systems, 1*(1), 57–64.

Chvatal, V., & Hammer, P. L. (1973). *Set-packing problems and threshold graphs. CORR, 73-21*. University of Waterloo.

Cohen, J. E. (1968). *Interval graphs and food webs: a finding and a problem. Document 17696-PR.* Santa Monica, CA: RAND Corporation.

Mathew, S., & Sunitha, M. S. (2009). Types of arcs in a fuzzy graph. *Information Sciences, 179*(11), 1760–1768. doi:10.1016/j.ins.2009.01.003

Mordeson, J. N., & Nair, P. S. (1996). Successor and source of (fuzzy) finite state machines and (fuzzy) directed graphs. *Information Sciences, 95*(1-2), 113–124. doi:10.1016/S0020-0255(96)00139-9

Mordeson, J. N., & Nair, P. S. (2000). *Fuzzy graphs and hypergraphs.* Physica-Verlag Heidelberg. doi:10.1007/978-3-7908-1854-3

Mordeson, J. N., & Peng, C. S. (1994). Operations on fuzzy graphs. *Information Sciences, 79*(3-4), 159–170. doi:10.1016/0020-0255(94)90116-3

Nagoorgani, A., & Hussain, R. J. (2009). Fuzzy effective distance k-dominating sets and their applications. International Journal of Algorithms, *Computing and Mathematics, 2*(3), 25–36.

Nagoorgani, A., & Latha, A. (2012). On irregular fuzzy graphs. *Applied Mathematical Sciences, 6*(11), 517–523.

Nagoorgani, A., & Radha, K. (2008). On regular fuzzy graphs. *The Journal of Physiological Sciences; JPS, 12,* 33–40.

Nair, P. S., & Cheng, S. C. (2001). Cliques and fuzzy cliques in fuzzy graphs. *IFSA World Congress and 20th NAFIPS International Conference, 4,* 2277 - 2280. 10.1109/NAFIPS.2001.944426

Ordman, E. T. (1985). Threshold coverings and resource allocation, *16th Southeastern Conference on Combinatorics, Graph Theory and Computing,* 99-113.

Rosenfeld, A. (1975). *Fuzzy graphs.* In L. A. Zadeh, K. S. Fu, & M. Shimura (Eds.), *Fuzzy Sets and their Applications* (pp. 77–95). New York: Academic Press.

Samanta, S., Akram, M., & Pal, M. (2015). -step fuzzy competition graphs. *Journal of Applied Mathematics and Computing, 47*(1-2), 461–472. doi:10.100712190-014-0785-2

Samanta, S., & Pal, M. (2011). Fuzzy tolerance graphs. *International Journal of Latest Trends in Mathematics, 1*(2), 57–67.

Samanta, S., & Pal, M. (2011). Fuzzy threshold graphs. *CIIT International Journal of Fuzzy Systems, 3*(12), 360–364.

Samanta, S., & Pal, M. (2013). Fuzzy *kp. Fuzzy Information and Engineering, 5*(2), 191–204. doi:10.100712543-013-0140-6

Samanta, S., & Pal, M. (2015). Fuzzy planar graphs. *IEEE Transactions on Fuzzy Systems, 23*(6), 1936–1942. doi:10.1109/TFUZZ.2014.2387875

Sunitha, M. S., & Vijaya Kumar, A. (2002). Complement of fuzzy graphs. *Indian Journal of Pure and Applied Mathematics, 33*(9), 1451–1464.

Zadeh, L. A. (1965). Fuzzy sets. *Information and Control, 8*(3), 338–353. doi:10.1016/S0019-9958(65)90241-X

Chapter 19
Fuzzy Graphs and Fuzzy Hypergraphs

Michael G. Voskoglou
Graduate T. E. I. of Western Greece, Greece

Tarasankar Pramanik
https://orcid.org/0000-0001-7582-2525
Khanpur Gangche High School, India

ABSTRACT

Relationship is the core building block of a network, and today's world advances through the complex networks. Graph theory deals with such problems more efficiently. But whenever vagueness or imprecision arises in such relationships, fuzzy graph theory helps. However, fuzzy hypergraphs are more advanced generalization of fuzzy graphs. Whenever there is a need to define multiary relationship rather than binary relationship, one can use fuzzy hypergraphs. In this chapter, interval-valued fuzzy hypergraph is discussed which is a generalization of fuzzy hypergraph. Several approaches to find shortest path between two given nodes in an interval-valued fuzzy graphs is described here. Many researchers have focused on fuzzy shortest path problem in a network due to its importance to many applications such as communications, routing, transportation, etc.

INTRODUCTION

Graph is a representation of relationship between objects. It can give a good idea of scope of relationship for any complex networking model. Graph has many variations such as directed graph, undirected graph, simple graph, pseudo graph, multi graph, finite graph, infinite graph, etc. In a directed graph the relation defined on V is not symmetric but in undirected graph the relation defined on V is symmetric. In a graph, loops may occur, that is, a vertex may have a relation to itself. Also, there may have more than one edges between two vertices, called parallel edges. Simple graphs have no multiple edges and loops at all. If in a graph, there are finite number of vertices and finite number of edges, then it is called finite graph otherwise, it is infinite graph. Most commonly, unless stated otherwise, graph means undirected

DOI: 10.4018/978-1-5225-9380-5.ch019

simple finite graph. In general, any mathematical problem involving points and connections among them can be called a graph and its pictorial representation may lead to a solution. Thus, graph as mathematical model of some problem can solve a graph-theoretic problem and then presents the solution of original problem.

Fuzzy graph is rather extension of (crisp) graph by introducing the concepts of fuzzy sets and fuzzy relations instead of (crisp) sets. The notion of fuzzy sets and fuzzy relations were first introduced by Rosenfeld (1975). After Rosenfeld, in 1977, Kaufmann (1973) introduced the notion of fuzzy hypergraphs.

FUZZY GRAPH

A **fuzzy graph** $\xi = \left(V, \sigma, \mu \right)$ is a triplet consisting of a non-empty set V together with a pair of functions $\sigma : V \rightarrow \left[0, 1 \right]$ and $\mu : V \times V \rightarrow \left[0, 1 \right]$ such that

$$\mu \left(xy \right) \le \min \left\{ \sigma \left(x \right), \sigma \left(y \right) \right\}$$

for all $x, y \in V$.

Here the two fuzzy sets σ and μ are called **fuzzy vertex set** and **fuzzy edge set** of ξ respectively. Clearly, μ is a fuzzy relation on σ.

The fuzzy subgraph of a fuzzy graph is a fuzzy graph whose fuzzy set is a subset of the fuzzy set of the given fuzzy graph.

A fuzzy graph $\xi = \left(V', \tau, \nu \right)$ is said to be a **partial fuzzy subgraph** of ξ if $\tau \subseteq \sigma$ and $\nu \subseteq \mu$.

The fuzzy graph $\xi = \left(V', \tau, \nu \right)$ is called a **fuzzy subgraph** of ξ if $\tau \left(x \right) \le \sigma \left(x \right)$ for all $x \in V'$ and $\nu \left(x, y \right) \le \mu \left(x, y \right)$ for all $x, y \in V'$ where $V' \subset V$.

A partial fuzzy subgraph $\xi' = \left(V', \tau, \nu \right)$ of ξ is said to span ξ if $\sigma = \tau$. This partial fuzzy subgraph ξ is called a spanning fuzzy subgraph of ξ.

An **underlying crisp graph** of a fuzzy graph $\xi = \left(V, \sigma, \mu \right)$ is a crisp graph $\xi' = \left(V, \sigma', \mu' \right)$ where

$$\sigma' = \left\{ u \in V \left(\xi \right) \middle| \sigma \left(u \right) \rangle 0 \right\}$$

and

$$\mu' = \left\{ \left(u, v \right) \middle| \mu \left(u, v \right) \rangle 0 \right\}.$$

A **path** P in a fuzzy graph $\xi = \left(V, \sigma, \mu \right)$ is a sequence of distinct vertices $v_1, v_2, \dots, v_n \left(n \ge 2 \right)$ such that

$$\mu \left(v_i, v_{i+1} \right) > 0, i = 1, 2, \dots, \left(n - 1 \right).$$

Here, $n-1$ is called the **length** of the path P. The **diameter** of $x, y \in V$, written $diam(x, y)$ is the length of the longest path joining x to y. A path P is a **cycle** if $v_1 = v_n$ and $n \geq 4$. That is $P = (V, \sigma, \mu)$ is a cycle in ξ if and only if (V, σ', μ') is a cycle in ξ'. A cycle $P = (V, \sigma, \mu)$ is a **fuzzy cycle** if it contains more than one weak edge (i.e., there is no unique $(x, y) \in \nu'$ such that

$$\mu(x, y) = \wedge \{\mu(u, v) : (u, v) \in \nu'\}).$$

Notice that, if a fuzzy graph is complete then the fuzzy graph is strong, but not vice versa. A fuzzy subgraph $\xi' = (V', \tau, \nu)$ of a fuzzy graph $\xi = (V, \sigma, \mu)$ is said to be a **fuzzy clique** if $(\xi')'$ is a clique and every cycle in ξ' is a fuzzy cycle.

The **union** of two fuzzy graphs $\xi_1 = (V_1, \sigma_1, \mu_1)$ and $\xi_2 = (V, \sigma_2, \mu_2)$ is denoted by

$$\xi_1 \cup \xi_2 = (V_1 \cup V_2, \sigma_1 \cup \sigma_2, \mu_1 \cup \mu_2),$$

where for all $x \in V_1 \cup V_2$,

$$(\sigma_1 \cup \sigma_2)(x) = \sigma_1(x) \wedge \sigma_2(x)$$

with $\sigma_1(x) = 0$ whenever $x \notin \sigma_1'$ and $\sigma_2(x) = 0$ whenever $x \notin \sigma_2'$, for all

$$(x, y) \in V_1 \cup V_2 \times V_1 \cup V_2, \ (\mu_1 \cup \mu_2)(x, y) = \mu_1(x, y) \wedge \mu_2(x, y)$$

with $\mu_1(x, y) = 0$ whenever $(x, y) \notin \mu_1'$ and $\mu_2(x, y) = 0$ whenever $(x, y) \notin \mu_2'$.

The strength of connectedness between two vertices u and ν is

$$\mu^\infty(u, v) = sup\{\mu^k(u, v) \mid k = 1, 2, \ldots\},$$

where

$$\mu^k(u, v) = sup\{\mu(u, u_1) \wedge \mu(u_1, u_2) \wedge \ldots \wedge \mu(u_{k-1}, \nu) \mid u_1, u_2, \ldots, u_{k-1} \in V\}.$$

In a fuzzy graph an arc (u, v) is said to be **strong arc** or **strong edge**, if $\mu(u, v) \geq \mu^\infty(u, v)$ and otherwise it is **weak**.

Fuzzy neighbourhood (Samanta & Pal, 2013) of a vertex ν of a fuzzy graph $\xi = (V, \sigma, \mu)$ is the fuzzy set $N(v) = (X_\nu, m_\nu)$, where

$$X_{\nu} = \left\{ u \,\middle|\, \mu\left(v,u\right)\right\rangle 0 \right\}$$

and $m_{\nu} : X_{\nu} \rightarrow \left[0,1\right]$ defined by $m_{\nu}\left(u\right) = \mu\left(v,u\right)$.

A **strong neighbourhood** (Nagoorgani & Vadivel, 2009) of a vertex u is a vertex v such that $\left(u,v\right)$ is strong edge.

A fuzzy graph ξ is said to be a **fuzzy star** (Vimala & Sathya, 2012) if every vertex of ξ has exactly one strong neighbour in ξ.

Directed fuzzy graphs (or simply fuzzy digraph) are the fuzzy graphs in which the fuzzy relations between edges are not necessarily symmetric. The definition of directed fuzzy graph is as follows:

Directed fuzzy graph (Mordeson & Nair, 1996) $\xi = \left(V,\sigma,\mu\right)$ is a non-empty set V together with a pair of functions $\sigma : V \rightarrow \left[0,1\right]$ and $\mu : V \times V \rightarrow \left[0,1\right]$ such that for all $x,y \in V$,

$$\mu\left(x,y\right)^{\rightarrow} \leq \sigma\left(x\right) \wedge \sigma\left(y\right).$$

Let $\xi = \left(V,\sigma,\mu\right)$ be a fuzzy graph, let x,y be two distinct vertices and let ξ be the partial fuzzy subgraph of ξ obtained by deleting the edge $\left(x,y\right)$. Then, $\xi' = \left(V,\sigma,\mu'\right)$, where $\mu'\left(x,y\right) = 0$ and $\mu' = \mu$ for all other pairs. Then the edge $\left(x,y\right)$ is said to be a **fuzzy bridge** in ξ if $\mu'^{\infty}\left(u,v\right) < \mu^{\infty}\left(u,v\right)$ for some $u,v \in \sigma'$.

The concept of fuzzy forests and fuzzy trees were first introduced by Rosenfeld

Like complete graph, the definition of complete fuzzy graph is given below.

A fuzzy graph $\xi = \left(V,\sigma,\mu\right)$ is **complete** if

$$\mu\left(u,v\right) = min\left\{\sigma\left(u\right),\sigma\left(v\right)\right\}$$

for all $u,v \in V$, where $\left(u,v\right)$ denotes the edge between the vertices u and v.

Theorem 1 (Rosenfeld, 1975)

Let $\xi = \left(V,\sigma,\mu\right)$ be a fuzzy graph. Then the following statements are equivalent.

1. $\left(x,y\right)$ *is a fuzzy bridge.*
2. $\mu'^{\infty}\left(x,y\right) < \mu\left(x,y\right)$
3. $\left(x,y\right)$ *is not the weakest edge of any cycle.*

Proof. If $\left(x,y\right)$ is not a fuzzy bridge, then

$$\mu'^{\infty}\left(x,y\right) = \mu^{\infty}\left(x,y\right) \geq \mu\left(x,y\right).$$

This shows that (ii) $\Rightarrow$ (i). Now, if (x,y) is the weakest edge of a cycle, then any path P involving edge (x,y) can be converted into a path P' not involving (x,y) but at least as strong as P, by using the rest of the cycle as a path from x to y. Thus, (x,y) cannot be a fuzzy bridge. Therefore, (i) $\Rightarrow$ (iii).

If $\mu'^{\infty}(x,y) \geq \mu(x,y)$, then there is a path from x to y not involving (x,y), that has strength $\geq \mu(x,y)$, and this path together with (x,y) forms a cycle of ξ in which (x,y) is a weakest edge.

Let w be any vertex and let ξ' be the partial fuzzy subgraph of ξ obtained by deleting the vertex w. That is, $\xi' = (V', \sigma', \mu')$ is the partial fuzzy subgraph of ξ such that $\sigma'(w) = 0, \sigma = \sigma'$ for all other vertices, $\mu'(w,z) = 0$ for all vertices z, and $\mu' = \mu$ for all other edges. This vertex w in ξ is called a **fuzzy cutvertex** if $\mu'^{\infty}(u,v) < \mu^{\infty}(u,v)$ for some u,v in V such that $u \neq w \neq v$. That is, the vertex w is a fuzzy cut-vertex if the strength of connectedness between some other pair of vertices reduces after deletion of the vertex w. In a fuzzy graph, a fuzzy subgraph ξ' is said to be a **block** if it does not contain any fuzzy cut-vertices. In a crisp graph, a block does not contain any bridges but in fuzzy graph, a block may have fuzzy bridges. Sometimes a block in fuzzy graph may be referred as fuzzy block.

A **maximum spanning tree** of a connected fuzzy graph (V, σ, μ) is a fuzzy spanning subgraph $T = (V, \sigma, \nu)$ of ξ, which is a tree, such that $\mu^{\infty}(u,v)$ is the strength of the unique strongest $u-v$ path in T for all $u, v \in \xi$. The next theorem characterizes the fuzzy cutvertices and fuzzy bridges of fuzzy graphs using maximum spanning trees.

Theorem 2 (Sunitha & Vijaykumar, 2005)

A vertex w of a fuzzy graph $\xi = (V, \sigma, \mu)$ is a fuzzy cut-vertex if and only if w is an internal vertex of every maximum spanning tree of ξ.

Proof. Let w be a fuzzy cut-vertex of a fuzzy graph $\xi = (V, \sigma, \mu)$. Then from the definition, it is obvious that w is on every strongest $u-v$ path for some distinct vertices u, v. Now each maximum spanning tree of ξ contains a unique strongest $u-v$ path and hence w is an internal vertex of every maximum spanning tree of ξ.

Conversely, let w be an internal vertex of every maximum spanning tree. Let T be a maximum spanning tree and let uw and wv be edges in T. Thus $u \to w \to v$ is a strongest $u-v$ path in T. If possible let, w is not a fuzzy cut-vertex. Then there exists at least one strongest $u-v$ path for every pair of vertices u, v which does not contain the fuzzy cut-vertex w. Assume one such $u-v$ path P which does not contain edges not in T. Now, let $\mu^{\infty}(u,v) = \mu(uw)$ in T. Then edges in P have strength $\geq \mu(uw)$. Therefore, if the edge uw is removed and the path P is added in T then another maximum spanning tree of ξ is formed in which w is an end vertex. This contradicts the assumption. Hence the result follows.

Theorem 3 (Mathew et al., 2018)

If $\xi = (V, \sigma, \mu)$ is a complete fuzzy graph, then for any edge $uv \in \mu$, $\mu^{\infty}(u, v) = \mu(u, v)$.

Proof. By definition,

$$\mu^2(u, v) = \vee_{z \in \sigma} \left\{ \mu(uz) \wedge \mu(zv) \right\}$$

$$\Rightarrow \vee \left\{ \sigma(u) \wedge \sigma(v) \wedge \sigma(z) \right\} = \sigma(u) \wedge \sigma(v) = \mu(uv).$$

Similarly, $\mu^3(u, v) = \mu(uv)$ and in the same way one can show that $\mu^k(u, v) = \mu(uv)$ for all positive integers k. Thus,

$$\mu^{\infty}(u, v) = \{ \mu^k(u, v) \vee \text{ for all integers} k \geq 1\} = \mu(uv).$$

Theorem 4 (Mathew et al., 2018)

If w is a common vertex of at least two fuzzy bridges, then w is a fuzzy cut-vertex.

Proof. Let $u_1 w$ and wu_2 be two fuzzy bridges. Then there exists u, v such that $u_1 w$ is on every strongest $u - v$ path. If w is distinct from u and v, then it follows that w is a fuzzy cut-vertex. Now, consider that one of the vertices u, v is w so that $u_1 w$ is on every strongest $u - w$ path or wu_2 is on every strongest $w - v$ path. Assume that w is not a fuzzy cut-vertex. Then between every two vertices there exists at least one strongest path not containing w. This path together with $u_1 w$ and wu_2 forms a cycle.

Therefore, there may arise two cases. First consider that u_1, w, u_2 is not a strongest path. Then clearly one of $u_1 w, wu_2$ or both become weakest edges of a cycle, which contradicts that $u_1 w$ and wu_2 are fuzzy bridges.

Second, suppose that u_1, w, u_2 is also a strongest path joining u_1 to u_2. Then

$$\mu^{\infty}(u_1, u_2) = \mu(u_1 w) \wedge \mu(wu_2),$$

the strength of P. Thus, edges of P are at least as strong as $\mu(u_1 w)$ and $\mu(wu_2)$, which implies that $u_1 w, wu_2$ are both weakest edges of a cycle, which again is a contradiction.

There are many variations in fuzzy graphs such as

1. Bipolar fuzzy graph,
2. Fuzzy intersection graph,
3. Fuzzy threshold graph,
4. Fuzzy planar graph,

5. Fuzzy competition graph,
6. Fuzzy tolerance graph,
7. Interval-valued fuzzy graph

These fuzzy graphs are being described here.

Bipolar Fuzzy Graph

In 1998, the concept of bipolar fuzzy set is introduced by Zhang (1998) as a generalization of fuzzy set. A bipolar fuzzy set is a generalization of Zadeh's fuzzy set. The range of the membership value of a bipolar fuzzy set is $\left[-1,1\right]$. In a bipolar fuzzy set, the membership value 0 of an element means that the element is not connected with the corresponding property, the membership value within $\left(0,1\right]$ of an element implies that the element satisfies the property with certain negotiations (higher the value indicates that there is lower amount of negotiations), and the negative membership value within $\left[-1,0\right)$ of an element means that the element satisfies the implicit counter-property to some extent.

Let V be not an empty set. A **bipolar fuzzy set (BFS)** (Lee, 2000; Zhang, 1998) B in V is characterized by

$$B = \{\left(V, \mu_A^P\left(v\right), \mu_A^N\left(v\right)\right) \mid v \in V\},$$

where $\mu_A^P : V \rightarrow \left[0,1\right]$ and $\mu_A^N : V \rightarrow \left[-1,0\right]$ are positive membership function and negative membership function respectively. The positive membership value $\mu_A^P\left(v\right)$ is used to denote the amount which the element v satisfies the property corresponding to a bipolar fuzzy set B, and the negative membership value $\mu_A^N\left(x\right)$ to denote the amount which the element v satisfies the implicit counter-property to some extent corresponding to a bipolar fuzzy set B.

A **bipolar fuzzy relation** (Zhang, 1994) $B = \left(\mu_B^P, \mu_B^N\right)$ on a non-empty set X is a mapping

$$B : X \times X \rightarrow \left[-1,1\right] \times \left[-1,1\right]$$

such that

$$\mu_B^P\left(u,v\right) \in \left[0,1\right] \, and \, \mu_B^N\left(u,v\right) \in \left[-1,0\right].$$

A **bipolar fuzzy graph** (BFG) (Akram & Dudek, 2011) ξ of a (crisp) graph $G = \left(V, E\right)$ is a pair $\xi = \left(V, A, B\right)$ where $A = \left(\mu_A^P, \mu_A^N\right)$ is a bipolar fuzzy set on V and $B = \left(\mu_B^P, \mu_B^N\right)$ is a bipolar fuzzy relation on $V \times V$ such that

$$\mu_B^P\left(u,v\right) \leq min\left\{\mu_A^P\left(u\right), \mu_A^P\left(v\right)\right\}$$

and

$$\mu_B^N \left(u, v \right) \leq min \left\{ \mu_A^N \left(u \right), \mu_A^N \left(v \right) \right\}$$

for all $\left(u, v \right) \in E$.

Let $\xi = \left(A, B \right)$ be a bipolar fuzzy graph on ξ. If all the vertices have the same open neighbourhood degree n, then ξ is called an n-**regular bipolar fuzzy graph**. The open neighbourhood degree of a vertex x in ξ is defined by

$$deg \left(x \right) = \left(deg^P \left(x \right), deg^N \left(x \right) \right),$$

where

$$deg^P \left(x \right) = \sum_{x \in V} \mu_A^P \left(x \right)$$

and

$$deg^N \left(x \right) = \sum_{x \in V} \mu_A^N \left(x \right).$$

Let $\xi = \left(A, B \right)$ be a regular bipolar fuzzy graph. The **order** of a regular bipolar fuzzy graph ξ is

$$O \left(G \right) = \left(\sum_{x \in V} \mu_A^P \left(x \right), \sum_{x \in V} \mu_A^N \left(x \right) \right).$$

The **size** of a regular bipolar fuzzy graph ξ

$$S \left(G \right) = \left(\sum_{(x,y) \in E} \mu_A^P \left(x, y \right), \sum_{(x,y) \in E} \mu_A^N \left(x, y \right) \right).$$

The fuzzy graph ξ is said to be a **totally regular bipolar fuzzy graph** if each vertex of ξ has same closed neighbourhood degree m. The closed neighbourhood degree of a vertex x is defined by

$$deg \left[x \right] = \left(deg^P \left[x \right], deg^N \left[x \right] \right),$$

where

$$deg^P \left[x \right] = deg^P \left(x \right) + \mu_A^P \left(x \right),$$

$$deg^{N}\left[x\right] = deg^{N}\left(x\right) + \mu_{A}^{N}\left(x\right).$$

This is to be noted that there is not inter-relationship between n-regular bipolar fuzzy graph and m-totally regular bipolar fuzzy graph.

There are some propositions regarding the order, size and regularity of bipolar fuzzy graphs.

Proposition 1 (Akram & Dudek, 2012)

The size of a n-regular bipolar fuzzy graph ξ is $\dfrac{nk}{2}$, where $\left|V\right| = k$.

Proof. Proof is obvious.

Proposition 2 (Akram & Dudek, 2012)

If ξ is a m-totally regular bipolar fuzzy graph ξ, then $2S\left(\xi\right) + O\left(\xi\right) = mk$, where $\left|V\right| = k$.

Proof. This is obvious to prove.

Proposition 3 (Akram & Dudek, 2012)

If ξ is both n-regular m-totally regular bipolar fuzzy graph ξ, then $O\left(G\right) = k\left(m - n\right)$, where $\left|V\right| = k$.

A bipolar fuzzy graph $\xi = \left(A, B\right)$ is said to be a **complete bipolar fuzzy graph** if the following conditions holds:

$$\mu_{B}^{P}\left(x, y\right) = min\left\{\mu_{A}^{P}\left(x\right), \mu_{A}^{P}\left(y\right)\right\}$$

and

$$\mu_{B}^{N}\left(x, y\right) = min\left\{\mu_{A}^{N}\left(x\right), \mu_{A}^{N}\left(y\right)\right\}$$

for all $x, y \in V$.

Theorem 5 (Akram & Dudek, 2012)

Every complete bipolar fuzzy graph is a totally regular bipolar fuzzy graph.

Theorem 6 (Akram & Dudek, 2012)

Let $\xi = \left(A, B\right)$ be a bipolar fuzzy graph of a graph ξ. Then $A = \left(\mu_{A}^{P}, \mu_{A}^{N}\right)$ is a constant function if and only if the following are equivalent:

1. ξ *is a regular bipolar fuzzy graph,*
2. ξ *is a totally regular bipolar fuzzy graph.*

Proof. Consider the bipolar fuzzy set $A = \left(\mu_A^P, \mu_A^N \right)$ is a constant function. Let $\mu_A^P(x) = c_1$ and $\mu_A^N(x) = c_2$ for all $x \in V$.

To prove (a) $\Rightarrow$ (b):

Consider the bipolar fuzzy graph ξ is n-regular bipolar fuzzy graph, then $deg^P(x) = n_1$ and $deg^N(x) = n_2$ for all $x \in V$. So

$$deg^P[x] = deg^P(x) + \mu_A^P(x),$$

$$deg^N[x] = deg^N(x) + \mu_A^N(x),$$

for all $x \in V$,

Thus,

$$deg^P[x] = n_1 + c_1, \; deg^N[x] = n_2 + c_2$$

for all $x \in V$.

Hence, ξ is a totally regular bipolar fuzzy graph.

Now, to prove (b) $\Rightarrow$ (a): Assume that ξ is a totally regular bipolar fuzzy graph, then

$$deg^P[x] = k_1, deg^N[x] = k_2$$

for all $x \in V$ or,

$$deg^P(x) + \mu_A^P(x) = k_1, deg^N(x) + \mu_A^N(x) = k_2$$

for all $x \in V$ or,

$$deg^P(x) + c_1 = k_1, deg^N(x) + c_2 = k_2$$

for all $x \in V$ or,

$$deg^P(x) = k_1 - c_1, deg^N(x) = k_2 - c_2$$

for all $x \in V$.

Thus, ξ is a regular bipolar fuzzy graph. Hence (a) and (b) are equivalent. The converse part is obvious.

Fuzzy Intersection Graph

In ordinary or classical concept for operations on sets, the **intersection graph** is generated by associating a vertex to each set in a family with a finite number of sets

$$S = \left\{ S_1, S_2, \ldots, S_n \right\}.$$

An edge between any of these two vertices exists if the intersection of the two sets corresponding to those two vertices is non-empty. That is, if any two sets are disjoint then the vertices corresponding them have no edge. This graph is denoted by $Z(S) = (V, E)$, where V is the set of vertices and E is the set of edges. Fuzzy intersection graph is a generalization of (crisp) intersection graph.

McAllister (1988) first introduces the fuzzy intersection graph. The definition of **fuzzy intersection graph** is given below.

Let

$$F = \left\{ A_1 = (X, m_1), A_2 = (X, m_2), \ldots, A_n = (X, m_n) \right\}$$

be a finite family of fuzzy sets defined on a set X and consider F as crisp vertex set $V = \left\{ v_1, v_2, \ldots, v_n \right\}$. Here, $m_1, m_2, \ldots, m_n$ are the membership functions for the fuzzy sets $A_1, A_2, \ldots, A_n$. The fuzzy intersection graph of F is the fuzzy graph $Int(F)$ where $\sigma : V \to [0,1]$ is defined by $\sigma(v_i) = h(A_i)$ and $\mu : V \times V \to [0,1]$ is defined by

$$\mu(v_i, v_j) = \{ h(A_i \cap A_j), if\, i \neq j\, 0, if\, i = j.$$

Fuzzy Threshold Graph

Multi-processor scheduling, bin packing, and the knapsack problem are the different variations of set-packing problems and are being very well studied problem in combinatorial optimization. These problems have large impact on design and analysis of fuzzy threshold graph. All of these problems involve packing items of different sizes into bins of finite capacities. Consider a parallel system consists of a set of independent processing units each of which has a set of time-sharable resources such as CPU, one or more disks, network controllers etc. Here all units have variable capacities as well as resources. Fuzzy threshold graph is defined as follows.

A fuzzy graph $\xi = (V, \sigma, \mu)$ is called a **fuzzy threshold graph** (Samanta & Pal, 2011a) if there exists a non-negative real number t such that $\sum_{u \in U} \sigma(u) \leq t$ if and only if $U \subseteq V$ is an independent set in G.

Figure 1. Example of a fuzzy threshold graph

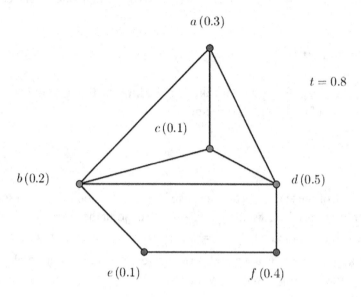

Table 1. The definition of membership function σ

Vertex	Membership Value	Vertex	Membership Value
a	0.3	d	0.5
b	0.7	e	0.1
c	0.2	f	0.4

Example 1: Let $\xi = (V, \sigma, \mu)$ be a fuzzy graph with vertex set $V = \{a, b, c, d, e, f\}$. The definition of membership functions σ, μ are given in Table 1 and 2 respectively.

Obviously the fuzzy graph ξ is a fuzzy threshold graph which is shown in Figure 1.

Now, definition of **fuzzy alternating 4-cycle** (Samanta & Pal, 2011a) is given.

Let $\xi = (V, \sigma, \mu)$ be a fuzzy graph where $V = \{a, b, c, d\}$. Then the four vertices a, b, c, d of ξ is said to form an alternating 4-cycle if

$$\mu(a, b) > 0, \mu(c, d) > 0, \mu(a, d) = 0$$

and $\mu(b, c) = 0$.

An example of fuzzy alternating 4-cycle is shown in Figure 2.

Theorem 7 (Samanta & Pal, 2011)

A fuzzy threshold graph does not have a strong fuzzy alternating 4-cycle.

Table 2. The definition of membership function μ

Vertex	Membership Value	Vertex	Membership Value
(a,b)	0.1	(b,e)	0.08
(a,c)	0.09	(c,d)	0.06
(a,d)	0.2	(d,f)	0.4
(b,c)	0.05	(e,f)	0.05
(b,d)	0.2		

Figure 2. Example of an alternating 4-cycle

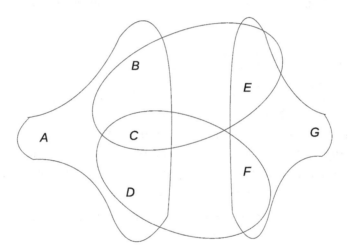

Proof. Let $\xi = (V, \sigma, \mu)$ be a fuzzy threshold graph. Let, if possible, ξ has a strong fuzzy alternating 4-cycle. So there exists vertices a, b, c, d with

$$\mu(a,b) > 0, \mu(c,d) > 0 \text{ and } \mu(a,d) = 0 = \mu(b,c).$$

As the graph is fuzzy threshold graph with threshold T, then

$$\sigma(a) + \sigma(b) > T, \sigma(c) + \sigma(d) > T$$

[as the a, b, c, d construct strong fuzzy alternating 4-cycle]

$$\sigma\left(a\right)+\sigma\left(d\right)\le T, \sigma\left(b\right)+\sigma\left(c\right)\le T$$

[as $\mu\left(a,d\right)=0=\mu\left(b,c\right)$].

These inequalities are inconsistent. Hence a,b,c,d does not construct a strong fuzzy alternating 4-cycle. So a fuzzy threshold graph does not have a strong fuzzy alternating 4-cycle.

A **fuzzy split graph** is a fuzzy graph in which the vertices can be partitioned into a fuzzy clique and a fuzzy independent set.

Theorem 8 (Samanta & Pal, 2011)

A fuzzy threshold graph is a fuzzy split graph.

Proof. Let $\xi=\left(V,\sigma,\mu\right)$ be a fuzzy threshold graph and K be the largest clique in ξ. If $\left(a,b\right)$ be a strong arc in $V-K\vee$, then by maximality of the K, there exists distinct vertices c,d in K such that $\mu\left(a,d\right)=0=\mu\left(b,c\right)$. These vertices create a strong fuzzy alternating 4-cycle. So contradiction arises. To avoid strong fuzzy alternating 4-cycle, $V-K\vee$ must be independent set. Therefore, ξ is a fuzzy split graph.

Theorem 9 (Samanta & Pal, 2011)

If ξ is a fuzzy threshold graph, then ξ can be constructed from the one vertex graph by repeatedly adding a fuzzy isolated vertex or a fuzzy dominating vertex.

Proof. It has already known that fuzzy threshold graph is a fuzzy split graph. Thus it is sufficient to show that the fuzzy graph ξ has a fuzzy isolated vertex or a fuzzy dominating vertex, if for such a vertex is removed, then the graph will be still fuzzy split graph. Let U be an independent set of the fuzzy threshold graph $\xi=\left(K,U\right)$. If U contains fuzzy isolated vertices only, then the result holds. If U has no fuzzy isolated vertices, then vertex $u\in U$ with smallest neighbourhood has some neighbour $v\in K$. As K is a fuzzy clique, so the vertex v is dominating vertex of ξ.

FUZZY PLANAR GRAPH

Day by day, the necessity of flyovers, subway tunnels, pipelines, metro lines increases due to demand in human kind. Number of crossing of routes increases the chance of accident. The cost of crossing of subways in underground is also high. But, the underground routes reduce the traffic jam. The system of routes without crossing is ideal for a city. But, lack of space and money often requires crossings of routes. %But due to lack of space, crossings are allowed somewhere. It is very true that the crossing between one congested and one non-congested route is better than the crossing between two congested routes.

It is true that, two congested crossing of routes is more safer than a congested and non-congested road crossing. The term "congested" has no specific meaning and measurement. To understand the exact load of a route we generally use the terms for routes like "congested", "very congested", "highly congested" routes, etc. These linguistic terms can be dealt in mathematics by giving some positive membership values and negative membership values in fuzzy sense.

In mathematical sense, strong route means highly congested route and weak route means low congested route. Thus crossing between a strong route and a weak route is better than the crossing between two strong routes.

That is, in city planning, crossing between strong routes and weak routes are allowed. The terms "strong route" and "weak route" lead strong edge and weak edge of a bipolar fuzzy graph respectively. And the approval of using the crossing between strong and weak edges lead to the concept of bipolar fuzzy planar graph. Jabbar et al. (2009) and Nirmala and Dhanabal (2012) introduced the concept of fuzzy planar graph. Recently, Samanta et al. (2014a) introduced fuzzy planar graph in a different way.

Theorem 10 (Samanta & Pal, 2015)

Let ξ be a fuzzy planar graph with fuzzy planarity value greater than 0.5 and considerable number c. The number of points of intersection between considerable edges in ξ is at most $\left\lceil \dfrac{1}{c} \right\rceil$ or $\dfrac{1}{c} - 1$ according as $\dfrac{1}{c}$ is not an integer or an integer respectively (here $\lceil x \rceil$ is greatest integer not exceeding x).

Theorem 11 (Samanta & Pal, 2015)

Let ξ be a fuzzy planar graph with fuzzy planarity value greater than 0.5. The number of points of intersection between effective edges in ξ is at most one.

FUZZY COMPETITION GRAPH

Fuzzy competition graph is a generalization of competition graph. This graph is related to fuzzy digraph. Fuzzy k-competition graph and m-step competition graph are the variations of fuzzy competition graph.

Fuzzy **out-neighbourhood** (Samanta et al., 2015) of a vertex $v \in V$ of a directed fuzzy graph $D^{\rightarrow} = \left(V, \sigma, \nu \right)$ is the fuzzy set

$$N^+\left(v\right) = \left(X_v^+, m_v^+ \right),$$

where

$$X_v^+ = \left\{ u \, \middle| \, \nu\left(v, u^{\rightarrow}\right) \rangle 0 \right\} \text{ and } m_v^+ : X_v^+ \rightarrow \left[0, 1 \right]$$

is defined by $m_v^+ = \nu\left(v, u^{\rightarrow}\right)$.

Fuzzy **in-neighbourhood** (Samanta et al., 2015) of a vertex $v \in V$ of a directed fuzzy graph $D^{\rightarrow} = \left(V, \sigma, \nu \right)$ is the fuzzy set

$$N^{-}\left(v\right)=\left(X_{v}^{-},m_{v}^{-}\right),$$

where

$$X_{v}^{-}=\left\{u\middle|\nu\left(u,v^{\rightarrow}\right)\right\rangle 0\right\}\text{ and }m_{v}^{-}:X_{v}^{-}\rightarrow\left[0,1\right]$$

is defined by $m_{v}^{-}=\nu\left(u,v^{\rightarrow}\right)$.

Fuzzy **neighbourhood** (Samanta et al., 2014c) of a vertex $v\in V$ of a fuzzy graph $\xi=\left(V,\sigma,\mu\right)$ is the fuzzy set

$$N\left(v\right)=\left(X_{v},m_{v}\right),$$

where

$$X_{v}=\left\{u\middle|\mu\left(u,v\right)\right\rangle 0\right\}\text{ and }m_{v}:X_{v}\rightarrow\left[0,1\right]$$

is defined by $m_{v}=\mu\left(u,v\right)$.

The m **-step fuzzy out-neighbourhood** (Samanta et al., 2015) of a vertex $v\in V$ of a directed fuzzy graph $D^{\rightarrow}=\left(V,\sigma,\nu\right)$ is the fuzzy set

$$N_{m}^{+}\left(v\right)=\left(X_{v}^{+},m_{v}^{+}\right),$$

where

$$X_{v}^{+}=\{u\middle|\mu_{m}^{\rightarrow}\left(v,u^{\rightarrow}\right)=min\left\{\nu\left(v,u_{1}^{\rightarrow}\right),\nu\left(u_{1},u_{2}^{\rightarrow}\right),...,\nu\left(u_{m},u^{\rightarrow}\right)\right\}\right\rangle 0,$$

$vu_{1}u_{2}...u_{m}u$ is a path from v to u} and $m_{v}^{+}:X_{v}^{+}\rightarrow\left[0,1\right]$ is defined by $m_{v}^{+}=\mu_{m}^{\rightarrow}\left(v,u^{\rightarrow}\right)$. If there is more than one fuzzy path of length m then we should take the path which has minimum membership value $\mu^{\rightarrow}_{m}\left(v,u^{\rightarrow}\right)$.

The definition of fuzzy competition graph is as follows.

The **fuzzy competition graph** of a fuzzy digraph $D^{\rightarrow}=\left(V,\sigma,\nu\right)$ is an undirected graph $C\left(D^{\rightarrow}\right)=\left(V,\sigma,\mu\right)$ which has the same fuzzy vertex set as in $D^{\rightarrow}$ and has a fuzzy edge between two vertices $u,v\in V$ in $C\left(D^{\rightarrow}\right)$ if and only if $N^{+}\left(u\right)\cap N^{+}\left(v\right)$ is non-empty fuzzy set in $D^{\rightarrow}$. The membership value of the edge $\left(u,v\right)$ in $C\left(D^{\rightarrow}\right)$ is

$$\mu\left(u, v\right) = \left(\sigma\left(u\right) \wedge \sigma\left(v\right)\right) h\left(N^+\left(u\right) \cap N^+\left(v\right)\right).$$

The m-**step fuzzy competition graph** (Samanta et al., 2015) of a digraph $D^\rightarrow = \left(V, \sigma, \nu\right)$ is denoted by $C_m\left(D^\rightarrow\right)$ and is defined by

$$C_m\left(D^\rightarrow\right) = \left(V, \sigma, \mu\right)$$

where

$$\mu\left(u, v\right) = \left(\sigma\left(u\right) \wedge \sigma\left(v\right)\right) h\left(N_m^+\left(u\right) \cap N_m^+\left(v\right)\right)$$

for all $u, v \in V$.

There is another variation of fuzzy competition graph, called p-competition fuzzy graph which is defined below.

Let p be a positive integer. The p-competition fuzzy graph $C^p\left(\xi\right) = \left(V, \sigma, \nu\right)$ of a fuzzy digraph $\xi^\rightarrow = \left(V, \sigma, \mu^\rightarrow\right)$ is an undirected fuzzy graph which has the same fuzzy vertex set as in $\xi^\rightarrow$ and has a fuzzy edge between two vertices x and y of V in $C^p\left(\xi^\rightarrow\right)$ if and only if

$$\left|supp\left(N^+\left(x\right) \cap N^+\left(y\right)\right)\right| \geq p.$$

The edge membership value of the edge $\left(x, y\right)$ in $C^p\left(\xi^\rightarrow\right)$ is

$$\nu\left(x, y\right) = \frac{n - p + 1}{n}\left[\sigma\left(x\right) \wedge \sigma\left(y\right)\right] h\left(N^+\left(x\right) \cap N^+\left(y\right)\right)$$

where

$$n = \left|supp\left(N^+\left(x\right) \cap N^+\left(y\right)\right)\right|.$$

Therefore, fuzzy p-competition graphs are graphs with edges between the vertices, if the vertices have exactly p number of common neighbourhoods. On the other hand, there is another variation of competition graph known as fuzzy k-competition graph (Pal et al., 2013) where, edges between two vertices exists if the minimum membership value of the common out-neighbourhoods of the vertices is more than positive real number k. Formal definition is given below.

Let k be a non-negative number. The fuzzy k-competition graph $C_k\left(\xi^{\rightarrow}\right) = \left(V, \sigma, \nu\right)$ of a fuzzy digraph $\xi^{\rightarrow} = \left(V, \sigma, \mu\right)$ is an undirected fuzzy graph which has the same vertex set as in $\xi^{\rightarrow}$ and has a fuzzy edge between two vertices $x, y \in V$ in $C_k\left(\xi^{\rightarrow}\right)$ if and only if

$$\left|N^+\left(x\right) \cap N^+\left(y\right)\right| > k \, .$$

The edge membership value between x and y in $C_k\left(\xi^{\rightarrow}\right)$ is

$$\nu\left(x, y\right) = \frac{k' - k}{k'}\left[\sigma\left(x\right) \wedge \sigma\left(y\right)\right] h\left(N^+\left(x\right) \cap N^+\left(y\right)\right)$$

where

$$k' = \left|N^+\left(x\right) \cap N^+\left(y\right)\right| \, .$$

Fuzzy tolerance of a fuzzy interval is denoted by T and is defined by an arbitrary fuzzy interval whose core length is a positive real number. If the real number is taken as L and $\left|i_k - i_{k-1}\right| = L$ where $i_k, i_{k-1} \in R$, a set of real numbers, then the fuzzy tolerance is a fuzzy set of the interval $\left[i_{k-1}, i_k\right]$.

Fuzzy tolerance graph is defined by Samanta and Pal in (2011b). They defined the fuzzy tolerance graph $G = \left(V, \sigma, \mu\right)$ as the fuzzy intersection graph of finite family of fuzzy intervals $I = \left\{I_1, I_2, \ldots, I_n\right\}$ on the real line along with tolerances $T = \left\{T_1, T_2, \ldots, T_n\right\}$ associated to each vertex of $\nu_i \in V$, where, $\sigma : V \rightarrow \left[0, 1\right]$ is defined by $\sigma\left(v_i\right) = h\left(I_i\right) = 1$ for all $v_i \in V$ and $\mu : V \times V \rightarrow \left[0, 1\right]$ is defined by

$$\mu\left(v_i, v_j\right) = \{1, if \, c\left(I_i \cap I_j\right) \geq \left\{c\left(T_i\right), c\left(T_j\right)\right\} \frac{s\left(I_i \cap I_j\right) - \min\left\{s(T_i, s\left(T_j\right)\right\}}{s\left(I_i \cap I_j\right)}$$

$$else \, if \, s\left(I_i \cap I_j\right) \geq \left\{s(T_i, s\left(T_j\right)\right\} 0, otherwise$$

Theorem 12 (Samanta & Pal, 2013)

Let $\xi^{\rightarrow} = \left(V, \sigma, \mu^{\rightarrow}\right)$ *be a fuzzy digraph. If* $N^+\left(x\right) \cap N^+\left(y\right)$ *contains single element of* $\xi^{\rightarrow}$, *then the edge* $\left(x, y\right)$ *of* $C\left(\xi^{\rightarrow}\right)$ *is strong if and only if*

$$\left|N^+\left(x\right) \cap N^+\left(y\right)\right| > 0.5 \, .$$

Proof. Here $\xi^{\rightarrow} = \left(V, \sigma, \mu^{\rightarrow}\right)$ is a fuzzy digraph. Let

$$N^+\left(x\right) \cap N^+\left(y\right) = \left\{\left(a, m\right)\right\},$$

where m is the membership value of the element a. Here

$$\left|N^+\left(x\right) \cap N^+\left(y\right)\right| = m = h\left(N^+\left(x\right) \cap N^+\left(y\right)\right).$$

So

$$\mu\left(x, y\right) = m \times \sigma\left(x\right) \wedge \sigma\left(y\right).$$

Hence, the edge $\left(x, y\right)$ in $C\left(\xi^{\rightarrow}\right)$ is strong if and only if $m > 0.5$.

If all the edges of a fuzzy digraph are strong, then all the edges of corresponding fuzzy competition graph may not be strong. This result is illustrated below. Let us consider two vertices x, y with

$$\sigma\left(x\right) = 0.3, \sigma\left(y\right) = 0.4$$

in a fuzzy digraph so that the vertices have a common prey z with $\sigma\left(z\right) = 0.2$. Let

$$\mu^{\rightarrow}\left(x, z\right) = 0.2, \mu^{\rightarrow}\left(y, z\right) = 0.15.$$

Clearly, the edges $\left(x, z\right)^{\rightarrow}$ and $\left(y, z\right)^{\rightarrow}$ are strong. But, the membership value of the edge $\left(x, y\right)$ in corresponding competition graph is $0.3 \times 0.15 = 0.045$. Hence, the edge is not strong as strength of the edge is $0.15 < 0.5$. But, if all the edges are strong of a fuzzy digraph, then the following result is obtained.

Theorem 13 (Samanta & Pal, 2013)

If all the edges of a fuzzy digraph $\xi^{\rightarrow} = \left(V, \sigma, \mu^{\rightarrow}\right)$ are strong, then

$$\frac{\mu\left(x, y\right)}{\left(\sigma\left(x\right) \wedge \sigma\left(y\right)\right)^2} > 0.5$$

for all edges $\left(x, y\right)$ in $C\left(\xi^{\rightarrow}\right)$.

Thus it is seen that if the height of intersection between two out-neighbourhoods of two vertices in a fuzzy digraph is greater than 0.5, the edge between the two vertices in corresponding fuzzy competition graph is strong. This result is not true in the corresponding fuzzy k-competition graph.

INTERVAL-VALUED FUZZY GRAPH

An **interval number** (Akram & Dudek, 2011) D is an interval $\left[a^-, a^+\right]$ with $0 \leq a^- \leq a^+ \leq 1$. For two interval numbers $D_1 = \left[a_1^-, a_1^+\right]$ and $D_2 = \left[a_2^-, a_2^+\right]$ the followings are defined:

1. $D_1 + D_2 = \left[a_1^-, a_1^+\right] + \left[a_2^-, a_2^+\right] = \left[a_1^- + a_2^- - a_1^- \cdot a_2^-, a_1^+ + a_2^+ - a_1^+ \cdot a_2^+\right],$

2. $min\left\{D_1, D_2\right\} = \left[min\left\{a_1^-, a_2^-\right\}, min\left\{a_1^+, a_2^+\right\}\right],$

3. $max\left\{D_1, D_2\right\} = \left[max\left\{a_1^-, a_2^-\right\}, max\left\{a_1^+, a_2^+\right\}\right],$

4. $D_1 \leq D_2 \Leftrightarrow a_1^- \leq a_2^-$ and $a_1^+ \leq a_2^+,$

5. $D_1 = D_2 \Leftrightarrow a_1^- = a_2^-$ and $a_1^+ = a_2^+,$

6. $D_1 < D_2 \Leftrightarrow D_1 \leq D_2$ but $D_1 \neq D_2,$

7. $kD_1 = \left[ka_1^-, ka_1^+\right],$ where $0 \leq k \leq 1.$

An interval-valued fuzzy set A on a set X is a mapping

$$\mu_A : X \rightarrow \left[0,1\right] \times \left[0,1\right],$$

called the membership function, i.e.

$$\mu_A\left(x\right) = \left[\mu_A^-\left(x\right), \mu_A^+\left(x\right)\right].$$

The support of A is

$$supp\left(A\right) = \{x \in X \mid \mu_A^-\left(x\right) \neq 0\}$$

and the core of A is

$$core\left(A\right) = \{x \in X \mid \mu_A^-\left(x\right) = 1\}.$$

The support length is $s\left(A\right) = \left|supp\left(A\right)\right|$ and the core length is $c\left(A\right) = \left|core\left(A\right)\right|$. The height of A is

$$h\left(A\right) = max\{\mu_A\left(x\right) \mid x \in X\} = \left[h^-\left(A\right), h^+\left(A\right)\right] = \left[max\left\{\mu_A^-\left(x\right)\right\}, max\left\{\mu_A^+\left(x\right)\right\}\right], \forall x \in X .$$

Let $F = \{A_1, A_2, \ldots, A_n\}$ be a finite family of interval-valued fuzzy subsets on a set X. The fuzzy intersection of two interval-valued fuzzy sets (IVFSs) A_1 and A_2 is an interval-valued fuzzy set defined by

$$A_1 \cap A_2 = \left\{ \left(x, \left[min\left\{ \mu_{A_1}^-(x), \mu_{A_2}^-(x) \right\}, min\left\{ \mu_{A_1}^+(x), \mu_{A_2}^+(x) \right\} \right] \right) : x \in X \right\}.$$

The fuzzy union of two IVFSs A_1 and A_2 is a IVFS defined by

$$A_1 \cup A_2 = \left\{ \left(x, \left[max\left\{ \mu_{A_1}^-(x), \mu_{A_2}^-(x) \right\}, max\left\{ \mu_{A_1}^+(x), \mu_{A_2}^+(x) \right\} \right] \right) : x \in X \right\}.$$

Here B is an interval-valued fuzzy relation on a set X, is denoted by

$$\mu_B : X \times X \Rightarrow [0,1] \times [0,1]$$

such that

$$\mu_B^-(x,y) \leq min\left\{ \mu_A^-(x), \mu_A^-(y) \right\} \quad \mu_B^+(x,y) \leq min\left\{ \mu_A^+(x), \mu_A^+(y) \right\}$$

An interval-valued fuzzy graph of a crisp graph $G = (V, E)$ is a graph $G = (V, A, B)$, where $A = [\mu_A^-, \mu_A^+]$ is an interval-valued fuzzy set on V and $B = [\mu_B^-, \mu_B^+]$ is an interval-valued fuzzy relation on E. An edge $(x,y), x, y \in V$ in an interval-valued fuzzy graph is said to be independent strong if

$$\mu_B^-(x,y) \geq \frac{1}{2} min\left\{ \mu_A^-(x), \mu_A^-(y) \right\}.$$

An interval-valued fuzzy digraph $G^\rightarrow = (V, A, B^\rightarrow)$ is an interval-valued fuzzy graph where the fuzzy relation $B^\rightarrow$ is antisymmetric.

An interval-valued fuzzy graph $G = (V, A, B)$ is said to be **complete interval-valued fuzzy graph** (Samanta & Pal, 2013). If

$$\mu^-(x,y) = min\left\{ \sigma^-(x), \sigma^-(y) \right\}$$

and

$$\mu^+(x,y) = min\left\{ \sigma^+(x), \sigma^+(y) \right\}, \forall x, y \in V.$$

An interval-valued fuzzy graph is said to be **bipartite** if the vertex set V can be partitioned into two sets V_1 and V_2 such that $\mu^-(u,v) = 0$ if $u,v \in V_1$ or $u,v \in V_2$ and $\mu^-(v_1, v_2) > 0$ if $v_1 \in V_1 (or V_2)$ and $v_2 \in V_2 (or V_1)$.

An interval-valued fuzzy graph $H = (V', A', B')$ is called an **interval-valued fuzzy subgraph** of the interval-valued fuzzy graph $G = (V, A, B)$ induced by V' if $V' \subseteq V$, $\mu^-_{A'}(x) = \mu^-_A(x)$ and $\mu^+_{A'}(x) = \mu^+_A(x)$ for all $x \in V'$ and $\mu^-_{B'}(x) = \mu^-_B(x)$ and $\mu^+_{B'}(x) = \mu^+_B(x)$ for all $x, y \in V'$.

An interval-valued fuzzy subgraph $H = (V', A', B')$ induced by A' is the **maximal interval-valued fuzzy subgraph** of $G = (V, A, B)$ which has the interval-valued fuzzy vertex set A' and the interval-valued fuzzy edge set B' be such that

$$\mu^-_{B'}(x,y) = \mu^-_{A'}(x) \wedge \mu^-_{A'}(y) \wedge \mu^-_B(x,y) \, and \, \mu^+_{B'}(x,y) = \mu^+_{A'}(x) \wedge \mu^+_{A'}(y) \wedge \mu^+_B(x,y)$$

for all $x, y \in V$. An interval-valued fuzzy graph $P = (V, A, B)$ is called an **interval-valued fuzzy cycle** if and only if it contains more than one weakest edge (i.e., there is no unique $(x, y) \in B'$ such that

$$\mu_B(x,y) = \wedge \{\mu_B(u,v) \mid (u,v) \in B'\})$$

and $P^* = (V, A^*, B^*)$ is a cycle.

An interval-valued fuzzy subgraph $H = (V', A', B')$ of $G = (V, A, B)$ is an **interval-valued fuzzy clique** if H^* is a clique and each cycle in H is an interval-valued fuzzy cycle.

FUZZY HYPERGRAPH

A hypergraph is a generalization of a graph in which an edge can connect any number of vertices. Formally, a hypergraph H^* is a pair $H^* = (X, E)$ where X is a set of elements called nodes or vertices and E is a set of non-empty subsets of X called hyperedges or edges. Therefore, E is a subset of $P(X) \setminus \{\phi\}$, where $P(X)$ is the power set (collection of all subsets) of X. While edges of a graph are pair of nodes or vertices, hyperedges are arbitrary sets of nodes and therefore contain arbitrary number of nodes. A directed hypergraph is a generalization of the concept of directed graph. It was first introduced by Byers and Waterman (1984) to represent functional dependencies in relational data base. A directed hypergraph is given by a set of nodes V and a set of pairs (T, h) (hyperedges) where T is a subset of V and h is a single node in V. The most obvious interpretation of a hyperedge (T, h) is that the information associated to h functionally depends on the information associated to nodes in T.

Sometimes hypergraphs can be defined by incidence matrix where each column represented by edge set and each row is represented by vertex set. Some useful definitions and notations (Akram & Dudek, 2013) related to hypergraphs are listed in Table 3.

Table 3. Notations and definitions related to hypergraphs

Symbol	Title	Definition
$H^* = \left(X, E \right)$	Crisp hypergraph	Generalization of graph where an edge can contain more than two vertices instead of only two vertices
$r \left(H^* \right)$	Rank of a hypergraph	Maximum number of nodes in one edge, i.e., $r \left(H^* \right) = \max_{j} \left| E_j \right|$
$s \left(H^* \right)$	Anti-rank of a hypergraph	Minimum number of nodes in one edge, i.e., $s \left(H^* \right) = \min_{j} \left| E_j \right|$
—	Uniform hypergraph	A hypergraph is said to be uniform if the rank and anti-rank of that hypergraph is equal, i.e., $r \left(H^* \right) = s \left(H^* \right)$
—	k-uniform hypergraph	A uniform hypergraph of rank k is called the k-uniform hypergraph. Hence a simple hypergraph is a 2-uniform hypergraph
$H^d = \left(X^d, E^d \right)$ $H^* = \left(X, E \right)$	Dual of a hypergraph	Dual of a hypergraph is obtained by interchanging of vertices and hyperedges in H^*

A hypergraph is useful in various combinatorial structures that generalize graphs. Directed hypergraph is an extension of directed graphs, and have often used in several areas such as a modelings and algorithmic tool. A brief introduction to directed hypergraphs is given by Gallo et al. (1993). Geotschel (1995) introduced the concept of fuzzy hypergraphs and Hebbian structures. Goetschel (1998) also explained the coloring of fuzzy hypergraphs. Intersecting fuzzy hypergraphs are defined by Goetschel and Voxman (1998).

Samanta and Pal have also worked on bipolar fuzzy hypergraphs and related fuzzy graphs in (Pramanik et al., 2016; Samanta & Pal, 2012; Pramanik et al., 2018; Pramanik et al., 2017; Samanta & Pal, 2012; Samanta et al., 2014b; 39, Samanta et al., 2016). Readers can found many recent works on [9, Dubois & Prade, 1983; Ghoseiri & Moghadam, 2008; Pramanik et al., 2016).

There are many real life problems such as communications, routing, transportation, etc., finding shortest path is very essential in research purposes. But finding shortest paths in hypergraphs are too more demanding in recent research areas. Hypergraphs can consider more complex networks such as protein-protein interaction network, social networks, information theory, publication data, collaborations, chemical processes, etc. Hypergraphs are learned to segment or classify the datas. In learning of distances between two nodes in a hypergraph, first create a normal graph by connecting nodes with weighted edges. Weight is the sum of the weights of hyperedges traversed in the shortest path. An example is given here.

Example 2: Suppose there are seven cities, say, A, B, C, D, E, F, G in a country. The hypergraph relation of these cities are given in Figure 3. A crisp graph can be constructed by specifying connections to each cities which are connected by roads such as

$$A \to B, A \to C, A \to D, B \to E, C \to E, C \to F, D \to F, E \to G, F \to G.$$

Figure 3. Example of a hypergraph

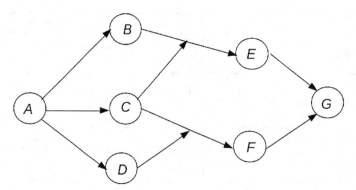

Figure 4. Graph representation of the hypergraph shown in Figure 3

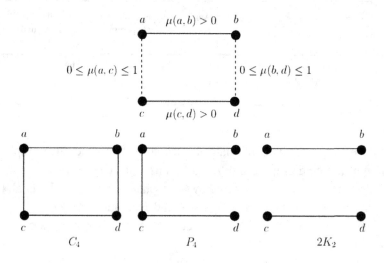

The corresponding graph is shown in Figure 4.

The population of G has demand of some products. The city A has a supplier to supply the products but wants the minimum cost of transportation to supply. Each roads have some cost of transportaion. Depending on many parameters (such as toll taxes, levies, security taxes, etc.) costs of transportation varies road by road and also time by time. So the costs can be treated as interval-valued trapezoidal fuzzy numbers where left end trapezoidal fuzzy number is the minimum cost and right end trapezoidal fuzzy number is the maximum cost to transport by that road. So this problem can be modelled as the interval-valued fuzzy hypernetwork with edge weights as trapezoidal fuzzy numbers.

Before going to be introduced with fuzzy hypergraph the formal definition of crisp hypergraph is given as follows:

Let $X = \{x_1, x_2, \ldots, x_n\}$ be a finite set. A hypergraph on X is a family $H = \{E_1, E_2, \ldots, E_m\}$ of subsets of X such that

$$E_i \neq \phi, \, (i = 1, 2, \ldots, m) \text{ and } \bigcup_{i=1}^{m} E_i = X.$$

Table 4. Fuzzy hypergraphs and its advancements

Author	Contribution	Year
Kaufmann (1973)	Introduced fuzzy hypergraphs	1977
Lee-Kwang and Lee (1995)	Generalization and redefinition of fuzzy hypergraphs	1995
Chen (1997)	Introduced interval-valued fuzzy hypergraphs	1997
Parvathi et al. (2009)	Introduced intuitionistic fuzzy hypergraphs	2009
Samanta and Pal (2012)	Introduced bipolar fuzzy hypergraphs	2012
Akram et al. (2013)	Developed certain properties on bipolar fuzzy hypergraphs	2013
Akram and Luqman (2017)	Introduced bipolar neutrosophic fuzzy hypergraphs	2017
Akram and Sarwar (2017a)	m -polar fuzzy hypergraphs	2017
Akram and Sarwar (2017b)	Discussed transversals of m -polar fuzzy hypergraphs with applications	2017
Akram and Shahzadi (2018)	Discussed new applications of m -polar fuzzy hypergraphs in decision-making problems	2018

The elements $x_1, x_2, \ldots, x_n$ are called the **vertices** and the sets $E_1, E_2, \ldots, E_m$ are called the **hyperedges** (or, simply **edges**) of the hypergraph.

A **simple hypergraph** is a hypergraph $H = \{E_1, E_2, \ldots, E_m\}$ such that $E_i \subset E_j \Rightarrow i = j$. A graph is a simple hypergraph each of whose edges has cardinality 2. An example of hypergraph is shown in Figure 3. In this example (X, H) is a hypergraph where,

$$X = \left\{ x_1, x_2, x_3, x_4, x_5, x_6, x_7, x_8 \right\}$$

is the set of vertices and

$$H = \left\{ \left\{ x_1, x_2, x_3 \right\}, \left\{ x_3, x_4, x_5 \right\}, \left\{ x_5, x_6, x_7 \right\}, \left\{ x_8 \right\} \right\}$$

is the set of hyperedges.

After Kaufmann, several extensions of fuzzy hypergraphs are found. A short survey of works on fuzzy hypergraphs are presented in Table 4.

FUZZY HYPERGRAPHS AND DEFINITIONS

Before going to be introduced with several properties of fuzzy graphs and fuzzy hypergraphs, the preliminary definitions are presented.

A fuzzy set A on a universal set X is defined by a mapping $m : X \to [0,1]$, which is called the membership function. A fuzzy set is denoted by $A = (X, m)$. A fuzzy graph (Rosenfeld, 1975) $G = (V, \sigma, \mu)$ is a non-empty set V together with a pair of functions $\sigma : V \to [0,1]$ and $\mu : V \times V \to [0,1]$

such that for all, where $\sigma(x)$ and $\mu(x,y)$ represent the membership values of the vertex x and of the edge (x,y) in G respectively. Fuzzy hypergraph is the mere extension of fuzzy graphs.

Let X be a finite set and let E be a family of non-empty fuzzy subsets of X such that

$$X = \cup\{\sup p(A) \mid A \in E\}.$$

Then the pair $H = (X,E)$ is called a **fuzzy hypergraph** on X. The basic notion of fuzzy hypergraph is given by Kaufmann (1973). Lee-Kwang and Lee (1995) have modified this definition and generalized this concept in the following way:

A fuzzy hypergraph $H = (X,E)$ is equipped with a finite set of fuzzy vertices $X = \{x_1, x_2, \ldots, x_n\}$ and the family of fuzzy subsets $E = \{E_1, E_2, \ldots, E_m\}$ of X with the following properties:

$$E_j = \{(x_i, \mu_j(x_i) \mid \mu_j(x_i)) 0\}, j = 1, 2, \ldots, m$$

$$E_j \neq \phi, j = 1, 2, \ldots, m$$

$$\cup_j supp(E_j) = X, j = 1, 2, \ldots, m$$

The hyperedges E_j is a fuzzy set of vertices. The membership value of the vertex x_i to the edge E_j is defined by $\mu_j(x_i)$.

A fuzzy hypergraph $H = (X,E)$ is called **simple** if E has no repeated fuzzy edges and whenever $\mu, \nu \in E$ and $\mu \subseteq \nu$, then $\mu = \nu$.

A fuzzy hypergraph $H = (X,E)$ is called **support simple**, if, whenever $\mu, \nu \in E$, $\mu \subseteq \nu$ and $supp(\mu) = supp(\nu)$, then $\mu = \nu$.

Suppose $H = (X,E)$ and $H' = (X',E')$ are two fuzzy hypergraphs. H is called a **partial fuzzy hypergraph** of H' if $E \subseteq E'$. If H is a partial fuzzy hypergraph of H', we write $H \subseteq H'$, If $H \subseteq H'$ and $E \subset E'$ then we write $H \subset H'$.

The **dual fuzzy hypergraph** H^d of a fuzzy hypergraph H is defined as $H^d = (X, E^d)$, where $X^d = \{e_1, e_2, \ldots, e_m\}$ is a finite set of vertices corresponding to $E = \{E_1, E_2, \ldots, E_m\}$ respectively and E^d is a finite set of hyperedges, $X_1, X_2, \ldots, X_n$ corresponding to $x_1, x_2, \ldots, x_n$ respectively, where

$$X_i = \{(e_j, \mu_i(e_j)) \mid \mu_i(e_j) = \mu_j(x_i)\}.$$

INTERVAL-VALUED FUZZY HYPERGRAPH

Let X be a finite set and let E be a family of non-empty interval-valued fuzzy subsets $E_1, E_2, \ldots, E_m$ of X such that $X = \cup\{supp(A) \mid A \in E\}$.

Then the pair $H = (X, E)$ is called an **interval-valued fuzzy hypergraph** on X.

An interval-valued fuzzy hypergraph $H = (X, E)$ is simple if $A = \left[\mu_A^-, \mu_A^+\right]$, $B = \left[\mu_B^-, \mu_B^+\right] \in E$ and $\mu_A^- \leq \mu_B^-$, $\mu_A^+ \leq \mu_B^+$ imply that $\mu_A^- = \mu_B^-$, $\mu_A^+ = \mu_B^+$. An interval-valued fuzzy hypergraph $H = (V, E)$ is **support simple** if $A = \left[\mu_A^-, \mu_A^+\right]$, $B = \left[\mu_B^-, \mu_B^+\right] \in E$, $supp(A) = supp(B)$ and $\mu_A^- \leq \mu_B^-$, $\mu_A^+ \leq \mu_B^+$ imply that $\mu_A^- = \mu_B^-$, $\mu_A^+ = \mu_B^+$. An interval-valued fuzzy hypergraph $H = (X, E)$ is **strongly support simple** if $A = \left[\mu_A^-, \mu_A^+\right]$, $B = \left[\mu_B^-, \mu_B^+\right] \in E$ and $supp(A) = supp(B)$ imply that $A = B$.

An interval-valued fuzzy set

$$A = \left[\mu_A^-, \mu_A^+\right] : X \to \left[0, 1\right]$$

is an elementary interval-valued fuzzy set if A is single valued on $supp(A)$. An **elementary interval-valued fuzzy hypergraph** (Akram & Alshehri, 2015) $H = (X, E)$ is an interval-valued fuzzy hypergraph whose edges are elementary.

Example 3: Let $X = \left\{x_1, x_2, x_3, x_4, x_5\right\}$ be a set of vertices. Consider a set E of interval-valued fuzzy sets E_1, E_2, E_3 such that

$$E_1 = \left\{\left(x_1, \left[0.3, 0.7\right]\right), \left(x_2, \left[0.5, 0.8\right]\right)\right\},$$

Figure 5. Example of an interval-valued fuzzy hypergraph

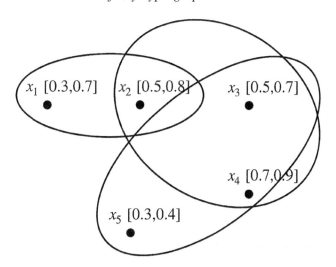

$$E_2 = \left\{ \left(x_2, [0.5, 0.8] \right), \left(x_3, [0.5, 0.7] \right), \left(x_4, [0.7, 0.9] \right) \right\},$$

$$E_3 = \left\{ \left(x_3, [0.5, 0.7] \right), \left(x_4, [0.7, 0.9] \right), \left(x_5, [0.3, 0.4] \right), \right\}.$$

Then the fuzzy graph $H = (X, E)$ is an interval-valued fuzzy graph. The graphical representation is shown in Figure 5.

A crisp-valued α-cut at β level hypergraph (Chen, 1997) $H_{\alpha, \beta}$ of an interval-valued fuzzy hypergraph H can be obtained by first cutting the interval-valued fuzzy hypergraph H at α-level to derive a fuzzy hypergraph H_α and then by applying β-cut to the derived fuzzy hypergraph H_α to obtain the α-cut at β level hypergraph $H_{\alpha, \beta}$ of H, where $\alpha \in [0, 1]$ and $\beta \in [0, 1]$. Let H be an interval-valued fuzzy hypergraph, $H = (X, E)$ and let E_j be a hyperedge in the interval-valued fuzzy hypergraph H, $E_j \in E$, where

$$E_j = \left\{ \left(x_1, [a_{11}, a_{12}] \right), \left(x_2, [a_{21}, a_{22}] \right), \ldots, \left(x_n, [a_{n1}, a_{n2}] \right) \right\}.$$

INTUITIONISTIC FUZZY HYPERGRAPHS

An **intuitionistic fuzzy set (IFS)** A in X is defined as an object of the form

$$A = \left\{ \left(x, \mu_A(x), \nu_A(x) \right) \mid x \in X \right\}$$

where the functions $\mu_A : X \rightarrow [0, 1]$ and $\nu_A : X \rightarrow [0, 1]$ define the degree of membership and the degree of non-membership of the element $x \in X$, respectively, and for every

$$x \in X, 0 \leq \mu_A(x) + \nu_A(x).$$

The **intuitionistic fuzzy hypergraph** H is an ordered pair $H = (X, E)$ where

$$X = \left\{ x_1, x_2, \ldots, x_n \right\}$$

is a finite set of vertices

$$E = \left\{ E_1, E_2, \ldots, E_m \right\}$$

is a family of intuitionistic fuzzy subsets of X

$$E_j = \left\{ \left(x_i, \mu_j \left(x_i \right), \nu_j \left(x_i \right) \right) : \mu_j \left(x_i \right), \nu_j \left(x_i \right) \geq 0 \, and \, \mu_j \left(x_i \right) + \nu_j \left(x_i \right) \leq 1 \right\}, \; j = 1, 2, \ldots, m$$

$$E_i \neq \phi, j = 1, 2, \ldots, m$$

$$\bigcup_j supp \left(E_j \right) = X, j = 1, 2, \ldots, m$$

The **dual intuitionistic fuzzy hypergraph** of an intuitionistic fuzzy hypergraph $H = \left(X, E \right)$ is denoted by $H^* = \left(E^*, \underline{X} \right)$ where, each e_i of $E^* = \left\{ e_1, e_2, \ldots, e_m \right\}$ is corresponding to E_i of

$$E = \left\{ E_1, E_2, \ldots, E_n \right\}$$

and

$$\underline{X} = \left\{ \underline{x}_1, \underline{x}_2, \ldots, \underline{x}_n \right\}$$

with

$$\underline{x}_j = \left\{ \left(e_j, \mu_i \left(e_j \right), \nu_i \left(e_j \right) \right) : \mu_i \left(e_j \right) = \mu_j \left(x_i \right), \nu_i \left(e_j \right) = \nu_j \left(x_i \right) \right\}.$$

CONCLUSION

Hypergraphs are the generalization of graphs in case of set of multi-ary relations. The hypergraph has been considered a useful tool for modeling system architectures or data structures and to represent a partition, covering and clustering in the area of circuit design. The concept of ordinary hypergraphs was extended to fuzzy theory. The interval-valued fuzzy sets constitute a generalization of Zadeh's fuzzy set theory. In this chapter, several types of fuzzy graphs and fuzzy hypergraphs are discussed. Researchers are keen to find several new types of fuzzy hypergraphs to consider more complex real world networks. There are till new scope to find various type of fuzzy hypergraphs and relationships among them.

ACKNOWLEDGMENT

We thank all the reviewers and editors for their valuable suggestion for improvement of this book chapter in all perspective.

REFERENCES

Akram, M., & Alshehri, N. O. (2015). Tempered interval-valued fuzzy hyeprgraphs. *U.P.B Sci. Bull, Series A, 77*(1), 39–47.

Akram, M., & Dudek, W. A. (2011). Interval-valued fuzzy graphs. *Computers & Mathematics with Applications (Oxford, England), 61*(2), 289–299. doi:10.1016/j.camwa.2010.11.004

Akram, M., & Dudek, W. A. (2012). Regular bipolar fuzzy graphs. *Neural Computing & Applications, 21*(1), 197–205. doi:10.100700521-011-0772-6

Akram, M., & Dudek, W. A. (2013). Intuitionistic fuzzy hypergraphs with applications. *Information Sciences, 218*, 182–193. doi:10.1016/j.ins.2012.06.024

Akram, M., Dudek, W. A., & Sarwa, S. (2013). Properties of bipolar fuzzy hypergraphs. *International Journal of Pure and Applied Mathematics, 31*, 141–160.

Akram, M., & Luqman, A. (2017). Bipolar neutrosophic hypergraphs with applications. *Journal of Intelligent & Fuzzy Systems, 33*(3), 1699–1713. doi:10.3233/JIFS-17228

Akram, M., & Sarwar, M. (2017a). Novel application of m-polar fuzzy hypergraphs. *Journal of Intelligent & Fuzzy Systems, 32*(3), 2747–2762. doi:10.3233/JIFS-16859

Akram, M., & Sarwar, M. (2017b). Transversals of m-polar fuzzy hypergraphs with applications. *Journal of Intelligent & Fuzzy Systems, 33*(1), 351–364. doi:10.3233/JIFS-161668

Akram, M., & Shahzadi, G. (2018). Hypergraphs in m-polar fuzzy environment. *Mathematics, 6*(2), 28. doi:10.3390/math6020028

Byers, T. H., & Waterman, M. S. (1984). Determining all optimal and near-optimal solutions when solving shortest path problems by dynamic programming. *Operations Research, 32*(6), 1381–1384. doi:10.1287/opre.32.6.1381

Chen, S. M. (1997). Interval-valued fuzzy hypergraph and fuzzy partition. *IEEE Transactions on Systems, Man, and Cybernetics. Part B, Cybernetics, 27*(4), 725–733. doi:10.1109/3477.604121 PMID:18255914

Dubois, D., & Prade, H. (1983). Ranking fuzzy numbers in the setting of possibility theory. *Information Sciences, 30*(3), 183–224. doi:10.1016/0020-0255(83)90025-7

Gallo, G., Longo, G., Nguyen, S., & Pallottino, S. (1993). Directed hypergraphs and applications. *Discrete Applied Mathematics, 40*(2-3), 177–201. doi:10.1016/0166-218X(93)90045-P

Ghoseiri, K., & Moghadam, A. R. J. (2008). Continuous fuzzy longest path problem in project networks. *Journal of Applied Sciences (Faisalabad), 8*(22), 4061–4069. doi:10.3923/jas.2008.4061.4069

Goetschel, R. H. Jr. (1995). Introduction to fuzzy hypergraphs and Hebbian structures. *Fuzzy Sets and Systems, 76*(1), 113–130. doi:10.1016/0165-0114(94)00381-G

Goetschel, R. H. Jr. (1998). Fuzzy colorings of fuzzy hypergraphs. *Fuzzy Sets and Systems, 94*(2), 185–204. doi:10.1016/S0165-0114(96)00256-4

Goetschel, R. H. Jr, & Voxman, W. (1998). Intersecting fuzzy hypergraphs. *Fuzzy Sets and Systems*, *99*(1), 81–96. doi:10.1016/S0165-0114(97)00005-5

Jabbar, N. A., Naoom, J. H., & Ouda, E. H. (2009). Fuzzy dual graph. *Journal of Al-Nahrain University*, *12*, 168–171.

Kaufmann, A. (1973). *Introduction la Thorie des Sous-Ensembles Flous Lusage des Ingnieurs (Fuzzy Sets Theory), Masson*. Paris: French.

Lee, K. M. (2000). Bipolar-valued fuzzy sets and their basic operations. *Proceedings of the International Conference*, 307-317.

Lee-Kwang, H., & Lee, K. M. (1995). Fuzzy hypergraph and fuzzy partition. *IEEE Transactions on Systems, Man, and Cybernetics*, *25*(1), 196–201. doi:10.1109/21.362951

Mathew, S., Mordeson, J. N., & Malik, D. S. (2018). *Fuzzy graph theory*. Springer International Publishing. doi:10.1007/978-3-319-71407-3

McAllister, M. L. N. (1988). Fuzzy intersection graphs. *Computers & Mathematics with Applications (Oxford, England)*, *15*(10), 871–886. doi:10.1016/0898-1221(88)90123-X

Mordeson, J. N., & Nair, P. S. (1996). Successor and source of (fuzzy) finite state machines and (fuzzy) directed graphs. *Information Sciences*, *95*(1-2), 113–124. doi:10.1016/S0020-0255(96)00139-9

Nagoorgani, A., & Vadivel, P. (2009). Relations between the parameters of independent domination and irredundance in fuzzy graph. *International Journal of Algorithms. Computing and Mathematics*, *2*(1), 15–19.

Nirmala, G., & Dhanabal, K. (2012). Special planar fuzzy graph configurations. *International Journal of Scientific and Research Publications*, *2*(7), 1–4.

Pal, M., Samanta, S., & Pal, A. (2013). Fuzzy *k*-competition graphs. In S*cience and Information Conference (SAI)*. IEEE-Xplore.

Parvathi, R., Thilagavathi, S., & Karunambigai, M. G. (2009). Intuitionistic fuzzy hypergraphs. *Cybernetics and Information Technologies*, *9*, 46–48.

Pramanik, T., Pal, M., & Mondal, S. (2016). Interval-valued fuzzy threshold graph. *Pacific Science Review A. Natural Science and Engineering*, *18*(1), 66–71.

Pramanik, T., Pal, M., Mondal, S., & Samanta, S. (2018). A study on bipolar fuzzy planar graph and its application in image shrinking. *Journal of Intelligent & Fuzzy Systems*, *34*(3), 1863–1874. doi:10.3233/JIFS-171209

Pramanik, T., Samanta, S., & Pal, M. (2016). Interval-valued fuzzy planar graphs. *International Journal of Machine Learning and Cybernetics*, *7*(4), 653–664. doi:10.100713042-014-0284-7

Pramanik, T., Samanta, S., & Pal, M. (2017). Fuzzy ϕ. *Soft Computing*, *21*(13), 3723–3734. doi:10.100700500-015-2026-5

Rosenfeld, A. (1975). Fuzzy graphs. New York: Academic Press.

Samanta, S., Akram, M., & Pal, M. (2015). -step fuzzy competition graphs. *Journal of Applied Mathematics and Computing*, *47*(1-2), 461–472. doi:10.100712190-014-0785-2

Samanta, S., & Pal, M. (2011a). Fuzzy threshold graphs. *CIIT International Journal of Fuzzy Systems*, *3*(12), 360–364.

Samanta, S., & Pal, M. (2011b). Fuzzy tolerance graphs. *International Journal of Latest Trends in Mathematics*, *1*(2), 57–67.

Samanta, S., & Pal, M. (2012). Bipolar fuzzy hypergraphs. *International Journal of Fuzzy Logic Systems*, *2*(1), 17–28. doi:10.5121/ijfls.2012.2103

Samanta, S., & Pal, M. (2012). Irregular bipolar fuzzy graphs. *International Journal of Applications of Fuzzy Sets*, *2*, 91–102.

Samanta, S., & Pal, M. (2013). Fuzzy k p. *Fuzzy Information and Engineering*, *5*(2), 191–204. doi:10.100712543-013-0140-6

Samanta, S., & Pal, M. (2015). Fuzzy planar graphs. *IEEE Transactions on Fuzzy Systems*, *23*(6), 1936–1942. doi:10.1109/TFUZZ.2014.2387875

Samanta, S., Pal, M., & Pal, A. (2014a). New concepts of fuzzy planar graph. *International Journal of Advanced Research in Artificial Intelligence*, *3*(1), 52–59. doi:10.14569/IJARAI.2014.030108

Samanta, S., Pal, M. & Pal, A. (2014b). Some more results on bipolar fuzzy sets and bipolar fuzzy intersection graphs. *The Journal of Fuzzy Mathematics*, *22*(2), 253-262.

Samanta, S., Pal, M., & Pal, A. (2014c). Some more results on fuzzy-competition graphs. *International Journal of Advanced Research in Artificial Intelligence*, *3*(1), 60–67. doi:10.14569/IJARAI.2014.030109

Samanta, S., Pramanik, T., & Pal, M. (2016). Fuzzy colouring of fuzzy graphs. *Afrika Matematika*, *27*(1-2), 37–50. doi:10.100713370-015-0317-8

Sunitha, M. S., & Vijayakumar, A. (2005). Blocks in fuzzy graphs. *The Journal of Fuzzy Mathematics*, *13*(1), 13–23.

Vimala, S., & Sathya, J. S. (2012). Connected point set domination of fuzzy graphs. *International Journal of Mathematics and Soft Computing*, *2*(2), 75–78. doi:10.26708/IJMSC.2012.2.2.10

Zhang, W. R. (1994, December). Bipolar fuzzy sets and relations: a computational framework for cognitive modeling and multiagent decision analysis. In *NAFIPS/IFIS/NASA'94. Proceedings of the First International Joint Conference of The North American Fuzzy Information Processing Society Biannual Conference. The Industrial Fuzzy Control and Intellige* (pp. 305-309). IEEE.

Zhang, W. R. (1998). Bipolar fuzzy sets. *Proceedings of FUZZ-IEEE*, 835–840.

Chapter 20
Energy of m–Polar Fuzzy Digraphs

Muhammad Akram
University of the Punjab, Pakistan

Danish Saleem
University of the Punjab, Pakistan

Ganesh Ghorai
Vidyasagar University, India

ABSTRACT

In this chapter, firstly some basic definitions like fuzzy graph, its adjacency matrix, eigenvalues, and its different types of energies are presented. Some upper bound and lower bound for the energy of this graph are also obtained. Then certain notions, including energy of m-polar fuzzy digraphs, Laplacian energy of m-polar fuzzy digraphs and signless Laplacian energy of m-polar fuzzy digraphs are presented. These concepts are illustrated with several example, and some of their properties are investigated.

INTRODUCTION

After the introduction of fuzzy sets by Zadeh (1965), fuzzy set theory has been applied in many research fields. Zhang (1994) presented the concept of bipolar fuzzy sets by depicting a positive degree of membership and negative degree of membership, which is an extension of fuzzy sets. This idea has been utilized in different ways, including preference modeling, multi-criteria decision inquiry and cooperative games. The positive degree of membership and negative degree of membership of bipolar fuzzy sets is lying in the interval [-1,1]. The degree of membership 0 of a component in a bipolar fuzzy set implies that the component is inapplicable to the property (K. M. Lee, 2000), the positive degree of membership (0,1] of a component argue that the component obviously satisfies the corresponding property and the negative degree of membership [-1,0) of a component argue that the component apparently satisfies the counter-property. Chen et al. (2014) introduced m-polar fuzzy sets as generalization of bipolar fuzzy

DOI: 10.4018/978-1-5225-9380-5.ch020

sets. He demonstrated that bipolar fuzzy sets and 2-polar fuzzy sets are cryptomorphic mathematical notions and that we can acquire briefly one from the corresponding one. The concept back of this is that "multipolar information" (not just bipolar information which correspond to two-valued logic) exists because the data for the real world problems is sometimes taken from n agents $(n \geq 2)$. For example, the exact degree of elecommunication safety of mankind is a point in $[0,1]^n$ ($n \approx 7 \times 10^9$) because different persons have been monitored different times. There are many other examples such as truth degrees of a logic formula which are based on n logic implication operators ($n \geq 2$), similarity degrees of two logic formulas which are based on n logic implication operators ($n \geq 2$), ordering results of a magazine, ordering results of a university, and inclusion degrees (accuracy measures, rough measures, approximation qualities, fuzziness measures, and decision preformation evaluations) of a rough set.

Graph theory was originated by Euler in 1736, when he solved the problem of the Königsberg's bridges, and presented his first paper on graph theory. The energy of a graph firstly proposed by Gutman (1978) and has a vast circle of applications in various fields, including computer science, physics, chemistry and other branch of mathematics. The upper and lower bound for the energy of a graph are discussed by several researchers (Brualdi, 2006; Gutman, 2001; Praba et al. 2014). The concept of fuzzy graphs was introduced by Kaufmann (1973) based on fuzzy relations of Zadeh (1971). Rosenfeld (1975) developed the structure of fuzzy graphs. Anjali and Mathew (2013) examined the energy of a fuzzy graph. Praba proposed the energy of an intuitionistic fuzzy graphs. Akram and Naz (2018) introduced the energy of Pythagorean fuzzy graphs. Naz et al. (2018) discussed some notions of energy in single-valued neutrosophic graphs. Naz et al. (2018a) proposed the notion of energy of bipolar fuzzy graphs. Akram and Waseem (2016) proposed the notion of m-polar fuzzy graphs and Ghorai and Pal (2015, 2016) described several operations and properties on it. Ghorai and Pal (2016a, 2016b) introduced m-polar fuzzy planar graphs, faces and dual of m-polar fuzzy planar graphs. Ghorai and Pal (2016c) also defined isomorphism and complement of m-polar fuzzy graphs. In this chapter, we present certain notions, including energy of m-polar fuzzy digraphs, Laplacian energy of m-polar fuzzy digraphs and signless Laplacian energy of m-polar fuzzy digraphs. We investigate some of their properties. For other notations, terminologies and applications not mentioned in the paper, the readers are referred to Akram (2019).

BACKGROUND

In this section, we review basic definitions which are helpful for later sections.

Definition 2.1 (Anjali and Mathew, 2013) An ordered pair of sets G = (V,E) where V is a nonempty finite set and E consisting of 2-element subsets of elements of V is called a graph. It is denoted by G = (V,E). V is called vertex and edge set respectively. The elements in V and E are called vertices and edges respectively.

Definition 2.2 A pair of the form $D = (V, E)$ is called directed graph, where V is a set whose elements are called vertices, nodes, or points and E is a set of ordered pairs of vertices, called arrows, directed edges, directed arcs, or directed lines.

Definition 2.3 (Gutman, 1978, 2001) Let $G = (V, E)$ be a graph. The energy of G is defined as the sum of its absolute eigenvalues, i.e.,

$$E(G) = \sum_{j=1}^{n} | \pi_j | .$$

Definition 2.4 (Pea and Rada, 2008) Let $D = (V, E)$ be a digraph. The energy of D is defined as the sum of the absolute real part of eigenvalues of D, i.e.,

$$E(D) = \sum_{j=1}^{n} | Re(\phi_j) | .$$

Below the definition of fuzzy graph is given as a generalization of crisp graph.

Definition 2.5 (Anjali and Mathew, 2013) A pair of the form $G = (\mu, \lambda)$ is said to be fuzzy graph of the crisp graph $G^* = (V, E)$, where μ is a fuzzy subset of V and λ is a fuzzy relation on μ, i.e.,

$$\lambda(z_j z_k) \leq \mu(z_j) \wedge \mu(z_k) \text{ for all } z_j, z_k \in V .$$

An example of a fuzzy graph is given in Figure 1.

G is said to be complete if

$$\lambda(z_j z_k) = \mu(z_j) \wedge \mu(z_k)$$

for all $z_j, z_k \in V$ and it is said to strong if

$$\lambda(z_j z_k) = \mu(z_j) \wedge \mu(z_k)$$

for all $z_j z_k \in E$.

Definition 2.6 (Anjali and Mathew, 2013) Let G be a fuzzy graph. The adjacency matrix $A(G)$ of G is a square matrix of order n defined by $A(G) = [a_{jk}]_{n \times n}$, where

$$a_{jk} = \begin{cases} \lambda(z_j z_k), & z_j z_k \in E, \\ 0, & otherwise. \end{cases}$$

Since researches or modelings on real world problems often involve multi-agent, multi-attribute, multi-object, multi-index, multi-polar information, uncertainty, or/and limit process, Chen et al (2014) introduced the notion of m-polar fuzzy set which is defined below.

Definition 2.7 (Chen et al., 2014) An m-polar fuzzy set in a non empty set Z is a mapping $U : Z \rightarrow [0,1]^m$. The degree of each element $z \in Z$ is defined as

$$U(z) = (P_1 \circ U(z), P_2 \circ U(z), \cdots, P_m \circ U(z)),$$

where

$$P_i \circ U : [0,1]^m \rightarrow [0,1]$$

is the i-th projection mapping. Note that $[0,1]^m$ (m-th power of $[0,1]$) is considered a poset with the point-wise order $\leq$, where m is an arbitrary ordinal number (we make an appointment that $m = n \mid n < m$ when $m > 0$), $\leq$ is defined by

$$z_1 \leq z_2 \Leftrightarrow P_i(z_1) \leq P_i(z_2)$$

for each

$$i \in m \ (z_1, z_2 \in [0,1]^m),$$

and $P_i : [0,1]^m \rightarrow [0,1]$ is the i-th projection mapping $(i \in m)$.

Definition 2.8 (Akram and Waseem, 2016) Let U be an m-polar fuzzy subset of a non-empty set Z. An m-polar fuzzy relation on U is an m-polar fuzzy subset V of $Z \times Z$ defined by the mapping $V : Z \times Z \rightarrow [0,1]^m$ such that for all $z_1, z_2 \in Z$,

$$P_i \circ V(z_1 z_2) \leq \inf\{P_i \circ U(z_1), P_i \circ U(z_2)\},$$

$1 \leq i \leq m$, where $P_i \circ U(z_1)$ denotes the i-th degree of membership of the vertex z_1 and $P_i \circ V(z_1 z_2)$ denotes the i-th degree of membership of the edge $z_1 z_2$.

ENERGY OF FUZZY GRAPHS

In this section, we present energy of fuzzy graphs which is discussed by Anjali and Mathew in [5].

Definition 3.1 (Anjali and Mathew, 2013) Let $G = (\mu, \lambda)$ be a fuzzy graph and A(G) be its adjacency matrix. The eigenvalues of A(G) are called eigenvalues of G. The spectrum of A(G) is called the spectrum of G. The set of all eigenvalues of G is called the spectrum of G. It is denoted by Spec(G).

Definition 3.2 (Anjali and Mathew, 2013) Let $G = (\mu, \lambda)$ be a fuzzy graph. The energy of G is defined as the sum of its absolute eigenvalues, i.e.,

$$E(G) = \sum_{j=1}^{n} | \pi_j | .$$

Example 3.3 Consider a fuzzy graph $G = (\mu, \lambda)$ on $Z = \{z_1, z_2, z_3, z_4, z_5\}$, as shown in Figure 1. The adjacency matrix of G is follows:

Figure 1. Fuzzy graph

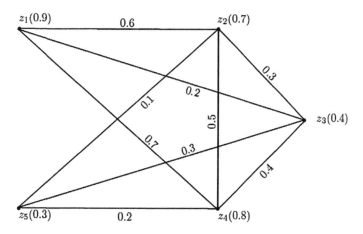

$$A(G) = \begin{pmatrix} 0 & 0.6 & 0.2 & 0.7 & 0 \\ 0.6 & 0 & 0.3 & 0.5 & 0.1 \\ 0.2 & 0.3 & 0 & 0.4 & 0.3 \\ 0.7 & 0.5 & 0.4 & 0 & 0.2 \\ 0 & 0.1 & 0.3 & 0.2 & 0 \end{pmatrix}.$$

$$Spec(G) = \{-0.7641, -0.4883, -0.3469, 0.1521, 1.4472\}.$$

The energy of fuzzy graph G is

$$E(G) = \sum_{j=1}^{n} |\pi_j|$$

$$= 3.1986.$$

Theorem 3.4 Let G be a fuzzy graph and $A(G)$ be its adjacency matrix. If $\pi_1 \geq \pi_2 \geq \cdots \geq \pi_2$ are the eigenvalues of $A(G)$. Then

1. $\sum_{j=1}^{n} \pi_j = 0,$

2. $\sum_{j=1}^{n} (\pi_j)^2 = 2 \sum_{1 \leq j < k \leq n} (\lambda(z_j z_k))^2.$

Proof.

1. Since the adjacency matrix of G is symmetric and has zero trace, its eigenvalues are real with sum equal to zero.
2. By trace properties of a matrix, we have

$$tr((A(G))^2) = \sum_{j=1}^{n} (\pi_j)^2,$$

where

$$tr((A(G))^2) = (0 + (\lambda(z_1 z_2))^2 + \cdots + (\lambda(z_1 z_n))^2)$$

$$+((\lambda(z_2 z_1))^2 + 0 + \cdots + (\lambda(z_2 z_n))^2)$$

$$\vdots$$

$$+((\lambda(z_n z_1))^2 + (\lambda(z_n z_2))^2 + \cdots + 0)$$

$$= 2 \sum_{1 \leq j < k \leq n} (\lambda(z_j z_k))^2.$$

Theorem 3.5 Let G be a fuzzy graph and $A(G)$ be its adjacency matrix. Then

$$\sqrt{2 \sum_{1 \leq j < k \leq n} (\lambda(z_j z_k))^2 + n(n-1) \mid det(A(\lambda(z_j z_k))) \mid^{\frac{2}{n}}} \leq E(A(G)) \leq \sqrt{2n \sum_{1 \leq j < k \leq n} (\lambda(z_j z_k))^2}.$$

Proof. Upper bound: Applying Cauchy-Schwarz inequality to the n numbers $(1, 1, \cdots, 1)$ and $(\mid \pi_1 \mid, \mid \pi_2 \mid, \cdots, \mid \pi_n \mid)$, we have

$$\sum_{j=1}^{n} \mid \pi_j \mid \leq \sqrt{n} \sqrt{\sum_{j=1}^{n} \mid \pi_j \mid^2} \tag{1}$$

$$\left(\sum_{j=1}^{n} \pi_j \right)^2 = \sum_{j=1}^{n} \mid \pi_j \mid^2 + 2 \sum_{1 \leq j < k \leq n} \pi_j \pi_k. \tag{2}$$

By comparing the coefficient of π^{n-2} in the characteristic polynomial

$$\prod_{j=1}^{n}(\pi - \pi_j) = \mid A(G) - \pi I \mid,$$

we have

$$\sum_{1 \leq j < k \leq n} \pi_j \pi_k = - \sum_{1 \leq j < k \leq n} (\lambda(z_j z_k))^2. \tag{3}$$

Substituting (3) in (2), we obtain

$$\sum_{j=1}^{n} \mid \pi_j \mid^2 = 2 \sum_{1 \leq j < k \leq n} (\lambda(z_j z_k))^2. \tag{4}$$

Substituting (4) in (1), we obtain

$$\sum_{j=1}^{n} \mid \pi_j \mid \leq \sqrt{n} \sqrt{2 \sum_{1 \leq j < k \leq n} (\lambda(z_j z_k))^2} = \sqrt{2n \sum_{1 \leq j < k \leq n} (\lambda(z_j z_k))^2}.$$

Therefore,

$$E(A(G)) \leq \sqrt{2n \sum_{1 \leq j < k \leq n} (\lambda(z_j z_k))^2}.$$

Lower bound:

$$(E(A(G)))^2 = (\sum_{j=1}^{n} \pi_j)^2 = \sum_{j=1}^{n} \mid \pi_j \mid^2 + 2 \sum_{1 \leq j < k \leq n} \pi_j \pi_k$$

$$= 2 \sum_{1 \leq j < k \leq n} (\lambda(z_j z_k))^2 + \frac{2n(n-1)}{2} AM\{\mid \pi_j \pi_k \mid\}.$$

Since

$$AM\{\mid \pi_j \pi_k \mid\} \geq GM\{\mid \pi_j \pi_k \mid\}, \ 1 \leq j < k \leq n,$$

Therefore,

$$E(A(G)) \geq \sqrt{2 \sum_{1 \leq j < k \leq n} (\lambda(z_j z_k))^2 + n(n-1)GM\{\mid \pi_j \pi_k \mid\}}.$$

Since

$$GM\{|\pi_j\pi_k|\} = (\prod_{1\le j<k\le n}|\pi_j\pi_k|)^{\frac{2}{n(n-1)}} = (\prod_{j=1}^{n}|\pi_j|^{n-1})^{\frac{2}{n(n-1)}}$$

$$= \left(\prod_{j=1}^{n}|\pi_j|\right)^{\frac{2}{n}}$$

$$= |det(A(\lambda(z_jz_k)))|^{\frac{2}{n}},$$

Thus,

$$E(A(G)) \ge \sqrt{2\sum_{1\le j<k\le n}(\lambda(z_jz_k))^2 + n(n-1)|det(A(\lambda(z_jz_k)))|^{\frac{2}{n}}}.$$

Hence,

$$\sqrt{2\sum_{1\le j<k\le n}(\lambda(z_jz_k))^2 + n(n-1)|det(A(\lambda(z_jz_k)))|^{\frac{2}{n}}} \le E(A(G)) \le \sqrt{2n\sum_{1\le j<k\le n}(\lambda(z_jz_k))^2}.$$

Theorem 3.6 Let G be a fuzzy graph and $A(G)$ be its adjacency matrix. If

$$n \le 2\sum_{1\le j<k\le n}(\lambda(z_jz_k))^2.$$

Then

$$E(A(G)) \le \frac{2\sum_{1\le j<k\le n}(\lambda(z_jz_k))^2}{n} + \sqrt{(n-1)\{2\sum_{1\le j<k\le n}(\lambda(z_jz_k))^2 - \left[\frac{2\sum_{1\le j<k\le n}(\lambda(z_jz_k))^2}{n}\right]^2\}}.$$

Proof. Take $B = [b_{jk}]_{n\times n}$ is a symmetric matrix with zero trace. Then

$$\pi_{\max} \ge \frac{2\sum_{1\le j<k\le n}b_{jk}}{n},$$

where, π_{max} is the maximum eigenvalue of B. If $A(G)$ is the adjacency matrix of a fuzzy graph G, then

$$\pi_{max} \geq \frac{2 \sum\limits_{1 \leq j < k \leq n} \lambda(z_j z_k)}{n},$$

where $\pi_1 \geq \pi_2 \geq \cdots \geq \pi_n$. Moreover, since

$$\sum_{j=1}^{n} (\pi_j)^2 = 2 \sum_{1 \leq j < k \leq n} (\lambda(z_j z_k))^2.$$

$$\sum_{j=2}^{n} (\pi_j)^2 = 2 \sum_{1 \leq j < k \leq n} (\lambda(z_j z_k))^2)^2 - \pi_1. \tag{5}$$

Applying Cauchy-Schwarz inequality to the vectors $(1,1,\cdots,1)$ and $(|\pi_1|,|\pi_2|,\cdots,|\pi_n|)$ with $n-1$ entries, we get

$$E(A(G)) - \pi_1 = \sum_{j=2}^{n} |\pi_j| \leq \sqrt{(n-1)\sum_{j=2}^{n} |\pi_j|^2}. \tag{6}$$

Substituting (5) in (6), we have

$$E(A(G)) - \pi_1 \leq \sqrt{(n-1)(2 \sum_{1 \leq j < k \leq n} (\lambda(z_j z_k))^2 - (\pi_1)^2)}.$$

$$E(A(G)) \leq \pi_1 + \sqrt{(n-1)(2 \sum_{1 \leq j < k \leq n} (\lambda(z_j z_k))^2 - (\pi_1)^2)}. \tag{7}$$

Since the function

$$F(u) = u + \sqrt{(n-1)(2 \sum_{1 \leq j < k \leq n} (\lambda(z_j z_k))^2 - u^2)}$$

decreases on the interval

$$(\sqrt{\frac{2 \sum\limits_{1 \leq j < k \leq n} (\lambda(z_j z_k))^2}{n}}, \sqrt{2 \sum_{1 \leq j < k \leq n} (\lambda(z_j z_k))^2}).$$

Also,

$$n \le 2 \sum_{1 \le j < k \le n} (\lambda(z_j z_k))^2, \; 1 \le \frac{2 \sum_{1 \le j < k \le n} (\lambda(z_j z_k))^2}{n}.$$

Thus,

$$\sqrt{\frac{2 \sum_{1 \le j < k \le n} (\lambda(z_j z_k))^2}{n}} \le \frac{2 \sum_{1 \le j < k \le n} (\lambda(z_j z_k))^2}{n} \le \frac{2 \sum_{1 \le j < k \le n} \lambda(z_j z_k)}{n} \le \pi_1 \le \sqrt{2 \sum_{1 \le j < k \le n} (\lambda(z_j z_k))^2}.$$

Therefore, (7) implies

$$E(A(G)) \le \frac{2 \sum_{1 \le j < k \le n} (\lambda(z_j z_k))^2}{n} + \sqrt{(n-1)\{2 \sum_{1 \le j < k \le n} (\lambda(z_j z_k))^2 - \left(\frac{2 \sum_{1 \le j < k \le n} (\lambda(z_j z_k))^2}{n}\right)^2\}}.$$

Theorem 3.7 Let G be a fuzzy graph on n vertices. Then

$$E(G) \le \frac{n}{2}(1 + \sqrt{n}).$$

Proof. Suppose that G be a fuzzy graph on $n$vertices.
If

$$n \le 2 \sum_{1 \le j < k \le n} (\lambda(z_j z_k))^2 = 2u,$$

then by routine calculus, we can easily show that

$$F(u) = \frac{2u}{n} + \sqrt{(n-1)(2u - (\frac{2u}{n})^2)}$$

is maximized when $u = \dfrac{n^2 + n\sqrt{n}}{4}$. Substituting this value of u in place of

$$u = 2 \sum_{1 \le j < k \le n} (\lambda(z_j z_k))^2$$

in Theorem 3.8, we must have

$$E(\lambda(z_j z_k)) \leq \frac{n}{2}(1 + \sqrt{n}).$$

Thus,

$$E(G) \leq \frac{n}{2}(1 + \sqrt{n}).$$

ENERGY OF *m*-POLAR FUZZY DIGRAPHS

In this section, energy, Laplacian energy and signless Laplacian energy of an m-polar fuzzy graph is introduced with examples. Some related results are also presented.

Definition 4.1 (Akram and N. Waseem, 2016) An m-polar fuzzy graph $G = (U, V)$ on an underlying set Z is a pair of functions $U : Z \to [0,1]^m$ and $V : Z \times Z \to [0,1]^m$ such that

$$V(z_1 z_2) \leq \inf\{U(z_1), U(z_2)\} \qquad \forall\, z_1, z_2 \in Z.$$

That is,

$$P_i \circ V(z_1 z_2) \leq \inf\{P_i \circ U(z_1), P_i \circ U(z_2)\},$$

$1 \leq i \leq m$, where U is an m-polar fuzzy vertex set of G and V is an m-polar fuzzy edge set of G. Note that

$$P_i \circ V(z_1 z_2) = 0\ \forall\, z_1 z_2 \in Z \times Z - E, \forall\, 1 \leq i \leq m.$$

We denote $P_i \circ U(z)$ by $\mu_U^{(i)}(z)$ and $P_i \circ V(z_1 z_2)$ by $\mu_V^{(i)}(z_1 z_2)$.

We now define the energy of m-polar fuzzy digraphs.

Definition 4.2 Let $D = (U, \vec{V})$ be an m-polar fuzzy digraph. The adjacency matrix $A(D)$ of D is a square matrix of order n defined by $A(D) = [a_{jk}]_{n \times n}$, where

$$a_{jk} = \begin{cases} \vec{V}(z_j z_k), & z_j z_k \in E, \quad i = 1, 2, \cdots, m, \\ 0, & otherwise. \end{cases}$$

Let $A_i(D)$ denote the adjacency matrix of i-th membership values of the vertices then

$$A_i(D) = [a_{jk}^{(i)}]_{n \times n}, i = 1, 2, \ldots, m,$$

is defined as,

$$a_{jk}^{(i)} = \begin{cases} \mu_{\overrightarrow{V}}^{(i)}(z_j z_k), & z_j z_k \in E, \\ 0, & otherwise. \end{cases}$$

Definition 4.3 The spectrum of m-polar fuzzy digraph D is denoted by $(T^{(1)}, T^{(2)}, \ldots, T^{(m)})$ where $T^{(i)}$ are the sets of eigenvalues of $A_i(D)$, respectively.

Definition 4.4 The energy of an m-polar fuzzy digraph G is defined as the $m-$tuple

$$E(D) = (E_1(D), E_2(D), \ldots, E_m(D))$$

where,

$$E_i(D) = E(A_i(D)) = \left(\sum_{j=1}^n | Re(\phi_j^{(i)}) | \right), i = 1, 2, \cdots, m.$$

Example 4.5 Consider a 5-polar fuzzy digraph $D = (U, \overrightarrow{V})$ on $Z = \{z_1, z_2, z_3, z_4, z_5\}$, as shown in Figure 2.

Table 1. 5-polar fuzzy set U

U	z_1	z_2	z_3	z_4	z_5
$\mu_U^{(1)}$	0.3	0.6	0.9	0.5	0.7
$\mu_U^{(2)}$	0.6	0.2	0.8	0.9	0.9
$\mu_U^{(3)}$	0.8	0.8	0.6	0.7	0.5
$\mu_U^{(4)}$	0.9	0.4	0.4	0.3	0.4
$\mu_U^{(5)}$	1	0.7	0.5	0.6	0.9

Table 2. 5-polar fuzzy relation $\overrightarrow{V}$

$\overrightarrow{V}$	$z_1 z_2$	$z_1 z_4$	$z_2 z_3$	$z_2 z_5$	$z_3 z_5$	$z_4 z_3$	$z_4 z_5$	$z_5 z_1$
$\mu_{\overrightarrow{V}}^{(1)}$	0.3	0.1	0.5	0.5	0.6	0.4	0.3	0.2
$\mu_{\overrightarrow{V}}^{(2)}$	0.2	0.4	0.1	0.1	0.7	0.7	0.8	0.5
$\mu_{\overrightarrow{V}}^{(3)}$	0.7	0.6	0.5	0.5	0.4	0.5	0.4	0.4
$\mu_{\overrightarrow{V}}^{(4)}$	0.3	0.2	0.3	0.2	0.2	0.1	0.2	0.3
$\mu_{\overrightarrow{V}}^{(5)}$	0.5	0.5	0.2	0.7	0.1	0.4	0.3	0.8

Figure 2. 5-polar fuzzy digraph

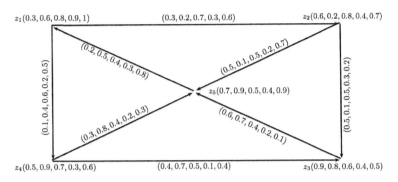

The adjacency matrix of D is follows as:

$$
A(D) = \begin{pmatrix}
(0,0,0,0,0) & (0.3,0.2,0.7,0.3,0.6) & (0,0,0,0,0) & (0.1,0.4,0.6,0.2,0.5) & (0,0,0,0,0) \\
(0,0,0,0,0) & (0,0,0,0,0) & (0.5,0.1,0.5,0.3,0.2) & (0,0,0,0,0) & (0.5,0.1,0.5,0.2,0.7) \\
(0,0,0,0,0) & (0,0,0,0,0) & (0,0,0,0,0) & (0,0,0,0,0) & (0.6,0.7,0.4,0.2,0.1) \\
(0,0,0,0,0) & (0,0,0,0,0) & (0.4,0.7,0.5,0.1,0.4) & (0,0,0,0,0) & (0.3,0.8,0.4,0.2,0.3) \\
(0.2,0.5,0.4,0.3,0.8) & (0,0,0,0,0) & (0,0,0,0,0) & (0,0,0,0,0) & (0,0,0,0,0)
\end{pmatrix}
$$

This matrix can be written in five different matrices as:

$$
A_1(D) = \begin{pmatrix}
0 & 0.3 & 0 & 0.1 & 0 \\
0 & 0 & 0.5 & 0 & 0.5 \\
0 & 0 & 0 & 0 & 0.6 \\
0 & 0 & 0.4 & 0 & 0.3 \\
0.2 & 0 & 0 & 0 & 0
\end{pmatrix}, \quad
A_2(D) = \begin{pmatrix}
0 & 0.2 & 0 & 0.4 & 0 \\
0 & 0 & 0.1 & 0 & 0.1 \\
0 & 0 & 0 & 0 & 0.7 \\
0 & 0 & 0.7 & 0 & 0.8 \\
0.5 & 0 & 0 & 0 & 0
\end{pmatrix}
$$

$$
A_3(D) = \begin{pmatrix}
0 & 0.7 & 0 & 0.6 & 0 \\
0 & 0 & 0.5 & 0 & 0.5 \\
0 & 0 & 0 & 0 & 0.4 \\
0 & 0 & 0.5 & 0 & 0.4 \\
0.4 & 0 & 0 & 0 & 0
\end{pmatrix}, \quad
A_4(D) = \begin{pmatrix}
0 & 0.3 & 0 & 0.2 & 0 \\
0 & 0 & 0.3 & 0 & 0.2 \\
0 & 0 & 0 & 0 & 0.2 \\
0 & 0 & 0.1 & 0 & 0.2 \\
0.3 & 0 & 0 & 0 & 0
\end{pmatrix}
$$

$$A_5(D) = \begin{pmatrix} 0 & 0.6 & 0 & 0.5 & 0 \\ 0 & 0 & 0.2 & 0 & 0.7 \\ 0 & 0 & 0 & 0 & 0.1 \\ 0 & 0 & 0.4 & 0 & 0.3 \\ 0.8 & 0 & 0 & 0 & 0 \end{pmatrix}.$$

Since,

$$Spec(A_1(D)) = \{0.4437, -0.0595 + 0.3933\iota, -0.0595 - 0.3933\iota, -0.3247, 0\},$$

$$Spec(A_2(D)) = \{0.6861, -0.1304 + 0.5855\iota, -0.1304 - 0.5855\iota, -0.4253, 0\},$$

$$Spec(A_3(D)) = \{0.7241, -0.1794 + 0.6009\iota, -0.1794 - 0.6009\iota, -0.3653, 0\},$$

$$Spec(A_4(D)) = \{0.3638, -0.0905 + 0.3018\iota, -0.0905 - 0.3018\iota, -0.1828, 0\},$$

$$Spec(A_5(D)) = \{0.7876, -0.3657 + 0.6674\iota, -0.3657 - 0.6674\iota, -0.0561, 0\}.$$

Therefore,

$$Spec(A(D)) = \{(0.4437, 0.6861, 0.7241, 0, 3638, 0.7876),$$
$$(-0.0595 + 0.3933\iota, -0.1304 + 0.5855\iota, -0.1794 + 0.6009\iota, -0.0905 + 0.3018\iota, -0.3657 + 0.6674\iota),$$
$$(-0.0595 - 0.3933\iota, -0.1304 - 0.5855\iota, -0.1794 - 0.6009\iota, -0.0905 - 0.3018\iota, -0.3657 - 0.6674\iota),$$
$$(-0.3247, -0.4253, -0.3653, -0.1828, -0.0561), (0, 0, 0, 0, 0)\}.$$

The energy of 5-polar fuzzy digraph is follows:

$$E(D) = (E(A_1(D)), E(A_2(D)), E(A_3(D)), E(A_4(D)), E(A_5(D)))$$

$$= \left(\sum_{j=1}^{n} |\phi_j^{(1)}|, \sum_{j=1}^{n} |\phi_j^{(2)}|, \sum_{j=1}^{n} |\phi_j^{(3)}|, \sum_{j=1}^{n} |\phi_j^{(4)}|, \sum_{j=1}^{n} |\phi_j^{(5)}| \right)$$

$$= (0.8874, 1.3722, 1.4482, 0.7276, 1.5751).$$

We now present the concept of Laplacian energy of m-polar fuzzy digraphs.

Definition 4.6 Let $D = (U, \overrightarrow{V})$ be an m-polar fuzzy digraph on n vertices. The out-degree matrix,

$$Deg^{out}(D) = (Deg^{out}(\mu_{\overrightarrow{V}}^{(i)}(z_j z_k))), \ i = 1, 2, \cdots, m,$$

of D is a $n \times n$ diagonal matrix defined as

$$d_{jk} = \begin{cases} d_D^{out}(z_j) & if \ j = k, \\ 0 & otherwise. \end{cases}$$

Definition 4.7 The Laplacian matrix of an m-polar fuzzy digraph $D = (U, \overrightarrow{V})$ is defined as

$$L(D) = (L(\mu_{\overrightarrow{V}}^{(i)}(z_j z_k))) = Deg^{out}(D) - A(D), \ i = 1, 2, \cdots, m,$$

Where $deg^{out}(D)$ is out-degree matrix and $A(D)$ is adjacency matrix of an m-polar fuzzy digraph $D = (U, \overrightarrow{V})$.

Definition 4.8 The spectrum of a Laplacian matrix of an m-polar fuzzy digraph $L(D)$ is defined as $(T_L^{(1)}, T_L^{(2)}, \cdots, T_L^{(m)})$, where $T_L^{(i)}$ are the sets of Laplacian eigenvalues of $L(\mu_{\overrightarrow{V}}^{(i)}(z_j z_k))$, respectively.

Theorem 4.9 Let $D = (U, \overrightarrow{V})$ be an m-polar fuzzy digraph and let L(D) be the Laplacian matrix of D. If

$$\phi_1^{(i)} \geq \phi_2^{(i)} \geq \cdots \geq \phi_n^{(i)}, \ i = 1, 2, \cdots, m,$$

are the eigenvalues of $L(\mu_{\overrightarrow{V}}^{(i)}(z_j z_k))$, respectively. Then

$$\sum_{j=1}^{n} Re(\phi_l^{(i)}) = \sum_{1 \leq j,k \leq n} \mu_{\overrightarrow{V}}^{(i)}(z_j z_k), \ i = 1, 2, \cdots, m.$$

Definition 4.10 The Laplacian energy of an m-polar fuzzy digraph $D = (U, \overrightarrow{V})$ is defined as

$$LE(D) = (LE(\mu_{\overrightarrow{V}}^{(1)}(z_j z_k)), LE(\mu_{\overrightarrow{V}}^{(2)}(z_j z_k)), \ldots, LE(\mu_{\overrightarrow{V}}^{(m)}(z_j z_k)))$$

$$= \left(\sum_{j=1}^{n} | \eta_j^{(i)} | \right), i = 1, 2, \cdots, m,$$

where

$$\eta_j^{(i)} = Re(\phi_j^{(i)}) - \frac{\sum_{1 \leq j,k \leq n} \mu_{\overrightarrow{V}}^{(i)}(z_j z_k)}{n}, \ i = 1, 2, \cdots, m.$$

Theorem 4.11 Let $D = (U, \vec{V})$ be an m-polar fuzzy digraph and let $L(D)$ be the Laplacian matrix of D. If

$$\phi_1^{(i)} \geq \phi_2^{(i)} \geq \cdots \geq \phi_n^{(i)}, \ i = 1, 2, \cdots, m,$$

are the eigenvalues of $L(\mu_{\vec{V}}^{(i)}(z_j z_k))$, respectively, and

$$\eta_j^{(i)} = Re(\phi_j^{(i)}) - \frac{\sum\limits_{1 \leq j, k \leq n} \mu_{\vec{V}}^{(i)}(z_j z_k)}{n}.$$

Then

$$\sum_{j=1}^{n} \eta_j^{(i)} = 0, i = 1, 2, \cdots, m.$$

Example 4.12 Consider an 3-polar fuzzy digraph $D = (U, \vec{V})$ on $Z = \{z_1, z_2, z_3, z_4, z_5, z_6\}$, as shown in Figure 3.

Table 3. 3-polar fuzzy set U

U	z_1	z_2	z_3	z_4	z_5	z_6
$\mu_U^{(1)}$	0.4	0.7	0.5	0.9	0.6	0.8
$\mu_U^{(2)}$	0.7	0.3	0.9	0.7	0.4	0.6
$\mu_U^{(3)}$	0.2	0.8	0.6	0.5	0.9	0.7

Table 4. 3-polar fuzzy relation $\vec{V}$

$\vec{V}$	$z_1 z_2$	$z_2 z_3$	$z_2 z_5$	$z_3 z_4$	$z_4 z_5$	$z_5 z_6$	$z_6 z_3$	$z_6 z_1$
$\mu_{\vec{V}}^{(1)}$	0.3	0.4	0.1	0.5	0.3	0.6	0.4	0.2
$\mu_{\vec{V}}^{(2)}$	0.2	0.3	0.3	0.6	0.2	0.4	0.3	0.5
$\mu_{\vec{V}}^{(3)}$	0.1	0.5	0.7	0.4	0.4	0.7	0.5	0.2

Figure 3. 3-polar fuzzy diagraph

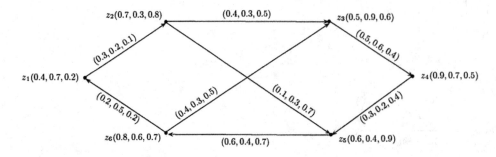

The adjacency matrix, degree matrix and Laplacian matrix of the m-polar fuzzy digraph shown in Figure 3 are as follows

$$A(D) = \begin{pmatrix} (0,0,0) & (0.3,0.2,0.1) & (0,0,0) & (0,0,0) & (0,0,0) & (0,0,0) \\ (0,0,0) & (0,0,0) & (0.4,0.3,0.5) & (0,0,0) & (0.1,0.3,0.7) & (0,0,0) \\ (0,0,0) & (0,0,0) & (0,0,0) & (0.5,0.6,0.4) & (0,0,0) & (0,0,0) \\ (0,0,0) & (0,0,0) & (0,0,0) & (0,0,0) & (0.3,0.2,0.4) & (0,0,0) \\ (0,0,0) & (0,0,0) & (0,0,0) & (0,0,0) & (0,0,0) & (0.6,0.4,0.7) \\ (0.2,0.5,0.2) & (0,0,0) & (0.4,0.3,0.5) & (0,0,0) & (0,0,0) & (0,0,0) \end{pmatrix}$$

$$Deg^{out}(D) = \begin{pmatrix} (0.3,0.2,0.1) & (0,0,0) & (0,0,0) & (0,0,0) & (0,0,0) & (0,0,0) \\ (0,0,0) & (0.5,0.6,1.2) & (0,0,0) & (0,0,0) & (0,0,0) & (0,0,0) \\ (0,0,0) & (0,0,0) & (0.5,0.6,0.4) & (0,0,0) & (0,0,0) & (0,0,0) \\ (0,0,0) & (0,0,0) & (0,0,0) & (0.3,0.2,0.4) & (0,0,0) & (0,0,0) \\ (0,0,0) & (0,0,0) & (0,0,0) & (0,0,0) & (0.6,0.4,0.7) & (0,0,0) \\ (0,0,0) & (0,0,0) & (0,0,0) & (0,0,0) & (0,0,0) & (0.6,0.8,0.7) \end{pmatrix}$$

$$L(D) = \begin{pmatrix} (0.3,0.2,0.1) & (-0.3,-0.2,-0.1) & (0,0,0) & (0,0,0) & (0,0,0) & (0,0,0) \\ (0,0,0) & (0.5,0.6,1.2) & (-0.4,-0.3,-0.5) & (0,0,0) & (-0.1,-0.3,-0.7) & (0,0,0) \\ (0,0,0) & (0,0,0) & (0.5,0.6,0.4) & (-0.5,-0.6,-0.4) & (0,0,0) & (0,0,0) \\ (0,0,0) & (0,0,0) & (0,0,0) & (0.3,0.2,0.4) & (-0.3,-0.2,-0.4) & (0,0,0) \\ (0,0,0) & (0,0,0) & (0,0,0) & (0,0,0) & (0.6,0.4,0.7) & (-0.6,-0.4,-0.7) \\ (-0.2,-0.5,-0.2) & (0,0,0) & (-0.4,-0.3,-0.5) & (0,0,0) & (0,0,0) & (0.6,0.8,0.7) \end{pmatrix}$$

The Laplacian spectrum of 3-polar fuzzy digraph $D = (U, \vec{V})$, is given below

$Laplacian\ Spec(\mu_{\vec{V}}^{(1)}(z_j z_k)) = \{0.9779, 0.0000, 0.5259 + 0.4000\iota, 0.5259 - 0.4000\iota, 0.3852 + 0.2076\iota, 0.3852 - 0.2076\iota\},$

$Laplacian\ Spec(\mu_{\vec{V}}^{(2)}(z_j z_k)) = \{0.9809, 0.0000, 0.5342 + 0.3050\iota, 0.5342 - 0.3050\iota, 0.3754 + 0.1195\iota, 0.3754 - 0.1195\iota\},$

$Laplacian \ Spec(\mu_{\overrightarrow{V}}^{(3)}(z_j z_k)) = \{0.0140, 0.1586, 0.5577 + 0.4366\iota, 0.5577 - 0.4366\iota, 1.2429, 0.9690\}.$

So,

$Laplacian \ spec(D) = \{(0.9779, 0.9809, 0.0140), (0.0000, 0.0000, 0.1586), (0.5259 + 0.4000\iota,$

$0.5342 + 0.3050\iota, 0.5577 + 0.4366\iota), \quad (0.5259 - 0.4000\iota, 0.5342 - 0.3050\iota, 0.5577 - 0.4366\iota),$

$(0.3852 + 0.2076\iota, 0.3754 + 0.1195\iota, 1.2429), (0.3852 - 0.2076\iota, 0.3754 - 0.1195\iota, 0.9690)\}.$

The Laplacian energy of a 3-polar fuzzy digraph $D = (U, \overrightarrow{V})$, shown in Figure 4.12 is

$$LE(D) = (LE(\mu_{\overrightarrow{V}}^{(1)}(z_j z_k)), \ LE(\mu_{\overrightarrow{V}}^{(2)}(z_j z_k)), \ LE(\mu_{\overrightarrow{V}}^{(3)}(z_j z_k)))$$

$$= (1.2593, 1.2985, 2.0905).$$

We now describe the notion of signless Laplacian energy of m-polar fuzzy digraphs.

Definition 4.13 The signless Laplacian matrix of an m-polar fuzzy digraph $D = (U, \overrightarrow{V})$ is defined as

$$L^+(D) = (L^+(\mu_{\overrightarrow{V}}^{(i)}(z_j z_k))) = Deg^{out}(D) - A(D), \ i = 1, 2, \cdots, m,$$

where $deg^{out}(D)$ is out-degree matrix and $A(D)$ is adjacency matrix of an m-polar fuzzy digraph $D = (U, \overrightarrow{V})$.

Definition 4.14 The spectrum of a signless Laplacian matrix of an m-polar fuzzy digraph $L^+(D)$ is defined as $(T_{L^+}^{(1)}, T_{L^+}^{(2)}, \cdots, T_{L^+}^{(m)})$, where $T_{L^+}^{(i)}$ are the sets of signless Laplacian eigenvalues of

$$L^+(\mu_{\overrightarrow{V}}^{(i)}(z_j z_k)), \ i = 1, 2, \cdots, m,$$

respectively.

Theorem 4.15 Let $D = (U, \overrightarrow{V})$ be an m-polar fuzzy digraph and let $L^+(D)$ be the signless Laplacian matrix of D. If

$$\phi_1^{(i)+} \geq \phi_2^{(i)+} \geq \cdots \geq \phi_n^{(i)+}$$

are the eigenvalues of

$$L^+(\mu_{\overrightarrow{v}}^{(i)}(z_j z_k)), \ \ i = 1, 2, \cdots, m,$$

respectively. Then

$$\sum_{j=1}^{n} Re(\phi_l^{(i)+}) = \sum_{1 \leq j,k \leq n} \mu_{\overrightarrow{V}}^{(i)}(z_j z_k), \ i = 1, 2, \cdots, m.$$

Definition 4.16 The signless Laplacian energy of an m-polar fuzzy digraph $D = (U, \overrightarrow{V})$ is defined as

$$LE^+(D) = (LE^+(\mu_{\overrightarrow{V}}^{(1)}(z_j z_k)), LE^+(\mu_{\overrightarrow{V}}^{(2)}(z_j z_k)), \cdots, LE^+(\mu_{\overrightarrow{V}}^{(m)}(z_j z_k))),$$

$$= \left(\sum_{j=1}^{n} \mid \eta_j^{(i)+} \mid \right),$$

where

$$\eta_j^{(i)+} = Re(\phi_j^{(i)+}) - \frac{\sum\limits_{1 \leq j,k \leq n} \mu_{\overrightarrow{V}}^{(i)}(z_j z_k)}{n}, \ i = 1, 2, \cdots, m.$$

Theorem 4.17 Let $D = (U, \overrightarrow{V})$ be an m-polar fuzzy digraph and let $L^+(D)$ be the signless Laplacian matrix of D. If

$$\phi_1^{(i)+} \geq \phi_2^{(i)+} \geq \cdots \geq \phi_n^{(i)+}, \ i = 1, 2, \cdots, m,$$

are the eigenvalues of $L^+(\mu_{\overrightarrow{V}}^{(i)}(z_j z_k))$, respectively, and

$$\eta_j^{(i)+} = Re(\phi_j^{(i)+}) - \frac{\sum\limits_{1 \leq j,k \leq n} \mu_{\overrightarrow{V}}^{(i)}(z_j z_k)}{n}, \ i = 1, 2, \cdots, m.$$

Then

$$\sum_{j=1}^{n} \eta_j^{(i)+} = 0, i = 1, 2, \cdots, m.$$

Example 4.18 Consider a 3-polar fuzzy digraph $D = (U, \overrightarrow{V})$ on $Z = \{z_1, z_2, z_3, z_4, z_5\}$, as shown in Figure 4.

Table 5. 3-polar fuzzy set U

U	z_1	z_2	z_3	z_4	z_5
$\mu_U^{(1)}$	0.7	0.9	0.6	0.5	0.4
$\mu_U^{(2)}$	0.9	0.8	0.4	0.6	0.3
$\mu_U^{(3)}$	0.6	0.4	0.5	0.7	0.9

Table 6. 3-polar fuzzy relation $\overrightarrow{V}$

$\overrightarrow{V}$	z_1z_2	z_2z_4	z_3z_4	z_4z_1	z_4z_5	z_5z_1
$\mu_{\overrightarrow{V}}^{(1)}$	0.6	0.1	0.4	0.5	0.3	0.2
$\mu_{\overrightarrow{V}}^{(2)}$	0.7	0.5	0.3	0.4	0.2	0.1
$\mu_{\overrightarrow{V}}^{(3)}$	0.3	0.3	0.4	0.2	0.6	0.5

Figure 4. 3-polar fuzzy diagraph

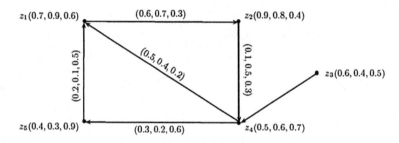

The adjacency matrix, out-degree matrix and signless Laplacian matrix of the 3-polar fuzzy digraph shown in Figure 4 are as follows

$$A(D) = \begin{pmatrix} (0,0,0) & (0.6,0.7,0.3) & (0,0,0) & (0,0,0) & (0,0,0) \\ (0,0,0) & (0,0,0) & (0,0,0) & (0.1,0.5,0.3) & (0,0,0) \\ (0,0,0) & (0,0,0) & (0,0,0) & (0.4,0.3,0.4) & (0,0,0) \\ (0.5,0.4,0.2) & (0,0,0) & (0,0,0) & (0,0,0) & (0.3,0.2,0.6) \\ (0.2,0.1,0.5) & (0,0,0) & (0,0,0) & (0,0,0) & (0,0,0) \end{pmatrix}$$

$$Deg^{out}(D) = \begin{pmatrix} (0.6,0.7,0.3) & (0,0,0) & (0,0,0) & (0,0,0) & (0,0,0) \\ (0,0,0) & (0.1,0.5,0.3) & (0,0,0) & (0,0,0) & (0,0,0) \\ (0,0,0) & (0,0,0) & (0.4,0.3,0.4) & (0,0,0) & (0,0,0) \\ (0,0,0) & (0,0,0) & (0,0,0) & (0.8,0.6,0.8) & (0,0,0) \\ (0,0,0) & (0,0,0) & (0,0,0) & (0,0,0) & (0.2,0.1,0.5) \end{pmatrix}$$

$$L^+(D) = \begin{pmatrix} (0.6,0.7,0.3) & (0.6,0.7,0.3) & (0,0,0) & (0,0,0) & (0,0,0) \\ (0,0,0) & (0.1,0.5,0.3) & (0,0,0) & (0.1,0.5,0.3) & (0,0,0) \\ (0,0,0) & (0,0,0) & (0.4,0.3,0.4) & (0.4,0.3,0.4) & (0,0,0) \\ (0.5,0.4,0.2) & (0,0,0) & (0,0,0) & (0.8,0.6,0.8) & (0.3,0.2,0.6) \\ (0.2,0.1,0.5) & (0,0,0) & (0,0,0) & (0,0,0) & (0.2,0.1,0.5) \end{pmatrix}$$

$$L^+ Spec(\mu_{\overrightarrow{V}}^{(1)}(z_j z_k)) = \{0.4000, 0.9284, 0.3401 + 0.1608\iota, 0.3401 - 0.1608\iota, 0.0913\},$$

$$L^+ Spec(\mu_{\overrightarrow{V}}^{(2)}(z_j z_k)) = \{0.3000, 1.1338, 0.3456 + 0.4583\iota, 0.3456 - 0.4583\iota, 0.0750\},$$

$$L^+ Spec(\mu_{\overrightarrow{V}}^{(3)}(z_j z_k)) = \{0.4000, 0.0587, 0.4362 + 0.3555\iota, 0.4362 - 0.3555\iota, 0.9689\}.$$

So,

$$L^+ spec(D) = \{(0.4000, 0.3000, 0.4000), (0.9284, 1.1338, 0.0587),$$

$$(0.3401 + 0.1608\iota, 0.3456 + 0.4583\iota, 0.4362 + 0.3555\iota),$$

$$(0.3401 - 0.1608\iota, 0.3456 - 0.4583\iota, 0.4362 - 0.3555\iota), (0.0913, 0.0750, 0.9689)\}.$$

The signless Laplacian energy of a 3-polar fuzzy digraph $D = (U, \overrightarrow{V})$, shown in Figure 4 is

$$LE^+(D) = \left(LE^+(\mu_{\overrightarrow{V}}^{(1)}(z_j z_k)), LE^+(\mu_{\overrightarrow{V}}^{(2)}(z_j z_k)), LE^+(\mu_{\overrightarrow{V}}^{(3)}(z_j z_k)) \right)$$

$$= (1.0169, 1.3876, 1.0178).$$

FUTURE RESEARCH DIRECTIONS

Graph theory has successfully employed to solve a wide range of problems encountered in diverse fields such as neural networks, transportation, expert systems, image capturing and network security. In past few years, a number of generalizations of graph theoretical concepts have introduced to model the impreciseness and uncertainties in network problems. The natural extension of this work are energy of m-polar fuzzy soft graphs, energy of m-polar rough graphs, energy of interval valued m-polar fuzzy graphs, etc.

CONCLUSION

An m-polar fuzzy model is a generalization of the fuzzy model. In this chapter, first of all a basic overview of energy of fuzzy graph is presented. Then, energy of m-polar fuzzy digraphs, Laplacian energy of m-polar fuzzy digraphs and signless Laplacian energy of m-polar fuzzy digraphs are introduced with several examples. Some of their properties are also investigated.

REFERENCES

Akram, M. (2011). Bipolar fuzzy graphs. *Information Sciences*, *181*(24), 5548–5564. doi:10.1016/j.ins.2011.07.037

Akram, M. (2013). Bipolar fuzzy graphs with applications. *Knowledge-Based Systems*, *39*, 1–8. doi:10.1016/j.knosys.2012.08.022

Akram, M., & Naz, S. (2018). Energy of Pythagorean fuzzy graphs with applications. *Mathematics*, *6*(8), 136. doi:10.3390/math6080136

Akram, M., & Waseem, N. (2016). Certain metrices in m-polar fuzzy graphs. *New Mathematics and Natural Computation*, *12*(2), 135–155. doi:10.1142/S1793005716500101

Akram, M., m-Polar Fuzzy Graphs, Studies in Fuzziness and Soft Computing, Springer,

371.   2019).

Anjali, N., & Mathew, S. (2013). Energy of a fuzzy graph. *Annals Fuzzy Maths and Informatics*, *6*(3), 455–465.

Brualdi, R. A. (2006). Energy of a graph. *Notes to AIM Workshop on Spectra of Families of Articles described by graphs, digraphs, and sign patterns*.

Chen, J., Li, S., Ma, S., & Wang, X. (2014). m- polar fuzzy sets: An extension of bipolar fuzzy sets. *The Scientific World Journal*, 1–8. PMID:25025087

Ghorai, G., & Pal, M. (2015). On some operations and density of m-polar fuzzy graphs, Pacific Science Review A. *Natural Science and Engineering*, *17*(1), 14–22.

Ghorai, G., & Pal, M. (2016). Some properties of m-polar fuzzy graphs, Paci_c Science Review A. *Natural Science and Engineering*, *18*(1), 38–46.

Ghorai, G., & Pal, M. (2016a). A study on m-polar fuzzy planar graphs. *International Journal of Computing Science and Mathematics*, *7*(3), 283–292. doi:10.1504/IJCSM.2016.077854

Ghorai, G., & Pal, M. (2016b). Faces and dual of m-polar fuzzy planar graphs. *Journal of Intelligent & Fuzzy Systems*, *31*(3), 2043–2049. doi:10.3233/JIFS-16433

Ghorai, G., & Pal, M. (2016c). Some isomorphic properties of m-polar fuzzy graphs with applica-tions. *SpringerPlus*, *5*(1), 1–21. doi:10.118640064-016-3783-z PMID:28066695

Gutman, I. (1978). The energy of a graph. *Ber Math Stat Sekt Forsch Graz, 103,* 1–22.

Gutman, I. (2001). *The energy of a graph: old and new results. In Algebraic Combinatorics and Applications* (pp. 196–211). Berlin: Springer.

Kaufmann, A. (1973). *Introduction a la Theorie des Sour-ensembles Flous.* Masson et Cie 1.

Lee, K. M. (2000). Bipolar-valied fuzzy sets and their basic operation. *Proceeding of the International Conference,* 307-317.

Mordeson, J. N., & Peng, C. S. (1994). Operations on fuzzy graphs. *Information Sciences, 79*(3), 159–170. doi:10.1016/0020-0255(94)90116-3

Naz, S., Akram, M., & Smarandache, F. (2018). Certain notions of energy in single-valued neutrosophic graphs. *Axioms, 7*(3), 1–30. doi:10.3390/axioms7030050

Naz, S., Ashraf, S., & Karaaslan, F. (2018a). Energy of a bipolar fuzzy graph and its applications in decision making. *Italian Journal of Pure and Applied Mathematics.*

Pea, I., & Rada, J. (2008). Energy of diagraph. *Linear and Multilinear Algebra, 56*(5), 565–579. doi:10.1080/03081080701482943

Praba, B., Chandrasekaran, V. M., & Deepa, G. (2014). Energy of an intuitionistic fuzzy graph. *International Journal of Pure and Applied Mathematics, 32,* 431–444.

Rosenfeld, A. (1975). *Fuzzy graphs.* In L. A. Zadeh, K. S. Fu, & M. Shimura (Eds.), *Fuzzy sets and their applications* (pp. 77–95). New York: Academic Press.

Zadeh, L. A. (1965). Fuzzy sets. *Information and Control, 8*(3), 338–353. doi:10.1016/S0019-9958(65)90241-X

Zadeh, L. A. (1971). Similarity relations and fuzzy oderings. *Information Sciences, 3*(2), 177–200. doi:10.1016/S0020-0255(71)80005-1

Zhang, W. R. (1994). Bipolar fuzzy sets and relations: a computational framework for cognitive modeling and multiagent decision analysis. *Proceedings of IEEE Conference,* 305-309.

Chapter 21
Bipolar Neutrosophic Cubic Graphs and Its Applications

C. Antony Crispin Sweety
Nirmala College for Women, India

K. Vaiyomathi
Nirmala College for Women, India

F. Nirmala Irudayam
Nirmala College for Women, India

ABSTRACT

The authors introduce neutrosophic cubic graphs and single-valued netrosophic Cubic graphs in bipolar setting and discuss some of their algebraic properties such as Cartesian product, composition, m-union, n-union, m-join, n-join. They also present a real time application of the defined model which depicts the main advantage of the same. Finally, the authors define a score function and present minimum spanning tree algorithm of an undirected bipolar single valued neutrosophic cubic graph with a numerical example.

INTRODUCTION

Most of the real time problems encounter one or the other forms of imprecise data. To deal with such types of data many authors have proposed various mathematical tools. And to name a few of these tools we have fuzzy sets, intuitionistic fuzzy sets, soft sets, vague set theory and so on. Though the fuzzy sets and intuitionistic fuzzy sets considered the hesitant degree along with membership and non-membership degree simultaneously, hesitant value cannot be a specific number always. Hence the development and generalization of fuzzy sets was not enough to deal with all types of uncertainties in real physical problems. Therefore, some more theories were required.

Neutrosophy, a newly born science, studies the origin, nature and scope of indeterminacies and their relations with various ideational spectra. The theory of neutrosophy make the concept intelligible, were the incidence of the application of a law, an idea, an axiom, a conceptual construction on an unclear,

DOI: 10.4018/978-1-5225-9380-5.ch021

imprecise, indeterminate phenomenon. Also the Neutrosophic set theory is applied in various fields such as topology, operations research, control theory, mechanics and in many more real life time problems. Neutrosophic theory is made more convenient by the introduction of Single Valued Neutrosophic Set, neutrosophic cubic sets, bipolar neutrosophic sets and so on. The Single Valued Neutrosophic sets are introduced to make the Neutrosophic theory more convenient to apply in real time physical and engineering problems. Neutrosophic cubic sets are very good tool by which are can manage vague data in a very effective way as compared to cubic sets and other models of fuzzy sets. Bipolar fuzzy sets are an extension of fuzzy sets whose membership degree range is [-1,1]. It is noted that positive information represents what is granted to be possible, while negative information represents what is considered to be impossible. In many domains, it is very convenient to deal with bipolar information.

On other hand graph is a convenient and attractive way of representing information in which the objects are represented by vertices and their relations by edges. When there is an uncertainty in describing an element or in its interconnections, fuzzy graph model and its extensions were designed. But Fuzzy graph theories and their extensions fail if the relation between the nodes in the problem is indeterminate. To overcome this neutrosophic graphs were developed. Many researchers applied the neutrosophic cubic sets to graph theory to make it more applicable and represent a problem physically in the form of matrices, diagrams etc., which is very easy to understand and deal with.

In this article, we propose the concept of neutrosophic cubic graph in bipolar setting and elucidate its various characterizations. We apply this defined model to real time problem. We also test its applicability based on present time and future prediction which is the key advantage of this model.

BACKGROUND

L. Zadeh (1965) introduced the concept of fuzzy sets by defining the degree of membership to deal with the data with uncertainties. To cope with the lack of non-membership degree K. T. Atanossov (1986) proposed the notion of intuitionistic fuzzy sets by associating the degree of non-membership in the concept of fuzzy set as an individual element. In addition to this Gau W. L and Buehrer D. J (1993) introduced vague sets. F. Smarandache (1999) introduced neutrosophic logic to handle and understand the indefinite information in a more effective way. F. Smaradache (2006) introduced neutrosophic sets as a generalization intuitionistic Fuzzy sets. Every element of a neutrosophic element has three grades of membership defined within the real non-standard interval]–0, 1+[. H. Wang and F. Smarandache (2010) defined single valued neutrosophic set which is a subclass of neutrosophic sets with three membership functions that are independent and their value defined in [0,1].

Fuzzy graph theory was designed by A. Rosenfeld (1975) and has been extended by many researchers. As a result, fuzzy hyper graphs by Mordeson, J.N. and Nair P. S(2001), intuitionistic fuzzy graph (2006) by M. G. Karunmbigai and Parvathi Rangasamy, strong intuitionistic fuzzy graph (2012) by M. Akram and Davvaz. M, interval- valued Fuzzy Planar Graphs by T. Pramanik Et. Al. are developed. S. Samanta and M. Pal studied Fuzzy Tolerence Graphs (2011), Fuzzy planar Graphs (2015) etc. Smarandache (2015) developed four kinds of neutrosophic graphs, two of them based on literal indeterminacy (I), named as I-edge neutrosophic graph and I-vertex neutrosophic graph. Many researchers studied deeply about these concepts, hence has gained more popularity due to its applications in many real world problems. Other two are based on (t,i,f) components namely the (t,i,f) – Edge neutrosophic graph and the (t,i,f)- vertex neutrosophic graph. Later on, third neutrosophic graph model was introduced by Broumi Et. al (2016) which allowed truth(t), indeterminacy(i) and falsity(f) values both to vertices and edges and studied some of its properties. Such a graph model is termed as single valued neutrosophic graph. Single valued neutrosophic graph theory is a conceptual framework which model and solves combinational problems that arise in many areas such as mathematics, physics, computer science and engineering. This advantage attracted many researchers and made them to pick various decisions making problems and expressed using cubic sets. Neutrosophic cubic sets (2017) introduced by Jun. Y.B., Smarandache. F. and Kim. C.S. have more compatibility, precision and flexibility as compared to previous defined fuzzy models.

W. R. Zhang (1998) discussed the concept of bipolar fuzzy sets as a generalization of fuzzy sets. Bipolar fuzzy sets are an extension of fuzzy sets whose membership degree range is [-1, 1]. In a bipolar fuzzy set, the membership degree 0 of an element means that the element is irrelevant to the corresponding property, the membership degree (0, 1] of an element indicates that the element somewhat satisfies the property, and the membership degree [-1,0) of an element indicates that the element somewhat satisfies the implicit counter-property. F. Smarandache Et. Al (2015) defined Bipolar Neutrosophic sets and their applications. M.Akram (2011) introduced bipolar fuzzy graphs and discussed its applications (2013). Bipolar fuzzy hypergraphs (2012), Irregular bipolar Fuzzy graphs (2012) was developed by S. Samanta and M. Pal. F. Smaradache Et. al. (2016) introduced bipolar single valued Neutrosophic Graphs.

BIOPOLAR NEUTROSOPHIC CUBIC GRAPHS

In this section, first we develop the model by introducing the idea of bipolar neutrosophic cubic graph, single valued neutrosophic cubic graph and fundamental algebraic operations of the same namely degree and order of bipolar neutrosophic cubic graphs union and join of bipolar neutrosophic cubic graphs, composition, Cartesian product, and some other results related with bipolar neutrosophic cubic graphs, strong bipolar neutrosophic cubic graph, complete bipolar neutrosophic cubic graph, the complement of strong bipolar neutrosophic cubic graph, regular bipolar neutrosophic cubic graph and illustrate them with many examples.

Definition

Let $G^* = (V, E)$ be a graph and $G(P,Q)$ is a bipolar neutrosophic cubic graph of G^*, if

$$P = (A, \lambda) = \left(T_A^+, T_\lambda^+ \right), \left(I_A^+, I_\lambda^+ \right), \left(F_A^+, F_\lambda^+ \right), \left(T_A^-, T_\lambda^- \right), \left(I_A^-, I_\lambda^- \right), \left(F_A^-, F_\lambda^- \right)$$

is the bipolar neutrosophic cubic set representation of vertex set V and

$$Q = (B, \mu) = \left(\left(T_B^+, T_\mu^+ \right), \left(I_B^+, I_\mu^+ \right), \left(F_B^+, F_\mu^+ \right), \left(T_B^-, T_\mu^- \right), \left(I_B^-, I_\mu^- \right), \left(F_B^-, F_\mu^- \right) \right)$$

is the bipolar neutrosophic cubic set representation of the edge set E such that

1.
$$T_B^+\left(u_iv_i\right) \le r\min\left\{T_A^+\left(u_i\right), T_A^+\left(v_i\right)\right\}, \quad T_\mu^+\left(u_iv_i\right) \ge r\max\left\{T_\mu^+\left(u_i\right), T_\mu^+\left(v_i\right)\right\}$$
$$T_B^-\left(u_iv_i\right) \ge r\max\left\{T_A^-\left(u_i\right), T_A^-\left(v_i\right)\right\}, \quad T_\mu^-\left(u_iv_i\right) \ge r\min\left\{T_\mu^-\left(u_i\right), T_\mu^-\left(v_i\right)\right\}$$

2.
$$I_B^+\left(u_iv_i\right) \le r\min\left\{I_A^+\left(u_i\right), I_A^+\left(v_i\right)\right\}, \quad I_\mu^+\left(u_iv_i\right) \ge r\max\left\{I_\mu^+\left(u_i\right), I_\mu^+\left(v_i\right)\right\}$$
$$I_B^-\left(u_iv_i\right) \ge r\max\left\{I_A^-\left(u_i\right), I_A^-\left(v_i\right)\right\}, \quad I_\mu^-\left(u_iv_i\right) \ge r\min\left\{I_\mu^-\left(u_i\right), I_\mu^-\left(v_i\right)\right\}$$

3.
$$F_B^+\left(u_iv_i\right) \le r\min\left\{F_A^+\left(u_i\right), F_A^+\left(v_i\right)\right\}, \quad F_\mu^+\left(u_iv_i\right) \ge r\max\left\{F_\mu^+\left(u_i\right), F_\mu^+\left(v_i\right)\right\}$$
$$F_B^-\left(u_iv_i\right) \ge r\max\left\{F_A^-\left(u_i\right), F_A^-\left(v_i\right)\right\}, \quad F_\mu^-\left(u_iv_i\right) \ge r\min\left\{F_\mu^-\left(u_i\right), F_\mu^-\left(v_i\right)\right\}$$

Let $G^* = \left(V, E\right)$ be a graph and $G(P,Q)$ is a bipolar neutrosophic cubic graph of G^*, if

$$P = \left(A, \lambda\right) = \left(T_A^+, T_\lambda^+\right), \left(I_A^+, I_\lambda^+\right), \left(F_A^+, F_\lambda^+\right), \left(T_A^-, T_\lambda^-\right), \left(I_A^-, I_\lambda^-\right), \left(F_A^-, F_\lambda^-\right)$$

is the bipolar neutrosophic cubic set representation of vertex set V and

$$Q = \left(B, \mu\right) = \left(\left(T_B^+, T_\mu^+\right), \left(I_B^+, I_\mu^+\right), \left(F_B^+, F_\mu^+\right), \left(T_B^-, T_\mu^-\right), \left(I_B^-, I_\mu^-\right), \left(F_B^-, F_\mu^-\right)\right).$$

And is the bipolar neutrosophic cubic set representation of the edge set E and λ and μ are bipolar neutrosophic sets.

Example

Let $G^* = \left(V, E\right)$ be a graph Where $V = \left\{a, b, c, d\right\}$ and $E = \left\{ab, bc, ac, ad, cd\right\}$ where P and Q are as follows:

$$p = \left\langle \begin{array}{l} \left[a, \left(\left[0.3, 0.6\right], 0.8\right), \left(\left[0.4, 0.6\right], 0.7\right), \left(\left[0.5, 0.6\right], 0.4\right), \right. \\ \left.\left(\left[-0.5, -0.4\right], -0.1\right), \left(\left[-0.6, -0.3\right], -0.1\right), \left(\left[-0.7, -0.6\right], -0.3\right)\right] \\ \left[b, \left(\left[0.1, 0.3\right], 0.5\right), \left(\left[0.2, 0.3\right], 0.6\right), \left(\left[0.6, 0.7\right], 0.8\right), \right. \\ \left.\left(\left[-0.8, -0.6\right], -0.2\right), \left(\left[-0.7, -0.3\right], -0.2\right), \left(\left[-0.9, -0.6\right], -0.4\right)\right] \\ \left[c, \left(\left[0.3, 0.5\right], 0.2\right), \left(\left[0.8, 0.9\right], 0.5\right), \left(\left[0.2, 0.4\right], 0.5\right), \right. \\ \left.\left(\left[-0.4, -0.3\right], -0.2\right), \left(\left[-0.6, -0.5\right], -0.8\right), \left(\left[-0.9, -0.8\right], -0.6\right)\right] \\ \left[d, \left(\left[0.9, 1\right], 0.7\right), \left(\left[0.6, 0.8\right], 0.1\right), \left(\left[0.4, 0.7\right], 0.1\right), \right. \\ \left.\left(\left[-0.8, -0.7\right], -0.5\right), \left(\left[-0.5, -0.2\right], -0.1\right), \left(\left[-0.3, -0.2\right], -0.1\right)\right] \end{array} \right\rangle$$

$$Q = \left\langle \begin{array}{l} \left[\begin{array}{l} ab, \big([0.1,0.3],0.8\big), \big([0.2,0.3],0.7\big), \big([0.6,0.7],0.4\big), \\ \big([-0.5,-0.4],-0.2\big), \big([-0.6,-0.3],-0.2\big), \big([-0.9,-0.6],-0.3\big) \end{array} \right] \\[10pt] \left[\begin{array}{l} ac, \big([0.3,0.5],0.8\big), \big([0.4,0.6],0.7\big), \big([0.5,0.6],0.3\big), \\ \big([-0.4,-0.3],-0.2\big), \big([-0.6,-0.3],-0.8\big), \big([-0.9,-0.8],-0.3\big) \end{array} \right] \\[10pt] \left[\begin{array}{l} ad, \big([0.3,0.6],0.8\big), \big([0.4,0.6],0.7\big), \big([0.5,0.7],0.1\big), \\ \big([-0.5,-0.1],-0.5\big), \big([-0.5,-0.2],-0.1\big), \big([-0.7,-0.4],-0.1\big) \end{array} \right] \\[10pt] \left[\begin{array}{l} bc, \big([0.1,3],0.5\big), \big([0.2,0.3],0.6\big), \big([0.6,0.7],0.5\big), \\ \big([-0.4,-0.3],-0.2\big), \big([-0.6,-0.3],-0.8\big), \big([-0.9,-0.7],-0.4\big) \end{array} \right] \\[10pt] \left[\begin{array}{l} bd, \big([0.1,0.3],0.7\big), \big([0.2,0.3],0.6\big), \big([0.6,0.7],0.1\big), \\ \big([-0.8,-0.6],-0.5\big), \big([-0.5,-0.2],-0.2\big), \big([-0.9,-0.6],-0.1\big) \end{array} \right] \\[10pt] \left[\begin{array}{l} cd, \big([0.3,0.5],0.7\big), \big([0.6,0.8],0.5\big), \big([0.4,0.7],0.1\big), \\ \big([-0.4,-0.3],-0.5\big), \big([-0.5,-0.2],-0.8\big), \big([-0.9,-0.8],-0.1\big) \end{array} \right] \end{array} \right\rangle$$

Figure 1.

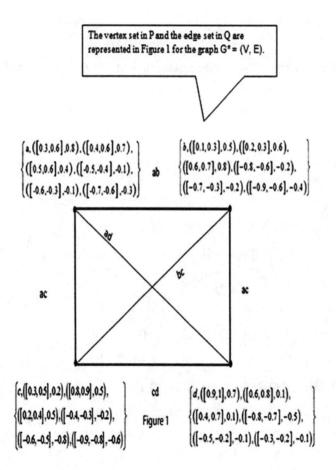

The vertex set in P and the edge set in Q are represented in Figure 1 for the graph G* = (V, E).

$\begin{bmatrix} a, ([0.3,0.6],0.8), ([0.4,0.6],0.7), \\ ([0.5,0.6],0.4), ([-0.5,-0.4],-0.1), \\ ([-0.6,-0.3],-0.1), ([-0.7,-0.6],-0.3) \end{bmatrix}$ ab $\begin{bmatrix} b, ([0.1,0.3],0.5), ([0.2,0.3],0.6), \\ ([0.6,0.7],0.8), ([-0.8,-0.6],-0.2), \\ ([-0.7,-0.3],-0.2), ([-0.9,-0.6],-0.4) \end{bmatrix}$

ad   bc   ac   ac

$\begin{bmatrix} c, ([0.3,0.5],0.2), ([0.8,0.9],0.5), \\ ([0.2,0.4],0.5), ([-0.4,-0.3],-0.2), \\ ([-0.6,-0.5],-0.8), ([-0.9,-0.8],-0.6) \end{bmatrix}$ cd

Figure 1

$\begin{bmatrix} d, ([0.9,1],0.7), ([0.6,0.8],0.1), \\ ([0.4,0.7],0.1), ([-0.8,-0.7],-0.5), \\ ([-0.5,-0.2],-0.1), ([-0.3,-0.2],-0.1) \end{bmatrix}$

Remark

1. If $n \geq 3$ in the vertex set and $n \geq 3$ in the set of edges then the graphs is a bipolar neutrosophic cubic polygon only when we join each vertex to the corresponding vertex through an edge.
2. If we have infinite elements in the vertex set and by joining the edge and every edge with each other we get a bipolar neutrosophic cubic curve.

Definition

Let $G(P,Q)$ be a bipolar neutrosophic cubic graph. The order of bipolar neutrosophic cubic graph is defined by

$$O(G) = \sum_{u \in v} \begin{bmatrix} \left(T_A^+, T_\lambda^+\right)(u), \left(I_A^+, I_\lambda^+\right)(u), \left(F_A^+, F_\lambda^+\right)(u), \\ \left(T_A^-, T_\lambda^-\right)(u), \left(I_A^-, I_\lambda^-\right)(u), \left(F_A^-, F_\lambda^-\right)(u) \end{bmatrix}$$

And the degree of a vertex u and G is defined by

$$\deg(u) = \sum_{uv \in E} \begin{bmatrix} \left(T_B^+, T_\mu^+\right)(uv), \left(I_B^+, I_\mu^+\right)(uv), \left(F_B^+, F_\mu^+\right)(uv), \\ \left(T_B^-, T_\mu^-\right)(uv), \left(I_B^-, I_\mu^-\right)(uv), \left(F_B^-, F_\mu^-\right)(uv) \end{bmatrix}$$

Example

In the above example, the order of a bipolar neutrosophic cubic graph is

$$\deg(a) = \begin{Bmatrix} \left([0.7, 1.4], 2.4\right), \left([1.0, 1.5], 2.1\right), \left([1.6, 2], 0.9\right) \\ \left([-1.4, -1.4], -1.1\right), \left([-1.7, -0.8], -1.1\right), \left([-2.5, -1.8], -0.7\right) \end{Bmatrix}$$

$$\deg(b) = \begin{Bmatrix} \left([0.3, 1.9], 2\right), \left([0.6, 0.9], 1.9\right), \left([1.8, 2.1], 1.0\right) \\ \left([-1.7, -1.3], -0.9\right), \left([-1.7, -0.8], -1.2\right), \left([-2.7, -2], -0.8\right) \end{Bmatrix}$$

$$\deg(c) = \begin{Bmatrix} \left([0.7, 1.3], 2\right), \left([1.2, 1.7], 1.8\right), \left([1.5, 2], 1.0\right) \\ \left([-1.2, -0.9], -0.9\right), \left([-1.7, -0.8], -2.4\right), \left([-2.7, -2.4], -0.8\right) \end{Bmatrix}$$

$$\deg(d) = \begin{Bmatrix} \left([0.7, 1.4], 2.2\right), \left([1.2, 1.7], 1.8\right), \left([1.5, 2.1], 0.3\right) \\ \left([-1.7, -1.3], -1.5\right), \left([-1.5, -0.6], -1.1\right), \left([-2.5, -1.8], -0.3\right) \end{Bmatrix}$$

Definition

Let $G_1 = \left(P_1, Q_1 \right)$ be a bipolar neutrosophic cubic graph of $G_1^* = \left(V_1, E_1 \right)$ and $G_2 = \left(P_2, Q_2 \right)$ be a bipolar neutrosophic cubic graph of $G_2^* = \left(V_2, E_2 \right)$. Then Cartesian product of G_1 and G_2 is denoted by

$$G_1 \times G_2 = \left(P_1 \times P_2, Q_1 \times Q_2 \right)$$

$$= \begin{pmatrix} \left(A_1^+, \lambda_1^+ \right) \times \left(A_2^+, \lambda_2^+ \right), \left(A_1^-, \lambda_1^- \right) \times \left(A_2^-, \lambda_2^- \right), \\ \left(B_1^+, \mu_1^+ \right) \times \left(B_2^+, \mu_2^+ \right), \left(B_1^-, \mu_1^- \right) \times \left(B_2^-, \mu_2^- \right) \end{pmatrix}.$$

$$= \begin{pmatrix} \left(A_1^+ \times A_2^+, \lambda_1^+ \times \lambda_2^+ \right), \left(A_1^- \times A_2^-, \lambda_1^- \times \lambda_2^- \right), \\ \left(B_1^+ \times B_2^+, \mu_1^+ \times \mu_2^+ \right), \left(B_1^- \times B_2^-, \mu_1^- \times \mu_2^- \right) \end{pmatrix}$$

$$G_1 \times G_2 = \left\langle \begin{matrix} \left(T_{A_1 \times A_2}^+, T_{\lambda_1 \times \lambda_2}^+ \right), \left(I_{A_1 \times A_2}^+, I_{\lambda_1 \times \lambda_2}^+ \right), \left(F_{A_1 \times A_2}^+, F_{\lambda_1 \times \lambda_2}^+ \right) \\ \left(T_{A_1 \times A_2}^-, T_{\lambda_1 \times \lambda_2}^- \right), \left(I_{A_1 \times A_2}^-, I_{\lambda_1 \times \lambda_2}^- \right), \left(F_{A_1 \times A_2}^-, F_{\lambda_1 \times \lambda_2}^- \right) \\ \left(T_{B_1 \times B_2}^+, T_{\mu_1 \times \mu_2}^+ \right), \left(I_{B_1 \times B_2}^+, I_{\mu_1 \times \mu_2}^+ \right), \left(F_{B_1 \times B_2}^+, F_{\mu_1 \times \mu_2}^+ \right) \\ \left(T_{B_1 \times B_2}^-, T_{\mu 1 \times \mu 2}^- \right), \left(I_{B_1 \times B_2}^-, I_{\mu_1 \times \mu_2}^- \right), \left(F_{B_1 \times B_2}^-, F_{\mu_1 \times \mu_2}^- \right) \end{matrix} \right\rangle$$

and is defined as follows:

1. $$\begin{pmatrix} T_{A_1 \times A_2}^+ \left(u, v \right) = r \min \left(T_{A_1}^+ \left(u \right), T_{A_2}^+ \left(v \right) \right), T_{\lambda_1 \times \lambda_2}^+ \left(u, v \right) = \max \left(T_{\lambda_1}^+ \left(u \right), T_{\lambda_2}^+ \left(v \right), \right) \\ T_{A_1 \times A_2}^- \left(u, v \right) = r \max \left(T_{A_1}^- \left(u \right), T_{A_2}^- \left(v \right) \right), T_{\lambda_1 \times \lambda_2}^- \left(u, v \right) = \min \left(T_{\lambda_1}^- \left(u \right), T_{\lambda_2}^- \left(v \right), \right) \end{pmatrix}$$

2. $$\begin{pmatrix} I_{A_1 \times A_2}^+ \left(u, v \right) = r \min \left(I_{A_1}^+ \left(u \right), I_{A_2}^+ \left(v \right) \right), I_{\lambda_1 \times \lambda_2}^+ \left(u, v \right) = \max \left(I_{\lambda_1}^+ \left(u \right), I_{\lambda_2}^+ \left(v \right), \right) \\ I_{A_1 \times A_2}^- \left(u, v \right) = r \max \left(I_{A_1}^- \left(u \right), I_{A_2}^- \left(v \right) \right), I_{\lambda_1 \times \lambda_2}^- \left(u, v \right) = \min \left(I_{\lambda_1}^- \left(u \right), I_{\lambda_2}^- \left(v \right) \right) \end{pmatrix}$$

3. $$\begin{pmatrix} F_{A_1 \times A_2}^+ \left(u, v \right) = r \max \left(F_{A_1}^+ \left(u \right), F_{A_2}^+ \left(v \right) \right), F_{\lambda_1 \times \lambda_2}^+ \left(u, v \right) = \min \left(F_{\lambda_1}^+ \left(u \right), F_{\lambda_2}^+ \left(v \right), \right) \\ F_{A_1 \times A_2}^- \left(u, v \right) = r \min \left(F_{A_1}^- \left(u \right), F_{A_2}^- \left(v \right) \right), F_{\lambda_1 \times \lambda_2}^- \left(u, v \right) = \max \left(F_{\lambda_1}^- \left(u \right), F_{\lambda_2}^- \left(v \right) \right) \end{pmatrix}$$

4. $$\begin{pmatrix} T_{B_1 \times B_2}^+ \left(\left(u, v_1 \right) \left(u, v_2 \right) \right) = r \min \left(T_{A_1}^+ \left(u \right), T_{B_2}^+ \left(v_1 v_2 \right) \right), \\ T_{B_1 \times B_2}^- \left(\left(u, v_1 \right) \left(u, v_2 \right) \right) = r \max \left(T_{A_1}^- \left(u \right), T_{B_2}^- \left(v_1 v_2 \right), \right) \\ T_{\mu_1 \times \mu_2}^+ \left(\left(u, v_1 \right) \left(u, v_2 \right) \right) = \max \left(T_{\lambda_1}^+ \left(u \right), T_{\mu_2}^- \left(v_1 v_2 \right) \right), \\ T_{\mu_1 \times \mu_2}^- \left(\left(u, v_1 \right) \left(u, v_2 \right) \right) = \min \left(T_{\lambda_1}^- \left(u \right), T_{\mu_2}^- \left(v_1 v_2 \right) \right) \end{pmatrix}$$

5. $$\begin{cases} I^+_{B_1 \times B_2}\left((u,v_1)(u,v_2)\right) = r\min\left(I^+_{A_1}(u), I^+_{B_2}(v_1v_2)\right), \\ I^-_{B_1 \times B_2}\left((u,v_1)(u,v_2)\right) = r\max\left(I^-_{A_1}(u), I^-_{B_2}(v_1v_2),\right) \\ I^+_{\mu_1 \times \mu_2}\left((u,v_1)(u,v_2)\right) = \max\left(I^+_{\lambda_1}(u), I^-_{\mu_2}(v_1v_2)\right), \\ I^-_{\mu_1 \times \mu_2}\left((u,v_1)(u,v_2)\right) = \min\left(I^-_{\lambda_1}(u), I^-_{\mu_2}(v_1v_2)\right) \end{cases}$$

6. $$\begin{cases} F^+_{B_1 \times B_2}\left((u,v_1)(u,v_2)\right) = r\max\left(I^+_{A_1}(u), I^+_{B_2}(v_1v_2)\right), \\ F^-_{B_1 \times B_2}\left((u,v_1)(u,v_2)\right) = r\max\left(I^-_{A_1}(u), I^-_{B_2}(v_1v_2),\right) \\ F^+_{\mu_1 \times \mu_2}\left((u,v_1)(u,v_2)\right) = \min\left(F^+_{\lambda_1}(u), F^-_{\mu_2}(v_1v_2)\right), \\ F^-_{\mu_1 \times \mu_2}\left((u,v_1)(u,v_2)\right) = \max\left(I^-_{\lambda_1}(u), I^-_{\mu_2}(v_1v_2)\right) \end{cases}$$

7. $$\begin{cases} T^+_{B_1 \times B_2}\left((u_1,v)(u_2,v)\right) = r\min\left(T^+_{B_1}(u_1,v_2), T^+_{\lambda_2}(v)\right), \\ T^-_{B_1 \times B_2}\left((u_1,v)(u_2,v)\right) = r\max\left(T^-_{B_1}(u_1,v_2), T^-_{\lambda_2}(v)\right) \\ T^+_{\mu_1 \times \mu_2}\left((u_1,v)(u_2,v)\right) = \max\left(T^+_{\mu_1}(u_1,v_2), T^+_{\lambda_2}(v)\right), \\ T^-_{\mu_1 \times \mu_2}\left((u_1,v)(u_2,v)\right) = \min\left(T^-_{\mu_1}(u_1,v_2), T^-_{\lambda_2}(v)\right) \end{cases}$$

8. $$\begin{cases} I^+_{B_1 \times B_2}\left((u_1,v)(u_2,v)\right) = r\min\left(I^+_{B_1}(u_1,v_2), T^+_{\lambda_2}(v)\right), \\ I^-_{B_1 \times B_2}\left((u_1,v)(u_2,v)\right) = r\max\left(I^-_{B_1}(u_1,v_2), I^-_{\lambda_2}(v)\right) \\ I^+_{\mu_1 \times \mu_2}\left((u_1,v)(u_2,v)\right) = \max\left(I^+_{\mu_1}(u_1,v_2), I^+_{\lambda_2}(v)\right), \\ I^-_{\mu_1 \times \mu_2}\left((u_1,v)(u_2,v)\right) = \min\left(I^-_{\mu_2}(u_1,v_2), I^-_{\lambda_2}(v)\right) \end{cases}$$

9. $$\begin{cases} F^+_{B_1 \times B_2}\left((u_1,v)(u_2,v)\right) = r\max\left(F^+_{B_1}(u_1,v_2), F^+_{\lambda_2}(v)\right), \\ F^-_{B_1 \times B_2}\left((u_1,v)(u_2,v)\right) = r\min\left(F^-_{B_1}(u_1,v_2), F^-_{\lambda_2}(v)\right) \\ F^+_{\mu_1 \times \mu_2}\left((u_1,v)(u_2,v)\right) = \min\left(F^+_{\mu_1}(u_1,v_2), F^+_{\lambda_2}(v)\right), \\ F^-_{\mu_1 \times \mu_2}\left((u_1,v)(u_2,v)\right) = \max\left(F^-_{\mu_1}(u_1,v_2), F^-_{\lambda_2}(v)\right) \end{cases}$$

Example

Let $G_1 = \left(P_1, Q_1\right)$ be a bipolar neutrosophic cubic graph of $G_1^* = \left(V_1, E_1\right)$ as shown in figure 2, Where $V_1 = \left\{a, b, c\right\}$, $E_1 = \left\{ab, bc, ac\right\}$.

$$P_1 = \left\langle \begin{array}{l} \left[a, \left([0.5,0.6],0.3 \right), \left([0.4,0.7]0.1 \right), \left([0.2,0.3],0.5 \right), \\ \left([-0.7,-0.6]-0.1 \right) \left([-0.4,-0.2],-0.5 \right), \left([-0.5,-0.4],-0.3 \right) \right] \\ \left[b, \left([0.1,0.2],0.4 \right), \left([0.7,0.3],0.9 \right), \left([0.2,0.4],0.1 \right), \\ \left([-0.3,-0.2],-0.1 \right), \left([-0.5,-0.3],-0.2 \right), \left([-0.7,-0.5],-0.3 \right) \right] \\ \left[c, \left([0.3,0.5],0.2 \right), \left([0.5,0.6],0.7 \right), \left([0.2,0.6],0.8 \right), \\ \left([-0.8,-0.7],-0.5 \right), \left([-0.7,-0.4],-0.3 \right), \left([-0.9,-0.4],-0.2 \right) \right] \end{array} \right\rangle$$

$$Q_1 = \left\langle \begin{array}{l} \left[ab, \left([0.1,0.2],0.4 \right), \left([0.4,0.7],0.9 \right), \left([0.2,0.3],0.5 \right), \\ \left([-0.3,-0.7],-0.1 \right), \left([-0.4,-0.2],-0.5 \right), \left([-0.7,-0.5],-0.3 \right) \right] \\ \left[bc, \left([0.1,0.2],0.4 \right), \left([0.5,0.6],0.9 \right), \left([0.2,0.4],0.8 \right), \\ \left([-0.3,-0.2],-0.5 \right), \left([-0.5,-0.3],-0.3 \right), \left([-0.9,-0.5],-0.1 \right) \right] \\ \left[ac, \left([0.3,0.5],0.3 \right), \left([0.4,0.6],0.7 \right), \left([0.2,0.3],0.8 \right), \\ \left([-0.7,-0.6],-0.5 \right), \left([-0.4,-0.2],-0.5 \right), \left([-0.9,-0.4],-0.2 \right) \right] \end{array} \right\rangle$$

And $G_2 = \left(P_2, Q_2 \right)$ be a bipolar neutrosophic cubic graph if $G_2^* = \left(V_2, E_2 \right)$ as shown in figure 3, where $V_2 = \left\{ x, y, z \right\}$ and $E_2 = \left\{ xy, yz, xz \right\}$

$$P_2 = \left\langle \begin{array}{l} \left[x, \left([0.6,0.7],0.3 \right), \left([0.2,0.5],0.6 \right), \left([0.3,0.4],0.7 \right), \\ \left([-0.4,-0.2],-0.1 \right), \left([-0.6,-0.4],-0.2 \right). \left([-0.8,-0.6],-0.4 \right) \right] \\ \left[y, \left([0.4,0.5],0.7 \right), \left([0.6,0.7],0.9 \right), \left([0.1,0.4],0.5 \right), \\ \left([-0.8,-0.7],-0.1 \right), \left([-0.5,-0.3],-0.2 \right), \left([-0.4,-0.2],-0.1 \right) \right] \\ \left[z, \left([0.3,0.5],0.1 \right), \left([0.9,1.0],0.4 \right), \left([0.5,0.8],0.9 \right), \\ \left([-0.6,-0.3],-0.2 \right), \left([-0.7,-0.5],-0.2 \right), \left([-0.5,-0.5],-0.1 \right) \right] \end{array} \right\rangle$$

$$Q_2 = \left\langle \begin{array}{l} \left[xy, \left([0.4,0.5],0.7 \right), \left([0.2,0.5],0.9 \right), \left([0.3,0.4],0.5 \right), \\ \left([-0.4,-0.2],-0.1 \right), \left([-0.5,-0.3],-0.2 \right), \left([-0.8,-0.6],-0.1 \right) \right] \\ \left[yz, \left([0.3,0.5],0.7 \right), \left([0.6,0.7],0.9 \right), \left([0.1,0.4],0.9 \right), \\ \left([-0.6,-0.3],-0.2 \right), \left([-0.5,-0.3],-0.2 \right), \left([-0.5,-0.4],-0.1 \right) \right] \\ \left[xz, \left([0.3,0.5],0.3 \right), \left([0.2,0.5],0.6 \right), \left([0.5,0.8],0.7 \right), \\ \left([-0.4,-0.2],-0.2 \right), \left([-0.6,-.4],-0.2 \right), \left([-0.8,-0.6],-0.1 \right) \right] \end{array} \right\rangle$$

$$Q_1 \times Q_2 = \left\langle \begin{array}{l} \left[\begin{array}{l} \big((a,x)(a,y)\big),\big([0.6,0.5],0.7\big),\big([0.2,0.5],0.9\big),\big([0.3,0.4],0.5\big), \\ \big([-0.4,-0.2]-0.1\big),\big([-0.4,-0.2],-0.5\big),\big([-0.8,-0.6],-0.1\big) \end{array}\right] \\[4pt] \left[\begin{array}{l} \big((a,y)(a,z)\big),\big([0.3,0.5],0.7\big),\big([0.4,0.7],0.9\big),\big([0.5,0.8],0.5\big), \\ \big([-0.6,-0.3],-0.2\big),\big([-0.4,-0.2],-0.5\big),\big([-0.5,-0.4],-0.1\big) \end{array}\right] \\[4pt] \left[\begin{array}{l} \big((a,z)(b,z)\big),\big([0.1,0.2],0.4\big),\big([0.4,0.7],0.9\big),\big([0.5,0.8],0.5\big), \\ \big([-0.3,-0.2],-0.2\big),\big([-0.4,-0.2],-0.5,\big),\big([-0.7,-0.5],-0.1\big) \end{array}\right] \\[4pt] \left[\begin{array}{l} \big((b,x)(b,z)\big),\big([0.1,0.2],0.4\big),\big([0.2,0.5],0.9\big),\big([0.5,0.8],0.1\big), \\ \big([-0.3,-0.2],-0.2\big),\big([-0.5,-0.3],-0.2\big),\big([-0.8,-0.6],-0.1\big) \end{array}\right] \\[4pt] \left[\begin{array}{l} \big((b,x)(b,y)\big),\big([0.1,0.2],0.7\big),\big([0.2,0.5],0.9\big),\big([0.3,0.4],0.1\big), \\ \big([-0.3,-0.2],-0.1\big),\big([-0.5,-0.3],-0.4\big),\big([-0.8,-0.6],-0.1\big) \end{array}\right] \\[4pt] \left[\begin{array}{l} \big((c,y)(c,z)\big),\big([0.3,0.5],0.7\big),\big([0.5,0.6],0.9\big),\big([0.5,0.8],0.5\big), \\ \big([-0.6,-0.3],-0.5\big),\big([-0.5,-0.3],-0.3\big),\big([-0.9,-0.5],-0.1\big) \end{array}\right] \\[4pt] \left[\begin{array}{l} \big((c,x)(c,z)\big),\big([0.3,0.5],0.3\big),\big([0.2,0.5],0.7\big),\big([0.5,0.8],0.7\big), \\ \big([-0.4,-0.2],-0.5\big),\big([-0.6,-0.4],-0.3\big),\big([-0.9,-0.6],-0.1\big) \end{array}\right] \\[4pt] \left[\begin{array}{l} \big((a,x)(c,x)\big),\big([0.3,0.5],0.3\big),\big([0.2,0.5],0.7\big),\big([0.3,0.6],0.5\big), \\ \big([-0.4,-0.2],-0.5\big),\big([-0.4,-0.2],-0.5\big),\big([-0.9,-0.6],-0.2\big) \end{array}\right] \end{array} \right\rangle$$

Proposition

The Cartesian product of two bipolar neutrosophic cubic graphs is again a bipolar neutrosophic cubic graph.

Proof

For $P_1 \times P_2$ the condition is obvious. Now we verify the conditions only for $Q_1 \times Q_2$, where

$$Q_1 \times Q_2 = \left\{ \begin{array}{l} \left(T^+_{B_1 \times B_2}, T^+_{\mu_1 \times \mu_2}\right), \left(I^+_{B_1 \times B_2}, I^+_{\mu_1 \times \mu_2}\right), \left(F^+_{B_1 \times B_2}, F^+_{\mu_1 \times \mu_2}\right), \\ \left(T^-_{B_1 \times B_2}, T^-_{\mu_1 \times \mu_2}\right), \left(I^-_{B_1 \times B_2}, I^-_{\mu_1 \times \mu_2}\right), \left(F^-_{B_1 \times B_2}, F^-_{\mu_1 \times \mu_2}\right) \end{array} \right\}$$

Then

Figure 2.

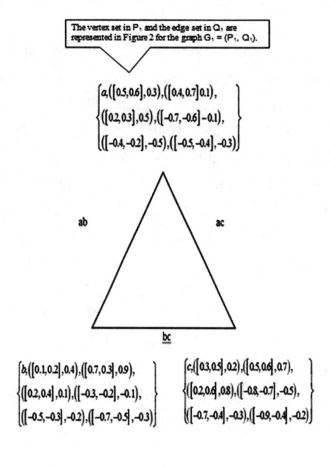

$$T_{B_1 \times B_2}^{+} \left((u, u_2), (u, v_2) \right) = r \min \left\{ \left(T_{A_1}^{+}(u), T_{B_2}^{+}(u_2 v_2) \right) \right\}$$

$$\leq r \min \left\{ \left(T_{A_1}^{+}(u) \right), \left(r \min \left(T_{A_2}^{+}(u_2), T_{A_2}^{+}(v_2) \right) \right) \right\}$$

$$= r \min \left\{ r \min \left(T_{A_1}^{+}(u), T_{A_2}^{+}(u_2) \right), r \min \left(T_{A_1}^{+}(u), T_{A_2}^{+}(v_2) \right) \right\}$$

$$= r \min \left\{ \left(\left(T_{A_1}^{+} \times T_{A_2}^{+} \right)(u v_2) \right), \left(\left(T_{A_1}^{+} \times T_{A_2}^{+} \right)(u v_2) \right) \right\}$$

$$T_{B_1 \times B_2}^{-} \left((u, u_2), (u, v_2) \right) = r \max \left\{ \left(T_{A_1}^{-}(u), T_{B_2}^{-}(u_2 v_2) \right) \right\}$$

$$\leq r \max \left\{ \left(T_{A_1}^{-}(u) \right), \left(r \max \left(T_{A_2}^{-}(u_2), T_{A_2}^{-}(v_2) \right) \right) \right\}$$

$$= r \max \left\{ r \max \left(T_{A_1}^{-}(u), T_{A_2}^{-}(u_2) \right), r \max \left(T_{A_1}^{-}(u), T_{A_2}^{-}(v_2) \right) \right\}$$

$$= r \max \left\{ \left(\left(T_{A_1}^{-} \times T_{A_2}^{-} \right)(u u_2) \right), \left(\left(T_{A_1}^{-} \times T_{A_2}^{-} \right)(u v_2) \right) \right\}$$

Figure 3.

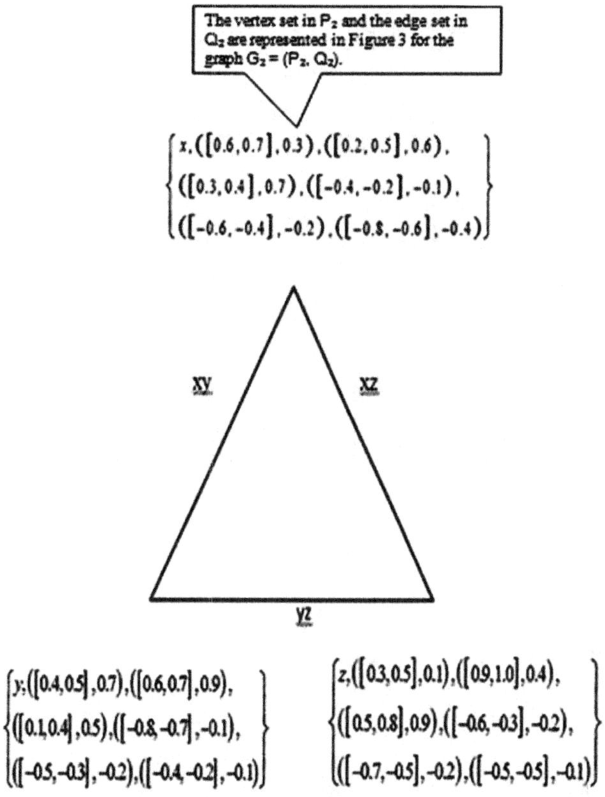

The vertex set in P_2 and the edge set in Q_2 are represented in Figure 3 for the graph $G_2 = (P_2, Q_2)$.

$$\left\{ \begin{array}{l} x, ([0.6, 0.7], 0.3), ([0.2, 0.5], 0.6), \\ ([0.3, 0.4], 0.7), ([-0.4, -0.2], -0.1), \\ ([-0.6, -0.4], -0.2), ([-0.8, -0.6], -0.4) \end{array} \right\}$$

XY   XZ

YZ

$$\left\{ \begin{array}{l} y, ([0.4, 0.5], 0.7), ([0.6, 0.7], 0.9), \\ ([0.1, 0.4], 0.5), ([-0.8, -0.7], -0.1), \\ ([-0.5, -0.3], -0.2), ([-0.4, -0.2], -0.1) \end{array} \right\}$$

$$\left\{ \begin{array}{l} z, ([0.3, 0.5], 0.1), ([0.9, 1.0], 0.4), \\ ([0.5, 0.8], 0.9), ([-0.6, -0.3], -0.2), \\ ([-0.7, -0.5], -0.2), ([-0.5, -0.5], -0.1) \end{array} \right\}$$

Figure 3 Bipolar Neutrosophic Cubic graph G_2

$$T^+_{\mu_1 \times \mu_2} \left((u, u_2), (u, v_2) \right) = r \max \left\{ \left(T^+_{\lambda_1}(u), T^+_{\mu_2}(u_2 v_2) \right) \right\}$$

$$\leq \max \left\{ \left(T^+_{\lambda_1}(u) \right), \left(\left(\max T^+_{\lambda_2}(u_2), T^+_{\lambda_2}(v_2) \right) \right) \right\}$$

$$= \max \left\{ \max \left(\left(T^+_{\lambda_1}(u), T^+_{\lambda_2}(u_2) \right) \right), r \min \left(T^+_{\lambda_1}(u), T^+_{\lambda_2}(v_2) \right) \right\}$$

$$= \max \left\{ \left(\left(T^+_{\lambda_1} \times T^+_{\lambda_2} \right)(u u_2) \right), \left(\left(T^+_{\lambda_1} \times T^+_{\lambda_2} \right)(u v_2) \right) \right\}$$

$$T^-_{\mu_1 \times \mu_2} \left((u, u_2), (u, v_2) \right) = \min \left\{ \left(T^-_{\lambda_1}(u), T^-_{\mu_2}(u_2 v_2) \right) \right\}$$

$$\geq \min \left\{ \left(T^-_{\lambda_1}(u) \right), \left(\min \left(T^-_{\lambda_2}(u_2), T^-_{\lambda_2}(v_2) \right) \right) \right\}$$

$$= \min \left\{ \min \left(\left(T^-_{\lambda_1}(u), T^-_{\lambda_2}(u_2) \right) \right), \min \left(T^-_{\lambda_1}(u), T^-_{\lambda_2}(v_2) \right) \right\}$$

$$= \min \left\{ \left(\left(T^-_{\lambda_1} \times T^-_{\lambda_2} \right)(u u_2) \right), \left(\left(T^-_{\lambda_1} \times T^-_{\lambda_2} \right)(u, v_2) \right) \right\}$$

$$I^+_{B_1 \times B_2}\left((u, u_2), (u, v_2)\right) = \min\left\{I^+_{A_1}(u), I^+_{B_2}(u_2 v_2)\right\}$$

$$\leq r\min\left\{\left(I^+_{A_1}(u)\right), \left(r\min\left(I^+_{A_1}(u_2), I^+_{A_2}(v_2)\right)\right)\right\}$$

$$= r\min\left\{r\min\left(I^+_{A_1}(u), I^+_{A2}(u_2)\right), r\min\left(I^+_{A_1}(u), I^+_{A_2}(v_2)\right)\right\}$$

$$= r\min\left\{\left(\left(I^+_{A_1} \times I^+_{A_2}\right)(uu_2)\right), \left(\left(I^+_{A_1} \times I^+_{A_2}\right)(u, v_2)\right)\right\}$$

$$I^-_{B_1 \times B_2}\left((u, u_2), (u, v_2)\right) = r\max\left\{\left(I^-_{A_1}(u), I^-_{B_2}(u_2 v_2)\right)\right\}$$

$$\geq r\max\left\{\left(I^-_{A_1}(u)\right), \left(r\max\left(I^-_{A_2}(u_2), I^-_{A_2}(v_2)\right)\right)\right\}$$

$$= r\max\left\{r\max\left(I^-_{A_1}(u), I^-_{A_2}(u_2)\right), r\max\left(I^-_{A_1}(u), I^-_{A_2}(v_2)\right)\right\}$$

$$= r\max\left\{\left(\left(I^-_{A_1} \times I^-_{A_2}\right)(u, u_2)\right), \left(\left(I^-_{A_1} \times I^-_{A_2}\right)(u, v_2)\right)\right\}$$

$$I^+_{\mu_1 \times \mu_2}\left((u, u_2), (u, v_2)\right) = \max\left\{\left(I^+_{\lambda_1}(u), I^+_{\mu_2}(u_2 v_2)\right)\right\}$$

$$\leq \max\left\{\left(I^+_{\lambda_1}(u)\right), \max\left(I^+_{\lambda_2}(u_2), I^+_{\lambda_2}(v_2)\right)\right\}$$

$$= \max\left\{\max\left(I^+_{\lambda_1}(u), I^+_{\lambda_2}(u_2)\right), \max\left(I^+_{\lambda_1}(u), I^+_{\lambda_2}(v_2)\right)\right\}$$

$$= \max\left\{\left(\left(I^+_{\lambda_1} \times I^+_{\lambda_2}\right)(u, u_2)\right), \left(\left(I^+_{\lambda_1} \times I^+_{\lambda_2}\right)(u, v_2)\right)\right\}$$

$$I^-_{\mu_1 \times \mu_2}\left((u, u_2), (u, v_2)\right) = \min\left\{\left(I^-_{\lambda_1}(u), I^-_{\mu_2}(u_2 v_2)\right)\right\}$$

$$\geq \min\left\{\left(I^-_{\lambda_1}(u)\right), \min\left(I^-_{\lambda_2}(u_2), I^-_{\lambda_2}(v_2)\right)\right\}$$

$$= \min\left\{\min\left(I^-_{\lambda_1}(u), I^-_{\lambda_2}(u_2)\right), \min\left(I^-_{\lambda_1}(u), I^-_{\lambda_2}(v_2)\right)\right\}$$

$$= \min\left\{\left(I^-_{\lambda_1} \times I^-_{\lambda_2}\right)(u, u_2), \left(I^-_{\lambda_1} \times I^-_{\lambda_2}\right)(u, v_2)\right\}$$

$$F^+_{B_1 \times B_2}\left((u, u_2), (u, v_2)\right) = r\max\left\{\left(F^+_{A_1}(u), F^+_{B_2}(u_2 v_2)\right)\right\}$$

$$\leq r\max\left\{\left(F^+_{A_1}(u)\right), r\max\left(F^+_{A_2}(u_2), F^-_{A_2}(v_2)\right)\right\}$$

$$= r\max\left\{r\max\left(F^+_{A_1}(u), F^+_{A_2}(u_2)\right), r\max\left(F^+_{A_1}(u), F^+_{A_2}(v_2)\right)\right\}$$

$$= r\max\left\{\left(\left(F^+_{A_1} \times T^+_{A_2}\right)(u, u_2)\right), \left(\left(F^+_{A_1} \times F^+_{A_2}\right)(u, v_2)\right)\right\}$$

$$F_{B_1 \times B_2}^{-}\left((u,u_2),(u,v_2)\right) = r\min\left\{\left(F_{A_1}^{-}(u), F_{B_2}^{-}(u_2 v_2)\right)\right\}$$

$$\geq r\min\left\{\left(F_{A_1}^{-}(u)\right), \left(r\min\left(F_{A_2}^{-}(u_2), F_{A_2}^{-}(v_2)\right)\right)\right\}$$

$$= r\min\left\{r\min\left(F_{A_1}^{-}(u), I_{A_2}^{-}(u_2)\right), r\min\left(F_{A_1}^{-}(u), F_{A_2}^{-}(v_2)\right)\right\}$$

$$= r\min\left\{\left(\left(F_{A_1}^{-} \times F_{A_2}^{-}\right)(u,u_2)\right), \left(\left(F_{A_1}^{-} \times F_{A_2}^{-}\right)(u,v_2)\right)\right\}$$

$$F_{\mu_1 \times \mu_2}^{+}\left((u,u_2),(u,v_2)\right) = \min\left\{\left(F_{\lambda_1}^{+}(u), F_{\mu_2}^{+}(u_2 v_2)\right)\right\}$$

$$\geq r\min\left\{\left(F_{\lambda_1}^{-}(u)\right), \min\left(F_{\lambda_2}^{-}(u_2), F_{\lambda_2}^{-}(v_2)\right)\right\}$$

$$= \min\left\{\min\left(F_{\lambda_1}^{-}(u), I_{\lambda_2}^{-}(u_2)\right), \min\left(F_{\lambda_1}^{-}(u), F_{\lambda_2}^{-}(v_2)\right)\right\}$$

$$= \min\left\{\left(F_{\lambda_1}^{-} \times F_{\lambda_2}^{-}\right)(u,u_2), \left(F_{\lambda_1}^{-} \times F_{\lambda_2}^{-}\right)(u,v_2)\right\}$$

$$F_{\mu_1 \times \mu_2}^{-}\left((u,u_2),(u,v_2)\right) = \max\left\{\left(F_{\lambda_1}^{-}(u), F_{\mu_2}^{-}(u_2 v_2)\right)\right\}$$

$$\geq \max\left\{\left(F_{\lambda_1}^{-}(u)\right), \min\left(F_{\lambda_2}^{-}(u_2), F_{\lambda_2}^{-}(v_2)\right)\right\}$$

$$= \max\left\{\max\left(F_{\lambda_1}^{-}(u), F_{\lambda_2}^{-}(u_2)\right), \max\left(F_{\lambda_1}^{-}(u), F_{\lambda_2}^{-}(v_2)\right)\right\}$$

$$= \max\left\{\left(\left(F_{\lambda_1}^{-} \times F_{\lambda_2}^{-}\right)(u,u_2)\right), \left(\left(F_{\lambda_1}^{-} \times F_{\lambda_2}^{-}\right)(u,v_2)\right)\right\}$$

Similarly we can prove it for $w \in V_2$ and $u_1, u_2 \in E_2$

Definition

Let $G_1 = \left(P_1, Q_1\right)$ and $G_2 = \left(P_2, Q_2\right)$ be two bipolar neutrosophic cubic graphs. The degree of a vertex in $G_1 \times G_2$ can be defined as follows for any $\left(u_1 \times u_2\right) \in v_1 \times v_2$

$$\deg\left(T_{A_1}^{+} \times T_{A_2}^{+}\right)(u_1, u_2)$$

$$= \sum_{(u_1 u_2)(v_1 v_2) \in E_2} r\max\left(T_{B_1}^{+} \times T_{B_2}^{+}\right)\left((u_1, u_2)(v_1, v_2)\right) = \sum_{u_1 = v_1 = u, u_2 v_2 \in E_2} r\max\left(T_{A_1}^{+}(u), T_{B_2}^{+}(u_2, v_2)\right)$$

$$+ \sum_{u_2 = v_2 = w, u_1 v_1 \in E} r\max\left(T_{A_2}^{+}(w), T_{B_1}^{+}(u_1, v_1)\right) + \sum_{u_1 = v_1 \in E, u_2 v_2 \in E_2} r\max\left(T_{B_2}^{+}(u_1, v_1), T_{B_2}^{+}(u_2, v_2)\right)$$

$$\deg\left(T_{A_1}^{-} \times T_{A_2}^{-}\right)\left(u_1, u_2\right)$$

$$= \sum_{\left(u_1 u_2\right)\left(v_1 v_2\right) \in E_2} r\min\left(T_{B_1}^{-} \times T_{B_2}^{-}\right)\left(\left(u_1, u_2\right)\left(v_1, v_2\right)\right) = \sum_{u_1 = v_1 = u, u_2 v_2 \in E_2} r\max\left(T_{A_1}^{-}\left(u\right), T_{B_2}^{-}\left(u_2, v_2\right)\right)$$

$$+ \sum_{u_2 = v_2 = w, u_1 v_1 \in E} r\max\left(T_{A_2}^{-}\left(w\right), T_{B_1}^{-}\left(u_1, v_1\right)\right) + \sum_{u_1 = v_1 \in E, u_2 v_2 \in E_2} r\max\left(T_{B_2}^{-}\left(u_1, v_1\right), T_{B_2}^{-}\left(u_2, v_2\right)\right)$$

$$\deg\left(T_{\lambda_1}^{+} \times T_{\lambda_2}^{+}\right)\left(u_1, u_2\right)$$

$$= \sum_{\left(u_1 u_2\right)\left(v_1 v_2\right) \in E_2} \min\left(T_{\mu_1}^{+} \times T_{\mu_2}^{+}\right)\left(\left(u_1, u_2\right)\left(v_1, v_2\right)\right) = \sum_{u_1 = v_1 = u, u_2 v_2 \in E_2} r\min\left(T_{\lambda_1}^{+}\left(u\right), T_{\mu_2}^{+}\left(u_2, v_2\right)\right)$$

$$+ \sum_{u_2 = v_2 = w, u_1 v_1 \in E} \min\left(T_{\lambda_2}^{+}\left(w\right), T_{\mu_1}^{+}\left(u_1, v_1\right)\right) + \sum_{u_1 = v_1 \in E, u_2 v_2 \in E_2} \min\left(T_{\mu_1}^{+}\left(u_1, v_1\right), T_{\mu_2}^{+}\left(u_2, v_2\right)\right)$$

$$\deg\left(T_{\lambda_1}^{-} \times T_{\lambda_2}^{-}\right)\left(u_1, u_2\right)$$

$$= \sum_{\left(u_1 u_2\right)\left(v_1 v_2\right) \in E_2} \max\left(T_{\mu_1}^{-} \times T_{\mu_2}^{-}\right)\left(\left(u_1, u_2\right)\left(v_1, v_2\right)\right) = \sum_{u_1 = v_1 = u, u_2 v_2 \in E_2} \max\left(T_{\lambda_1}^{-}\left(u\right), T_{\mu_2}^{-}\left(u_2, v_2\right)\right)$$

$$+ \sum_{u_2 = v_2 = w, u_1 v_1 \in E} \max\left(T_{\lambda_2}^{-}\left(w\right), T_{\mu_1}^{-}\left(u_1, v_1\right)\right) + \sum_{u_1 = v_1 \in E, u_2 v_2 \in E_2} \max\left(T_{\mu_1}^{-}\left(u_1, v_1\right), T_{\mu_2}^{-}\left(u_2, v_2\right)\right)$$

$$\deg\left(I_{A_1}^{+} \times I_{A_2}^{+}\right)\left(u_1, u_2\right)$$

$$= \sum_{\left(u_1 u_2\right)\left(v_1 v_2\right) \in E_2} r\max\left(I_{B_1}^{+} \times I_{B_2}^{+}\right)\left(\left(u_1, u_2\right)\left(v_1, v_2\right)\right) = \sum_{u_1 = v_1 = u, u_2 v_2 \in E_2} r\max\left(I_{B_1}^{+}\left(u\right), I_{B_2}^{+}\left(u_2, v_2\right)\right)$$

$$+ \sum_{u_2 = v_2 = w, u_1 v_1 \in E} r\max\left(I_{A_2}^{+}\left(w\right), I_{B_1}^{+}\left(u_1, v_1\right)\right) + \sum_{u_1 = v_1 \in E, u_2 v_2 \in E_2} r\max\left(T_{B_1}^{+}\left(u_1, v_1\right), T_{B_2}^{+}\left(u_2, v_2\right)\right)$$

$$\deg\left(I_{A_1}^{-} \times I_{A_2}^{-}\right)\left(u_1, u_2\right)$$

$$= \sum_{\left(u_1 u_2\right)\left(v_1 v_2\right) \in E_2} r\min\left(I_{B_1}^{-} \times I_{B_2}^{-}\right)\left(\left(u_1, u_2\right)\left(v_1, v_2\right)\right) = \sum_{u_1 = v_1 = u, u_2 v_2 \in E_2} r\max\left(I_{B_1}^{-}\left(u\right), I_{B_2}^{-}\left(u_2, v_2\right)\right)$$

$$+ \sum_{u_2 = v_2 = w, u_1 v_1 \in E} r\max\left(I_{A_2}^{-}\left(w\right), I_{B_1}^{-}\left(u_1, v_1\right)\right) + \sum_{u_1 = v_1 \in E, u_2 v_2 \in E_2} r\max\left(T_{B_1}^{-}\left(u_1, v_1\right), T_{B_2}^{-}\left(u_2, v_2\right)\right)$$

$$\deg\left(I_{\lambda_1}^{+} \times I_{\lambda_2}^{+}\right)\left(u_1, u_2\right)$$

$$= \sum_{\left(u_1 u_2\right)\left(v_1 v_2\right) \in E_2} \min\left(I_{\mu_1}^{+} \times I_{\mu_2}^{+}\right)\left(\left(u_1, u_2\right)\left(v_1, v_2\right)\right) = \sum_{u_1 = v_1 = u, u_2 v_2 \in E_2} \min\left(I_{\lambda_1}^{+}\left(u\right), I_{\mu_2}^{+}\left(u_1, v_2\right)\right)$$

$$+ \sum_{u_2 = v_2 = w, u_1 v_1 \in E} \min\left(I_{\lambda_2}^{+}\left(w\right), I_{\mu_1}^{+}\left(u_1, v_1\right)\right) + \sum_{u_1 = v_1 \in E, u_2 v_2 \in E_2} \min\left(I_{\mu_1}^{+}\left(u_1, v_1\right), I_{\mu_2}^{+}\left(u_2, v_2\right)\right)$$

$$\deg\left(I_{\lambda_1}^- \times I_{\lambda_2}^-\right)\left(u_1, u_2\right)$$

$$= \sum_{\left(u_1 u_2\right)\left(v_1 v_2\right)\in E_2} \max\left(I_{\mu_1}^- \times I_{\mu_2}^-\right)\left(\left(u_1, u_2\right)\left(v_1, v_2\right)\right) = \sum_{u_1=v_1=u, u_2 v_2 \in E_2} \max\left(I_{\lambda_1}^-\left(u\right), I_{\mu_2}^-\left(u_1, v_2\right)\right)$$

$$+ \sum_{u_2=v_2=w, u_1 v_1 \in E} \max\left(I_{\lambda_2}^-\left(w\right), I_{\mu_1}^-\left(u_1, v_1\right)\right) + \sum_{u_1=v_1 \in E, u_2 v_2 \in E_2} \min\left(I_{\mu_1}^-\left(u_1, v_1\right), I_{\mu_2}^-\left(u_2, v_2\right)\right)$$

$$\deg\left(F_{A_1}^+ \times F_{A_2}^+\right)\left(u_1, u_2\right)$$

$$= \sum_{\left(u_1 u_2\right)\left(v_1 v_2\right)\in E_2} r\min\left(F_{B_1}^+ \times F_{B_2}^+\right)\left(\left(u_1, u_2\right)\left(v_1, v_2\right)\right) = \sum_{u_1=v_1=u, u_2 v_2 \in E_2} r\min\left(F_{\lambda_1}^+\left(u\right), F_{\mu_2}^+\left(u_2, v_2\right)\right)$$

$$+ \sum_{u_2=v_2=w, u_1 v_1 \in E} r\min\left(F_{\lambda_2}^+\left(w\right), F_{\mu_1}^+\left(u_1, v_1\right)\right) + \sum_{u_1=v_1 \in E, u_2 v_2 \in E_2} r\min\left(F_{\mu_1}^+\left(u_1, v_1\right), F_{\mu_2}^+\left(u_2, v_2\right)\right)$$

$$\deg\left(F_{A_1}^- \times F_{A_2}^-\right)\left(u_1, u_2\right)$$

$$= \sum_{\left(u_1 u_2\right)\left(v_1 v_2\right)\in E_2} r\max\left(F_{B_1}^- \times F_{B_2}^-\right)\left(\left(u_1, u_2\right)\left(v_1, v_2\right)\right) = \sum_{u_1=v_1=u, u_2 v_2 \in E_2} r\min\left(F_{\lambda_1}^-\left(u\right), F_{\mu_2}^-\left(u_2, v_2\right)\right)$$

$$+ \sum_{u_2=v_2=w, u_1 v_1 \in E} r\max\left(F_{\lambda_2}^-\left(w\right), F_{\mu_1}^-\left(u_1, v_1\right)\right) + \sum_{u_1=v_1 \in E, u_2 v_2 \in E_2} r\max\left(F_{\mu_1}^-\left(u_1, v_1\right), F_{\mu_2}^-\left(u_2, v_2\right)\right)$$

$$\deg\left(F_{\lambda_1}^+ \times F_{\lambda_2}^+\right)\left(u_1, u_2\right)$$

$$= \sum_{\left(u_1 u_2\right)\left(v_1 v_2\right)\in E_2} \max\left(F_{\mu_1}^+ \times I_{\mu_2}^+\right)\left(\left(u_1, u_2\right)\left(v_1, v_2\right)\right) = \sum_{u_1=v_1=u, u_2 v_2 \in E_2} \max\left(F_{\lambda_1}^+\left(u\right), F_{\mu_2}^+\left(u_1, v_2\right)\right)$$

$$+ \sum_{u_2=v_2=w, u_1 v_1 \in E} \max\left(F_{\lambda_2}^+\left(w\right), F_{\mu_1}^+\left(u_1, v_1\right)\right) + \sum_{u_1=v_1 \in E, u_2 v_2 \in E_2} \max\left(F_{\mu_1}^+\left(u_1, v_1\right), F_{\mu_2}^+\left(u_2, v_2\right)\right)$$

$$\deg\left(F_{\lambda_1}^- \times F_{\lambda_2}^-\right)\left(u_1, u_2\right)$$

$$= \sum_{\left(u_1 u_2\right)\left(v_1 v_2\right)\in E_2} \min\left(F_{\mu_1}^- \times I_{\mu_2}^-\right)\left(\left(u_1, u_2\right)\left(v_1, v_2\right)\right) = \sum_{u_1=v_1=u, u_2 v_2 \in E_2} \min\left(F_{\lambda_1}^-\left(u\right), F_{\mu_2}^+\left(u_1, v_2\right)\right)$$

$$+ \sum_{u_2=v_2=w, u_1 v_1 \in E} \min\left(F_{\lambda_2}^-\left(w\right), F_{\mu_1}^-\left(u_1, v_1\right)\right) + \sum_{u_1=v_1 \in E, u_2 v_2 \in E_2} \min\left(F_{\mu_1}^-\left(u_1, v_1\right), F_{\mu_2}^-\left(u_2, v_2\right)\right)$$

Definition

Let $G_1 = \left(P_1, Q_1\right)$ be a bipolar neutrosophic cubic graph of $G_1^* = \left(V_1^*, E_1^*\right)$ and $G_2 = \left(P_2, Q_2\right)$ be a bipolar neutrosophic cubic graph of $G_2^* = \left(V_2^*, E_2^*\right)$. Then the composition of G_1 and G_2 is denoted by $G_2[G_2]$ and defined as follows:

$$G_1\big[G_2\big] = \big(P_1,Q_1\big)\big[\big(P_2,Q_2\big)\big]$$
$$= \big\{P_1\big[P_2\big], Q_1\big[Q_2\big]\big\}$$
$$= \big\{\big(A_1,\lambda_1\big)\big[\big(A_2,\lambda_2\big)\big],\big(B_1,\mu_1\big)\big[\big(B_2,\mu_2\big)\big]\big\}$$
$$= \big\{\big(A_1\big[A_2\big],\lambda_1\big[\lambda_2\big]\big),\big(B_1\big[B_2\big],\mu_1\big[\mu_2\big]\big)\big\}$$

$$=\left[\begin{array}{l}\left\langle\begin{array}{l}\big(\big(T^+_{A_1}oT^+_{A_2}\big),\big(T^+_{\lambda_1}oT^+_{\lambda_2}\big)\big),\big(\big(I^+_{A_1}oI^+_{A_2}\big),\big(I^+_{\lambda_1}oI^+_{\lambda_2}\big)\big),\big(\big(F^+_{A_1}oF^+_{A_2}\big),\big(F^+_{\lambda_1}oF^+_{\lambda_2}\big)\big),\\ \big(\big(T^-_{A_1}oT^-_{A_2}\big),\big(T^-_{\lambda_1}oT^-_{\lambda_2}\big)\big),\big(\big(I^-_{A_1}oI^-_{A_2}\big),\big(I^-_{\lambda_1}oI^-_{\lambda_2}\big)\big),\big(\big(F^-_{A_1}oF^-_{A_2}\big),\big(F^-_{\lambda_1}oF^-_{\lambda_2}\big)\big)\end{array}\right\rangle\\ \left\langle\begin{array}{l}\big(\big(T^+_{B_1}oT^+_{B_2}\big),\big(T^+_{\mu_1 1}oT^+_{\mu_2}\big)\big),\big(\big(I^+_{B_1}oI^+_{B_2}\big),\big(I^+_{\mu_1}oI^+_{\mu_2}\big)\big),\big(\big(F^+_{B_1}oF^+_{B_2}\big),\big(F^+_{\mu_1}oF^+_{\mu_2}\big)\big),\\ \big(\big(T^-_{B_1}oT^-_{B_2}\big),\big(T^-_{\mu_1}oT^-_{\mu_2}\big)\big),\big(\big(I^-_{B_1}oI^-_{B_2}\big),\big(I^-_{\lambda_1}oI^-_{\lambda_2}\big)\big),\big(\big(F^-_{B_1}oF^-_{B_2}\big),\big(F^-_{\mu_1}oF^-_{\mu_2}\big)\big)\end{array}\right\rangle\end{array}\right]$$

1. $\forall\big(u,v\big)\in\big(v_1,v_2\big)=V$

$$\big(T^+_{A_1}oT^+_{A_2}\big)(u,v)=r\min\big(T^+_{A_1}(u),T^+_{A_2}(v)\big),\big(T^+_{\lambda_1}oT^+_{\lambda_2}\big)(u,v)=\max\big(T^+_{\lambda_1}(u),T^+_{\lambda_2}(v)\big)$$
$$\big(T^-_{A_1}oT^-_{A_2}\big)(u,v)=r\max\big(T^-_{A_1}(u),T^-_{A_2}(v)\big),\big(T^-_{\lambda_1}oT^-_{\lambda_2}\big)(u,v)=\min\big(T^-_{\lambda_1}(u),T^-_{\lambda_2}(v)\big)$$

$$\big(I^+_{A_1}oI^+_{A_2}\big)(u,v)=r\min\big(I^+_{A_1}(u),I^+_{\lambda_2}(v)\big),\big(I^+_{\lambda_1}oI^+_{\lambda_2}\big)(u,v)=\max\big(I^+_{\lambda_1}(u),I^+_{\lambda_2}(v)\big)$$
$$\big(I^-_{A_1}oI^-_{A_2}\big)(u,v)=r\max\big(I^-_{A_1}(u),I^-_{\lambda_2}(v)\big),\big(I^-_{\lambda_1}oI^-_{\lambda_2}\big)(u,v)=\min\big(I^-_{\lambda_1}(u),I^-_{\lambda_2}(v)\big)$$

$$\big(F^+_{A_1}oF^+_{A_2}\big)(u,v)=r\max\big(F^+_{A_1}(u),F^+_{\lambda_2}(v)\big),\big(F^+_{\lambda_1}oF^+_{\lambda_2}\big)(u,v)=\max\big(F^+_{\lambda_1}(u),F^+_{\lambda_2}(v)\big)$$
$$\big(F^-_{A_1}oF^-_{A_2}\big)(u,v)=r\min\big(F^-_{A_1}(u),F^-_{\lambda_2}(v)\big),\big(F^-_{\lambda_1}oF^-_{\lambda_2}\big)(u,v)=\min\big(F^-_{\lambda_1}(u),F^-_{\lambda_2}(v)\big)$$

2. $\forall u\in V_1$ and $v_1v_2\in E$

$$\big(T^+_{B_1}oT^+_{B_2}\big)\big((u,v_1)(u,v_2)\big)=r\min\big(T^+_{A_1}(u),T^+_{B_2}(v_1v_2)\big),$$
$$\big(T^+_{\mu_1}oT^+_{\mu_2}\big)\big((u,v_1)(u,v_2)\big)=\max\big(T^+_{\lambda_1}(u),T^+_{\mu_2}(v_1v_2)\big)$$

$$\big(T^-_{B_1}oT^-_{B_2}\big)\big((u,v_1)(u,v_2)\big)=r\max\big(T^-_{A_1}(u),T^-_{B_2}(v_1v_2)\big),$$
$$\big(T^-_{\mu_1}oT^-_{\mu_2}\big)\big((u,v_1)(u,v_2)\big)=\min\big(T^-_{\lambda_1}(u),T^-_{\mu_2}(v_1v_2)\big)$$

$$\left(F_{B_1}^+ o F_{B_2}^+\right)\left((u,v_1)(u,v_2)\right) = r\max\left(F_{A_1}^+(u), F_{B_2}^+(v_1v_2)\right),$$
$$\left(F_{\mu_1}^+ o F_{\mu_2}^+\right)\left((u,v_1)(u,v_2)\right) = \min\left(F_{\lambda_1}^+(u), F_{\mu_2}^+(v_1v_2)\right)$$

$$\left(F_{B_1}^- o F_{B_2}^-\right)\left((u,v_1)(u,v_2)\right) = r\max\left(F_{A_1}^-(u), F_{B_2}^-(v_1v_2)\right),$$
$$\left(F_{\mu_1}^- o F_{\mu_2}^-\right)\left((u,v_1)(u,v_2)\right) = \min\left(F_{\lambda_1}^-(u), F_{\mu_2}^-(v_1v_2)\right)$$

3. $v \in V_2$ and $u_1 u_2 \in E_1$

$$\left(T_{B_1}^+ o T_{B_2}^+\right)\left((u_1,v)(u_2,v)\right) = r\min\left(T_{B_1}^+(u_1u_2), T_{A_2}^+(v)\right),$$
$$\left(T_{\mu_1}^+ o T_{\mu_2}^+\right)\left((u_1,v)(u_2,v)\right) = \max\left(T_{\mu_1}^+(u_1,u_2), T_{\lambda_1}^+(v)\right)$$

$$\left(T_{B_1}^- o T_{B_2}^-\right)\left((u_1,v)(u_2,v)\right) = r\max\left(T_{B_1}^-(u_1,u_2), T_{A_2}^-(v)\right),$$
$$\left(T_{\mu_1}^- o T_{\mu_2}^-\right)\left((u_1,v)(u_2,v)\right) = \min\left(T_{\mu_1}^-(u_1,u_2), T_{\lambda_2}^-(v)\right)$$

$$\left(I_{B_1}^+ o I_{B_2}^+\right)\left((u_1,v)(u_2,v)\right) = r\min\left(I_{B_1}^+(u_1u_2), I_{A_2}^+(v)\right),$$
$$\left(I_{\mu_1}^+ o I_{\mu_2}^+\right)\left((u_1,v)(u_2,v)\right) = \max\left(I_{\mu_1}^+(u_1,u_2), I_{\lambda_2}^+(v)\right)$$

$$\left(I_{B_1}^- o I_{B_2}^-\right)\left((u_1,v)(u_2,v)\right) = r\max\left(I_{B_1}^-(u_1,u_2), I_{A_2}^-(v)\right),$$
$$\left(I_{\mu_1}^- o I_{\mu_2}^-\right)\left((u_1,v)(u_2,v)\right) = \min\left(I_{\mu_1}^-(u_1,u_2), I_{\lambda_2}^-(v)\right)$$

$$\left(F_{B_1}^+ o F_{B_2}^+\right)\left((u_1,v)(u_2,v)\right) = r\max\left(F_{B_1}^+(u_1u_2), F_{A_2}^+(v)\right),$$
$$\left(F_{\mu_1}^+ o F_{\mu_2}^+\right)\left((u_1,v)(u_2,v)\right) = \min\left(F_{\mu_1}^+(u_1,u_2), F_{\lambda_2}^+(v)\right)$$

$$\left(F_{B_1}^- o F_{B_2}^-\right)\left((u_1,v)(u_2,v)\right) = r\min\left(F_{B_1}^-(u_1,u_2), F_{A_2}^-(v)\right),$$
$$\left(F_{\mu_1}^- o F_{\mu_2}^-\right)\left((u_1,v)(u_2,v)\right) = \max\left(F_{\mu_1}^-(u_1,u_2), F_{\lambda_2}^-(v)\right)$$

4. $\forall (u_1 v_1)(u_2 v_2) \in E^O - E$

$$\left(T_{B_1}^+ o T_{B_2}^+\right)\left((u_1,v_1)(u_2,v_2)\right) = r\min\left(T_{A_2}^+(v_1), T_{A_2}^+(v_2), T_{B_1}^+(u_1,v_2)\right)$$
$$\left(T_{\mu_1}^+ o T_{\mu_2}^+\right)\left((u_1,v_1)(u_2,v_2)\right) = \max\left(T_{\lambda_2}^+(v_1), T_{\lambda_2}^+(v_2), T_{\mu_1}^+(u_1,v_2)\right)$$

$$\left(T_{B_1}^- o T_{B_2}^-\right)\left((u_1,v_1)(u_2,v_2)\right) = r\max\left(T_{A_2}^-(v_1), T_{A_2}^-(v_2), T_{B_1}^-(u_1,v_2)\right)$$

$$\left(T_{\mu_1}^- o T_{\mu_2}^-\right)\left((u_1,v_1)(u_2,v_2)\right) = \min\left(T_{\lambda_2}^-(u_1,u_2), T_{\lambda_2}^-(v_2), T_{\mu_1}^+(u_1,v_2)\right)$$

$$\left(I_{B_1}^+ o I_{B_2}^+\right)\left((u_1,v_1)(u_2,v_2)\right) = r\min\left(I_{A_2}^+(v_1), I_{A_2}^+(v_2), I_{B_1}^+(u_1,v_2)\right)$$

$$\left(I_{\mu_1}^+ o I_{\mu_2}^+\right)\left((u_1,v_1)(u_2,v_2)\right) = \max\left(I_{\lambda_2}^+(v_1), I_{\lambda_2}^+(v_2), I_{\mu_1}^+(u_1,v_2)\right)$$

$$\left(I_{B_1}^- o I_{B_2}^-\right)\left((u_1,v_1)(u_2,v_2)\right) = r\max\left(I_{A_2}^-(v_1), I_{A_2}^-(v_2), I_{B_1}^-(u_1,v_2)\right)$$

$$\left(I_{\mu_1}^- o I_{\mu_2}^-\right)\left((u_1,v_1)(u_2,v_2)\right) = \min\left(I_{\lambda_2}^-(u_1,u_2), I_{\lambda_2}^-(v_2), I_{\mu_1}^+(u_1,v_2)\right)$$

$$\left(F_{B_1}^+ o F_{B_2}^+\right)\left((u_1,v_1)(u_2,v_2)\right) = r\max\left(F_{A_2}^+(v_1), F_{A_2}^+(v_2), F_{B_1}^+(u_1,v_2)\right)$$

$$\left(F_{\mu_1}^+ o F_{\mu_2}^+\right)\left((u_1,v_1)(u_2,v_2)\right) = \max\left(F_{\lambda_2}^+(v_1), F_{\lambda_2}^+(v_2), F_{\mu_1}^+(u_1,v_2)\right)$$

$$\left(F_{B_1}^- o F_{B_2}^-\right)\left((u_1,v_1)(u_2,v_2)\right) = r\max\left(F_{A_2}^-(v_1), F_{A2}^-(v_2), F_{B_1}^-(u_1,v_2)\right)$$

$$\left(F_{\mu_1}^- o F_{\mu_2}^-\right)\left((u_1,v_1)(u_2,v_2)\right) = \min\left(F_{\lambda_2}^-(u_1,u_2), F_{\lambda_2}^-(v_2), F_{\mu_1}^+(u_1,v_2)\right)$$

Example

Let $G_1^* = (V_1, E_1)$ and $G_2^* = (V_2, E_2)$ be two bipolar neutrosophic cubic graphs, where $V_1 = (a,b)$ and $V_2 = (c,d)$. Suppose P_1 and P_2 be the bipolar neutrosophic cubic set representations of V_1 And V_2. Also Q_1 and Q_2 be the bipolar neutrosophic cubic set representations of E_1 and E_2 defined as follows:

$$P_1 = \left\langle \begin{matrix} a, \left([0.5,0.6], 0.2\right), \left([0.1,0.2], 0.4\right), \left([0.8,0.9], 0.2\right), \\ \left([-0.8,-0.7], -0.6\right), \left([-0.7,-0.3], -0.1\right), \left([-0.9,-0.6], -0.3\right) \\ b, \left([0.4,0.5], 0.4\right), \left([0.2,0.3], 0.1\right), \left([0.5,0.6], 0.6\right), \\ \left([-0.5,-0.2], -0.1\right), \left([-0.3,-0.2], -0.2\right), \left([-0.6,-0.5], -0.2\right) \end{matrix} \right\rangle$$

$$Q_1 = \left\langle \begin{matrix} ab, \left([0.4,0.5], 0.4\right), \left([0.1,0.2], 0.4\right), \left([0.8,0.9], 0.2\right), \\ \left([-0.5,-0.2], -0.6\right), \left([-0.3,-0.2], -0.1\right), \left([-0.9,-0.6], -0.2\right) \end{matrix} \right\rangle$$

$$P_2 = \left\langle \begin{bmatrix} c, \big([0.6, 0.8], 0.4\big), \big([0.7, 0.8], 0.8\big), \big([0.2, 0.3], 0.5\big), \\ \big([-0.8, -0.6], -0.2\big), \big([-0.9, -0.6], -0.5\big), \big([-0.4, -0.2], -0.1\big) \end{bmatrix} \\ \begin{bmatrix} d, \big([0.3, 0.4], 0.7\big), \big([0.6, 0.7], 0.5\big), \big([0.9, 1.0], 0.9\big), \\ \big([-0.5, -0.4], -0.1\big), \big([-0.6, -0.5], -0.4\big), \big([-0.3, -0.2], -0.1\big) \end{bmatrix} \right\rangle$$

$$Q_2 = \left\langle \begin{bmatrix} cd, \big([0.3, 0.4], 0.7\big), \big([0.6, 0.7], 0.8\big), \big([0.9, 1.0], 0.9\big), \\ \big([-0.5, -0.4], -0.1\big), \big([-0.6, -0.5], -0.4\big), \big([-0.3, -0.2], -0.1\big) \end{bmatrix} \right\rangle$$

Figure 4.

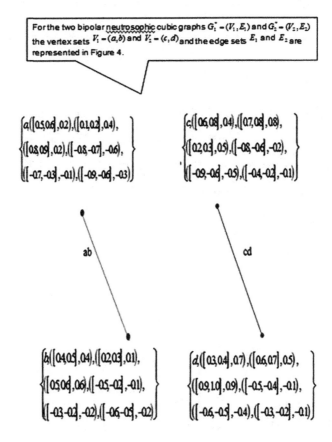

For the two bipolar neutrosophic cubic graphs $G_1^* = (V_1, E_1)$ and $G_2^* = (V_2, E_2)$ the vertex sets $V_1 = (a, b)$ and $V_2 = (c, d)$ and the edge sets E_1 and E_2 are represented in Figure 4.

The composition of two bipolar neutrosophic cubic graphs G_1 and G_2 is again a bipolar neutrosophic cubic graph, where

$$P_1[P_2] = \left\langle \begin{array}{l} \left[\begin{array}{l} (a,c), \left([0.5,0.6], 0.4 \right), \left([0.1,0.2], 0.8 \right), \left([0.2,0.3], 0.5 \right), \\ \left([-0.8,-0.6], -0.6 \right), \left([-0.3,-0.2], -0.1 \right), \left([-0.9,-0.6], -0.2 \right) \end{array} \right] \\ \left[\begin{array}{l} (a,d), \left([0.3,0.4], 0.7 \right), \left([0.1,0.2], 0.5 \right), \left([0.8,0.9], 0.9 \right), \\ \left([-0.5,-0.4], -0.6 \right), \left([-0.6,-0.3], -0.4 \right), \left([-0.9,-0.6], -0.1 \right) \end{array} \right] \\ \left[\begin{array}{l} (b,c), \left([0.4,0.5], 0.4 \right), \left([0.2,0.3], 0.8 \right), \left([0.5,0.6], 0.5 \right), \\ \left([-0.5,-0.2], -0.2 \right), \left([-0.3,-0.2], -0.5 \right), \left([-0.6,-0.5], -0.2 \right) \end{array} \right] \\ \left[\begin{array}{l} (b,d), \left([0.3,0.4], 0.7 \right), \left([0.2,0.3], 0.5 \right), \left([0.9,1.0], 0.6 \right), \\ \left([-0.5,-0.2], -0.1 \right), \left([-0.3,-0.2], -0.4 \right), \left([-0.6,-0.5], -0.1 \right) \end{array} \right] \end{array} \right\rangle$$

$$Q_1[Q_2] = \left\langle \begin{array}{l} \left[\begin{array}{l} ((a,c)(a,d)), \left([0.3,0.4], 0.7 \right), \left([0.1,0.2], 0.8 \right), \left([0.8,0.9], 0.5 \right), \\ \left([-0.5,-0.4], -0.6 \right), \left([-0.3,-0.2], -0.4 \right), \left([-0.9,-0.6], -0.1 \right) \end{array} \right] \\ \left[\begin{array}{l} ((a,d)(b,d)), \left([0.3,0.4], 0.7 \right), \left([0.2,0.3], 0.5 \right), \left([0.9,1.0], 0.6 \right), \\ \left([-0.5,-0.2], -0.6 \right), \left([-0.3,-0.2], -0.4 \right), \left([-0.9,-0.6], -0.1 \right) \end{array} \right] \\ \left[\begin{array}{l} ((b,d)(b,c)), \left([0.3,0.4], 0.7 \right), \left([0.2,0.3], 0.8 \right), \left([0.9,1.0], 0.5 \right), \\ \left([-0.5,-0.2], -0.2 \right), \left([-0.3,-0.2], -0.5 \right), \left([-0.6,-0.5], -0.1 \right) \end{array} \right] \\ \left[\begin{array}{l} ((b,c)(a,c)), \left([0.4,0.5], 0.4 \right), \left([0.2,0.3], 0.8 \right), \left([0.5,0.6], 0.5 \right), \\ \left([-0.5,-0.2], -0.6 \right), \left([-0.3,-0.2], -0.5 \right), \left([-0.9,-0.6], -0.2 \right) \end{array} \right] \\ \left[\begin{array}{l} ((a,c)(b,d)), \left([0.3,0.4], 0.7 \right), \left([0.1,0.2], 0.8 \right), \left([0.9,1.0], 0.5 \right), \\ \left([-0.5,-0.2], -0.6 \right), \left([-0.3,-0.2], -0.4 \right), \left([-0.9,-0.6], -0.1 \right) \end{array} \right] \\ \left[\begin{array}{l} ((a,d)(b,c)), \left([0.3,0.4], 0.7 \right), \left([0.1,0.2], 0.8 \right), \left([0.8,0.9], 0.5 \right), \\ \left([-0.5,-0.2], -0.6 \right), \left([-0.3,-0.2], -0.5 \right), \left([-0.9,-0.6], -0.1 \right) \end{array} \right] \end{array} \right\rangle$$

Definition: $G_1 = (P_1, Q_1)$ and $G_2 = (P_2, Q_2)$ be two bipolar neutrosophic cubic graphs of the graph G_1^* and G_2^* respectively. Then M- union is denoted by $G_1 U_M G_2$ and is defined as

Figure 5.

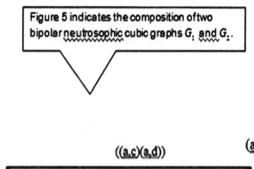

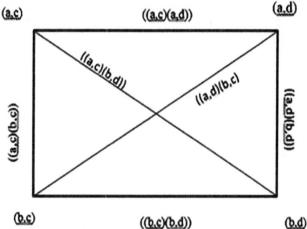

Composition of G_1 and G_2

$$G_1 U_M G_2 = \left\{ (P_1, Q_1) U_M (P_2, Q_2) \right\} = \left\{ P_1 U_M P_2, Q_1 U_M Q_2 \right\}$$

$$= \begin{cases} \left\langle \begin{array}{l} ((T_{A_1}^+ \cup_M T_{A_2}^+), (T_{\lambda_1}^+ \cup_M T_{\lambda_2}^+)), ((I_{A_1}^+ \cup_M I_{A_2}^+), (I_{\lambda_1}^+ \cup_M I_{\lambda_2}^+)), ((F_{A_1}^+ \cup_M F_{A_2}^+), (F_{\lambda_1}^+ \cup_M F_{\lambda_2}^+)) \\ ((T_{A_1}^- \cup_M T_{A_2}^-), (T_{\lambda_1}^- \cup_M T_{\lambda_2}^-)), ((I_{A_1}^- \cup_M I_{A_2}^-), (I_{\lambda_1}^- \cup_M I_{\lambda_2}^-)), ((F_{A_1}^- \cup_M F_{A_2}^-), (F_{\lambda_1}^- \cup_M F_{\lambda_2}^-)) \end{array} \right\rangle \\ \left\langle \begin{array}{l} ((T_{B_1}^+ \cup_M T_{B_2}^+), (T_{\mu_1}^+ \cup_M T_{\mu_2}^+)), ((I_{B_1}^+ \cup_M I_{B_2}^+), (I_{\mu_1}^+ \cup_M I_{\mu_2}^+)), ((F_{B_1}^+ \cup_M F_{B_2}^+), (F_{\mu_1}^+ \cup_M F_{\mu_2}^+)) \\ ((T_{B_1}^- \cup_M T_{B_2}^-), (T_{\mu_1}^- \cup_M T_{\mu_2}^-)), ((I_{B_1}^- \cup_M I_{B_2}^-), (I_{\mu_1}^- \cup_M I_{\mu_2}^-)), ((F_{B_1}^- \cup_M F_{B_2}^-), (F_{\mu_1}^- \cup_M F_{\mu_2}^-)) \end{array} \right\rangle \end{cases}$$

where

$$(T_{A_1}^+ \cup_M T_{A_2}^+)(u) = \begin{cases} T_{A_1}^+(u), & , if \quad u \in v_1 - v_2 \\ T_{A_2}^+(u), & , if \quad u \in v_2 - v_1 \\ r \max \left\{ T_{A_1}^+(u), T_{A_2}^+(u) \right\}, & if \quad u \in v_1 \cap v_2 \end{cases}$$

$$(T_{A_1}^- \cup_M T_{A_2}^-)(u) = \begin{cases} T_{A_1}^-(u), & ,if \quad u \in v_1 - v_2 \\ T_{A_2}^-(u), & ,if \quad u \in v_2 - v_1 \\ r\min\left\{T_{A_1}^-(u), T_{A_2}^-(u)\right\}, & ,if \quad u \in v_1 \cap v_2 \end{cases}$$

$$(T_{\lambda_1}^+ \cup_M T_{\lambda_2}^+)(u) = \begin{cases} T_{\lambda_1}^+(u), & ,if \quad u \in v_1 - v_2 \\ T_{\lambda_2}^+(u), & ,if \quad u \in v_2 - v_1 \\ \max\left\{T_{\lambda_1}^+(u), T_{\lambda_2}^+(u)\right\}, & ,if \quad u \in v_1 \cap v_2 \end{cases}$$

$$(T_{\lambda_1}^- \cup_M T_{\lambda_2}^-)(u) = \begin{cases} T_{\lambda_1}^-(u), & ,if \quad u \in v_1 - v_2 \\ T_{\lambda_2}^-(u), & ,if \quad u \in v_2 - v_1 \\ \min\left\{T_{\lambda_1}^-(u), T_{\lambda_2}^-(u)\right\}, & ,if \quad u \in v_1 \cap v_2 \end{cases}$$

$$(I_{A_1}^+ \cup_M I_{A_2}^+)(u) = \begin{cases} I_{A_1}^+(u), & ,if \quad u \in v_1 - v_2 \\ I_{A_2}^+(u), & ,if \quad u \in v_2 - v_1 \\ r\max\left\{T_{A_1}^+(u), T_{A_2}^+(u)\right\}, & ,if \quad u \in v_1 \cap v_2 \end{cases}$$

$$(I_{A_1}^- \cup_M I_{A_2}^-)(u) = \begin{cases} I_{A_1}^-(u), & ,if \quad u \in v_1 - v_2 \\ I_{A_2}^-(u), & ,if \quad u \in v_2 - v_1 \\ r\min\left\{I_{A_1}^-(u), I_{A_2}^-(u)\right\}, & ,if \quad u \in v_1 \cap v_2 \end{cases}$$

$$(I_{\lambda_1}^+ \cup_M I_{\lambda_2}^+)(u) = \begin{cases} I_{\lambda_1}^+(u), & ,if \quad u \in v_1 - v_2 \\ I_{\lambda_2}^+(u), & ,if \quad u \in v_2 - v_1 \\ \max\left\{I_{\lambda_1}^+(u), I_{\lambda_2}^+(u)\right\}, & ,if \quad u \in v_1 \cap v_2 \end{cases}$$

$$(I_{\lambda_1}^- \cup_M I_{\lambda_2}^-)(u) = \begin{cases} I_{\lambda_1}^-(u), & ,if \quad u \in v_1 - v_2 \\ I_{\lambda_2}^-(u), & ,if \quad u \in v_2 - v_1 \\ \min\left\{I_{\lambda_1}^-(u), I_{\lambda_2}^-(u)\right\}, & ,if \quad u \in v_1 \cap v_2 \end{cases}$$

$$(F_{A_1}^+ \cup_M F_{A_2}^+)(u) = \begin{cases} F_{A_1}^+(u), & , if \quad u \in v_1 - v_2 \\ F_{A_2}^+(u), & , if \quad u \in v_2 - v_1 \\ r\max\left\{F_{A_1}^+(u), F_{A_2}^+(u)\right\}, if \quad u \in v_1 \cap v_2 \end{cases}$$

$$(F_{A_1}^- \cup_M F_{A_2}^-)(u) = \begin{cases} F_{A_1}^-(u), & , if \quad u \in v_1 - v_2 \\ F_{A_2}^-(u), & , if \quad u \in v_2 - v_1 \\ r\min\left\{F_{A_1}^-(u), F_{A_2}^-(u)\right\}, if \quad u \in v_1 \cap v_2 \end{cases}$$

$$(F_{\lambda_1}^+ \cup_M F_{\lambda_2}^+)(u) = \begin{cases} F_{\lambda_1}^+(u), & , if \quad u \in v_1 - v_2 \\ F_{\lambda_2}^+(u), & , if \quad u \in v_2 - v_1 \\ \max\left\{F_{\lambda_1}^+(u), F_{\lambda_2}^+(u)\right\}, if \quad u \in v_1 \cap v_2 \end{cases}$$

$$(F_{\lambda_1}^- \cup_M F_{\lambda_2}^-)(u) = \begin{cases} F_{\lambda_1}^-(u), & , if \quad u \in v_1 - v_2 \\ F_{\lambda_2}^-(u), & , if \quad u \in v_2 - v_1 \\ \min\left\{F_{\lambda_1}^-(u), F_{\lambda_2}^-(u)\right\}, if \quad u \in v_1 \cap v_2 \end{cases}$$

$$(T_{B_1}^+ \cup_M T_{B_2}^+)(u_2 v_2) = \begin{cases} T_{B_1}^+(u_2 v_2), & , if \quad u_2 v_2 \in v_1 - v_2 \\ T_{B_2}^+(u_2 v_2), & , if \quad u_2 v_2 \in v_2 - v_1 \\ r\max\left\{T_{B_1}^+(u_2 v_2), T_{B_2}^+(u_2 v_2)\right\}, if \quad u_2 v_2 \in E_1 \cap E_2 \end{cases}$$

$$(T_{B_1}^- \cup_M T_{B_2}^-)(u_2 v_2) = \begin{cases} T_{B_1}^-(u_2 v_2), & , if \quad u_2 v_2 \in v_1 - v_2 \\ T_{B_2}^-(u_2 v_2), & , if \quad u_2 v_2 \in v_2 - v_1 \\ r\min\left\{T_{B_1}^-(u_2 v_2), T_{B_2}^-(u_2 v_2)\right\}, if \quad u_2 v_2 \in E_1 \cap E_2 \end{cases}$$

$$(T_{\mu_1}^+ \cup_M T_{\mu_2}^+)(u_2 v_2) = \begin{cases} T_{\mu_1}^+(u_2 v_2), & , if \quad u_2 v_2 \in v_1 - v_2 \\ T_{\mu_2}^+(u_2 v_2), & , if \quad u_2 v_2 \in v_2 - v_1 \\ \max\left\{T_{\mu_1}^+(u_2 v_2), T_{\mu_2}^+(u_2 v_2)\right\}, if \quad u_2 v_2 \in E_1 \cap E_2 \end{cases}$$

$$
(T_{\mu_1}^- \cup_M T_{\mu_2}^-)(u_2 v_2) = \begin{cases} T_{\mu_1}^-(u_2 v_2), & , if \quad u_2 v_2 \in v_1 - v_2 \\ T_{\mu_2}^-(u_2 v_2), & , if \quad u_2 v_2 \in v_2 - v_1 \\ \min\left\{ T_{\mu_1}^-(u_2 v_2), T_{\mu_2}^-(u_2 v_2) \right\} & , if \quad u_2 v_2 \in E_1 \cap E_2 \end{cases}
$$

$$
(I_{B_1}^+ \cup_M I_{B_2}^+)(u_2 v_2) = \begin{cases} I_{B_1}^+(u_2 v_2), & , if \quad u_2 v_2 \in v_1 - v_2 \\ I_{B_2}^+(u_2 v_2), & , if \quad u_2 v_2 \in v_2 - v_1 \\ r \max\left\{ I_{B_1}^+(u_2 v_2), I_{B_2}^+(u_2 v_2) \right\} & , if \quad u_2 v_2 \in E_1 \cap E_2 \end{cases}
$$

$$
(I_{B_1}^- \cup_M I_{B_2}^-)(u_2 v_2) = \begin{cases} I_{B_1}^-(u_2 v_2), & , if \quad u_2 v_2 \in v_1 - v_2 \\ I_{B_2}^-(u_2 v_2), & , if \quad u_2 v_2 \in v_2 - v_1 \\ r \min\left\{ I_{B_1}^-(u_2 v_2), I_{B_2}^-(u_2 v_2) \right\} & , if \quad u_2 v_2 \in E_1 \cap E_2 \end{cases}
$$

$$
(I_{\mu_1}^+ \cup_M I_{\mu_2}^+)(u_2 v_2) = \begin{cases} I_{\mu_1}^+(u_2 v_2), & , if \quad u_2 v_2 \in v_1 - v_2 \\ I_{\mu_2}^+(u_2 v_2), & , if \quad u_2 v_2 \in v_2 - v_1 \\ \max\left\{ I_{\mu_1}^+(u_2 v_2), I_{\mu_2}^+(u_2 v_2) \right\} & , if \quad u_2 v_2 \in E_1 \cap E_2 \end{cases}
$$

$$
(I_{\mu_1}^- \cup_M I_{\mu_2}^-)(u_2 v_2) = \begin{cases} I_{\mu_1}^-(u_2 v_2), & , if \quad u_2 v_2 \in v_1 - v_2 \\ I_{\mu_2}^-(u_2 v_2), & , if \quad u_2 v_2 \in v_2 - v_1 \\ \min\left\{ I_{\mu_1}^-(u_2 v_2), I_{\mu_2}^-(u_2 v_2) \right\} & , if \quad u_2 v_2 \in E_1 \cap E_2 \end{cases}
$$

$$
(F_{B_1}^+ \cup_M F_{B_2}^+)(u_2 v_2) = \begin{cases} F_{B_1}^+(u_2 v_2), & , if \quad u_2 v_2 \in v_1 - v_2 \\ F_{B_2}^+(u_2 v_2), & , if \quad u_2 v_2 \in v_2 - v_1 \\ r \max\left\{ F_{B_1}^+(u_2 v_2), F_{B_2}^+(u_2 v_2) \right\} & , if \quad u_2 v_2 \in E_1 \cap E_2 \end{cases}
$$

$$
(F_{B_1}^- \cup_M F_{B_2}^-)(u_2 v_2) = \begin{cases} F_{B_1}^-(u_2 v_2), & , if \quad u_2 v_2 \in v_1 - v_2 \\ F_{B_2}^-(u_2 v_2), & , if \quad u_2 v_2 \in v_2 - v_1 \\ r \min\left\{ F_{B_1}^-(u_2 v_2), F_{B_2}^-(u_2 v_2) \right\} & , if \quad u_2 v_2 \in E_1 \cap E_2 \end{cases}
$$

$$(F_{\mu_1}^+ \cup_M F_{\mu_2}^+)(u_2 v_2) = \begin{cases} F_{\mu_1}^+(u_2 v_2), & ,if \quad u_2 v_2 \in v_1 - v_2 \\ F_{\mu_2}^+(u_2 v_2), & ,if \quad u_2 v_2 \in v_2 - v_1 \\ \max\left\{F_{\mu_1}^+(u_2 v_2), F_{\mu_2}^+(u_2 v_2)\right\} & ,if \quad u_2 v_2 \in E_1 \cap E_2 \end{cases}$$

$$(F_{\mu_1}^- \cup_M F_{\mu_2}^-)(u_2 v_2) = \begin{cases} F_{\mu_1}^-(u_2 v_2), & ,if \quad u_2 v_2 \in v_1 - v_2 \\ F_{\mu_2}^-(u_2 v_2), & ,if \quad u_2 v_2 \in v_2 - v_1 \\ \min\left\{F_{\mu_1}^-(u_2 v_2), F_{\mu_2}^-(u_2 v_2)\right\} & ,if \quad u_2 v_2 \in E_1 \cap E_2 \end{cases}$$

And the N-union is denoted by $G_1 \cup_N G_2$ and is defined as follows:

$$G_1 U_N G_2 = \left\{(P_1, Q_1) U_N (P_2, Q_2)\right\} = \left\{P_1 U_N P_2, Q_1 U_N Q_2\right\}$$

$$\begin{bmatrix} \left\langle \begin{matrix} ((T_{A_1}^+ \cup_N T_{A_2}^+),(T_{\lambda_1}^+ \cup_N T_{\lambda_2}^+)),((I_{A_1}^+ \cup_N I_{A_2}^+),(I_{\lambda_1}^+ \cup_N I_{\lambda_2}^+)),((F_{A_1}^+ \cup_N F_{A_2}^+),(F_{\lambda_1}^+ \cup_N F_{\lambda_2}^+)) \\ ((T_{A_1}^- \cup_N T_{A_2}^-),(T_{\lambda_1}^- \cup_N T_{\lambda_2}^-)),((I_{A_1}^- \cup_N I_{A_2}^-),(I_{\lambda_1}^- \cup_N I_{\lambda_2}^-)),((F_{A_1}^- \cup_N F_{A_2}^-),(F_{\lambda_1}^- \cup_N F_{\lambda_2}^-)) \end{matrix} \right\rangle \\ \left\langle \begin{matrix} ((T_{B_1}^+ \cup_N T_{B_2}^+),(T_{\mu_1}^+ \cup_N T_{\mu_2}^+)),((I_{B_1}^+ \cup_N I_{B_2}^+),(I_{\mu_1}^+ \cup_N I_{\mu_2}^+)),((F_{B_1}^+ \cup_N F_{B_2}^+),(F_{\mu_1}^+ \cup_N F_{\mu_2}^+)) \\ ((T_{B_1}^- \cup_N T_{B_2}^-),(T_{\mu_1}^- \cup_N T_{\mu_2}^-)),((I_{B_1}^- \cup_N I_{B_2}^-),(I_{\mu_1}^- \cup_N I_{\mu_2}^-)),((F_{B_1}^- \cup_N F_{B_2}^-),(F_{\mu_1}^- \cup_N F_{\mu_2}^-)) \end{matrix} \right\rangle \end{bmatrix}$$

$$(T_{A_1}^+ \cup_N T_{A_2}^+)(u) = \begin{cases} T_{A_1}^+(u), & ,if \quad u \in v_1 - v_2 \\ T_{A_2}^+(u), & ,if \quad u \in v_2 - v_1 \\ r\max\left\{T_{A_1}^+(u), T_{A_2}^+(u)\right\} & ,if \quad u \in v_1 \cap v_2 \end{cases}$$

$$(T_{A_1}^- \cup_N T_{A_2}^-)(u) = \begin{cases} T_{A_1}^-(u), & ,if \quad u \in v_1 - v_2 \\ T_{A_2}^-(u), & ,if \quad u \in v_2 - v_1 \\ r\min\left\{T_{A_1}^-(u), T_{A_2}^-(u)\right\} & ,if \quad u \in v_1 \cap v_2 \end{cases}$$

$$(T_{\lambda_1}^+ \cup_N T_{\lambda_2}^+)(u) = \begin{cases} T_{\lambda_1}^+(u), & ,if \quad u \in v_1 - v_2 \\ T_{\lambda_2}^+(u), & ,if \quad u \in v_2 - v_1 \\ \min\left\{T_{\lambda_1}^+(u), T_{\lambda_2}^+(u)\right\} & ,if \quad u \in v_1 \cap v_2 \end{cases}$$

$$(T_{\lambda_1}^- \cup_N T_{\lambda_2}^-)(u) = \begin{cases} T_{\lambda_1}^-(u), & , if \quad u \in v_1 - v_2 \\ T_{\lambda_2}^-(u), & , if \quad u \in v_2 - v_1 \\ \max\left\{ T_{\lambda_1}^-(u), T_{\lambda_2}^-(u) \right\}, & if \quad u \in v_1 \cap v_2 \end{cases}$$

$$(I_{A_1}^+ \cup_N I_{A_2}^+)(u) = \begin{cases} I_{A_1}^+(u), & , if \quad u \in v_1 - v_2 \\ I_{A_2}^+(u), & , if \quad u \in v_2 - v_1 \\ r\max\left\{ T_{A_1}^+(u), T_{A_2}^+(u) \right\}, & , if \quad u \in v_1 \cap v_2 \end{cases}$$

$$(I_{A_1}^- \cup_N I_{A_2}^-)(u) = \begin{cases} I_{A_1}^-(u), & , if \quad u \in v_1 - v_2 \\ I_{A_2}^-(u), & , if \quad u \in v_2 - v_1 \\ r\min\left\{ I_{A_1}^-(u), I_{A_2}^-(u) \right\}, & , if \quad u \in v_1 \cap v_2 \end{cases}$$

$$(I_{\lambda_1}^+ \cup_N I_{\lambda_2}^+)(u) = \begin{cases} I_{\lambda_1}^+(u), & , if \quad u \in v_1 - v_2 \\ I_{\lambda_2}^+(u), & , if \quad u \in v_2 - v_1 \\ \min\left\{ I_{\lambda_1}^+(u), I_{\lambda_2}^+(u) \right\}, & , if \quad u \in v_1 \cap v_2 \end{cases}$$

$$(I_{\lambda_1}^- \cup_N I_{\lambda_2}^-)(u) = \begin{cases} I_{\lambda_1}^-(u), & , if \quad u \in v_1 - v_2 \\ I_{\lambda_2}^-(u), & , if \quad u \in v_2 - v_1 \\ \max\left\{ I_{\lambda_1}^-(u), I_{\lambda_2}^-(u) \right\}, & , if \quad u \in v_1 \cap v_2 \end{cases}$$

$$(F_{A_1}^+ \cup_N F_{A_2}^+)(u) = \begin{cases} F_{A_1}^+(u), & , if \quad u \in v_1 - v_2 \\ F_{A_2}^+(u), & , if \quad u \in v_2 - v_1 \\ r\max\left\{ F_{A_1}^+(u), F_{A_2}^+(u) \right\}, & , if \quad u \in v_1 \cap v_2 \end{cases}$$

$$(F_{A_1}^- \cup_N F_{A_2}^-)(u) = \begin{cases} F_{A_1}^-(u), & , if \quad u \in v_1 - v_2 \\ F_{A_2}^-(u), & , if \quad u \in v_2 - v_1 \\ r\min\left\{ F_{A_1}^-(u), F_{A_2}^-(u) \right\}, & , if \quad u \in v_1 \cap v_2 \end{cases}$$

$$(F_{\lambda_1}^+ \cup_N F_{\lambda_2}^+)(u) = \begin{cases} F_{\lambda_1}^+(u), & ,if \quad u \in v_1 - v_2 \\ F_{\lambda_2}^+(u), & ,if \quad u \in v_2 - v_1 \\ \min\left\{F_{\lambda_1}^+(u), F_{\lambda_2}^+(u)\right\} & ,if \quad u \in v_1 \cap v_2 \end{cases}$$

$$(F_{\lambda_1}^- \cup_N F_{\lambda_2}^-)(u) = \begin{cases} F_{\lambda_1}^-(u), & ,if \quad u \in v_1 - v_2 \\ F_{\lambda_2}^-(u), & ,if \quad u \in v_2 - v_1 \\ \max\left\{F_{\lambda_1}^-(u), F_{\lambda_2}^-(u)\right\} & ,if \quad u \in v_1 \cap v_2 \end{cases}$$

$$(T_{B_1}^+ \cup_N T_{B_2}^+)(u_2 v_2) = \begin{cases} T_{B_1}^+(u_2 v_2), & ,if \quad u_2 v_2 \in v_1 - v_2 \\ T_{B_2}^+(u_2 v_2), & ,if \quad u_2 v_2 \in v_2 - v_1 \\ r\max\left\{T_{B_1}^+(u_2 v_2), T_{B_2}^+(u_2 v_2)\right\} & ,if \quad u_2 v_2 \in E_1 \cap E_2 \end{cases}$$

$$(T_{B_1}^- \cup_N T_{B_2}^-)(u_2 v_2) = \begin{cases} T_{B_1}^-(u_2 v_2), & ,if \quad u_2 v_2 \in v_1 - v_2 \\ T_{B_2}^-(u_2 v_2), & ,if \quad u_2 v_2 \in v_2 - v_1 \\ r\min\left\{T_{B_1}^-(u_2 v_2), T_{B_2}^-(u_2 v_2)\right\} & ,if \quad u_2 v_2 \in E_1 \cap E_2 \end{cases}$$

$$(T_{\mu_1}^+ \cup_N T_{\mu_2}^+)(u_2 v_2) = \begin{cases} T_{\mu_1}^+(u_2 v_2), & ,if \quad u_2 v_2 \in v_1 - v_2 \\ T_{\mu_2}^+(u_2 v_2), & ,if \quad u_2 v_2 \in v_2 - v_1 \\ \min\left\{T_{\mu_1}^+(u_2 v_2), T_{\mu_2}^+(u_2 v_2)\right\} & ,if \quad u_2 v_2 \in E_1 \cap E_2 \end{cases}$$

$$(T_{\mu_1}^- \cup_N T_{\mu_2}^-)(u_2 v_2) = \begin{cases} T_{\mu_1}^-(u_2 v_2), & ,if \quad u_2 v_2 \in v_1 - v_2 \\ T_{\mu_2}^-(u_2 v_2), & ,if \quad u_2 v_2 \in v_2 - v_1 \\ \max\left\{T_{\mu_1}^-(u_2 v_2), T_{\mu_2}^-(u_2 v_2)\right\} & ,if \quad u_2 v_2 \in E_1 \cap E_2 \end{cases}$$

$$(I_{B_1}^+ \cup_N I_{B_2}^+)(u_2 v_2) = \begin{cases} I_{B_1}^+(u_2 v_2), & ,if \quad u_2 v_2 \in v_1 - v_2 \\ I_{B_2}^+(u_2 v_2), & ,if \quad u_2 v_2 \in v_2 - v_1 \\ r\max\left\{I_{B_1}^+(u_2 v_2), I_{B_2}^+(u_2 v_2)\right\} & ,if \quad u_2 v_2 \in E_1 \cap E_2 \end{cases}$$

$$(I^-_{B_1} \cup_N I^-_{B_2})(u_2 v_2) = \begin{cases} I^-_{B_1}(u_2 v_2), & ,if \quad u_2 v_2 \in v_1 - v_2 \\ I^-_{B_2}(u_2 v_2), & ,if \quad u_2 v_2 \in v_2 - v_1 \\ r\min\left\{ I^-_{B_1}(u_2 v_2), I^-_{B_2}(u_2 v_2) \right\} & ,if \quad u_2 v_2 \in E_1 \cap E_2 \end{cases}$$

$$(I^+_{\mu_1} \cup_N I^+_{\mu_2})(u_2 v_2) = \begin{cases} I^+_{\mu_1}(u_2 v_2), & ,if \quad u_2 v_2 \in v_1 - v_2 \\ I^+_{\mu_2}(u_2 v_2), & ,if \quad u_2 v_2 \in v_2 - v_1 \\ \min\left\{ I^+_{\mu_1}(u_2 v_2), I^+_{\mu_2}(u_2 v_2) \right\} & ,if \quad u_2 v_2 \in E_1 \cap E_2 \end{cases}$$

$$(I^-_{\mu_1} \cup_N I^-_{\mu_2})(u_2 v_2) = \begin{cases} I^-_{\mu_1}(u_2 v_2), & ,if \quad u_2 v_2 \in v_1 - v_2 \\ I^-_{\mu_2}(u_2 v_2), & ,if \quad u_2 v_2 \in v_2 - v_1 \\ \max\left\{ I^-_{\mu_1}(u_2 v_2), I^-_{\mu_2}(u_2 v_2) \right\} & ,if \quad u_2 v_2 \in E_1 \cap E_2 \end{cases}$$

$$(F^+_{B_1} \cup_N F^+_{B_2})(u_2 v_2) = \begin{cases} F^+_{B_1}(u_2 v_2), & ,if \quad u_2 v_2 \in v_1 - v_2 \\ F^+_{B_2}(u_2 v_2), & ,if \quad u_2 v_2 \in v_2 - v_1 \\ r\max\left\{ F^+_{B_1}(u_2 v_2), F^+_{B_2}(u_2 v_2) \right\} & ,if \quad u_2 v_2 \in E_1 \cap E_2 \end{cases}$$

$$(F^-_{B_1} \cup_N F^-_{B_2})(u_2 v_2) = \begin{cases} F^-_{B_1}(u_2 v_2), & ,if \quad u_2 v_2 \in v_1 - v_2 \\ F^-_{B_2}(u_2 v_2), & ,if \quad u_2 v_2 \in v_2 - v_1 \\ r\min\left\{ F^-_{B_1}(u_2 v_2), F^-_{B_2}(u_2 v_2) \right\} & ,if \quad u_2 v_2 \in E_1 \cap E_2 \end{cases}$$

$$(F^+_{\mu_1} \cup_N F^+_{\mu_2})(u_2 v_2) = \begin{cases} F^+_{\mu_1}(u_2 v_2), & ,if \quad u_2 v_2 \in v_1 - v_2 \\ F^+_{\mu_2}(u_2 v_2), & ,if \quad u_2 v_2 \in v_2 - v_1 \\ \min\left\{ F^+_{\mu_1}(u_2 v_2), F^+_{\mu_2}(u_2 v_2) \right\} & ,if \quad u_2 v_2 \in E_1 \cap E_2 \end{cases}$$

$$(F^-_{\mu_1} \cup_N F^-_{\mu_2})(u_2 v_2) = \begin{cases} F^-_{\mu_1}(u_2 v_2), & ,if \quad u_2 v_2 \in v_1 - v_2 \\ F^-_{\mu_2}(u_2 v_2), & ,if \quad u_2 v_2 \in v_2 - v_1 \\ \max\left\{ F^-_{\mu_1}(u_2 v_2), F^-_{\mu_2}(u_2 v_2) \right\} & ,if \quad u_2 v_2 \in E_1 \cap E_2 \end{cases}$$

Example

Let us consider the two bipolar neutrosophic cubic graphs as $G_1 = (P_1, Q_1)$ and $G_2 = (P_2, Q_2)$

$$P_1 = \left\langle \begin{cases} \begin{aligned} & a, \big([0.4, 0.7]\, 0.1\big), \big([0.5, 0.6], 0.3\big), \big([0.1, 0.2], 0.4\big), \\ & \big([-0.4, -0.2], -0.5\big), \big([-0.8, -0.7], -0.5\big), \big([-0.5, -0.4], -0.3\big) \end{aligned} \\ \begin{aligned} & b, \big([0.2, 0.3], 0.5\big), \big([0.2, 0.4], 0.1\big), \big([0.7, 0.3], 0.9\big), \\ & \big([-0.7, -0.4], -0.3\big), \big([-0.7, -0.5], -0.3\big), \big([-0.5, -0.3], -0.2\big) \end{aligned} \\ \begin{aligned} & c, \big([0.2, 0.6], 0.8\big), \big([0.3, 0.5], 0.2\big), \big([0.5, 0.6], 0.7\big), \\ & \big([-0.7, -0.6] - 0.1\big), \big([-0.3, -0.2], -0.1\big), \big([-0.9, -0.4], -0.2\big) \end{aligned} \end{cases} \right\rangle$$

$$Q_1 = \left\langle \begin{cases} \begin{aligned} & ab, \big([0.2, 0.3], 0.5\big), \big([0.2, 0.4], 0.3\big), \big([0.7, 0.3], 0.4\big), \\ & \big([-0.4, -0.2], -0.5\big), \big([-0.7, -0.5], -0.5\big), \big([-0.5, -0.4], -0.2\big) \end{aligned} \\ \begin{aligned} & bc, \big([0.2, 0.3], 0.8\big), \big([0.2, 0.4], 0.2\big), \big([0.7, 0.6], 0.7\big), \\ & \big([-0.7, -0.4], -0.3\big), \big([-0.3, -0.2], -0.3\big), \big([-0.9, -0.4], -0.2\big) \end{aligned} \\ \begin{aligned} & ac, \big([0.2, 0.6], 0.8\big), \big([0.3, 0.5], 0.3\big), \big([0.5, 0.6], 0.4\big), \\ & \big([-0.4, -0.2], -0.5\big), \big([-0.3, -0.2], -0.5\big), \big([-0.9, -0.4], -0.2\big) \end{aligned} \end{cases} \right\rangle$$

$$P_2 = \left\langle \begin{cases} \begin{aligned} & a, \big([0.2, 0.6]\, 0.8\big), \big([0.3, 0.7], 0.8\big), \big([0.5, 0.6], 0.9\big), \\ & \big([-0.6, -0.4], -0.1\big), \big([-0.5, -0.4], -0.3\big), \big([-0.9, -0.8], -0.5\big) \end{aligned} \\ \begin{aligned} & b, \big([0.3, 0.6]\, 0.1\big), \big([0.4, 0.7], 0.9\big), \big([0.3, 0.5], 0.6\big), \\ & \big([-0.4, -0.3], -0.1\big), \big([-0.8, -0.6], -0.2\big), \big([-0.7, -0.5], -0.1\big) \end{aligned} \\ \begin{aligned} & c, \big([0.4, 0.7]\, 0.9\big), \big([0.5, 0.7], 0.8\big), \big([0.2, 0.4], 0.6\big), \\ & \big([-0.5, -0.3], -0.2\big), \big([-0.6, -0.4], -0.1\big), \big([-0.3, -0.2], -0.1\big) \end{aligned} \end{cases} \right\rangle$$

$$Q_2 = \left\langle \begin{cases} \begin{aligned} & ab, \big([0.2, 0.6]\, 0.8\big), \big([0.3, 0.7], 0.9\big), \big([0.3, 0.5], 0.9\big), \\ & \big([-0.4, -0.3], -0.1\big), \big([-0.5, -0.4], -0.3\big), \big([-0.9, -0.8], -0.1\big) \end{aligned} \\ \begin{aligned} & bc, \big([0.3, 0.6]\, 0.9\big), \big([0.4, 0.7], 0.9\big), \big([0.3, 0.5], 0.6\big), \\ & \big([-0.4, -0.3], -0.2\big), \big([-0.6, -0.4], -0.2\big), \big([-0.7, -0.5], -0.1\big) \end{aligned} \\ \begin{aligned} & ac, \big([0.2, 0.6]\, 0.9\big), \big([0.3, 0.7], 0.8\big), \big([0.5, 0.6], 0.6\big), \\ & \big([-0.5, -0.3], -0.2\big), \big([-0.5, -0.4], -0.3\big), \big([-0.9, -0.8], -0.1\big) \end{aligned} \end{cases} \right\rangle$$

Here M-union of the bipolar neutrosophic cubic graph $G_1 U_M G_2$ as follows:

$$P_1 U_M P_2 = \left\langle \begin{array}{l} \left[a, \left([0.4, 0.7], 0.8 \right), \left([0.5, 0.7], 0.8 \right), \left([0.5, 0.6], 0.9 \right), \right. \\ \left. \left([-0.6, -0.4], -0.5 \right), \left([-0.8, -0.7], -0.5 \right), \left([-0.9, -0.8], -0.5 \right) \right] \\ \left[b, \left([0.3, 0.6], 0.5 \right), \left([0.4, 0.7], 0.9 \right), \left([0.7, 0.5], 0.9 \right), \right. \\ \left. \left([-0.7, -0.4], -0.3 \right), \left([-0.8, -0.6], -0.3 \right), \left([-0.7, -0.5], -0.2 \right) \right] \\ \left[c, \left([0.4, 0.7], 0.9 \right), \left([0.5, 0.7], 0.8 \right), \left([0.5, 0.6], 0.7 \right), \right. \\ \left. \left([-0.7, -0.6], -0.2 \right), \left([-0.6, -0.4], -0.1 \right), \left([-0.9, -0.4], -0.2 \right) \right] \end{array} \right\rangle$$

$$Q_1 U_M Q_2 = \left\langle \begin{array}{l} \left[ab, \left([0.2, 0.6], 0.8 \right), \left([0.3, 0.7], 0.9 \right), \left([0.7, 0.5], 0.9 \right), \right. \\ \left. \left([-0.4, -0.3], -0.5 \right), \left([-0.7, -0.5], -0.5 \right), \left([-0.9, -0.8], -0.2 \right) \right] \\ \left[bc, \left([0.3, 0.6], 0.9 \right), \left([0.4, 0.7], 0.9 \right), \left([0.7, 0.6], 0.7 \right), \right. \\ \left. \left([-0.7, -0.4], -0.3 \right), \left([-0.6, -0.4], -0.3 \right), \left([-0.9, -0.5], -0.2 \right) \right] \\ \left[ac, \left([0.2, 0.6], 0.9 \right), \left([0.3, 0.7], 0.8 \right), \left([0.5, 0.6], 0.6 \right), \right. \\ \left. \left([-0.5, -0.3], -0.5 \right), \left([-0.5, -0.4], -0.5 \right), \left([-0.9, -0.8], -0.2 \right) \right] \end{array} \right\rangle$$

Here N-union of the bipolar neutrosophic cubic graph $G_1 U_N G_2$ as follows:

$$P_1 U_N P_2 = \left\langle \begin{array}{l} \left[a, \left([0.4, 0.7], 0.1 \right), \left([0.5, 0.7], 0.3 \right), \left([0.5, 0.6], 0.4 \right), \right. \\ \left. \left([-0.6, -0.4], -0.1 \right), \left([-0.8, -0.7], -0.3 \right), \left([-0.9, -0.8], -0.3 \right) \right] \\ \left[b, \left([0.3, 0.6], 0.1 \right), \left([0.4, 0.7], 0.1 \right), \left([0.7, 0.5], 0.6 \right), \right. \\ \left. \left([-0.7, -0.4], -0.1 \right), \left([-0.8, -0.6], -0.2 \right), \left([-0.7, -0.5], -0.1 \right) \right] \\ \left[c, \left([0.4, 0.7], 0.8 \right), \left([0.5, 0.7], 0.2 \right), \left([0.5, 0.6], 0.6 \right), \right. \\ \left. \left([-0.7, -0.6], -0.1 \right), \left([-0.6, -0.4], -0.1 \right), \left([-0.9, -0.4], -0.1 \right) \right] \end{array} \right\rangle$$

$$Q_1 U_N Q_2 = \left\langle \begin{array}{l} \left[ab, \left([0.2, 0.6], 0.5 \right), \left([0.3, 0.7], 0.3 \right), \left([0.7, 0.5], 0.4 \right), \right. \\ \left. \left([-0.4, -0.3], -0.1 \right), \left([-0.7, -0.5], -0.3 \right), \left([-0.9, -0.8], -0.1 \right) \right] \\ \left[bc, \left([0.3, 0.6], 0.8 \right), \left([0.4, 0.7], 0.2 \right), \left([0.7, 0.6], 0.6 \right), \right. \\ \left. \left([-0.7, -0.4], -0.2 \right), \left([-0.6, -0.4], -0.2 \right), \left([-0.9, -0.5], -0.1 \right) \right] \\ \left[ac, \left([0.2, 0.6], 0.8 \right), \left([0.3, 0.7], 0.3 \right), \left([0.5, 0.6], 0.4 \right), \right. \\ \left. \left([-0.5, -0.3], -0.2 \right), \left([-0.5, -0.4], -0.3 \right), \left([-0.9, -0.8], -0.1 \right) \right] \end{array} \right\rangle$$

Proposition

The M-union of the two bipolar neutrosophic cubic graphs is again a neutrosophic cubic graphs.

Definition

Let $G_1 = (P_1, Q_1)$ and $G_2 = (P_2, Q_2)$ be two bipolar neutrosophic cubic graphs of the graphs G_1^* and G_2^* respectively, then M-join is denoted by $G_1 +_M G_2$ and is defined as follows:

$$G_1 +_M G_2 = (P_1, Q_1) +_M (P_2, Q_2) = (P_1 +_M P_2, Q_1 +_M Q_2)$$

$$= \begin{bmatrix} \left\langle \begin{array}{l} ((T_{A_1}^+ +_N T_{A_2}^+),(T_{\lambda_1}^+ +_N T_{\lambda_2}^+)),((I_{A_1}^+ +_M I_{A_2}^+),(I_{\lambda_1}^+ +_M I_{\lambda_2}^+)),((F_{A_1}^+ +_M F_{A_2}^+),(F_{\lambda_1}^+ +_M F_{\lambda_2}^+)) \\ ((T_{A_1}^- +_M T_{A_2}^-),(T_{\lambda_1}^- +_M T_{\lambda_2}^-)),((I_{A_1}^- +_M I_{A_2}^-),(I_{\lambda_1}^- +_M I_{\lambda_2}^-)),((F_{A_1}^- +_M F_{A_2}^-),(F_{\lambda_1}^- +_M F_{\lambda_2}^-)) \end{array} \right\rangle \\ \left\langle \begin{array}{l} ((T_{B_1}^+ +_M T_{B_2}^+),(T_{\mu_1}^+ +_M T_{\mu_2}^+)),((I_{B_1}^+ +_M I_{B_2}^+),(I_{\mu_1}^+ +_M I_{\mu_2}^+)),((F_{B_1}^+ +_M F_{B_2}^+),(F_{\mu_1}^+ +_M F_{\mu_2}^+)) \\ ((T_{B_1}^- +_M T_{B_2}^-),(T_{\mu_1}^- +_M T_{\mu_2}^-)),((I_{B_1}^- +_M I_{B_2}^-),(I_{\mu_1}^- +_M I_{\mu_2}^-)),((F_{B_1}^- +_M F_{B_2}^-),(F_{\mu_1}^- +_M F_{\mu_2}^-)) \end{array} \right\rangle \end{bmatrix}$$

where

1. if $u \in v_1 \cap v_2$

$$(T_{A_1}^+ +_M T_{A_2}^+)(u) = (T_{A_1}^+ \cup_M T_{A_2}^+)(u) \quad (T_{\lambda_1}^+ +_M T_{\lambda_2}^+)(u) = (T_{\lambda_1}^+ \cup_M T_{\lambda_2}^+)(u)$$
$$(T_{A_1}^- +_M T_{A_2}^-)(u) = (T_{A_1}^- \cup_M T_{A_2}^-)(u) \quad (T_{\lambda_1}^- +_M T_{\lambda_2}^-)(u) = (T_{\lambda_1}^- \cup_M T_{\lambda_2}^-)(u)$$

$$(I_{A_1}^+ +_M I_{A_2}^+)(u) = (I_{A_1}^+ \cup_M I_{A_2}^+)(u) \quad (I_{\lambda_1}^+ +_M I_{\lambda_2}^+)(u) = (I_{\lambda_1}^+ \cup_M I_{\lambda_2}^+)(u)$$
$$(I_{A_1}^- +_M I_{A_2}^-)(u) = (I_{A_1}^- \cup_M I_{A_2}^-)(u) \quad (I_{\lambda_1}^- +_M I_{\lambda_2}^-)(u) = (I_{\lambda_1}^- \cup_M I_{\lambda_2}^-)(u)$$

$$(F_{A_1}^+ +_M F_{A_2}^+)(u) = (F_{A_1}^+ \cup_M F_{A_2}^+)(u) \quad (F_{\lambda_1}^+ +_M F_{\lambda_2}^+)(u) = (F_{\lambda_1}^+ \cup_M F_{\lambda_2}^+)(u)$$
$$(F_{A_1}^- +_M F_{A_2}^-)(u) = (F_{A_1}^- \cup_M F_{A_2}^-)(u) \quad (F_{\lambda_1}^- +_M F_{\lambda_2}^-)(u) = (F_{\lambda_1}^- \cup_M F_{\lambda_2}^-)(u)$$

2. if $u \in E_1 \cup E_2$

$$(T_{B_1}^+ +_M T_{B_2}^+)(uv) = (T_{B_1}^+ \cup_M T_{B_2}^+)(uv) \quad (T_{\mu_1}^+ +_M T_{\mu_2}^+)(u) = (T_{\mu_1}^+ \cup_M T_{\mu_2}^+)(uv)$$
$$(T_{B_1}^- +_M T_{B_2}^-)(uv) = (T_{B_1}^- \cup_M T_{B_2}^-)(uv) \quad (T_{\mu_1}^- +_M T_{\mu_2}^-)(u) = (T_{\mu_1}^- \cup_M T_{\mu_2}^-)(uv)$$

$$(I_{B_1}^+ +_M I_{B_2}^+)(uv) = (I_{B_1}^+ \cup_M I_{B_2}^+)(uv) \quad (I_{\mu_1}^+ +_M I_{\mu_2}^+)(u) = (I_{\mu_1}^+ \cup_M I_{\mu_2}^+)(uv)$$
$$(I_{B_1}^- +_M I_{B_2}^-)(uv) = (I_{B_1}^- \cup_M I_{B_2}^-)(uv) \quad (I_{\mu_1}^- +_M I_{\mu_2}^-)(u) = (I_{\mu_1}^- \cup_M I_{\mu_2}^-)(uv)$$

$$(F_{B_1}^+ +_M F_{B_2}^+)(uv) = (F_{B_1}^+ \cup_M F_{B_2}^+)(uv) \quad (F_{\mu_1}^+ +_M F_{\mu_2}^+)(u) = (F_{\mu_1}^+ \cup_M F_{\mu_2}^+)(uv)$$

$$(F_{B_1}^- +_M F_{B_2}^-)(uv) = (F_{B_1}^- \cup_M F_{B_2}^-)(uv) \quad (F_{\mu_1}^- +_M F_{\mu_2}^-)(u) = (F_{\mu_1}^- \cup_M F_{\mu_2}^-)(uv)$$

3. if $uv \in E^*$, where E^* = The set of all edges joining the vertices of v_1 & v_2 .

$$(T_{B_1}^+ +_M T_{B_2}^+)(uv) = r\min\left\{T_{A_1}^+(u), T_{A_2}^+(v)\right\} \quad (T_{\mu_1}^+ +_M T_{\mu_2}^+)(u) = \min\left\{T_{\lambda_1}^+(u), T_{\lambda_2}^+(v)\right\}$$

$$(T_{B_1}^- +_M T_{B_2}^-)(uv) = r\max\left\{T_{A_1}^-(u), T_{A_2}^-(v)\right\} \quad (T_{\mu_1}^- +_M T_{\mu_2}^-)(u) = \max\left\{T_{\lambda_1}^-(u), T_{\lambda_2}^-(v)\right\}$$

$$(I_{B_1}^+ +_M I_{B_2}^+)(uv) = r\min\left\{I_{A_1}^+(u), I_{A_2}^+(v)\right\} \quad (I_{\mu_1}^+ +_M I_{\mu_2}^+)(u) = \min\left\{I_{\lambda_1}^+(u), I_{\lambda_2}^+(v)\right\}$$

$$(I_{B_1}^- +_M I_{B_2}^-)(uv) = r\max\left\{I_{A_1}^-(u), I_{A_2}^-(v)\right\} \quad (I_{\mu_1}^- +_M I_{\mu_2}^-)(u) = \max\left\{I_{\lambda_1}^-(u), I_{\lambda_2}^-(v)\right\}$$

$$(F_{B_1}^+ +_M F_{B_2}^+)(uv) = r\min\left\{F_{A_1}^+(u), F_{A_2}^+(v)\right\} \quad (F_{\mu_1}^+ +_M F_{\mu_2}^+)(u) = \min\left\{F_{\lambda_1}^+(u), F_{\lambda_2}^+(v)\right\}$$

$$(F_{B_1}^- +_M F_{B_2}^-)(uv) = r\max\left\{F_{A_1}^-(u), F_{A_2}^-(v)\right\} \quad (F_{\mu_1}^- +_M F_{\mu_2}^-)(u) = \max\left\{F_{\lambda_1}^-(u), F_{\lambda_2}^-(v)\right\}$$

Definition

Let $G_1 = (P_1, Q_1)$ and $G_2 = (P_2, Q_2)$ be two bipolar neutrosophic cubic graphs of the graphs G_1^* and G_2^* respectively, then N-join is denoted by $G_1 +_N G_2$ and is defined as follows:

$$G_1 +_N G_2 = (P_1, Q_1) +_N (P_2, Q_2) = (P_1 +_N P_2, Q_1 +_N Q_2)$$

$$= \left\{ \begin{array}{l} \left\langle \begin{array}{l} ((T_{A_1}^+ +_N T_{A_2}^+),(T_{\lambda_1}^+ +_N T_{\lambda_2}^+)),((I_{A_1}^+ +_N I_{A_2}^+),(I_{\lambda_1}^+ +_N I_{\lambda_2}^+)),((F_{A_1}^+ +_N F_{A_2}^+),(F_{\lambda_1}^+ +_N F_{\lambda_2}^+)) \\ ((T_{A_1}^- +_N T_{A_2}^-),(T_{\lambda_1}^- +_N T_{\lambda_2}^-)),((I_{A_1}^- +_N I_{A_2}^-),(I_{\lambda_1}^- +_N I_{\lambda_2}^-)),((F_{A_1}^- +_N F_{A_2}^-),(F_{\lambda_1}^- +_N F_{\lambda_2}^-)) \end{array} \right\rangle \\ \left\langle \begin{array}{l} ((T_{B_1}^+ +_N T_{B_2}^+),(T_{\mu_1}^+ +_N T_{\mu_2}^+)),((I_{B_1}^+ +_N I_{B_2}^+),(I_{\mu_1}^+ +_N I_{\mu_2}^+)),((F_{B_1}^+ +_N F_{B_2}^+),(F_{\mu_1}^+ +_N F_{\mu_2}^+)) \\ ((T_{B_1}^- +_N T_{B_2}^-),(T_{\mu_1}^- +_N T_{\mu_2}^-)),((I_{B_1}^- +_N I_{B_2}^-),(I_{\mu_1}^- +_N I_{\mu_2}^-)),((F_{B_1}^- +_N F_{B_2}^-),(F_{\mu_1}^- +_N F_{\mu_2}^-)) \end{array} \right\rangle \end{array} \right\}$$

1. if $u \in v_1 \cap v_2$

$$(T_{A_1}^+ +_N T_{A_2}^+)(u) = (T_{A_1}^+ \cup_N T_{A_2}^+)(u) \quad (T_{\lambda_1}^+ +_N T_{\lambda_2}^+)(u) = (T_{\lambda_1}^+ \cup_N T_{\lambda_2}^+)(u)$$

$$(T_{A_1}^- +_N T_{A_2}^-)(u) = (T_{A_1}^- \cup_N T_{A_2}^-)(u) \quad (T_{\lambda_1}^- +_N T_{\lambda_2}^-)(u) = (T_{\lambda_1}^- \cup_N T_{\lambda_2}^-)(u)$$

$$(I_{A_1}^+ +_N I_{A_2}^+)(u) = (I_{A_1}^+ \cup_N I_{A_2}^+)(u) \quad (I_{\lambda_1}^+ +_N I_{\lambda_2}^+)(u) = (I_{\lambda_1}^+ \cup_N I_{\lambda_2}^+)(u)$$

$$(I_{A_1}^- +_N I_{A_2}^-)(u) = (I_{A_1}^- \cup_N I_{A_2}^-)(u) \quad (I_{\lambda_1}^- +_N I_{\lambda_2}^-)(u) = (I_{\lambda_1}^- \cup_N I_{\lambda_2}^-)(u)$$

$$(F_{A_1}^+ +_N F_{A_2}^+)(u) = (F_{A_1}^+ \cup_N F_{A_2}^+)(u) \quad (F_{\lambda_1}^+ +_N F_{\lambda_2}^+)(u) = (F_{\lambda_1}^+ \cup_N F_{\lambda_2}^+)(u)$$

$$(F_{A_1}^- +_N F_{A_2}^-)(u) = (F_{A_1}^- \cup_N F_{A_2}^-)(u) \quad (F_{\lambda_1}^- +_N F_{\lambda_2}^-)(u) = (F_{\lambda_1}^- \cup_N F_{\lambda_2}^-)(u)$$

2. if $u \in E_1 \cup E_2$

$$(T_{B_1}^+ +_N T_{B_2}^+)(uv) = (T_{B_1}^+ \cup_N T_{B_2}^+)(uv) \quad (T_{\mu_1}^+ +_N T_{\mu_2}^+)(u) = (T_{\mu_1}^+ \cup_N T_{\mu_2}^+)(uv)$$

$$(T_{B_1}^- +_N T_{B_2}^-)(uv) = (T_{B_1}^- \cup_N T_{B_2}^-)(uv) \quad (T_{\mu_1}^- +_N T_{\mu_2}^-)(u) = (T_{\mu_1}^- \cup_N T_{\mu_2}^-)(uv)$$

$$(I_{B_1}^+ +_N I_{B_2}^+)(uv) = (I_{B_1}^+ \cup_N I_{B_2}^+)(uv) \quad (I_{\mu_1}^+ +_N I_{\mu_2}^+)(u) = (I_{\mu_1}^+ \cup_N I_{\mu_2}^+)(uv)$$

$$(I_{B_1}^- +_N I_{B_2}^-)(uv) = (I_{B_1}^- \cup_N I_{B_2}^-)(uv) \quad (I_{\mu_1}^- +_N I_{\mu_2}^-)(u) = (I_{\mu_1}^- \cup_N I_{\mu_2}^-)(uv)$$

$$(F_{B_1}^+ +_N F_{B_2}^+)(uv) = (F_{B_1}^+ \cup_N F_{B_2}^+)(uv) \quad (F_{\mu_1}^+ +_N F_{\mu_2}^+)(u) = (F_{\mu_1}^+ \cup_N F_{\mu_2}^+)(uv)$$

$$(F_{B_1}^- +_N F_{B_2}^-)(uv) = (F_{B_1}^- \cup_N F_{B_2}^-)(uv) \quad (F_{\mu_1}^- +_N F_{\mu_2}^-)(u) = (F_{\mu_1}^- \cup_N F_{\mu_2}^-)(uv)$$

3. if $uv \in E^*$, where $E^* =$ The set of all edges joining the vertices of v_1 & v_2.

$$(T_{B_1}^+ +_N T_{B_2}^+)(uv) = r\min\left\{T_{A_1}^+(u), T_{A_2}^+(v)\right\} \quad (T_{\mu_1}^+ +_N T_{\mu_2}^+)(u) = \max\left\{T_{\lambda_1}^+(u), T_{\lambda_2}^+(v)\right\}$$

$$(T_{B_1}^- +_N T_{B_2}^-)(uv) = r\max\left\{T_{A_1}^-(u), T_{A_2}^-(v)\right\} \quad (T_{\mu_1}^- +_N T_{\mu_2}^-)(u) = \min\left\{T_{\lambda_1}^-(u), T_{\lambda_2}^-(v)\right\}$$

$$(I_{B_1}^+ +_N I_{B_2}^+)(uv) = r\min\left\{I_{A_1}^+(u), I_{A_2}^+(v)\right\} \quad (I_{\mu_1}^+ +_N I_{\mu_2}^+)(u) = \max\left\{I_{\lambda_1}^+(u), I_{\lambda_2}^+(v)\right\}$$

$$(I_{B_1}^- +_N I_{B_2}^-)(uv) = r\max\left\{I_{A_1}^-(u), I_{A_2}^-(v)\right\} \quad (I_{\mu_1}^- +_N I_{\mu_2}^-)(u) = \min\left\{I_{\lambda_1}^-(u), I_{\lambda_2}^-(v)\right\}$$

$$(F_{B_1}^+ +_N F_{B_2}^+)(uv) = r\min\left\{F_{A_1}^+(u), F_{A_2}^+(v)\right\} \quad (F_{\mu_1}^+ +_N F_{\mu_2}^+)(u) = \max\left\{F_{\lambda_1}^+(u), F_{\lambda_2}^+(v)\right\}$$

$$(F_{B_1}^- +_N F_{B_2}^-)(uv) = r\max\left\{F_{A_1}^-(u), F_{A_2}^-(v)\right\} \quad (F_{\mu_1}^- +_N F_{\mu_2}^-)(u) = \min\left\{F_{\lambda_1}^-(u), F_{\lambda_2}^-(v)\right\}$$

Proposition

The M-join and N-join of two bipolar neutrosophic cubic graphs is again a neutrosophic cubic graph.

APPLICATIONS OF BIOPOLAR NEUTROSOPHIC CUBIC GRAPHS

Since our aim is to address the uncertainties associated with every problem, in this chapter we present the real time applications of bipolar neutrosophic cubic Sets.

Numerical Example 1

Let us consider three teams of software experts represented by the vertex set $V=\{A,B,C\}$. And let the truth-value denote the completion of the project with less effort, the indeterminacy-value denote the situation in which the team puts a moderate effort to complete the project, the false value denotes the heavy competition between the teams.

Let the vertex is given as follows:

$$P = \left\langle \begin{array}{c} \{A,([0.4,0.5],0.3),([0.7,0.8],0.4),([0.3,0.5],0.6), \\ ([-0.6,-0.5],-0.2),([-0.4,-0.3],-0.1),([-0.9,-0.8],-0.7)\} \\ \{B,([0.2,0.4],0.6),([0.5,0.7],0.8),([0.3,0.6],0.7),([-0.4,-0.3],-0.2), \\ ([-0.6,-0.5],-0.4),([-0.8,-0.5],-0.3)\} \\ \{C,([0.5,0.6],0.1),([0.9,0.8],0.4),([0.6,0.7],0.2), \\ ([-0.5,-0.4],-0.2),([-0.7,-0.5],-0.4),([-0.6,-0.5],-0.3)\} \end{array} \right\rangle$$

where the fixed single-valued membership indicates the success strength and strategies of team in future and interval- valued membership indicate the success strength and strategies of teams at present. So on the basis of P we get a set of edges Q defined as follows:

$$Q = \left\langle \begin{array}{c} \{AB,([0.2,0.4],0.6),([0.5,0.7],0.8),([0.3,0.6],0.6), \\ ([-0.4,-0.3],-0.2),([-0.4,-0.3],-0.4),([-0.9,-0.8],-0.3) \} \\ \{BC,([0.2,0.4],0.6),([0.5,0.7],0.8),([0.6,0.7],0.2),([-0.4,-0.3],-0.2), \\ ([-0.6,-0.5],-0.4),([-0.8,-0.5],-0.3)\} \\ \{AC,([0.4,0.5],0.3),([0.7,0.8],0.4),([0.6,0.7],0.2), \\ ([-0.5,-0.4],-0.2),([-0.4,-0.3],-0.4),([-0.9,-0.8],-0.3)\} \end{array} \right\rangle$$

Finally, we see that the strength of one team strongly affect its success with other teams.

$$\text{Order (G)} = \begin{array}{c} \{([1.1,1.5],1.0),([2.1,2.3],1.6),([1.2,1.8],1.5),([-1.5,-1.2],-0.6), \\ ([-1.7,-1.3],-0.9),([-2.3,-1.8],-1.3)\} \end{array}$$

$$\text{degree(A)} = \begin{array}{c} \{([0.6,0.9],0.9),([1.2,1.5],1.2),([0.9,1.3],0.8),([-0.9,-0.7],-0.4), \\ ([-0.8,-0.6],-0.8),([-1.8,-1.6],-0.6)\} \end{array}$$

$$\text{degree(B)} = \begin{array}{l} \{\left(\left[0.4,0.8\right],1.2\right),\left(\left[1.0,1.4\right],1.6\right),\left(\left[0.9,1.3\right],0.8\right),\left(\left[-0.8,,-0.6\right],-0.5\right), \\ \left(\left[-1.0,-0.8\right],-0.8\right),\left(\left[-1.7,-1.3\right],-0.6\right)\} \end{array}$$

$$\text{degree(C)} = \begin{array}{l} \{\left(\left[0.6.0.9\right],0.9\right),\left(\left[1.2,1.5\right],1.2\right),\left(\left[1.2,1.4\right],0.4\right),\left(\left[-0.9,-0.7\right],-0.4\right), \\ \left(\left[-1.0,-0.8\right],-0.8\right),\left(\left[-1.7,-1.3\right],-0.6\right)\} \end{array}$$

The order of G represents the overall effort of above given teams A, B and C. Degree of A represents the effect of other team on A through an edge with the team A. The minimum degree of A is 0 when it cannot be with any other.

Figure 6.

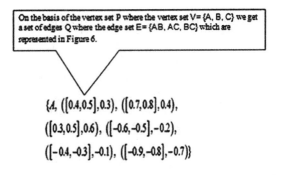

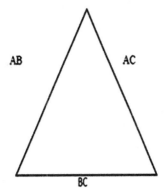

Numerical Example 2

Let us consider an interview process and we evaluate the overall performance of the candidates. The important criteria considered are technical skill, communication skill, work experience, creativity and originality perspective, organizational knowledge, work ethic. The above said criteria are taken in the form of interval valued on future prediction and single-valued number from the unit interval [0,1] to represent present type.

$$
P = \left\langle
\begin{array}{l}
\{A, ([0.4,0.5],0.3), ([0.7,0.8],0.4), ([0.3,0.5],0.6), \\
([-0.6,-0.5],-0.2), ([-0.4,-0.3],-0.1), ([-0.9,-0.8],-0.7)\} \\
\{B, ([0.2,0.4],0.6), ([0.5,0.7],0.8), ([0.3,0.6],0.7), \\
([-0.4,-0.3],-0.2), ([-0.6,-0.5],-0.4), ([-0.8,-0.5],-0.3)\} \\
\{C, ([0.5,0.6],0.1), ([0.9,0.8],0.4), ([0.6,0.7],0.2), \\
([-0.5,-0.4],-0.2), ([-0.7,-0.5],-0.4), ([-0.6,-0.5],-0.3)\} \\
\{D, ([0.6,0.7],0.3), ([0.4,0.7],0.8), ([0.5,0.6],0.2), \\
([-0.7,-0.6],-0.2), ([-0.4,-0.2],-0.1), ([-0.6,-0.5],-0.4) \\
\{E, ([0.4,0.5],0.2), ([0.6,0.8],0.9), ([0.3,0.5],0.1), \\
([-0.5,-0.3],-0.2), ([-0.6,-0.4],-0.8), ([-0.3,-0.2],-0.1)\} \\
\{F, ([0.4,0.8],0.3), ([0.5,0.7],0.4), ([0.3,0.5],0.1), \\
([-0.6,-0.4],-0.2), ([-0.7,-0.5],-0.3), ([-0.4,-0.2],-0.1)\}
\end{array}
\right\rangle
$$

To find the combined effect of all these factors we are need to use bipolar neutrosophic cubic sets for the edges. Where the edge

$$
\{AB, ([0.2,0.4],0.6), ([0.5,0.7],0.8), ([0.3,0.6],0.6), ([-0.4,-0.3],-0.2), \\
([-0.4,-0.3],-0.4), ([-0.9,-0.8],-0.3)\}
$$

denotes the combined result of technical skill, communication skill in the interview. similarly, the other combinations of combined results can also be represented.

Now, if we are interest to find which factor are more effective to the overall performance, we may use the score and accuracy of the bipolar neutrosophic cubic sets, which will give us the closer view of the factors.

Figure 7.

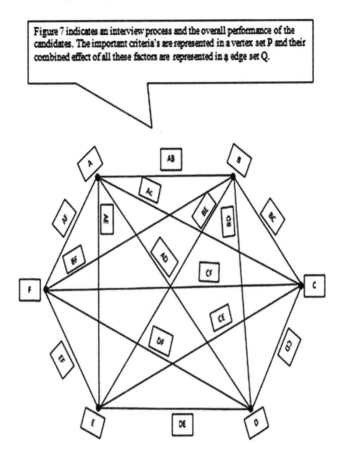

Figure 7 indicates an interview process and the overall performance of the candidates. The important criteria's are represented in a vertex set P and their combined effect of all these factors are represented in a edge set Q.

Bipolar Neutrosophic Cubic Graphs and Minimum Spanning Tree Algorithm

In this section, we first define a score function and present a minimum spanning tree problem and discuss it on a graph with bipolar single-valued neutrosophic cubic edge weight.

Definition: Let A be a bipolar single-valued neutrosophic cubic set, we define a new score function as follows:

$$S(A) = \frac{1}{15}\left\{\begin{array}{l} \left(T^+_{A_L} + T^+_{A_U}\right) + \left(I^+_{A_L} + I^+_{A_U}\right) + \left(1 - \left(F^+_{A_L} + F^+_{A_U}\right)\right) + \left(T^+_\lambda + I^+_\lambda - F^+_\lambda\right) \\ + \left(1 - \left(T^-_{A_L} + T^-_{A_U}\right)\right) + \left(1 - \left(I^-_{A_L} + I^-_{A_U}\right)\right) + \left(F^-_{A_L} + F^-_{A_U}\right) + \left(F^-_\lambda + I^-_\lambda - T^-_\lambda\right) \end{array}\right\}$$

Minimum Spanning Tree Algorithm:

Step 1: Input bipolar single-valued neutrosophic cubic adjacency matrix A.

Step 2: Convert thebipolar single-valued neutrosophic cubic matrix into a score matrix $\left[S_{ij}\right]_{n \times n}$ Using the defined score function.

Step 3: Iterate step 4 and step 5 until all (n-1) elements of matrix of S are either marked to 0 or all the nonzero ($\neq 0$) elements of the matrix are marked.

Step 4: Find the cost of the corresponding edge that is the minimum element in $\left[S_{ij}\right]_{n \times n}$ row wise or column wise.

Step 5: Set $\left[S_{ij}\right] = 0$ if the corresponding e_{ij} of the chosen S_{ij} produces a cycle with the previous marked entries of the score matrix S else mark S_{ij}.

Step 6: Draw the tree T including only the marked elements from the S and it will be the computed minimum spanning tree of the graph G.

Step 7: Stop

Numerical Example

Consider the graph $G=(V,E)$ where V represents the vertices and E represents the edge of the graph. The edges are represented with bipolar single-valued neutrosophic cubic weight. Here we have 6 vertices and 9 edges.

Figure 8 represents the Undirected bipolar single-valued neutrosophic cubic graph.

(I)
$$\left(\left[0.2, 0.4\right], 0.6\right), \left(\left[0.7, 0.5\right], 0.4\right), \left(\left[0.3, 0.5\right], 0.5\right), \left(\left[-0.3, -0.2\right], -0.1\right),$$
$$\left(\left[-0.2, -0.1\right], -0.3\right), \left(\left[-0.8, -0.6\right], -0.4\right)$$

(II)
$$\left(\left[0.6, 0.7\right], 0.3\right), \left(\left[0.4, 0.5\right], 0.2\right), \left(\left[0.1, 0.3\right], 0.2\right), \left(\left[-0.7, -0.5\right], -0.2\right),$$
$$\left(\left[-0.5, -0.3\right], -0.1\right), \left(\left[-0.6, -0.4\right], -0.2\right)$$

Figure 8.

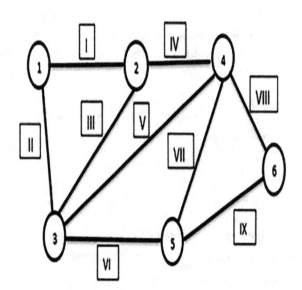

(III) $\Big(\big[0.4,0.5\big],0.1\Big),\Big(\big[0.7,0.8\big],0.2\Big),\Big(\big[0.2,0.5\big],0.1\Big),\Big(\big[-0.9,-0.8\big],-0.6\Big),$
$\Big(\big[-0.3,-0.2\big],-0.1\Big),\Big(\big[-0.6,-0.4\big],-0.2\Big)$

(IV) $\Big(\big[0.9,0.1\big],0.5\Big),\Big(\big[0.7,0.8\big],0.4\Big),\Big(\big[0.2,0.5\big],0.1\Big),\Big(\big[-0.6,-0.3\big],-0.2\Big),$
$\Big(\big[-0.5,-0.4\big],-0.3\Big),\Big(\big[-0.8,-0.6\big],-0.3\Big)$

(V) $\Big(\big[0.6,0.8\big],0.2\Big),\Big(\big[0.3,0.5\big],0.2\Big),\Big(\big[0.5,0.6\big],0.4\Big),\Big(\big[-0.5,-0.4\big],-0.1\Big),$
$\Big(\big[-0.6,-0.4\big],-0.3\Big),\Big(\big[-0.7,-0.6\big],-0.1\Big)$

(VI) $\Big(\big[0.8,0.9\big],0.2\Big),\Big(\big[0.4,0.6\big],0.1\Big),\Big(\big[0.4,0.5\big],0.2\Big),\Big(\big[-0.5,-0.3\big],-0.1\Big),$
$\Big(\big[-0.7,-0.6\big],-0.3\Big),\Big(\big[-0.4,-0.3\big],-0.2\Big)$

(VII) $\Big(\big[0.6,0.7\big],0.5\Big),\Big(\big[0.2,0.6\big],0.1\Big),\Big(\big[0.3,0.4\big],0.2\Big),\Big(\big[-0.6,-0.5\big],-0.4\Big),$
$\Big(\big[-0.8,-0.7\big],-0.5\Big),\Big(\big[-0.5,-0.4\big],-0.3\Big)$

(VIII) $\Big(\big[0.5,0.7\big],0.3\Big),\Big(\big[0.5,0.6\big],0.2\Big),\Big(\big[0.3,0.4\big],0.2\Big),\Big(\big[-0.6,-0.5\big],-0.1\Big),$
$\Big(\big[-0.8,-0.5\big],-0.3\Big),\Big(\big[-0.3,-0.2\big],-0.1\Big)$

(IX) $\Big(\big[0.8,0.9\big],0.6\Big),\Big(\big[0.3,0.4\big],0.1\Big),\Big(\big[0.2,0.5\big],0.1\Big),\Big(\big[-0.4,-0.3\big],-0.2\Big),$
$\Big(\big[-0.5,-0.4\big],-0.1\Big),\Big(\big[-0.8,-0.9\big],-0.5\Big)$

Using the Score function, we get the score matrix S (A) as follows:

SCORE MATRIX

$$S(A) = \begin{bmatrix} 0 & 0.22 & 0.4 & 0 & 0 & 0 \\ 0.22 & 0 & 0.4267 & 0.4733 & 0 & 0 \\ 0.4 & 0.4267 & 0 & 0.2933 & 0.3933 & 0 \\ 0 & 0.4733 & 0.2933 & 0 & 0.4067 & 0.4333 \\ 0 & 0 & 0.3933 & 0.4067 & 0 & 0.3467 \\ 0 & 0 & 0 & 0.4333 & 0.3467 & 0 \end{bmatrix}$$

The minimum entry is 0.22 and is selected in score matrix and the corresponding edge (1, 2) is highlighted by red color in Figure 9.

Figure 9 represents the bipolar single-valued neutrosophic cubic graph where the edge (1, 2) is highlighted.

In the score matrix, the next non zero minimum entry is 0.2933. It is marked and the corresponding edge (3, 4) highlighted by red color in Figure 10.

Figure 10 represents bipolar single-valued neutrosophic cubic graph where the edge (3, 4) is highlighted.

In the score matrix, the next non zero minimum entry is 0.3467. It is marked and the corresponding edge (5,6) highlighted by red color in Figure 11.

Figure 11 represents the bipolar single-valued neutrosophic cubic graph where the edge (5, 6) is highlighted.

In the score matrix, the next non zero minimum entry is 0.3933. It is marked and the corresponding edge (3, 5) highlighted by red color in Figure-12.

Figure 12 represents the bipolar single-valued neutrosophic cubic graph where the edge (3, 5) is highlighted.

Figure 9.

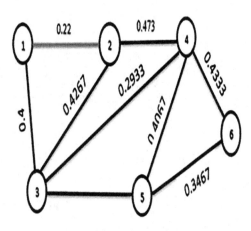

Figure 10.

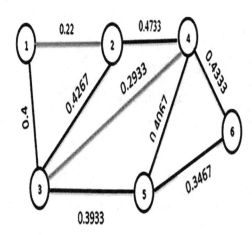

Figure 11.

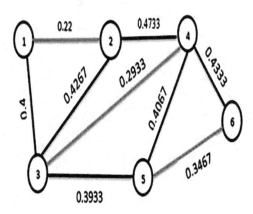

Figure 12.

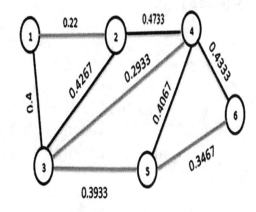

Figure 13.

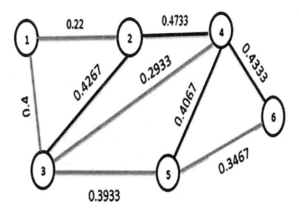

In the score matrix, the next non zero minimum entry is 0.4. It is marked and the corresponding edge (1, 3) highlighted by red color in Figure 13.

Figure 13 represents the bipolar single-valued neutrosophic cubic graph where the edge (1, 3) is highlighted.

In the score matrix, the next non zero minimum entry is 0.4067 but it produces a cycle and hence we delete and mark it as zero instead of 0.4067.

Figure 14.

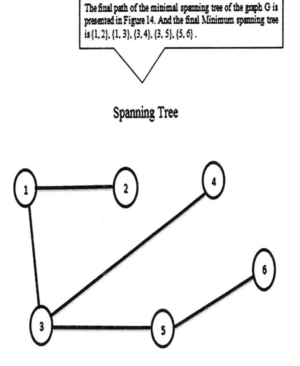

Spanning Tree

Similarly, next minimum entries 0.4267, 0.4333 and 0.4733 will also produces a cycle. So we delete and mark it as zero instead.

$$
S(A) = \begin{bmatrix}
0 & 0.22 & 0.4 & 0 & 0 & 0 \\
0.22 & 0 & \cancel{0.4267} & \cancel{0.4733} & 0 & 0 \\
0.4 & \cancel{0.4267} & 0 & 0.2933 & 0.3933 & 0 \\
0 & \cancel{0.4733} & 0.2933 & 0 & \cancel{0.4067} & \cancel{0.4333} \\
0 & 0 & 0.3933 & \cancel{0.4067} & 0 & 0.3467 \\
0 & 0 & 0 & \cancel{0.4333} & 0.3467 & 0
\end{bmatrix}
$$

The above score matrix $S(A)$ represents the bipolar single-valued neutrosophic cubic score matrix indicating the deleted edges.

After the above steps the final path of the minimal spanning tree of the graph G is presented in Figure 14. The Crisp Minimum Cost Spanning Tree is 3 and the final Minimum spanning tree is

{1,2}, {1,3}, {3,4}, {3,5}, {5,6}.

CONCLUSION

Graph theory has vital application in the field of science, and technology and helps to represent communication network, organization of data, computational flow. The concept of bipolar neutrosophic cubic set is generalization of smarandache's neutrosophic set theory which is more flexible, compatible and precise compared to the classical and fuzzy generalizations. we have introduced the concept of bipolar neutrosophic cubic graphs and described its algebraic properties. The applications of the bipolar neutrosophic cubic graphs are also discussed. As an application of bipolar single-valued neutrosophic cubic graphs we defined the minimum spanning tree problem. The concept of bipolar neutrosophic graphs can be applied in various fields of computer science: expert systems, database theory, neural networks, signal processing, artificial intelligence, pattern recognition, robotics, computer networks, medical science and engineering. This work can be hybridized with rough graphs, soft graphs, hypergraphs soft graphs.

REFERENCES

Akram, M. (2011). Bipolar fuzzy graphs. *Information Sciences*, *181*(24), 5548–5564. doi:10.1016/j.ins.2011.07.037

Akram, M. (2013). Bipolar fuzzy graphs with applications. *Knowledge-Based Systems*, *39*, 1–8. doi:10.1016/j.knosys.2012.08.022

Akram, M., & Davvaz, M. (2012). Strong intuitionistic fuzzy graphs. *Filomat*, *26*(1), 177–196. doi:10.2298/FIL1201177A

Atanassov, K. T. (1986). Intuitionistic fuzzy sets. *Fuzzy Sets and Systems, 20*(1), 87–96. doi:10.1016/S0165-0114(86)80034-3

Broumi, S., Bakali, A., Talea, M., & Smarandache, F. (2016). Isolated Single Valued Neutrosophic Graphs. *Neutrosophic Sets and Systems, 11*, 74–78.

Broumi, S., Bakali, A., Talea, M., & Smarandache, F. (2017). Generalized Bipolar Neutrosophic Graphs of Type 1. *20th International Conference on Information Fusion,* 1714-1720.

Broumi, S., Bakali, A., Talea, M., Smarandache, F., & Rao, V. V. (2018). Interval Complex Neutrosophic Graph of Type 1. *Neutrosophic Operational Research, 3*, 87–106.

Broumi, S., Bakali, A., Talea, M., Smarandache, F., & Verma, R. (2017). Computing Minimum Spanning Tree in Interval Valued Bipolar Neutrosophic Environment. *International Journal of Modeling and Optimization, 7*(5), 300–304. doi:10.7763/IJMO.2017.V7.602

Broumi, S., Talea, M., Bakali, A., & Smarandache, F. (2016). Bipolar Single Valued Neutrosophic Graphs. *New Ternds in Neutrosophic Theory and Appications,* 203-221.

8. Broumi, S., Talea, M., Bakali, A., & Smarandache, F. (2016). On bipolar single valued neutrosophic graphs. *Journal of New Theory, 11*, 84–102.

Broumi, S., Talea, M., Bakkali, A., & Smarandache, F. (2016). Single Valued Neutrosophic Graph. *Journal of New Theory, 10*, 68–101.

Bustince, H., & Burilloit, P. (1996). Vague sets are intuitionistic fuzzy sets. *Fuzzy Sets and Systems, 79*(30), 403–405. doi:10.1016/0165-0114(95)00154-9

Deli, M., & Smarandache, F. (2015). Bipolar neutrosophic sets and their application based on multi-criteria decision making problems. *Advanced Mechatronic Systems (ICAMechS),* 249-254.

Gau, W. L., & Buehrer, D. J. (1993). Vague sets. *IEEE Transactions on Systems, Man, and Cybernetics, 23*(2), 610–614. doi:10.1109/21.229476

Hai-Long, She, & Yanhonge, & Xiuwu. (2015). On single valued neutrosophic relations. *Journal of Intelligent & Fuzzy Systems,* 1–12.

Jun, Y. B., Smarandache, F., & Kim, C. S. (2017). Neutrosophic cubic sets. *New Mathematics and Natural Computation, 13*(01), 41–54. doi:10.1142/S1793005717500041

Mordeson, J. N., & Nair, P. S. (2001). *Fuzzy Graphs and Fuzzy Hypergraphs.* Heidelberg, Germany: Physica Verlag.

Parvathi, R., & Karunambigai, M. G. (2006). Intuitionistic Fuzzy Graphs. *Computational Intelligence, Theory and Applications,* 139-150.

Pramanik, T., Samanta, M., & Pal, S. (2016). Interval-valued fuzzy planar graphs. *International Journal of Machine Learning and Cybernetics, 7*(4), 653–664. doi:10.100713042-014-0284-7

Rosenfeld, A. (1975). *Fuzzy graphs in Fuzzy sets and their applications.* Academic Press.

Samanta, S., & Pal. (2011). Fuzzy Tolerance Graphs. *International Journal of Latest Trends in Mathematics.*

Samanta, S., & Pal. (2012). *Irregular bipolar fuzzy graphs.* arXiv preprint arXiv, 1209, 1682.

Samanta, S., & Pal, M. (2012). Bipolar fuzzy hypergraphs. *International Journal of Fuzzy Logic System,* 2(1), 17–28. doi:10.5121/ijfls.2012.2103

Samanta, S., & Pal, M. (2015). Fuzzy planar graphs. *IEEE Transactions on Fuzzy Systems,* 23(6), 936–1942. doi:10.1109/TFUZZ.2014.2387875

Smarandache, F. (1999). *A unifying field in logics. neutrosophy: neutrosophic probability, set and logic.* Rehoboth: American Research Press.

Smarandache, F. (2006). Neutrosophic set - a generalization of the intuitionistic fuzzy set, Granular Computing. *IEEE International Conference,* 38 – 42.

Smarandache, F. (2015). *Types of Neutrosophic Graphs and Neutrosophic Algebraic Structures together with their Applications in Technology, seminar, Universitatea Transilvania din Brasov.* Brasov, Romania: Facultatea de Design de Produs si Mediu.

Wang, H., Smarandache, F., Zhang, Y. Q., & Sunderraman, R. (2010). Single-valued neutrosophic sets. *Multisspace and Multistruct,* 4, 410–413.

Zadeh, L. A. (1965). Fuzzy sets. *Information and Control,* 8(3), 338–353. doi:10.1016/S0019-9958(65)90241-X

Zhang, W. R. (1998). Bipolar fuzzy sets. *IEEE International Conference on fuzzy systems,* 835-840.

Compilation of References

(2016). Ediz, On the reduced first Zagreb index of graphs. *Pacific J. Appl. Math.*, *8*(2), 99–102.

(2018e). V.R. Kulli General reduced second Zagreb index of certain networks. *International Journal of Current Research in Life Sciences*, *7*(11), 2827–2833.

371. 2019).

8.   Broumi, S., Talea, M., Bakali, A., & Smarandache, F. (2016). On bipolar single valued neutrosophic graphs. *Journal of New Theory*, *11*, 84–102.

Abbasi, A., Jalili, M., & Sadeghi-Niaraki, A. (2018). Influence of network-based structural and power diversity on research performance. *Scientometrics*, *117*(1), 579–590. doi:10.100711192-018-2879-3

Abhishek & Germina. (2012). On Set valued graphs. *J. Fuzzy Set Valued Analysis*. doi: . doi:10.5899/2012/jfsva-00127

Abhishek, K. (2007). *New directions in the theory of set-valuations of graphs* (Ph. D Thesis). Kannur University, Kannur, India.

Abhishek, K. (2009). *New Directions in the Theory of Set-Valuations of Graphs* (Ph.D. Thesis). Kannur University, Kannur, Kerala, India.

Abhishek, K. (2013). Set-valued graphs-II. *J. Fuzzy Set Valued Anal.*, *2013*, 1–16. doi:10.5899/2013/jfsva-00149

Abhishek, K. (2015). A note on set-indexed graphs. *J. Discrete Math. Sci. Cryptography*, *18*(1-2), 31–40. doi:10.1080 /09720529.2013.867637

Abhishek, K. (2015). Set-valued graphs: A survey. *J. Discrete Math. Sci. Cryptography*, *18*(1-2), 55–80. doi:10.1080/ 09720529.2014.894306

Acharya, Abhishek, & Germina. (2012). Hypergraphs of minimal arc bases in a digraph. *J. Combin. Inform. Syst. Sci*, *37*(2-4), 329–341.

Acharya, B. D. (1983). Set-Valuations and their Applications. MRI Lecture Notes in Applied Math., 2, 1-17.

Acharya, B. D., Germina, K. A., & Paul, V. (2010). Linear hypergraph set-indexers of a graph. *Int. Math. Forum*, *5*(68), 3359-3370.

Acharya, B. D., Germina, K. A., Princy, K. L., & Rao, S. B. (2008). On set-valuations of graphs. In Labelling of Discrete Structures and Applications. Narosa Pub. House.

Acharya, Germina, & Paul. (2010). Linear hypergraph set-indexres of graph. *Int. Math. Forum, 5*(68), 3359–3370.

Acharya, Rao, & Walika. (2003). *Energy of signed digraphs.* Lecture Notes, Group Discussion on Energy of a Graph. Karnatak University.

Acharya. (1982). *Two structural criteria for voltage graphs satisfying Kirchoff's voltage law.* Res. Rep. No. HCS/DST/409/7/82-83.

Acharya. (1983). On d-sequential graphs. *J. Math. Phys. Sci., 17*(1) 21–35.

Acharya, B. D. (1983). *Set-valuations and their applications. MRI Lecture Notes in Applied Mathematics, No.2.* Allahabad: The Mehta Research Institute of Mathematics and Mathematical Physics.

Acharya, B. D. (2001). Set-indexers of a graph and set-graceful graphs. *Bull. Allahabad Math. Soc., 16,* 1–23.

Acharya, B. D. (2010). K_s is the only Eulerian set-sequential graph. *Bulletin of the Calcutta Mathematical Society, 102*(5), 465–470.

Acharya, B. D. (2012). Set-valuations of a signed digraph. *J. Combin. Inform. Syst. Sci., 37*(2-4), 145–167.

Acharya, B. D. (2012). Set-valuations of signed digraphs. *J. Combin. Inform. System Sci., 37*(2-4), 145–167.

Acharya, B. D., & Germina, K. A. (2010). Unigeodesic Graphs and Related Recent Notions, International Journal of Algorithms. *Computing and Mathematics, 3*(1), 89–92.

Acharya, B. D., & Germina, K. A. (2011). Distance compatible set-labeling of graphs. *Indian J. Math. Comp. Sci. Jhs., 1,* 49–54.

Acharya, B. D., & Germina, K. A. (2013). Set-valuations of graphs and their applications: A survey. *Ann. Pure Appl. Math., 4*(1), 8–42.

Acharya, B. D., Germina, K. A., Princy, K. L., & Rao, S. B. (2008). Graph labellings, embedding and NP-completeness theorems. *J. Combin. Math. Combin. Computing, 67,* 163–180.

Acharya, B. D., Germina, K. A., Princy, K. L., & Rao, S. B. (2008). On set-valuations of graphs. In *Proc. II Group Discussion, Labeling of Discrete Structures and Their Applications.* Narosa Publishing House.

Acharya, B. D., Germina, K. A., Princy, K. L., & Rao, S. B. (2009). On set-valuations of graphs: Embedding and NP-completeness theorems for set-graceful, topologically set-graceful and set-sequential graphs. *J. Discrete Math. Sci. Cryptog., 12*(4), 481–487. doi:10.1080/09720529.2009.10698249

Acharya, B. D., Germina, K. A., Princy, K. L., & Rao, S. B. (2012). Topologically set-graceful graphs. *J. Combin. Inform. Syst. Sci., 37*(2-4), 309–328.

Acharya, B. D., & Hegde, S. M. (1985). Set-sequential graphs. *National Academy Science Letters, 8*(12), 387–390.

Acharya, B. D., Rao, S. B., & Arumugam, S. (2008). Embeddings and NP-Complete Problems for Graceful Graphs. In B. D. Acharya, S. Arumugam, & A. Rosa (Eds.), *Labeling of Discrete Structures and Applications* (pp. 57–62). New Delhi: Narosa Publishing House.

Acharya, & Germina,, & Joy. (2011). Set-valuation of digraphs. *Global J. Pure Appl. Math., 7*(3), 237–243.

Acharya, & Germina, & Joy. (2011). Topogenic graphs. *Advanced Stud. Contemp. Math., 21*(2), 139–159.

Acharya, & Germina, & Koshy. (2012). A creative survey on complementary distance pattern uniform sets of vertices in a graph. *J. Combin. Inform. Syst. Sci, 37*(1), 291–308.

Acharya, & Germina, Abhishek, & Slater. (2012). Some new results on set-graceful and set-sequential graphs. *J. Combin. Inform. Syst. Sci, 37*(2-4), 239–249.

Acharya, & Germina, Abhishek, & Slater. (2012). Some new results on set-graceful and set-sequential graphs. *J. Combin. Inform. Syst. Sci., 37*(2-4), 39–249.

Acharya, Germina, & Ahbishek, Rao, & Zaslavsky. (2009). Point- and arc-reaching sets of vertices in a digraph. *Indian J. Math., 51*(3), 597–609.

Adamic, L. (1999). *The Small World Web*. Retrieved from http: //www. parc. xerox. com/istl / groups / i a /www / small world paper. html

Adiga, C., & Rakshith, B. R. (2016). On spectra of variants of the corona of two graphs and some new equienergetic graphs. *Discussiones Mathematicae. Graph Theory, 36*(1), 127–140. doi:10.7151/dmgt.1850

Adiono, T., Fuada, S., Saputo, R. A., & Luthfi, M. (2018). Internet access over visible light. *Proceedings of the 2018 IEEE International Conference on Consumers Electronics*. Retrieved from https://ieeexplore.ieee.org/abstract/document/8552123

Afinogenov, Gr. (2013). Andrei Ershov and the Soviet Information Age. *Kritika: Explorations in Russian and Eurasian History, 14*(3), 561–584. doi:10.1353/kri.2013.0046

Aggarwal, A. A. (2003). Internetalization of end-users. *Journal of Organizational and End User Computing, 15*(1), 54–56. doi:10.4018/joeuc.2003010104

Aiqun, Z. (2018). An IT capability approach to informatization construction of higher education Institutions. *Procedia Computer Science, 131*, 683–690. doi:10.1016/j.procs.2018.04.312

Akiyama, J., Avis, D., Chavtal, V., & Era, H. (1981). Balancing signed graphs. *Discrete Applied Mathematics, 3*(4), 227–233. doi:10.1016/0166-218X(81)90001-9

Akram, M., m-Polar Fuzzy Graphs, Studies in Fuzziness and Soft Computing, Springer,

Akram, M. (2011). Bipolar fuzzy graphs. *Information Sciences, 181*(24), 5548–5564. doi:10.1016/j.ins.2011.07.037

Akram, M. (2013). Bipolar fuzzy graphs with applications. *Knowledge-Based Systems, 39*, 1–8. doi:10.1016/j.knosys.2012.08.022

Akram, M., & Alshehri, N. O. (2015). Tempered interval-valued fuzzy hyeprgraphs. *U.P.B Sci. Bull, Series A, 77*(1), 39–47.

Akram, M., & Davvaz, M. (2012). Strong intuitionistic fuzzy graphs. *Filomat, 26*(1), 177–196. doi:10.2298/FIL1201177A

Akram, M., & Dudek, W. A. (2011). Interval-valued fuzzy graphs. *Computers & Mathematics with Applications (Oxford, England), 61*(2), 289–299. doi:10.1016/j.camwa.2010.11.004

Akram, M., & Dudek, W. A. (2012). Regular bipolar fuzzy graphs. *Neural Computing & Applications, 21*(1), 197–205. doi:10.100700521-011-0772-6

Akram, M., & Dudek, W. A. (2013). Intuitionistic fuzzy hypergraphs with applications. *Information Sciences, 218*, 182–193. doi:10.1016/j.ins.2012.06.024

Akram, M., Dudek, W. A., & Sarwa, S. (2013). Properties of bipolar fuzzy hypergraphs. *International Journal of Pure and Applied Mathematics, 31*, 141–160.

Akram, M., & Luqman, A. (2017). Bipolar neutrosophic hypergraphs with applications. *Journal of Intelligent & Fuzzy Systems, 33*(3), 1699–1713. doi:10.3233/JIFS-17228

Akram, M., & Naz, S. (2018). Energy of Pythagorean fuzzy graphs with applications. *Mathematics, 6*(8), 136. doi:10.3390/math6080136

Akram, M., & Sarwar, M. (2017a). Novel application of m-polar fuzzy hypergraphs. *Journal of Intelligent & Fuzzy Systems, 32*(3), 2747–2762. doi:10.3233/JIFS-16859

Akram, M., & Sarwar, M. (2017b). Transversals of m-polar fuzzy hypergraphs with applications. *Journal of Intelligent & Fuzzy Systems, 33*(1), 351–364. doi:10.3233/JIFS-161668

Akram, M., & Shahzadi, G. (2018). Hypergraphs in m-polar fuzzy environment. *Mathematics, 6*(2), 28. doi:10.3390/math6020028

Akram, M., & Waseem, N. (2016). Certain metrices in m-polar fuzzy graphs. *New Mathematics and Natural Computation, 12*(2), 135–155. doi:10.1142/S1793005716500101

al Liu, L. G. (2007). Weighted network properties of Chinese nature science basic research. *Physica A, 377*(1), 302–314. doi:10.1016/j.physa.2006.11.011

al Liu, X. (2005). Co-authorship networks in the digital library research community. *Information Processing & Management, 41*(6), 1462–1480. doi:10.1016/j.ipm.2005.03.012

Albert, R., & Barabási, A. L. (2002). The statistical mechanics of complex networks. *Reviews of Modern Physics, 74*(1), 47–97. doi:10.1103/RevModPhys.74.47

Albertson, M. O. (1997). The irregularity of a graph. *Ars Combinatoria, 46*, 219–225.

Al-Hawary, T. (2011). Complete fuzzy graphs. *International J Math Combin, 4*, 26–34.

Ali, W., Shamsuddin, S. M., & Ismail, A. S. (2011). A survey of web caching and prefetching. *Int. J. Advance. Soft Comput. Appl, 3*(1), 18–44.

Allan, R. B., Laskar, R., & Hedetniemi, S. T. (1984). A note on total domination. *Discrete Mathematics, 49*(1), 7–13. doi:10.1016/0012-365X(84)90145-6

Alrawi, K. (2017). The internet and international marketing. *Competitiveness Review, 17*(4), 222–233. doi:10.1108/10595420710844316

Amanathulla & Pal. (2017c). $L(3,2,1)$- and $L(4,3,2,1)$-labeling problems on interval graphs. *AKCE International Journal of Graphs and Combinatorics, 14*, 205-215.

Amanathulla, & Pal. (2016a). - and -labeling problems on circular-arc graphs. *International Journal of Soft Computing, 11*(6), 343-350.

Amanathulla, & Pal. (2016b). - and -labeling problems on circular-arc graphs. *International Journal of Control Theory and Applications, 9*(34), 869-884.

Amanathulla, S., & Pal, M. (2017a). -labeling problems on interval graphs. *International Journal of Control Theory and Applications, 10*(1), 467–479.

Amanathulla, S., & Pal, M. (2017b). -labeling problems on permutation graphs. *Transylvanian Review, 25*(14), 3939–3953.

Amanathulla, S., & Pal, M. (2017v). Labeling problems on circular-arc graphs. *Far East Journal of Mathematical Sciences, 102*(6), 1279–1300. doi:10.17654/MS102061279

Amanathulla, S., & Pal, M. (2018). Surjective $L(2,1)$. *Journal of Intelligent & Fuzzy Systems*, *35*(1), 739–748. doi:10.3233/JIFS-171176

Amanathulla, S., Sahoo, S., & Pal, M. (n.d.). Labeling numbers of square of paths, complete graphs and complete bipartite graphs. *Journal of Intelligent & Fuzzy Systems*. doi:10.3233/JIFS-172195

Amihud, Y. (2002). Illiquidity and stock returns: Cross-section and time-series effects *. *Journal of Financial Markets*, *5*(1), 31–56. doi:10.1016/S1386-4181(01)00024-6

Anandavally, T. M. K. (2013). A characterisation of 2-uniform IASI graphs. *Int. J. of Contemp. Math. Sci.*, *8*(10), 459–462.

Ananthakumar, R., & Germina, K. A. (2011). Distance pattern distinguishing sets in a graph. *Advanced Stud. Contemp. Math*, *21*(1), 107–114.

Andrasfai. (1991). *Flows, Matrices*. Adam Hilger.

Angeltveit, V., & Mckay, B. D. (2017). *R(5, 5) ≤ 48*. arXiv:1703.08768

Anjali, N., & Mathew, S. (2013). Energy of a fuzzy graph. *Annals Fuzzy Maths and Informatics*, *6*(3), 455–465.

An, M., & Xiong, L. (2015). Some results on the difference of the Zagreb indices of a graph. *Bulletin of the Australian Mathematical Society*, *92*(2), 117–186. doi:10.1017/S0004972715000386

An, M., & Xiong, L. (2018). Some results on the inverse sum indeg index of a graph. *Information Processing Letters*, *134*, 42–46. doi:10.1016/j.ipl.2018.02.006

Apache Hadoop. (n.d.). Retrieved October 28, 2018, from http://hadoop.apache.org/

Apostol, T. M. (1989). *Introduction to analytic number theory*. New York: Springer-Verlag.

Asinowski, A., Cohen, E., Golumbic, M. C., Limouzy, V., Lipshteyn, M., & Stern, M. (2012). Vertex intersection graphs of paths on a grid. *Journal of Graph Algorithms and Applications*, *16*(2), 129–150. doi:10.7155/jgaa.00253

Atanassov, K. T. (1986). Intuitionistic fuzzy sets. *Fuzzy Sets and Systems*, *20*(1), 87–96. doi:10.1016/S0165-0114(86)80034-3

Aumann, Y., Lewenstein, M., Melamud, O., Pinter, R. Y., & Yakhini, Z. (2005). Dotted interval graphs and high throughput genotyping. *Proceedings of the Sixteenth Annual ACM-SIAM Symposium on Discrete Algorithms*, 339–348.

Aumann, Y., Lewenstein, M., Melamud, O., Pinter, R. Y., & Yakhini, Z. (2012). Dotted interval graphs. *ACM Transactions on Algorithms*, *8*(2).

Bae, J., & Kim, S. (2014). Identifying and ranking influential spreaders in complex networks by neighborhood coreness. *Physica A*, *395*, 549–559. doi:10.1016/j.physa.2013.10.047

Bai, B., Wang, L., Han, Z., Chen, W., & Svensson, T. (2016). Caching based socially-aware D2D communications in wireless content delivery networks: A hypergraph framework. *IEEE Wireless Communications*, *23*(4), 74–81. doi:10.1109/MWC.2016.7553029

Baig, A. Q., Nadeem, M., & Gao, W. (2018). Revan and hyper Revan indices of octahedral and icosahedral networks. *Applied Mathematics and Nonlinear Sciences*, *3*(1), 33–40. doi:10.21042/AMNS.2018.1.00004

Balakrishnan, R. (2004). The energy of a graph. *Linear Algebra and Its Applications*, *387*, 287–295. doi:10.1016/j.laa.2004.02.038

Balakrishnan, R., & Paulraja, P. (1983). Powers of chordal graphs. *Journal of the Australian Mathematical Society, 35*(2), 211–217. doi:10.1017/S1446788700025696

Barman, S. C., Mondal, S., & Pal, M. (2009). An efficient algorithm to find next-to-shortest path on permutation graphs. *Journal of Applied Mathematics and Computing, 31*(1), 369–384. doi:10.100712190-008-0218-1

Barman, S. C., Mondal, S., & Pal, M. (2010a). A linear time algorithm to construct a tree 4-spanner on trapezoid graphs. *International Journal of Computer Mathematics, 87*(4), 743–755. doi:10.1080/00207160802037880

Barman, S. C., Pal, M., & Mondal, S. (2007). An efficient algorithm to find next-to-shortest path on trapezoid graphs. *Advances in Applied Mathematical Analysis, 2*(2), 97–107.

Barman, S. C., Pal, M., & Mondal, S. (2010b). The k-neighbourhood-covering problem on interval graphs. *International Journal of Computer Mathematics, 87*(9), 1918–1935. doi:10.1080/00207160802676570

Barsegyan, A. A., Kupriyanov, M. S., Stepanenko, V. V., & Holod, I. I. (2008). *Technologies for data analysis: data mining, visual mining, text mining, OLAP* (p. 384). Saint Petersburg: BHV-Petersburg.

Basavanagoud, B., Barangi, A. P., & Hosamani, S. M. (2018). First neighbourhood Zagreb index of some nanostructures. *Proceedings of IAM, 7*(2), 178-193.

Basavanagoud, B., & Chitra, E. (2018). On leap Zagreb indices of some nanostructures. *Malaya Journal of Matematik, 6*(4), 816–822.

Basavanagoud, B., & Desai, V. R. (2016). Forgotten topological index and hyper Zagreb index of generalized transformation graphs. *Bull. Math. Sci. Appl., 14*, 1–6.

Basavanagoud, B., & Jakkannavar, P. (2018). Computing first leap Zagreb index of some nanostructures. *Int. J. Math. And. Appl., 6*(2-B), 141–150.

Basavanagoud, B., & Jakkannavar, P. (2018). Kulli-Basava indices of graphs. *Inter. J. Appl. Engg. Research, 14*(1), 325–342.

Basavanagoud, B., & Patil, S. (2016). A note on hyper Zagreb index of graph operation. *Iran. J. Math. Chem., 7*, 89–92.

Basavanagoud, B., & Patil, S. (2017). The Hyper-Zagreb index of four operations on graphs. *Mathematical Sciences Letters, 6*(2), 193–198. doi:10.18576/msl/060212

Battista, G. D., Eades, P., Tamassia, R., & Tollis, I. G. (1998). *Graph drawing: algorithms for the visualization of graphs.* Prentice Hall PTR.

Bavelas, A. (1948). A mathematical model for groups' structure. *Applied Anthropology, 7*(3), 16-30. Available in: http://www.jstor.org/stable/44135428

Bavelas, A. (1948). A mathematical model for group structures. *Applied Anthropology, 7*, 16–30.

Bavelas, A. (1950). Communication patterns in task oriented groups. *The Journal of the Acoustical Society of America, 22*(6), 725–730. doi:10.1121/1.1906679

Beauchamp, M. A. (1965). An improved index of centrality. *Behavioral Science, 10*(2), 161–163. doi:10.1002/bs.3830100205 PMID:14284290

Bebarta & Venkatesh. (2016). A *Low Complexity FLANN Architecture for Forecasting Stock Time Series Data Training with MetaHeuristic Firefly Algorithm.* Springer.

Bell, J., Deans, K., Ibbotson, P., & Sinkovics, R. (2001). Towards the "internetalization" of international marketing education. *Marketing Education Review, 22*(9), 69–79. doi:10.1080/10528008.2001.11488758

Benoumhani, M. (2006). The Number of topologies on a finite set. *J. Integer Sequences, 9*(2), 9–10.

Bera, D., Pal, M., & Pal, T. K. (2002). An efficient algorithm to generate all maximal cliques on trapezoid graphs. *International Journal of Computer Mathematics, 79*(10), 1057–1065. doi:10.1080/00207160212707

Bera, D., Pal, M., & Pal, T. K. (2003a). An efficient algorithm for finding all hinge vertices on trapezoid graphs. *Theory of Computing Systems, 36*(1), 17–27. doi:10.100700224-002-1004-3

Bera, D., Pal, M., & Pal, T. K. (2003b). An optimal PRAM algorithm for a spanning tree on trapezoid graphs. *Journal of Applied Mathematics and Computing, 12*(1–2), 21–29. doi:10.1007/BF02936178

Bera, D., Pal, M., & Pal, T. K. (2001). A parallel algorithm for computing all hinge vertices on interval graphs. *Korean Journal of Computational & Applied Mathematics, 8*(2), 295–309.

Berge, C. (1973). *Graphs and Hypergraphs*. Amsterdam: North Holland.

Berge, C. (1979). *Graphs and hypergraphs*. Amsterdam: North-Holland.

Berge, C. (1989). *Hypergraphs*. Amsterdam: North-Holland.

Berge, C. (1989). *Hypergraphs: Combinatorics of finite sets*. Amsterdam: North-Holland.

Berge, C. (2001). *Theory of graphs*. Dover Pub.

Berger, D. S., Sitaraman, R. K., & Harchol-Balter, M. (2017). AdaptSize: Orchestrating the Hot Object Memory Cache in a Content Delivery Network. In NSDI (pp. 483-498). Academic Press.

Bertossi, A. A., & Bonuccelli, M. A. (1995). Code assignment for hidden terminal interference avoidance in multi-hop packet radio networks. *IEEE/ACM Transactions on Networking, 3*(4), 441–449. doi:10.1109/90.413218

Bhadoria, R. S., Chaudhari, N. S., & Vidanagama, V. T. N. (2017). Analyzing the role of interfaces in enterprise service bus: A middleware epitome for service-oriented systems. *Computer Standards & Interfaces, 56*, 146–155.

Bhargav, T. N., & Ahlborn, T. J. (1968). On topological spaces associated with digraphs. *Acta Math. Acad. Scient. Hungar, 19*(1-2), 47–52. doi:10.1007/BF01894678

Bhutani, K. R. (1989). On automorphism of fuzzy graphs. *Pattern Recognition Letters, 9*(3), 159–162. doi:10.1016/0167-8655(89)90049-4

Bhutani, K. R., Moderson, J., & Rosenfeld, A. (2004). On degrees of end nodes and cut nodes in fuzzy graphs. *Iranian Journal of Fuzzy Systems, 1*(1), 57–64.

Billera, L. J. (1971). On the composition and decomposition of clutters. *Journal of Combinatorial Theory, 11B*, 243–245.

Bindhu, K. (2010). *Advanced Studies on Labeling of graphs and Hypergraphs and Related Topics* (Ph.D. Thesis). Kannur University, Kannur, Kerala, India.

Bindhu, K., & Germina, K.A. (2010). (k,r)-Arithmetic dcsl labeling of graphs. *Int. Math. Forum, 5*(45), 2237–2247.

Bindhu,, K., & Germina, K.A. (2010). Distance compatible set-labeling index of graphs. *Int. J. Contemp. Math. Sci., 5*(19), 911–919.

Bindhu,, K., Germina, K.A., & Joy. (2010). On Sequential topogenic graphs. *Int. J. Contemp. Math. Sci., 5*(36), 1799–1805.

Bloom, G. S., & Golomb, S. W. (1977). Applications of numbered undirected graphs. *Proceedings of the IEEE, 65*(4), 562–570. doi:10.1109/PROC.1977.10517

Bogart, K. P., Jacobson, M. S., Langley, L. J., & Mcmorris, F. R. (2001). Tolerance orders and bipartite unit tolerance graphs. *Discrete Mathematics, 226*(1), 35–50. doi:10.1016/S0012-365X(00)00124-2

Boginski, V., Sergiy, B., & Pardalos, P. M. (2006). Mining market data: A network approach. *Computers & Operations Research, 33*(11), 3171–3184. doi:10.1016/j.cor.2005.01.027

Bollabás, B. (1998). *Modern graph theory* (International Edition). Springer.

Bollobas, B., Jayawardene, C. J., Sheng, Y. J., Ru, H. Y., Rousseau, C. C., & Min, Z. K. (2000). On a Conjecture Involving Cycle-Complete Graph Ramsey Numbers. *The Australasian Journal of Combinatorics, 22*, 63–71.

Bolnokina, E. V., Oleinikova, S. A., & Kravets, O. J. (2019). Determination of optimal composition of team of executors for multistage service system. *International Journal on Information Technologies and Security, 11*(1), 51-58.

Bomze, I. M., Budinich, M., Pardalos, P. M., & Pelillo, M. (1999). The maximum clique problem. In D.-Z. Du & P. M. Pardalos (Eds.), *Handbook of combinatorial optimization* (pp. 1–74). Dordrecht: Kluwer Academic Publishers. doi:10.1007/978-1-4757-3023-4_1

Bonacich, P. (1972). Factoring and weighing approaches to status scores and clique identification. *The Journal of Mathematical Sociology, 2*(1), 113–120. doi:10.1080/0022250X.1972.9989806

Bonacich, P. (1987). Power and centrality: A family of measures. *American Journal of Sociology, 92*(5), 1170–1182. doi:10.1086/228631

Bonacich, P. (2007). Some unique properties of eigenvector centrality. *Social Networks, 29*(4), 555–564. doi:10.1016/j.socnet.2007.04.002

Bonacich, P., & Lloyd, P. (2001). Eigenvector-like measures of centrality for asymmetric relations. *Social Networks, 23*(3), 191–201. doi:10.1016/S0378-8733(01)00038-7

Bonanno, Vandewalle, & Mantegna. (2000). Taxonomy of stock market indices. *Physical Review E, 62*, R7615.

Bondy, J. A., & Murty, U. S. R. (1976). Graph Theory with Applications. Macmillan.

Bondy, J. A., & Chvátal, V. (1976). A method in graph theory. *Discrete Mathematics, 15*(2), 111–135. doi:10.1016/0012-365X(76)90078-9

Bondy, J. A., & Murty, U. S. R. (2008). *Graph theory*. Springer.

Borgatti, S. (2005). Centrality and network flow. *Social Networks, 27*(1), 55–71. doi:10.1016/j.socnet.2004.11.008

Borgatti, S., & Everett, R. (2006). A graph-theoretic perspective on centrality. *Social Networks, 28*(4), 466–484. doi:10.1016/j.socnet.2005.11.005

Brandes, U. (2001). A faster algorithm for betweenness centrality. *Journal of Mathematical Sociology, 25*(2), 163-177.

Brandstädt, A., Le, V. B., & Spinrad, J. P. (1999). *Graph classes: A survey*. Philadelphia: SIAM.

Brankov, V., Stevanovic, D., & Gutman, I. (2004). Equienergetic chemical trees. *Journal of the Serbian Chemical Society, 69*(7), 549–553. doi:10.2298/JSC0407549B

Brigham, R. C., & Dutton, R. D. (1990). Factor domination in graphs. *Discrete Mathematics, 86*(1-3), 127–136. doi:10.1016/0012-365X(90)90355-L

Britto Antony Xavier, G., Suresh, E., & Gutman, I. (2014). Counting relations for general Zagreb indices, Kragujevac, J. *Math., 38*, 95–103.

Broberg, J., Buyya, R., & Tari, Z. (2009). MetaCDN: Harnessing 'Storage Clouds' for high performance content delivery. *Journal of Network and Computer Applications*, *32*(5), 1012–1022. doi:10.1016/j.jnca.2009.03.004

Broumi, S., Talea, M., Bakali, A., & Smarandache, F. (2016). Bipolar Single Valued Neutrosophic Graphs. *New Ternds in Neutrosophic Theory and Appications*, 203-221.

Broumi, S., Bakali, A., Talea, M., & Smarandache, F. (2016). Isolated Single Valued Neutrosophic Graphs. *Neutrosophic Sets and Systems*, *11*, 74–78.

Broumi, S., Bakali, A., Talea, M., & Smarandache, F. (2017). Generalized Bipolar Neutrosophic Graphs of Type 1. *20th International Conference on Information Fusion*, 1714-1720.

Broumi, S., Bakali, A., Talea, M., Smarandache, F., & Rao, V. V. (2018). Interval Complex Neutrosophic Graph of Type 1. *Neutrosophic Operational Research*, *3*, 87–106.

Broumi, S., Bakali, A., Talea, M., Smarandache, F., & Verma, R. (2017). Computing Minimum Spanning Tree in Interval Valued Bipolar Neutrosophic Environment. *International Journal of Modeling and Optimization*, *7*(5), 300–304. doi:10.7763/IJMO.2017.V7.602

Broumi, S., Talea, M., Bakkali, A., & Smarandache, F. (2016). Single Valued Neutrosophic Graph. *Journal of New Theory*, *10*, 68–101.

Brualdi, R. A. (2006). Energy of a graph. *Notes to AIM Workshop on Spectra of Families of Articles described by graphs, digraphs, and sign patterns.*

Bruun, J., & Brewe, E. (2013). Talking and learning physics: Predicting future grades from network measures and Force Concept Inventory pretest scores. *Physical Review Physics Education Research*, *9*(2), 020109. doi:10.1103/PhysRevSTPER.9.020109

Buckley, F., & Harary, F. (1990). Distances in Graphs. Addison–Wesley.

Buckley, F. (1981). Iterated line graphs. *Congr. Numer.*, *33*, 390–394.

Buckley, F. (1993). The size of iterated line graphs. *Graph Theory Notes, New York*, *25*, 33–36.

Bulancev, V.Y., Zaripov, I.A., Kuren'kov, D.B., Lukashenko, S.N., & Miroshkin A.A. (2008). "Shine and poverty" of tax planning for imports. *International Banking.*

Buneman, P. (1974). A characterisation of rigid circuit graphs. *Discrete Mathematics*, *9*(3), 205–212. doi:10.1016/0012-365X(74)90002-8

Burger, A. P., & Vuuren, V. J. H. (2004). Ramsey numbers in Complete Balanced Multipartite Graphs. Part II: Size Numbers. *Discrete Mathematics*, *283*(1-3), 45–49. doi:10.1016/j.disc.2004.02.003

Burton, D. M. (2007). *Elementary number theory.* New Delhi: Tata McGraw-Hill Inc.

Bustince, H., & Burilloit, P. (1996). Vague sets are intuitionistic fuzzy sets. *Fuzzy Sets and Systems*, *79*(30), 403–405. doi:10.1016/0165-0114(95)00154-9

Byers, T. H., & Waterman, M. S. (1984). Determining all optimal and near-optimal solutions when solving shortest path problems by dynamic programming. *Operations Research*, *32*(6), 1381–1384. doi:10.1287/opre.32.6.1381

Calamoneri, T. (2014). The $L(h,k)$-labeling problem, an updated survey and annotated bibliography. *The Computer Journal*, *54*(8), 1–54.

Calamoneri, T., Caminiti, S., Petreschi, R., & Olariu, S. (2009). On the $L(h, k)$. *Networks*, *53*(1), 27–34. doi:10.1002/net.20257

Cancan & Aldemir. (2017). *On ve-degree and ev-degree Zagreb indices of titania nanotubes*. Academic Press.

Caparossi, G., Hansen, P., & Vukičević, D. (2010). Comparing Zagreb indices of cyclic graphs. *MATCH Commun. Math. Comput. Chem*, *63*, 44–451.

Cárdenas-Tapia, M., Klingler-Kaufman, C., & Rivas-Tovar, L. A. (2012, October). *Structural analysis of a knowledge network through graph theory*. Paper presented at the XVII International Accounting and Administration Congress, School of Accounting and Administration at National Autonomous University of Mexico, Mexico City. Available in: http://congreso.investiga.fca.unam.mx/es/congreso_xvii.php

Carlisle, M. C., & Lloyd, E. L. (1995). On the k-coloring of intervals. *Discrete Applied Mathematics*, *59*(3), 225–235. doi:10.1016/0166-218X(95)80003-M

Cartwright, D., & Zander, A. (Eds.). (1953). *Group dynamics*. Londres: Tavistock.

Ceballos, H. G., Fangmeyer, J. J. Jr, Galeano, N., Juarez, E., & Cantu-Ortiz, F. J. (2017). Impelling research productivity and impact through collaboration: A scientometric case study of knowledge management. *Knowledge Management Research and Practice*, *15*(3), 346–355. doi:10.105741275-017-0064-8

Central Intelligence Agency. (2019). *The World Factbook*. Available in: https://www.cia.gov/library/publications/resources/the-world-factbook/

Chalopin, J., Gonçalves, D., & Ochem, P. (2007). Planar graphs are in 1-STRING. *Proceedings of the Eighteenth Annual ACM-SIAM Symposium on Discrete Algorithms*, 609–617.

Chandran, L. S., & Sivadasan, N. (2006, August 15). *Geometric representation of graphs in low dimension*. Academic Press. doi:10.1007/11809678_42

Chandran, L. S., & Sivadasan, N. (2007). Boxicity and treewidth. *Journal of Combinatorial Theory Series B*, *97*(5), 733–744. doi:10.1016/j.jctb.2006.12.004

Chang, G. J., & Kuo, D. (1996). The $L(2,1)$. *SIAM Journal on Discrete Mathematics*, *9*(2), 309–316. doi:10.1137/S0895480193245339

Chartrand, G., & Lesniak, L. (1996). *Graphs and digraphs*. CRC Press.

Chartrand, G., & Zhang, P. (2005). *Introduction to graph theory*. McGraw-Hill Inc.

Cheah, F., & Corneil, D. G. (1996). On the structure of trapezoid graphs. *Discrete Applied Mathematics*, *66*(2), 109–133. doi:10.1016/0166-218X(94)00158-A

Chellali, M., Hynes, T. W., Hedetniemi, S. T., & Lewis, T. W. (2017). On ve-degrees and ev-degrees in graphs. *Discrete Mathematics*, *340*(2), 31–38. doi:10.1016/j.disc.2016.07.008

Chen, C. M. (2006). CiteSpace II: Detecting and visualizing emerging trends and transient patterns in scientific literature. *Journal of the American Society for Information Science and Technology*, *57*(3), 359377. doi:10.1002/asi.20317

Chen, E., Yang, L., & Yuan, H. (2007). Improved algorithms for largest cardinality 2-interval pattern problem. *Journal of Combinatorial Optimization*, *13*(3), 263–275. doi:10.100710878-006-9030-8

Cheng, F. F., Huang, Y. W., Tsaih, D. C., & Wu, C. S. (2019). Trend analysis of co-authorship network in Library Hi Tech. *Library Hi Tech*, *37*(1), 43–56. doi:10.1108/LHT-11-2017-0241

Chen, H., & Deng, H. (2018). The inverse sum indeg index of graphs with some given parameters. *Discrete Mathematics, Algorithms, and Applications, 10*(1). doi:10.1142/S1793830918500064

Chen, J., Li, S., Ma, S., & Wang, X. (2014). m- polar fuzzy sets: An extension of bipolar fuzzy sets. *The Scientific World Journal*, 1–8. PMID:25025087

Chen, S. M. (1997). Interval-valued fuzzy hypergraph and fuzzy partition. *IEEE Transactions on Systems, Man, and Cybernetics. Part B, Cybernetics, 27*(4), 725–733. doi:10.1109/3477.604121 PMID:18255914

Che, Z., & Chen, Z. (2016). Lower and upper bounds of the forgotten topological index. *MATCH Commun. Math. Comput. Chem., 76*, 635–648.

Chithra, K. P., Sudev, N. K., & Germina, K. A. (2014). A study on the sparing number of the corona of certain graphs, *Res. Review. Discrete Math. Structure (London, England), 1*(2), 5–15.

Chithra, K. P., Sudev, N. K., & Germina, K. A. (2014). The sparing number of the Cartesian product of certain graphs. *Commun. Math. Appl., 5*(1), 23–30.

Chithra, K. P., Sudev, N. K., & Germina, K. A. (2015). On the sparing number of the edge-corona of graphs. *International Journal of Computers and Applications, 118*(1), 1–5. doi:10.5120/20706-3025

Christoforidis, G., Stykas, V. A., & Kasso, T. (2018). Simulated comparison of push/pull production with committed and non-committed automated guided vehicles. *Proceedings of the 32nd International conference on Information Technologies - InfoTech-2018*.

Chuang, H. Y., Lee, E., Liu, Y. T., Lee, D., & Ideker, T. (2007). Network-based classification of breast cancer metastasis. *Molecular Systems Biology, 3*, 140–150. doi:10.1038/msb4100180 PMID:17940530

Chvatal, V., & Hammer, P. L. (1973). *Set-packing problems and threshold graphs. CORR, 73-21*. University of Waterloo.

Cielen, D., Meysman, A., & Ali, M. (2016). Introducing data science: big data, machine learning, and more, using Python tools. Manning Publications Co.

Clarivate. (2019). *Master Journal List. Scope Notes: Science Citation Index Expanded*. Available in: http://mjl.clarivate.com.uchile.idm.oclc.org/scope/scope_scie/

Clarke, J., & Holton, D. A. (1991). *A first look at graph theory*. Singapore: World Scientific Pub.

Cockayne, E. J., & Hedetniemi, S. T. (1974). Independence graphs. *Proc. Fifth. S.E. conf. on Combinotories, Graph Theory and Computing, Utilities Mathematics*, 471-479.

Cockayne, E. J., Dawqes, R. M., & Hedetniemi, S. T. (1980). Total domination in graphs. *Networks, 10*(3), 211–219. doi:10.1002/net.3230100304

Cohen, J. E. (1968). *Interval graphs and food webs: a finding and a problem. Document 17696-PR*. Santa Monica, CA: RAND Corporation.

Coles, N. (2001). Analyzing serious crime groups as social network. *British Journal of Criminology, 41*, 580–594. doi:10.1093/bjc/41.4.580

CORFO. (1962). *Twenty years of labor 1939 - 1959*. Santiago de Chile: Corporation for the Promotion of Production. Available in: http://www.memoriachilena.cl/602/w3-article-9037.html

Cormen, T. H., Leiserson, C. E., Rivest, R. L., & Stein, C. (2009). Introduction to algorithms. MIT Press.

Corneil, D., Olariu, S., & Stewart, L. (1997). Asteroidal triple-free graphs. *SIAM Journal on Discrete Mathematics*, *10*(3), 399–430. doi:10.1137/S0895480193250125

Cosyn, E. (2002). Coarsening a knowledge structure. *Journal of Mathematical Psychology*, *46*(2), 123–139. doi:10.1006/jmps.2001.1376

Coulson, C. A. (1940). On the calculation of the energy in unsaturated hydrocarbon molecules. *Proceedings of the Cambridge Philosophical Society*, *36*(2), 201–203. doi:10.1017/S0305004100017175

Crochemore, M., Hermelin, D., Landau, G. M., Rawitz, D., & Vialette, S. (2008). Approximating the 2-interval pattern problem. *Theoretical Computer Science*, *395*(2), 283–297. doi:10.1016/j.tcs.2008.01.007

Cvetkovic, D. M., Doob, M., & Sachs, H. (1980). *Spectra of Graphs*. New York: Academic Press.

Cvetkovic, D. M., & Gutman, I. (1986). The computer system GRAPH: A useful tool in chemical graph theory. *Journal of Computational Chemistry*, *7*, 640–644. doi:10.1002/jcc.540070505

Da Fonseca, C. M., & Stevanović, D. (2014). Further properties of the second Zagreb index. *MATCH Commun. Math. Comput. Chem.*, *72*, 655–668.

Dagan, I., Golumbic, M. C., & Pinter, R. Y. (1988). Trapezoid graphs and their coloring. *Discrete Applied Mathematics*, *21*(1), 35–46. doi:10.1016/0166-218X(88)90032-7

Daneshvar-Rouyendegh, B., & Feryal-Can, G. (2012). Selection of working area for industrial engineering students. *Procedia: Social and Behavioral Sciences*, *31*, 15–19. doi:10.1016/j.sbspro.2011.12.008

Das, K. C. (2010). Atom bond connectivity index of graphs. *Discrete Applied Mathematics*, *158*(11), 1181–1188. doi:10.1016/j.dam.2010.03.006

Das, K. C. (2010). On geometric-arithmetic index of graphs. *MATCH Commun. Math. Comput. Chem.*, *64*, 619–630.

Das, K. C., Das, S., & Zhou, B. (2016). Sum connectivity index of a graph. *Frontiers of Mathematics in China*, *11*(1), 47–54. doi:10.100711464-015-0470-2

Das, K. C., & Gutman, I. (2004). Some properties of the second Zagreb index. *MATCH Commun. Math. Comput. Chem.*, *52*, 103–112.

Das, K. C., Gutman, I., & Furtula, B. (2011). On atom bond connectivity index. *Chemical Physics Letters*, *511*(4-6), 452–454. doi:10.1016/j.cplett.2011.06.049

Das, K. C., Gutman, I., & Horoldagva, B. (2012). Comparing Zagreb indices and coindices of trees. *MATCH Commun. Math. Comput. Chem.*, *67*, 189–198.

Das, K. C., Gutman, I., & Zhou, B. (2009). New upper bounds on Zagreb indices. *Journal of Mathematical Chemistry*, *46*(2), 514–521. doi:10.100710910-008-9475-3

Das, K. C., & Trinajstić, N. (2010). Comparison between first geometric-arithmetic index and atom-bond connectivity index. *Chemical Physics Letters*, *497*(1-3), 149–151. doi:10.1016/j.cplett.2010.07.097

Das, K. C., Xu, K., & Gutman, I. (2013). On Zagreb and Harary indices. *MATCH Commun. Math. Comput. Chem.*, *70*, 301–314.

Das, K., Samanta, S., & Pal, M. (2018). Study on centrality measures in social networks: A survey. *Social Network Analysis and Mining*, 1–11.

DataStax Enterprise Graph Super-Powering Your Data Relationships. (n.d.). Retrieved October 28, 2018, from https://www.datastax.com/products/datastax-enterprise-graph/

De, Nayeem, & Pal. (2016). F-index of some graph operations. *Discrete Math. Algorithms Appl., 8.*

deBruijn, N. G. (1959). Generalization of Polya's fundamental theorem in enumerative combinatorial analysis. *Indagationes Mathematicae, 21*, 59–69. doi:10.1016/S1385-7258(59)50008-6

Delgado-Saab, I., Carapaica, I., & Ortiz-Sosa, D. (2016). Software tool for problem analysis whose abstraction is based on graphs. *Tekhné, 1*(10). Available in: http://revistasenlinea.saber.ucab.edu.ve/temas/index.php/tekhne/article/view/3002

Deli, M., & Smarandache, F. (2015). Bipolar neutrosophic sets and their application based on multi-criteria decision making problems. *Advanced Mechatronic Systems (ICAMechS),* 249-254.

De, N. (2016). F-index of bridge and chain graphs. *Malay. J. Fund. Appl. Sci., 12*, 109–113.

Deng, H., Saralab, D., Ayyaswamy, S. K., & Balachandran, S. (2016). The Zagreb indices of four operations on graphs. *Applied Mathematics and Computation, 275*, 422–431. doi:10.1016/j.amc.2015.11.058

Deng, X., Hell, P., & Huang, J. (1996). Linear-time representation algorithms for proper circular-arc graphs and proper interval graphs. *SIAM Journal on Computing, 25*(2), 390–403. doi:10.1137/S0097539792269095

Deo, N. (1974). *Graph theory with application to engineering and computer science.* Delhi: Prentice Hall of India Pvt. Ltd.

Deza & Sikirić. (2008). Geometry of Chemical Graphs: Polycycles and Two-Faced Maps. Encyc. Math. Appl., 119.

Dirac, G. A. (1961). On rigid circuit graphs. *Abhandlungen aus dem Mathematischen Seminar der Universität Hamburg, 25*(1), 71–76. doi:10.1007/BF02992776

Dordick, H. S., & Wang, G. (1993). *Information society: A retrospective view.* Sage Publications, Inc.

Doslic, T., Furtula, B., Graovać, A., Gutman, I., Moradi, S., & Yarahmadi, Z. (2011). On vertex degree based molecular structure descriptors. *MATCH Commun. Math. Comput. Chem., 66*, 613–626.

Dragomir, S. S. (2003). A survey on Cauchy-Buyakorsky-Schwarz type discrete inequalities. *Journal of Inequalities in Pure and Applied Mathematics, 4*(3), 63.

Dreżewski, R., Sepielak, J., & Filipkowski, W. (2015). The application of social network analysis algorithms in a system supporting money laundering detection. *Information Sciences, 295*, 18–32. doi:10.1016/j.ins.2014.10.015

Dubois, D., & Prade, H. (1983). Ranking fuzzy numbers in the setting of possibility theory. *Information Sciences, 30*(3), 183–224. doi:10.1016/0020-0255(83)90025-7

Duchet, P. (1984). Classical perfect graphs: an introduction with emphasis on triangulated and interval graphs. In *North-Holland mathematics studies* (Vol. 88, pp. 67–96). North-Holland.

Dugundji, J. (1966). *Topology.* Boston: Allyn and Bacon.

Ediz, S. (2015). Maximum chemical trees of the second reverse Zagreb index. *Pacific J. Appl. Math., 7*(4), 291–295.

Ediz, S. (2017). A new tool for QSPR researches: Ve-degree Randić index. *Celal Bayar University Journal of Science, 13*(3), 615–618.

Ediz, S. (2017). Predicting some physicochemical properties of octane isomers: A topological approach using ve-degree and ve-degree Zagreb indices. *International Journal of System Science and Applied Mathematics, 2*, 87–92.

Ediz, S. (2018). Maximal graphs of the first reverse Zagreb beta index. *TWMS J. Appl. Eng. Math*, *8*, 306–310.

Ediz, S. (2018). On ve-degree molecular topological properties of silicate and oxygen networks, Int. J. *Computing Science and Mathematics*, *9*(1), 1–12.

Ediz, S., & Cancan, M. (2016). Reverse Zagreb indices of Cartesian product of graphs. *International Journal of Mathematics and Computer Science*, *11*(1), 51–58.

Edmonds, J., & Fulkerson, D. R. (1970). Bottleneck extrema. *J. Comb. Theory*, *8*(3), 299–306. doi:10.1016/S0021-9800(70)80083-7

Ehrlich, G., Even, S., & Tarjan, R. E. (1976). Intersection graphs of curves in the plane. *Journal of Combinatorial Theory Series B*, *21*(1), 8–20. doi:10.1016/0095-8956(76)90022-8

Eliasi, M., Irammanesh, A., & Gutman, I. (2012). Multiplicative versions of first Zagreb index. *MATCH Commun. Math. Comput. Chem.*, *68*, 217–230.

Eliasi, M., & Taeri, B. (2009). Four new sums of graphs and their Wiener indices. *Discrete Applied Mathematics*, *157*(4), 794–803. doi:10.1016/j.dam.2008.07.001

Escobar, A. (2005). Welcome to Cyberia. Notes for an anthropology of the cyberculture. *Revista de Estudios Sociales*, (22): 15–35. doi:10.7440/res22.2005.01

Estrada, E. (2000). Characterization of 3D molecular structure. *Chemical Physics Letters*, *319*(5-6), 713–718. doi:10.1016/S0009-2614(00)00158-5

Estrada, E., & Rodriguez-Velazquez, J. A. (2005). Subgraph centrality in complex networks. *Physical Review*, *71*, 056103. PMID:16089598

Estrada, E., Torres, L., Rodriguez, L., & Gutman, I. (1998). An atom bond connectivity index: Modeling the enthalpy of formation of alkanes. *Indian Journal of Chemistry*, *37A*, 849–855.

Evans, J. W., Harary, F., & Lynn, M. S. (1967). On Computer enumeration of finite topologies. *Communications of the ACM*, *10*(5), 295–298. doi:10.1145/363282.363311

Everett, M. G., & Borgatti, S. P. (1999). The centrality of groups and classes. *The Journal of Mathematical Sociology*, *23*(3), 181–201. doi:10.1080/0022250X.1999.9990219

Evers, H. D., Gerke, S., & Menkoff, T. (2010). Knowledge clusters and knowledge hubs: Designing epistemic landscapes for development. *Journal of Knowledge Management*, *14*(5), 678–689.

Exoo, G. (1998). A lower bound for r(5, 5). *Journal of Graph Theory*, *13*(1), 97–98. doi:10.1002/jgt.3190130113

Fabri, J. (1982). *Automatic storage optimization*. Academic Press.

Fajtolowicz, S. (1988). On conjectures of Grafitti. *Discrete Mathematics*, *72*(1-3), 113–118. doi:10.1016/0012-365X(88)90199-9

Falahati Nezhad, F., & Azari, M. (2016). Bounds on the hyper Zagreb index. *J. Appl. Math. Inform.*, *34*(3_4), 319–330. doi:10.14317/jami.2016.319

Falahati-Nezhad, F., & Azari, M. (2016). The inverse sum indeg index of some nanotubes. *Studia Ubb Chemia*, *LXI*(1), 63–70.

Falahati-Nezhad, F., Azari, M., & Došlić, T. (2017). Sharp bounds on the inverse sum indeg index. *Discrete Applied Mathematics*, *217*(2), 185–195. doi:10.1016/j.dam.2016.09.014

Fath-Tabar, G. H. (2011). Old and new Zagreb indices of graphs. *MATCH Commun. Math. Comput. Chem., 65*, 79–84.

Feng, L., Zhang, P., Liu, H., Liu, W., Liu, M., & Hu, Y. (2017). Spectral conditions for some graphical properties. *Linear Algebra and Its Applications, 524*, 182–198. doi:10.1016/j.laa.2017.03.006

Festinger, L. (1949). The analysis of sociograms using matrix algebra. *Human Relations, 2*(2), 153–158. doi:10.1177/001872674900200205

Fiedler, M., & Nikiforov, V. (2010). Spectral radius and Hamiltonicity of graphs. *Linear Algebra and Its Applications, 432*(9), 2170–2173. doi:10.1016/j.laa.2009.01.005

Fiksel, J. (1980). Dynamic evolution of societal networks. *The Journal of Mathematical Sociology, 7*(1), 27–46. doi:10.1080/0022250X.1980.9989897

Finck, H. J., & Grohmann, G. (1965). Vollstandiges Produkt, chromatische Zahl und charakteristisches Polynom regular Graphen I. *Wiss. Z. TH Ilmenau, 11*, 1–3.

Firdous, S., Nazeer, W., & Farahani, M. R. (2018). Mathematical properties and computations of Banhatti indices for a nanostructure "Toroidal Polyhex Network". *Asian Journal of Nanoscience and Materials, 1*, 43–47.

Flotow, C. (1995). On powers of m-trapezoid graphs. *Discrete Applied Mathematics, 63*(2), 187–192. doi:10.1016/0166-218X(95)00062-V

Fomin, F. V., Gaspers, S., Golovach, P., Suchan, K., Szeider, S., van Leeuwen, E. J., … Villanger, Y. (2012). k-gap interval graphs. In D. Fernández-Baca (Ed.), LATIN 2012: Theoretical Informatics (pp. 350–361). Springer Berlin Heidelberg.

François, K. (2007). *Biological Network. Complex Systems and Interdisciplinary Science – Volume 3*. World Scientific Publishing Company.

Freeman, L. C. (1977). A Set of Measures of Centrality Based on Betweenness. *Sociometry, 40*(1), 35–41. doi:10.2307/3033543

Freeman, L. C. (1978). Centrality in social networks conceptual clarification. *Social Networks, 1*(3), 215–239. doi:10.1016/0378-8733(78)90021-7

Freeman, L. C. (1984). Turning a profit from mathematics: The case of social networks. *The Journal of Mathematical Sociology, 10*(3-4), 343–360. doi:10.1080/0022250X.1984.9989975

Freeman, L. C., Borgatti, S. P., & White, D. R. (1991). Centrality in valued graphs: A measure of betweenness based on network flow. *Social Networks, 13*(2), 141–154. doi:10.1016/0378-8733(91)90017-N

Fulkerson, D., & Gross, O. (1965). Incidence matrices and interval graphs. *Pacific Journal of Mathematics, 15*(3), 835–855. doi:10.2140/pjm.1965.15.835

Furtula, B., Graovac, A., & Vukičević, D. (2009). Atom bond connectivity index of trees. *Discrete Applied Mathematics, 157*(13), 2828–2835. doi:10.1016/j.dam.2009.03.004

Furtula, B., & Gutman, I. (2015). A forgotten topological index. *Journal of Mathematical Chemistry, 53*(4), 1184–1190. doi:10.100710910-015-0480-z

Furtula, B., Gutman, I., & Ediz, S. (2014). On difference of Zagreb indices. *Discrete Applied Mathematics, 178*, 83–88. doi:10.1016/j.dam.2014.06.011

Galinier, P., Habib, M., & Paul, C. (1995). Chordal graphs and their clique graphs. In M. Nagl (Ed.), *Graph-Theoretic Concepts in Computer Science* (pp. 358–371). Springer Berlin Heidelberg.

Gallian. (2011). A Dynamic survey of graph labelling. *Electronic J. Combin., Dynamic Surveys.*

Gallian, J. A. (2018). A dynamic survey of graph labelling. *The Journal of Combinatorics*, DS-6.

Gallo, G., Longo, G., Nguyen, S., & Pallottino, S. (1993). Directed hypergraphs and applications. *Discrete Applied Mathematics*, *40*(2-3), 177–201. doi:10.1016/0166-218X(93)90045-P

Gambardella, A., & Giarratana, M. (2010). Organizational attributes and the distribution of rewards in a region: Managerial firms vs. knowledge clusters. *Organization Science*, *21*(2), 573–586.

Gan, L., Hou, H., & Liu, B. (2011). Some results on atom bond connectivity index of graphs. *MATCH Commun. Math. Comput. Chem.*, *66*, 669–680.

Gao, W., Farahani, M. R., Siddiqui, M. K., & Jamil, M. K. (2016). On the first and second Zagreb and first and second hyper Zagreb indices of carbon nanocones CNC_k [n]. *Journal of Computational and Theoretical Nanoscience*, *13*(10), 7475–7482. doi:10.1166/jctn.2016.5742

Gao, W., Jamil, M. K., & Farahani, M. R. (2017). The hyper-Zagreb index and some graph operations. *J. Appl. Math. Comput.*, *54*(1-2), 263–275. doi:10.100712190-016-1008-9

Gao, W., Jamil, M. K., Javed, A., Farahani, M. R., & Imran, M. (2018). Inverse sum indeg index of the line graphs of subdivision graphs of some chemical structures. *U. P. B. Sci. Bull. Series B*, *80*(3), 97–104.

Gao, W., Muzaffar, B., Nazeer, W., & Banhatti, K. (2017). K-Banhatti and K-hyper Banhatti Indices of Dominating David Derived Network. *Open J. Math. Anal.*, *1*(1), 13–24. doi:10.30538/psrp-oma2017.0002

Gao, W., Siddiqui, M. K., Imran, M., Jamil, M. K., & Farahani, M. R. (2016). Forgotten topological index of chemical structure in drugs. *Saudi Pharmaceutical Journal*, *24*(3), 258–264. doi:10.1016/j.jsps.2016.04.012 PMID:27275112

Gao, W., Yonuas, M., Farooq, A., Virk, A. R., & Nazeer, W. (2018). Some reverse degree based topological indices and polynomials of dendrimers. *Mathematics*, *6*(10), 214. doi:10.3390/math6100214

Garey, M. R., & Johnson, D. S. (1990). *Computers and Intractability; A Guide to the Theory of NP-Completeness*. New York, NY: W. H. Freeman & Co.

Garfield, E. (1987). Little Science, Big Science…and beyond gathers together the major works of Derek de Solla Price. *Essays of an Information Scientist*, *10*(11), 72. Available in: http://www.garfield.library.upenn.edu/essays/v10p072y1987.pdf

Gau, W. L., & Buehrer, D. J. (1993). Vague sets. *IEEE Transactions on Systems, Man, and Cybernetics*, *23*(2), 610–614. doi:10.1109/21.229476

Gavril, F. (1974). The intersection graphs of subtrees in trees are exactly the chordal graphs. *Journal of Combinatorial Theory Series B*, *16*(1), 47–56. doi:10.1016/0095-8956(74)90094-X

Germina & Jose. (2011). Distance pattern distinguishable trees. *Int. Math. Forum, 12*, 591–604.

Germina & Jose. (2012). A creative review on distance pattern distinguishing sets in a graph. *J. Combin. Inform. Syst. Sci*, *37*(2-4), 267–278.

Germina & Joseph. (2011). Some general results on distance pattern distinguishable graphs. *Int. J. Contemp. Math. Sci.*, *6*, 713–720.

Germina & Joy. (2009). Topogenic graphs: II. Embeddings. *Indian J. Math.*, *51*(3), 645–661.

Germina & Joy. (2010). Enumeration of graphical realization. *Int. J. Algorithms Comput. Math.*, *3*(1), 31–46.

Germina & Koshy. (2009). Independent complementary distance pattern uniform graphs. *J. Math. Combin.*, *4*, 63–74.

Germina & Koshy. (2010). Complementary distance pattern uniform graphs. *Int. J. Contemp. Math. Sci.*, *5*(55), 2745–2751.

Germina & Koshy. (2011). New perspectives on CDPU graphs. *General Math. Notes*, *4*(1), 90–98.

Germina & Kurian. (2011). Bi-distance pattern uniform number. *Int. Math. Forum, 7*(27), 1303–1308.

Germina & Kurian. (2012). Bi-distance pattern uniform number. *Int. Math. Forum, 7*(27), 1303–1308.

Germina & Marykutty. (2012). Open distance pattern coloring of a graph. *J. Fuzzy Set Valued Analysis.* doi:10.5899/2012/jfsva-00144

Germina K.A., (2011). *Set-valuations of a graph and applications.* Final Technical Report, DST Grant-In-Aid Project No.SR/S4/277/05, The Dept. of Science and Technology (DST), Govt. of India.

Germina, & Kurian. (2013). Bi-distance pattern uniform graphs. *Proc. Jangjeon Math. Soc.*, *16*(1), 87–90.

Germina, Joseph, & Jose. (2010). Distance neighbourhood pattern matrices. *European J. Pure Appl. Math.*, *3*(4), 748–764.

Germina, Joy, & Thomas. (2010). On Sequential topogenic graphs. *Proc. Int. Conf. on Mathematics and Computer Sci.,* (5-6), 131–134.

Germina, K. A. (2010). Distance pattern distinguishing sets in a graph. *Int. Math. Forum, 5*(34), 1697–1704.

Germina, K. A. (2011). *Set-valuations of Graphs and Applications.* Technical Report, DST Grant-In-Aid Project No. SR/S4/277/05. Department of Science and Technology (DST), Govt. of India.

Germina, K. A., & Sudev, N. K. (2013). On weakly uniform integer additive set-indexers of graphs. *Int. Math. Forum, 8*(37), 1827-1834. DOI: 10.12988/imf.2013.310188

Germina, & Bindhu, K. (2010). On bitopological graphs. *Int. J. Algorithms Comput. Math.*, *4*(1), 71–79.

Germina, K. A. (2012). Uniform Distance-compatible set-labelings of graphs. *J. Combin. Inform. Syst. Sci*, *37*(2-4), 179–188.

Germina, K. A. (2013). Out set-magic digraphs. *J. Discrete Math. Sci. Cryptog.*, *16*(1), 45–59. doi:10.1080/09720529.2013.778458

Germina, K. A., Abhishek, K., & Princy, K. L. (2008). Further results on set-valued graphs. *J. Discrete Math. Sci. Cryptog, 11*(5), 559–566. doi:10.1080/09720529.2008.10698208

Germina, K. A., & Anandavally, T. M. K. (2012). Integer additive set-indexers of a graph: Sum square graphs. *J. Combin. Inform. System Sci.*, *37*(2-4), 345–358.

Ghorai, G., & Pal, M. (2015). On some operations and density of m-polar fuzzy graphs, Pacific Science Review A. *Natural Science and Engineering*, *17*(1), 14–22.

Ghorai, G., & Pal, M. (2016). Some properties of m-polar fuzzy graphs, Paci_c Science Review A. *Natural Science and Engineering*, *18*(1), 38–46.

Ghorai, G., & Pal, M. (2016a). A study on m-polar fuzzy planar graphs. *International Journal of Computing Science and Mathematics*, *7*(3), 283–292. doi:10.1504/IJCSM.2016.077854

Ghorai, G., & Pal, M. (2016b). Faces and dual of m-polar fuzzy planar graphs. *Journal of Intelligent & Fuzzy Systems*, *31*(3), 2043–2049. doi:10.3233/JIFS-16433

Ghorai, G., & Pal, M. (2016c). Some isomorphic properties of m-polar fuzzy graphs with applica-tions. *SpringerPlus*, *5*(1), 1–21. doi:10.118640064-016-3783-z PMID:28066695

Ghoseiri, K., & Moghadam, A. R. J. (2008). Continuous fuzzy longest path problem in project networks. *Journal of Applied Sciences (Faisalabad)*, *8*(22), 4061–4069. doi:10.3923/jas.2008.4061.4069

Ghosh, S., & Pal, A. (2018). Exact algorithm for -labeling of cartesian product between complete bipartite graph and path. *ACES Conference*.

Ghosh, S., & Pal, A. (2016). $L(3,1)$-labeling of some simple graphs. *Advanced Modeling and Optimization*, *18*(2), 243–248.

Ghosh, S., Paul, S., & Pal, A. (2017). -Labeling of cartesian product of complete bipartite graph and path. *Journal of Informatics and Mathematical Sciences*, *9*(3), 685–698.

Ghouilo-Houri, A. (1962). *Characterisation des graphs non orientes dont on peut orienter les arretes de maniere a obtenir le graphe d'une relation d'ordre* (Vol. 254). Paris: C. R. Acad. Sci.

Gilmore, P. C., & Hoffman, A. J. (1964). A characterization of comparability graphs and of interval graphs. *Canadian Journal of Mathematics*, *16*, 539–548. doi:10.4153/CJM-1964-055-5

Gilsing, V., Nooteboom, B., Vanhaverbeke, W., Duysters, G., & van der Oord, A. (2008). Network embeddedness and the exploration of novel technologies: Technological distance, betweenness centrality and density. *Research Policy*, *37*(10), 1717–1731. doi:10.1016/j.respol.2008.08.010

Giuliani, E. (2005). *The Structure of Cluster Knowledge Networks Uneven, not Pervasive and Collective*. DRUID Working Papers 05-11, DRUID, Copenhagen Business School, Department of Industrial Economics and Strategy/Aalborg University, Department of Business Studies. Available in: https://ideas.repec.org/p/aal/abbswp/05-11.html

Godsil, C., & Royle, G. (2001). *Algebraic Graph Theory*. New York: Springer. doi:10.1007/978-1-4613-0163-9

Goetschel, R. H. Jr. (1995). Introduction to fuzzy hypergraphs and Hebbian structures. *Fuzzy Sets and Systems*, *76*(1), 113–130. doi:10.1016/0165-0114(94)00381-G

Goetschel, R. H. Jr. (1998). Fuzzy colorings of fuzzy hypergraphs. *Fuzzy Sets and Systems*, *94*(2), 185–204. doi:10.1016/S0165-0114(96)00256-4

Goetschel, R. H. Jr, & Voxman, W. (1998). Intersecting fuzzy hypergraphs. *Fuzzy Sets and Systems*, *99*(1), 81–96. doi:10.1016/S0165-0114(97)00005-5

Golberg, D. E. (1989). *Genetic algorithms in search, optimization, and machine learning*. Reading, MA: Addison Wesley.

Golomb, S. W. (1972). How to number a graph. In R. C. Read (Ed.), *Graph Theory and Computing* (pp. 13–22). Academic Press.

Golomb, S. W. (1972). How to Number a Graph. In R. C. Read (Ed.), *Graph Theory and Computing* (pp. 23–37). New York: Academic Press. doi:10.1016/B978-1-4832-3187-7.50008-8

Golomb, S. W. (1974). The Largest graceful subgraph of the complete graph. *The American Mathematical Monthly*, *81*(5), 499–501. doi:10.1080/00029890.1974.11993597

Golumbic, M. C. (1980). *Algorithmic graph theory and perfect graphs* (Vol. 57). Academic Press.

Golumbic. (2004). *Algorithmic graph theory and perfect graphs* (2nd ed.). Elsevier.

Golumbic, M. C. (1980). *Algorithmic Graph Theory and Perfect Graphs*. San Diego, CA: Academic Press.

Golumbic, M. C., Monma, C. L., & Trotter, W. T. (1984). Tolerance graphs. *Discrete Applied Mathematics, 9*(2), 157–170. doi:10.1016/0166-218X(84)90016-7

Golumbic, M. C., & Trenk, A. N. (2004). *Tolerance Graphs*. Cambridge, UK: Cambridge University Press.

Goncalves, D. (2008). On the $L(d,1)$-labelinng of graphs. *Discrete Mathematics, 308*, 1405–1414.

Goncharov, V. N. (2018). Informatization of society: social and economic aspects of development. *Proceedings of the 8th International scientific Conference Informatization of society: Socio-Economic, Socio-Cultural and International Aspects.*

Gong, S., Li, X., Xu, G., Gutman, I., & Furtula, B. (2015). Borderenergetic graphs. *MATCH Commun. Math. Comput. Chem., 74*, 321–332.

Graovac, A., Ghorbani, M., & Hosseinzadeh, M. A. (2011). Computing fifth geometric-arithmetic index of nanostar dendrimers. *Journal of Mathematical Nanoscience, 1*(1), 33–42.

Griggs, J. R., & Yeh, R. K. (1992). Labeling graphs with a condition at distance two. *SIAM Journal on Discrete Mathematics, 5*(4), 586–595. doi:10.1137/0405048

Griggs, J., & West, D. (1980). Extremal Values of the Interval Number of a Graph. *SIAM Journal on Algebraic Discrete Methods, 1*(1), 1–7. doi:10.1137/0601001

Gross, J., & Yellen, J. (1999). *Graph theory and its applications*. CRC Press.

Grunspan, D. Z., Wiggins, B. L., & Goodreau, S. M. (2014). Understanding Classrooms through Social Network Analysis: A Primer for Social Network Analysis in Education Research. *CBE Life Sciences Education, 13*(2), 167–178. doi:10.1187/cbe.13-08-0162 PMID:26086650

Guimera, R. (2005) The worldwide air transportation network: anomalous centrality, community structure, and cities global roles. *Proceedings of the National Academy of Sciences.* 10.1073/pnas.0407994102

Guo, L., Wang, S., Kang, L., Li, Q., Chen, G., & Li, C. (2014). A method of manufacture resource informatization in cloud manufacturing. *Journal of Software Engineering, 8*(1), 32-40. Doi:10.3923/jse.2014.32.40

Gutiérrez-Jarpa, G., Laporte, G., Marianov, V., & Moccia, L. (2017). Multi-objective rapid transit network design with modal competition: The case of Concepción, Chile. *Computers & Operations Research, 78*, 27–43. doi:10.1016/j.cor.2016.08.009

Gutiérrez-Jarpa, G., Obreque, C., Laporte, G., & Marianov, V. (2013). Rapid transit network design for optimal cost and origin–destination demand capture. *Computers & Operations Research, 40*(12), 3000–3009. doi:10.1016/j.cor.2013.06.013

Gutman, I., Milovanović, E., & Milovanović, I. (2018). Beyond the Zagreb indices. *AKCE International Journal of Graph and Combinatorics.*

Gutman, Milovanović, & Milovanović. (2018). Beyond the Zagreb indices. *AKCE International Journal of Graphs and Combinatorics.* .2018.05.002 doi:10.1016/jakcej

Gutman, I. (1977). Acyclic systems with extremal Huckel π-electron energy. *Theoretica Chimica Acta, 45*(2), 79–87. doi:10.1007/BF00552542

Gutman, I. (1978). The energy of a graph. *Ber Math Stat Sekt Forsch Graz, 103*, 1–22.

Gutman, I. (1978). The energy of a graph. *Ber. Math. Stat. Sekt. Forschungszentrum Graz, 103*, 1–22.

Gutman, I. (1999). Hyperenergetic molecular graphs. *Journal of the Serbian Chemical Society, 64*, 199–205.

Gutman, I. (2001). *The energy of a graph: old and new results. In Algebraic Combinatorics and Applications* (pp. 196–211). Berlin: Springer.

Gutman, I. (2001). The energy of a graph: old and new results. In A. Betten, A. Kohnert, R. Laue, & A. Wassermann (Eds.), *Algebraic Combinatorics and Applications* (pp. 196–211). Berlin: Springer. doi:10.1007/978-3-642-59448-9_13

Gutman, I. (2013). Degree based topological indices. *Croatica Chemica Acta, 86*(4), 351–361. doi:10.5562/cca2294

Gutman, I. (2015). Open problems for equienergetic graphs. *Iranain J. Math. Chem., 6*, 185–187.

Gutman, I. (2017). On hyper Zagreb index and coindex, Bull. Acad. Sebre Sci. Arts. *Cl. Sci. Math. Natur., 150*, 1–8.

Gutman, I., & Das, K. C. (2004). The first Zagreb index 30 years after. *MATCH Commun. Math. Comput. Chem., 50*, 83–92.

Gutman, I., & Furtula, B. (Eds.). (2008). *Recent Results in the theory of Randić index*. Kragujevac: University Kragujevac.

Gutman, I., Furtula, B., & Elphick, C. (2014). Three new/old vertex degree based topological indices. *MATCH Commun. Math. Comput. Chem., 72*, 617–682.

Gutman, I., Furtula, B., Vukićević, Ž. K., & Popivoda, G. (2015). On Zagreb indices and Coindices. *MATCH Commun. Math. Comput. Chem., 74*, 5–16.

Gutman, I., Hou, Y., Walikar, H. B., Ramane, H. S., & Hampiholi, P. R. (2000). No Huckel graph is hyperenergetic. *Journal of the Serbian Chemical Society, 65*(11), 799–801. doi:10.2298/JSC0011799G

Gutman, I., Kulli, V. R., Chaluvaraju, B., & Boregowda, H. S. (2017). On Banhatti and Zagreb indices. *Journal of the International Mathematical Virtual Institute, 7*, 53–67.

Gutman, I., Matejić, M., Milovanović, E., & Milovanović, I. (2020). Lower bounds for inverse sum indeg index of graphs. *Kragujevac J. Math., 44*(4), 551–562.

Gutman, I., & Polansky, O. E. (1986). *Mathematical Concepts in Organic Chemistry*. Berlin: Springer. doi:10.1007/978-3-642-70982-1

Gutman, I., & Rodr. (2019). Linear and non-linear inequalities on the inverse sum indeg index. *Discrete Applied Mathematics, 258*, 123–134. doi:10.1016/j.dam.2018.10.041

Gutman, I., Ruščić, B., Trinajstić, N., & Wilcox, C. F. Jr. (1975). Graph theory and molecular orbitals. XII. Acyclic polyenes. *The Journal of Chemical Physics, 62*(9), 3399–3405. doi:10.1063/1.430994

Gutman, I., Togan, M., Yurtlas, A., Cavik, A. S., & Cangul, I. N. (2018). Inverse problem for sigma index. *MATCH Commun. Math. Comput. Chem., 79*, 491–508.

Gutman, I., & Trinajstić, N. (1972). Graph theory and molecular orbitals. Total π-electron energy of alternant hydrocarbons. *Chemical Physics Letters, 17*(4), 535–538. doi:10.1016/0009-2614(72)85099-1

Haemers, W. H. (2008). Strongly regular graphs with maximal energy. *Linear Algebra and Its Applications, 429*(11-12), 2719–2723. doi:10.1016/j.laa.2008.03.024

Haemers, W. H., & Xiang, Q. (2010). Strongly regular graphs with parameters $(4m^4, 2m^4 + m^2, m^4 + m^2, m^4 + m^2)$ exist for all $m > 1$. *European Journal of Combinatorics, 31*, 1553–1559. doi:10.1016/j.ejc.2009.07.009

Hage, P., & Harary, F. (1983). *Structural models in anthropology*. Cambridge, UK: Cambridge University Press.

Hage, P., & Harary, F. (1995). Ecentricity and centrality in networks. *Social Networks*, *17*(1), 57–63. doi:10.1016/0378-8733(94)00248-9

Hai-Long, She, & Yanhonge, & Xiuwu. (2015). On single valued neutrosophic relations. *Journal of Intelligent & Fuzzy Systems*, 1–12.

Hall, M. (1986). *Combinatorial Theory*. New York: Wiley.

Hammack, R., Imrich, W., & Clavzar, S. (2011). *Handbook of product graphs*. CRC Press.

Harary & Norman. (1953). *Graph Theory as a Mathematical Model in Social Science*. University of Michigan Institute of Social Research.

Harary, F. (1953). On the notion of balance of a signed graph. *The Michigan Mathematical Journal*, *2*(2), 143–146.

Harary, F. (1967). *Graph Theory and Theoretical Physics*. London: Academic Press.

Harary, F. (1972). *Graph Theory*. Reading, MA: Addison–Wesley.

Harary, F., Kabell, J. A., & McMorris, F. R. (1982). Bipartite intersection graphs. *Commentationes Mathematicae Universitatis Carolinae*, *23*(4), 739–745.

Harary, F., & Norman, R. Z. (1953). *Graph theory as a mathematical model in social science*. Ann Arbor, MI: Institute for Social Science.

Harary, F., Norman, R. Z., & Cartwright, D. W. (1965). *Structural Models: An Introduction to the Theory of Directed Graphs*. New York: Wiley.

Harary, F., & Palmer, E. M. (1973). *Graphical enumeration*. Academic Press Inc.

Hartman, I. B.-A., Newman, I., & Ziv, R. (1991). On grid intersection graphs. *Discrete Mathematics*, *87*(1), 41–52. doi:10.1016/0012-365X(91)90069-E

Hasani, M. (2017). Study of inverse sum indeg index. *Journal of Mathematical Nanoscience*, *7*(2), 103–109.

Havet, F., Reed, B., & Sereni, J. S. (2008). L(2, 1)-labeling of graphs. *Proceedings 19th Annual ACM-SIAM Symposium on Discrete Algorithms, SODA 2008*, 621-630.

Hedetniemi, S. T., & Laskar, R. (1984). *Connected domination in graphs*. In B. Bollobas (Ed.), *Graph theory and Combinatories* (pp. 209–218). London: Academic Press.

Hegde, S. M. (1989). *Numbered Graphs and their Applications* (Ph. D thesis). University of Delhi, Delhi.

Hegde, S. M. (1991). On set-valuations of graphs. *National Academy Science Letters*, *14*(4), 181–182.

Hegde, S. M. (1993). On k-sequential graphs. *National Academy Science Letters*, *16*(11-12), 299–301.

Hegde, S. M. (2008). On set-labelings of graphs. In B. D. Acharya, S. Arumugam, & A. Rosa (Eds.), *Labelings of Discrete Structures and Applications* (pp. 97–107). New Delhi, India: Narosa Publishing House.

Hegde, S. M. (2009). Set colorings of graphs. *European Journal of Combinatorics*, *30*(4), 986–995. doi:10.1016/j.ejc.2008.06.005

Hell, P., & Huang, J. (2004). Interval bigraphs and circular arc graphs. *Journal of Graph Theory*, *46*(4), 313–327. doi:10.1002/jgt.20006

Henn, S., & Bathelt, H. (2018). Cross-local knowledge fertilization, cluster emergence, and the generation of buzz. *Industrial and Corporate Change*, *27*(3), 449–466. doi:10.1093/icc/dtx036

Hermelin, D., Mestre, J., & Rawitz, D. (2014). Optimization problems in dotted interval graphs. *Discrete Applied Mathematics*, *174*, 66–72. doi:10.1016/j.dam.2014.04.014

Horne, J. A., & Smith, J. C. (2005). Dynamic programming algorithms for the conditional covering problem on path and extended star graphs. *Networks*, *46*(4), 177–185. doi:10.1002/net.20086

Horoldagva, B., Buyantoglok, L., & Dorjsembe, S. (2017). Difference of Zagreb indices and reduced second Zagreb index of cyclic graphs with cut edges. *MATCH Commun. Math. Comput. Chem.*, *78*, 337–350.

Hosamani, S. M., & Gutman, I. (2014). Zagreb indices of transformation graphs and total transformation graphs. *Applied Mathematics and Computation*, *247*, 1156–1160. doi:10.1016/j.amc.2014.09.080

Hota, M. (2005). *Sequential and parallel algorithms on some problems of trapezoid graph*. Midnapore, India: Vidyasagar University.

Hota, M., Pal, M., & Pal, T. K. (1999). An efficient algorithm to generate all maximal independent sets on trapezoid graphs. *International Journal of Computer Mathematics*, *70*(4), 587–599. doi:10.1080/00207169908804777

Hota, M., Pal, M., & Pal, T. K. (2001). An efficient algorithm for finding a maximum weight k-independent set on trapezoid graphs. *Computational Optimization and Applications*, *18*(1), 49–62. doi:10.1023/A:1008791627588

Hota, M., Pal, M., & Pal, T. K. (2004). Optimal sequential and parallel algorithms to compute all cut vertices on trapezoid graphs. *Computational Optimization and Applications*, *27*(1), 95–113. doi:10.1023/B:COAP.0000004982.13444.bc

Hou, Y., & Gutman, I. (2001). Hyperenergetic line graphs. *MATCH Commun. Math. Comput. Chem.*, *43*, 29–39.

Huang. (2005). *Graph-based Analysis for E-commerce Recommendation* (PhD dissertation). The University of Arizona.

Huckel, E. (1931). Quantentheoretische Beitrage zum Benzolproblem. *Zeitschrift fur Physik*, *70*, 204–286. doi:10.1007/BF01339530

Huggins, R. (2008). The evolution of knowledge clusters. Progress and policy. *Economic Development Quarterly*, *22*(4), 277–289. doi:10.1177/0891242408323196

Ilić, A., & Stevanović, D. (2011). On Comparing Zagreb indices. *MATCH Commun. Math. Comput. Chem.*, *66*, 681–687.

Indulal, G., & Vijayakumar, A. (2006). On a pair of equienergetic graphs. *MATCH Commun. Math. Comput. Chem.*, *55*, 83–90.

Indulal, G., & Vijayakumar, A. (2007). Energies of some non-regular graphs. *Journal of Mathematical Chemistry*, *42*(3), 377–386. doi:10.100710910-006-9108-7

Jabbar, N. A., Naoom, J. H., & Ouda, E. H. (2009). Fuzzy dual graph. *Journal of Al-Nahrain University*, *12*, 168–171.

Jaeger, F., & Payen, C. (1972). Relations du type Nordhaus Gaddum pour le nombre d' absorption d'un graphs simple. *C.R. Acad. Sci. Paris*, *274*, 728–730.

Jayawardene, C. J. (2018). On a Ramsey problem involving the 3-Pan Graph. *Annals of Pure and Applied Mathematics*, *16*(2), 437–441.

Jayawardene, C. J., & Hewage, T. (2017a). On a Ramsey problem involving quadrilaterals. *Annals of Pure and Applied Mathematics*, *13*(2), 297–304. doi:10.22457/apam.v13n2a16

Jayawardene, C. J., Rousseau, C. C., & Harboth, H. (2017b). On Path Convex Ramsey Numbers. *Journal of Graph Theory*, *86*(3), 286–294. doi:10.1002/jgt.22126

Jayawardene, C. J., & Samerasekara, L. (2017c). Size Ramsey numbers for C_3 versus all graphs G up to 4 vertices. *National Science Foundation*, *45*(1), 67–72. doi:10.4038/jnsfsr.v45i1.8039

Jayawardene, C. J., & Samerasekara, L. (2017d). Size Multipartite Ramsey numbers for K_4-e versus all graphs G up to 4 vertices. *Annals of Pure and Applied Mathematics*, *13*(1), 9–26. doi:10.22457/apam.v13n1a2

Jayaweera, I. M. L. N., Perera, K. K. K. R., & Munasinghe, J. (2017). Centrality Measures to Identify Traffic Congestion on Road Networks: A Case Study of Sri Lanka. *IOSR Journal of Mathematics*, *13*(2), 13–19. doi:10.9790/5728-1302011319

Jeong, H., Mason, S. P., Barabási, A.-L., & Oltvai, Z. N. (2001). Lethality and centrality in protein networks. *Nature*, *411*(6833), 41–42. doi:10.1038/35075138 PMID:11333967

Jin, X. T., & Yeh, R. K. (2004). Graph distance-dependent labeling related to code assignment in compute networks. *Naval Research Logistics*, *51*, 159–164.

Johnson. (2010). *Molecular Graph Theory, A Masters Project*. Worcester Polytechnic Institute.

Jonas, K. (1993). *Graph coloring analogues with a condition at distance two:* $L(2,1)$ *-labelings and list* λ *-labelings* (Ph.D. Thesis). University of South Carolina, Columbia, SC.

Jose, B. K. (2009). Open distance pattern uniform graphs. *Int. J. Math. Combin.*, *3*, 103–115.

Jose, B. K. (2009). Open distance-pattern uniform graphs. *Int. J. Math. Combin.*, *3*, 103–115.

Joshi, K. D. (1983). *Introduction to General Topology*. New Delhi: New Age International.

Joshi, K. D. (2003). *Applied discrete structures*. New Delhi: New Age International.

Joy & Germina. (2010). Topogenic set-indexers: Extended abstract. *Proc. National Workshop on Graph Theory Applied to Chemistry*, 112–114.

Joy & Germina. (2012). On gracefully topogenic graphs. *J. Combin. Inform. Syst. Sci.*, *37*, 279–289.

Joy. (2011). *A Study on Topologies Arising from Graphs and Digraphs* (Ph.D. Thesis). Kannur University, Kannur, Kerala, India.

Joyce, K. E., Laurienti, P. J., Burdette, J. H., & Hayasaka, S. (2010). A New Measure of Centrality for Brain Networks. *PLoS One*, *5*(8), 12200. doi:10.1371/journal.pone.0012200 PMID:20808943

Jun, Y. B., Smarandache, F., & Kim, C. S. (2017). Neutrosophic cubic sets. *New Mathematics and Natural Computation*, *13*(01), 41–54. doi:10.1142/S1793005717500041

Kalbfleisch, J. G., Stanton, R. G., & Horton, J. D. (1971). On covering sets and error-correcting codes. *Journal of Combinatorial Theory Series A*, *11*(3), 233–250. doi:10.1016/0097-3165(71)90051-3

Kandan, P., Chandrasekaran, E., & Priyadharshini, M. (2018). The Revan weighted szeged index of graphs. *Journal of Emerging Technologies and Innovative Research*, *5*(9), 358–366.

Kandan, P., & Joseph Kennedy, A. (2018). Reverse Zagreb indices of corona product of graphs. *Malaya Journal of Matematik*, *6*(4), 720–714. doi:10.26637/MJM0604/0003

Kasyanov, V. N., & Evstigneev, V. A. (2003). *Graphs in programming: processing, visualization and application*. SPb.: BHV-Petersburg.

Kasyanov, V. N., & Kasyanova, E. V. (2013). Information visualisation based on graph models. *Enterprise Information Systems*, 7(2), 187–197. doi:10.1080/17517575.2012.743188

Katz, L. (1953). A new status index derived from sociometric analysis. *Psychometrika*, 18(1), 39–43. doi:10.1007/BF02289026

Kaufmann, A. (1973). *Introduction a la Theorie des Sour-ensembles Flous*. Masson et Cie 1.

Kaufmann, A. (1973). *Introduction la Thorie des Sous-Ensembles Flous Lusage des Ingnieurs (Fuzzy Sets Theory), Masson*. Paris: French.

Kerr, A., & Waddington, J. (2013). E-Communications: An aspect of union renewal or merely doing thing electronically? *British Journal of Industrial Relations*, 52(4), 658–681. doi:10.1111/bjir.12010

Khalifeh, M. H., Yousefi-Azari, H., & Ashrafi, A. R. (2009). The first and second Zagreb indices of some graph operations. *Discrete Applied Mathematics*, 157(4), 804–811. doi:10.1016/j.dam.2008.06.015

Khan, N., Pal, M., & Pal, A. (2012a). Labeling of cactus graphs. *Mapana Journal of Science*, 11(4), 15–42. doi:10.12723/mjs.23.2

Khan, N., Pal, M., & Pal, A. (2012b). -Abeling of cactus graphs. *Communications and Network*, 4(01), 18–29. doi:10.4236/cn.2012.41003

Kharchenko, V. (2018). Big Data and Internet of Things for safety critical applications: Challenges, methodology and industry cases. *International Journal on Information Technologies and Security*, 10(4), 3-16.

Kim, K., & Chung, Y. (2018). Overview of journal metrics. *Science Editing*, 5(1), 16–20. doi:10.6087/kcse.112

Kitsak, M., Gallos, L. K., Havlin, S., Liljeros, F., Muchnik, L., Stanley, H. E., & Makse, H. A. (2010). Identification of influential spreaders in complex networks. *Nature Physics*, 6(11), 888–893. doi:10.1038/nphys1746

Kluver, R. (2004). Globalization, informatization, and intercultural communication. In Fr. E. Jandt (Ed.), Intercultural Communication: A Global Reader (pp. 425-437). SAGE Publications.

Knoke, D., & Laumann, E. O. (2012). Social organization in national politics areas. Exploration of some structural hypothesis. In *Social network analysis. Origins, theories and applications*. Madrid: Centro de Investigaciones Sociológicas.

Koolen, J. H., & Moulton, V. (2001). Maximal energy graphs. *Advances in Applied Mathematics*, 26(1), 47–52. doi:10.1006/aama.2000.0705

Koolen, J. H., & Moulton, V. (2003). Maximal energy bipartite graphs. *Graphs and Combinatorics*, 19(1), 131–135. doi:10.100700373-002-0487-7

Koschutzki, D. (2005) Centrality Indices. Analysis: Methodological Foundations, 3418, 16–61. doi:10.1007/978-3-540-31955-9_3

Koschützki, D., & Schreiber, F. (2004) Comparison of centralities for biological networks. *German Conference on Bioinformatics, 53*, 199-206.

Koschutzki, D., & Schreiber, F. (2008). Centrality analysis methods for biological networks and their application to gene regulatory networks. *Gene Regulation and Systems Biology*, 2, 193–201. doi:10.4137/GRSB.S702 PMID:19787083

Koshy & Germina. (2009). Independent complementary distance pattern uniform graphs. *Int. J. Math. Combin.*, (4), 63–74.

Koshy & Germina. (2010). Distance pattern segregated graphs. *Proc. Int. Conf. on Mathematics and Computer Science, 5-6*, 135-138.

Koshy & Germina. (2010). M-Complementary distance uniform matrix of a graph. *Int. Math. Forum, 5*(45), 2225–2235.

Koshy, B. (2012). *Labelings of Graphs and Hypergraphs* (Ph.D. Thesis). Kannur University, Kannur, Kerala, India.

Kral, D., & Skrekovski, R. (2003). A theory about channel assignment problem. *SIAM Journal on Discrete Mathematics, 16*(3), 426–437. doi:10.1137/S0895480101399449

Kravets, O. J., Choporov, O. N., & Bolnokin, V. E. (2018). Mathematical models and algorithmization of monitoring control an affiliated network in maintenance service distributed organization. *Quality – Access to Success, 19*(167), 68-72.

Kravets, O. J., Kosorukov, O., & Utyusheva, L. (2017). Specific features of graphical-analytical representation of a process model of managing innovation activity in the framework of software development for import substitution projects. *Journal of Theoretical and Applied Information Technology, 95*(14), 3337-3351.

Krishnamoorthy, V. (1966). On the number of topologies of finite sets. *The American Mathematical Monthly, 73*(2), 154–157.

Kulikov, S. S. (2016). Working with MySQL, MS SQL Server and Oracle in the examples of EPAM Systems. Minsk: BOFF.

Kulli, Chaluvaraju, & Boregowda. (n.d.). *On sum connectivity Banhatti index of some nanostructures.* (submitted)

Kulli, Stone, Wang, & Wei. (2017). Generalized multiplicative indices of polycyclic aromatic hydrocarbons and benzenoid systems. *Z. Naturforsch, 72*(6), 573-576.

Kulli. (2017). Geometric-arithmetic reverse and sum connectivity reverse indices of silicate and hexagonal networks. *International Journal of current Research in Science and Technology, 3*(10), 29-33.

Kulli. (2017). On the product connectivity reverse index of silicate and hexagonal networks. *International Journal of Mathematics and its Applications, 5*(4-B), 175-179.

Kulli. (2017). Revan indices of oxide and honeycomb networks. *International Journal of Mathematics and its Applications, 5*(4-E), 663-667.

Kulli. (2017a). Computing topological indices of dendrimer nanostars. *International Journal of Mathematics and Its Applications, 5*(3-A), 163-169.

Kulli. (2018). Hyper Revan indices and their polynomials of silicate networks. *International Journal of Mathematics and its Applications, 4*(3), 17-21.

Kulli, V. R. (2016). Multiplicative connectivity indices of certain nanotubes. *Annals of Pure and Applied Mathematics, 12*(2), 169–176. doi:10.22457/apam.v12n2a8

Kulli, V. R. (2016). On K hyper-Banhatti indices and coindices of graphs. *International Research Journal of Pure Algebra, 6*(5), 300–304.

Kulli, V. R. (2017). Computation of some Gourava indices of titania nanotubes. *Intern. J. Fuzzy Mathematical Archive, 12*(2), 75–81.

Kulli, V. R. (2017). General fifth M-Zagreb indices and fifth M-Zagreb polynomials of PAMAM dendrimers. *Intern. J. Fuzzy Mathematical Archive, 13*(1), 99–103.

Kulli, V. R. (2017). On hyper-Gourava indices and coindices. *International Journal of Mathematical Archive, 8*(12), 116–120.

Kulli, V. R. (2017). On the product connectivity Revan index of certain nanotubes. *J. Comp. and Math. Sci., 8*(10), 562–567.

Kulli, V. R. (2017). On the sum connectivity Gourava index. *International Journal of Mathematical Archive, 8*(6), 211–217.

Kulli, V. R. (2017). On the sum connectivity reverse index of oxide and honeycomb networks. *J. Comp. and Math. Sci., 8*(9), 408–413.

Kulli, V. R. (2017). The Gourava indices and coindices of graphs. *Annals of Pure and Applied Mathematics, 14*(1), 33–38. doi:10.22457/apam.v14n1a4

Kulli, V. R. (2017). The product connectivity Gourava index. *J. Comp. and Math. Sci., 8*(6), 235–242.

Kulli, V. R. (2017). The sum connectivity Revan index of silicate and hexagonal networks. *Annals of Pure and Applied Mathematics, 14*(3), 401–406. doi:10.22457/apam.v14n3a6

Kulli, V. R. (2017b). General topological indices of tetrameric 1,3-adamantane. *International Journal of Current Research in Science and Technology, 3*(8), 26–33.

Kulli, V. R. (2017c). General topological indices of some dendrimer nanostars. *Journal of Global Research in Mathematical Archives, 14*(11), 83–90.

Kulli, V. R. (2017d). Edge version of F-index, general sum connectivity index of certain nanotubes. *Annals of Pure and Applied Mathematics, 14*(3), 449–455. doi:10.22457/apam.v14n3a11

Kulli, V. R. (2018). Atom bond connectivity reverse and product connectivity reverse indices of oxide and honeycomb networks. *Intern. J. Fuzzy Mathematical Archive, 15*(1), 1–5.

Kulli, V. R. (2018). Computation of F-reverse and modified reverse indices of some nanostructures. *Annals of Pure and Applied Mathematics, 18*(1), 37–43.

Kulli, V. R. (2018). Computation of some minus indices of titania nanotubes. *International Journal of Current Research in Science and Technology, 4*(12), 9–13.

Kulli, V. R. (2018). Computing Banhatti indices of networks. *International Journal of Advances in Mathematics, 2018*(1), 31–40.

Kulli, V. R. (2018). Computing the F-Revan and modified Revan indices of certain nanostructures. *J. Comp. and Math. Sci., 9*(10), 1326–1333.

Kulli, V. R. (2018). Computing the F-ve-degree index and its polynomial of dominating oxide and regular triangulate oxide networks. *Intern. J. Fuzzy Mathematical Archive, 16*(1), 1–6.

Kulli, V. R. (2018). Leap hyper-Zagreb indices and their polynomials of certain graphs. *International Journal of Current Research in Life Sciences, 7*(10), 2783–2791.

Kulli, V. R. (2018). Minus leap and square leap indices and their polynomials of some special graphs. *Int. Res. J. Pure Algebra, 8*(11), 54–60.

Kulli, V. R. (2018). *Multiplicative Connectivity Indices of Nanostructures*. LAP Lambert Academic Publishing.

Kulli, V. R. (2018). On the square ve-degree index and its polynomial of certain network. *Journal of Global Research in Mathematical Archives, 5*(10), 1–4.

Kulli, V. R. (2018). On ve-degree indices and their polynomials of dominating oxide networks. *Annals of Pure and Applied Mathematics, 18*(1), 1–7. doi:10.22457/apam.v13n1a1

Kulli, V. R. (2018). Product connectivity leap index and ABC leap index of helm graphs. *Annals of Pure and Applied Mathematics, 18*(2), 183–193.

Kulli, V. R. (2018). Revan indices and their polynomials of certain rhombus networks. *International Journal of Current Research in Life Sciences, 7*(5), 2110–2116.

Kulli, V. R. (2018). Square reverse index and its polynomial of certain networks. *International Journal of Mathematical Archive, 9*(10), 27–33.

Kulli, V. R. (2018a). On KV indices and their polynomials of two families of dendrimers. *International Journal of Current Research in Life Sciences, 7*(9), 2739–2744.

Kulli, V. R. (2018b). Multiplicative connectivity indices of dendrimers. *Journal of Mathematics and Informatics, 15*, 1–7.

Kulli, V. R. (2018c). Dakshayani indices. *Annals of Pure and Applied Mathematics, 18*(2), 139–146.

Kulli, V. R. (2018d). On reduced Zagreb indices of polycyclic aromatic hydrocarbons and benzenoid systems. *Annals of Pure and Applied Mathematics, 18*(1), 73–78.

Kulli, V. R. (2018f). Reduced second hyper-Zagreb index and its polynomial of certain silicate networks. *Journal of Mathematics and Informatics, 14*, 11–16. doi:10.22457/jmi.v14a2

Kulli, V. R. (2018g). Computing reduced connectivity indices of certain nanotubes. *Journal of Chemistry and Chemical Sciences, 8*(11), 1174–1180. doi:10.29055/jccs/688

Kulli, V. R. (2018h). Reverse Zagreb and reverse hyper-Zagreb indices and their polynomials of rhombus silicate networks. *Annals of Pure and Applied Mathematics, 16*(1), 47–51. doi:10.22457/apam.v16n1a6

Kulli, V. R. (2019). Sum connectivity leap index and geometric-arithmetic leap index of certain windmill graphs. *J. Global Research in Mathematical Archives, 6*(1), 15–20.

Kulli, V. R. (2019a). On hyper KV and square KV indices and their polynomials of certain families of dendrimers. *J. Comp. and Math. Sci., 10*(2), 279–186.

Kulli, V. R. (2019b). On connectivity KV indices of certain families of dendrimers. *International Journal of Mathematical Archive, 10*(2), 14–17.

Kulli, V. R., Chaluvaraju, B., & Baregowda, H. S. (2017). K-Banhatti and K hyper-Banhatti indices of windmill graphs. *South East Asian J. of Math. and Math. Sci, 13*(1), 11–18.

Kulli, V. R., Chaluvaraju, B., & Boregowda, H. S. (2017). Connectivity Banhatti indices for certain families of benzenoid systems. *Journal of Ultra Chemistry, 13*(4), 81–87. doi:10.22147/juc/130402

Kulli, V. R., Chaluvaraju, B., & Boregowda, H. S. (2019). On sum connectivity Banhatti index of a graph. *Discussiones Mathematicae. Graph Theory, 39*, 205–217.

Kulli, V. R., & New, K. (2017). Banhatti topological indices. *Intern.J.Fuzzy Mathematical Archive, 12*(1), 29–37.

Kulli, V. R., & On, K. (2016). Banhatti indices of graphs. *J. Comp. and Math. Sci., 7*, 213–218.

Kyungsik, K., Soo Yong, K., & Deock-Ho, H. (2007). Characteristic of Networks in Financial Markets. *Computer Physics Communications, 177*(1-2), 184–185. doi:10.1016/j.cpc.2007.02.037

Lacasa, L., Luque, B., Ballesteros, F., Luque, J., & Nuño, J. C. (2008). From time series to Complex networks: The visibility graph. *Proceedings of the National Academy of Sciences of the United States of America, 105*(13), 4972–4975. doi:10.1073/pnas.0709247105 PMID:18362361

Lanktree, C., & Briere, J. (1991, January). *Early data on the trauma symptom checklist for children (TSC-C)*. Paper presented at the meeting of the American Professional Society on the Abuse of Children, San Diego, CA.

Lee, K. M. (2000). Bipolar-valied fuzzy sets and their basic operation. *Proceeding of the International Conference*, 307-317.

Lee, K. M. (2000). Bipolar-valued fuzzy sets and their basic operations. *Proceedings of the International Conference*, 307-317.

Lee-Kwang, H., & Lee, K. M. (1995). Fuzzy hypergraph and fuzzy partition. *IEEE Transactions on Systems, Man, and Cybernetics*, *25*(1), 196–201. doi:10.1109/21.362951

Lekkeikerker, C., & Boland, J. (1962). Representation of a finite graph by a set of intervals on the real line. *Fundamenta Mathematicae*, *51*, 45–64. doi:10.4064/fm-51-1-45-64

Li, R. (2019). *The hyper-Zagreb index and some Hamiltonian properties of graphs*. Academic Press.

Li. (2019). *The hyper-Zagreb index and some Hamiltonian properties of graphs*. Academic Press.

Li, M., Wei, W., Wang, L., & Qi, X. (2018). Approach to evaluating accounting informatization based on entropy in intuitionistic fuzzy environment. *Entropy (Basel, Switzerland)*, *20*(6), 476. doi:10.3390/e20060476

Liu, Y. (2015). Identify influential spreaders in complex networks: The role of neighborhood. *Physica A, 452*, 289–298.

Liu, B., & You, Z. (2011). A survey on comparing Zagreb indices, A. Ilić, D. Stevanović, On Comparing Zagreb indices. *MATCH Commun. Math. Comput. Chem.*, *66*, 581–593.

Liu, C. L. (1968). *Introduction to combinatorial mathematics*. McGraw-Hill.

Liu, J. G., Ren, Z. M., & Guo, Q. (2014). Ranking the spreading influence in complex networks. *Physica A*, *392*(18), 4154–4159. doi:10.1016/j.physa.2013.04.037

Liu, J., & Liu, B. (2008). Note on a pair of equienergetic graphs. *MATCH Commun. Math. Comput. Chem.*, *59*, 275–278.

Liu, M., & Liu, B. (2010). Some properties of the first general Zagreb index. *Australas J. Combin.*, *47*, 285–294.

Li, X., & Gutman, I. (2006). *Mathematical Aspects of Randić-type Molecular Structure Descriptors*. Kragujevac: University Kragujevac.

Li, X., Li, Y., & Shi, Y. (2010). Note on the energy of regular graphs. *Linear Algebra and Its Applications*, *432*(5), 1144–1146. doi:10.1016/j.laa.2009.10.023

Li, X., Shi, Y., & Gutman, I. (2012). *Graph Energy*. New York: Springer. doi:10.1007/978-1-4614-4220-2

Li, X., & Zhao, H. (2004). Trees with the first three smallest and largest generalised topological indices. *MATCH Commun. Math. Comput. Chem.*, *50*, 57–62.

Li, X., & Zheng, J. (2005). A unified approach to the external trees for different indices. *MATCH Commun. Math. Comput. Chem.*, *54*, 195–208.

Llić, A., & Milosavljević, N. (2017). The weighted vertex PI index. *Mathematical and Computer Modelling*, *57*, 393–406.

Lozares, C. (1996). La teoría de las redes sociales. *Papers*, *48*, 103–126. doi:10.5565/rev/papers/v48n0.1814

Luo, Z. (2016). Applications on hyper Zagreb index of generalized hierarchical product graphs. *Journal of Computational and Theoretical Nanoscience*, *13*(10), 7355–7361. doi:10.1166/jctn.2016.5726

Maddah-Ali, M. A., & Niesen, U. (2015). Decentralized coded caching attains order-optimal memory-rate tradeoff. *IEEE/ACM Transactions on Networking*, *23*(4), 1029–1040. doi:10.1109/TNET.2014.2317316

Malinin, L., & Malinina, N. (2009). Graph isomorphism in theorems and algorithms. Moscow: Librocom.

Mandal, S., & Pal, M. (2006a). A sequential algorithm to solve next-to-shortest path problem on circular-arc graphs. *The Journal of Physiological Sciences; JPS, 10*, 201–217.

Mandal, S., & Pal, M. (2006b). Maximum weight independent set of circular-arc graph and its application. *Journal of Applied Mathematics and Computing, 22*(3), 161–174. doi:10.1007/BF02832044

Mandal, S., & Pal, M. (2007). An optimal algorithm to compute all hinge vertices on circular-arc graphs. *Arab Journal of Mathematics and Mathematical Sciences, 1*(1), 16–27.

Mandere, E. (2009). *Financial networks and their applications to the stock market* (Thesis). Graduate College of Bowling Green State University.

Mangili, M., Martignon, F., & Capone, A. (2016). Performance analysis of Content-Centric and Content-Delivery networks with evolving object popularity. *Computer Networks, 94*, 80–98. doi:10.1016/j.comnet.2015.11.019

Maskell, P. (2014). Accessing remote knowledge - the roles of trade fairs, pipelines, crowdsourcing and listening posts. *Journal of Economic Geography, 14*(5), 883–902. doi:10.1093/jeg/lbu002

Ma, T. H., & Spinrad, J. P. (1994). On the 2-chain subgraph cover and related problems. *Journal of Algorithms, 17*(2), 251–268. doi:10.1006/jagm.1994.1034

Mathew, S., Mordeson, J. N., & Malik, D. S. (2018). *Fuzzy graph theory*. Springer International Publishing. doi:10.1007/978-3-319-71407-3

Mathew, S., & Sunitha, M. S. (2009). Types of arcs in a fuzzy graph. *Information Sciences, 179*(11), 1760–1768. doi:10.1016/j.ins.2009.01.003

Maynard, H. B., & Zandin, K. B. (Eds.). (2001). *Maynard's industrial engineering handbook* (5th ed.). New York: McGraw-Hill.

McAllister, M. L. N. (1988). Fuzzy intersection graphs. *Computers & Mathematics with Applications (Oxford, England), 15*(10), 871–886. doi:10.1016/0898-1221(88)90123-X

McClelland, B. J. (1971). Properties of latent roots of a matrix: The estimation of π-electron energies. *The Journal of Chemical Physics, 54*(2), 640–643. doi:10.1063/1.1674889

McConnell, R. M. (2003). Linear-time recognition of circular-arc graphs. *Algorithmica, 37*(2), 93–147. doi:10.100700453-003-1032-7

McKee, T., & McMorris, F. (1999). *Topics in intersection graph theory*. Retrieved from https://epubs.siam.org/doi/abs/10.1137/1.9780898719802

McKee, T. A., & McMorris, F. R. (1999). *Topics in intersection graph theory*. SIAM.

Mehmet, M., & Wijesekera, D. (2013). *Detecting the evolution of money laundering schemes*. Paper presented at Ninth Annual IFIP WG 11.9 Digital Forensics, Orlando, FL.

Mehmet, M., Wijesekera, D., & Buchholtz, M. F. (2014). Money laundering detection framework to link the disparate and evolving schemes. *Journal of Digital Forensics, Security and Law, 8*(3), 40–70.

Mehta, A. R., & Vijayakumar, G. R. (2008). A note on ternary sequences of strings of 0 and 1. *AKCE Int. J. Graphs Combin., 5*(2), 175–179.

Melnyk, V. V. (2018). Information management as a factor of innovative society development. *Humanities Bulletin of Zaporizhzhe State Engineering Academy, 74*(74), 39–47. doi:10.30839/2072-7941.2018.149651

Metsidik, M., Zhang, W., & Duan, F. (2010). Hyper- and reverse-Wiener indices of F-sums of graphs. *Discrete Applied Mathematics*, *158*(13), 1433–1440. doi:10.1016/j.dam.2010.04.003

Milićević, A., & Nikolić, S. (2004). On variable Zagreb indices. *Croatica Chemica Acta*, *77*, 97–101.

Milovanović, E., Matejić, M., & Milovanović, I. (2019). Some new upper bounds for the hyper-Zagreb index. *Discrete Mathematics Letters*, *1*, 30–35.

Mishra, B. K. (n.d.). *Molecular (graph) characteristics of some hydrocarbons through graph theory*. Retrieved from http://citeseerx.ist.psu.edu/viewdoc/download?doi=10.1.1.604.8277&rep=rep1&type=pdf

Mogharrab, M., & Fath-Tabar, G. H. (2011). Some bound on GA_1 index of graphs. *MATCH Commun. Math. Comput. Chem*, *65*, 33–38.

Mohammadi, M., & Mukhtar, M. (2018). Service-Oriented Architecture and process modelling. *Proceedings of the 32nd International Conference on Information Technologies – InfoTech-2018*.

Mollard, M., Payan, C., & Shixin, S. (1987). *Graceful problems*. In Seventh Hungarian Colloquium on Finite and Infinite Combinatorics, Budapest, Hungary.

Mollard, M., & Payan, C. (1989). On two conjectures about set-graceful graphs. *European Journal of Combinatorics*, *10*(2), 185–187. doi:10.1016/S0195-6698(89)80047-2

Mondal, De, & Pal. (2018). *On neighbourhood index of product of graphs*. arXivi1805.05273v1

Mondal, S., Pal, M., & Pal, T. K. (1999). An optimal algorithm for finding depth-first spanning tree on permutation graphs. *Korean Journal of Computational & Applied Mathematics*, *6*(3), 493–500. doi:10.1007/BF03009943

Mondal, S., Pal, M., & Pal, T. K. (2002a). An optimal algorithm for solving all-pairs shortest paths on trapezoid graphs. *International Journal of Computational Engineering Science*, *03*(02), 103–116. doi:10.1142/S1465876302000575

Mondal, S., Pal, M., & Pal, T. K. (2002b). An optimal algorithm to solve 2-neighbourhood covering problem on interval graphs. *International Journal of Computer Mathematics*, *79*(2), 189–204. doi:10.1080/00207160211921

Mondal, S., Pal, M., & Pal, T. K. (2003a). Optimal sequential and parallel algorithms to compute a Steiner tree on permutation graphs. *International Journal of Computer Mathematics*, *80*(8), 937–943. doi:10.1080/0020716031000112330

Mondal, S., Pal, M., & Pal, T. K. (2003b). An optimal algorithm to solve the all-pairs shortest paths problem on permutation graph. *Journal of Mathematical Modelling and Algorithms*, *2*(1), 57–65. doi:10.1023/A:1023695531209

Mordeson, J. N., & Nair, P. S. (1996). Successor and source of (fuzzy) finite state machines and (fuzzy) directed graphs. *Information Sciences*, *95*(1-2), 113–124. doi:10.1016/S0020-0255(96)00139-9

Mordeson, J. N., & Nair, P. S. (2000). *Fuzzy graphs and hypergraphs*. Physica-Verlag Heidelberg. doi:10.1007/978-3-7908-1854-3

Mordeson, J. N., & Nair, P. S. (2001). *Fuzzy Graphs and Fuzzy Hypergraphs*. Heidelberg, Germany: Physica Verlag.

Mordeson, J. N., & Peng, C. S. (1994). Operations on fuzzy graphs. *Information Sciences*, *79*(3-4), 159–170. doi:10.1016/0020-0255(94)90116-3

Moreno, J. L. (1953). Who Shall Survive? New York: Beacon House Inc. Beacon, N,Y. Library of Congress Catalog Card No. 53-7284

Müller, H. (1997). Recognizing interval digraphs and interval bigraphs in polynomial time. *Discrete Applied Mathematics*, *78*(1), 189–205. doi:10.1016/S0166-218X(97)00027-9

Naduvath, S. (2017). A study on modular sumset labelling of graphs. *Discrete Mathematics, Algorithms, and Applications, 9*(1), 1–16. doi:10.1142/S1793830917500392

Naduvath, S., & Augustine, G. (2018). *An introduction of sumset valued graphs.* Beau Bassin, Mauritius: Lambert Academic Publ.

Nagoorgani, A., & Hussain, R. J. (2009). Fuzzy effective distance *k*-dominating sets and their applications. International Journal of Algorithms, *Computing and Mathematics, 2*(3), 25–36.

Nagoorgani, A., & Latha, A. (2012). On irregular fuzzy graphs. *Applied Mathematical Sciences, 6*(11), 517–523.

Nagoorgani, A., & Radha, K. (2008). On regular fuzzy graphs. *The Journal of Physiological Sciences; JPS, 12*, 33–40.

Nagoorgani, A., & Vadivel, P. (2009). Relations between the parameters of independent domination and irredundance in fuzzy graph. *International Journal of Algorithms. Computing and Mathematics, 2*(1), 15–19.

Nair, P. S., & Cheng, S. C. (2001). Cliques and fuzzy cliques in fuzzy graphs. *IFSA World Congress and 20th NAFIPS International Conference, 4*, 2277 - 2280. 10.1109/NAFIPS.2001.944426

Naji, A. M., & Soner, N. D. (2018). The first leap Zagreb index of some graph operations. *Int. J. Appl. Graph Theory, 2*(1), 7–18.

Naji, A. M., Soner, N. D., & Gutman, I. (2017). On leap Zagreb indices of graphs. *Commun. Comb. Optim., 2*(2), 99–107.

Narumi, H., & Katayama, M. (1984). Simple topological index, A newly devised index characterizing the topological nature of structural isomers of saturated hydrocarbons. *Mem. Fac. Engin. Hokkaido Univ., 16*, 209–214.

Nathanson, M. B. (1996). *Additive number theory: Inverse problems and geometry of sumsets.* New York: Springer.

National Accreditation Commission - Chile. (2014). *Exempt Resolution Nº639 of Postgraduate Accreditation: Doctorate in Engineering Sciences, Industrial Engineering given by the University of Santiago, Chile. Santiago de Chile.* Chile: National Accreditation Commission.

National Accreditation Commission - Chile. (2016). *Exempt Resolution Nº852 of Postgraduate Accreditation: Doctorate in Engineering Sciences, Industrial Engineering and Transport given by the Pontifical Catholic University of Chile. Santiago de Chile.* National Accreditation Commission.

National Accreditation Commission - Chile. (2018). *Exempt Resolution Nº1023 of Postgraduate Accreditation: Doctorate in Industrial Engineering and Operations Research given by the Adolfo Ibáñez University. Santiago de Chile.* Chile: National Accreditation Commission.

National Accreditation Commission - Chile. (2019). *Advanced Search for Accreditations.* Retrieved from: https://www.cnachile.cl/Paginas/buscador-avanzado.aspx

National Historical Museum of Chile. (2009). *Earthquakes in Chile.* Santiago de Chile: Origo Editions.

Naz, S., Ashraf, S., & Karaaslan, F. (2018a). Energy of a bipolar fuzzy graph and its applications in decision making. *Italian Journal of Pure and Applied Mathematics.*

Naz, S., Akram, M., & Smarandache, F. (2018). Certain notions of energy in single-valued neutrosophic graphs. *Axioms, 7*(3), 1–30. doi:10.3390/axioms7030050

Neo4j – Platform for connected data. (n.d.). Retrieved October 28, 2018, from https://neo4j.com/ PL/SQL

Newman, M. (2003). The structure and function of complex networks. *SIAM Review, 45*(2), 167–256. doi:10.1137/S003614450342480

Nieminen, J. (1974). On the centrality in a graph. *Scandinavian Journal of Psychology, 15*(1), 322–336. doi:10.1111/j.1467-9450.1974.tb00598.x PMID:4453827

Nikolić, S., Kovaćević, G., Milićević, A., & Trinajstić, N. (2003). The Zagreb indices 30 years after. *Croat. Chem. Acta CCACAA, 76*(2), 113–124.

Nilsson, M. (2019). Proximity and the trust formation process. *European Planning Studies, 27*(5), 841–861. doi:10.10 80/09654313.2019.1575338

Nirmala, G., & Dhanabal, K. (2012). Special planar fuzzy graph configurations. *International Journal of Scientific and Research Publications, 2*(7), 1–4.

Olariu, S. (1991). An optimal greedy heuristic to color interval graphs. *Information Processing Letters, 37*(1), 21–25. doi:10.1016/0020-0190(91)90245-D

Olawumi, O., Väänänen, A., Haataja, U., & Toivanen, P. (2017). Security issues in smart homes and mobile health system: Threat analysis, possible countermeasures and lessons learned. *International Journal on Information Technologies and Security, 9*(1), 31–52.

Onagh, B. N. (2017). The Harmonic index of subdivision graphs. *Transactions on Combinatorics, 6*(4), 15–27.

Onagh, B. N. (2018). The harmonic index of edge-semitotal graphs, total graphs and related sums. *Kragujevac J. Math., 42*(2), 217–228. doi:10.5937/KgJMath1802217O

Opsahl, T., Agneessens, F., & Skvoretz, J. (2010). Node centrality in weighted networks. Generalizing degree and shortest paths. *Social Networks, 32*(3), 245–251. doi:10.1016/j.socnet.2010.03.006

Ordman, E. T. (1985). Threshold coverings and resource allocation, *16th Southeastern Conference on Combinatorics, Graph Theory and Computing*, 99-113.

Ore, O. (1962). *Theory of Graphs*. American Math. Soc. Colloquium Pub.

Ore, O. (1982). Theory of Graphs. *Amer. Math. Scc. Collog. Publ., 38*, 206-212.

Ore, O. (1962). *Theory of Graphs, Amer. Math. Soc. Colloq. Publ* (Vol. 38). Providence, RI: Amer. Math. Soc.

Pach & Tóth. (2002). Recognizing string graphs is decidable. *Discrete & Computational Geometry, 28*(4), 593–606. doi:10.100700454-002-2891-4

Pal, M. (1995a). *Some sequential and parallel algorithms on interval graphs* (Ph. D Thesis). Indian Institute of Technology, Kharagpur, India.

Pal, M. (1996a). An efficient parallel algorithm for computing a maximum-weight independent set of a permutation graph. *Proc.: 6th National Seminar on Theoretical Computer Science, Banasthali Vidyapith, Rajasthan, India*, 276–285.

Pal, M., Samanta, S., & Pal, A. (2013). Fuzzy *k*-competition graphs. In S*cience and Information Conference (SAI)*. IEEE-Xplore.

Pal, M. (1998a). A parallel algorithm to generate all maximal independent sets on permutation graphs. *International Journal of Computer Mathematics, 67*(3–4), 261–274. doi:10.1080/00207169808804664

Pal, M. (1998b). Efficient algorithms to compute all articulation points of a permutation graph. *Korean Journal of Computational & Applied Mathematics, 5*(1), 141–152. doi:10.1007/BF03008943

Pal, M., & Bhattacharjee, G. (1996b). An optimal parallel algorithm to color an interval graph. *Parallel Processing Letters, 6*(4), 439–449.

Pal, M., & Bhattacharjee, G. (1997a). An optimal parallel algorithm for all-pairs shortest paths on unweighted interval graphs. *Nordic Journal of Computing, 4*(4), 342–356.

Pal, M., & Bhattacharjee, G. P. (1995b). Optimal sequential and parallel algorithms for computing the diameter and the center of an interval graph. *International Journal of Computer Mathematics, 59*(1–2), 1–13. doi:10.1080/00207169508804449

Pal, M., & Bhattacharjee, G. P. (1996c). A sequential algorithm for finding a maximum weight k-independent set on interval graphs. *International Journal of Computer Mathematics, 60*(3–4), 205–214. doi:10.1080/00207169608804486

Pal, M., & Bhattacharjee, G. P. (1997b). A data structure on interval graphs and its applications. *Journal of Circuits, Systems, and Computers, 07*(03), 165–175. doi:10.1142/S0218126697000127

Parvathi, R., & Karunambigai, M. G. (2006). Intuitionistic Fuzzy Graphs. *Computational Intelligence, Theory and Applications,* 139-150.

Parvathi, R., Thilagavathi, S., & Karunambigai, M. G. (2009). Intuitionistic fuzzy hypergraphs. *Cybernetics and Information Technologies, 9*, 46–48.

Patil, G. H., & Chaudhary, M. S. (1995). A recursive determination of topologies on finite sets. *Indian Journal of Pure and Applied Mathematics, 26*(2), 143–148.

Patrimonio Consultores. (2009). *History of the Production Development Corporation 1939 - 2009.* Santiago de Chile: Patrimonio Consultores. Available in: http://repositoriodigital.corfo.cl/bitstream/handle/11373/7229/HISTORIA%20 CORFO%20FINAL.pdf

Pattabiraman, K. (2018). Inverse Sum Indeg index of graphs. *AKCE International Journal of Graph and Combinatorics, 15*(2), 155-167.

Pattabiraman, K., & Vijayaragavan, M. (2017). Hyper Zagreb indices and its coindices of graphs. *Bull. Int. Math. Virt. Inst., 7*, 31–41.

Paul & Germina. (2012). On linear hypergraph set indexers of a graph: Cyclomatic number, cyclicity and 2-colorability. *J. Combin. Inform. Syst. Sci., 37*(2 4), 227–237.

Paul, Pal, & Pal. (n.d.a). $L(0,1)$-labeling of trapezoid graphs. *International Journal of Applied Computational Mathematics.* DOI doi:10.100740819-017-0372-y

Paul, Pal, & Pal. (n.d.b). $L(2,1)$-labeling of trapezoid graphs. *Communicated.*

Paul, V. (2012). *Labeling and set-indexing hypergraphs of a graph and related areas* (PhD Thesis). Kannur Univ., Kannur, India.

Paul, V. (2012). *Labeling of Graphs and Set-Indexing Hypergraphs and Related Topics* (Ph.D. Thesis). Kannur University, Kannur, Kerala, India.

Paul, S., Pal, M., & Pal, A. (2013). An efficient algorithm to solve $L(0,1)$-labeling problem on interval graphs. *Advanced Modelling and Optimization, 15*(1), 31–43.

Paul, S., Pal, M., & Pal, A. (2014). Labeling of circular-arc graph. *Annals of Pure and Applied Mathematics, 5*(2), 208–219.

Paul, S., Pal, M., & Pal, A. (2015). Labeling of interval graphs. *Journal of Applied Mathematics and Computing, 49*(1), 419–432. doi:10.100712190-014-0846-6

Paul, S., Pal, M., & Pal, A. (2015). Labeling of permutation and bipartite permutation graphs. *Mathematical in Computer Science, 9*(1), 113–123. doi:10.100711786-014-0180-2

Paul, S., Pal, M., & Pal, A. (2015). Labeling of permutation graphs. *Journal of Mathematical Modelling and Algorithms*, *14*(4), 469–479. doi:10.100710852-015-9280-5

Paul, V., & Germina, K. A. (2011). On 3-uniform linear hypergraph set indexers of a graph. *Int. J. Contemp. Math. Sci.*, *6*(18), 861–868.

Paul, V., & Germina, K. A. (2011). On uniform linear hypergraph set-indexers of graphs. *Int. J. Contemp. Math. Sci.*, *6*(18), 861–868.

Paul, V., & Germina, K. A. (2012). On Structural properties of 3-uniform linear hypergraph set indexers of a graph. *Advances Theor. Appl. Math.*, *7*(1), 95–104.

Pea, I., & Rada, J. (2008). Energy of diagraph. *Linear and Multilinear Algebra*, *56*(5), 565–579. doi:10.1080/03081080701482943

Peay, E. R. (1982). Structural models with qualitative values. *The Journal of Mathematical Sociology*, *8*(2), 161–192. doi:10.1080/0022250X.1982.9989921

Pinar, B. S., Günther, B., & Fazleena, J. (2016). A Novel Framework for Achieving Sustainable Value Creation through Industrial Engineering Principles. *Procedia CIRP*, *40*, 516–523. doi:10.1016/j.procir.2016.01.126

Pnueli, A., Lempel, A., & Even, S. (1971). Transitive orientation of graphs and identification of permutation graphs. *Canadian Journal of Mathematics*, *23*(1), 160–175. doi:10.4153/CJM-1971-016-5

Praba, B., Chandrasekaran, V. M., & Deepa, G. (2014). Energy of an intuitionistic fuzzy graph. *International Journal of Pure and Applied Mathematics*, *32*, 431–444.

Pramanik, T., Pal, M., & Mondal, S. (2016). Interval-valued fuzzy threshold graph. *Pacific Science Review A. Natural Science and Engineering*, *18*(1), 66–71.

Pramanik, T., Pal, M., Mondal, S., & Samanta, S. (2018). A study on bipolar fuzzy planar graph and its application in image shrinking. *Journal of Intelligent & Fuzzy Systems*, *34*(3), 1863–1874. doi:10.3233/JIFS-171209

Pramanik, T., Samanta, S., & Pal, M. (2016). Interval-valued fuzzy planar graphs. *International Journal of Machine Learning and Cybernetics*, *7*(4), 653–664. doi:10.100713042-014-0284-7

Pramanik, T., Samanta, S., & Pal, M. (2017). Fuzzy ϕ . *Soft Computing*, *21*(13), 3723–3734. doi:10.100700500-015-2026-5

Pranav, P., & Chirag, P. (2013). Various Graphs and Their Applications in Real World. International Journal of Engineering Research and Technology, 2(12).

Prathap, G. (2018). Eugene Garfield: From the metrics of science to the science of metrics. *Scientometrics*, *114*(2), 637–650. doi:10.100711192-017-2525-5

Prathap, G. (2018b). A bibliometric tale of two cities: Hong Kong and Singapore. *Scientometrics*, *117*(3), 2169–2175. doi:10.100711192-018-2927-z

Price, D. (1976). A general theory of bibliometric and other cumulative advantage processes. *Journal of the Association for Information Science*, *27*, 292–306. doi:10.1002/asi.4630270505

Price, D. (1986). *Little science, big science – and beyond*. Columbia University Press.

Price, D. J. S. (1963). *Little Science, Big Science*. Columbia University Press.

Princy, K. L. (2007). *Some studies on set-valuations of graph-embedding and NP-completeness* (Ph. D Thesis). Kannur Univ., Kannur, India.

Princy, K. L. (2007). *Some Studies on Set-Valuations of Graphs–Embedding and NP-Completeness* (Ph.D. Thesis). Kannur University, Kannur, Kerala, India.

Rafajlovski, G., & Digalovski, M. (2018). Controlling induction motors. *Proceedings of the 32nd International conference on Information Technologies – InfoTech-2018*.

Ramalingam, G., & Rangan, C. P. (1988). A unified approach to domination problems on interval graphs. *Information Processing Letters*, *27*(5), 271–274.

Ramane, H. S., Basavanagoud, B., & Yalnaik, A. S. (2016). Harmonic Status Index og Graphs. *Bulletin of Mathematical Sciences and Applications*, *17*, 24–32. doi:10.18052/www.scipress.com/BMSA.17.24

Ramane, H. S., Gutman, I., Walikar, H. B., & Halkarni, S. B. (2004). Another class of equienergetic graphs. *Kragujevac J. Math.*, *26*, 15–17.

Ramane, H. S., & Jummannaver, R. B. (2016). Note on forgotten topological index of chemical structure in drugs. *Applied Mathematics and Nonlinear Sciences*, *1*(2), 369–374. doi:10.21042/AMNS.2016.2.00032

Ramane, H. S., & Walikar, H. B. (2007). Construction of equienergetic graphs. *MATCH Commun. Math. Comput. Chem.*, *57*, 203–210.

Ramane, H. S., Walikar, H. B., & Gutman, I. (2009). Equienergetic graphs. *J. Comb. Math. Comb. Comput.*, *69*, 165–173.

Ramane, H. S., Walikar, H. B., Rao, S. B., Acharya, B. D., Hampiholi, P. R., Jog, S. R., & Gutman, I. (2004). Equienergetic graphs. *Kragujevac J. Math.*, *26*, 5–13.

Ramane, H. S., Walikar, H. B., Rao, S. B., Acharya, B. D., Hampiholi, P. R., Jog, S. R., & Gutman, I. (2005). Spectra and energies of iterated line graphs of regular graphs. *Applied Mathematics Letters*, *18*(6), 679–682. doi:10.1016/j.aml.2004.04.012

Rana, A., Pal, A., & Pal, M. (2011c). Efficient algorithms to solve k-domination problem on permutation graphs. In Y. Wu (Ed.), High Performance Networking, Computing, and Communication Systems (pp. 327–334). Springer Berlin Heidelberg.

Rana, A. (2011). Pal, A., & Pal, M. (2011a). *The 2-neighbourhood covering problem on permutation graphs. Advanced Modelling and Optimization*, *13*(3), 463–476.

Randić, M. (1975). On characterization of molecular branching. *Journal of the American Chemical Society*, *97*(23), 6609–6615. doi:10.1021/ja00856a001

Ranjini, P. S., Lokesha, V., & Cangül, I. N. (2011). On the Zagreb indices of the line graphs of the subdivision graphs. *Applied Mathematics and Computation*, *218*(3), 699–702. doi:10.1016/j.amc.2011.03.125

Ranjini, P. S., Lokesha, V., & Usha, A. (2013). Relation between phenylene and hexagonal squeeze using harmonic index. *Int. J. Graph Theory*, *1*, 116–121.

Rao, S. B., & Germina, K. A. (2011). Graph labelings and complexity problems: A review. In P. Panigrahi & S. B. Rao (Eds.), *Graph Theory Research Directions*. New Delhi: Narosa Pub. House.

Raychaudhuri, A. (1987). On powers of interval and unit interval graphs. *Congr. Numer.*, *4*, 235–242.

Rehan, Haran, Chauhan, & Grover. (2013). Visualizing the Indian stock market: A complexnetworks approach. *International Journal of Advances in Engineering and Technology, 6*(3), 1348–1354.

Réti, T. (2012). On the relationships between the first and second Zagreb indices. *MATCH Commun. Math. Comput. Chem., 68*, 169–188.

Roberts, F. S. (1969). On the boxicity and cubicity of a graph. *Recent Progresses in Combinatorics*, 301–310.

Roberts, F. S. (1978). *Graph theory and its practical applications to problems of society*. Philadelphia: SIAM.

Rodrigues, M., Moreira, A., Azevedo, E., Neves, M., Sadok, D., Callado, A., & Souza, V. (2013). On learning how to plan content delivery networks. In *Proceedings of the 46th Annual Simulation Symposium* (p. 13). Society for Computer Simulation International.

Rodriguez, J. A., Estrada, E., & Gutierrez, A. (2006). Functional centrality in graphs. *Linear and Multilinear Algebra, 55*(3), 293–302. doi:10.1080/03081080601002221

Romansky, R. (2017a). A survey of digital world opportunities and challenges for user's privacy. *International Journal on Information Technologies and Security, 9*(4), 97-112.

Romansky, R., & Noninska, I. (2017c). Stochastic investigation of secure access to the resources of a corporative system. *International Journal of Scientific & Engineering Research, 8*(1), 578-584.

Romansky, R., & Parvanova, E. (2010). Formalization and discrete modelling of the information servicing in distributed learning environment. *Communication & Cognition, 43*(1&2), 1-15.

Romansky, R. (2017b). *Information servicing in distributed learning environments. Formalization and model investigation*. Saarbrüken, Germany: LAP LAMBERT Academic Publishing.

Romansky, R., & Kirilov, K. (2018). Architectural design and modelling of a web based application for GDPR clarification. *AIP Conference Proceedings (American Institute of Physics), Proceedings of the 44th International Conference on Applications of Mathematics in Engineering and Economics (AMEE'18)*. 10.1063/1.5082121

Romansky, R., & Noninska, I. (2016). Discrete formalization and investigation of secure access to corporative resources. *International Journal of Engineering Research and Management, 3*(5), 97–101.

Rosa, A. (1967). *On certain valuations of the vertices of a graph, in Theory of Graphs (Int. Symposium, Rome, July 1966)*. Gordon and Breach.

Rosa, A. (1967). On certain valuation of the vertices of a graph. In *Theory of Graphs*. New York: Gordon and Breach.

Rose, D., Tarjan, R., & Lueker, G. (1976). Algorithmic aspects of vertex elimination on graphs. *SIAM Journal on Computing, 5*(2), 266–283. doi:10.1137/0205021

Rosemann, M. (2013) The Internet of Things: new digital capital in the hands of customers. *Business Transformation Journal, 9*, 6-15. Retrieved from http://eprints.qut.edu.au/66451/

Rosenfeld, A. (1975). *Fuzzy graphs in Fuzzy sets and their applications*. Academic Press.

Rosenfeld, A. (1975). *Fuzzy graphs*. In L. A. Zadeh, K. S. Fu, & M. Shimura (Eds.), *Fuzzy Sets and their Applications* (pp. 77–95). New York: Academic Press.

Rosenfeld, A. (1975). Fuzzy graphs. New York: Academic Press.

Rousa, I. K. (2008). *Sumsets and structures*. Budapest: Alfréd Rényi Institute of Mathematics.

Saavedra, I. (1985). Philosophy, science, technology and development. *Ambiente y Desarrollo, 1*(3), 43-54. Available in: http://www.cipmachile.com/web/200.75.6.169/RAD/1984-85/3_Saavedra.pdf

Sabidussi, G. (1966). The centrality index of a graph. *Psychometrika, 31*(4), 581-603.

Sachs, H. (1962). Uber selbstkomplementare Graphen. *Publicationes Mathematicae (Debrecen), 9*, 270–288.

Sachs, H. (1967). Uber Teiler, Faktoren und charakteristiche Polynome Von Graphen, Teil II. *Wiss. Z. TH Ilmenau, 13*, 405–412.

Saha, A., & Pal, M. (2003). Maximum weight k-independent set problem on permutation graphs. *International Journal of Computer Mathematics, 80*(12), 1477–1487. doi:10.1080/00207160310001614972

Saha, A., Pal, M., & Pal, T. K. (2005a). An optimal parallel algorithm to construct a tree 3-spanner on interval graphs. *International Journal of Computer Mathematics, 82*(3), 259–274. doi:10.1080/00207160412331286851

Saha, A., Pal, M., & Pal, T. K. (2005b). An optimal parallel algorithm for solving all-pairs shortest paths problem on circular-arc graphs. *Journal of Applied Mathematics and Computing, 17*(1), 1–23.

Saha, A., Pal, M., & Pal, T. K. (2005c). An efficient PRAM algorithm for maximum-weight independent set on permutation graphs. *Journal of Applied Mathematics and Computing, 19*(1–2), 77–92. doi:10.1007/BF02935789

Saha, A., Pal, M., & Pal, T. K. (2007). Selection of programme slots of television channels for giving advertisement: A graph theoretic approach. *Information Sciences, 177*(12), 2480–2492. doi:10.1016/j.ins.2007.01.015

Sahin, B., & Ediz, S. (2018). On ev-degree and ve-degree topological indices. *Iranian J. Math. Chem., 9*(4), 263–277.

Sahoo, J., Salahuddin, M., Glitho, R., Elbiaze, H., & Ajib, W. (2016). A Survey on Replica Server Placement Algorithms for Content Delivery Networks. *IEEE Communications Surveys and Tutorials*.

Samanta, S., & Pal. (2011). Fuzzy Tolerance Graphs. *International Journal of Latest Trends in Mathematics*.

Samanta, S., & Pal. (2012). *Irregular bipolar fuzzy graphs*. arXiv preprint arXiv, 1209, 1682.

Samanta, S., Pal, M. & Pal, A. (2014b). Some more results on bipolar fuzzy sets and bipolar fuzzy intersection graphs. *The Journal of Fuzzy Mathematics, 22*(2), 253-262.

Samanta, S., Akram, M., & Pal, M. (2015). -step fuzzy competition graphs. *Journal of Applied Mathematics and Computing, 47*(1-2), 461–472. doi:10.100712190-014-0785-2

Samanta, S., & Pal, M. (2011). Fuzzy threshold graphs. *CIIT International Journal of Fuzzy Systems, 3*(12), 360–364.

Samanta, S., & Pal, M. (2011). Fuzzy tolerance graphs. *International Journal of Latest Trends in Mathematics, 1*(2), 57–67.

Samanta, S., & Pal, M. (2012). Bipolar fuzzy hypergraphs. *International Journal of Fuzzy Logic Systems, 2*(1), 17–28. doi:10.5121/ijfls.2012.2103

Samanta, S., & Pal, M. (2012). Irregular bipolar fuzzy graphs. *International Journal of Applications of Fuzzy Sets, 2*, 91–102.

Samanta, S., & Pal, M. (2013). Fuzzy kp. *Fuzzy Information and Engineering, 5*(2), 191–204. doi:10.100712543-013-0140-6

Samanta, S., & Pal, M. (2015). Fuzzy planar graphs. *IEEE Transactions on Fuzzy Systems, 23*(6), 1936–1942. doi:10.1109/TFUZZ.2014.2387875

Samanta, S., Pal, M., & Pal, A. (2014a). New concepts of fuzzy planar graph. *International Journal of Advanced Research in Artificial Intelligence, 3*(1), 52–59. doi:10.14569/IJARAI.2014.030108

Samanta, S., Pal, M., & Pal, A. (2014c). Some more results on fuzzy-competition graphs. *International Journal of Advanced Research in Artificial Intelligence, 3*(1), 60–67. doi:10.14569/IJARAI.2014.030109

Samanta, S., Pramanik, T., & Pal, M. (2016). Fuzzy colouring of fuzzy graphs. *Afrika Matematika, 27*(1-2), 37–50. doi:10.100713370-015-0317-8

Sampathkumar, E., & Germina, K. A. (2010). k-Transitive digraphs and topologies. *Int. Math. Forum, 15*(63), 111–3119.

Sampathkumar, E. (1989). The global domination number of a graph. *J. Math. Phys. Sci., 23*, 377–385.

Sampathkumar, E. (2006). Generalized graph structures. *Bull. Kerala Math. Assoc., 3*(2), 67–123.

Sampathkumar, E., & Kulkarni, K. H. (1973). Transitive digraphs and topologies on a set. *J. Karnatak Univ. Sci., 18*, 266–273.

Sampathkumar, E., Neelagiri, P. S., & Venkatachalam, C. V. (1988). Odd and even colorings of a graph. *J. Karnatak Univ. Sci., 33*, 128–133.

Sampathkumar, E., & Pushpalatha, L. (1996). Strong, weak domination and domination balance in a graph. *Discrete Mathematics, 161*(1-3), 235–242. doi:10.1016/0012-365X(95)00231-K

Sampathkumar, E., & Walikar, H. B. (1979). The connected domination number of a graph. *J. Math. Phys. Sci, 13*(6).

Sarala, D., Deng, H., Natarajan, C., & Ayyaswamy, S.K. (2019). F index of graphs based on four new operations related to the strong product. *AKCE International Journal of Graphs and Combinatorics*. doi:10.1016/j.akcej.2018.07.003

Sarkar, D., Rakesh, N., & Mishra, K. K. (2016). Content delivery networks: Insights and recent advancement. In *Parallel, Distributed and Grid Computing (PDGC), 2016 Fourth International Conference on* (pp. 1-5). IEEE.

SAS Anti-Money Laundering (SAS AML). (n.d.). Retrieved October 28, 2018, from http://www.tadviser.ru/index.php/%D0%9F%D1%80%D0%BE%D0%B4%D1%83%D0%BA%D1%82:SAS_Anti-Money_Laundering_(SAS_AML)

Schadt, E. E., Lamb, J., Yang, X., Zhu, J., Edwards, S., Guhathakurta, D., ... Lusis, A. J. (2005). An integrative genomics approach to infer causal associations between gene expression and disease. *Nature Genetics, 37*(7), 710–717. doi:10.1038/ng1589 PMID:15965475

Scheinerman, E. R. (1985). Characterizing intersection classes of graphs. *Discrete Mathematics, 55*(2), 185–193. doi:10.1016/0012-365X(85)90047-0

Scheinerman, E. R., & West, D. B. (1983). The interval number of a planar graph: Three intervals suffice. *Journal of Combinatorial Theory Series B, 35*(3), 224–239. doi:10.1016/0095-8956(83)90050-3

Sedlacek, J. (1976). Some properties of magic graphs, in Graphs, Hypergraphs and Block Systems. *Proc. Conf.*, 247–253.

Sedlacek, J. (1976). On magic graphs. *Mathematica Slovaca, 26*(4), 329–335.

Sedlar, J., Stevanović, D., & Vasilyev, A. (2015). On the inverse sum indeg index. *Discrete Applied Mathematics, 184*, 202–212. doi:10.1016/j.dam.2014.11.013

Sergeeva, M. G., Karavanova, L. Z., Bereznatskaya, M. A., Klychkov, K. E., Loktionova, T. E., & Chauzova, V. A. (2018). Socialization of a personality under the conditions of globalization and informatization of the society. *Revista Espacios, 39*(21), 28. Retrieved from http://www.revistaespacios.com/a18v39n21/a18v39n21p28.pdf

Shanmugam, K., Golrezaei, N., Dimakis, A. G., Molisch, A. F., & Caire, G. (2013). Femtocaching: Wireless content delivery through distributed caching helpers. *IEEE Transactions on Information Theory*, *59*(12), 8402–8413. doi:10.1109/TIT.2013.2281606

Shao, Z., & Deng, F. (2016). Correcting the number of borderenergetic graphs of order 10. *MATCH Commun. Math. Comput. Chem.*, *75*, 263–266.

Sharp, H. Jr. (1968). Cardinality of finite topologies. *J. Combin. Theory*, *5*(1), 82–86. doi:10.1016/S0021-9800(68)80031-6

Shaw, M. E. (1954). Group structure and the behavior of individuals in small groups. *The Journal of Psychology*, *38*(1), 139–149. doi:10.1080/00223980.1954.9712925

Shiladhar, Naji, & Soner. (n.d.). Computation of leap Zagreb indices of some windmill graphs. *Int. J. Math. And. Appl.*, *6*(2-B, 183-191.

Shiladhar, P., Naji, A. M., & Soner, N. D. (2018). Leap Zagreb indices of some wheel related graphs. *J. Comp. and Math. Sci.*, *9*(3), 221–231.

Shimbel, A. (1953). Structural parameters of communication networks. *The Bulletin of Mathematical Biophysics*, *15*(4), 501–507. doi:10.1007/BF02476438

Shirdel, G. H., Rezapour, H., & Sayadi, A. M. (2013). The hyper-Zagreb index of graph operations. *Iranian J. Math. Chem.*, *4*(2), 213–220.

Shumway & Stoffer. (n.d.). *Time Series Analysis and Its Applications.* Springer-Verlag.

Silyutin, D. S. (2004). *Genetic search of graph isomorphism controlled by multi-agent system* (Doctoral dissertation). Taganrog State University of Radio Engineering.

Singh, G. S. (1998). A note on labelings of graphs. *Graphs and Combinatorics*, *14*, 201–207.

Škraba, A., Stanovov, V., Semenkin, E., KoloŽvari, A., & Kofjač, D. (2018). Development of algorithm for combination of cloud services for speech control of cyber-physical systems. *International Journal on Information Technologies and Security*, *10*(1), 73-82.

Smarandache, F. (1999). *A unifying field in logics. neutrosophy: neutrosophic probability, set and logic.* Rehoboth: American Research Press.

Smarandache, F. (2006). Neutrosophic set - a generalization of the intuitionistic fuzzy set, Granular Computing. *IEEE International Conference*, 38 – 42.

Smarandache, F. (2015). *Types of Neutrosophic Graphs and Neutrosophic Algebraic Structures together with their Applications in Technology, seminar, Universitatea Transilvania din Brasov.* Brasov, Romania: Facultatea de Design de Produs si Mediu.

Solow, R. M. (1957). Technical Change and the Aggregate Production Function. *The Review of Economics and Statistics*, *39*(3), 312-320. Available in: http://www.jstor.org/stable/1926047

Sparrow, M. K. (1991). The application of network analysis to criminal intelligence: An assessment of the prospects. *Social Networks*, *13*(3), 251–274. doi:10.1016/0378-8733(91)90008-H

Spinak, E. (1998). Indicadores cienciométricos. *Ciência da Informação*, *27*(2), 141-148. Available in: http://revista.ibict.br/ciinf/article/view/795

Spinrad, J. (1985). On comparability and permutation graphs. *SIAM Journal on Computing, 14*(3), 658–670. doi:10.1137/0214048

Srinuvasu, M. A., & Prasad, D. D. (2016). Class-Oriented Model Graph Design Based on Abstract Syntax Tree. International Journal of Computer Science and Communication, 7(2).

Srivastav, M. K., & Nath, A. (2017). Content Delivery and Management System. *Proc. of IEEE, International Conference on Information, Communication, Instrumentation and Control (ICICIC 2017).* 10.1109/ICOMICON.2017.8279058

Srivastav, M.K., & Nath, A. (2016). Web Content Management System. *International Journal of Innovative Research in Advanced Engineering, 3*(3), 51-56.

Steiner, G. (1996). The recognition of indifference digraphs and generalized semiorders. *Journal of Graph Theory, 21*(2), 235–241.

Stephen, D. (1968). Topology on finite sets. *The American Mathematical Monthly, 75,* 739–741.

Stephenson, K., & Zelen, M. (1989). Rethinking centrality: Methods and examples. *Social Networks, 11*(1), 1–37. doi:10.1016/0378-8733(89)90016-6

Stevanovic, D. (2005). Energy and NEPS of graphs. *Linear and Multilinear Algebra, 53*(1), 67–74. doi:10.1080/0308 1080410001714705

Stevanovic, D., & Stankovic, I. (2005). Remarks on hyperenergetic circulant graphs. *Linear Algebra and Its Applications, 400,* 345–348. doi:10.1016/j.laa.2005.01.001

Stocker, V., Smaragdakis, G., Lehr, W., & Bauer, S. (2017). The growing complexity of content delivery networks: Challenges and implications for the Internet ecosystem. *Telecommunications Policy, 41*(10), 1003–1016. doi:10.1016/j.telpol.2017.02.004

Stoll R.R., (1979). *Set theory and Logic.* Dover pub.

Sudev, N. K. (2015). *Set-valuations of discrete structures and their applications* (PhD Thesis). Kannur University, Kannur, India.

Sudev, N.K., & Germina, K.A. (2015). On integer additive set-sequential graphs. *Int. J. Math. Combin., 2015*(3), 125-133.

Sudev, N. K., Ashraf, P. K., & Germina, K. A. (2019).Integer additive set-valuations of signed graphs. *TWMS J. Appl. Engg. Math.*

Sudev, N. K., Chithra, K. P., & Germina, K. A. (2015). On integer additive set-filter graphs. *J. Abst. Comput. Math., 3,* 8–15.

Sudev, N. K., & Germina, K. A. (2014). A Characterisation of strong integer additive set indexers of graphs. *Commun. Math. Appl., 5*(3), 101–110.

Sudev, N. K., & Germina, K. A. (2014). A characterization of weak integer additive set-indexers of graphs. *J. Fuzzy Set Valued Anal., 2014,* 1–7. doi:10.5899/2014/jfsva-00189

Sudev, N. K., & Germina, K. A. (2014). A note on the sparing number of graphs. *Adv. Appl. Discrete Math., 14*(1), 51–65.

Sudev, N. K., & Germina, K. A. (2014). A study of semi-arithmetic integer additive set-indexers of graphs. *Int. J. Math. Sci. Eng. Appl., 8*(III), 157–165.

Sudev, N. K., & Germina, K. A. (2014). A study on semi-arithmetic set-indexers of graphs. *Int. J. Math. Sci. Eng. Appl., 8*(III), 157–165.

Sudev, N. K., & Germina, K. A. (2014). Associated graphs of certain arithmetic IASI graphs. *Int. J. Math. Soft Comput.*, *4*(2), 71–80.

Sudev, N. K., & Germina, K. A. (2014). Further studies on the sparing number of graphs, *TechS Vidya e-J. Res.*, *2*, 25–36.

Sudev, N. K., & Germina, K. A. (2014). On integer additive set-indexers of graphs. *Int. J. Math. Sci. Eng. Appl.*, *8*(II), 11–22.

Sudev, N. K., & Germina, K. A. (2014). On the sparing number of certain graph structures. *Annals Pure Appl. Math.*, *6*(2), 140–149.

Sudev, N. K., & Germina, K. A. (2014). Weak integer additive set-indexers of certain graph products. *J. Inform. Math. Sci.*, *6*(1), 35–43.

Sudev, N. K., & Germina, K. A. (2014). *Weak integer additive set-indexers of graph operations*, Global J. Math. Sci. Theory. *Practical*, *6*(1), 25–36.

Sudev, N. K., & Germina, K. A. (2015). A note on the sparing number on the sieve graphs of certain graphs, *Appl. Mathematical Notes*, *15*, 29–37.

Sudev, N. K., & Germina, K. A. (2015). A study on the nourishing number of graphs and graph power. *Math.*, *3*, 29–39. doi:10.3390/math3010029

Sudev, N. K., & Germina, K. A. (2015). A study on topogenic integer additive set-labelled graphs. *J. Adv. Res. Pure Math.*, *7*(3), 15–22. doi:10.5373/jarpm.2230.121314

Sudev, N. K., & Germina, K. A. (2015). A study on topological integer additive set-labelling of graphs. *Electron. J. Graph Theory Appl.*, *3*(1), 70–84. doi:10.5614/ejgta.2015.3.1.8

Sudev, N. K., & Germina, K. A. (2015). Integer additive set-valuations of signed graphs. *Carpathian Math. Publ.*, *7*(2), 236–246.

Sudev, N. K., & Germina, K. A. (2015). On certain arithmetic integer additive set-indexers of graphs. *Discrete Mathematics, Algorithms, and Applications*, *7*(3), 1–15. doi:10.1142/S1793830915500251

Sudev, N. K., & Germina, K. A. (2015). On the hypergraphs associated with certain integer additive set-labelled graphs. *J. Adv. Res. Appl. Math.*, *7*(4), 23–33. doi:10.5373/jaram.2287.021015

Sudev, N. K., & Germina, K. A. (2015). Some new results on strong integer additive set indexers of graphs. *Discrete Mathematics, Algorithms, and Applications*, *7*(1), 1–11. doi:10.1142/S1793830914500657

Sudev, N. K., & Germina, K. A. (2015). Some new results on weak integer additive set-labelling of graphs. *International Journal of Computers and Applications*, *128*(5), 1–5. doi:10.5120/ijca2015906514

Sudev, N. K., & Germina, K. A. (2015). Switched signed graphs of integer additive set-valued signed graphs. *Discrete Mathematics, Algorithms, and Applications*, *9*(4), 1–10. doi:10.1142/S1793830917500434

Sudev, N. K., & Germina, K. A. (2015). Weak integer additive set indexes of certain graph classes. *J. Discrete Math. Sci. Cryptography*, *18*(2-3), 117–128. doi:10.1080/09720529.2014.962866

Sudev, N. K., & Germina, K. A. (2017). A study on prime arithmetic integer additive set-indexers of graphs. *Proyecciones J. Math.*, *36*(2), 195–208. doi:10.4067/S0716-09172017000200195

Sudev, N. K., & Germina, K. A. (2018). Arithmetic integer additive set-indexers of certain graph operations. *J. Inform. Math. Sci.*, *10*(1-2), 321–332. doi:10.26713/jims.v10i1+&+2.617

Sudev, N. K., & Germina, K. A. (2019). A study on integer additive set-graceful graphs. *Southeast Asian Bulletin of Mathematics*.

Sudev, N. K., Germina, K. A., & Chithra, K. P. (2014). Weak set-labelling number of certain integer additive set-labelled graphs. *International Journal of Computers and Applications*, *114*(2), 1–6. doi:10.5120/19947-1772

Sunitha, M. S., & Vijaya Kumar, A. (2002). Complement of fuzzy graphs. *Indian Journal of Pure and Applied Mathematics*, *33*(9), 1451–1464.

Sunitha, M. S., & Vijayakumar, A. (2005). Blocks in fuzzy graphs. *The Journal of Fuzzy Mathematics*, *13*(1), 13–23.

Syafrizal, S., & Baskoro, E. T. (2012). Lower bounds of the size multipartite Ramsey numbers. *The 5th Mathematics*, *AIP Conf. Proc.*, *1450*, 259-261.

Syafrizal, S., Baskoro, E. T., & Uttunggadewa, S. (2005). The size multipartite Ramsey number for paths. *Journal Combin. Math. Combin. Comput.*, *55*, 103–107.

Tallman, S., Jenkins, M., Henry, N., & Pinch, S. (2004). Knowledge, clusters, and competitive advantage. *Academy of Management Review*, *29*(2), 258–271. doi:10.5465/amr.2004.12736089

Tapscott, D. (2014). *The digital economy anniversary edition: Rethinking promise and peril in the age of networked intelligence* (2nd ed.). McGraw-Hill Education.

Teichert, H.-M., & Sonntag, M. (2004). Competition hypergraphs. *Discrete Applied Mathematics*, *143*(1-3), 324–329. doi:10.1016/j.dam.2004.02.010

The Eurasian Group on Combating Money Laundering and Financing of Terrorism (EAG-1). (n.d.). *Glossary of definitions of the Eurasian Group on AML/CFT*. Retrieved October 28, 2018, from https://eurasiangroup.org/en/glossary

The Eurasian Group on Combating Money Laundering and Financing of Terrorism (EAG-2). (n.d.). *Typologies research topics*. Retrieved October 28, 2018, from https://eurasiangroup.org/en/typologies-research-topics

The Financial Action Task Force (FATF). (2006). *Trade-based money laundering*. Retrieved October 28, 2018, from http://www.fatf-gafi.org/publications/methodsandtrends/documents/trade-basedmoneylaundering.html

The Financial Action Task Force (FATF). (2007). *Laundering the proceeds of VAT carousel fraud*. Retrieved October 28, 2018, from http://www.fatf-gafi.org/documents/documents/launderingtheproceedsofvat carouselfraudreport.html

The Financial Action Task Force (FATF). (2008). *Best practices on trade based money laundering*. Retrieved October 28, 2018, from http://www.fatf-gafi.org/media/fatf/documents/recommendations/BPP%20Trade%20Based%20Money%20 Laundering%202012%20COVER.pdf

Thomas, B. K. (2009). *Advanced studies on labelling of graphs and hypergraphs and related areas* (PhD Thesis). Kannur University, Kannur, India.

Tiskin, A. (2015). Fast distance multiplication of unit-Monge matrices. *Algorithmica*, *71*(4), 859–888. doi:10.100700453-013-9830-z

Todeschini, R., Ballabio, D., & Consonni, V. (2010). Noval molecular descriptors based on functions of new vertex degrees. In I. Gutman & B. Furtula (Eds.), *Novel molecular structure descriptors – Theory and Applications I* (pp. 73–100). Kragujevac: Univ. Kragujevac.

Todeschini, R., & Consonni, V. (2009). *Molecular Descriptors for Chemoinformatics*. Weinheim: Wiley-VCH. doi:10.1002/9783527628766

Todeschini, R., & Consonni, V. (2010). New local vertex invariants and descriptors based on functions of vertex degrees. *MATCH Commun. Math. Comput. Chem.*, *64*, 359–372.

Trinajstic, N. (1988). *MATH/CHEM/COMP 1987* (R. C. Lacher, Ed.; Vol. 83). Amsterdam: Elsevier.

Trinajstić, N. (2011). *Chemical graph theory* (2nd ed.). CRC Press.

Trinajstić, N., Nikolić, S., Miličvić, A., & Gutman, I. (2010). On Zagreb indices. *Kem. Ind.*, *59*, 577–589.

TsinghuaUniversity.(2018).*QiuYongvisitsChiletopromotecooperationbetweenChinaandChileonHigherEducation*.Available in: http://news.tsinghua.edu.cn/publish/thunewsen/9670/2018/20181219103403591578325/20181219103403591578325_.html

Tucker, A. (1971). Matrix characterizations of circular-arc graphs. *Pacific Journal of Mathematics*, *39*(2), 535–545.

Tucker, A. (1974). Structure theorems for some circular-arc graphs. *Discrete Mathematics*, *7*(1), 167–195. doi:10.1016/S0012-365X(74)80027-0

Turro & Angew. (1986). Geometric and Topological thinking in Organic Chemistry. *Chem. Int. Ed. Engl*, *25*, 882.

Tutorial. (n.d.). Retrieved October 28, 2018, from https://www.tutorialspoint.com/plsql/

van Rooij, A. C. M., & Wilf, H. (1965). The interchange graph of a finite graph. *Acta Mathematica Hungarica*, *16*(3–4), 263–269. doi:10.1007/BF01904834

Vega, M. A., Benítez, G. J., & Yévenes, S. A. (2005). Generic strategies for regional development in the plastics industry supply chain. *Horizontes Empresariales*, *4*(1), 31-45. Available in: http://revistas.ubiobio.cl/index.php/HHEE/article/view/2068

Vega, A., & Salinas, C. M. (2017). Scientific Production Analysis in Public Affairs of Chile and Peru. Challenges for a Better Public Management. *Lex*, *15*(20), 463–478. doi:10.21503/lex.v15i20.1451

Venketesh, P., & Venkatesan, R. (2009). A survey on applications of neural networks and evolutionarytechniques in web caching. *IETE Technical Review*, *26*(3), 171–180. doi:10.4103/0256-4602.50701

Vijayakumar, G. R. (2011). A *note on set-graceful graphs*. arXiv:1101.2729

Vijayakumar, G. R. (2007). A note on set-magic graphs. In *Proc. II Group Discussion, Labelling of Discrete Structures and their Applications*. Narosa Publishing House.

Vijayakumar, G. R. (2007). Set-magic labellings of infinite graphs. In *Proc. II Group Discussion, Labelling of Discrete Structures and their Applications*. Narosa Publishing House.

Vimala, S., & Sathya, J. S. (2012). Connected point set domination of fuzzy graphs. *International Journal of Mathematics and Soft Computing*, *2*(2), 75–78. doi:10.26708/IJMSC.2012.2.2.10

Voloshin, V. I. (2009). *Introduction to graph and hypergraph theory*. New York: Nova Science Pub.

Von Landesberger, T., Kuijper, A., Schreck, T., Kohlhammer, J., van Wijk, J. J., Fekete, J. D., & Fellner, D. W. (2011). Visual analysis of large graphs: state-of-the-art and future research challenges. Computer Graphics Forum, 30(6), 1719-1749.

Vorobiev, E. G., Petrenko, S. A., Kovaleva, I. V., & Abrosimov, I. K. (2017). Organization of the entrusted calculations in crucial objects of informatization under uncertainty. *Proceedings of the 2017 XX IEEE International Conference on Soft Computing and Measurements*. 10.1109/SCM.2017.7970566

Vukičević, D., & Furtula, B. (2009). Topological index on the ratios of geometrical and arithmetical means of end-vertex degrees of edges. *Journal of Mathematical Chemistry*, *46*(4), 1369–1376. doi:10.100710910-009-9520-x

Vukičević, D., & Gašperov, M. (2010). Bond additive Modeling 1. Adriatic indices. *Croatica Chemica Acta*, *83*, 243–260.

Vukičević, D., Li, Q., Sedlar, J., & Došlić, T. (2018). Lanzhou index. *MATCH Commun. Math. Comput. Chem.*, *8*, 863–876.

Walikar & Ramane. (2005). Energy of trees with edge independence number two. *Proc. Nat. Acad. Sci., India*, *75*(A), 137 – 140.

Walikar, H. B., Gutman, I., Hampiholi, P. R., & Ramane, H. S. (2001). Nonhyperenergetic graphs. *Graph Theory Notes, New York*, *51*, 14–16.

Walikar, H. B., Ramane, H. S., Gutman, I., & Halkarni, S. B. (2007). On equienergetic graphs and molecular graphs. *Kragujevac J. Sci.*, *29*, 73–84.

Walikar, H. B., Ramane, H. S., & Hampiholi, P. R. (1999). On the energy of a graph. In R. Balakrishnan, H. M. Mulder, & A. Vijaykumar (Eds.), *Graph Connections* (pp. 120–123). New Delhi: Allied Publishers.

Walikar, H. B., Ramane, H. S., & Hampiholi, P. R. Energy of trees with edge independence number three. In *Proc. Nat. Conf. Math. Comput. Models*. Allied Publishers.

Walikar, H. B., Ramane, H. S., & Jog, S. R. (2008). On an open problem of R. Balakrishnan and the energy of products of graphs. *Graph Theory Notes, New York*, *55*, 41–44.

Wang, J. (2017). A novel weight neighborhood centrality algorithm for identifying influential spreaders in complex networks. *Physica A*, (17), 30121-8.

Wang, B., & Xing, H. Y. (2011). The application of cloud computing in education informatization. *Proceedings of the 2011 International Conference on Computer Science and Service System*. 10.1109/CSSS.2011.5973921

Wang, H., Smarandache, F., Zhang, Y. Q., & Sunderraman, R. (2010). Single-valued neutrosophic sets. *Multisspace and Multistruct*, *4*, 410–413.

Wang, S., Zhou, B., & Trinajstić, N. (2011). On the sum connectivity index. *Filomat*, *25*(3), 29–42. doi:10.2298/FIL1103029W

Wassermann, S., & Faust, K. (1994). *Social Network Analysis: Methods and Applications*. Cambridge, UK: Cambridge University Press. doi:10.1017/CBO9780511815478

Watts, D., & Strogatz, S. (1998). Collective Dynamics of 'Small-World' Networks. *Nature*, *393*(6), 440–442. doi:10.1038/30918 PMID:9623998

Wei-Qiang, H., Xin-Tian, Z., & Shuang, Y. (2009). A Network Analysis of the Chinese Stock Market. *Physica*, *388*(14), 2956–2964. doi:10.1016/j.physa.2009.03.028

West, D. B. (2002). *Introduction to graph Theory*. Prentice Hall India.

White, D. R., & Borgatti, S. P. (1994). Betweenness centrality measures for directed graphs. *Social Networks*, *16*(4), 335–346. doi:10.1016/0378-8733(94)90015-9

Wiener, H. (1947). Structural determination of parattin boiling points. *Journal of the American Chemical Society*, *69*(1), 17–20. doi:10.1021/ja01193a005 PMID:20291038

WoS. (2019). *Web of Science Core Collection. Advanced Search*. Available in: https://clarivate.com/products/web-of-science/web-science-form/web-science-core-collection/

Wuchty, S., & Stadler, P. F. (2003). Centers of complex networks. *Journal of Theoretical Biology, 223*(1), 45–53. doi:10.1016/S0022-5193(03)00071-7 PMID:12782116

Xing, R., Zhou, B., & Du, Z. (2010). Further results on atom bond connectivity index of trees, discrete. *Applications of Mathematics, 158*, 1536–1545.

Xing, R., Zhou, B., & Tranajstić, N. (2010). Sum connectivity index of molecular trees. *Journal of Mathematical Chemistry, 48*(3), 583–591. doi:10.100710910-010-9693-3

Xu, K., Tang, K., Liu, H., & Wang, J. (2015). The Zagreb indices of bipartite graphs with more edges. *J. Appl. Math. and Informatics, 33*(3), 365–377. doi:10.14317/jami.2015.365

Xu, L., & Hou, Y. (2007). Equienergetic bipartite graphs. *MATCH Commun. Math. Comput. Chem., 57*, 363–370.

Yang, L., Li, K., & Huang, H. (2018). A new network model for extracting text keywords. *Scientometrics, 116*(1), 339–361. doi:10.100711192-018-2743-5

Yeh, R. K. (1990). *Labeling graphs with a condition at distance two* (Ph.D Thesis). University of South Carolina, Columbia, SC.

Yu, C.-W., & Chen, G.-H. (1993). Generate all maximal independent sets in permutation graphs. *International Journal of Computer Mathematics, 47*(1–2), 1–8. doi:10.1080/00207169308804157

Yu, X., & Xue, Y. (2016). Smart grids: A cyber-physical systems perspective. *Proceedings of the IEEE, 104*(5), 1058–1070. doi:10.1109/JPROC.2015.2503119

Zadeh, L. A. (1965). Fuzzy sets. *Information and Control, 8*(3), 338–353. doi:10.1016/S0019-9958(65)90241-X

Zadeh, L. A. (1971). Similarity relations and fuzzy oderings. *Information Sciences, 3*(2), 177–200. doi:10.1016/S0020-0255(71)80005-1

Zaslavsky, T. (1981). Characterizations of signed graphs. *Journal of Graph Theory, 5*(4), 401–406. doi:10.1002/jgt.3190050409

Zaslavsky, T. (1982). Signed graphs. *Discrete Applied Mathematics, 4*(1), 47–74. doi:10.1016/0166-218X(82)90033-6

Zaslavsky, T. (2010). *Balance and clustering in signed graphs*. Hyderabad, India: C R Rao Adv. Instt. of Mathematics Statistics and Computer Science.

Zaslavsky, T. (2012). Signed graphs and geometry. *J. Combin. Inform. System Sci., 37*(2-4), 95–143.

Zeng, A., & Zhang, C. J. (2013). Ranking spreaders by decomposing complex networks. *Physics Letters. [Part A], 377*(14), 1031–1035. doi:10.1016/j.physleta.2013.02.039

Zhang, J. (2018). Research on evaluation model of the effective EFL teaching in the era of informatization. *Journal of Language Teaching and Research, 9*(4), 738-745.

Zhang, W. R. (1994, December). Bipolar fuzzy sets and relations: a computational framework for cognitive modeling and multiagent decision analysis. In *NAFIPS/IFIS/NASA'94. Proceedings of the First International Joint Conference of The North American Fuzzy Information Processing Society Biannual Conference. The Industrial Fuzzy Control and Intellige* (pp. 305-309). IEEE.

Zhang. (2009). *Stock market network topology analysis based on a minimum spanning Tree approach* (Thesis). Graduate College of Bowling Green State University.

Zhang, W. R. (1994). Bipolar fuzzy sets and relations: a computational framework for cognitive modeling and multiagent decision analysis. *Proceedings of IEEE Conference*, 305-309.

Zhang, W. R. (1998). Bipolar fuzzy sets. *IEEE International Conference on fuzzy systems*, 835-840.

Zhang, W. R. (1998). Bipolar fuzzy sets. *Proceedings of FUZZ-IEEE*, 835–840.

Zhang, Z., Zhang, J., & Lu, X. (2005). The relation of matching with inverse degree of a graph. *Discrete Mathematics*, *301*(2-3), 243–246. doi:10.1016/j.disc.2003.01.001

Zhivko, M. (2018). New economy: synergy of informatization and global civil society. *Journal of European Economy*, *17*(1), 34-55.

Zhou, B. (2004). Energy of graphs. *MATCH Commun. Math. Comput. Chem.*, *51*, 111–118.

Zhou, B. (2004). Zagreb indices. *MATCH Commun. Math. Comput. Chem.*, *52*, 113–118.

Zhou, B. (2007). Remarks on Zagreb indices. *MATCH Commun. Math. Comput. Chem.*, *57*, 597–616.

Zhou, B., & Gutman, I. (2005). Further properties of Zagreb indices. *MATCH Commun. Math. Comput. Chem.*, *54*, 233–239.

Zhou, B., & Ramane, H. S. (2008). On upper bounds for energy of bipartite graphs. *Indian Journal of Pure and Applied Mathematics*, *39*, 483–490.

Zhou, B., & Trinajstic, N. (2009). On a novel connectivity index. *Journal of Mathematical Chemistry*, *46*(4), 1252–1270. doi:10.100710910-008-9515-z

About the Contributors

Madhumangal Pal is currently a Professor of Applied Mathematics, Vidyasagar University, India. He has received several awards for his research and contribution in the academic performance. Prof. Pal has successfully guided 31 research scholars for Ph.D. degrees and has published more than 290 articles in international and national journals. His specializations include algorithmic graph theory, fuzzy graph theory, inventory control, fuzzy game theory, fuzzy matrices, genetic algorithms and parallel algorithms. Prof. Pal is the Editor-in-Chief of the journals "Journal of Physical Sciences" and "Annals of Pure and Applied Mathematics" and member of the several editorial Boards of the Indian and foreign journals. He is also a reviewer of several international journals. Prof. Pal is the author of the eight books written for undergraduate and postgraduate students published from India and abroad. He organized several national seminars/ conferences/workshops. Also, he has visited London, China, Malaysia, Thailand, Hong Kong and Bangladesh to participated, delivered invited talks, chaired seminars, conferences, winter school, refresher course, etc.

* * *

Muhammad Akram received MSc degrees in Mathematics and Computer Science, MPhil in Computational Mathematics and PhD in Fuzzy Mathematics. He is currently a Professor in the Department of Mathematics at the University of the Punjab, Lahore, Pakistan, where he has been serving as a PhD supervisor of more than 10 students. Dr. Akram's research interests include numerical solutions of parabolic PDEs, fuzzy graphs, fuzzy algebras, and fuzzy decision support systems. He has published 6 monographs and 275 research articles in international peer-reviewed journals. He has served as editorial board member of 10 international academic journals and as reviewer of 120 International journals, including Mathematical Reviews and Zentralblatt Math.

Sk Amanathulla, M.Sc, PhD is an expert in graph theory, computational mathematics, discrete mathematics, graph labeling, algorithm design, etc.

Juan Manuel Arjona-Fuentes has a PhD in Business Management and Quantitative Methods, Universidad de Córdoba, Spain. He works as Statistics and Operations Research Professor in the Department of Quantitative Methods and is a Member of the Research Team of the Doctoral Program in Data Science, Universidad Loyola Andalucia (Spain). ORCID: https://orcid.org/0000-0002-9694-3358.

Germina K. Augusthy is an Associate Professor in the Department of Mathematics, Central University of Kerala, Karsargod, Kerala, India. She is a well-known researcher in the field of Discrete Mathematics, especially in algebraic graph theory, labeling of graphs, spectral graph theory and signed and gain graph theory. She has supervised 19 Ph. D theses, 12 M. Phil thesis and many M. Sc projects. She published many research articles in both international/national journals and a very active researcher in imparting invited talks and plenary talks at various International/national conference.

Robin Bhadoria has obtained his Ph.D. degree in the month of January 2018 from Discipline of Computer Science & Engineering at Indian Institute of Technology (IIT) Indore, Madhya Pradesh, India. He also finished his Master of Technology (M.Tech) and Bachelor of Engineering (B. Eng.) in CSE from Rajiv Gandhi Technological University (RGTU) of Madhya Pradesh, India in 2011 and 2008 respectively. He received Gold Medal at M.Tech. His current area of Interests are Big Data Analytics, Service-Oriented Architecture, Internet of Things and Wireless Sensor Network. He has published more than 50 articles in IEEE, Elsevier, and Springer that also includes the book chapters. Presently, a professional member for - IEEE (USA), IAENG (Hong-Kong), Internet Society, Virginia (USA), IACSIT (Singapore).

Ankur Bharali is currently working as Assistant Professor in the Department of Mathematics, Dibrugarh University, India. He has obtained his master degree from Indian Institute of Technology Guwahati in Mathematics and Computing and awarded PhD. in Mathematics by Dibrugarh University. He has more 10 years of teaching experience in Post Graduate level. His areas of interests are Graph Theory and Complex Networks. He has published more than 25 research articles in reputed journals and edited books.

Ankur Bharali is an assistant professor in the Department of Mathematics, Dibrugarh University, India.

Jibonjyoti Buragohain is a research scholar in the Department of Mathematics, Dibrugarh University, India.

Kousik Das is a research scholar in the Department of Mathematics, Vidyasagar University. He is working on Social Network Analysis and related issues.

Amitav Doley is an Assistant Professor in the Department of Mathematics at DHSK College.

Ganesh Ghorai is working as an Assistant Professor in the Department of Applied Mathematics, Vidyasagar University from 2012. He has completed his M. Sc from IIT Bombay in 2012 and Ph. D degree from Vidyasagar University in 2017. His research interests includes fuzzy graph theory with application and their generalization, fuzzy algebra, fuzzy functional analysis, etc. He has published 24 research articles in highly reputed Internationally journals.

F. Nirmala Irudayam is an Assistant Professor at Nirmala College For Women. She has completed her doctorate form Nirmala College for Women, Bharathiar University, Coimbatore. She authored over 26 publications in the field of topology, neutrosophy and soft sets.

Kavikumar Jacob, Associate Professor, earned his M.Sc in 1999 and Ph.D in 2005 from Annamalai University, India. Since 2006 he has been at Department of mathematics and Statistics, Faculty of Science, Technology and Human Development, Universiti Tun Hussein Onn Malaysia, Malaysia. He has over nine years of teaching and research experience. He started his career as a Senior Lecturer in Mathematics at Vellore Institute of Technology, India in 2004. His current research interests include bipolar quantum logic gates and finite switchboard state machines. He has successfully supervised three PhD students. Dr. Kavikumar Jacob has given many presentations and invited talks in fuzzy mathematics in Malaysia and in India. He has published about 40 research papers. He is actively involved in collaborative research work from University of Malaya, Malaysia and Gyeongsang National University, South Korea. He is a member in several professional bodies such as American Mathematical Society (AMS) and Society for Industrial and Applied Mathematics.

Chula Jayawardene is currently a Senior Lecturer at the Department of Mathematics, University of Colombo.

Johan Kok is a police officer by profession and a recognised researcher in the area of graph theory. He has published more than a hundred articles in recognised international research journals. He is very active in academia as a member of professional academic societies and reviewer/editor to many recognised research journals.

Rupkumar Mahapatra is a research scholar in the Department of Mathematics, Vidyasagar University. He is working on fuzzy graph theory and related issues.

Natalia Miloslavskaya after graduating NRNU MEPhI works at the MEPhI. She was a Vice Dean on International Affairs of the MEPhI's Information Security Faculty. She has 25 years of experience in the field of information security. At present, her research interests lie in information security management systems and network security of different types of systems, for example, of the open systems (such as the Internet and Intranets) and the automated banking systems. She does research on security solutions (in particular SIEM, NGFW, IDPS, security scanners, VPN), services and policies. She lectures at MEPhI for students and masters and at the retraining courses for the Russian banking specialists and supervises graduates/post-graduates. She wrote or co-authored 62 textbooks for her original educational courses in MEPhI. She actively participated in the development of the "Business Continuity and Information Security Maintenance" Master's degree programme, in the implementation of which she is very deeply involved. 48 times she was an international conference PC member, co-chair, and chair. She has more than 310 publications (in Russian and English). For the third time, she is an IFIP TC11 WG11.8 Vice Chair on Information Security Education. Among her main awards are the highest internal IFIP Silver Core 2013 Award, INFOFORUM 2015 Silver Dagger for the strengthening of international cooperation in the field of information security, 2016 Best Reviewer Award from SDIWC, and the Certificate of Outstanding Contribution in Reviewing from Future Generation Computing Systems (Elsevier, 2017).

Sudev Naduvath is currently working as Associate Professor in the Department of Mathematics, Christ University, Bangalore, INDIA. He has a teaching experience of over fifteen years at the collegiate level and research experience of over eight years. He has authored about ninety research papers in various reputed international research journals and delivered lectures in many conferences. He has several reputed research collaborators within and outside India. He is an active member of many national and international professional societies and is included in the editorial board of more than ten research journals and in referee panel of more than forty-five journals. He is also a reviewer of prestigious Mathematical Reviews, ZbMATH, Computing Reviews and MAA Reviews.

Andrey Nikiforov is a post-graduate student after graduating NRNU MEPhI works at the MEPhI. His research areas are information security, big data technologies and standardization.

Kirill Plaksiy is a post-graduate student after graduating NRNU MEPhI works at the MEPhI. His research areas are the graph theory and information and financial security.

Tarasankar Pramanik received his Bachelor of Science degree with honours in Mathematics in 2006 from Tamralipta Mahavidyalaya, Purba Medinipur, West Bengal, India and Master of Science degree in Mathematics in 2008 from Vidyasagar University, West Bengal, India. He has ranked 201 in UGC of Council of Scientific and Industrial Research, Human Resource Developement Group, India in 2009. He has qualified GATE, conducted by Indian Institute of Technology, Kharagpur in 2009. He has completed B. Ed. degree from Netaji Subhas Open University in 2015. He is an Assistant Teacher of a Govt. Sponsored High School since 2009. He has completed his Ph. D. from the Department of Applied Mathematics, Vidyasagar University in 2018. His research interest includes fuzzy graphs and their applications in real world. He has participated and presented scientific works in many national and international seminars/conferences/workshops.

Harishchandra S. Ramane obtained his B.Sc., M.Sc. and Ph.D. from Karnatak University, Dharwad, India. Currently he is Professor of Mathematics at Karnatak University, Dharwad. He has 25 years of teaching and research experience. His area of interest includes Graph theory, Spectral graph theory, Topological indices. He has published several research articles and delivered invited talks in conferences. Eleven students completed Ph.D. under his guidance.

Radi Romansky is a full professor in Technical University of Sofia, Bulgaria; Doctor of Science in Informatics and Computer Science, Vice Rector. He has over 195 scientific publications and 19 published monographies, books and manuals. Participant in 33 scientific research projects in the field of computer systems and technologies, e-learning, etc. Full member of the European Network of Excellence on High Performance and Embedded Architectures and Compilation – HiPEAC. Member of the International Editorial Board of scientific journals (Bulgaria, India, Slovakia, USA, etc.), chairman of the Organizing and Program committee of International Conference on Information Technologies. Scientific areas: Computer systems and architectures, Computer modelling, Information technologies, Personal data protection, etc.

Danish Saleem received her M.Phil. degree in Mathematics and M.Sc. degree in Mathematics from the University of the Punjab, Lahore. His research interest includes extensions of fuzzy graphs and their applications. He has published 05 research articles in scientific journals.

Sovan Samanta has completed his post-doctoral degree from Hanyang University, South Korea. Currently, he is working as an Assistant Professor in the Department of Mathematics, Tamralipta Mahavidyalaya (Vidyasagar University), India. He has teaching and research experiences from Indian Institute of Information Technology Nagpur. He is working on Graph Theory, Social Network Analysis, Fuzzy Systems, etc. Still now, he has published more than 50 articles in different reputed journals including 22 SCIE/SCI journals.

E Sampathkumar, MSc PhD, is a retired professor of Mathematics. Dr. Sampathkumar's research interest includes graph theory, coloring, and domination. Dr. Sampathkumar has published more than 100 papers and was a founding member of Ramanujan Mathematical Society, Academy of Discrete Mathematics and a Managing Editor of the *Journal of Ramanujan Mathematical Society*.

C. Antony Crispin Sweety holds a Ph. D in mathematics from Nirmala College for Women, Bharathiar University, Coimbatore. She has authored over 10 publications in the field of Neutrosophy, Rough Sets, Soft Sets, Topology and Granular Computing. Her current area of research is Neutrosphic Topology, Neutrosphic Operation Research, Neutrosophic Graph theory and Neutrosophic Controls.

Alexander Tolstoy after graduating NRNU MEPhI works at MEPhI. He was a Vice Dean on Educational Work of the MEPhI's Information Security Faculty. He has 30 years of experience in the field of information security. At present, his research interests lie in information security management systems and network security of different types of systems, for example, of the open systems (such as the Internet and Intranets) and the automated banking systems. He does research on security policies and security education. He lectures at MEPhI for students and masters and at the retraining courses for the Russian banking specialists and supervises graduates/post-graduates. He wrote or co-authored 62 textbooks for his original educational courses in MEPhI. He actively participated in the development of the "Business Continuity and Information Security Maintenance" Master's degree programme, in the implementation of which he is very deeply involved. He has more than 300 publications (in Russian and English).

K. Vaiyomathi is an Assistant Professor at Nirmala College for Women, Coimbatore. She has completed her M. Phil and currently pursing Ph. D in Mathematics. Her current area of research is Neutrosophy, Operation Research, Graph Theory and Topology.

Alejandro Vega-Muñoz has a Ph.D. in Business Studies, Universidad Antonio de Nebrija (Spain), M.Sc. in Industrial Engineering, Universidad de Concepción (Chile), B.Sc. in Port Engineering, Universidad Católica de la Santísima Concepción (Chile), and B.Sc. in Education, Universidad de Antofagasta (Chile). His research interests are focused on sectoral economics, international relations, economic geography and their relationship with economic development studies. He currently works as a Researcher in Universidad San Sebastián (Chile), Foundation 'School of International Affairs' (Chile), PIIE Corporation (Chile), and Universidad Nacional de la Patagonia San Juan Bosco (Argentina). Member of SOCHIGEO (Sociedad Chilena de Ciencias Geográficas), and Professor at HEIs in Chile, Spain, Argentina, and Peru. ORCID: http://orcid.org/0000-0002-9427-2044.

Michael G. Voskoglou (B.Sc., M.Sc., M.Phil., Ph.D. in Mathematics) is an Emeritus Professor of Mathematical Sciences at the School of Technological Applications of the Graduate Technological Educational Institute of Western Greece. He is the author of 9 books (in Greek and in English language) and of more than 300 papers published in reputed journals and proceedings of conferences of 25 countries in 5 continents, with many references from other researchers.

Index

A

Acentric Factor 92, 100-101

Adjacency Matrix 121, 267, 269, 274, 355, 359, 377-379, 469, 471-474, 476-477, 479-481, 483, 485-486, 488, 529, 537

Anti-Money Laundering And Combating Financing Of Terrorism 297

Arithmetic Sumset Labeling 249

B

Bibliometrics 415

Bipartite Fuzzy Graph 419, 422-423

Bipolar Fuzzy Graph 437, 442-447, 443-447, 451, 491

Bipolar Neutrosophic Cubic Edge 492

Bipolar Neutrosophic Cubic Vertex 492

Bradford Core 415

C

Cache Scheme 386

Cache Server 386-395, 386-389, 391-395

CDPU Graphs 196, 204

Centrality 357, 368, 371-373, 375-385, 375-380, 382, 399, 403, 412, 414-415, 415

Chemistry 66, 68, 83-85, 88, 90-91, 134, 206, 267, 293-295, 354, 359, 369, 470, 546

Chordal Graphs 24-25, 27, 42, 51, 55, 60-62

Circular-Arc Graphs 24, 27, 39, 54, 61, 63, 65, 135-139, 135-137, 139, 149-150, 153, 155, 157, 168

Client-Server Mode 386

Communication Infrastructure 320, 326, 330, 333

Complete Fuzzy Graph 419, 422, 440, 442

Connected Domination 1, 17, 22-23

Content Delivery Network (CDN) 386

D

Dakshayani Indices 66, 71, 88

Dcsl Graphs 196-197

Degree Of A Vertex 51, 71, 92-93, 93, 106, 419, 421, 444, 497, 505

Design And Analysis Of Algorithms 24

Digital World 320-321, 321, 350-351, 350

Discrete Modelling 320, 336, 352

Disk Graphs 24, 27

Distance-Compatible Set-Labeling 171, 196

Domination On Graphs 1

Domination Parameters 1

DRMS 345-346, 349, 353

E

Eigen Values 359, 469

Energy Of M-Polar Fuzzy Digraphs 469, 479, 486, 490

Entropy 92, 100, 102, 323, 351

Equienergetic Graphs 267, 285-286, 289-296, 289-293

F

Fifth Zagreb Indices 66, 69

First Zagreb Index 67-68, 70-71, 80, 83, 85, 92-95, 100, 102, 104, 107, 113, 115, 275-276

Forgotten Topological Index 68, 82, 84, 89, 120, 134, 537, 546

Frequency Assignment 135-136, 135

Fuzzy Competition Graph 427, 433-435, 443, 451-453, 455

Fuzzy Graphs 419-420, 422-427, 435-437, 435, 437, 439-443, 445, 459, 461-462, 465-470, 465, 470, 472, 489-491, 489, 494, 534-536

Fuzzy Hypergraph 437, 458, 460, 462-467, 462-465

Fuzzy Intersection Graph 437, 442, 447, 454

Fuzzy Logic 386, 391, 468, 536

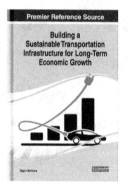

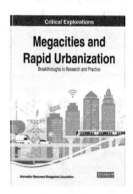
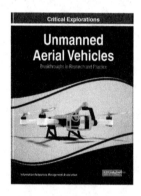
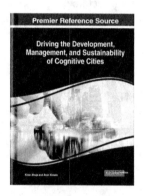

Ensure Quality Research is Introduced to the Academic Community

Become an IGI Global Reviewer for Authored Book Projects

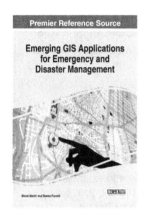

The overall success of an authored book project is dependent on quality and timely reviews.

In this competitive age of scholarly publishing, constructive and timely feedback significantly expedites the turnaround time of manuscripts from submission to acceptance, allowing the publication and discovery of forward-thinking research at a much more expeditious rate. Several IGI Global authored book projects are currently seeking highly-qualified experts in the field to fill vacancies on their respective editorial review boards:

Applications and Inquiries may be sent to:
development@igi-global.com

Applicants must have a doctorate (or an equivalent degree) as well as publishing and reviewing experience. Reviewers are asked to complete the open-ended evaluation questions with as much detail as possible in a timely, collegial, and constructive manner. All reviewers' tenures run for one-year terms on the editorial review boards and are expected to complete at least three reviews per term. Upon successful completion of this term, reviewers can be considered for an additional term.

If you have a colleague that may be interested in this opportunity, we encourage you to share this information with them.

Printed in the United States
By Bookmasters